Lecture Notes in Computer Science   10828

Commenced Publication in 1973
Founding and Former Series Editors:
Gerhard Goos, Juris Hartmanis, and Jan van Leeuwen

More information about this series at http://www.springer.com/series/7409

Jian Pei · Yannis Manolopoulos
Shazia Sadiq · Jianxin Li (Eds.)

Database Systems
for Advanced Applications

23rd International Conference, DASFAA 2018
Gold Coast, QLD, Australia, May 21–24, 2018
Proceedings, Part II

 Springer

Editors
Jian Pei
Simon Fraser University
Burnaby, BC
Canada

Shazia Sadiq
University of Queensland
Brisbane, QLD
Australia

Yannis Manolopoulos
Aristotle University of Thessaloniki
Thessaloniki
Greece

Jianxin Li
University of Western Australia
Crawley, WA
Australia

ISSN 0302-9743   ISSN 1611-3349 (electronic)
Lecture Notes in Computer Science
ISBN 978-3-319-91457-2   ISBN 978-3-319-91458-9 (eBook)
https://doi.org/10.1007/978-3-319-91458-9

Library of Congress Control Number: 2018943431

LNCS Sublibrary: SL3 – Information Systems and Applications, incl. Internet/Web, and HCI

This Springer imprint is published by the registered company Springer Nature Switzerland AG
The registered company address is: Gewerbestrasse 11, 6330 Cham, Switzerland

Preface

It is our great pleasure to present the proceedings of the 23rd International Conference on Database Systems for Advanced Applications (DASFAA). DASFAA 2018 is an annual international database conference, which showcases state-of-the-art R&D activities in database systems and their applications. It provides a forum for technical presentations and discussions among database researchers, developers, and users from academia, business, and industry.

DASFAA 2018 was held on the Gold Coast, Australia, during May 21–24, 2018. The Gold Coast is a coastal city in the state of Queensland, 66 km (41 mi) from the state capital Brisbane. With a population of 638,090 (2016), the Gold Coast is the sixth largest city in Australia. It is a major tourist destination with its sunny subtropical climate and has become widely known for its surfing beaches, high-rise-dominated skyline, theme parks, nightlife, and rainforest hinterland. It is also the major film production hub for Queensland. The Gold Coast will host the 2018 Commonwealth Games.

This year we introduced a Senior Program Committee (SPC) at DASFAA. The SPC comprised 12 distinguished leaders in the area of database systems and advanced applications: Amr El Abbadi, UC Santa Barbara, USA; K. Selcuk Candan, Arizona State University, USA; Lei Chen, Hong Kong University of Science and Technology, Hong Kong; Chengfei Liu, Swinburne University of Technology, Australia; Nikos Mamoulis, University of Ioannina/University of Hong Kong, Hong Kong; Kyuseok Shim, Seoul National University, Korea; Michalis Vazirgiannis, Ecole Polytechnique Paris, France; Xiaokui Xiao, Nanyang Technological University, Singapore; Xiaochun Yang, Northeastern University, China; Jeffrey Xu Yu, Chinese University of Hong Kong, Hong Kong; Xiaofang Zhou, University of Queensland, Australia; and Aoying Zhou, East China Normal University, China. We are grateful for the role played by the SPC and acknowledge that the SPC provided a significant level of support and expert advice in the efficient paper-reviewing process that resulted in an excellent selection of papers.

We received 360 submissions, each of which was assigned to at least three Program Committee (PC) members and one SPC member. The thoughtful discussion on each paper by the PC with facilitation and meta-review provided by the SPC resulted in the selection of 83 full research papers (acceptance ration of 23%). In addition, we included 21 short papers, six industry papers, and eight demo papers in the program. This year the dominant topics for the selected papers included learning models, graph and network data processing, and social network analysis, followed by text and data mining, recommendation, data quality and crowd sourcing, and trajectory and stream data. Selected papers also included topics relating to network embedding, sequence and temporal data processing, RDF and knowledge graphs, security and privacy, medical data mining, query processing and optimization, search and information retrieval, multimedia data processing, and distributed computing. Last but not least, the conference program included keynote presentations by Dr. C. Mohan (IBM Almaden Research Center, San Jose, USA), Prof. Xuemin Lin (UNSW, Sydney, Australia), and Prof. Yongsheng Gao (Griffith University, Brisbane, Australia).

Four workshops were selected by the workshop co-chairs to be held in conjunction with DASFAA 2018: the 5th International Workshop on Big Data Management and Service (BDMS 2018); the 5th International Symposium on Semantic Computing and Personalization (SeCoP 2018); the Second International Workshop on Graph Data Management and Analysis (GDMA 2018); and the Third Workshop on Big Data Quality Management (BDQM 2018). The workshop papers are included in a separate volume of the proceedings also published by Springer in its *Lecture Notes in Computer Science* series.

We are grateful to the general chairs, Yanchun Zhang, Victoria University, and Rao Kotagiri, University of Melbourne, all SPC members, PC members and external reviewers who contributed their time and expertise to the DASFAA 2018 paper reviewing process. We would like to thank all the members of the Organizing Committee, and many volunteers, for their great support in the conference organization. Special thanks go to the DASFAA 2018 local Organizing Committee chair, Junhu Wang (Griffith University), for his tireless work before and during the conference. Many thanks to the authors who submitted their papers to the conference. Lastly we acknowledge the generous financial support from Griffith University, Destination Gold Coast, and Springer.

March 2018

Shazia Sadiq
Jian Pei
Yannis Manolopoulos

Organization

General Co-chairs

Yanchun Zhang   Victoria University, Australia
Rao Kotagiri   University of Melbourne, Australia

Program Committee Co-chairs

Jian Pei   Simon Fraser University, Canada
Yannis Manolopoulos   Aristotle University of Thessaloniki, Greece
Shazia Sadiq   The University of Queensland, Australia

Industrial/Practitioners Track Co-chairs

Yu Zheng   Urban Computing Group, Microsoft Research, China
Qing Liu   Data61, CSIRO, Australia

Demo Track Co-chairs

Sebasitian Link   University of Auckland, New Zealand
Chaoyi Pang   NIT, Zhejiang University, China

Workshop Co-chairs

Chengfei Liu   Swinburne University of Technology, Australia
Lei Zou   Peking University, China

Tutorial Chair

Yoshiharu Ishikawa   Nagoya University, Japan

PhD Consortium Chair

Zhiguo Gong   University of Macau, China

Panel Co-chairs

Sean Wang   Fudan University, China
Sven Hartman   Clausthal University of Technology, Germany

Proceedings Chair

Jianxin Li   The University of Western Australia, Australia

Publicity Co-chairs

Ji Zhang   University of Southern Queensland, Australia
Xin Wang   Tianjin University, China
Shuo Shang   KAUST, Saudi Arabia

Local Organization Co-chairs

Junhu Wang   Griffith University, Australia
Bela Stantic   Griffith University, Australia
Alan Liew   Griffith University, Australia

DASFAA Liaison officer

Kyuseok Shim   Seoul National University, South Korea

Web Master

Xuguang Ren   Griffith University, Australia

Senior Program Committee Members

Amr El Abbadi   UC Santa Barbara, USA
K. Selcuk Candan   Arizona State University, USA
Lei Chen   Hong Kong University of Science and Technology,
                                  SAR China
Chengfei Liu   Swinburne University of Technology, Australia
Nikos Mamoulis   University of Ioannina/University of Hong Kong,
                                  SAR China
Kyuseok Shim   Seoul National University, Korea
Michalis Vazirgiannis   Ecole Polytechnique Paris, France
Xiaokui Xiao   Nanyang Technological University, Singapore
Xiaochun Yang   Northeastern University, China
Jeffrey Xu Yu   Chinese University of Hong Kong, SAR China
Xiaofang Zhou   University of Queensland, Australia
Aoying Zhou   East China Normal University, China

Program Committee

Alberto Abello   Universitat Politècnica de Catalunya, Spain
Akhil Arora   Indian Institute of Technology, India
Jie Bao   Independent

Zhifeng Bao	RMIT University, Australia
Ladjel Bellatreche	LIAS/ENSMA, France
K. Selcuk Candan	Arizona State University, USA
Huiping Cao	New Mexico State University, USA
Barbara Catania	DIBRIS-University of Genoa, Italy
Lei Chen	The Hong Kong University of Science and Technology, SAR China
Reynold Cheng	The University of Hong Kong, SAR China
Lingyang Chu	Simon Fraser University, Canada
Gao Cong	Nanyang Technological University, Singapore
Antonio Corral	University of Almeria, Spain
Bn Cui	Peking University, China
Ernesto Damiani	University of Milan, Italy
Lars Dannecker	SAP SE
Hasan Davulcu	Arizona State University, USA
Gianluca Demartini	The University of Queensland, Australia
Ugur Demiryurek	University of Southern California, USA
Curtis Dyreson	Utah State University, USA
Amr El Abbadi	University of California, USA
Elena Ferrari	University of Insubria, Italy
Yanjie Fu	Missouri University of Science and Technology, USA
Johann Gamper	Free University of Bozen-Bolzano, Italy
Hong Gao	Harbin Institute of Technology, China
Yunjun Gao	Zhejiang University, China
Neil Gong	Iowa State University, USA
Le Gruenwald	The University of Oklahoma, USA
Jingrui He	Arizona State University, USA
Juhua Hu	Simon Fraser University, Canada
Wen Hua	The University of Queensland, Australia
Helen Zi Huang	The University of Queensland, Australia
Nguyen Quoc Viet Hung	Griffith University, Australia
Yoshihara Ishikawa	Nagoya University, Japan
Md Saiful Islam	Griffith University, Australia
Cheqing Jin	East China Normal University, China
Alekh Jindal	Microsoft
Ioannis Karydis	Ionian University, Greece
Latifur Khan	UTD
Jinha Kim	Oracle Labs
Anne Laurent	LIRMM - UM
Young-Koo Lee	Kyung Hee University, South Korea
Guoliang Li	Tsinghua University, China
Jianxin Li	University of Western Australia, Australia
Zhixu Li	Soochow University, China
Xiang Lian	Kent State University, USA
Chengfei Liu	Swinburne University of Technology, Australia
Guanfeng Liu	Soochow University, China

Qing Liu	CSIRO, Australia
Eric Lo	The Chinese University of Hong Kong, SAR China
Hua Lu	Aalborg University, Denmark
Nikos Mamoulis	University of Ioannina, Greece
Yannis Manolopoulos	Aristotle University of Thessaloniki, Greece
Mikolaj Morzy	Poznan University of Technology, Poland
Kyriakos Mouratidis	Singapore Management University, Singapore
Parth Nagarkar	New Mexico State University, USA
Yunmook Nah	Dankook University, South Korea
Sarana Yi Nutanong	City University of Hong Kong, SAR China
Kjetil Nørvåg	Norwegian University of Science and Technology, Norway
Vincent Oria	NJIT
Dhaval Patel	IBM
Jian Pei	Simon Fraser University, Canada
Ruggero G. Pensa	University of Turin, Italy
Dieter Pfoser	George Mason University, USA
Evaggelia Pitoura	University of Ioannina, Greece
Silvestro Poccia	University of Turin, Italy
Weixiong Rao	Tongji University, China
Simon Razniewski	Max Planck Institute for Informatics, Germany
Matthias Renz	George Mason University, USA
Oscar Romero	Universitat Politècnica de Catalunya, Spain
Florin Rusu	University of California, USA
Shazia Sadiq	The University of Queensland, Australia
Simonas Saltenis	Aalborg University, Denmark
Maria Luisa Sapino	University of Turin, Italy
Claudio Schifanella	University of Turin, Italy
Shuo Shang	KAUST, Saudi Arabia
Hengtao Shen	University of Science and Technology of China, China
Yanyan Shen	Shanghai Jiao Tong University, China
Kyuseok Shim	Seoul National University, South Korea
Alkis Simitsis	HP (Hewlett Packard) Lab, USA
Shaoxu Song	Tsinghua University, China
Yangqiu Song	The Hong Kong University of Science and Technology, SAR China
Nan Tang	Qatar Computing Research Institute, Qatar
Christian Thomsen	Aalborg University, Denmark
Hanghang Tong	Arizona State University, USA
Yongxin Tong	Beihang University, China
Ismail Hakki Toroslu	Middle East Technical University, Turkey
Efthymia Tsamoura	University of Oxford, UK
Vincent S. Tseng	National Chiao Tung University, Taiwan
Theodoros Tzouramanis	University of the Aegean, Greece
Panos Vassiliadis	University of Ioannina, Greece
Michalis Vazirgiannis	AUEB, Greece

Sabrina De Capitani Vimercati	University of Milan, Italy
Bin Wang	NEU, China
Jianmin Wang	Tsinghua University, China
Wei Wang	National University of Singapore
Xin Wang	Tianjin University, China
John Wu	Berkeley Lab, USA
Xiaokui Xiao	Nanyang Technological University, Singapore
Xike Xie	University of Science and Technology of China, China
Jianlian Xu	Hong Kong Baptist University, SAR China
Xiaochun Yang	Northeastern University, China
Yu Yang	Simon Fraser University, Canada
Hongzhi Yin	The University of Queensland, Australia
Man Lung Yiu	The Hong Kong Polytechnic University, SAR China
Ge Yu	Northeastern University, China
Jeffrey Xu Yu	The Chinese University of Hong Kong, SAR China
Yi Yu	National Institute of Informatics, Japan
Ye Yuan	NEU, China
Fuzheng Zhang	Microsoft
Wenjie Zhang	The University of New South Wales, Australia
Ying Zhang	University of Technology, Australia
Zhengjie Zhang	Yitu Technology
Bolong Zheng	Aalborg University, Denmark
Kai Zheng	University of Science and Technology of China, China
Yu Zheng	JD Finance
Aoying Zhou	East China Normal University, China
Xiangmin Zhou	RMIT University, Australia
Xiaofang Zhou	The University of Queensland, Australia
Yongluan Zhou	University of Copenhagen, Denmark
Hengshu Zhu	Baidu Inc.
Yuanyuan Zhu	Wuhan University, China
Andreas Zuefle	George Mason University, USA

Additional Reviewers

Al-Baghdadi, Ahmed
Alserafi, Ayman
Alves Peixoto, Douglas
Anisetti, Marco
Ardagna, Claudio
Askar, Ahmed
Banerjee, Prithu
Behrens, Hans

Bellandi, Valerio
Benkrid, Soumia
Berkani, Nabila
Bilalli, Besim
Bioglio, Livio
Cao, Xin
Casagranda, Paolo
Castelltort, Arnaud

Ceh Varela, Edgar
Ceravolo, Paolo
Chen, Chen
Chen, Jinpeng
Chen, Lu
Chen, Xiaoshuang
Chen, Xilun
Cheng, Yu

Chondrogiannis, Theodoros
Cong, Zicun
Cui, Yufei
Dellal, Ibrahim
Dian, Ouyang
Du, Boxin
Du, Dawei
Du, Xingzhong
Feng, Kaiyu
Feng, Shi
Feng, Xing
Frey, Christian
Fu, Xiaoyi
Galhotra, Sainyam
Galicia Auyon, Jorge
Gan, Junhao
Garg, Yash
Gianini, Gabriele
Gkountouna, Olga
Gong, Qixu
Gu, Yu
Guo, Long
Guo, Shangwei
Guo, Tao
Gurukar, Saket
Hao, Yifan
Hewasinghage, Moditha
Hu, Jiafeng
Hu, Xia
Huang, Jun
Huang, Shengyu
Huang, Xiangdong
Huang, Zhipeng
Imani, Maryam
Jovanovic, Petar
Kang, Jian
Kang, Rong
Kefalas, Pavlos
Khan, Hina
Khouri, Selma
Lai, Longbin
Lei, Mingtao
Li, Guorong
Li, Hangyu
Li, Huan

Li, Huayu
Li, Jingjing
Li, Liangyue
Li, Lin
Li, Mao-Lin
Li, Pengfei
Li, Xiaodong
Li, Xinsheng
Li, Xiucheng
Liang, Yuan
Liu, Qing
Liu, Sicong
Liu, Weiwei
Liu, Wu
Liu, Yiding
Luo, Siqiang
Ma, Chenhao
Ma, Yujing
Mao, Jiali
Mattheis, Sebastian
Mesmoudi, Amin
Moscato, Vincenzo
Munir, Rana Faisal
Mustafa, Ahmad
Nadal, Sergi
Nei, Wendy
Nelakurthi, Arun
Nelakurthi, Arun Reddy
Nie, Tiezheng
Pande, Shiladitya
Pang, Junbiao
Paraskevopoulos, Pavlos
Peng, Hao
Peng, Jinglin
Pham, Nguyen Tuan Anh
Pham, Tuan Anh
Piantadosi, Gabriele
Qin, Chengjie
Qin, Dong
Rai, Niranjan
Rakthanmanon, Thanawin
Ren, Weilong
Roukh, Amine
Ruan, Sijie
Sarwar, Raheem
Shan, Caihua

Shao, Yingxia
Sharma, Vishal
Song, Shaoxu
Sperlì, Giancarlo
Su, Li
Sun, Haiqi
Tang, Bo
Tao, Hemeng
Tiakas, Eleftherios
Tzouramanis, Theodoros
Vachery, Jithin
Varga, Jovan
Vassilakopoulos, Michael
Wang, Hanchen
Wang, Hongwei
Wang, Kai
Wang, Li
Wang, Qinyong
Wang, Shuhui
Wang, Sibo
Wang, Wei
Wang, Weiqing
Wang, Zhefeng
Wen, Lijie
Xiao, Chuan
Xu, Cheng
Xu, Jianqiu
Xu, Tong
Xu, Wenjian
Xu, Xing
Xue, Zhe
Yan, Jing
Yavanoğlu, Uraz
Zhang, Ce
Zhang, Fan
Zhang, Jilian
Zhang, Liming
Zhang, Pengfei
Zhang, Si
Zhao, Kaiqi
Zhao, Weijie
Zhou, Qinghai
Zhou, Yao
Zhu, Lei
Zhu, Zichen

Contents – Part II

Query Processing and Optimizations

Data Quality and Crowdsourcing

Learning Models

Multimedia Data Processing

Distributed Computing

Contents – Part I

Graph and Network Data Processing

Sequence and Temporal Data Processing

Trajectory and Streaming Data

RDF and Knowledge Graphs

Text and Data Mining

Medical Data Mining

Personalized Prescription for Comorbidity

Lu Wang[1], Wei Zhang[1], Xiaofeng He[1(✉)], and Hongyuan Zha[2]

[1] School of Computer Science and Software Engineering,
East China Normal University, Shanghai, China
`joywanglulu@163.com`, `zhangwei.thu2011@gmail.com`, `xfhe@sei.ecnu.edu.cn`
[2] Georgia Institute of Technology, Atlanta, USA
`zha@cc.gatech.edu`

Abstract. Personalized medicine (PM) aiming at tailoring medical treatment to individual patient is critical in guiding precision prescription. An important challenge for PM is comorbidity due to the complex interrelation of diseases, medications and individual characteristics of the patient. To address this, we study the problem of PM for comorbidity and propose a neural network framework Deep Personalized Prescription for Comorbidity (PPC). PPC exploits multi-source information from massive electronic medical records (EMRs), such as demographic information and laboratory indicators, to support personalized prescription. Patient-level, disease-level and drug-level representations are simultaneously learned and fused with a trilinear method to achieve personalized prescription for comorbidity. Experiments on a publicly real world EMRs dataset demonstrate PPC outperforms state-of-the-art works.

Keywords: Personalized prescription · Deep learning
Multi-source fusion · Comorbidity

1 Introduction

Restricted by the traditional care delivery models, many doctors still prescribe therapies based on their own experience and population averages, which causes inefficient care for significant portions of patients [1]. As reported from the literature, 75% patients on average take ineffective cancer drugs and 70% patients take ineffective Alzheimer's drugs [2]. Personalized medicine (PM) which tailors the medical treatment to individual patient is promising to guide precision prescription [3]. An extremely important challenge for PM is comorbidity. Comorbidity stands for two or more complex disease conditions in the same patient and has complex interrelation of diseases, medications and individual characteristics of the patient [4,5]. Some researches show comorbidity is reported in 35% to 80% of all ill people [6,7]. In the United States, about 80% of medicare costs are caused by patients with 4 or more chronic diseases [8]. Recently with the availability of massive electronic medical records (EMRs), exploring the healthcare data has great potential to support intelligent personalized prescription for comorbidity.

© Springer International Publishing AG, part of Springer Nature 2018
J. Pei et al. (Eds.): DASFAA 2018, LNCS 10828, pp. 3–19, 2018.
https://doi.org/10.1007/978-3-319-91458-9_1

Researches about prescription based on EMRs are mainly divided into pattern-based and model-based approaches. Pattern-based methods recommend prescriptions by measuring the similarities among records of patients [9,10]. These methods are challenging to learn the relation of patients' information (e.g., disease, demographic information, lab information, etc) and medications. Model-based methods include decision-theoretic methods [11] and statistical methods [12]. But these methods only focus on one specific disease. Recently, two deep models are proposed to learn a nonlinear mapping from multiple diseases to multiple drugs based on EMRs [13,14], and achieve significant improvements. Without considering patient-specific information, these deep methods recommend constant-treatment for patients with same diseases. However, it is not in line with real situations. As shown in Table 1, the two patients are with the same diseases. Due to the different physiologic states, they take different treatments.

Table 1. The difference and intersection treatments of two patients with same diseases.

Diagnosis	Intersection treatments	Difference treatments
Pure hypercholesterolemia, Intermediate coronary syndrome, Hypertension NOS, Coronary atherosclerosis of native coronary artery	Meperidine, Neostigmine, Phenylephrine HCl, Ranitidine, Oxycodone-Acetaminophen, Metoclopramide, Calcium Gluconate, Glycopyrrolate, Magnesium Sulfate, Milk of Magnesia, Nitroglycerin, Aspirin EC, Acetaminophen, Sucralfate, Bisacodyl, Docusate Sodium, Potassium Chloride, Furosemide, Morphine Sulfate, Aspirin, Metoprolol	Propofol, Vancomycin HCl, Ibuprofen, Midazolam HCl, Chlorpheniramine Maleate, Hydrochlorothiazide, Hespan, Nitroprusside Sodium, Ondansetron, Diphenhydramine HCl
		CefazoLIN, Insulin Human Regular, Propofol, Docusate Sodium, Dextrose 50%, Insulin, Simvastatin, Sodium Chloride 0.9% Flush

There are two important issues remained in the aforementioned methods. (1) Non-personalized medicine. Existing methods for comorbidity ignore massive individual characteristics of the patient, such as demographic and laboratory information, which fail to recommend patient-specific prescription. (2) Lack of medical knowledge. Medical knowledge can guide us to learn a more effective and interpretable model. Furthermore, learning different "weights" of multiple diseases for comorbidity patients is also a difficult issue [15].

To tackle these issues, we integrate multi-source patient-specific information to learn patient-level representation. The representations and severities of multiple diseases are learned by employing medical knowledge and attention mechanism. The main contributions of this paper can be summarized as follows:

- To obtain the interdependencies among diseases, medications and individual characteristics of the patient, we design a deep learning model to integrate multi-source information to learn the patient-level, disease-level and drug-level representations simultaneously, and fuse them with a trilinear method. (for comorbidity challenge)

- Patient-level representation is learned based on multiple patient-specific information, such as demographic and laboratory information (for issue 1). Disease-level representation is obtained by medical ontologies, where an attention mechanism is used to learn the different severities of multiple diseases (for issue 2).
- We evaluate our method over a real world EMRs MIMIC-3 and show that it outperforms state-of-the-art approaches for prescription.

The rest of this paper is organized as follows. We summarize the related work in Sect. 2. The proposed method is presented in Sect. 3. Experimental results and analysis are introduced in Sect. 4. We conclude our work in Sect. 5.

2   Related Work

Computational methods that leverage EMRs to support healthcare begin to draw attention in recent years. To learn good representations of diagnosis and prescription, several models from the fields, such as image processing and machine translation, are also leveraged to represent medical ontology.

Diagnosis is first handled by neural networks in 1989 [16]. Recently, deep models such as multi-layer perceptron (MLP) and recurrent neural networks (RNN) are applied to diagnose life-threatening diseases. Lipton et al. are the first to apply long short-term memory (LSTM) [17] to multi-label diagnoses, which takes the clinical variables as input to predict the diagnosis in intensive care unit setting [18]. A gated recurrent unit (GRU) [19] model is used to early detect heart failure with the row value of patients' records [20]. However, for the distinct tasks and different input, these methods can not be directly applied to prescription.

Prescription settled by pattern-based methods is to identify the treatments based on the similarities among records of patients [9,10,21]. As for model-based studies, Cheerla and Gevaert [12] use SVM to recommend proper treatments for pan-cancer patients with microRNA. Concurrently, Bajor and Lasko use a GRU model to predict the total medications for multiple diagnosis records of a patient to check the EMRs records [13]. However, the disease representations learned by Bajor et al. are not well aligned to the medical knowledge [13]. Zhang et al. also design a deep learning model LEAP to predict safe prescription with the input of multiple diseases [14]. Bajor's method and LEAP are established as state-of-the-art approaches, but they ignore the patient-specific information. These approaches are not effective for personalized prescription in comorbidity for: (1) due to the complex and abstruse correlation among multiple diseases, it is hard to measure their similarities; (2) ignoring the individual information of patients, the methods may recommend the same medications for patients with the same disease. As shown in Fig. 1, it is not in line with the real situation.

Neural Attention Model is designed for solving neural machine translation tasks which cause a bottleneck by using a fixed-length vector to represent a sentence [22]. To predict a target word, attention model automatically focuses on the related words in the source sentence. Recently, it is applied to image processing [23], dialog systems [24], machine translation [22] and popularity prediction [25]. Retain [26] is the pioneer work to apply attention mechanism to healthcare, which considers the historical visit records of patients in a reverse time to learn attentions of different visits.

Distributed Representation for language is proposed to predict the neighbors of a word using a simple neural network such as Skip-gram and Continuous Bag-of-Words (CBOW) [27]. In medical domain, Riccardo et al. propose an unsupervised method to learn the patients representations using a three-layer stack of denoising autoencoders [28]. To improve the interpretation of representations, GRAM employs an attention mechanism based on the hierarchical medical ontology to learn the representation of diseases and drugs [29]. However, GRAM overlooks the severity of diseases when the patients suffer from multiple diseases. Indeed, these works mainly focus on learning representation instead of prescription.

This paper extends prescription methods in a number of important dimensions, including: (1) a deep learning model to learn the patient-level, disease-level, drug-level representations simultaneously from multi-source information of EMRs to achieve patient-specific prescription for comorbidity, and (2) an effective representation of comorbidity learned by hierarchical disease ontologies and a neural attention model.

3   Personalized Prescription for Comorbidity

In this section, we first define the notations of medical ontology and EMRs data, followed by an overview of our approach. Then we introduce the detailed components of learning disease, patient and drug representations, and a fusion method to integrate these representations for personalized prescription.

3.1   Preliminaries

Considering a set of N patients $\mathbb{P} = \{p_1, p_2, ..., p_n, ..., p_N\}$, a patient p_n is specified by his or her patient-specific information P_n (demographic and laboratory information), diagnosis information D_n and medication information Y_n, where $P_n = \{p_n^{age}, p_n^{heartrate}, ...\}$, $D_n = \{d_1^n, d_2^n, ..., d_i^n, ..., d_I^n\}$, $Y_n = \{y_1^n, y_2^n, ..., y_k^n, ..., y_K^n\}$. d_i^n denotes the i-th disease in D_n and $y_k^n \in \{0, 1\}$ denotes whether a medication in the k-th medicine class treated for the patient p_n. G is a directed acyclic graph (DAG) of disease (coded in ICD-9) ontology[1]. We only focus on three main levels of ICD-9 ontology (1-digit nodes, 3-digit nodes and leaf-nodes) in this paper to ensure good generalization. Also, the three levels are often used to identify

---

[1] http://bioportal.bioontology.org/ontologies/ICD9CM.

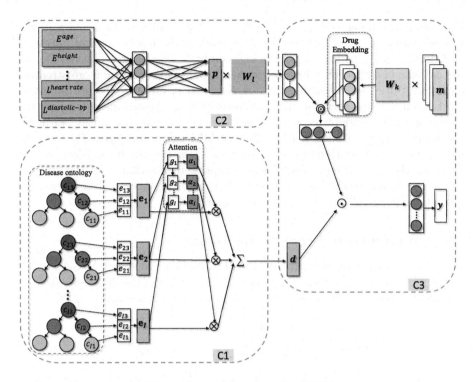

Fig. 1. General framework of PPC.

pharmacological subgroups. There is a hyponymy relation between the high level and low level nodes in G, where the leaf-node c_{i1} (in level-1) represents the i-th disease d_i^n, and the non-leaf nodes c_{i2} (in level-2), c_{i3} (in level-3) show a concept generalized from their child-nodes. Inspired by [29], each node in the three levels is associated with a basic embedding, where e_{ij} represents the basic embedding of node c_{ij} in j-th level.

PROBLEM DEFINITION (Personalized Prescription for Comorbidity.)

For a patient p_n, given his or her patient-specific information P_n and diseases D_n, where $P_n = \{p_n^{age}, p_n^{heartrate}, ...\}, D_n = \{d_1^n, d_2^n, ..., d_I^n\}$, the problem is to predict the personalized treatment Y_n ($Y_n = \{y_1^n, y_2^n, ..., y_K^n\}$) for the patient.

3.2 Algorithm Overview

As shown in Fig. 1, our approach is a deep learning model which includes three main components: **C1:** learning to represent the diagnosis, **C2:** learning to represent the patient, **C3:** fusing representations with a trilinear method.

PPC employs the hierarchical structure of disease ontology in knowledge graph G to learn a interpretable representation of disease d_i^n. It first finds a path from the leaf-node c_{i1} to the highest level node c_{i3} in G. Then, PPC concatenates the basic embedding vectors $e_{ij} \in \mathbb{R}^m$ $(j = 1, 2, 3)$ of the three nodes as the representation $\hat{e}_i \in \mathbb{R}^{3m}$ of disease d_i^n. Combing the information of ancestors and children helps to learn a robust and comprehensive representation. Due to the patients in comorbidity have more than two diseases, playing more attention on severity diseases is beneficial to alleviate the symptoms. We use attention mechanism [22] to learn the different severities of diseases and represent the diagnosis of the patient as a single embedding d^n. Simultaneously, we learn the patient representation using a 2-layer MLP with the input P_n, and the medication representation m_k $(k = 1, 2, ..., K)$ is learned by a 1-layer MLP. To learn the interdependencies among diseases, medications and the patient, a trilinear fusion method is adapted to integrate the three representations to predict the personalized treatments for comorbidity.

3.3 C1: Learning to Represent the Diagnosis

Diagnostic information in EMRs consists of the patients' diseases. Medical ontology in this paper is used to facilitate the representation of the diagnosis. We first concatenate the three basic embeddings into a single embedding $\hat{e}_i \in \mathbb{R}^{3m}$:

$$\hat{e}_i = [e_{i1}, e_{i2}, e_{i3}], \tag{1}$$

$$e_{i1} = W_{emb1} c_{i1}, \quad e_{i2} = W_{emb2} c_{i2}, \quad e_{i3} = W_{emb3} c_{i3},$$

where $\hat{e}_i$ is the embedding of disease d_i^n, $c_{ij} \in \mathbb{R}^D$ is the one-hot representation of node c_{ij} $(j = 1, 2, 3)$, W_{emb1}, W_{emb2} and $W_{emb3} \in \mathbb{R}^{m \times D}$ are the embedding matrixes corresponding to c_{i1}, c_{i2} and c_{i3} respectively.

Then, we use the convex combination of multiple diseases to represent the diagnosis of the patient:

$$d^n = \sum_{i=1}^{I} \alpha_i \hat{e}_i, \quad \sum_{i=1}^{I} \alpha_i = 1, \alpha_i \geq 0 \qquad \text{for } d_i^n \in D_n, \tag{2}$$

where I is the number of diseases in D_n. α_i is the attention weight of the disease d_i^n, which also indicates the severity of d_i^n for the patient. The scalar α_i is generated as follows,

$$\alpha_i = \frac{\exp(f(\hat{e}_i))}{\sum_{j=1}^{I} \exp(f(\hat{e}_j))}. \tag{3}$$

Using a 1-layer GRU and a 1-layer MLP, we obtain $f(\hat{e}_i)$ as follows,

$$(g_1, ..., g_i, .., g_I) = \text{GRU}(\hat{e}_1, ..., \hat{e}_i, ..., \hat{e}_I), \tag{4}$$

$$h_i = w_k^\mathrm{T} g_i + b_k, \tag{5}$$

$$f(\hat{\mathbf{e}}_i) = \tanh(\mathbf{w}_i[h_1, \ h_2, ..., \ h_I]^\mathrm{T} + b_i), \qquad for \ \ i = 1, 2, ..., I, \qquad (6)$$

where $\mathbf{g}_i \in \mathbb{R}^p$ is the hidden layer of GRU[2]. h_i is the hidden layer of MLP and $\mathbf{w}_k \in \mathbb{R}^p$, b_k, $\mathbf{w}_i \in \mathbb{R}^I$, b_i are parameters to learn. The GRU layer learns the attentions of diseases separately, while the MLP learns the attentions of diseases jointly.

The final representation of diseases $\mathbf{d}_n \in \mathbb{R}^m$ can also be calculated by $C_n \in \mathbb{R}^{3D \times I}$ as shown in Eq. (8), where $\boldsymbol{\alpha} \in \mathbb{R}^I$ is the attention vector. As shown in Eq. (9), $\mathbf{W}_{emb} \in \mathbb{R}^{m \times 3D}$ is the concatenation embedding matrix of disease ontologies in the three levels. Overall, we represent the diagnosis of patients by employing the hierarchical structure of disease ontologies in knowledge graph G and learning different severities of multiple diseases.

$$\mathbf{d}_n = \mathbf{W}_{emb}(C_n \boldsymbol{\alpha}) \qquad (7)$$

$$C_n = [\hat{\mathbf{c}}_1, \hat{\mathbf{c}}_2, ..., \hat{\mathbf{c}}_I], \qquad \text{where } \hat{\mathbf{c}}_i = [c_{i1}, c_{i2}, c_{i3}], \quad i = 1, 2, ..., I \qquad (8)$$

$$\mathbf{W}_{emb} = [\mathbf{W}_{emb1}, \mathbf{W}_{emb2}, \mathbf{W}_{emb3}]. \qquad (9)$$

3.4   C2: Learning to Represent the Patient

Demographic and laboratory information belongs to patient-specific indicators. Demographic information consists of age, gender, height, weight, language, ethnicity, etc. Laboratory indicators include blood pressure, temperature, blood oxygen saturation, etc. The patient-specific information is important to the design of therapeutic regimen and dosage.

The demographic information is denoted as E:

$$E = \{E^{age}, E^{height}, ..., E^{weight}\},$$

and the laboratory indicators are denoted as L:

$$L = \{L^{blood-pressure}, L^{temperature}, ..., L^{ph}\}.$$

Each element in E and L indicates a variable of P_n. Let $\hat{\boldsymbol{p}}_n$ be the intermediate representation of patients where the discrete variables are represented as one-hot codes, and the continuous variables keep invariant.

We use a 2-layer MLP to learn the patient representation:

$$\mathbf{h}_z = f(\mathbf{W}_z \hat{\boldsymbol{p}}_n + \mathbf{b}_z), \qquad (10)$$

$$\mathbf{p}_n = f(\mathbf{W}_u \mathbf{h}_z + \mathbf{b}_u), \qquad (11)$$

where $\mathbf{W}_z$ and $\mathbf{b}_z$ are the parameters of first layer, $\mathbf{W}_u$ and $\mathbf{b}_u$ are parameters of second layer, f is the activation function ReLUs, and $\mathbf{p}_n \in \mathbb{R}^n$ is the final representation of the patient.

---

[2] We have also examined LSTM and other activation functions to learn to represent diagnosis, but they have less efficiency and worse performance.

3.5   C3: Fusing Representations with Trilinear Method

We propose a trilinear fusion method to integrate different sources of informa-tion. The input of the trilinear fusion method consists of three types of variables: diagnosis C_n, patient-specific information $\mathbf{p}_n$ and candidate medications $\mathbf{m}_k$ $(k = 1, 2, ..., K)$, where $\mathbf{m}_k \in \mathbb{R}^K$ is the one-hot representation of the medicine. C_n is the concatenation of one-hot representations of diseases as shown in Eq. (8). The trilinear fusion method characterizes such a specific treatment event by con-sidering the interdependencies among medications, the patient and diagnosis. Assume $h_{k,n}$ is the index of the probability of the medication m_k recommended for the patient p_n, and the probability is shown in Eq. (13). The trilinear method is described as follows,

$$h_{k,n} = (\mathbf{W}_{emb}(C_n\boldsymbol{\alpha}))^{\mathrm{T}}(\mathbf{W}_m\mathbf{m}_k \odot \mathbf{W}_l\mathbf{p}_n), \tag{12}$$

where $\odot$ denotes the element-wise multiplication and $\boldsymbol{\alpha}$, $\mathbf{W}_{emb}$, $\mathbf{W}_m \in \mathbb{R}^{m \times K}$, $\mathbf{W}_l \in \mathbb{R}^{m \times n}$ are parameters to learn. To predict whether to recommend drug m_k for patient p_n, we use a sigmoid function to predict the probability of rec-ommending m_k as follows:

$$f_{k,n} = \frac{1}{1 + e^{-h_{k,n}}}. \tag{13}$$

3.6   Objective Optimization

To solve this multi-label problem, we optimize the loss function of the K labels simultaneously:

$$Loss = \frac{1}{N}\frac{1}{K}\sum_{n=1}^{N}\sum_{k=1}^{K} l(f_k(\mathbf{C}_n, \mathbf{m}_k, \mathbf{p}_n), y_{k,n}), \tag{14}$$

$$l(f_{k,n}(\mathbf{C}_n, \mathbf{m}_k, \mathbf{p}_n), y_{k,n}) = -(1 - y_{k,n}) * log(1 - f_{k,n}) - y_{k,n} * log(f_{k,n}), \tag{15}$$

where $l(f_{k,n}(\mathbf{C}_n, \mathbf{m}_k, \mathbf{p}_n), y_{k,n})$ is the cross-entropy loss, N is the number of patients in training set. If we believe the solutions with small parameters are more general, we may optionally add a l1-penalty term, which will often make the parameters be nonzero in only a few states to prevent overfitting[3].

4   Experiment

In this section, we conduct experiments to evaluate our proposed method. We first report the dataset and models for comparison, followed by quantitative and qualitative measurements. Quantitative measurements include the common multi-label metrics and mean Jaccard. Qualitative measurements focus on how well the presented method solves the issues mentioned in Sect. 1, such as person-alized prescription analysis, the interpretable representation of diseases analysis and the effect of the diseases' severities learned by attention mechanism.

---

[3] We have examined both l1-norm and l2-norm, and find their performance are similar.

4.1   Dataset Description

The experiments are conducted on a public EMRs dataset MIMIC-3 [30]. MIMIC-3 contains 43K patients in critical care units during 2001 and 2012. There are 6,695 distinct diseases and 4,127 drugs in MIMIC-3. The median number of diseases of each record is 9 (Q1–Q3:6–15). Following the procedure adopted in [13], we extract the top 1,000 most medications and top 2,000 most diseases (ICD-9 codes) in the first 24 h after the admission of patients. Because the patient states always change after 24 h and the first 24 h are the most critical time of the patient. These medications and diseases cover 85.4% of all medication records and 95.3% of all disease records. The medications in patient's diagnosis records are coded in NDC[4]. To obtain the hierarchical information of medications, we map the medication code from NDC into the third level of ATC[5] using the public tool[6]. ATC is another medication code which is hierarchically structured by anatomic and therapeutic classes. Finally, we obtain 180 ATC codes, which is also the number of labels in our multi-label classification task.

For learning the patient representation, we choose 8 demographic features: gender, age, weight, height, religion, language, marital status and ethnicity and 11 clinical variables (followed by the physician's suggestion): diastolic blood pressure, Glasgow coma scale, blood glucose, systolic blood pressure fraction of inspired O2, heart rate, pH, respiratory rate, blood oxygen saturation, body temperature, and urine output. These variables are first rescaled to z-scores, then rescaled to [0,1]. We extract the results of clinical variables in the first 24 h after the patients admitted to the intensive care unit. We further fill the missing values by sampling them from the clinically normal interval as defined by clinical physicians. It is reasonable because clinicians often think the variables are norm and do not measure them [18]. For good generalization, we remove the records with more than 10 missing variables. Finally, we obtain 39,260 patients, and randomly divide the dataset for training, validation and testing by the ratio of 80/10/10.

We use the common metrics of multi-label, which contains micro under the ROC curve (micro-AUC), macro under the ROC curve (macro-AUC), label ranking average precision score and label ranking loss to promise fair and honest evaluation [31,32]. Also, we use mean Jaccard to measure the combination of recommended drugs as [14]. Initial PPC and PPC are our proposed methods, while the others are baselines. We describe these methods in detail as follows:

- **Popularity-20 (POP-20):** This is a patten-based method, which considers the top-k most frequent medications prescribed for each disease as predictions. We set K to be 20 for its best performance on validation dataset.
- **Random Forest (RF):** This is a classical machine learning method for multi-label problem. To reduce the massive computation, we use scalar to represent the different diseases, and train the model with 180 independent

---

[4] http://www.fda.gov/Drugs/DevelopmentApprovalProcess/.
[5] http://www.whocc.no/atc/structure and principles/.
[6] https://www.nlm.nih.gov/research/umls/rxnorm/.

forests, each forest is trained with the total diseases and predict one of the 180 treatments.

- **LM** [13]: This is a non-personalized prescription method, which uses a 3-level-MLP to recommend treatments for patients. The goal of LM is to check the errors and omissions in EMRs. The input is historical diseases of the patient in the EMRs records. The output is a single vector which is used to predict the medications treated for the historical diseases of the patient. To test the performance of LM, we use the current diseases of a patient as input and predict the medications for current diseases.
- **LG** [13]: This model is with the same setting as LM. But it uses a GRU model instead of MLP.
- **LEAP** [14]: LEAP uses a MLP framework to train a multi-label model which uses multiple diseases to predict multiple medications and considers the dependence of medications.
- **Knowledge-based LM (LMK):** We extend LM by incorporating hierarchical structure of disease ontology. The results of LMK can be utilized to test the effectiveness of considering medical ontology.
- **Personalized-infor-based LM (LMKF):** We further extend LMK by concatenating demographic information of patients, clinical measurements and diseases together as input. The results of LMKF can be used to verify the benefit of considering the patient-specific information.
- **initial-PPC (i-PPC):** It is with the same setting as our model, except using GRAM [29] to learn the representation of diseases.
- **PPC:** This is the model proposed in this paper. We aim at comparing it with other methods to demonstrate its advantages in multi-aspects. The basic embeddings $e_{i,j}$ of i-PPC and PPC are both randomly initialized.

The main goals of this section is to answer the following core questions, which guide the design of the experiments.

1. **Prediction Accuracy:** Can patient-specific information support more accurate prescription than other non-personalized prescription? (for issue 1)
2. **Ablation Study:** What is the contribution of each factor (diagnosis, patient, medicine information) to PPC?
3. **Embedding Analysis:** Does medical knowledge help to learn a better representation? (for issue 2)
4. **Attention Analysis:** How well does PPC learn the different severities among diseases?

4.2 Prediction Accuracy

Table 2 shows the performance of aforementioned methods on MIMIC-3. LMK outperforms LM by 2%–5.8%. This result shows combing medical knowledge to learn representations of diseases is significant to improve the accuracy of prescription. LMKF outperforms LMK by 0.1%–2.4%, which verifies the precision treatment is benefit from patient-specific information. Moreover, PPC consistently outperforms other baselines. For non-personalized deep models, such as

LEAP, LG and LM, PPC achieves 0.7%–4.1% improvement, because patient-specific information can help to prescribe more effective medications by identifying different physiologic states and characteristics of patients. It also outperforms LMKF by 0.2%–4.2% because the trilinear fusion method endows PPC with ability of learning rich and integrated representations based on different sources of information. Compared to i-PPC, PPC also achieves better improvements, because learning the different severities can help PPC pay more attention to important diseases.

Table 2. Performance comparisons on test sets for comorbidity prescription (%).

Method	Micro-AUC	Macro-AUC	Label ranking avg. precision	Label ranking loss	Jaccard
POP-20	76.2	55.8	52.7	40.8	37.8
RF	88.3	71.8	60.2	9.8	38.5
LM	89.2	73.2	62.8	9.4	36.6
LG	91.7	77.3	67.0	7.8	39.3
LEAP	92.0	78.9	67.5	7.6	40.8
LMK	92.1	79.0	68.2	7.4	40.1
LMKF	92.2	81.4	68.3	7.1	40.5
i-PPC	92.7	81.0	68.6	7.0	41.3
PPC	**93.1**	**83.0**	**69.9**	**6.90**	**44.7**

As for the other baselines, POP-20 is not effective due to its incapability of learning relation between multiple diseases and medications. RF works poor than deep models, because it fails to learn high-level representations of diseases.

4.3 Ablation Study

We conduct ablation study here to verify the contributions of the three types of information employed in this study. More specifically, we denote PPC-m, PPC-d and PPC-p as the variants of PPC by removing medical information, diagnosis information, and patient-specific information respectively. As the results presented in Table 3, all the information makes a positive contribution to precision treatment, where the contribution of diagnosis information is the most significant.

4.4 Embedding Analysis

To evaluate the effectiveness of disease representations learned by PPC, we use t-SNE [33] to visualize the final embeddings of 2000 diseases in our experiments. As shown in Fig. 2, different colors correspond to different categories of diseases

Table 3. Factor contribution analysis for PPC (%).

Method	Micro-AUC	Macro-AUC	Label ranking avg. precision	Label ranking loss	Jaccard
PPC-m	92.9	82.7	68.5	7.3	44.1
PPC-d	89.5	70.8	61.1	9.8	37.7
PPC-p	92.3	81.3	68.6	7.1	43.2
PPC	**93.1**	**83.0**	**69.9**	**6.90**	**44.7**

in the highest level of G. The names of the categories are represented aside the color-bar. The result shows that the embeddings of diseases in different categories can be roughly separated. In addition, we randomly select two impact point sets in Fig. 2, where the blue digits indicate the leaf-ICD-9 codes. The result shows that the codes are indeed related to their neighbors. However, the most related codes are not with the shortest distance because of the insufficient data. In deed, training the embeddings always need sufficient data, for example, training Skip-gram requires large amount of documents.

4.5   Attention Analysis

The attentions of the diseases can be explained intuitively using a randomly chosen case. Case 1: a patient with 13 diseases and 39 drugs. As mentioned in Sect. 1, learning different "weights" of multiple diseases is still a significant problem to be well addressed. In this section, we validate the availability of diseases' attentions using the domain knowledge and the amount of medications.

Analysis Based on the Domain Knowledge: As verified by a doctor, this is a patient with two main diseases: Parkinson and Chronic airway obstruction. More specifically, the patient is with diseases and symptoms such as: Parkinson, Chronic airway obstruction, depression, constipation, eye infections, esophagitis, indigestion, pneumonia, respiratory failure and congestive heart failures. About 1/3 of Parkinson patients suffer from severe depression and may cause constipation, abnormal gastrointestinal motility, and some eye diseases. Therefore, part of these symptoms and diseases may be caused by Parkinson's disease. In addition, Parkinson's patients are difficult to clean up the sputum, who easily infect pneumonia. Chronic airway obstruction which is unrelated to Parkinson's disease, may cause pulmonary heart disease and lung inflammation. Thus in this case, the patient also suffer from pneumonia, respiratory failure and congestive heart failure. Overall, Parkinson's disease and Chronic airway obstruction are the main diseases in this case and most of the other diseases are complications. As shown in Fig. 3, Parkinson's disease achieves the most attention ($\alpha = 0.12$), while Chronic airway obstruction obtains the third ($\alpha = 0.094$).

Analysis Based on the Amount of Medications: As shown in Fig. 4, to validate our results, we choose level-1 ATC codes to represent the medications. The

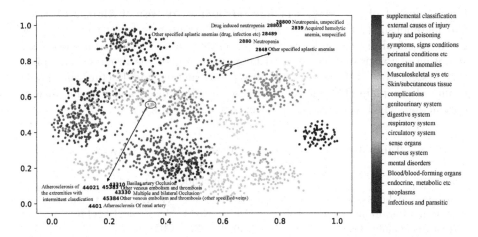

Fig. 2. The visualization (t-SNE, 2-D) of diseases' embeddings learned by PPC. Different colors correspond to different categories of disease in the highest level of G. The name of categories are represented aside the color-bar. The blue digits indicate the leaf-ICD-9 codes (diseases) of two randomly point sets. (Color figure online)

Fig. 3. Attentions learned from a comorbidity patient. Each rectangle represents a disease of this patient. The different color shades shows the volume of the attention of the disease. (DHF: Diastolic heart failure Acute on chronic, UPE: Unspecified pleural effusion, CHF: Congestive heart failure, PD: Parkinson's disease, HYPS: Hyposmolality and/or hyponatremia, HYPO: Hyperpotassemia, LLR: Lymphoid leukemia in remission, AU: Anemia, unspecified, POU: Pneumonia, organism unspecified, EH: Essential hypertension, UAFA: Upper arm and forearm Other cellulitis and abscess, CAO: Chronic airway obstruction, ARF: Acute respiratory failure.)

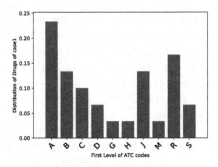

Fig. 4. Distribution of the number of drugs in this case. Abscissa represents the highest level of drug codes (ATC) of this case. The descriptions of partial codes are: A: Alimentary tract and metabolism, R: Respiratory system, J: Antiinfectives for systemic use, B: Blood and blood forming organs. C: Cardiovascular system.

largest amount of the drugs is mainly targeted for disease of Alimentary tract and metabolism (A). As mentioned before, Parkinson is most likely to cause gastrointestinal disease and constipation. So that these drugs are prescribed for symptom of Parkinson. The second largest amount of the drug is for respiratory system (R), such as Chronic airway obstruction and Pneumonia. These results also explain that the main diseases are Parkinson and Chronic airway obstruction, which is in line with our experiment results.

Table 4. Prescriptions for two patients with same diseases.

Diagnosis	Methods	Recommended treatments
Secondary malignant neoplasm of brain and spine, breast malignancy, Other convulsions, Secondary malignant neoplasm of lung, Hypertension Cerebral edema	PPC_{p1}	B05C, B05X, A10A, C08C, A02B, N03A, N02A, N02B, C02D, A12C, A06A, C03A, C03C
	$LEAP_{p1}$	B05C, B05X, A02B, N03A, N02A, N02B, A12C, A06A
	LG_{p1}	B05C, B05X, A10A, A02B, N02A, N02B, A12C, A06A
	PPC_{p2}	B05C, B05X, A10A, C08C, A02B, N03A, A04A, N02A, N02B, C02D, A12C, A06A
	$LEAP_{p2}$	B05C, B05X, A02B, N03A, N02A, N02B, A12C, A06A
	LG_{p2}	B05C, B05X, A10A, A02B, N02A, N02B, A12C, A06A

4.6   Personalized Prescription Analysis

With the subjectively examining performance on 30 randomly selected cases, we find the favorably performs of PPC comparing against other baselines. We choose one of these cases for analysis. In Fig. 5, we show 2 patients with same diseases, where the diseases and mediations recommend by 3 prescription methods are shown in Table 4. For the first patient, PPC_{p_1} recommends a set of medications with 78.6% coverage, where p_i $(i = 1, 2)$ represents the i-th patient. The recommendation coverage of $LEAP_{p_1}$ and LG_{p_1} are both 42.9%. For the second patient, PPC_{p_2} recommends a set of medications with 100% coverage. In contrast, the coverage of $LEAP_{p_2}$ and LG_{p_2} are 88.9% and 77.8%. The case is also the evidence of patient-specific medications. Due to the different physiologic states of patients, the mediations which the patients need are changed. In this case, the first patient is with systolic blood pressure 142 mmHg, while the second patient is 117 mmHg. Considering the patient-specific information, PPC recommends the drugs for p1 with C08C, C02D, C03A, C03C, which targets hypertension. For p2, these drugs were largely reduced. However, as shown

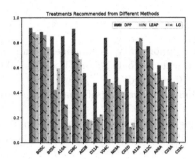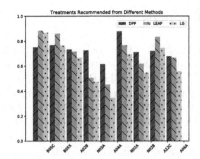

Fig. 5. Medication predictions confidence of two patients with same diseases. ATC codes on abscissa axis represents the prescriptions of doctors, where the hight of the bar indicates the prediction confidence of the three methods. We predict the medicine with the confidence $>=0.5$.

in Table 4, LEAP and LG that only consider diseases for prescription always recommend the same drugs for the patients with same diseases and ignore the hypertension states of the patients.

5   Conclusion

In order to solve the challenge and issues of personalized prescription for comorbidity, we propose an end-to-end deep learning model PPC. PPC integrates different sources of information to jointly learn representations of patients, diseases and medications and fuses them with a trilinear method to realize personalized prescription. Multiple patient-specific information is exploited to learn patient-level representation, and medical knowledge is combined to learn disease-level representation where an attention mechanism is used to learn different severities of comorbidity. Exploiting multi-source patient-specific information, PPC can recommend customized treatments which may be different for patients even having same diseases but different physiologic states, which achieves better results. Furthermore, PPC learns a good representation of disease and discriminates different severities of multiple diseases of comorbidity patients well. In the future, we will study how to solve the scalability issue for fuller set of medications.

Acknowledgements. This work was partially supported by the National Key Research and Development Program of China under Grant No. 2016YFB1000904, NSFC (61702190), NSFC-Zhejiang (U1609220), Shanghai Sailing Program (17YF1404500) and SHMEC (16CG24).

References

1. PMC2017: The personalized medicine opportunity, challenges, and the future (2017). http://www.personalizedmedicinecoalition.org/Userfiles/PMC-Corporate/file/The-Personalized-Medicine-Report1.pdf
2. Spear, B., Heath-Chiozzi, M., Huff, J.: Clinical application of pharmacogenetics. Trends Mol. Med. **7**(5), 201–204 (2001)
3. Berwick, D.M., Finkelstein, J.A.: Preparing medical students for the continual improvement of health and health care: Abraham flexner and the new public interest. Acad. Med. S56–S65 (2010)
4. Munoz, E., Rosner, F., Friedman, R., Sterman, H., Goldstein, J., Wise, L.: Financial risk, hospital cost, and complications and comorbidities in medical non-complications and comorbidity-stratified diagnosis-related groups. Am. J. Med. **84**(5), 933–939 (1988)
5. Jakovljević, M., Reiner, Ž., Miličić, D., Crnčević, Ž.: Comorbidity, multimorbidity and personalized psychosomatic medicine: epigenetics rolling on the horizon. Psychiatr. Danub. **22**(2), 184–189 (2010)
6. Taylor, A.W., Price, K., Gill, T.K., Adams, R., Pilkington, R., Carrangis, N., Shi, Z., Wilson, D.: Multimorbidity-not just an older person's issue. Results from an Australian biomedical study. BMC Public Health **10**(1), 718 (2010)
7. Bonavita, V., De Simone, R.: Towards a definition of comorbidity in the light of clinical complexity. Neurol. Sci. **29**(1), 99–102 (2008)
8. Valderas, J.M., Starfield, B., Sibbald, B., Salisbury, C., Roland, M.: Defining comorbidity: implications for understanding health and health services. Ann. Family Med. **7**(4), 357–363 (2009)
9. Sun, L., Liu, C., Guo, C., Xiong, H., Xie, Y.: Data-driven automatic treatment regimen development and recommendation. In: KDD, pp. 1865–1874 (2016)
10. Hu, J., Perer, A., Wang, F.: Data driven analytics for personalized healthcare. In: Weaver, C.A., Ball, M.J., Kim, G.R., Kiel, J.M. (eds.) Healthcare Information Management Systems. HI, pp. 529–554. Springer, Cham (2016). https://doi.org/10.1007/978-3-319-20765-0_31
11. Lusted, L.B.: Introduction to medical decision making. Am. J. Phys. Med. Rehabil. **49**(5), 322 (1970)
12. Cheerla, N., Gevaert, O.: Microrna based pan-cancer diagnosis and treatment recommendation. BMC Bioinform. **18**(1), 32 (2017)
13. Bajor, J.M., Lasko, T.A.: Predicting medications from diagnostic codes with recurrent neural networks. In: ICLR (2017)
14. Zhang, Y., Chen, R., Tang, J., Stewart, W.F., Sun, J.: Leap: learning to prescribe effective and safe treatment combinations for multimorbidity. In: KDD, pp. 1315–1324 (2017)
15. Jakovljevi, M., Ostoji, L.: Comorbidity and multimorbidity in medicine today: challenges and opportunities for bringing separated branches of medicine closer to each other. Psychiatr Danub **25**, 18–28 (2013)
16. Hart, A., Wyatt, J.: Connectionist models in medicine: an investigation of their potential. In: Hunter, J., Cookson, J., Wyatt, J. (eds.) AIME 89, pp. 115–124. Springer, Heidelberg (1989). https://doi.org/10.1007/978-3-642-93437-7_15
17. Hochreiter, S., Schmidhuber, J.: Long short-term memory. Neural Comput. **9**(8), 1735–1780 (1997)
18. Lipton, Z.C., Kale, D.C., Elkan, C., Wetzell, R.: Learning to diagnose with LSTM recurrent neural networks. In: ICLR (2016)

19. Chung, J., Gulcehre, C., Cho, K.H., Bengio, Y.: Empirical evaluation of gated recurrent neural networks on sequence modeling. arXiv preprint arXiv:1412.3555 (2014)
20. Choi, E., Bahadori, M.T., Schuetz, A., Stewart, W.F., Sun, J.: Doctor AI: predicting clinical events via recurrent neural networks. In: Machine Learning for Healthcare Conference, pp. 301–318 (2016)
21. Zhang, P., Wang, F., Hu, J., Sorrentino, R.: Towards personalized medicine: leveraging patient similarity and drug similarity analytics. In: AMIA Joint Summits on Translational Science Proceedings, p. 132 (2014)
22. Bahdanau, D., Cho, K., Bengio, Y.: Neural machine translation by jointly learning to align and translate. arXiv preprint arXiv:1409.0473 (2014)
23. Xu, K., Ba, J., Kiros, R., Cho, K., Courville, A., Salakhudinov, R., Zemel, R., Bengio, Y.: Show, attend and tell: neural image caption generation with visual attention. In: ICML, pp. 2048–2057 (2015)
24. Yang, L., Ai, Q., Guo, J., Croft, W.B.: aNMM: ranking short answer texts with attention-based neural matching model. In: CIKM, pp. 287–296 (2016)
25. Zhang, W., Wang, W., Wang, J., Zha, H.: User-guided hierarchical attention network for multi-modal social image popularity prediction. In: WWW (2018)
26. Choi, E., Bahadori, M.T., Sun, J., Kulas, J., Schuetz, A., Stewart, W.: Retain: an interpretable predictive model for healthcare using reverse time attention mechanism. In: NIPS, pp. 3504–3512 (2016)
27. Rumelhart, D.E., Hinton, G.E., McClelland, J.L., et al.: A general framework for parallel distributed processing. Parallel Distrib. Process.: Explor. Microstruct. Cogn. 1, 45–76 (1986)
28. Riccardo, M., Li, L., Kidd, B.A., Dudley, J.T.: Deep patient: an unsupervised representation to predict the future of patients from the electronic health records. Sci. Rep. 6, 26094 (2016)
29. Choi, E., Bahadori, M.T., Song, L., Stewart, W.F., Sun, J.: GRAM: graph-based attention model for healthcare representation learning. In: KDD, pp. 787–795 (2017)
30. Johnson, A.E.W., Pollard, T.J., Shen, L., Lehman, L.H., Feng, M., Ghassemi, M., Moody, B., Szolovits, P., Celi, L.A., Mark, R.G.: MIMIC-III, a freely accessible critical care database. Sci. Data 3, 160035 EP (2016)
31. Zhang, M.L., Zhou, Z.H.: A review on multi-label learning algorithms. TKDE 26(8), 1819–1837 (2014)
32. Zhang, W., Wang, L., Yan, J., Wang, X., Zha, H.: Deep extreme multi-label learning. arXiv preprint arXiv:1704.03718 (2017)
33. Maaten, L.V.D., Hinton, G.: Viualizing data using T-SNE. J. Mach. Learn. Res. 9, 2579–2605 (2008)

Modeling Patient Visit Using Electronic Medical Records for Cost Profile Estimation

Kangzhi Zhao[1]([✉]), Yong Zhang[1], Zihao Wang[1], Hongzhi Yin[2],
Xiaofang Zhou[2], Jin Wang[3], and Chunxiao Xing[1]

[1] RIIT, TNList, Department of Computer Science and Technology,
Institute of Internet Industry, Tsinghua University, Beijing 100084, China
{zkz15,wzh17}@mails.tsinghua.edu.cn, {zhangyong05,xingcx}@tsinghua.edu.cn
[2] The University of Queensland, Brisbane, Australia
h.yin1@uq.edu.au, zxf@itee.uq.edu.au
[3] Computer Science Department, University of California,
Los Angeles, Los Angeles, USA
jinwang@cs.ucla.edu

Abstract. Estimating health care cost of patients provides promising opportunities for better management and treatment to medical providers and patients. Existing clinical approaches only focus on patient's demographics and historical diagnoses but ignore ample information from clinical records. In this paper, we formulate the problem of patient's cost profile estimation and use Electronic Medical Records (EMRs) to model patient visit for better estimating future health care cost. The performance of traditional learning based methods suffered from the sparseness and high dimensionality of EMR dataset. To address these challenges, we propose Patient Visit Probabilistic Generative Model (PVPGM) to describe a patient's historical visits in EMR. With the help of PVPGM, we can not only learn a latent patient condition in a low dimensional space from sparse and missing data but also hierarchically organize the high dimensional EMR features. The model finally estimates the patient's health care cost through combining the effects learned both from the latent patient condition and the generative process of medical procedure. We evaluate the proposed model on a large collection of real-world EMR dataset with 836,033 medical visits from over 50,000 patients. Experimental results demonstrate the effectiveness of our model.

Keywords: Electronic medical records · Cost profile estimation
Health care data mining · Probabilistic generative model

1 Introduction

The growth rate of health care cost is an alarming problem in many countries all over the world. The cost of health care in the United States is steadily rising, making it the most expensive in the world[1]. Studies show that advanced analytics

---

[1] https://www.cms.gov/.

© Springer International Publishing AG, part of Springer Nature 2018
J. Pei et al. (Eds.): DASFAA 2018, LNCS 10828, pp. 20–36, 2018.
https://doi.org/10.1007/978-3-319-91458-9_2

can help slow this upward trend [12], among which accurate estimation of health care cost serves as the first step. In clinical practice, a more expensive treatment regimen at the onset of diseases could lower the patient's total cost in the next 10 years. Doctors need to predict the future cost of a patient in order to give a more reasonable and economical treatment. For health insurance companies, there is a significant need to look beyond applicants' future cost, especially for those with chronic diseases.

Therefore, it is important to estimate patient's health care cost aiming at better management and treatment for both medical providers and patients. Traditionally, previous approaches in this field are mainly based on demographics and diagnosis data. The Diagnosis-Related Groups (DRGs) classifies patients into groups by their age, sex and history diagnosis to infer patient's current problem and their potential health care cost [1,7,10]. Moturu et al. [21] utilized different classification algorithms to predict future high cost patients. Recently, with the development of health informatics, a large amount of data from medical institute become available. These Electronic Medical Records (EMR) document patient visits, including demographic information (birth date, gender, etc.), diagnosis (historic and current), treatment (medication and surgery), laboratory results and clinical notes, providing an opportunity for researchers to develop data-driven models for health care data analysis.

In this paper, we study the problem of estimating future health care cost of a patient from his history data in EMR. More specifically, given a series of hospital visit data of a patient, the goal of this study is to estimate the high cost risk of future visits. This approach could help doctors identify potential high cost patients in the near future, who could be enrolled in case-management or enroll-management program and get early intervention. The health insurance companies can also benefit from them to develop personalized contracts. However, there are some obstacles for making analysis on EMR due to its intricate nature. Generally speaking, there are two aspects of challenges:

Sparse and Missing Data. There are thousands of possible items in the clinical laboratory, only a few of them could be done in each patient visit. Even for a certain kind of disease, a patient can take a small bundle of medical tests in hundreds of related tests, leading to a highly sparse dataset. There are 566 different medical test features in our dataset, but each patient visit only contains 13.92 medical tests (2.46%) on average. 70.37% of the 836,033 patient visits have less than 10 medical tests. The sparse dataset will make it difficult in constructing the model and definitely deteriorate learning performance. In clinical trails, Wood et al. [26] proposed multiple imputation methods to handle missing data, but little statistical imputation could handle our highly sparse dataset with guaranteed bias.

Confounding Effect and High Dimensionality. Confounding arises when a variable (confounder) is not connecting the exposure with the outcome but associated with them [22]. Using EMRs often has confounding effects [20]. For medical cost prediction problems, risk score features based on DRGs and indicators of chronic conditions can be extracted [8]. However, a spurious association may be generated when we study on a small bundle of risk factors and the high cost risk of patients. On the other hand, there are 746 different features in our EMR dataset containing detailed information of patient visits. But modeling of the high dimensional features without domain knowledge is still an open question.

To address these challenges, we propose the Patient Visit Probabilistic Generative Model (PVPGM) to describe the generative process of the high dimensional features of a patient visit. For handling sparse and missing data, PVPGM learns a latent patient condition in a low dimensional space from the sparse and missing medical test data. For handling the confounding effect and high dimensionality, PVPGM hierarchically organizes the high dimensional features of a patient visit including demographic information, diagnosis, treatment and laboratory results, and then describes the generative process by logistic functions. With the help of such a generative model, we can predict the high cost risk of a patient as well as expose the risk factors. To the best of our knowledge, it is the first paper to study the cost prediction problem in EMR data mining. We evaluate the proposed model on a large collection of the real-world electronic medical records from a famous cardiovascular hospital in China and the experiment results demonstrate the effectiveness of our model.

To summarize, our work contributes on the following aspects:

- We study the problem of estimating patients' cost, which has been found widespread application prospect. To the best of our knowledge, no previous work has studied this problem in machine learning perspective.
- We propose a probabilistic generative model, PVPGM, to solve this problem. Our model aims at addressing two common challenges in EMR data mining: it first projects the high dimensional sparse data into a low dimensional space and then models the features of a patient visit from electronic medical records to estimate the patient's cost profile.
- We evaluate our proposed model on a large scale real-world dataset, including 836,033 medical visits from over 50,000 patients. Experimental results show that our model outperforms baseline methods significantly.

2   Problem Definition

In this section, we introduce and define related concepts and formulate our cost profile estimation problem.

2.1   Definition

As we mentioned in the above section, our model jointly describes patient visits in the generative process. Formally, we define a patient visit for a patient first.

Definition 1 *(Patient Visit). A patient visit n for a patient is defined as a set* $s_n = \{x, \iota, d, t, c, \tau\}$*, where* $x = \{(m_l, r_l)\}$ *is the medical test set of* (*item, result*) *pairs, demographical information* ι*, diagnosis* d*, medical treatment* t*, health care cost* c *and time stamp* τ*.*

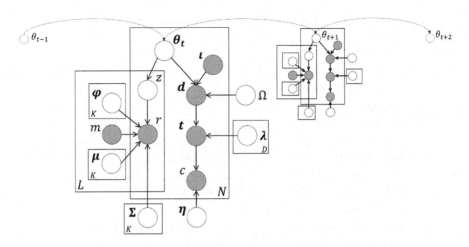

Fig. 1. Patient visit probabilistic generative model

A patient visit is the behavior and corresponding record of the inpatient or outpatient documented by hospital information system. We organize all possible information in a clinical record that EMR provides to us. A patient may visit a hospital multiple times, so we use a patient visit sequence $S = (s_i)$ to represent a patient where $\tau_i < \tau_j$ if $i < j$. Then we define the concept of patient condition as below.

Definition 2 *(Patient Condition). The patient condition $\psi = (\iota, \theta)$ denotes one particular patient's current condition, which is defined as a pair of demographical information ι and intrinsic condition θ.*

The patient condition for one patient visit includes the extrinsic condition (the demographic information such as age, gender) and intrinsic one. θ is a latent low dimensional probability vector learned from patient's medical test. Each dimension of the latent intrinsic condition denotes the weight of the generative distribution since we adopt a mixture of the generative process to the medical tests. Similar to the above, we further define a condition sequence $\Psi = (\psi_i)$ for a particular patient to describe the development of his/her disease, where ψ_i is earlier than ψ_j if $i < j$. Furthermore, we define the concept related to our object problem as follows.

Definition 3 *(Cost Profile). The cost profile is the description of one patient's health care cost sequence $C = (c_i)$. For a particular patient visit i, the m-period cost profile is the average cost of the future m patient visits:*

$$CP(i, m) = \frac{1}{m} \sum_{k=0}^{m-1} c_{i+k}$$

Table 1. Variable descriptions

Notation	Description
N	The number of all patient visits
L	The number of medical tests
K	The number of latent variables
D	The number of different diagnoses
ι	The demographical information of the patient on patient visit n, which indicates patient's current extrinsic condition
θ	The latent variable of mixture generative process of medical tests, which indicates patient's current intrinsic condition
d	The diagnose code vector of patient visit n
Ω	The coefficient matrix $(\omega_1^T, ..., \omega_D^T)^T$ of multinomial logistic regression to generate d from patient condition ι and θ
t	The treatment vector of patient visit n, including medication and surgery
λ	The distribution parameter for each treatments
c	The label node indicating whether the STCP of patient visit n is in the high group
η	The coefficient for cost c
z	The distribution of medical test l w.r.t. patient visit n
m	The type of medical test l
r	The result of medical test l w.r.t. patient visit n
μ, σ	The parameters for Gaussian distributions to sample continuous medical tests r
φ	The parameter for multinomial distribution to sample categorical medical tests r

The cost profile of patient visit i includes two aspects: short-term description $STCP = CP(i, 1)$ and m periods long-term description $LTCP(m) = CP(i, m)$. Based on the above definitions, we further define our prediction task as follows.

2.2   Cost Profile Estimation Problem

The problem we address is to forecast cost profile based on the patient visit sequence. The input of cost profile estimation includes a set of patients $V = \{v_n\}$ and a patient visit sequence $S_n(\tau_0) = (..., s_{i_0-1}, s_{i_0})$ for each patient v_n where $\tau_{i_0} \leq \tau_0 < \tau_{i_0+1}$. Our goal of cost profile estimation is for each patient v_n and $i_0 + 1$ patient visit, determining if $STCP_n$ is in the high group. The threshold will be tuned to verify the robustness of our approach.

3   Model

In this section, we describe our proposed Patient Visit Probabilistic Generative Model, PVPGM to estimate the cost profile of a patient. The model includes the generative process of patient visit and patient condition so as to produce the unknown condition of next period and corresponding cost profile.

3.1   Assumptions

With the support of our clinical experts, we have the following assumptions based on the medical insights.

Assumption 1. *The health care cost of patients with non-communicable disease (NCD) mainly depends on the progression of disease.*

Non-communicable diseases (NCDs) tend to be of long duration, generally slow progression such as cardiovascular disease (CVD) and diabetes mellitus[2]. Our problem focuses on NCDs whose health care cost of inpatient and outpatient is more predictable than other diseases such as surgery injuries. We can learn patient condition from medical tests to infer the progression of disease and make a more accurate prediction about patient's future cost.

Assumption 2. *A diagnosis is made by obtaining patient condition and physicians give out the treatment based on the diagnosis.*

In order to make a diagnose, physicians analyze patient's extrinsic and intrinsic condition from inquiry and medical tests based on their own experience. Then, the corresponding treatment is given and the health care cost is incurred.

3.2   Patient Visit Probabilistic Generative Model

Based on the assumptions, we propose our Patient Visit Probabilistic Generative Model, PVPGM, as shown in Fig. 1. Table 1 describes the variables of our model.

According to our problem definition, we have a patient visit sequence $S(\tau_0) = (..., s_{i_0-1}, s_{i_0})$ for each patient. We use c_n to denote whether STCP of patient visit n is in the high group. $c_n = 1$ represents the positive result and $c_n = -1$ is the negative result. Our target is to map the input visit sequence to the label of patient visit $i_0 + 1$, i.e., $f : S(\tau_0) \mapsto c_{i_0+1}$. We can build a classification model to map the sequence to a target label. In order to handle the challenges of our dataset, in each time slice, PVPGM models the medical procedure and extracts low dimensional features from the sparse medical tests.

Medical Procedure. The right part of each time slice in Fig. 1 represents the generative process of medical procedure. Assumption 2 describes our intuitive insights. According to the model, for each patient visit n, the diagnose $\boldsymbol{d}_n$ is

---

[2] http://www.who.int.

generated from a multinomial logistic regression Ω based on the patient condition $\psi_n = (\iota_n, \theta_n)$. We concatenate ψ_n to a condition vector

$$\vartheta_n = \iota_n \bigoplus \theta_n = (\iota_{n1}, ..., \iota_{nI}, \theta_{n1}, ..., \theta_{nK})$$

Then the treatment t is generated from a d_n-specific multinomial distribution λ_{d_n}. Finally c is generated from logistic regression based on the treatment.

Mixture Generative Process for Medical Tests. To extract low dimensional latent feature θ from the sparse medical tests dataset, we use the hierarchical generative architecture which is similar to the Latent Dirichlet Allocation (LDA) model [2]. Left part of our model shown in Fig. 1 denotes the design. Marginalized out z, we have

$$p(r_{n,l}|\boldsymbol{\theta_n}, \boldsymbol{\mu_l}, \boldsymbol{\sigma}, \boldsymbol{\Phi_l}) = \sum_{k=1}^{K} \theta_{n,k} \cdot p(r_{n,l}|\mu_{l,k}, \sigma_k, \varphi_{l,k}) = \sum_{k=1}^{K} \theta_{n,k} \cdot \Pi_{l,k}(r_{n,l}) \quad (1)$$

---

1 **foreach** *patient visit n* **do**
2   Draw a diagnosis d_n from multinomial logistic regression based on patient condition $\psi = (\iota, \boldsymbol{\theta})$, i.e., $d_n \sim multi(softmax(\Omega \cdot (\iota \bigoplus \theta)))$;
3   **foreach** *treatment i* **do**
4   Draw a treatment $t_{n,i}$ from d_n-specific multinomial distribution, i.e., $t_{n,i} \sim multi(\lambda_{d_n})$;
5   **end**
6   Draw the cost c_n from the treatment $\boldsymbol{\eta}$ i.e., $c_n \sim multi(sigmoid(\boldsymbol{\eta}^T \cdot \boldsymbol{t_n}))$;
7   **foreach** *medical test l* **do**
8   Draw a hidden pattern $k = z_{n,l}$ from multinomial distribution, i.e., $z_{n,l} \sim multi(\theta_n)$;
9   Draw $r_{n,l}$ from mixture distribution $\Pi_{l,k}$ which is shown in Eq. (2);
10   **end**
11 **end**

Algorithm 1. Probabilistic generative process

Unlike LDA, there are two kinds of generative target r: numerical and categorical. Similar to Liu et at. [19], we generate numerical variables from Gaussian distributions and categorical variable from multinomial distributions. Thus the k-specific distribution $\Pi_{l,k}$ is defined as

$$\Pi_{l,k}(r_{n,l}) = p(r_{n,l}|\mu_{l,k}, \sigma_k, \varphi_{l,k})$$

$$= \begin{cases} \dfrac{1}{(2\pi\sigma_k^2)^{\frac{1}{2}}} e^{-\frac{(r_{n,l}-\mu_{l,k})^2}{2\sigma_k^2}} & r_{n,l} \text{ is numerical.} \\ \varphi_{l,k,r_{n,l}} & r_{n,l} \text{ is caterorical.} \end{cases} \quad (2)$$

Input: a feature matrix X, learning rate α, weight decay term α'
Output: parameter configuration Ξ

1 Calculate λ according to Eq. (13);
2 Initialize $\theta, \Omega, \eta, \mu, \sigma, \varphi$ randomly according to the constrains of Eq. (17);
3 **repeat**
4 | Calculate $p^{(i)}(k|n, l)$ according to Eq. (7);
5 | Update $\theta, \varphi, \mu, \sigma$ according to Eq. (8)-(11);
6 | Update η according to Eq. (14);
7 | Calculate $p(d_n = d|\theta_n, \Omega)$ according to Eq. (15);
8 | Update Ω according to Eq. (16);
9 **until** *convergence*;

Algorithm 2. Learning algorithm

The whole generative process is shown in Algorithm 1. We define

$$\Xi = \{\theta, \Omega, \lambda, \eta, \mu, \sigma, \varphi\} \tag{3}$$

as our parameter configuration. By adopting the chain rule of probability, the log-likelihood objective function can be obtained as follows:

$$
\begin{aligned}
O(\Xi) &= \sum_n \log p(r_n, d_n, t_n, c_n | \Xi) \\
&= \sum_n \sum_{l_1} I(n, l_1) \log(\sum_{k=1}^{K} \theta_{n,k} \cdot \frac{1}{(2\pi\sigma_k^2)^{\frac{1}{2}}} e^{-\frac{(r_{n,l_1} - \mu_{l_1,k})^2}{2\sigma_k^2}}) \\
&+ \sum_n \sum_{l_2} I(n, l_2) \log(\sum_{k=1}^{K} \theta_{n,k} \cdot \varphi_{l_2,k,r_{n,l_2}}) \\
&+ \sum_n (c_n \log \frac{1}{1 + e^{-\eta^T \cdot t_n}} + (1 - c_n) \log (\frac{e^{-\eta^T \cdot t_n}}{1 + e^{-\eta^T \cdot t_n}})) \\
&+ \sum_n \sum_t \log \lambda_{d_n, t_{n,t}} + \sum_n \log \frac{e^{\omega_{d_n}^T \cdot \vartheta_n}}{\sum_j e^{\omega_j^T \cdot \vartheta_n}}
\end{aligned}
\tag{4}
$$

where l_1 denotes the continuous medical tests and l_2 denotes the categorical tests.

$$
I(n, l) = \begin{cases} 0 & r_{n,l} \text{ is missing.} \\ 1 & r_{n,l} \text{ is not missing.} \end{cases}
$$

is an indicator function of our dataset.

3.3 Model Learning

From the above, we show the maximum-likelihood estimation problem of PVPGM as follows:

$$\Xi^* = \arg\max_{\Xi} O(\Xi)$$

$$s.t. \quad \sum_k \theta_{n,k} = 1, \sum_x \phi_{l,k,x} = 1, \sum_t \lambda_{d,t} = 1 \tag{5}$$

In this paper, we use iterative approximation method to learn the parameters because Eq. (5) does not have the closed-form solution. This algorithm separates the objective function into two parts and maximizes them iteratively. For the first part, we fix the parameters of medical procedure $\{\Omega, \boldsymbol{\lambda}, \boldsymbol{\eta}\}$ and learn the remaining mixture generative process. We have

$$O_1(\boldsymbol{\theta}, \varphi, \mu, \sigma) = \sum_n \sum_{l_1} I(n, l_1) \log(\sum_{k=1}^K \theta_{n,k} \cdot \frac{1}{(2\pi\sigma_k^2)^{\frac{1}{2}}} e^{-\frac{(r_{n,l_1} - \mu_{l_1,k})^2}{2\sigma_k^2}})$$

$$+ \sum_n \sum_{l_2} I(n, l_2) \log(\sum_{k=1}^K \theta_{n,k} \cdot \varphi_{l_2,k,r_{n,l_2}}) \tag{6}$$

We can adopt Eq. (7) to get the lower bound of Eq. (6) by Jensen's inequality, then maximize and update the lower bound iteratively to generate the global maximum. We have the following update functions.

$$p^{(i)}(k|n, l) = \frac{\theta_{n,k}^{(i)} \cdot \Pi_{l,k}(r_{n,l})}{\sum_{k'=1}^K \theta_{n,k'}^{(i)} \cdot \Pi_{l,k'}(r_{n,l})} \tag{7}$$

$$\mu_{l_1,k}^{(i+1)} = \frac{\sum_n I(n, l_1) p^{(i)}(k|n, l_1) r_{n,l_1}}{\sum_n I(n, l_1) p^{(i)}(k|n, l_1)} \tag{8}$$

$$\sigma_k^{(i+1)2} = \frac{\sum_n \sum_{l_1} I(n, l_1) p^{(i)}(k|n, l_1)(r_{n,l_1} - \mu_{l_1,k}^{(i+1)})^2}{\sum_n I(n, l_1) p^{(i)}(k|n, l_1)} \tag{9}$$

$$\varphi_{k,l_2,r}^{(i+1)} = \frac{\sum_n I(n, l_2) 1_r(r_{n,l_2}) p^{(i)}(k|n, l_2)}{\sum_n \sum_r I(n, l_2) 1_r(r_{n,l_2}) p^{(i)}(k|n, l_2)} \tag{10}$$

$$\theta_{n,k}^{(i+1)} = \frac{\sum_{l_1} I(n, l_1) p^{(i)}(k|n, l_1) + \sum_{l_2} I(n, l_2) p^{(i)}(k|n, l_2)}{\sum_{l_1} I(n, l_1) + \sum_{l_2} I(n, l_2)} \tag{11}$$

where $1_x(y)$ indicates whether x is equal to y.

For the second part, we fix the parameters of mixture generative process $\{\boldsymbol{\theta}, \varphi, \mu, \sigma\}$ and learn the medical procedure. We have

$$O_2(\Omega, \boldsymbol{\lambda}, \boldsymbol{\eta}) = \sum_n (c_n \log \frac{1}{1 + e^{-\boldsymbol{\eta}^T \cdot t_n}} + (1 - c_n) \log(\frac{e^{-\boldsymbol{\eta}^T \cdot t_n}}{1 + e^{-\boldsymbol{\eta}^T \cdot t_n}}))$$

$$+ \sum_n \sum_t \log \lambda_{d_n, t_{n,t}} + \sum_n \log \frac{e^{\omega_{d_n}^T \cdot \vartheta_n}}{\sum_j e^{\omega_j^T \cdot \vartheta_n}} \tag{12}$$

We have closed-form solution for $\lambda_{d,t}$:

$$\lambda_{d,t} = \frac{\sum_n \sum_{t'} 1_d(d_n) 1_t(t_{n,t'})}{\sum_t \sum_n \sum_{t'} 1_d(d_n) 1_t(t_{n,t'})} \tag{13}$$

We adopt gradient ascent algorithm to update the remaining parameters of O_2 iteratively. Following are the update functions.

$$\eta_t^{(i+1)} = \eta_t^{(i)} + \alpha \sum_n (c_n - sigmoid(\boldsymbol{\eta}^{(i)^T} \cdot \boldsymbol{t}_n))t_{n,t} \tag{14}$$

$$p(d_n = d|\boldsymbol{\theta}_n, \Omega) = \sum_n \frac{e^{\omega_d^{(i)^T} \cdot \boldsymbol{\vartheta}_n}}{\sum_{d'=1}^{D} e^{\omega_{d'}^{(i)^T} \cdot \boldsymbol{\vartheta}_n}} \tag{15}$$

$$\omega_{d,k}^{(i+1)} = \omega_{d,k}^{(i)} + \alpha(\sum_n (1_d(d_n) - p(d_n = d|\boldsymbol{\theta}_n, \Omega))\theta_{n,k} + \alpha'\omega_{d,k}^{(i)}) \tag{16}$$

where α is the learning rate with the gradient and α' is the weight decay term. Algorithm 2 summarizes the learning algorithm.

3.4 Cost Profile Estimation

With the generative model above, a patient's diagnose code $\boldsymbol{d}$, treatment $\boldsymbol{t}$ and health care cost c can be generated from his condition ψ in a certain patient visit n. In order to estimate cost profile of this patient, we learn the sequence of his history condition. Figure 2 shows the procedure of our cost estimation. Instead of directly producing c_{i_0+1} (or equivalently s_{i_0+1}), we first learn a probabilistic graphical model which hierarchically organizes patients' medical features. Then the condition vectors for different time slices of one patient can be used to predict the condition of next visit. Finally c_{i_0+1} can be drawn from the inference of probabilistic graph.

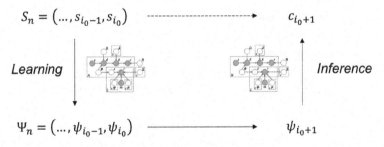

Fig. 2. The procedure of cost estimation

With the model parameters and patient condition vector $\boldsymbol{\vartheta}_n$ for each time slice, we assume the patient condition sequence $\Psi = (\psi_i)$ has Markov property. So we can learn the transition matrix by solving the following optimization problem

$$X^* = \arg\min_X \sum_{i \leq i_0} \|\psi_i - X\psi_{i-1}\| \tag{17}$$

where $\tau_{i_0} \leq \tau_0$ is the last patient visit before τ_0.

Then we can calculate next condition vector by $\psi_{i_0+1} = X\psi_{i_0}$ and generate the corresponding cost profile.

4   Experiments

In this section, we comprehensively evaluate our PVPGM model on a real-world dataset and compared it with multiple state-of-the-art methods. Firstly, we introduce the experiment setup, including the description of our EMR dataset, evaluation metrics and baseline models. Then we present the experimental result and illustrate the effectiveness of our proposed approach. We also discuss the effect of some important parameter values in this section.

4.1   Experiment Setup

We use a collection of real medical records from a leading hospital in cardiovascular medicine, cardiovascular surgery and geriatrics in Beijing. Our dataset is described in Table 2. It contains over 836,033 patient visits from over 50,624 patients. Each record consists of 566 medical tests with 96 demographical information and 102 treatment columns. On average each EMR record contains 13.92 different medical tests (2.46% of all medical tests), which indicates our dataset has a serious problem with feature sparsity.

Table 2. Description of EMR dataset

Items	Numbers
#patient visits	836,033
#patients	50,624
#demographics	96
#treatments	102
#medical tests	566
Sparse ratio of medical tests	2.46%
Time span	Jan 2003 to Dec 2015

We randomly picked 80% of the patients as training set and the rest for testing, then adopt 5-fold cross validation during training. We evaluate the proposed model in terms of precision, recall and F-Measure. In order to better evaluate the classifier on imbalanced data, we also compare the ROC curve with different baseline methods to validate its effectiveness. These metrics are widely used in data mining studies. Receiver Operating Characteristic (ROC) curve is extensively adopted to evaluate the imbalanced dataset especially in clinical practice [9].

4.2   Baseline Methods

We select the following methods as baselines for the cost estimation problem and compare our proposed method with them:

- **DTC.** Decision Tree Classifier (DTC) [3] learns simple decision rules inferred from the data features, which will generate a data-driven model similar to the rule-based DRGs models. Diagnosis-Related Group (DRG) is a widely used classification model to estimate patient's health care cost [1,7] as we mentioned in Sect. 1. We adopt DTC to illustrate the effectiveness of DRGs.
- **NI+SVM.** Normal Imputation (NI) is employed on our dataset. In medical tests, it is reasonable that doctors would not check the irrelevant or unnecessary medical lab tests when they are in the reference range. Normal imputation is a widely used method in clinical practice [26]. We treat medical records as features and LIBSVM [5] is employed as the classification model for health care cost estimation.
- **ALS+SVM.** Alternative Least Square (ALS) collaborative filtering [13] is an algorithm based imputation method. ASL features are employed as the input of SVM classification model.
- **PGM.** A traditional probabilistic generative model is used as classification model. It is a part of PVPGM and has the same architecture with the medical procedure mentioned in Sect. 3.2. We employ gradient descent algorithm to learn the parameters in PGM [16] and set the learning rate parameter as 0.1.
- **PGM+PCA.** PCA [11] converts high dimensional data into a set of principal components. We adopt PCA to solve the data sparsity problem and employ it as the input of our PGM classifier.

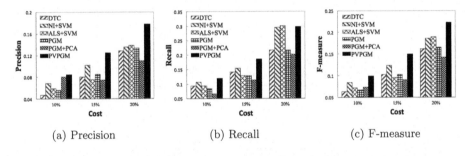

(a) Precision   (b) Recall   (c) F-measure

Fig. 3. Model comparison on precision, recall and F-measure

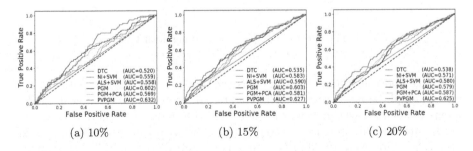

(a) 10%   (b) 15%   (c) 20%

Fig. 4. Model comparison on ROC for cost threshold 10%, 15%, 20%

4.3   Performance Comparison

We now compare the performance with baseline models to evaluate the effectiveness of our model in the health care cost estimation. We set the threshold of the high group for patient's health care cost to the top 10%, 15% and 20%. We also empirically set the compression parameter of latent variables (dimension of latent variable/origin) as 0.05.

Figure 3 shows the results of precision, recall and F-measure of different models with the cost threshold as 10%. Figure 4 demonstrates the ROC curves on different cost thresholds of health care cost estimation task. From these results, we can see that PVPGM achieves the best aggregate performance in all the three thresholds. Generally, PVPGM models the medical procedure and lab tests of a patient visit at the same time, which captures the dependence and constraint between medical features and latent condition variable of patients. It improves the precision without lowering recall, thus results in the best F_1 score.

DTC is used to simulate DRGs model to learn decision rules from patient's demographics and diagnosis. NI+SVM and ALS+SVM adopt the rule-based or statistical method to impute missing data in medical test. Compared with DTC, NI+SVM and ALS+SVM include patient's condition information and get a better performance. However, noise is introduced in such imputation methods and they also suffer from high dimensionality and feature sparsity.

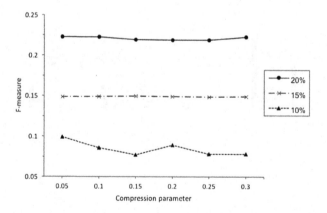

Fig. 5. The effect of compression parameter in PVPGM

Unlike above methods, PGM models the medical procedure of patient visit with a graph-based method. Such traditional probabilistic graphical model adopts domain knowledge from the medical field to reduce the parameter space, which improves the precision without hurting the recall substantially compared with DTC. However, PGM+PCA does not improve the overall performance because the sparse modeling and classification process are separated into two steps. While our PVPGM learns the patient's hidden condition and features from the medical process at the same time. Such integrating methods estimate the parameters better and outperform all the baselines.

4.4 Sensitivity Analysis

We also investigate the impact of the latent variable dimensionality. In this study, our PVPGM embeds high dimensional and sparse medical tests into a low dimensional space representing patient's hidden condition. So we focus on the compression parameter of latent variable dimensionality (latent variables/origin). Figure 5 displays the F-measure of PVPGM to the compression parameter. We can see that the overall performance of our model does not change much in terms of the dimensionality of latent variables when the cost threshold is set as 15% or 20%. Lower dimension is better for 10% cost probably because a larger latent space will increase the number of training parameters and thus hurt the generalization of the classifier. There is also a trade-off between effectiveness and training time. So in our experiment, we choose the compression parameter as 0.05.

5   Related Work

In this section, we review related work about cost estimation problem using EMR dataset in health care data mining and sparse modeling problem using probabilistic graphical models.

Cost Estimation Problem in EMR Data Mining. Recently modeling electronic medical records for prediction has attracted the interests of researchers from various areas. There is a long stream of studies for clinical problems in machine learning perspective [23,27]. Various medical models are developed based on the specific clinical problems such as patient phenotype identification [18,24], potential complications of diseases [6,29] and risk profiling [17]. For health care cost estimation problem, current studies mainly focus on statistical and economic models [1,21] but few researchers in this domain adopt medical test features from EMR dataset. To the best of our knowledge, none of the previous work estimates cost profile of a patient from a medical model.

Probabilistic Graphical Models in Sparse Modeling. Due to the interpretability and good generalization ability, lots of papers in health care adopt probabilistic generative models to analyze EMR dataset. Liu et al. [18] employ the generative model to capture the relationship of medical events. Some studies [4,28] design dynamic latent variables to model patient's behavior over time. Sparse modeling is an important problem in machine learning tasks. Lee et al. [15] and Zhang et al. [37] extensively study the sparse coding problem. Probabilistic graphical models are widely adopted in sparse modeling. The most important applications include probabilistic topic models [30–35]. Sparse conditional random field [39], sparse generalized linear models [14] and sparse factor models [29,36] utilize probabilistic graphical structure to solve the sparse problem. CNN based solution [25] and considering syntactic similarity issues [38] are also good ways to address data sparsity. Inspired by the medical assumptions, we describe a distribution over the observed patient visit by recovering latent random variables from a probabilistic graphical structure.

6   Conclusion

In this paper, we study the problem of using a large volume of electronic medical records to estimate the cost profile of a patient. We propose Patient Visit Probabilistic Generative Model (PVPGM) to model patient visit for better estimating future health care cost. Our model learns the latent patient condition from the sparse dataset and integrates it into the generative process of medical procedure. We validate our model on a large collection of real-world electronic medical records from a famous cardiovascular hospital in China and the experiment results show the effectiveness of our model.

For the future work, we will continuously work with our medical group to reveal the explanations of our model. The mixture generative process of high dimensional medical tests can be regarded as clusters of lab items, which may help mine the possible medical phenotypes. The sensitivity of demographics can reveal a lot of risk factors on health economics. Our estimation architecture can also be used for the intelligent diagnosis or personalized treatment.

Acknowledgment. This work was supported by NSFC (91646202), the National High-tech R&D Program of China (SS2015AA020102), NSSFC (15CTQ028), Research/Project 2017YB142 supported by Ministry of Education of The People's Republic of China, the 1000-Talent program and Tsinghua Fudaoyuan Research Fund.

References

1. Ash, A.S., Ellis, R.P., Pope, G.C., Ayanian, J.Z., Bates, D.W., Burstin, H., Iezzoni, L.I., MacKay, E., Yu, W.: Using diagnoses to describe populations and predict costs. Health Care Financ. Rev. **21**(3), 7 (2000)
2. Blei, D.M., Ng, A.Y., Jordan, M.I.: Latent dirichlet allocation. J. Mach. Learn. Res. **3**(Jan), 993–1022 (2003)
3. Breiman, L., Friedman, J., Stone, C.J., Olshen, R.A.: Classification and Regression Trees. CRC Press, Boca Raton (1984)
4. Caballero Barajas, K.L., Akella, R.: Dynamically modeling patient's health state from electronic medical records: a time series approach. In: KDD, pp. 69–78 (2015)
5. Chang, C., Lin, C.: LIBSVM: a library for support vector machines. ACM TIST **2**(3), 27:1–27:27 (2011)
6. Feld, S.I., Cobian, A.G., Tevis, S.E., Kennedy, G.D., Craven, M.: Modeling the temporal evolution of postoperative complications. In: AMIA (2016)
7. Fetter, R.B., Shin, Y., Freeman, J.L., Averill, R.F., Thompson, J.D.: Case mix definition by diagnosis-related groups. Med. Care **18**(2), i–53 (1980)
8. Fleishman, J.A., Cohen, J.W.: Using information on clinical conditions to predict high-cost patients. Health Serv. Res. **45**(2), 532–552 (2010)

9. Hajian-Tilaki, K.: Receiver operating characteristic (ROC) curve analysis for medical diagnostic test evaluation. Casp. J. Internal Med. **4**(2), 627 (2013)
10. Horn, S.D., Bulkley, G., Sharkey, P.D., Chambers, A.F., Horn, R.A., Schramm, C.J.: Interhospital differences in severity of illness: problems for prospective payment based on diagnosis-related groups (DRGs). N. Engl. J. Med. **313**(1), 20–24 (1985)
11. Jolliffe, I.T.: Principal component analysis and factor analysis. In: Jolliffe, I.T. (ed.) Principal Component Analysis, pp. 115–128. Springer, New York (1986). https://doi.org/10.1007/978-1-4757-1904-8_7
12. Koh, H.C., Tan, G., et al.: Data mining applications in healthcare. J. Healthc. Inf. Manag. **19**(2), 65 (2011)
13. Koren, Y., Bell, R.M., Volinsky, C.: Matrix factorization techniques for recommender systems. IEEE Comput. **42**(8), 30–37 (2009)
14. Krishnapuram, B., Carin, L., Figueiredo, M.A.T., Hartemink, A.J.: Sparse multinomial logistic regression: fast algorithms and generalization bounds. IEEE Trans. Pattern Anal. Mach. Intell. **27**(6), 957–968 (2005)
15. Lee, H., Battle, A., Raina, R., Ng, A.Y.: Efficient sparse coding algorithms. In: NIPS, pp. 801–808 (2006)
16. Lee, J.D., Hastie, T.J.: Learning the structure of mixed graphical models. J. Comput. Graph. Stat. **24**(1), 230–253 (2015)
17. Lin, Y.K., Chen, H., Brown, R.A., Li, S.H., Yang, H.J.: Healthcare predictive analytics for risk profiling in chronic care: a Bayesian multitask learning approach. MIS Q. **41**(2), 473–495 (2017)
18. Liu, C., Wang, F., Hu, J., Xiong, H.: Temporal phenotyping from longitudinal electronic health records: a graph based framework. In: KDD, pp. 705–714 (2015)
19. Liu, L., Tang, J., Cheng, Y., Agrawal, A., Liao, W.K., Choudhary, A.: Mining diabetes complication and treatment patterns for clinical decision support. In: CIKM, pp. 279–288 (2013)
20. Moher, D., Jones, A., Cook, D.J., Jadad, A.R., Moher, M., Tugwell, P., Klassen, T.P., et al.: Does quality of reports of randomised trials affect estimates of intervention efficacy reported in meta-analyses? Lancet **352**(9128), 609–613 (1998)
21. Moturu, S.T., Johnson, W.G., Liu, H.: Predicting future high-cost patients: a real-world risk modeling application. In: BIBM, pp. 202–208. IEEE (2007)
22. Pearl, J.: Causality. Cambridge University Press, Cambridge (2009)
23. Shickel, B., Tighe, P., Bihorac, A., Rashidi, P.: Deep EHR: a survey of recent advances on deep learning techniques for electronic health record (EHR) analysis. arXiv preprint arXiv:1706.03446 (2017)
24. Shivade, C., Raghavan, P., Fosler-Lussier, E., Embi, P.J., Elhadad, N., Johnson, S.B., Lai, A.M.: A review of approaches to identifying patient phenotype cohorts using electronic health records. J. Am. Med. Inform. Assoc. **21**(2), 221–230 (2013)
25. Wang, J., Wang, Z., Zhang, D., Yan, J.: Combining knowledge with deep convolutional neural networks for short text classification. In: IJCAI, pp. 2915–2921 (2017)
26. Wood, A.M., White, I.R., Thompson, S.G.: Are missing outcome data adequately handled? A review of published randomized controlled trials in major medical journals. Clin. Trials **1**(4), 368–376 (2004)
27. Yadav, P., Steinbach, M., Kumar, V., Simon, G.: Mining electronic health records: a survey. arXiv preprint arXiv:1702.03222 (2017)
28. Yang, S., Khot, T., Kersting, K., Natarajan, S.: Learning continuous-time Bayesian networks in relational domains: a non-parametric approach. In: AAAI, pp. 2265–2271 (2016)

29. Yang, Y., Luyten, W., Liu, L., Moens, M.F., Tang, J., Li, J.: Forecasting potential diabetes complications. In: AAAI, pp. 313–319 (2014)
30. Yin, H., Cui, B.: Spatio-Temporal Recommendation in Social Media. Springer Briefs in Computer Science. Springer, Singapore (2016). https://doi.org/10.1007/978-981-10-0748-4
31. Yin, H., Cui, B., Zhou, X., Wang, W., Huang, Z., Sadiq, S.W.: Joint modeling of user check-in behaviors for real-time point-of-interest recommendation. ACM Trans. Inf. Syst. **35**(2), 11:1–11:44 (2016)
32. Yin, H., Hu, Z., Zhou, X., Wang, H., Zheng, K., Hung, N.Q.V., Sadiq, S.W.: Discovering interpretable geo-social communities for user behavior prediction. In: ICDE, pp. 942–953. IEEE Computer Society (2016)
33. Yin, H., Wang, W., Wang, H., Chen, L., Zhou, X.: Spatial-aware hierarchical collaborative deep learning for POI recommendation. IEEE Trans. Knowl. Data Eng. **29**(11), 2537–2551 (2017)
34. Yin, H., Zhou, X., Cui, B., Wang, H., Zheng, K., Hung, N.Q.V.: Adapting to user interest drift for POI recommendation. IEEE Trans. Knowl. Data Eng. **28**(10), 2566–2581 (2016)
35. Yin, H., Zhou, X., Shao, Y., Wang, H., Sadiq, S.W.: Joint modeling of user check-in behaviors for point-of-interest recommendation. In: CIKM, pp. 1631–1640. ACM (2015)
36. Yoshida, R., West, M.: Bayesian learning in sparse graphical factor models via variational mean-field annealing. J. Mach. Learn. Res. **11**, 1771–1798 (2010)
37. Zhang, X., Yu, Y., White, M., Huang, R., Schuurmans, D.: Convex sparse coding, subspace learning, and semi-supervised extensions. In: AAAI (2011)
38. Zhang, Y., Li, X., Wang, J., Zhang, Y., Xing, C., Yuan, X.: An efficient framework for exact set similarity search using tree structure indexes. In: ICDE, pp. 759–770 (2017)
39. Zhong, P., Wang, R.: Learning sparse crfs for feature selection and classification of hyperspectral imagery. IEEE Trans. Geosci. Remote Sens. **46**(12), 4186–4197 (2008)

Learning the Representation of Medical Features for Clinical Pathway Analysis

Xiao Xu, Ying Wang, Tao Jin[(✉)], and Jianmin Wang

School of Software, Tsinghua University, Beijing 100084, China
{xu-x14,wang-yin17}@mails.tsinghua.edu.cn, jintao05@gmail.com,
jimwang@tsinghua.edu.cn

Abstract. Clinical Pathway (CP) represents the best practice of treatment process management for inpatients with specific diagnosis, and a treatment process can be divided into several stages, usually in units of days. With the explosion of medical data, CP analysis is receiving increasing attention, which provides important support for CP design and optimization. However, these data-driven researches often suffer from the high complexity of medical data, so that a proper representation of medical features is necessary. Most of existing representation learning methods in healthcare domain focus on outpatient data, which get weak performance and interpretability when adopted for CP analysis. In this paper, we propose a new representation, RoMCP, which can capture both diagnosis information and temporal relations between days. The learned diagnosis embedding grasps the key factors of the disease, and each day embedding is determined by the diagnosis together with the preorder days. We evaluate RoMCP on real-world dataset with 538K inpatient visits for several typical CP analysis tasks. Our method demonstrates significant improvement on performance and interpretation.

1 Introduction

With the significant improvement of living standards, the conflict between the quality healthcare demand and the financial pressure by governments is rising. Clinical Pathway (CP) is one of the most important tools to manage inpatient's[1] treatment process with better therapeutic effect and less economical cost. CP refers to a set of defined treatment activities during the healthcare process. In populous China, more than one thousand diseases have been managed by CPs for standardized treatment and regulated expenditure. Table 1 shows an example of Chinese CP about Intracerebral Hemorrhage (ICH). At present, most of CPs are designed by domain experts based on the clinical guidelines and their clinical experiences. However, these static CPs can be hardly adopted in practice for hospitals with different resources and patients under various conditions.

---

[1] A patient who stays in a hospital while receiving medical care or treatment. In general, it takes up more resources and cost compared to outpatients.

© Springer International Publishing AG, part of Springer Nature 2018
J. Pei et al. (Eds.): DASFAA 2018, LNCS 10828, pp. 37–52, 2018.
https://doi.org/10.1007/978-3-319-91458-9_3

During the past decade, large amounts of healthcare data, covering patients' whole treatment process, have been accumulated. Data-driven methods that discover and analyze execution careflows from the historical data are becoming an important support for CP (re)design and application. Faced with the complex medical data, the representation of medical features strongly affects the performance of these CP analysis methods. Process mining technologies pay attention to the temporal information in data [29]. They regard each treatment process as an event sequence (medical activities with timestamps), and extract the process model from massive sequences. The resultant models can be also used for conformance checking and enhancement. However, because of the high dimension of medical data, process mining methods usually generate spaghetti-like graph models which are difficult for understanding and executing. Another common representation of medical data for CP-related tasks is conducted through aggregation way. Each treatment process is mapped to a vector space by summing over the number of each medical activity. Various statistical models for CP, such as topic modeling and clustering, are performed on the aggregation representation. This representation always ignores temporal information, however, this ignored information is critical in CP.

Table 1. The National CP of ICH released by Ministry of Health of China.

Examination: (1) Blood, Urine, Stool routine examination (2) Hepatorenal function, Electrolyte, Blood glucose, Blood lipids, Cardiac enzymes, Coagulation function, Infectious Disease screening (3) Brain CT, Chest X-ray, ECG	
Medication: Mannitol, Glycerol fructose, Furosemidum, Antihypertensive drugs, Antibacterial drugs, Laxative, Electrolyte drugs	
Stage1 (Day1)	**Long term medical order:** (1) Neurology nursing routine (2) Level I care (3) Normal diet (4) Keep the bed (5) Observing vital signs
	Temporary medical order: (1) Blood, Urine, Stool routine examination (2) Hepatorenal function, Electrolyte, Blood glucose, Blood lipids, Cardiac enzymes, Coagulation function, Infectious disease screening (3) Brain CT, Chest X-ray, ECG (4) *when necessary:* Brain MRI, CTA, MRA or DSA
Stage2 (Day2)	**Long term medical order:** (1) Neurology nursing routine (2) Level I care (3) Normal diet (4) Keep the bed (5) Observing vital signs (6) Basic drugs
	Temporary medical order: (1) Re-examination for abnormal laboratory values (2) *when necessary:* Re-examination CT
...	......
Stage6 (Day8–14)	**Long term medical order:** (1) Discharge with drugs

Recently, inspired by word2vec [22, 23], medical concept embedding has been widely studied. It achieved significant performance improvement for various healthcare tasks. However, most of these methods focused on the outpatients, rather than the inpatients. We would face with the following challenges when adopt these methods on inpatients:

- **Semi-temporality.** For an inpatient, one visit represents a treatment process for a diagnosis that composed by several stages (usually counted in days). The medical activities between different stages are in sequence, while the medical activities in one stage are unordered[2]. In general, the temporal relation between stages for inpatient is stronger than that between visits for outpatient[3].
- **Numerical sensitivity.** The dosage of medications, the count of procedures and the quantity of medical consumables are crucial for inpatient treatment. Most of existing representations for outpatient ignore the numeric feature.
- **Interpretability.** The interpretability of the representation is important in healthcare domain. Thus, an inpatient-oriented representation is essential for CP analysis.

By taking into account all challenges mentioned above, we propose a three-layer neural network to learn the representations of medical features for CP analysis (RoMCP). The key principle of RoMCP is regarding one visit as a day-level sequence around its disease (corresponding to a diagnosis code). It stems from the view of CP that each diagnosis contains a set of core activities and each hospitalized day of a visit is based on both the diagnosis and the preorder days. Thus, in our method we first map the medical activities in one day to a fixed-length vector. We use the normalized real value of each activity for the input layer, instead of the one-hot coding (one if the activity happened and zero otherwise) which is common in previous methods. Inspired by doc2vec [17], we then map each diagnosis to a unique vector and use the temporal information between days to derive the representation. It means that given a visit with several days, the diagnosis vector and day vectors are concatenated to predict the next day. Each diagnosis vector is shared across all the visits that have the same diagnosis. It plays a role like a memory that remember the important topics of the diagnosis. Both disease vectors and day vectors are fed by stochastic gradient descent.

To summarize, we make the following primary contributions:

- In this paper, the representation learning of medical features for CP analysis is investigated. A concise three-layer network RoMCP is brought forward to address this issue, without any expert knowledge.

---

[2] Doctors usually make a prescription with multiple events together. There is no strict temporal relations between these events.

[3] Outpatient contains several visits. It is common that the events between sequential visits are quite different, due to the different diagnosis.

– RoMCP incorporates both diagnosis information and temporal information effectively into the embedding process. In particular, each diagnosis is mapped to a diagnosis embedding that can capture the key topics of the diagnosis; each day embedding is derived from the combination of the disease embedding and the preorder days' embeddings.
– We evaluate our approach on real-world data through three typical CP analysis tasks, and the experimental results show that our method outperforms the baselines. We also validate the interpretability of the learned representation in a case study with medical expert.

The remainders are organized as follows: In Sect. 2, we briefly discuss some related work. Then, details about our method are presented in Sect. 3. In Sect. 4, we demonstrate the experimental results conducted on a real world dataset. Finally, we conclude our study and prospect our future work in Sect. 5.

2   Related Work

In this section, we first discuss some related work on CP analysis, and then review some relevant studies about representation learning in healthcare.

2.1   CP Analysis

CP can be seen as a process management in healthcare. It defines the most common actions that represent best practice for most patients most of the time. Recently, various researches have been conducted on CP analysis. They can be divided into two categories:

Process mining, which focuses on discovering and verifying the temporal relation between medical activities. Traditional process mining technologies, such as Heuristic Miner, Fuzzy Miner and Hidden Markov models, were used to extract the execution process from medical log data [21,26,27]. In [3–5,29], researchers adopted the discovered process model for anomalies detection. However, due to the complexity and diversity of treatment behaviors in CPs are far higher than that of common business processes, these methods can hardly generate interpretable results without prior knowledge from medical experts [28].

Topic modeling, which can effectively identify different clinical phenotypes based on aggregation features of medical data. In [14–16], Huang et al. incorporated LDA and its variants into the clinical pattern discovery. In [30–32], LDA and process mining was combined to generate topic-based CP models. However, they weakened the temporal relations among medical activities.

2.2   Representation Learning in Healthcare

In the past few years, word embedding has achieved remarkable results in text mining [22,23]. The basic principle is that words with similar contexts should have similar meanings. Inspired by this idea, many research efforts are attracted to the representation learning of medical features for various healthcare applications. According to the date type, the related work can be classified into two categories:

Medical Free-text, including medical articles, doctor's advice and so on. Follow the idea of word2vec, a medical term was represented by its adjacent terms [11,12].

Structured Longitudinal Medical Data, such as Electronic Health Record (EHR) and claims data. It contains various treatment information over time. In [11,18,24], the medical events that happened in a short period of time were treated as the context in word2vec. The learned embeddings were useful across medical informatics tasks, such as heart failure prediction and readmission prediction. Instead of using the relative positions of medical events in timeline, Zhu et al. [34] adopted the actual timestamps in Electronic Health Record (EHR) for window size selection that acute condition medical concepts were assigned short context and vice versa. They used a convolutional neural network (CNN) framework to measure the patient similarities. However, these methods take a simple randomized policy for the medical events with the same timestamp. If the sizes of these events are imbalanced, the context would be varies widely. Considering the semi-temporality of medical data, Choi et al. proposed Med2Vec [7], which learned both medical code and visit representations with high interpretability from EHR datasets. They incorporated medical ontology to enhance the representation performance in [8]. There are also many studies leverage recurrent neural networks (RNN) to learn medical representation for various prediction problems, like risk [25], heart failure [9,10], diagnosis [19,20] and medication category [6]. Among all of the above studies about representation learning, [7] is the most related work to ours. While it differs from our work in several aspects:

- We are dealing with inpatient from the perspective of CP, whose temporal relation between days are stronger than that between visits of outpatient.
- The work in [7] put the diagnosis codes (represent the disease) of each visit together with the medication and procedure codes. While we regard the diagnosis as an independent vector, because the diagnosis has a global impact on all days for an inpatient.
- We take the numeric feature into account, instead of using the one-hot input.

3   Methodology

In this section, we start by introducing some definitions of CP and the related notations. Then we give a brief review of word2vec. Finally, we describe our RoMCP for representation learning of medical features.

3.1   Preliminary

Definition 1 (Visit). *A visit for an inpatient refers to a treatment process from admission to discharge. We denote the set of visits as $\mathcal{V} = \{V_1, V_2, \ldots, V_N\}$, where N is the size of visits in our dataset.*

Definition 2 (Diagnosis). *Each visit has a diagnosis[4]. We define all the diagnosis in our dataset as* $\mathcal{G} = \{g_1, g_2, \ldots, g_{|\mathcal{G}|}\}$ *with size* $|\mathcal{G}|$. *The diagnosis of visit* V_n *is denoted as* $g^{(n)} \in \mathcal{G}$.

Definition 3 (Day). *From the viewpoint of CP, a visit can be divided into several stages. In most cases, the stages are defined by days. We also use day as the basic unit in this study. Thus, a visit is composed by a day sequence* $\{d_{n,1}, d_{n,2}, \ldots, d_{n,T_n}\}$, *where* T_n *is the size of day in* V_n. *We denote the day set in our dataset as* $\mathcal{D}$ *with size* $|\mathcal{D}| = \sum_{n=1}^{N} T_n$.

Definition 4 (Activity). *Each day contains a set of medical activities. We define all the unique activities in the entire dataset as* $\mathcal{A} = \{a_1, a_2, \ldots, a_{|\mathcal{A}|}\}$ *with size* $|\mathcal{A}|$. *Each day* $d_{n,t}$, *containing a subset of medical activities* $(d_{n,t} \subseteq \mathcal{A})$, *is denoted by a real-valued vector* $x_{n,t} \in \mathbb{R}^{|\mathcal{A}|}$. *The i-element of* $x_{n,t}$ *represents the normalized value of* a_i *in* $d_{n,t}$.

3.2   Brief Review of Word2vec

Word2vec assumes that a word can be represented by the context words. It contains two models: continuous bag-of-word (CBOW) and Skip-gram model.

Given a word sequence $\{w_{i-k}, \ldots, w_{i-1}, w_i, w_{i+1}, \ldots, w_{i+k}\}$, the objective of CBOW is to maximize the following average log probability:

$$\frac{1}{T} \sum_{i=1}^{T} \log P(w_i | w_{i-k}, \ldots, w_{i-1}, w_{i+1}, \ldots, w_{i+k}) \tag{1}$$

where T is the corpus size, k is the size of the context window, and w_i is the target word. The conditional probability is defined by the softmax function as follows:

$$P(w_i | w_{i-k}, \ldots, w_{i-1}, w_{i+1}, \ldots, w_{i+k}) = \frac{\exp(\nu_i \cdot \nu_k)}{\sum_{w \in \mathcal{W}} \exp(\nu_w \cdot \nu_k)} \tag{2}$$

where $\mathcal{W}$ is the vocabulary, ν_i is the embedding of the target word w_i, and ν_k is the average of all the context word embeddings.

For skip-gram model, the target word is used to predict the context words. Both of the two models ignore the word order in the context.

3.3   RoMCP

Our proposed method aims to learning the CP-oriented representation of medical features. Figure 1 shows the framework of RoMCP, and two types of embeddings can be found in it:

---

[4] Some visits may have more than one diagnosis. While for CP, we only concern the first diagnosis, which largely determines the treatment strategy.

- **Diagnosis Embedding.** A CP is designed for a specific diagnosis. We map each diagnosis to a vector of dimension p that represents the most important treatment topics.
- **Day Embedding.** CP for inpatient is a day-based management strategy for treatment process. Therefore, we would like to map each day of dimension $|\mathcal{A}|$ to a vector of dimension q.

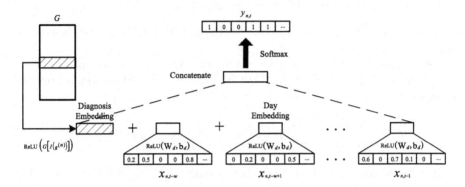

Fig. 1. The overview of RoMCP. Each day is derived from the diagnosis embedding together with the preorder days' embeddings.

Inspired by word2vec [22] and doc2vec [17], we use both diagnosis information and temporal relation between days to learn the representation. Formally, given the context window size w, we try to learn the two embeddings by maximizing the probability as follow:

$$\mathcal{L} = \frac{1}{N} \sum_{n=1}^{N} \frac{1}{T_n} \sum_{t=1}^{T_n} P(E_d(d_{n,t}) \mid E_g(g^{(n)}), E_d(d_{n,t-w}), \ldots, E_d(d_{n,t-1})) \qquad (3)$$

where E_g and E_d refers to the embedding function for diagnosis and day respectively.

Specifically, suppose that the set of diagnosis $\mathcal{G}$ are mapped to the matrix $G \in \mathbb{R}^{|\mathcal{G}| \times p}$, so that $E_g(g^{(n)})$ for diagnosis $g^{(n)}$ is represented by a row in G:

$$E_g(g^{(n)}) = \text{ReLU}(G[I(g^{(n)})]) \qquad (4)$$

where $G[i]$ is the i-th row in G, $I(g^{(n)})$ is the index of $g^{(n)}$ in $\mathcal{G}$, and ReLU is the rectified linear unit defined as $\text{ReLU}(s) = \max(0, s)$.

For day embedding, we convert a day $d_{n,t}$ represented by real value vector $x_{n,t}$ to a vector with dimension q as follows:

$$E_d(d_{n,t}) = \text{ReLU}(W_d \cdot x_{n,t} + b_d) \qquad (5)$$

where $W_d \in \mathbb{R}^{q \times |\mathcal{A}|}$ is the weight matrix, $b_d \in \mathbb{R}^q$ is the bias vector.

To predict a target day $d_{n,t}$, we concatenate the diagnosis embedding $E_g(g^{(n)})$ and the preorder days' embedding $E_d(d_{n,t-w}), \ldots, E_d(d_{n,t-1})$ that within the context window. Similar to Med2Vec [7], we employ a binary vector $y_{n,t} \in \{0,1\}^{|\mathcal{A}|}$ (the i-th element is 1 if $d_{n,t}$ contains a_i) as the predict label for $d_{n,t}$, which shows better performance than $x_{n,t}$ and $E_d(d_{n,t})$. Therefore, a softmax classifier is used to produce the prediction result:

$$\hat{y}_{n,t} = \text{Softmax}(W_s \cdot Concat[E_g(g^{(n)}), E_d(d_{n,t-w}), \ldots, E_d(d_{n,t-1})] + b_s) \quad (6)$$

where $W_s \in \mathbb{R}^{|\mathcal{A}| \times (p+q*w)}$ and $b_s \in \mathbb{R}^{|\mathcal{A}|}$ is the weight matrix and bias vector for softmax.

Cross entropy between the label vector $y_{n,t}$ and our prediction vector $\hat{y}_{n,t}$ is used as the objective function. That is, Eq. 3 is achieved by minimizing the follow loss:

$$\mathcal{L} = -\frac{1}{N}\sum_{n=1}^{N}\frac{1}{T_n}\sum_{t=2}^{T_n}(y_{n,t}^\top \log\hat{y}_{n,t} + (1 - y_{n,t})^\top \log(1 - \hat{y}_{n,t})) \quad (7)$$

It is worth mentioning that we predict each day in a visit without the first day. Because the first day, corresponding to inpatients' admission, usually contains some activities that used to make a definite diagnosis by doctors. These activities are strongly related to the individuals, so that it is unreasonable to predict them by only diagnosis information. For other days (except the first day) whose preorder days' size less than the window size w, we use zero vector to complete the Concat operation in Eq. 6. For example, given $d_{n,2}$ with $w = 3$, the Eq. 6 would be $\hat{y}_{n,2} = \text{Softmax}(W_s \cdot Concat[E_g(g^{(n)}), \mathbf{0}, \mathbf{0}, E_d(d_{n,1})] + b_s)$.

To summarize, there are 5 categories parameters to be learned, including G, W_d, b_d, W_s and b_s. We train the network by back-propagation and stochastic gradient descent. The computational complexity is as follows:

$$\mathcal{O}(|\mathcal{D}||\mathcal{A}|(p + qw)) \quad (8)$$

4   Experiments

In this section, experiments demonstrate the effectiveness and efficiency of the proposed method in medical feature representation for different CP analysis tasks. We begin with the description of dataset and the experimental setting. Then we describe the three CP analysis tasks for evaluation, and show the detailed results. Finally, we give a computational complexity discussion and a case study. The source code is available at https://github.com/wuyuxiaobi/RoMCP.

4.1   Dataset

We use a real-world claims data in the experiments, and the detail statistics are summarized in Table 2. It was collected from the New Rural Cooperative Medical

System (NRCMS) [18] of a Chinese city, which covers the medicare of 2 million rural residents. The data consists of 538,945 visits, and each of them corresponds to a diagnosis code that follows the ICD-10[5]. There are totally 4,496,230 recorded days in these visits. Here, a **recorded day** refers to a day in a visit that has at least one medical activity claim record. Therefore, the size of recorded days of a visit may be not equal to its length-of-stay (**LOS**, the actual hospitalized days of a visit from admission to discharge). The day sequence discussed in Sect. 3 refers the recorded days. We excluded the visits that meet the following criteria: (1) a visit has less than two recorded days; (2) a visit whose size of recorded days is less than half of the LOS. In the following discussion, we use day to refer to recorded day.

Table 2. Statistics of our dataset

# of visits	538,945	# of medical codes	4,216
# of days	4,496,230	# of disease codes	7,232
Avg. # of activities per day	14.32	Avg. of length of stay	8.75
Max. # of activities per day	182	Max. of length of stay	360

4.2   Baseline Methods

To demonstrate the performance of our proposed model for representation learning, we compared it with the following methods:

- **Raw vector model (RVM).** Each day is represented by a vector that concatenating the real-value vector of activity $(x_{n,t})$ and the one-hot vector of diagnosis (with dimension 7,232). It can be seen as the baseline without any embedding method.
- **Stacked autoencoder (SAE).** We first train a three-layer stacked autoencoder [2] for each day by minimizing the reconstruction error of the raw vector that mentioned in RVM. Then the learned representation is used as the initial value for further representation learning by incorporating temporal relations (similar to Eqs. 6 and 7, without our diagnosis embedding).
- **Med2Vec.** Med2Vec is a skip-gram based algorithm for representation learning of medical features, which takes both code-level and visit-level information into account. Since in Med2Vec the visits are temporally ordered and the medical codes within a visit are unordered, we use the day concept in our dataset as the visit concept in Med2Vec to train the day representation. In addition, Med2Vec takes binary vector as the input, so that we fed it by concatenating the binary vector of activity $(y_{n,t})$ and the one-hot vector of diagnosis for each day.

---

[5] It refers to the 10th revision of the International Statistical Classification of Diseases and Related Health Problems that listed by the World Health Organization. In our dataset, an Chinese version is used for NRCMS.

4.3 Experimental Setting

We randomly divided the dataset into training, validation and testing set in a 0.7:0.1:0.2 ratio. The validation set is used to tune the hyper-parameters: diagnosis embedding size $p \in \{100, 150, 200, 250, 300\}$, day embedding size $q \in \{100, 150, 200, 250, 300\}$, context window size $w \in \{1, 2, 3, 4\}$, regularization coefficient $L_2 \in \{0.1, 0.01, 0.001, 0.0001\}$, and dropout rate $\epsilon \in \{0.0, 0.2, 0.5, 0.8\}$. The optimal value selected for RoMCP training is $p = 150$, $q = 200$, $w = 2$, $L_2 = 0.001$, and $\epsilon = 0.2$. For the compared methods, we use the same parameters (without p). Specially, Med2Vec is a skip-gram model that the window size is half of ours. All the methods are implemented with TensorFlow 1.3.0[6]. We train each models for 10 epochs, and Adadelta [33] with a mini-batch of 1,024 days is used.

4.4 Evaluation

An important evaluation criteria for representation learning method is that if the learned low-dimensional embedding can capture the core structures underlying the high-dimensional input data, which are useful for different machine learning tasks. Therefore, we designed three typical CP analysis tasks with quantitative measurements to evaluate the representation quality.

- **Diagnosis Clustering.** This task aims to evaluate the capability of diagnosis embedding for capturing key topics of the disease. We use the hierarchy of ICD-10 to group the 7,232 diagnosis into 432 categories (e.g. I63.902 is mapped to I63). It is worth mentioning that the diagnosis embedding of compared methods can be extracted from the weight matrix (the part corresponded to diagnosis vector) between input layer and day embedding [7]. We apply K-means for the clustering (implemented by scikit-learn 0.19.0[7]), and take normalized mutual information (NMI) as the measurement [1].
- **Next Day Recommendation.** CP provides activity recommendation for next day. In RoMCP and SAE, we mimic the prediction procedure to train the representation. Thus, the prediction results can be used to evaluate the recommendation performance. Specifically, given the target next day $d_{n,t}$, we calculate $Recall@k$ for the correctly predicted medical activities in top k of $\hat{y}_{d,t}$ as the measurement.

$$Recall@k = \frac{|A_{n,t}|}{\min(k, |d_{n,t}|)} \qquad (9)$$

where $|d_{n,t}|$ refers to the number of medical activities that actually occurred in $d_{n,t}$, $A_{n,t}$ is the intersection between $d_{n,t}$ and the top k of $\hat{y}_{d,t}$. In the experiments, we vary k from 5 to 30. In addition, a variance of $Recall@k$ with adaptive k, which is denoted as $Recall@A = \frac{|A_{n,t}|}{|d_{n,t}|}$, is also used as the measurement.

---

[6] https://www.tensorflow.org/.
[7] http://scikit-learn.org/stable/.

Since Med2Vec is a skip-gram model, we use the learned representation to train another next day prediction model for this task. Similarly, for RVM, we use the raw vector to train the prediction model.

- **LOS Prediction.** LOS is one of the most important indicators for inpatients management. Patients with long LOS stand for higher expenditure, longer beds occupancy and lower resource turnover ratio [13]. In this experiment, given the beginning 1/2/3 days of a visit, we train a logistic regression (implemented by scikit-learn 0.19.0) to predict that whether the inpatient would be discharged in one week. Area Under the ROC Curve (AUC) is used to measure the binary classification.

4.5   Results

Diagnosis Clustering. Table 3 reports the clustering performance for all the methods excepts RVM, which has no diagnosis embedding. We can observe that our proposed method, RoMCP outperforms the baselines. SAE achieves the similar performance to RoMCP, and Med2Vec shows the weakest conformity. The significant performance gap between Med2Vec and the other two methods may stem from the different utilization of daily temporal relations. RoMCP and SAE use the exact ordering of days to train the representation, while Med2Vec is a skip-gram model that only considering the context window. Therefore, RoMCP and SAE can fully take all the temporal information into consideration. In addition, Med2Vec takes binary vector as the input, which ignores the importance of the numerical features for disease.

Next Day Recommendation. Table 4 shows the experimental results with *Recall@k* measurements. It is observed that the performance of all the methods improves as *k* increases. Compared to RVM, the three embedding methods show improvement, which demonstrates the advantage of representation learning in

Table 3. Diagnosis clustering measured by NMI

	SAE	Med2Vec	RoMCP
NMI	0.5409	0.3029	**0.5557**

Table 4. Next day recommendation measured by Recall

Model	Recall@A	Recall@k						
		5	10	15	20	25	30	50
RVM	0.5613	0.3354	0.5477	0.6532	0.6935	0.7277	0.7484	0.8103
SAE	0.5772	0.3787	0.5752	0.6602	0.7028	0.7252	0.7432	0.7863
Med2Vec	0.6179	0.3829	0.6149	0.6996	0.7351	0.7585	0.7766	0.8268
RoMCP	**0.7684**	**0.4555**	**0.7031**	**0.7965**	**0.8393**	**0.8522**	**0.8674**	**0.9050**

Table 5. LOS prediction measured by AUC

Model	# of beginning days		
	1	2	3
RVM	0.7626	0.7833	**0.7962**
SAE	0.7460	0.7559	0.7637
Med2Vec	0.7327	0.7283	0.7328
RoMCP	**0.7650**	**0.7888**	0.7894

next day recommendation task. RoMCP consistently shows the best performance across various k. Since in this task the most remarkable difference of RoMCP is the independent diagnosis embedding, it suggests that our diagnosis embedding can better grasp the key topics of the disease.

LOS Prediction. The results for LOS prediction are summarized in Table 5. Since the type and number of the medical activities used in the beginning several days in part reflect the patient's condition and treatment schedule, in most cases the longer given days contribute more to AUC metric. As can be seen from the table, our method get the best performance when given the beginning 1/2 days. Unlike the next day recommendation task, it is surprising that RVM achieves a competitive performance. A possible reason is that RVM contains more complete information about the past days, especially the entire numerical information, which are important for LOS prediction. It can be also inferred from the weak performance of Med2Vec, which takes binary vector as the input.

4.6   Computational Complexity Study

We show the execution time of the three representation learning methods in Fig. 2, and all the experiments were conducted on a machine with 24 cores of Intel(R) Xeon(R) CPU E5-2620 v3 @2.40 GHz, 256 GB memory and one Nvidia K40m Tesla cards. As can be seen from the Table, all the three methods show linear time complexity on different proportions of data. SAE takes the longest time due to the reconstruction procedure (one epoch for reconstruction, and one epoch for representation learning). In Med2Vec, the code-level loss dominates the complexity, so that it takes longer time than RoMCP. It is worth mentioning that although the structure of RoMCP is similar to doc2vec [17], we map each diagnosis rather than each visit to a vector space, so that we have linear time complexity according to the number of training instances.

4.7   Case Study

In this subsection, we select a typical visit of ICH (whose Chinese official CP is shown in Fig. 1) to demonstrate the interpretability of our embedding.

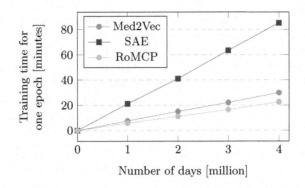

Fig. 2. Running time of different size of data.

Table 6. Top medical activities of ICH discovered by RoMCP. All of them have been classified into the four categories (we extract biochemical assay from examination as an independent category). The number in bracket represents the ranking of the medical activity in Eq. 10.

Category	Medical activities
Medication	(2) Glycerin fructose, (3) Vitamin C injection, (4) Sodium chloride, (5) Glucose, (7) Glucose injection, (7) Vitamin B6 injection, (10) Vitamin B6, (11) Tranexamic acid, (12) Vitamin C, (16) Glucose and sodium chloride, (19) Potassium chloride, (36) Mannitol, (41) Aminocaproic acid, (42) Pantoprazole, (43) Omeprazole
Biochemical assay	(6) Seromucoid, (8) Cystatin, (13) Plasma fibrinogen, (15) Thrombin time, (22) Plasma prothrombin time, (23) GGT, (24) Serum globulin, (28) ALT, (30) Chloride, (31) APTT, (32) TBA, (33) Sodium, (34) ABO blood subgroup, (35) ALP, (36) Serum uric acid, (37) Inorganic phosphorus, (40) Apo AI, (42) 5'-NT, (43) TBil, (44) AST, (45) Apo B, (48) Treponema pallidum specific antibody
Nursing	(14) Infusion, (15) Urinary intubation, (25) Absorbing oxygen cylinders, (26) Oxygen Inhalation, (27) Retention intubation, (29) oral care, (39) II-level nursing, (47) Intensive care unit, (50) Arterial and venous intubation
Examination	(1) Ambulatory electroencephalogram, (18) Blood oxygen saturation monitoring, (20) Electrocardiogram monitoring (ECG), (21) Brain CT

For each diagnosis, we can derive the most important medical activities (top k) from the diagnosis embedding and softmax weight matrix as follows:

$$argsort(W_s[0:p] \cdot E_g(g^{(n)})[0:k] \tag{10}$$

where argsort returns the top k activities with the largest values.

Table 6 shows the top 50 medical activities with the strongest value in Eq. 10. Compared to the official CP (as shown in Fig. 1), all these medical activities are critical for ICH treatment, such as brain computed tomography (CT), glycein fructose and various biochemical examinations.

Table 7 shows parts of a visit of ICH. Given the first day (column 1) contains some basic medications, nursing and admission examinations (biochemical assay), the prediction (column 3) derived from the day embedding includes two new examinations (brain CT and ECG) and the drugs in-use with high ranking, while no biochemical assay that already happened. This prediction conforms to the truth of the next day (column 2) with $Recall@A = 0.9565$. It demonstrates that RoMCP can effectively grasp the temporal relations between days.

Table 7. A demo visit for the case study. Total # refers to the number of medical activities in the day. BA refers to biomedical assay. For the category with many activities, we just list the number of the activities in it.

Day1-True	Day2-True	Day2-Predict
Total #: 58	**Total #**: 23	**Total #**: 23
BA: 36	**BA**: 0	**BA**: <u>0</u>
Nursing: 11	**Nursing**: 11 = Day1	**Nursing**: 10 = Day2-True (*except* vein intubation)
Medication: 11	**Medication**: 11 = Day1	**Medication**: 11 = Day2-True
Examination: 0	**Examination**: Brain CT	**Examination**: <u>Brain CT</u> (Rank 5), ECG (Rank 22)

5   Conclusion

In this paper, we present a representation learning approach, RoMCP, that embeddings medical features for CP analysis. RoMCP mimics the practice of CP by incorporating both diagnosis information and temporal relations so that it can effectively tackles the key challenges in inpatient medical data, including semi-temporality, numerical sensitivity and interpretability. The core topics of a disease are captured by the diagnosis embedding, and each day embedding is derived from the diagnosis embedding together with the preorder days. Meanwhile, the method is scalable and without depending on expert knowledge. Experimental results on a large real-world claims dataset prove the effectiveness and efficiency of the proposed RoMCP for CP analysis tasks.

A number of open problems need to be solved to allow further development of the proposed representation learning work. One direction is to introduce examination data, such as laboratory tests results and radiological examination reports. These data can help us to identify the patient status, which are useful for the representation learning. Another direction is to explore an adaptive strategy to determine the context window size for better usage of the temporal relations.

Acknowledgments. This work was supported by The National Key Technology R&D Program (No. 2015BAH14F02), and Project 61325008 (Mining and Management of Large Scale Process Data) supported by NSFC.

References

1. Andrews, N.O., Fox, E.A.: Recent Developments in Document Clustering (2007)
2. Bengio, Y., Lamblin, P., Popovici, D., Larochelle, H.: Greedy layer-wise training of deep networks. In: Proceedings of NIPS, pp. 153–160 (2007)
3. Binder, M., et al.: On analyzing process compliance in skin cancer treatment: an experience report from the evidence-based medical compliance cluster (EBMC^2). In: Ralyté, J., Franch, X., Brinkkemper, S., Wrycza, S. (eds.) CAiSE 2012. LNCS, vol. 7328, pp. 398–413. Springer, Heidelberg (2012). https://doi.org/10.1007/978-3-642-31095-9_26
4. Bouarfa, L., Dankelman, J.: Workflow mining and outlier detection from clinical activity logs. J. Biomed. Inform. **45**(6), 1185–1190 (2012)
5. Caron, F., Vanthienen, J., Vanhaecht, K., Van Limbergen, E., De Weerdt, J., Baesens, B.: Monitoring care processes in the gynecologic oncology department. Comput. Biol. Med. **44**, 88–96 (2014)
6. Choi, E., Bahadori, M.T., Schuetz, A., Stewart, W.F., Sun, J.: Doctor AI: predicting clinical events via recurrent neural networks. In: Proceedings of MLHC, pp. 301–318 (2016)
7. Choi, E., Bahadori, M.T., Searles, E., Coffey, C., Thompson, M., Bost, J., Tejedor-Sojo, J., Sun, J.: Multi-layer representation learning for medical concepts. In: Proceedings of KDD, pp. 1495–1504. ACM (2016)
8. Choi, E., Bahadori, M.T., Song, L., Stewart, W.F., Sun, J.: GRAM: graph-based attention model for healthcare representation learning. In: Proceedings of KDD, pp. 787–795. ACM (2017)
9. Choi, E., Bahadori, M.T., Sun, J., Kulas, J., Schuetz, A., Stewart, W.: Retain: An interpretable predictive model for healthcare using reverse time attention mechanism. In: Proceedings of NIPS, pp. 3504–3512 (2016)
10. Choi, E., Schuetz, A., Stewart, W.F., Sun, J.: Using recurrent neural network models for early detection of heart failure onset. J. Am. Med. Inform. Assoc. **24**(2), 361–370 (2016)
11. Choi, Y., Chiu, C.Y.I., Sontag, D.: Learning low-dimensional representations of medical concepts. In: AMIA Summits on Translational Science Proceedings 2016, p. 41 (2016)
12. De Vine, L., Zuccon, G., Koopman, B., Sitbon, L., Bruza, P.: Medical semantic similarity with a neural language model. In: Proceedings of CIKM, pp. 1819–1822. ACM (2014)
13. Harutyunyan, H., Khachatrian, H., Kale, D.C., Galstyan, A.: Multitask learning and benchmarking with clinical time series data. arXiv preprint arXiv:1703.07771 (2017)
14. Huang, Z., Dong, W., Ji, L., Gan, C., Lu, X., Duan, H.: Discovery of clinical pathway patterns from event logs using probabilistic topic models. J. Biomed. Inform. **47**, 39–57 (2014)
15. Huang, Z., Dong, W., Ji, L., He, C., Duan, H.: Incorporating comorbidities into latent treatment pattern mining for clinical pathways. J. Biomed. Inform. **59**, 227–239 (2016)
16. Huang, Z., Lu, X., Duan, H.: Latent treatment pattern discovery for clinical processes. J. Med. Syst. **37**(2), 1–10 (2013)
17. Le, Q., Mikolov, T.: Distributed representations of sentences and documents. In: Proceedings of ICML, pp. 1188–1196 (2014)

18. Li, C., Hou, Y., Sun, M., Lu, J., Wang, Y., Li, X., Chang, F., Hao, M.: An evaluation of China's new rural cooperative medical system: achievements and inadequacies from policy goals. BMC Public Health **15**(1), 1079 (2015)

19. Lipton, Z.C., Kale, D.C., Elkan, C., Wetzell, R.: Learning to diagnose with LSTM recurrent neural networks. arXiv preprint arXiv:1511.03677 (2015)

20. Ma, F., Chitta, R., Zhou, J., You, Q., Sun, T., Gao, J.: Dipole: diagnosis prediction in healthcare via attention-based bidirectional recurrent neural networks. In: Proceedings of KDD, pp. 1903–1911. ACM (2017)

21. Mans, R.S., Schonenberg, M.H., Song, M., van der Aalst, W.M.P., Bakker, P.J.M.: Application of process mining in healthcare – a case study in a Dutch hospital. In: Fred, A., Filipe, J., Gamboa, H. (eds.) BIOSTEC 2008. CCIS, vol. 25, pp. 425–438. Springer, Heidelberg (2008). https://doi.org/10.1007/978-3-540-92219-3_32

22. Mikolov, T., Chen, K., Corrado, G., Dean, J.: Efficient estimation of word representations in vector space. arXiv preprint arXiv:1301.3781 (2013)

23. Mikolov, T., Sutskever, I., Chen, K., Corrado, G.S., Dean, J.: Distributed representations of words and phrases and their compositionality. In: Proceedings of NIPS, pp. 3111–3119 (2013)

24. Nguyen, P., Tran, T., Wickramasinghe, N., Venkatesh, S.: Deepr: a convolutional net for medical records. J. Biomed. Health Inf. **21**(1), 22–30 (2017)

25. Pham, T., Tran, T., Phung, D., Venkatesh, S.: DeepCare: a deep dynamic memory model for predictive medicine. In: Bailey, J., Khan, L., Washio, T., Dobbie, G., Huang, J.Z., Wang, R. (eds.) PAKDD 2016. LNCS (LNAI), vol. 9652, pp. 30–41. Springer, Cham (2016). https://doi.org/10.1007/978-3-319-31750-2_3

26. Poelmans, J., Dedene, G., Verheyden, G., Van der Mussele, H., Viaene, S., Peters, E.: Combining business process and data discovery techniques for analyzing and improving integrated care pathways. In: Perner, P. (ed.) ICDM 2010. LNCS (LNAI), vol. 6171, pp. 505–517. Springer, Heidelberg (2010). https://doi.org/10.1007/978-3-642-14400-4_39

27. Prodel, M., Augusto, V., Xie, X., Jouaneton, B., Lamarsalle, L.: Discovery of patient pathways from a national hospital database using process mining and integer linear programming. In: T-ASE, pp. 1409–1414. IEEE (2015)

28. Rojas, E., Munoz-Gama, J., Sepúlveda, M., Capurro, D.: Process mining in healthcare: a literature review. J. Biomed. Inform. **61**, 224–236 (2016)

29. Rovani, M., Maggi, F.M., de Leoni, M., van der Aalst, W.M.: Declarative process mining in healthcare. Expert Syst. Appl. **42**(23), 9236–9251 (2015)

30. Xu, X., Jin, T., Wang, J.: Summarizing patient daily activities for clinical pathway mining. In: Proceedings of Healthcom, pp. 1–6. IEEE (2016)

31. Xu, X., Jin, T., Wei, Z., Lv, C., Wang, J.: TCPM: topic-based clinical pathway mining. In: Proceedings of CHASE, pp. 292–301. IEEE (2016)

32. Xu, X., Jin, T., Wei, Z., Wang, J.: Incorporating domain knowledge into clinical goal discovering for clinical pathway mining. In: Proceedings of BHI, pp. 261–264. IEEE (2017)

33. Zeiler, M.D.: ADADELTA: an adaptive learning rate method. arXiv preprint arXiv:1212.5701 (2012)

34. Zhu, Z., Yin, C., Qian, B., Cheng, Y., Wei, J., Wang, F.: Measuring patient similarities via a deep architecture with medical concept embedding. In: Proceedings of ICDM, pp. 749–758. IEEE (2016)

Domain Supervised Deep Learning Framework for Detecting Chinese Diabetes-Related Topics

Xinhuan Chen[✉], Yong Zhang, Kangzhi Zhao,
Qingcheng Hu, and Chunxiao Xing

Research Institute of Information Technology, Tsinghua National Laboratory
for Information Science and Technology, Department of Computer Science
and Technology, Institute of Internet Industry, Tsinghua University,
Beijing 100084, China
{xh-chen13,zkz15}@mails.tsinghua.edu.cn,
{zhangyong05,xingcx}@tsinghua.edu.cn

Abstract. As millions of people are diagnosed with diabetes every year in China, many diabetes-related websites in Chinese provide news and articles. However, most of the online articles are uncategorized or lack a clear or unified topic, users often cannot find their topics of interest effectively and efficiently. The problem of health text classification on Chinese websites cannot be easily addressed by applying existing approaches, which have been used for English documents, in a straightforward manner. To address this problem and meet users' demand for diabetes-related information needs, we propose a Chinese domain lexicon, adopt some professional diabetes topic explanations as domain knowledge and incorporate them into deep learning approach to form our topic classification framework. Our experiments using real datasets showed that the framework significantly achieved a higher effectiveness and accuracy in categorizing diabetes-related topics than most of the state-of-the-art benchmark approaches. Our experimental analysis also revealed that some health websites provided some incorrect or misleading category information.

Keywords: Domain knowledge · Stacked Denoising Autoencoders
Healthcare · Chinese

1 Introduction

Diabetes is a common chronic disease. According to the survey by the International Diabetes Federation, 109.6 million people were diagnosed with diabetes and more than 13 million people died of diabetes in China in 2015 [9]. To meet the increasing demand for health information and knowledge, many diabetes-related websites in Chinese provide various resources and services, including health news, articles, discussion forums, and online patient communities.

© Springer International Publishing AG, part of Springer Nature 2018
J. Pei et al. (Eds.): DASFAA 2018, LNCS 10828, pp. 53–71, 2018.
https://doi.org/10.1007/978-3-319-91458-9_4

The documents of the news, articles and discussion posts in China have two key characteristics. First, the quality of the documents is uneven in various authors, for example healthcare professionals and experts produce many high-quality health news and articles based on their domain knowledge, while a lot of low-quality and uncategorized documents are generated by patients and small website editors. Second, the classification standards of topics of the documents differ in various sources, the websites provided by the authoritative and official institutions have more clear and unified standards for a certain disease than the forums and communities websites. Even though there is the same classification standard for different websites, the results of classification for a certain disease may be different. This situation is especially obvious in developing countries that have less healthcare experts and official institutions. As a result, most of the news and articles are uncategorized and lack a clear or unified topics, making it very time-consuming and overwhelming for users to browse and search for information about specific topics from the search engines (e.g., Google) or special health websites (e.g., tnbz.com). While automatically categorizing topics out of uncategorized health-related articles could be very useful for various types of health information users, including patients and their family members, health-care professionals (e.g., physicians and nurses), and researchers. Especially, it can help newly diagnosed patients find valuable educational materials for self-management of their diseases and health conditions more effectively.

Many approaches and methods have been proposed for categorizing topics in English articles in various application domains, including the healthcare and medical domain [6,33]. Unfortunately, the work of health-related topic classification of online Chinese articles has been comparatively unexplored. The problem of health-related topic classification of online Chinese articles is quite challenging. It cannot be easily addressed by applying existing approaches and methods, which have been used for English articles, in a straightforward manner due to the difference between Chinese and English, the lack of Chinese medical lexicons, and the special characteristics of Chinese online health articles. First, Chinese is based on ideographic writing systems, whose structure and grammar are quite different from those of English, which is based on alphabetic [24]. For instance, since there is no space between Chinese characters, it is more difficult to parse Chinese sentences into unambiguous word segments. Second, many prior studies adopt a standard medical knowledge base, UMLS (Unified Medical Language System) [2] provided by U.S. National Library of Medicine, to extract medical terms and features when categorizing English articles. Unfortunately, there has not been a standard Chinese medical knowledge base available. Third, there has been a lack of widely adopted standards for categorizing online Chinese health articles. Because of the lack of diet, medicine related classification, ICD-10 (International Classification of Diseases 10th version) is not enough for healthcare research. Consequently, the topic categories of many online Chinese articles are misleading.

To address these challenges, we develop a Chinese domain lexicon, adopt some professional diabetes topic explanations and incorporate them into deep learning approach to form our topic classification framework. Deep learning [7] is a promising machine learning approach. Among many different deep learning architectures, RNN based word embedding model [14] is a promising text classification method with a large scale corpus training. But with limited online Chinese articles available, we introduce domain knowledge to help improve performance of classification. In order to represent the domain knowledge better, we tried many models and finally choose Stacked Denoising Autoencoders (SDA) [28] model because its loss function is easy to be modified [11]. In this research, We develop deep learning based models for text classification application that identify topics related to diabetes on Chinese online articles. The main contributions of this work are summarized as follow:

- We develop a Chinese domain lexicon, adopt some professional diabetes topic explanations as domain knowledge and incorporate them into deep learning approach to form our topic classification framework.
- We propose domain supervised SDA that incorporates domain knowledge into the process of model training, which makes use of limited auxiliary domain knowledge to help improve performance of topic classification.
- Our experiments show that our methods significantly outperform the state-of-the-art text classification techniques for diabetes-related articles including basic deep learning model SDA and reveal some health websites provide some incorrect or misleading category information.

The remainder of this paper is organized as follows. We present a review of literature on deep learning and healthcare text classification, and then describe our topic classification framework, and two SDA-based deep learning models. Next, we report on our experiments and discuss the results in detail. Finally, we conclude this paper.

2 Related Work

In this section, we review related work about text classification methods, especially the deep learning methods, and their applications in health-related domains.

Text classification is a set of important and well-developed classification methods for categorizing the growing number of electronic documents worldwide. Text classification has been widely used in natural language processing and information retrieval applications, including Web page classification, spam filtering, email routing, genre classification, readability assessment, and sentiment analysis [3]. Deep learning is a latest advanced representation-learning approach [15]. The various deep learning architecture are widely used in the processing of images, audios, videos and text documents [4,23,31,32]. Images and sounds can be easily represented as input feature vectors, deep learning is a natural technique for image processing and speech recognition applications. However, it is difficult to find appropriate feature vectors to represent text. In other words, when representing text documents, the feature vectors are usually based on bag-of-words or other feature extraction methods, which often cause information loss. Word2vector [14] is a promising embedding model for computing vector representations of text documents, but it cannot receive good performance without a large scale corpus training. Only a limited number of studies have employed the deep learning techniques to process small text corpus. Nolle et al. applied denoising autoencoders to deal with anomaly detection in noisy business process event text logs [17]. Wang et al. proposed SDA based model to learn more effective text representation for tag recommendation [29]. As a result, it remains a major challenge to use deep learning techniques in text classification applications with small text corpus. In this study, we identify and combine feature vectors by using different methods to represent text in SDA-based model with better classification performance.

On the other hand, some researchers [27,30] have used domain knowledge to improve the classification performance. Sinha and Zhao [27] compared the performance of seven classification methods with and without incorporating domain knowledge. They found that incorporation of domain knowledge significantly improves classification performance. Thus we introduce auxiliary domain knowledge to help further enhance the classification performance with limited online Chinese articles available.

As a data mining and text mining approach, classification has been used to categorize electronic medical documents (e.g., medical literatures and clinical records) and Web documents into meaningful topics in the health-related domains. [15] reviewed the recent literature on deep learning technologies for health-related domain. Some studies often rely on a standard domain lexicon during the process of document classification. Liu et al. [12] assessed the effectiveness of several traditional learning models against a number of performance metrics based on a relational graph database of clinical entities. Sibunruang and Polpinij [25] leveraged an ontology of Cancer Technical Term Net to select keyword-based features, where this ontology is used as a lexicon. Nie et al. [16] proposed a novel deep learning scheme to infer the possible diseases given the questions of health seekers, they obtained features of bag-of-words with the aid of MetaMap tool. However, there has not been a standard medical lexicon available in Chinese, not to mention a medical ontology. This adds more difficulty for the task of categorizing medical and health-related documents in Chinese.

We focus on articles about diabetes because it is one of the most widely studied chronic diseases and there are many diabetic patients in China. Simon et al. [26] applied association rule mining for EMR to discover sets of risk factors and their corresponding subpopulations that represent patients at particularly high risk of developing diabetes. In this research, we develop a Chinese domain lexicon and some professional diabetes-related topic explanations and incorporate them into the deep learning framework to identify topics related to diabetes on Chinese Web articles.

3   The Topic Classification Framework

Our framework includes four stages: data collection, data preprocessing, model training, and topic classification.

3.1   Data Collection

We developed a Web crawler to fetch and download online articles from Chinese health websites. We used the keyword "diabetes" to locate diabetes-related pages. Figure 1 presents an example of collected pages in our dataset. Text parsers were used to extract various fields from the pages including article ID, URL, title, article source, posted time, and article body as shown in Fig. 1. Navigation paths (e.g., Home Page > Diabetes Information > Child Diabetes) that provide category information were also extracted. Note that not all websites contain navigation paths. We chose the pages with navigation paths as the training and testing data in the experiments. Additionally, some pages contain tags or keywords (e.g., "Child Diabetes", "Diabetes Complications") that highlight the focuses of the content. These tags and keywords were extracted as well. We focus on the article body of the pages and refer to them as articles in the following sections. From all the navigation paths in the Web pages, we concluded 26 distinct diabetes-related terms and compiled definitions and explanations for these 26 terms based on classical books for Medicine and Diabetes [35]. Based on which, we created a professional diabetes vocabulary.

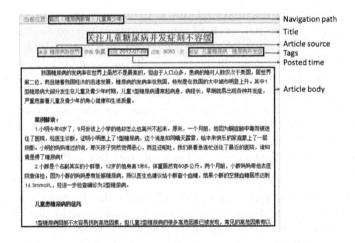

Fig. 1. An example of a diabetes-related Web page

For example, the term 低血糖 (Hypoglycemia) is defined as 低血糖反应是最常见 的反应。常见于胰岛素过量、注射胰岛素后未按时进餐或活动量过大所致 (Hypoglycemia is low blood sugar.

It is one of the most common symptoms of diabetes, often caused by insulin overdose, missing meals after insulin injections, or excessive physical activities.). In this study we used this professional diabetes vocabulary and their definitions and explanations as the **domain knowledge** for helping the diabetes-related topic classification.

3.2   Data Preprocessing

Data preprocessing prepares the data for the model training stage in our framework. Data preprocessing consists of five steps: lexicon creation, word segmentation, topic category map construction, article annotation, and feature extraction.

Lexicon Creation. A domain lexicon, which contains terms in a specific domain, is usually used for feature extraction and word segmentation in text mining applications. Unfortunately, there has not been a standard domain lexicon in Chinese for diabetes. To build this lexicon, we combined entries from Diabetes Dictionary App [18], which was the easy available commercial mobile application for Chinese diabetic patients, and the extracted tags from diabetes-related Web pages to obtain a relatively complete Chinese diabetes lexicon. The resulting lexicon contains 1,065 terms related to diabetes including its medication, treatment, care, and prevention.

Word Segmentation. We used a Chinese word segmentation tool, ICTCLAS to remove stop words and perform word segmentation for the content bodies of Web pages.

Topic Category Map Construction. Our diabetes topic category map is a tree structure with nested levels of topic categories related to diabetes. The top level consists of main topic categories, each of which is broken down to lower-levels of sub-categories. The map was built based on both the professional vocabulary (with the 26 terms) and the navigation paths extracted from the web pages in the data collection stage. Because most navigation paths of the web pages comprised non-professional, layperson terms, we made a semantic mapping between the professional vocabulary and the extracted navigation paths with the help of domain experts. Figure 2 presents the resulting topic category map with six main categories on the top level and sub-category levels. Because of the space limit, some lower-level sub-categories are not shown on the map.

Article Annotation. This is an optional step in this framework and can be skipped if all the collected Web pages contain navigation paths that can be used as ground truth, namely category topic labels. For pages without navigation paths, they should be manually annotated based on the diabetes topic category map for the model training and testing purposes.

Feature Extraction. The features used as input in the deep learning based model are then extracted from the articles. In prior studies [34], the bag-of-words (BOW) approach has often been employed to generate features for text data in deep learning models. In this approach, each distinct word in an article is treated

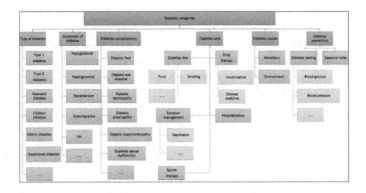

Fig. 2. Diabetes topic category tree

as a feature. However, this approach can result in a large but sparse feature vector, significantly affecting the performance of deep learning models [21]. To reduce the dimensionality of the feature vectors, we combined 1,065 diabetes-related terms from our Chinese diabetes domain lexicon and 123 feature values using the TFIDF method [22]. The first 1,065 features were binary, indicating whether the corresponding term appears in the article or not. The remaining 123 features were calculated by obtaining the N greatest mean weights from TFIDF output using a certain threshold. As a result, each article was represented as a feature vector of 1,188 dimensions. We will demonstrate the effectiveness of the combined features in the experiments.

3.3   Model Training and Topic Classification

The SDA-based models will be described in detail in the next section. The feature vectors representing the articles serve as the inputs to the deep learning based models. For each article, the models produce 26 probability values, each of which corresponds to one category node on the topic category map.

In the topic classification stage, the primary topic category of the article is identified against the topic category map. For each article, this primary topic category is mapped to one of the six main category nodes on the first level of the tree as shown in Fig. 2.

4   The SDA-Based Models

We propose two Stacked Denoising Autoencoders(SDA) based models for the topic classification. Specifically, the first model is called TSSDA that incorporates topic of corresponding input data for supervision into the SDA model, the second model is called DSSDA and incorporates the domain knowledge supervision into the SDA model. Figures 3 and 4 present the architecture of our TSSDA and DSSDA models.

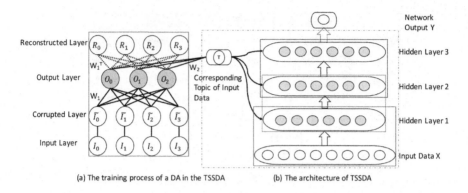

Fig. 3. The training process and architecture of TSSDA

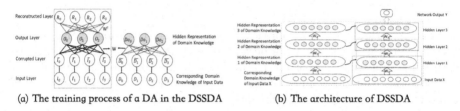

Fig. 4. The training process and architecture of DSSDA

4.1   Topic Supervised Stacked Denoising Autoencoders

Our task is to classify the diabetes-related articles to the suitable topics, a simple idea is to guide the unsupervised pre-training process with topic supervision so that the trained hidden layers of SDA are topic-specialized and improve performance of topic classification. The idea is similar with the SDA with sentiment supervision [11].

Figure 3 shows how we incorporate topic supervision into the SDA. We add a fully connected logistic regression layer and use softmax function to predict the topic of input data. Thus output layer in a DA (a stochastic version of autoencoder [8] that randomly sets some values of the input to zero in corrupted layer, the stochastic corruption process forces the output layer to discover more robust features and avoid simply learning the identity) is not only used for reconstructing the input, but also predicting the topic label. In this way, the TSSDA is encouraged to extract topic-specific features better in the classification task. This is what we expected will be demonstrated in the experiments.

$$Loss(I, R) = -\sum_j (I_j log R_j + (1 - I_j) log(1 - R_j)) \quad (1)$$

In Fig. 3(a), W_1 is the weights of the links between the corrupted layer and the output layer, W_2 is the weights of the links between the output layer and the topic of the input data. Mathematically, the probability that an output data O is a member of a topic t can be written as:

$$P(T = t|O) = softmax(OW_2 + c) = \frac{e^{OW_2^{(t)}+c^{(t)}}}{\sum_{j=1}^{tN} e^{OW_2^{(j)}+c^{(j)}}} \qquad (2)$$

where c is node biases of logistic regression layer, tN is the number of topics in our diabetes-related articles classification. It is very common to use the negative log-likelihood as the loss function in the case of muticlass logistic regression. Thus we define the loss function of topic supervision as follow:

$$loss(T) = -logP(T = t|O) \qquad (3)$$

By combining the reconstruction loss (Eq. 1) and topic supervision loss, we obtain the new loss function of the DA in TSSDA as follow:

$$loss_{TSSDA} = loss(I, R) + \lambda loss(T) \qquad (4)$$

where λ is used to control the balance of the two loss function between 0 to 1. We still use stochastic gradient descent to minimize the new loss function under all training data.

4.2   Domain Supervised Stacked Denoising Autoencoders

If incorporating topic supervision in the SDA model introduced in previous section is effective, a more novel idea is to combine domain knowledge (e.g. diabetes topics's professional definitions and explanations) into the process of topic classification using the SDA, enforcing the hidden representation of input data more domain-favorable.

As depicted in Fig. 4(a), the network models in the left corresponding domain knowledge of the input data are **shared parameters** with the DA in the right that models the input data and will output the hidden representation of domain knowledge whose dimensions are the same as the number of dimension of output layer in the right DA. The process mathematically expresses as follow:

$$p(O_k = 1|\widetilde{I}) = sigmoid(b_k + \sum_j (\widetilde{I}_j W_{jk})) \qquad (5)$$

$$p(Do_k = 1|\widetilde{D}) = sigmoid(b_k + \sum_j (\widetilde{D}_j W_{jk})) \qquad (6)$$

where Do_k and $\widetilde{D}_j$ are the states for the k-th node in hidden representation layer of domain knowledge and the j-th node in the corrupted layer of domain knowledge, respectively.

In order to measure the degree of approximation between the hidden representation layer of the domain knowledge and the output layer of the input data, we apply cross-entropy again to express the domain supervision loss as follow:

$$loss(Do, O) = -\sum_j (Do_j logO_j + (1 - Do_j)log(1 - O_j)) \qquad (7)$$

where the distributions of Do and O can be obtained from Eqs. 5 and 6. We naturely obtain the new loss function of the DA in DSSDA through combining the reconstruction loss and domain supervision loss as follow:

$$loss_{DSSDA} = loss(I, R) + \eta loss(Do, O) \tag{8}$$

where η controls the balance of the two loss function between 0 to 1. We use stochastic gradient descent to minimize the new loss function under all training data. Thus we stack the domain supervised DA as the building blocks as shown in Fig. 4(b).

At the end, for each article, the SDA-based models select the category with the maximum probability value. All parameters obtained in the SDA-based models training process are used to test the performance of the model in the experiments.

5   Experiments

We conducted the experiments to evaluate the performance of the SDA-based models for categorizing diabetes-related topics. In this section, we report the results and findings from these experiments.

5.1   The Datasets

We collected all pages posted between July 2010 and September 2013 from two dedicated Chinese health websites for diabetes. A summary of the two datasets is shown in the Dataset 1 column and Dataset 2 column in Table 1. Some pages contain tags such as "High Satiety" (高饱腹感) and "Diet Control" (饮食控制). There are in total 912 distinct tags in Dataset 1 and Dataset 2.

Table 1. Dataset statistics summary

Dataset	Dataset 1	Dataset 2	Dataset 3
Website	tnbz.com	zzcxhg.com	Combined
No. of articles	3936	15682	1000
No. of all tags	6933	7023	–
No. of distinct tags	888	49	–

To evaluate our models' performance for uncategorized Web pages, we created Dataset 3 by randomly selecting 1,000 Web pages out of Datasets 1 and 2 (see the Dataset 3 column in Table 1). We removed the navigation paths in the original pages and manually annotated the 1,000 articles. Two graduate students with diabetes knowledge annotated the data (with 0.88 inter-coder reliability); and the remaining disagreement was resolved by in-person discussions between the two students.

5.2   Evaluation Metrics and Benchmarks

The evaluation metrics of the experiments are *precision, recall, F-measure* and *accuracy*. We selected the following six techniques as the benchmarks.

- **SDA.** This model uses only one SDA without supervision in unsupervised pre-traing stage and takes only the articles from the websites as the input.
- **DBN.** The training process of DBN (Deep Belief Network [8]) is very similar with the SDA, because both model invole the unsupervised layer-wise pre-training followed by supervised fine-tuning. The main difference is that the SDA use DAs instead of RBMs. This model uses only one DBN module and takes only the articles from the websites as the input.
- **SVM** (Support Vector Machine). SVM is one of the most popular and effective classification techniques, it has been adopted in medical research [1].
- **GNB** (Gaussian Naive Bayes). GNB [5] is a simple probabilistic classifier implementing Bayes' theorem and has been shown to perform superior in some text classification tasks.
- **PE** (Perceptron). Perceptron [20] is one of the first artificial neural networks and suitable for supervised classification.
- **DT** (Decision Trees). DT [19] is a non-parametric supervised learning method for classification.

5.3   Comparison Results

To compare the effectiveness of the DSSDA and TSSDA with the benchmarks, we conducted 10-fold cross validation on each of the three datasets. We also report the p-values of the t-tests for comparing between DSSDA and each of the benchmarks for F-measure (only consider F-measure here because it is a comprehensive metric incorporating both Precision and Recall) and accuracy, respectively. The values of the default parameters in the models are presented in Table 2. In the training stage, we set the learning rates for the training and tuning processes to be 0.1, and the number of iterations to be 100 and 200, respectively.

Table 2. The values of the default parameters in the models

Parameters	DBN	SDA	TSSDA	DSSDA
Dimensions of input	1188	1188	1188	1188
Dimensions of domain topic explanation input	–	–	–	1188
No. of hidden layers	3	3	3	3
Ratio of input values to zero in each DA's corrupted layer	–	0.05	0.05	0.05
Dimensions of hidden layers	594	594	594	594
Dimensions of output	26	26	26	26
Control factor of in the loss function	–	–	0.05	0.05

In our implementation, the deep learning related models contain three hidden layers with 594 dimensions (i.e., half of the number of input dimensions) in each layer. In Tables 3 through 5, the metric values that are significantly less than those of the DSSDA model are highlighted with asterisks. The last four columns report the average values of metrics over the 10-fold cross validations.

Table 3 summarizes the results of the performance comparison among proposed models and the benchmarks over Dataset 1. Compared with the benchmark models, the DSSDA model is significantly better than all benchmarks including the TSSDA in F-measure and accuracy. As a result, the supplementary information provided by the domain knowledge helps enhance the performance of the deep learning. Table 4 presents the performance comparison results over Dataset 2. It shows that the performance is worse than that over Dataset 1. This is because the distribution in dataset 2 is more uneven, which results in many indistinguishable class boundaries. For example, there are no articles in 7 out of the 26 (27%) categories in this dataset. Since the model always outputs 26 probabilities values, it may assign some articles into categories that do not exist in the dataset. However, even though the DSSDA's F-measure and accuracy are lower than those in Dataset 1, these two measures are still significantly greater than those of the benchmarks. Table 5 shows the results from Dataset 3. It can be seen that when the category distributed is more even (with no missing data in any category), all classification models' performances are better, and that the DSSDA is even significantly more effective and accurate than the TSSDA model.

Table 3. Performance comparison over Dataset 1

Method	Precision%	Recall%	F-measure%	Accuracy%
DSSDA	71.40	64.91	67.92	78.52
TSSDA	68.80	64.57	**66.52***	**77.76***
SDA	67.39	64.60	**65.90***	**77.28***
DBN	62.67	62.93	**62.73*****	**74.78*****
SVM	79.65	51.07	**62.14*****	**75.14*****
GNB	42.00	50.60	**45.88*****	**53.33*****
PE	64.50	59.10	**61.47*****	**73.69*****
DT	51.97	51.77	**51.83*****	**67.28*****

*: $p < 0.05$, ***: $p < 0.001$

In summary, the DSSDA model is significantly more effective and accurate in categorizing topics in diabetes-related Chinese Web pages than mainstream classification models, including SVM, GNB, PE, and DT. In most cases, the DSSDA, TSSDA and SDA models perform betther than the DBN model in terms of F-measure and accuracy. Therefore, the SDA-based model is more suitable for categorizing topics by using domain knowledge in diabetes-related Chinese Web pages.

Table 4. Performance comparison over Dataset 2

Method	Precision%	Recall%	F-measure%	Accuracy%
DSSDA	46.36	38.83	42.16	56.58
TSSDA	43.79	37.94	**40.60***	**55.63****
SDA	44.47	38.47	**41.18***	**55.55***
DBN	38.07	36.60	**37.31*****	**52.96*****
SVM	58.33	27.55	**37.29*****	**43.32*****
GNB	27.73	30.55	**29.04*****	**39.97*****
PE	38.50	36.40	**37.29*****	**54.28***
DT	33.25	31.72	**32.45*****	**49.21*****

*: $p < 0.05$, **: $p < 0.005$, ***: $p < 0.001$

Table 5. Performance comparison over Dataset 3

Method	Precision%	Recall%	F-measure%	Accuracy%
DSSDA	55.95	59.74	57.28	64.3
TSSDA	50.63	55.66	**52.46*****	**60.4****
SDA	51.52	55.11	**52.77****	**60.2****
DBN	49.45	53.89	**51.16****	**61.1***
SVM	61.05	33.16	**42.62****	**52.3***
GNB	36.74	44.54	**39.69*****	**47.1****
PE	52.80	53.70	**52.72***	**60.8***
DT	45.70	49.47	**47.23*****	**56.0****

*: $p < 0.05$, **: $p < 0.005$, ***: $p < 0.001$

5.4 Extra Comparison Results Using English Datasets

In order to demonstrate the effectiveness of domain supervision SDA model on text dataset in other language, we select the small number of English datasets to do some extra experiments. We investigated some diabetes-related research using English text and obtained the dataset of diabetes online community of the American Diabetes Association from the paper [13].

We selected three types topics: gestational diabetes, type 1 diabetes and type 2 diabetes. After randomly selecting 300 English articles of each topic from the dataset, we got a small English dataset with 900 articles in total. Professional diabetes topics explanations of three topics as domain knowledge are achieved by applying **UMLS** to extract related text. We used TFIDF method to extract features for each English articles and then conducted 10-fold cross validation on the dataset. The Table 6 shows the performance among proposed SDA-based models and the SDA model.

Compared with the basic SDA models, our DSSDA model is significantly better than basic SDA including the TSSDA in F-measure and accuracy. As a

Table 6. Performance comparison over English Dataset

Method	Precision%	Recall%	F-measure%	Accuracy%
DSSDA	89.25	89.46	89.35	89.22
TSSDA	87.42	87.61	**87.51***	**87.33***
SDA	87.88	88.14	**88.01***	**87.78***

*: $p < 0.05$

result, the supplementary information provided by the domain knowledge helps enhance the performance in our proposed model.

5.5 The Effects of the Parameter Values

In order to find best performance of our SDA-base models, we conducted additional experiments to examine the effects of the parameters on model performance. Especially, we focused on the effects of the dimensions of hidden layers, the number of hidden layers, the ratio of corrupted input values and the control factors of the loss function in the SDA-based models. We used the default values for other parameters (e.g., learning rate). Dataset 3 with smallest articles was used for these experiments.

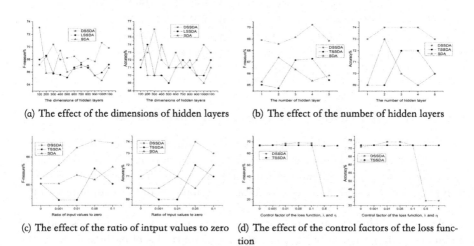

(a) The effect of the dimensions of hidden layers   (b) The effect of the number of hidden layers

(c) The effect of the ratio of intput values to zero   (d) The effect of the control factors of the loss function

Fig. 5. The effects of parameter values

Figure 5(a) shows the performance of the SDA-base models in terms of the number of nodes (dimensions) of the hidden layers. As we can see from the Figure, as the dimension of hidden layers increases, the performance of the models goes up first and then goes down. This indicates that for the SDA-based models, having hidden units about the half of the number of input dimensions is sufficient to train the model.

Figure 5(b) displays the F-measures and accuracy of the DSSDA as a function of the number of layers. As the number of layers in the DSSDA model increases, the DSSDA's effectiveness increases until it reaches its top at the point for four layers. The right of the Fig. 5(b) presents the changes in models' effectiveness in response to the changes in the number of layers showing a tendency of a convex peak. There is a trade-off between the performance and training time, thus we chose to use three layers in the SDA-based models.

Figure 5(c) shows the performance of the SDA-base models in terms of the ratio of input values to zero in each DA's corrupted layer. As we can see from the Figure, as the ratio of input values to zero increases, the performance of the models goes up first, then goes down and reaches top at the point 0.05. Thus we chose to use this point to train the models.

We report the performance of DSSDA and TSSDA among different control factors of the loss functions in the Fig. 5(d). As the control factors of the loss function in the DSSDA increases, the DSSDA's performance increases until it reaches its top at the point 0.05 and then goes down fast. However, the performance of TSSDA is steady for the changes in the control factors of the loss function in the TSSDA. Therefore, we set the control factors as 0.05 for training the models.

These results also show that the performance of DSSDA is better than TSSDA and SDA in general.

5.6   The Effects of the Feature Vectors

We also investigated the effects of different feature vectors on our model. In this study, we firstly applied paragraph2vector (a word2vector based model for documents representation) [10] to do preliminary experiment, but it had poor performance for the lack of large scale corpus. Finally we used two types of features: 1,065 binary features from the diabetes lexicon and the 123 features by using the TFIDF method. We call them BOW (bag of words) and TFIDF, respectively. The combined features, which was what we used in the above experiments, is called B-TFIDF. To demonstrate the value of B-TFIDF, we compared it with the BOW and TFIDF feature vectors over Dataset 3. The performance of the DSSDA model by using the three different feature vectors are shown in Table 7. Our results demonstrate that the combined feature vectors, B-TFIDF, yields the best performance in terms of F-measure and accuracy.

5.7   Causes of Misclassification Errors

To further investigate and analyze the causes of misclassification errors, we selected some samples misclassified by the DSSDA model. We randomly selected 100 articles out of Dataset 1 and 15% of these articles were misclassified by the

Table 7. Performance using different feature vectors

Method	Precision%	Recall%	F-measure%	Accuracy%
B-TFIDF	55.95	59.74	57.28	64.3
BOW	52.37	57.58	**54.29***	**62.9**
TFIDF	36.75	40.63	**38.26****	**47.1****

*: $p < 0.05$; **: $p < 0.005$

DSSDA model. We conducted an indepth analysis of the 15 articles and identified three possible causes for the misclassification errors: **wrong label**, **multiple topics**, and **model error**.

- **Wrong label.** We found that 7% of the articles were "misclassified" by the DSSDA because of their navigation paths, which were used as topic category labels in our experiments, were actually wrong. In other words, the DSSDA model assigned correct topics based on the articles' contents and the topic category map. For example, the article's category label provided by the health website is Glucose Tests, while the article actually describes the diabetes complications.
- **Multiple topics.** Because our model selected only one primary topic category for each article, an article may be misclassified if it contains multiple topics. We found that 3% of the articles were misclassified due to this reason. For example, one article covers two topics: Diabetes Diet and Diabetes Complications; and the original website used only Diabetes Diet as the topic label. Since our model outputs Diabetes Complications as the topic category, it was treated as a kind of misclassification error.
- **Model error.** For the remaining 5% of the misclassified articles, the model simply did not capture the main point of the articles. For instance, "Hyperglycemia" was mistakenly assigned by the DSSDA model as the topic category regarding glucose testing.

In short, if the training data are correctly labeled, the DSSDA model will achieve better performance in accuracy. The cases of **wrong label** and **multiple topics** show that some health websites provide some categorized articles but the category information is incorrect or misleading. There is also a lack of widely accepted standard for categorizing diabetes-related topics. These issues are also the reasons that we propose this SDA-based framework. We provide an effective approach (and standard) to categorize diabetes-related topics from online Chinese articles.

6 Conclusion

In this study, we propose a deep learning based framework for categorizing diabetes-related topics on Chinese health websites. Our experiments using real data show that the DSSDA model significantly outperforms several state-of-the-art benchmark classification methods without the large scale corpus training.

The contributions of our research are summarized as follow. First, we develop a Chinese domain lexicon, adopt some professional diabetes topic explanations and incorporate them into deep learning approach to form our topic classification framework. Second, we propose domain supervised SDA that incorporates domain knowledge into the process of model training, which makes use of limited auxiliary domain knowledge.

One limitation of our model is that the training process is time-consuming, affecting the model's scalability for large datasets.

Our future work will be done in two directions: (a) seek more collaboration with physicians and medical professionals to refine the topic category map; and (b) extend and apply our framework to other chronicle diseases and even other domains.

Acknowledgments. This work was supported by NSFC (91646202), NSSFC (15CTQ028), the National High-tech R&D Program of China (SS2015AA020102), Research/Project 2017YB142 supported by Ministry of Education of The People's Republic of China, the 1000-Talent program and Tsinghua University Initiative Scientific Research Program.

References

1. Adeva, J.G., Atxa, J.P., Carrillo, M.U., Zengotitabengoa, E.A.: Automatic text classification to support systematic reviews in medicine. Expert Syst. Appl. **41**(4), 1498–1508 (2014)
2. Bodenreider, O.: The unified medical language system (UMLS): integrating biomedical terminology. Nucleic Acids Res. **32**(Suppl. 1), 267–270 (2004)
3. Bollegala, D., Mu, T., Goulermas, J.Y.: Cross-domain sentiment classification using sentiment sensitive embeddings. TKDE **28**(2), 398–410 (2016)
4. Charalampous, K., Gasteratos, A.: A tensor-based deep learning framework. Image Vis. Comput. **32**(11), 916–929 (2014)
5. Cheeseman, P., Kelly, J., Self, M., Stutz, J., Taylor, W., Freeman, D.: Autoclass: a Bayesian classification system. In: Readings in Knowledge Acquisition and Learning, pp. 431–441 (1993)
6. Chen, X., Zhang, Y., Xing, C., Liu, X., Chen, H.: Diabetes-related topic detection in chinese health websites using deep learning. In: Zheng, X., Zeng, D., Chen, H., Zhang, Y., Xing, C., Neill, D.B. (eds.) ICSH 2014. LNCS, vol. 8549, pp. 13–24. Springer, Cham (2014). https://doi.org/10.1007/978-3-319-08416-9_2
7. Hinton, G., Osindero, S., Teh, Y.W.: A fast learning algorithm for deep belief nets. Neural Comput. **18**(7), 1527–1554 (2006)
8. Hinton, G.E., Salakhutdinov, R.R.: Reducing the dimensionality of data with neural networks. Science **313**(5786), 504–507 (2006)
9. IDF: IDF Diabetes Atlas, 7th edn. (2016). http://www.diabetesatlas.org/component/attachments/?task=download&id=116
10. Le, Q., Mikolov, T.: Distributed representations of sentences and documents. In: ICML, pp. 1188–1196 (2014)

11. Liu, B., Huang, M., Sun, J., Zhu, X.: Incorporating domain and sentiment supervision in representation learning for domain adaptation. In: IJCAI, pp. 1277–1283 (2015)
12. Liu, W., Sweeney, H.J., Chung, B., Glance, D.G.: Constructing consumer-oriented medical terminology from the web a supervised classifier ensemble approach. In: Pham, D.-N., Park, S.-B. (eds.) PRICAI 2014. LNCS (LNAI), vol. 8862, pp. 770–781. Springer, Cham (2014). https://doi.org/10.1007/978-3-319-13560-1_61
13. Liu, X., Chen, H.: AZDrugMiner: an information extraction system for mining patient-reported adverse drug events in online patient forums. In: Zeng, D., Yang, C.C., Tseng, V.S., Xing, C., Chen, H., Wang, F.-Y., Zheng, X. (eds.) ICSH 2013. LNCS, vol. 8040, pp. 134–150. Springer, Heidelberg (2013). https://doi.org/10.1007/978-3-642-39844-5_16
14. Mikolov, T., Sutskever, I., Chen, K., Corrado, G.S., Dean, J.: Distributed representations of words and phrases and their compositionality. In: NIPS, pp. 3111–3119 (2013)
15. Miotto, R., Wang, F., Wang, S., Jiang, X., Dudley, J.T.: Deep learning for healthcare: review, opportunities and challenges. Brief. Bioinform. **1**, 11 (2017). bbx044
16. Nie, L., Wang, M., Zhang, L., Yan, S., Zhang, B., Chua, T.S.: Disease inference from health-related questions via sparse deep learning. TKDE **27**(8), 2107–2119 (2015)
17. Nolle, T., Seeliger, A., Mühlhäuser, M.: Unsupervised anomaly detection in noisy business process event logs using denoising autoencoders. In: Calders, T., Ceci, M., Malerba, D. (eds.) DS 2016. LNCS (LNAI), vol. 9956, pp. 442–456. Springer, Cham (2016). https://doi.org/10.1007/978-3-319-46307-0_28
18. Omesoft: Diabetes dictionary. http://shouji.baidu.com/software/item?docid=1018036888&from=as
19. Quinlan, J.R.: Induction of decision trees. Mach. Learn. **1**(1), 81–106 (1986)
20. Rosenblatt, F.: The perceptron, a perceiving and recognizing automaton Project Para. Cornell Aeronautical Laboratory (1957)
21. Salakhutdinov, R., Hinton, G.: Semantic hashing. Int. J. Approx. Reason. **50**(7), 969–978 (2009)
22. Salton, G., McGill, M.J.: Introduction to Modern Information Retrieval (1986)
23. Sarikaya, R., Hinton, G.E., Deoras, A.: Application of deep belief networks for natural language understanding. IEEE/ACM Trans. Audio Speech Lang. Process. (TASLP) **22**(4), 778–784 (2014)
24. Schmitt, B.H., Pan, Y., Tavassoli, N.T.: Language and consumer memory: the impact of linguistic differences between Chinese and English. J. Consum. Res. **21**, 419–431 (1994)
25. Sibunruang, C., Polpinij, J.: Ontology-based text classification for filtering cholangiocarcinoma documents from PubMed. In: Ślęzak, D., Tan, A.-H., Peters, J.F., Schwabe, L. (eds.) BIH 2014. LNCS (LNAI), vol. 8609, pp. 266–277. Springer, Cham (2014). https://doi.org/10.1007/978-3-319-09891-3_25
26. Simon, G.J., Caraballo, P.J., Therneau, T.M., Cha, S.S., Castro, M.R., Li, P.W.: Extending association rule summarization techniques to assess risk of diabetes mellitus. TKDE **27**(1), 130–141 (2015)
27. Sinha, A.P., Zhao, H.: Incorporating domain knowledge into data mining classifiers: an application in indirect lending. Decis. Support Syst. **46**(1), 287–299 (2008)
28. Vincent, P., Larochelle, H., Bengio, Y., Manzagol, P.A.: Extracting and composing robust features with denoising autoencoders. In: ICML, pp. 1096–1103. ACM (2008)

29. Wang, H., Shi, X., Yeung, D.Y.: Relational stacked denoising autoencoder for tag recommendation. In: AAAI, pp. 3052–3058 (2015)
30. Wang, J., Wang, Z., Zhang, D., Yan, J.: Combining knowledge with deep convolutional neural networks for short text classification. In: IJCAI, pp. 2915–2921 (2017)
31. Wu, Z., Jiang, Y.G., Wang, J., Pu, J., Xue, X.: Exploring inter-feature and inter-class relationships with deep neural networks for video classification. In: MM, pp. 167–176. ACM (2014)
32. Xu, W., Sun, H., Deng, C., Tan, Y.: Variational autoencoder for semi-supervised text classification. In: AAAI, pp. 3358–3364 (2017)
33. Yang, H., Kundakcioglu, E., Li, J., Wu, T., Mitchell, J.R., Hara, A.K., Pavlicek, W., Hu, L.S., Silva, A.C., Zwart, C.M., et al.: Healthcare intelligence: turning data into knowledge. IEEE Intell. Syst. **29**(3), 54–68 (2014)
34. Yin, W., Schütze, H.: Deep learning embeddings for discontinuous linguistic units. arXiv preprint arXiv:1312.5129 (2013)
35. Yin, X.: Diabetology. Shanghai Scientific and Technical Publishers (2003)

Security and Privacy

Publishing Graph Node Strength Histogram with Edge Differential Privacy

Qing Qian[1], Zhixu Li[1], Pengpeng Zhao[1], Wei Chen[1],
Hongzhi Yin[2], and Lei Zhao[1(✉)]

[1] School of Computer Science and Technology, Soochow University, Suzhou, China
qqian@stu.suda.edu.cn, {zhixuli,ppzhao,zhaol}@suda.edu.cn,
wchzhg@gmail.com
[2] The School of Information Technology and Electrical Engineering Brisbane,
The University of Queensland, St Lucia, Australia
db.hongzhi@gmail.com

Abstract. Protecting the private graph data while releasing accurate estimate of the data is one of the most challenging problems in data privacy. Node strength combines the topological information with the weight distribution of the weighted graph in a natural way. Since an edge in graph data oftentimes represents relationship between two nodes, edge-differential privacy (edge-DP) can protect relationship between two entities from being disclosed. In this paper, we investigate the problem of publishing the node strength histogram of a private graph under edge-DP. We propose two clustering approaches based on sequence-aware and local density to aggregate histogram. Our experimental study demonstrates that our approaches can greatly reduce the error of approximating the true node strength histogram.

Keywords: Differential privacy · Node strength
Histogram publishing

1 Introduction

Many kinds of private data can be well represented by graph data, e.g., social network activities [1–4], communication patterns [5], and disease transmission [6]. Due to the sensitivity of these valuable network data, they cannot be directly exploited by analysts. Even if the private data is anonymized, publishing graph data has risk to reveal sensitive information of an individual [7,8]. To tackle the problem, Differential Privacy (DP) has been widely used to design sanitization mechanisms for publishing information of graphs [9]. Based on the original description of differential privacy, it guarantees that the change of one record will not significantly affect the output distribution of an analysis procedure. This model is very effective for releasing data in the form of histogram, since the magnitude of the statistical noise is often dominated by random variation in the data [10].

© Springer International Publishing AG, part of Springer Nature 2018
J. Pei et al. (Eds.): DASFAA 2018, LNCS 10828, pp. 75–91, 2018.
https://doi.org/10.1007/978-3-319-91458-9_5

In the context of graph data, two interpretations of differential privacy have been proposed: *edge* and *node* differential privacy [11]. Intuitively, edge differential privacy ensures that the algorithm's output does not reveal the inclusion or removal of a particular edge in the graph, and node differential privacy hides the inclusion or removal of a node together with all its adjacent edges. In edge-DP, two graphs are neighboring if they differ on a single edge. The meaning of an edge in the graph could connote friendship, email exchange, etc. [12] assumes that edge represents a sensitive relationship that should be kept private. For some applications, edge-DP seems to be a reasonable privacy standard. For example, consider the study of [13], in which they analyze a graph derived from the email communication among students and faculty of a large university. What makes this dataset sensitive is that it reveals who emails whom, and edge-DP can protect email relationships from being disclosed.

Numerous recent studies [10,12,14,15] release differentially private graphs under edge-DP and node-DP. Many of them focus on publishing the node degree of graphs [12,14,15]. However, the degree distribution of a graph cannot fully represent the importance of nodes. Bitcoin OTC [16] is a who-trusts-whom network of people who trade using Bitcoin on a platform. The edge weight is the distrust level of a transaction and the node degree is the number of transactions. In this case, node degree cannot represent a user's reputation. And an individual edge does not provide a general picture of all transactions. Instead, we can use node strength to measure the importance of nodes. Node strength has combined the topological information with the weight distribution of the network in a natural way. The quantity measures the strength of nodes in terms of the total weight of their connections. A more significant measure of the network properties in terms of the actual weights is obtained by studying the node strength.

In this paper, we investigate the problem of publishing the node strength histogram of a weighted graph under edge differential privacy. Given a graph $G = (V, E, W)$, the goal is to release a node strength histogram that approximates the true distribution of G as much as possible while satisfying edge differential privacy. A key challenge is that the true distribution is too sparse. If the maximum node strength is far greater than the number of nodes, the values of a larger number of bins will be zero. This means that when the query interval is large, the noise accumulation will lead to the low accuracy of query results. To address this problem, we propose the following solutions. First, we reduce the number of bins of histogram by setting a upper bound t for edge weights. Note that, given a larger t, more weight information can be preserved. Meanwhile, a lower t means that the distribution of node strength can be less sparse. Then we generate a node strength histogram of the weight bounded graph. Furthermore, We privately learn a partitioning of the bins B and replace the value of each bin with the mean of its group's sum. Finally, to reduce the errors caused by noise, transformation and aggregation, we use exponential mechanism to output the accurate estimate of node strength histogram. We also prove that publishing the node strength histogram under edge-DP has a sensitivity of 4.

Our main contributions can be summarized as follows.

1. To the best of our knowledge, we are the first to study the problem of releasing node strength histogram on weighted graph under edge differential privacy.
2. We propose two efficient clustering approaches, sequence-aware and density-based, to group histogram under differential privacy, and design a low-sensitivity quality function to obtain a trade-off between the error of noise and aggregation.
3. We have conducted extensive experiments on four real world datasets. The experimental results demonstrate that our proposed mechanisms have significant improvement over the baseline systems. We also perform the introspective analysis to investigate the impacts of weight bound t and aggregation strategies B.

The rest of this paper is organized as follows. In Sect. 2, we give the problem definition, discuss ϵ-differential privacy and its application on graphs. Section 3 introduces our proposed approaches. Experimental results are given in Sect. 4. We discuss related work in Sect. 5, and conclude in Sect. 6.

2 Preliminaries

2.1 Node Strength

A weighted and finite graph $G = (V, E, W)$ is defined by a set of nodes $V = \{v_1, v_2, \ldots, v_n\}$, a set of edges $E = \{e_{ij} | i, j \in V\}$, and $W = \{w_{ij} | w_{ij} = value(e_{ij})\}$, where $value(e_{ij})$ represents the connection weight of edge e_{ij}. In this paper, for simplicity of presentation, we assume that $n = |V|$, the number of nodes of the input graph G, is publicly known.

We generally extend the degree of a node v_i to the sum of weights of edges that connect v_i, denoted by node strength, which is given as follows:

$$s_i = \sum_{j \in \Gamma(i)} w_{ij}$$

where $\Gamma(i)$ is the set of adjacent nodes of node v_i.

2.2 Differential Privacy

The notion of (ϵ, δ)-differential privacy [17] is defined based on the concept of *neighboring databases*. Two datasets G and G' are defined as neighboring datasets if they only differ in one record.

Definition 1 ((ϵ, δ)-Differential Privacy). *A randomized algorithm $\mathcal{A}$ is (ϵ, δ)-differential privacy if for all events S in the output space, and for any two neighboring databases G and G', we have*

$$Pr[\mathcal{A}(G) \in S] \leq \exp(\epsilon) \times Pr[\mathcal{A}(G') \in S] + \delta$$

where $S \subseteq Range(\mathcal{A})$.

When $\delta = 0$, the algorithm is ϵ-differential privacy. The parameter ϵ refers to the privacy budget and a smaller ϵ represents a stronger privacy level. For queries that produce numerical outputs, the differential privacy can be satisfied by adding appropriately random noise to the answer. The noise depends on the query's sensitivity.

Definition 2 (Global Sensitivity). *For a query $f : G \to \mathbb{R}^d$, and neighboring databases G and G', the l_1-global sensitivity of f is defined as:*

$$\triangle f = \max_{G \simeq G'} ||f(G) - f(G')||_1$$

where the L1 distance $||x||_1$ is the sum of the absolute values of each element of the vector x.

While there are many approaches to achieve differential privacy, the best known and most-widely used two for this purpose are the Laplace mechanism [17] and the exponential mechanism [18].

Laplace Mechanism. For a query $f : G \to \mathbb{R}$ over a database G, we use Laplace mechanism to satisfy ϵ-DP, noise $Lap(\triangle f/\epsilon)$ is added to the output of the query.

$$\mathcal{A}(G) = f(G) + Lap(\frac{\triangle f}{\epsilon})^d$$

and $Pr[Lap(\beta) = x] = \frac{1}{2\beta}e^{-|x|/\beta}$.

Exponential Mechanism. Given a quality score $u(G, H_i)$ for outputting H_i on input G and a randomized algorithm $\mathcal{A}$, we have

$$\mathcal{A}(G, H_i) = \left\{ H_i : |Pr[H_i \in \mathcal{H}] \propto exp(\frac{\epsilon u(G, H_i)}{2 \triangle u}) \right\}$$

where $\triangle u = max_{\forall H_i, G, G'}|u(G, H_i) - u(G', H_i)|$ is the global sensitivity of the quality function. The sampling probability for each $H_i \in H$ is determined based on a user-specified quality function u.

Composition Properties. Differential privacy satisfies sequential composition and transformation invariance [19,20]. If an algorithm $\mathcal{A}$ runs t randomized algorithm $\mathcal{A}_1, \mathcal{A}_2, \ldots, \mathcal{A}_t$, each of which is (ϵ_i, δ_i)-differential privacy, and applies an arbitrary algorithm g to their result, i.e., $\mathcal{A}(G) = g(\mathcal{A}_1(G), \ldots, \mathcal{A}_t(G))$, then $\mathcal{A}$ is $(\sum_i \epsilon_i, \sum_i \delta_i)$-differential privacy.

3   Proposed Approaches

A graph is t-bounded if the weight of each edge is no more than t. Given an input graph G, we transform G to a t-bounded graph G_t, where the weight of each edge is defined as follows:

$$w_{ij} = \begin{cases} t & if\ value(e_{ij}) > t \\ value(e_{ij}) & otherwise \end{cases}$$

During the transformation, we use m to denote the number of edges whose weights are larger than t. Obviously, t is an important parameter to our study. On the one hand, given a larger t, we can preserve more weight information, but the number of bins in the node strength histogram will increase and more noise will be added. On the other hand, although we can obtain a dense distribution of node strength with a smaller t, more weight information will be lost.

Next, based on the graph G_t, we can generate a node strength histogram $hist(G_t) = \{h_1, h_2, \ldots h_n\}$, where h_i is the number of nodes whose node strength are i. Note that, according to the following Lemma 1, the global sensitivity of releasing $hist(G_t)$, which is defined as $\triangle hist = ||hist(G_t) - hist(G_t')||_1$, is no more than 4.

Lemma 1. *For any $G \simeq G'$ that differ in one edge, we have*

$$||hist(G_t) - hist(G_t')||_1 \leq 4$$

Proof. Assume, without loss of generality, that $G' = (V', E')$ has an additional edge e^+ compared to $G = (V, E)$, i.e. $V' = V$, $E' = E \cup \{e^+\}$. For arbitrary nodes v_i and v_j, assuming that $e^+ = (v_i, v_j)$, so we have $s_i = \sum_{k \in \Gamma(i)} w_{ik}$ and $s_i' = s_i + w(e^+)$. Similarly, $s_j = \sum_{k \in \Gamma(j)} w_{jk}$ and $s_j' = s_j + w(e^+)$. Obviously, only the node strength of v_i and v_j has changed. And one change in node strength of arbitrary node can cause two changes in the histogram, thus the change of one edge induces a difference of at most 4 in the histogram.

3.1   Sequence-Aware Clustering

The standard solution to differential privacy is to add Laplace noise directly to each bin in the original histogram, which will cause a large magnitude of noise. One way to reduce the effect of noise is to merge the adjacent bins into a group and use an average to estimate each bin within a group [21,22]. However, this will lead to extra approximate error. In this section, we introduce a sequence-aware clustering (SC) algorithm which groups neighboring bins with close values into the same bucket. As shown in Algorithm 1, the algorithm takes the node strength histogram $H_t = \{h_1, h_2, \ldots, h_n\}$ as input as well as ϵ, where the parameter ϵ represents the privacy budget. For each given k, we partition H_t into k disjoint sets, e.g., $B = \{b_1, b_2, \ldots, b_k\}$, and each $b_i \in B$ is a group of contiguous bins in H_t.

ALGORITHM 1. Sequence-aware Clustering algorithm

Input: Original histogram H_t, number of buckets k, privacy budget ϵ, and the number of iterations N

Output: The clustered buckets B.

1 Choose k candidates $c_1, ..., c_k$ randomly from the histogram H_t
2 **repeat**
3   **for** *each* $h_i \in H_t$ **do**
4   compute the cumulative distance $dist(c_l, h_i)$, $dist(h_i, c_r)$ between h_i and two neighboring candidates c_l, c_r
5   **end**
6   Associate each h_i with the nearest candidate c_j, and partition H_t into k sets $b_1, ..., b_k$
7   **for** $1 \leq j \leq k$ **do**
8          $sum'(b_j) = sum(b_j) + Lap(\frac{\triangle SC}{\epsilon})$, $num'(b_j) = num(b_j) + Lap(\frac{\triangle SC}{\epsilon})$
9   compute the mean of the bins in b_j, $mean'(b_j) = sum'(b_j)/num'(b_j)$
10   set the nearest bin from $mean'(b_j)$ as the new candidate c_j
11   **end**
12 **until** *the number of iterations reaches* N;
13 **return** B

Algorithm 1 has two steps. First, it randomly selects k bins as initial candidates $c_1, \ldots, c_k$. Then for each bin in H_t, the algorithm determines its associated candidate according to the sequence-aware cumulative distance. The sequence-aware cumulative distance $dist(c_l, h_i)$ and $dist(h_i, c_r)$ between h_i and two neighboring candidates c_l, c_r is defined as:

$$dist(c_l, h_i) = \frac{\sum_{h_k \in (c_l, h_i]} |h_k - c_l|}{i - p}$$

$$dist(h_i, c_r) = \frac{\sum_{h_k \in [h_i, c_r)} |h_k - c_r|}{q - i}$$

where p and q is the position of c_l and c_r in the histogram H_t respectively.

If $dist(c_l, h_i) < dist(h_i, c_r)$, then we associate all bins between c_l and h_i to c_l. Otherwise, the bins between h_i and c_r are associated to c_r. The results of this step is k sets $\{b_1, \ldots, b_k\}$.

Second, we update the candidate c_j according to the mean of all bins in the set b_j. Although computing the nearest mean of any one bin would break privacy, as mentioned in [23], to compute an average among an unknown set b_j can be equivalent to compute the $sum(b_j)$ and divide by $num(b_j)$, where $sum(b_j)$ is the sum of the bins, and $num(b_j)$ is the number of the bins in the set b_j. Thus, the computation only needs to expose the approximate cardinalities of the b_j, instead of the sets themselves. And the k candidate means provide a differential private approximation to the mean in updating step. In our case, the denominators $num(b_j)$ will not change, and the numerators $sum(b_j)$ has sensitivity at most 2. Therefore, the global sensitivity of SC is not more than 2. We can compute a new set of candidate means by dividing the approximate sum of the bins $sum'(b_j)$ by the approximation to the cardinality $num'(b_j)$.

Lemma 2. *For any* $G \simeq G'$ *that differ in one edge, we have*

$$||SC(hist(G_T)) - SC(hist(G'_T))||_1 \leq 2$$

Proof. We follow the idea and notations in the proof of Lemma 1. Since neighboring databases G and G' only differ at one edge, which induces a difference of at most 4 in the histogram. And there must be at most 4 bins where $h_i \neq h_i'$. Assume, without loss of generality, an additional edge causes the bin h_{x1}, h_{x2} to add by 1 and h_{x3}, h_{x4} to minus 1. In the worst case, $sum(b_j)$ can change by 2, which can only happen when $h_{x1}, h_{x2} \in b_j$ and $h_{x3}, h_{x4} \notin b_j$.

ALGORITHM 2. Density-based Clustering algorithm

Input: Original histogram H_t, bucket counts k and privacy budget ϵ
Output: The clustered buckets B.
1 **for** *each* $h_i \in H_t$ **do**
2 | compute the distance with neighboring bin h_j as $d(i,j)$
3 | $\tilde{d}(i,j) = d(i,j) + Lap(\frac{\triangle DC}{\epsilon})$
4 **end**
5 **for** *each* h_i **do**
6 | compute the local density ρ_i
7 **end**
8 Select the top k bins with the highest local density as the cluster centers
9 Associate each h_i with the nearest cluster center, and partition H_t into k sets $b_1, ..., b_k$
10 **return** B

The updating rule of the clustering is iterated until the convergence condition is satisfied, or a fixed number of iterations is reached. In our approach, we fix the number of iterations into N. Then we can use the noise distribution $Lap(\triangle SC * N/\epsilon)$ to satisfy ϵ-differential privacy. Finally, we obtain a set of disjoint buckets $B = \{b_1, b_2, \ldots, b_k\}$.

3.2 Density-Based Clustering

The sequence-aware clustering method preserves ϵ-differential privacy and reduces the error of noise and approximation. However, its initial candidates are randomly selected, which cannot guarantee a stable result. In this section, we discuss another approach that adopts a different strategy, which is based on the observation that the cluster centers has a smaller fluctuation among the adjacent bins. If a bin has more neighboring bins whose values are similar to its, then the more suitable it is to serve as a cluster center. And we consider that this bin has a higher local density.

In Algorithm 2, we present the details of our proposed density-based clustering (DC) method. First, for each bin h_i, we compute its local density ρ_i, which is measured by the cumulative average distance between h_i and its neighboring bin h_j. The cumulative average distance is

$$d(i,j) = \sum_{|j-i|<\xi, l\in[i,j]} |h_l - avg(i,j)|$$

where $avg(i,j) = \frac{\sum_{|j-i|\leq\xi, l\in[i,j]} h_l}{|j-i|}$.

Note that directly computing the cumulative average distance $d(i,j)$ between any two bins cannot be done based on their true values, otherwise differential privacy will be violated. For this reason, we obtain the noisy $\widetilde{d}(i,j)$ by the Laplace mechanism using privacy parameter ϵ. Since the histogram has 4 changed bins, and Lemma 3 below shows that the total sensitivity of the computation of $d(i,j)$ is at most 6.

Lemma 3. *For any $G \simeq G'$ that differ in one edge, $d(i,j)$ is the cumulative average distance in $hist(G)$ and $d'(i,j)$ is the cumulative average distance in $hist(G')$, we have*

$$||d(i,j) - d'(i,j)||_1 \leq 6$$

Proof. Assume, without loss of generality, let $i < j$, we use $avg(i,j) = \frac{\sum_{l=i}^{j} x_l}{j-i}$ to denote the average value of bins within $[x_i, x_j]$, then $d(i,j) = \sum_{t=i}^{j} |x_t - avg(i,j)|$. Following the idea and notations in the proof of Lemma 1, the change of one edge induces a difference of at most 4 in the histogram. In the worst case, there are two bins between x_i and x_j having a difference of 2, so we have $\sum_{t=i}^{j} |x_t - x'_t| \leq 4$. And the change of sum of the bins between x_i and x_j is at most 2, i.e., $|\sum_{m=i}^{j} x_m - \sum_{m=i}^{j} x'_m| \leq 2$.

Thus, we have

$$||d(i,j) - d'(i,j)||_1 = \left| \sum_{l=i}^{j} |x_l - avg(i,j)| - \sum_{l=i}^{j} |x'_l - avg'(i,j)| \right|$$

$$\leq \left| \sum_{l=i}^{j} (x_l - avg(i,j)) - \sum_{l=i}^{j} (x'_l - avg'(i,j)) \right|$$

$$\leq \left| \sum_{l=i}^{j} (x_l - x'_l) - \sum_{l=i}^{j} (avg(i,j) - avg'(i,j)) \right|$$

$$\leq \sum_{l=i}^{j} |x_l - x'_l| + \sum_{l=i}^{j} |avg(i,j) - avg'(i,j)|$$

$$\leq \sum_{l=i}^{j} |x_l - x'_l| + (j-i) * |\sum_{m=i}^{j} x_m - \sum_{m=i}^{j} x'_m|/(j-i)$$

$$\leq \sum_{l=i}^{j} |x_l - x'_l| + |\sum_{m=i}^{j} x_m - \sum_{m=i}^{j} x'_m|$$

$$\leq 4 + 2 \leq 6$$

In fact, the proposed DC method is sensitive only to the relative magnitude of $d(i,j)$. Therefore, the global sensitivity of DC is not more than 6. Then we the local density ρ_i of h_i is defined as

$$\rho_i = \sum_j \exp(-\sqrt{\frac{\widetilde{d}(i,j)/|j-i|}{d_c}})$$

where d_c is a threshold value.

Basically, ρ_i is equivalent to the degree of proximity between the value of the neighboring bins and h_i. And for a bin with a larger local density, the probability of being selected as a cluster center should be higher. Thus given the buckets size k, we choose the top k bins with the maximum density as the cluster centers greedily. After the cluster centers have been found, each remaining bin is assigned to the nearest bucket.

ALGORITHM 3. t-Bounded-Buckets-Hist algorithm

Input: A weighted graph $G = (V, E, W)$, privacy budget ϵ, candidates T and K
Output: A histogram $\widetilde{H}$ satisfying the differential privacy
1 $\epsilon = \epsilon_1 + \epsilon_2 + \epsilon_3$
2 **for** *each $t_i \in T$* **do**
3 $\quad$ **for** *each $k_j \in K$* **do**
4 $\quad\quad$ | $\quad$ clustering a partition B_{ij} of node strength histogram of G_{t_i} with privacy budget ϵ_3
5 $\quad$ **end**
6 **end**
7 Calculate $cost(G, t_i, B_{ij}, \epsilon_1)$ for each $(t_i, B_{ij}) \in T \times B$,
8 Select (t', B') with probability proportional to $\exp(\frac{-\epsilon_2 cost(G, t_i, B_{ij}, \epsilon_1)}{2\triangle cost})$
9 Calculate average count h_k for every bin in bucket b_k
10 Add noise to counts as $\widetilde{h}_i = h_k + Lap(\frac{\triangle hist}{\epsilon_1 |b_k|})$
11 **return** $\widetilde{H}$

3.3   Finding the Least Cost Partition

When publishing node strength histogram, the proposed clustering strategies with a smaller bucket size k means less noise error is added to the histogram. However, it can lead to more approximate error. Similarly, a larger t can preserve more weight information, but the number of bins in the node strength histogram will increase and more noise will be added.

In this section, we propose a strategy which features a sophisticated evaluation of the trade-off between the approximation error due to clustering and the Laplace error due to Laplace noise injected. To do so, we design a low-sensitivity quality function to select the optimal group strategy B and weight bound t simultaneously, and generate an accurate publication of the node strength histogram H. The main framework is summarized in Algorithm 3.

The t-Bounded-Buckets-Hist method is a ϵ-differential privacy algorithm that takes as input a weighted graph G, transformation parameters T and candidates K, represented as the set of weight bounds and the number of buckets. The output is an estimate $\widetilde{H}$ of original node strength histogram H. To ensure that the overall algorithm satisfies ϵ-differential privacy, we split the total ϵ budget into ϵ_1, ϵ_2 and ϵ_3 such that $\epsilon = \epsilon_1 + \epsilon_2 + \epsilon_3$, and use these three portions of the budget on the respective stages of the algorithm.

The first step obtains a clustering result $B = \{b_1, b_2, \ldots, b_k\}$ of the histogram. We propose two novel differential privacy algorithms SC and DC that use ϵ_3 budget to partition H. Then for the partition B, the second step derives noisy estimates of the bucket counts. In the last step we derive $\widetilde{H} = \{\widetilde{h}_1, \widetilde{h}_2, \ldots, \widetilde{h}_n\}$, which is the differential private estimate of H, and $\widetilde{h}_i = h_i + f_i$. The counts $\sum_{j\in b_k} h_j$ of each bucket b_k spread uniformly amongst each bin of b_k. After playing this, the resulting estimate for h_i is:

$$\widetilde{h}_i = \frac{\sum_{j\in b_k} h_j}{|b_k|} + \frac{F_k}{|b_k|}$$

where each h_i is near the mean of the bucket $\frac{\sum_{j\in b_k} h_j}{|b_k|}$, F_k is the noise added to bucket b_k and $F_k \sim Laplace(\triangle hist/\epsilon)$.

Since the scale of F_k is fixed, larger buckets can have less noise per individual $\widetilde{h}_j$. However, larger buckets can have more approximation error. To select an optimal partition result for histogram, we design a cost function similar to that used in [22], which has three components. The first component captures the number of edges whose weights are changed after transformation. The intuition is that each of such edge will cause two changes on node strength, which results in four changes in corresponding histogram before and after transformation. Given a partition B_j and a transformation parameter t_i, the cost of this step is

$$cost_{tran}(G, t_i, B_j) = 2 * |\{v | e \in E(v), value(e) > t_i\}|$$

where $value(e)$ gives the weight of edge e in the graph and $E(v)$ contains all edges adjacent to node v.

The second component captures the error due to the aggregation step, which approximates each bin h_i in the bucket by the mean value $\frac{\sum_{s\in b_k} h_s}{|b_k|}$. And the cost of the aggregation step is

$$cost_{agg}(G, t_i, B_j) = \sum_{b_k \in B_j} \sum_{s\in b_k} (|h_s - \frac{\sum_{s\in b_k} h_s}{|b_k|}|)$$

where h_s denotes the number of nodes with node strength s in $hist(G_{t_i})$.

The third component captures the error due to the added noises produced by employing Laplace mechanism (with a budget of ϵ_1). Recall the resulting estimate for h_i is $\widetilde{h}_i = \frac{\sum_{s\in b_k} h_s}{|b_k|} + \frac{F_k}{|b_k|}$, where h_i is in the bucket b_k. The cost of this step is

$$cost_{noise}(G, t_i, B_j) = \sum_{b_k \in B_j} \sum_{s\in b_k} \frac{F_i}{|b_k|} = \sum_{b_k \in B_j} \frac{\triangle hist}{\epsilon_1} = \frac{4k}{\epsilon_1}$$

Combining the three components, the proposed cost function is

$$cost(G, t_i, B_j) = cost_{tran} + cost_{agg} + cost_{noise}$$

To apply the exponential mechanism to select an optimal (t, B) pair, we need to have a upper bound on the global sensitivity of $cost(G, t_i, B_j)$ given before. The following lemma shows that the global sensitivity is bounded by 12.

Lemma 4. *For any $G \simeq G'$ that differ in one edge, we have*

$$|cost(G, t_i, B_j) - cost(G', t_i, B_j)| \leq 12$$

Proof. We have $cost_{proj}(G, t_i) = 4 * |\{e|e \in E, value(e) > t_i\}|$, and based on the proof of Lemma 1, $|cost_{proj}(G, t_i) - cost_{proj}(G', t_i)| \leq 4$. $cost_{noise}(G, t_i, B_j) = 2k/\epsilon_2$ and $k = |B_j|$, we can see $cost_{noise}(G, t_i, B_j)$ is independent of the input dataset and thus does not change between two neighboring graphs, so we have $|cost_{noise}(G', t_i, B_j) - cost_{noise}(G, t_i, B_j)| = 0$. Since $cost(G, t_i) = cost_{noise} + cost_{proj} + cost_{group}$, we only need to analyze the part of $cost_{group}$. We have

$$\sum_{b_k \in B_j} \sum_{s \in b_k} \left(\frac{\sum_{s \in b_k} h_s}{|b_k|} - \frac{\sum_{s \in b_k} h'_s}{|b_k|} \right) = \sum_{b_k \in B_j} |\sum_{s \in b_k} h_s - \sum_{s \in b_k} h'_s| \leq \sum_{b_k \in B_j} \sum_{s \in b_k} |h_s - h'_s| = 4$$

Thus, we have

$$|cost_{group}(G, t_i, B_j) - cost_{group}(G', t_i, B_j)|$$

$$= \left| \sum_{b_k \in B_j} \sum_{s \in b_k} \left(|h_s - \frac{\sum_{s \in b_k} h_s}{|b_k|}| \right) \right|$$

$$\leq \sum_{b_k \in B_j} \sum_{s \in b_k} \left| \left(|h_s - \frac{\sum_{s \in b_k} h_s}{|b_k|}| - |h'_s - \frac{\sum_{s \in b_k} h'_s}{|b_k|}| \right) \right|$$

$$\leq \sum_{b_k \in B_j} \sum_{s \in b_k} \left| \left(h_s - \frac{\sum_{s \in b_k} h_s}{|b_k|} \right) - \left(h'_s - \frac{\sum_{s \in b_k} h'_s}{|b_k|} \right) \right|$$

$$\leq \sum_{b_k \in B_j} \sum_{s \in b_k} \left| (h_s - h'_s) - \left(\frac{\sum_{s \in b_k} h_s}{|b_k|} - \frac{\sum_{s \in b_k} h'_s}{|b_k|} \right) \right|$$

$$\leq \sum_{b_k \in B_j} \sum_{s \in b_k} |h_s - h'_s| + \sum_{b_k \in B_j} \sum_{s \in b_k} \left| \frac{\sum_{s \in b_k} h_s}{|b_k|} - \frac{\sum_{s \in b_k} h'_s}{|b_k|} \right|$$

$$\leq 4 + 4 \leq 8$$

Finally, we give the privacy guarantee of *t-Bounded-Buckets-Hist* as well as the proof in the following.

Lemma 5. *t-Bounded-Buckets-Hist in Algorithm 3 satisfies $(\epsilon_1 + \epsilon_2 + \epsilon_3)$-edge-differential privacy.*

Proof. In Algorithm 3, the step of clustering (Line 4) uses Laplace mechanism with privacy budget ϵ_3. The step of Lines 8 uses exponential mechanism with privacy budget ϵ_2. The step of publishing histogram with aggregated counts(Line 10) uses Laplace mechanism and satisfy edge-differential privacy for ϵ_1. By the composition theorem and transformation invariance, this algorithm satisfies $(\epsilon_1 + \epsilon_2 + \epsilon_3)$-edge-differential privacy.

4   Experiment

In this section, we report experimental results comparing our proposed approaches with approaches NoiseFirst and StructureFirst proposed in [21], and analyse how different aspects of our proposed approaches affect the utility.

4.1   Datasets and Settings

Our experiments are based on 4 real-world datasets downloaded from [16], as shown in Table 1. These datasets are from different domains: (i) USairports dataset denotes the network of passenger flights between airports in the United States. (ii) Facebook is a social networks dataset, and we add random weight values to each edge of the unweighted graph. (iii) Bitcoin OTC trust network is a who-trusts-whom network of people who trade using Bitcoin on a platform called Bitcoin OTC. This is the first explicit weighted signed directed network available for research. (iv) Enron dataset is an email network obtained from a dataset of around a million emails. Nodes of the network are email addresses and if an address i sent at least one email to address j, the graph contains an undirected edge from i to j. We add random weight values to each edge of the unweighted graph, which indicates the number of email exchanges between nodes. Table 1 illustrates the properties of the datasets such as maximum edge weight $weight_{max}$ and maximum node strength ns_{max}.

Table 1. Information about datasets

Graph	$|V|$	$|E|$	$weight_{max}$	ns_{max}
USairports	755	4660	53	1700
Facebook	4039	88234	10	5794
Bitcoin OTC	5881	35592	20	874
Enron	36692	183831	25	17844

Following previous works [12,14,15], we use L1 error and Kolmogorov-Smirnov distance (KS-distance) as the utility metrics.

We use the L1 distance [15] between the published distribution and the true distribution (or L1 error) to evaluate different approaches. More formally, the L1 distance between any two distributions d and d' with length n can be computed by $||d - d'||_1 = \sum_{i=0}^{n-1} |d_i - d'_i|$. Some techniques may publish a distribution with size smaller than n. We follow the same procedure in [24] to pad d with 0 if its size is less than n for comparison.

In addition to the L1 error, we also employ the KS-distance used in [12] to evaluate the published distribution. Given two distributions d and d', the KS-distance between d and d' is used to test the closeness between them and is defined as: $KS(d, d') = max_i |CDF_d(i) - CDF_{d'}(i)|$, where $CDF_d(i)$ is the value of cumulative distribution function on node strength i from distribution d.

We compare our proposed methods t-Sequence-aware-Hist and t-Density-based-Hist, against two state-of-the-art methods NoiseFirst and StructrueFirst for answering a given set of range queries. And we modify NoiseFirst and StructrueFirst algorithms to publish node strength histogram under edge-DP.

We evaluate the approximation result on privacy budget $\epsilon \in [0.1, 2.0]$ in Sect. 4.2, where each privacy budget ϵ is divide into $\epsilon_1 = 0.8\epsilon$, $\epsilon_2 = 0.1\epsilon$ and $\epsilon_3 = 0.1\epsilon$ in our proposed approaches. All results published by exponential mechanism are the averages from 50 runs. When partitioning the histogram, the t-Sequence-aware-Hist runs 10 times for relieving the uncertainty of the random chosen of initial cluster center.

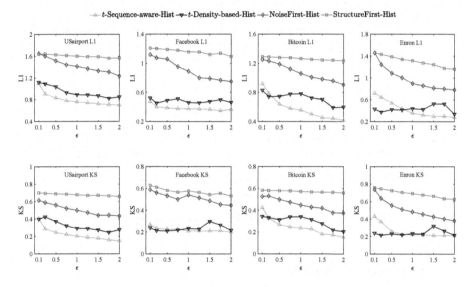

Fig. 1. The L1 error and KS distance of approaches on different datasets

4.2 Evaluating t-Sequence-aware-Hist and t-Density-based-Hist

Figure 1 compares the quality of the resulting node strength histograms of our two proposed methods to those of variant of NoiseFirst and StructureFirst. The upper half is L1 error and the lower half is KS distance.

The experiment results show that StructureFirst generally performs the worst, followed by NoiseFirst. And our two proposed methods perform significantly better than StructureFirst both in L1 error and KS distance. t-Sequence-aware-Hist results in quite accurate node strength histograms, especially when $\epsilon > 1$. For the KS distance, we can see that t-Sequence-aware-Hist performs almost identically with t-Density-based-Hist, and they both perform noticeably better than StructureFirst and NoiseFirst on the four datasets.

The reason that t-Density-based-Hist performs not as well as t-Sequence-aware-Hist in terms of some datsets is because density-based clustering is a greedy strategy that can only guarantee the local optimal. Meanwhile, sequence-aware clustering is more accurate for the randomness is reduced by iteration. Overall, both t-Sequence-aware-Hist and t-Density-based-Hist can achieve advanced performance for publishing the node strength distribution under edge-DP.

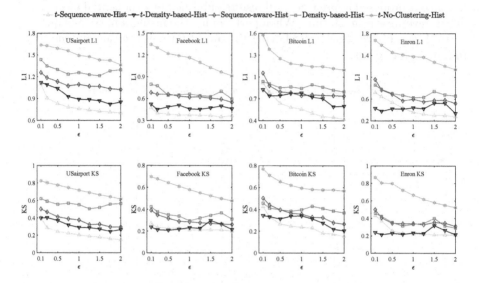

Fig. 2. Comparison of weight bound and clustering components on different datasets

4.3   Introspective Analysis

In this section, we perform the introspective analysis, to understand how different aspects of our approaches affect the utility.

In this analysis, we would like to know how the use of weight bound and aggregation affect the performance of our approach. Figure 2 illustrates the result, the upper half is L1 error and the lower half is KS distance. t-No-Clustering-Hist restrict the upper bound of weight but does not use aggregation strategy. Sequence-aware-Hist uses the Sequence-aware aggregation strategy without limiting the weight bound. Density-based-Hist is similar to Sequence-aware-Hist.

From the results, we can see that the t-No-Clustering-Hist performs the worst both in L1 error and KS distance, which indicates that the aggregation strategy significantly impacts the accuracy. And for all the four datasets, t-Sequence-aware-Hist and t-Density-based-Hist both outperform the Sequence-aware-Hist and Density-based-Hist, demonstrating the effectiveness of restricting the bound of edge weight.

5   Related Work

DP on Graph Data. Applying differential privacy to the graph has been studied extensively. Two concepts of differential privacy in graph data have been defined: edge differential privacy [10,12,25,26] and node differential privacy [14, 15]. Nissim et al. [25] conducted the first differential privacy research into graph data. They showed how to evaluate the number of triangles in a social network with edge differential privacy and showed how to calibrate the noise for subgraph counts accordingly. Hay et al. [12] translated the language of differential privacy to the graph context, and gave the formal definitions of edge and node differential privacy. It provided solutions to releasing the degree sequence of a sensitive graph, and use some sophisticated post-process techniques to reduce noise. Zhang et al. [10] claimed that the isomorphic graph could be used to generate accurate query answers if one could find an isomorphic graph with proper statistical properties that is similar to the original graph.

Histogram Publishing Under DP. Numerous recent studies [21,22,27–29] have been studied to publish histogram under differential privacy. One promising line of research is [21,22,29] based on the idea of aggregation bins for more accurate histogram. Xu et al. [21] proposed the NoiseFirst and StructureFirst algorithms. NoiseFirst formed clusters by applying the non-private optimal histogram construction technique over a noisy histogram. StructureFirst used a cost function based on L2 to solve a different optimization problem. DAWA [22] was most similar to StructureFirst, and aimed to minimize the sum of errors caused by the noise and errors caused by applying dynamic programming to select a configuration. P-HP [29] used the exponential mechanism to recursively bisect each interval into subintervals. Publishing node strength histogram under differential privacy has the challenge that the histogram is distributed very sparsely. To address this problem, we set a upper bound of edge weights, and propose two clustering strategies for aggregating the bins of histogram to achieve an accurate publication.

6   Conclusion

In this paper, we discuss the necessity of node strength, and study how to publish node strength histograms while satisfying edge-DP. We employ a transformation method to restrict edge weight and limit the length of node strength histogram on the t-bounded graph. Based on the transformation, we propose two clustering approaches, sequence-aware and density-based clustering, and experimentally compare with existing studies for publishing node strength histogram. The experimental results show that our proposed approaches have significant improvement over the state-of-the-art methods. In the future, we plan to improve the effect furthermore by using node-DP.

Acknowledgements. This work was supported by the National Natural Science Foundation of China under Grant Nos. 61572335, 61572336, 61472263, 61402312 and 61402313, the Natural Science Foundation of Jiangsu Province of China under Grant No. BK20151223, and Collaborative Innovation Center of Novel Software Technology and Industrialization, Jiangsu, China.

References

1. Yin, H., Cui, B., Zhou, X., Wang, W., Huang, Z., Sadiq, S.W.: Joint modeling of user check-in behaviors for real-time point-of-interest recommendation. ACM Trans. Inf. Syst. **35**(2), 11:1–11:44 (2016)
2. Yin, H., Chen, H., Sun, X., Wang, H., Wang, Y., Nguyen, Q.V.H.: SPTF: a scalable probabilistic tensor factorization model for semantic-aware behavior prediction. In: ICDM, pp. 585–594 (2017)
3. Yin, H., Cui, B.: Spatio-Temporal Recommendation in Social Media. Springer Briefs in Computer Science. Springer, Singapore (2016). https://doi.org/10.1007/978-981-10-0748-4
4. Yin, H., Zhou, X., Cui, B., Wang, H., Zheng, K., Hung, N.Q.V.: Adapting to user interest drift for POI recommendation. IEEE Trans. Knowl. Data Eng. **28**(10), 2566–2581 (2016)
5. Zhao, Y., Zhang, Z., Wang, Y., Liu, J.: Robust mobile spamming detection via graph patterns. In: ICPR, pp. 983–986 (2012)
6. Thedchanamoorthy, G., Piraveenan, M., Uddin, S., Senanayake, U.: Influence of vaccination strategies and topology on the herd immunity of complex networks. Soc. Netw. Anal. Min. **4**(1), 213 (2014)
7. Backstrom, L., Dwork, C., Kleinberg, J.M.: Wherefore art thou r3579x?: anonymized social networks, hidden patterns, and structural steganography. In: WWW, pp. 181–190 (2007)
8. Narayanan, A., Shmatikov, V.: De-anonymizing social networks. In: Symposium on Security and Privacy, pp. 173–187 (2009)
9. Proserpio, D., Goldberg, S., McSherry, F.: Calibrating data to sensitivity in private data analysis. PVLDB **7**, 637–648 (2014)
10. Zhang, J., Cormode, G., Procopiuc, C.M., Srivastava, D., Xiao, X.: Private release of graph statistics using ladder functions. In: SIGMOD, pp. 731–745 (2015)
11. Zhu, T., Li, G., Zhou, W., Yu, P.S.: Differentially private data publishing and analysis: a survey. IEEE Trans. Knowl. Data Eng. **29**, 1619–1638 (2017)
12. Hay, M., Li, C., Miklau, G., Jensen, D.D.: Accurate estimation of the degree distribution of private networks. In: ICDM, pp. 169–178 (2009)
13. Kossinets, G., Watts, D.J.: Empirical analysis of an evolving social network. Science **311**(5757), 88–90 (2006)
14. Day, W., Li, N., Lyu, M.: Publishing graph degree distribution with node differential privacy. In: SIGMOD, pp. 123–138 (2016)
15. Kasiviswanathan, S.P., Nissim, K., Raskhodnikova, S., Smith, A.: Analyzing graphs with node differential privacy. In: Sahai, A. (ed.) TCC 2013. LNCS, vol. 7785, pp. 457–476. Springer, Heidelberg (2013). https://doi.org/10.1007/978-3-642-36594-2_26
16. Stanford large network dataset collection. http://snap.stanford.edu/data/
17. Dwork, C.: Differential privacy. In: ICALP, pp. 1–12 (2006)
18. McSherry, F., Talwar, K.: Mechanism design via differential privacy. In: FOCS, pp. 94–103 (2007)

19. McSherry, F.: Privacy integrated queries: an extensible platform for privacy-preserving data analysis. In: SIGMOD, pp. 19–30 (2009)
20. Kifer, D., Lin, B.: Towards an axiomatization of statistical privacy and utility. In: SIGMOD, pp. 147–158 (2010)
21. Xu, J., Zhang, Z., Xiao, X., Yang, Y., Yu, G., Winslett, M.: Differentially private histogram publication. VLDB **22**(6), 797–822 (2013)
22. Li, C., Hay, M., Miklau, G., Wang, Y.: A data- and workload-aware query answering algorithm for range queries under differential privacy. PVLDB **7**(5), 341–352 (2014)
23. Dwork, C.: A firm foundation for private data analysis. Commun. ACM **54**(1), 86–95 (2011)
24. Raskhodnikova, S., Smith, A.D.: Efficient lipschitz extensions for high-dimensional graph statistics and node private degree distributions. CoRR, abs/1504.07912 (2015)
25. Nissim, K., Raskhodnikova, S., Smith, A.D.: Smooth sensitivity and sampling in private data analysis. In: STOC, pp. 75–84 (2007)
26. Sala, A., Zhao, X., Wilson, C., Zheng, H., Zhao, B.Y.: Sharing graphs using differentially private graph models. In: SIGCOMM, pp. 81–98 (2011)
27. Zhang, X., Chen, R., Xu, J., Meng, X., Xie, Y.: Towards accurate histogram publication under differential privacy. In: ICDM, pp. 587–595 (2014)
28. Qardaji, W.H., Yang, W., Li, N.: Understanding hierarchical methods for differentially private histograms. PVLDB **6**(14), 1954–1965 (2013)
29. Ács, G., Castelluccia, C., Chen, R.: Differentially private histogram publishing through lossy compression. In: ICDM, pp. 1–10 (2012)

PrivTS: Differentially Private Frequent Time-Constrained Sequential Pattern Mining

Yanhui Li[1,3(✉)], Guoren Wang[2], Ye Yuan[1], Xin Cao[3(✉)], Long Yuan[3], and Xuemin Lin[3]

[1] School of Computing Science and Engineering,
Northeastern University, Shenyang, China
lyhneu5068223280163.com
[2] School of Computing Science and Technology,
Beijing Institute of Technology, Beijing, China
[3] School of Computing Science and Engineering,
The University of New South Wales, Sydney, Australia
xin.cao@unsw.edu.au

Abstract. In this paper, we address the problem of mining *time-constrained sequential patterns* under the differential privacy framework. The mining of time-constrained sequential patterns from the sequence dataset has been widely studied, in which the transition time between adjacent items should not be too large to form frequent sequential patterns. A wide spectrum of applications can greatly benefit from such patterns, such as movement behavior analysis, targeted advertising, and POI recommendation. Improper releasing and use of such patterns could jeopardize the individually's privacy, which motivates us to apply differential privacy to mining such patterns. It is a challenging task due to the inherent sequentiality and high complexity. Towards this end, we propose a two-phase algorithm *PrivTS*, which consists of *sample-based filtering* and *count refining* modules. The former takes advantage of an improved sparse vector technique to retrieve a set of potentially frequent sequential patterns. Utilizing this information, the latter computes their noisy supports and detects the final frequent patterns. Extensive experiments conducted on real-world datasets demonstrate that our approach maintains high utility while providing privacy guarantees.

1 Introduction

Frequent sequential pattern mining (FSM) is a fundamental task in data mining. Given a collection of input sequences, FSM aims to find all subsequences that occur in the input sequences more frequently than a user-specified threshold. Finding all sequential patterns usually returns overwhelming number of patterns, which limits the utility of the detected patterns. Hence, the time elapsed between adjacent items are often taken into account to obtain the frequent

© Springer International Publishing AG, part of Springer Nature 2018
J. Pei et al. (Eds.): DASFAA 2018, LNCS 10828, pp. 92–111, 2018.
https://doi.org/10.1007/978-3-319-91458-9_6

time-constrained sequential patterns [2,20], which can benefit a wide spectrum of important practical applications, such as web usage analysis [2] and disease diagnosis [20].

Motivation. Despite valuable insights the discovery of such frequent time-constrained sequential patterns could potentially provide, if the data is sensitive, releasing these frequent patterns may pose considerable threats to individual's privacy. In fact, the malicious adversaries may exploit these information for nefarious purposes such as stalking, spamming, and inferring political/religious affiliations or alternative lifestyles. To illustrate, let's consider the following example.

Example 1. Figure 1 shows a trajectory dataset D consisting of 5 users $\{o_1, o_2, \ldots, o_5\}$ and 3 places $\{p_1, p_2, p_3\}$. Let the threshold $\sigma = 3$, and the maximum time gap constraint $\triangle t = 60$. The sequential pattern $p_1 \rightarrow p_2 \rightarrow p_3$ becomes frequent since it appears in the trajectories of o_1, o_2, and o_4. Meanwhile, both transitions $p_1 \rightarrow p_2$ and $p_2 \rightarrow p_3$ occur in no more than 60 min. Such a **time-constrained sequential pattern** clearly reveals a common behavior that people visiting p_1 and p_2 would like to visit p_3 within an hour. When an adversary has all knowledge about D except the trajectory of o_4, to mine frequent patterns, he can derive that $p_1 \rightarrow p_2 \rightarrow p_3$ is a pattern with support 2. Combining the released information, he can infer that $p_1 \rightarrow p_2 \rightarrow p_3$ must appear in the trajectory of o_4. If this pattern is "school $\rightarrow$ hospital $\rightarrow$ hospital", the adversary can infer that o_4 may work at school and suffers from serious health problem with high probability. This inference violates the privacy of o_4.

Object	Trajectory Database D
o_1	<(**p_2**,0), (**p_1**,10), (**p_2**,30), (**p_3**,40)>
o_2	<(**p_1**,0), (**p_2**,30), (**p_1**,360), (**p_2**,400), (**p_3**,420)>
o_3	<(**p_2**,0), (**p_3**,30)>
o_4	<(**p_1**,0), (**p_1**,120), (**p_3**,140), (**p_2**,150), (**p_3**,180)>
o_5	<(**p_1**,0), (**p_2**,80), (**p_3**,120), (**p_1**,210)>

* *The timestamps are in minute.*

* *Bold elements match the pattern $p_1 \rightarrow p_2 \rightarrow p_3$.*

Fig. 1. An example of a time-constrained sequence pattern $p_1 \rightarrow p_2 \rightarrow p_3 (\sigma = 3, \triangle t = 60)$

Differential privacy [11] has become one popular paradigm that can be used to provide strong privacy guarantees. It ensures that the output of an algorithm is insensitive to the change of any record. Only recently, several techniques [4,27,28] have been proposed to mine FSM under this model. It would seem attractive to adapt those techniques to address the problem of frequent time-constrained sequential pattern mining. Unfortunately, they all have some drawbacks. The technique proposed in [4] is only applicable to mining consecutive patterns. The state-of-the-art algorithm PFS^2 [27,28] fails to satisfy the specific requirements of mining frequent time-constrained sequential patterns because: (i) The sequence shrinking strategy of PFS^2 could result in the loss of some frequent patterns, or the violation of the time constraint. (ii) The candidate set generation based on the downward closure property [1] is only a subset of real candidate set. To our best knowledge, none of existing work is able to privately mine time-constrained sequential patterns.

Our solution. Motivated by the facts above, we propose a novel differentially private time-constrained sequential pattern mining algorithm *PrivTS*. We observe that, mining the sequential patterns directly from the data incurs excessive noise due to the large number of generated candidate sequential patterns. Therefore, we propose to use two phases to mine the patterns privately: (i) **Sample-based filtering**. All potentially frequent time-constrained sequential patterns are identified at this stage. The algorithm does not compute the accurate supports of the patterns, but only knows that their supports are probably above the threshold; (ii) **Count refining**. Using the information obtained from the first phase, this phase computes the noisy support of each identified sequential pattern, and discovers the final frequent patterns with privacy guarantees.

In the first phase, we observe that in real dataset the number of frequent sequential patterns is much smaller than the number of candidate patterns. Inspired by this, our *PrivTS* algorithm makes use of the advanced sparse vector technique [7,13] to effectively filter out unpromising candidate patterns. The nice property of this technique is that the information disclosure affecting differential privacy occurs only for patterns above the threshold; negative answers do not consume the privacy budget. Therefore, the final impact of the perturbation noise is reduced significantly.

Our solution in the second phase is inspired by the power of the *group-based scheme* [25], which renders the sensitivity on the new counts up to constant, irrelevant to the maximum number of the original counts affected by a sequence. We group the identified sequential patterns into disjoint groups, and set their noisy counts as the averaging count in each group with adding noise. Unfortunately, the determination of the group strategy is a challenge. To address this issue, we leverage an effective grouping technique with low privacy cost based on *sampling*. Grouping introduces the *approximate error*, which may considerably balance off the benefits from the reduced Laplace noise, resulting in non-effective for low sampling rate scenarios. To tackle this problem, a greedy-based counting is applied.

Contributions. In designing our solution to the problem of privately frequent time-constrained sequential patterns mining, our contributions can be summarized as below:

(i) This is the first work to study the problem of differentially private frequent time-constrained pattern mining.

(ii) We propose an algorithm *PrivTS* consists of two phases: *sample-based filtering* and *count refining*, which has high data utility while satisfying ϵ-differential privacy.

(iii) Through formal privacy analysis, we show that our proposed algorithm guarantees ϵ-differential privacy.

(vi) Extensive experiments demonstrate that our algorithm can privately find frequent time-constrained sequential patterns with high data utility.

2 Preliminaries

Let $I = \{L_1, L_2, \ldots, L_{|I|}\}$ be the universe of items, where $|I|$ is the size of the universe. Formally, a sequence S of length $|S|$ is an ordered list of items $S = L_1 \to L_2 \to \ldots \to L_{|S|}$, where $\forall i \in [1, |S|]$, $L_i \in I$. A sequence S is called a k-sequence if $|S| = k$. Due to time continuity, the transition time between two consecutive items should not too large. Below, we introduce the concepts of *T-sequence* and *containment*.

Definition 1 *(Time-constrained sequential pattern, T-sequence). A length-k T-sequence S has the form $T_k = L_1 \xrightarrow{\Delta t} L_2 \xrightarrow{\Delta t} \ldots \xrightarrow{\Delta t} L_k$ where Δt is the maximum transition time between any two consecutive items.*

Definition 2 *(Containment). Given an input sequence $S \leq (L_1, t_1)(L_2, t_2)$ $\ldots (L_l, t_l)$ and a T_k-sequence $T_k = L_1 \xrightarrow{\Delta t} L_2 \xrightarrow{\Delta t} \ldots \xrightarrow{\Delta t} L_k (k \leq l)$, S contains T_k (denoted as $T_k \sqsubseteq S$) if there exist integers $1 < w_1 < w_2 < \ldots < w_k \leq l$ such that: (1) $\forall 1 \leq i \leq k, L_{w_i} = L_i$; and (2) $\forall 1 \leq i \leq k-1, 0 < t_{w_{i+1}} - t_{w_i} \leq \Delta t$.*

A sequential dataset D of size $|D|$ is composed of a multiset of sequences $D = \{S_1, S_2, \ldots, S_{|D|}\}$. Each input sequence represents an individual's record. The support of a T-sequence is the number of input sequences containing T. Given the user-specified minimum support threshold σ, a T-sequence is called frequent if its support is no less than this threshold.

To ensure privacy protection, we require that mining frequent T-sequences should be performed with an algorithm that satisfies ϵ-differential privacy, which is defined based on the concept of *neighboring datasets*. Two datasets D and D' are referred to as *neighboring* if we can obtain D' from D by removing or adding a sequence, donated by $D \ominus D'$.

Definition 3 *(ϵ-Differential Privacy, ϵ-DP). A randomized algorithm $\mathcal{A}$ is ϵ-differential privacy if for any pair of neighboring datasets D and D', and for any subset of output $S \subseteq Range(\mathcal{A})$,*

$$Pr(\mathcal{A}(D) \in S) \leq e^\epsilon Pr(\mathcal{A}(D') \in S) \tag{1}$$

The most common mechanism for achieving DP is *Laplace mechanism* [12]. Given an analysis task with a numeric output $f \colon D \rightarrow \mathbb{R}^d$ and a privacy budget ϵ, the Laplace mechanism injects into f random Laplace noise of scale $\frac{\Delta f}{\epsilon}$, where Δf is called the *sensitivity* of the function f. In particular, $\Delta f = \max_{D \ominus D'} \| f(D) - f(D') \|_1$, where $\| . \|_1$ represents the L_1 norm.

For an analysis task with a categorical output (e.g., an item), injecting random noise no longer yields meaningful results. The exponential mechanism [18] tackles this problem by performing random perturbations during the selection of the output. More specifically, it draws a sample from the distribution on the output domain R which assigns each possible output $r \in R$ a probability mass proportion to $exp(\frac{\epsilon q(D,r)}{2 \Delta q})$. Here, Δq is the sensitivity of the utility function $q(D,r)$, i.e., $\Delta q = \max_{\forall r, D \ominus D'} |q(D,r) - q(D',r)|$. Intuitively, a high scoring output r is exponentially more likely to be chosen.

In some cases, $q(D,r)$ satisfies the condition that when the input dataset is changed from D to D', the changes of all quality values are one-directional, i.e., $\forall_{D \ominus D'}[(\exists_{r1} q(D,r_1) < q(D',r_1)) \rightarrow (\forall_{r2} q(D,r_2) < q(D',r_2))]$. Then one can remove the factor of $1/2$ in the exponent of $exp(\frac{\epsilon q(D,r)}{2 \Delta q})$ and return r with probability proportional to $exp(\frac{\epsilon q(D,r)}{\Delta q})$, instead of $exp(\frac{\epsilon q(D,r)}{2 \Delta q})$. This improves the accuracy of results.

Two composition properties are extensively used to ensure the overall privacy, known as sequential and parallel compositions.

Theorem 1 *(Sequential Composition [19]). Let $\mathcal{A}_i, \ldots, \mathcal{A}_m$ be m algorithms, each provides ϵ_i-DP. A sequential of algorithms $\mathcal{A}_i(D)$ over the dataset D provides $(\sum_i \epsilon_i)$-DP.*

Theorem 2 *(Parallel Composition [19]). Let $\mathcal{A}_i, \ldots, \mathcal{A}_m$ be m algorithms, each provides ϵ_i-DP. Then, a sequential of $\mathcal{A}_i(DS_i)$ over disjoint subsets DS_i of dataset D provides $(\max_i \epsilon_i)$-DP.*

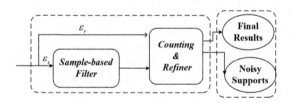

Fig. 2. Overview of PrivTS

3   Overview of PrivTS

To privately mine frequent T-sequences, we present a two-phase algorithm *PrivTS*. An overview of our algorithmic framework is illustrated in Fig. 2. In the first phase, a sampled-based filter is utilized to mine potentially frequent

T-sequences in ascending order of length. At this stage, we focus on whether the support of a T-sequence exceeds the threshold, rather than its accurate support. In the second phase, we design a counting refiner to derive the noisy count of each potentially frequent T-sequence and retrieve the final frequent results with private guarantees.

Our strategy presents several advantages. First, the use of a sample-based filter allows us to achieve high utility for mining potentially frequent sequential patterns. Second, the use of the sample-based filter helps us to significantly reduce the space of computing noisy supports of the frequent patterns. In fact, rather than considering the entire universe of patterns, we focus only on those potentially frequent patterns. Third, we refine the count of all the mined potentially frequent patterns, to derive their noisy supports. In this way, the sensitivity of the counting query can be controlled, further leading to a smaller amount of perturbation noise in reporting their noisy supports. We introduce the two phases in details in the subsequent two sections.

4   Phase 1: Sample-Based Filtering

In this section, we first introduce advanced sparse vector technique. Then, we formally present the sample-based filter algorithm and its theoretical privacy analysis. Table 1 summarizes some important notation.

Table 1. Summary of notation

Symbol	Meaning
D/D_s	Input dataset/sampled dataset
Q	Query that ask for all considering pattern counts
C_k	Candidate set of T_k-sequences
TS_k/FS_k	Set of potentially frequent/frequent T_k-sequences
g/Q_g	Group strategy/Count query given g
$\mathbf{o}/\mathbf{o_g}$	Result vector of Q/Q_g on D
γ_{min}	Minimum sampling rate
TS_s/TS_d	Set of TS_i whose sampling rate is larger/smaller than γ_{min}
$\|A\|/\widetilde{A}$	Size of A/ Noisy count of A

4.1   Sparse Vector Algorithm

The sparse vector algorithm (SVT) [7,13] was widely adopted to release κ count queries that are above the given threshold σ. For a count query q, it outputs either $\bot$ (negative response) or $\top$ (positive response). This algorithm works in two steps: (1) perturb the threshold σ by injecting Laplace noise $Lap(2/\epsilon)$ to get the noisy threshold $\widetilde{\sigma}$. (2) calculate the noisy count of each query q by injecting

Laplace noise $Lap(2\kappa\Delta/\epsilon)$ and compares it with $\tilde{\sigma}$. Output $\perp$ if $\tilde{q} < \tilde{\sigma}$ and $\top$ otherwise. Note that each query q has sensitivity bounded by Δ. Let K denote the algorithm.

$$K(D) = \begin{cases} q(D) + Lap(\frac{2\kappa\Delta}{\epsilon}) & q(D) + Lap(\frac{2\kappa\Delta}{\epsilon}) \geq \tilde{\sigma} \\ \perp & \text{otherwise} \end{cases} \qquad (2)$$

The algorithm divides the remaining privacy budget $\frac{2}{\epsilon}$ into κ count queries. After outputting κ positive responses it halts. This technique guarantees that privacy only degrades with (a) the maximum sensitivity of any one query, and (b) the number of positive responses output by the algorithm. Hence, any number of below threshold queries can be answered without compromising privacy. In that case, the amount of injected noise is proportionate to the number of above threshold queries, irrelevant to the number of all queries. This allows us to answer the query more accurately.

4.2 Sample-Based Filtering

In general, the amount of noise required for counting frequent T_k-sequences is proportionate to the number of candidate T_k-sequences. Despite this negative results, in real datasets, the number of real frequent sequences is much smaller than the number of candidate sequences. Thus, if we could filter unpromising candidates, the amount of noise required by differential privacy can be significantly reduced, which considerably improves the utility of the results. Intuitively, the sample from within a statistical population can be used to estimate characteristics of the whole population. Thus, for most T-sequences, it is sufficient to estimate if they are frequent based on a small part of the dataset. In this way, we can further filter the unpromising T-sequences.

To this end, we propose a sampled-based filter approach. Given the candidate T_k-sequences and a small sample dataset D_k randomly drawn from the original datasets, the local supports of candidate T_k-sequences in the sample dataset are used to estimate whether they are potentially frequent. Due to the privacy requirement, we have to inject noise to the local support of candidate T_k-sequences. To make estimation more accurately, an improved sparse vector technique is leveraged.

Algorithm 1 shows the steps of our sample-based filtering approach, which follows the general framework of the Apriori-based. It involves four main inputs: D (the sequence dataset), σ (the threshold), ϵ_s (the privacy budget), and Δt (the time constraint). The output of the algorithm are potentially frequent T-sequences TS. More specifically, the algorithm starts by extracting some statistic information, after which it performs the initialization (lines 1–2). The subsequent part of the algorithm consists of a number of iterations (lines 3–14). In each iteration, it first generates the candidate T_k-sequences and leverage binary estimation method to estimate the number of frequent T_k-sequences (lines 4–5). Then, the algorithm uses the advanced SVT technique to discover frequent T_k-sequences (lines 6–14). During this process, to make more accurate estimation, the threshold for the sample dataset is relaxed (line 6).

Algorithm 1. Sample-based Filter Algorithm(D, ϵ_s, σ, Δt)

---

1 $L_f \leftarrow$ compute_max_frequent_sequence_length(D, σ, Δt, ϵ_1);

2 randomly partition datasets $D = \{D_1, D_2, ..., D_{L_f}\}$;

3 **for** k *from 1 to* L_f **do**

4 $\quad$ generate the candidate T_k-sequences C_k;

5 $\quad$ $c_k \leftarrow$ binary_estimation(D_k, C_k, $\frac{\epsilon_2}{2}$);

6 $\quad$ $\sigma_k = \frac{|D_k|}{|D|} \times \sigma$, $\widetilde{\sigma_k} = \sigma_k + \text{Lap}(\frac{1}{\epsilon_t})$, count=0;

7 $\quad$ **for** *Each candidate* $q_i \in C_k$ **do**

8 $\quad\quad$ $v_i = Lap(\frac{c_k}{\epsilon_p})$;

9 $\quad\quad$ **if** $(q_i(D_k) + v_i) \geq \widetilde{\sigma}$ **then**

10 $\quad\quad\quad$ output $a_i = \top$;

11 $\quad\quad\quad$ $TS_k = TS_k \cup q_i$;

12 $\quad\quad\quad$ count=count+1, **Abort** if count$\geq c_k$.

13 $\quad\quad$ **else**

14 $\quad\quad\quad$ output $a_i = \bot$

15 $\quad$ $TS = TS \cup TS_k$;

4.3 Candidate Generating

During the process of mining frequent sequences, the downward closure property [1] is extensively used for generating candidate sequences. It states that a sequence is frequent iff all its subsequences are frequent. Thus, the candidate k-sequences are those sequences whose $(k-1)$-subsequences are all frequent. For mining frequent T-sequences, this property is no longer applicable. This is because the temporal continuity constraint is not exerted on any two items but only on two consecutive items. Thus, the candidate T-sequences generated according to this property is only a subset of real candidate set. For example, suppose for a dataset D, its frequent T_3-sequences are $TS_3 = \{L_1 \rightarrow L_2 \rightarrow L_3, L_2 \rightarrow L_3 \rightarrow L_1\}$. If the algorithm has identified $TS_2 = \{L_1 \rightarrow L_2, L_2 \rightarrow L_3, L_3 \rightarrow L_1, L_1 \rightarrow L_3\}$, by applying the downward closure property, we can only obtain $C_3 = \{L_1 \rightarrow L_2 \rightarrow L_3\}$. As a result, the frequent pattern $L_2 \rightarrow L_3 \rightarrow L_1$ is lost. To address this problem, we propose a *modified downward closure property* to generate candidate T-sequences, formulated as below:

Modified Downward Closure Property: Assume that two frequent T_{k-1}-sequences $T_{c1} = \{L_{i_1} \rightarrow L_{i_2} \rightarrow \ldots \rightarrow L_{i_{k-1}}\}$ and $T_{c2} = \{L_{j_1} \rightarrow L_{j_2} \rightarrow \ldots \rightarrow L_{j_{k-1}}\}$, they can be used to generate candidate T_k-sequences T_{c3} if $\forall n \in [2, k-1]$, $L_{i_n} = L_{j_{n-1}}$ or $L_{j_n} = L_{i_{n-1}}$. If $\forall n \in [2, k-1]$ and $L_{i_n} = L_{j_{n-1}}$, $T_{c3} = \{L_{i_1} \rightarrow L_{i_2} \rightarrow \ldots \rightarrow L_{i_{k-1}} \rightarrow L_{j_{k-1}}\}$; else if $L_{j_n} = L_{i_{n-1}}$, $T_{c3} = \{L_{j_1} \rightarrow L_{j_2} \rightarrow \ldots \rightarrow L_{j_{k-1}} \rightarrow L_{i_{k-1}}\}$.

Lemma 1. *Any frequent T_k-sequences must be in the candidate set C_k generated by applying the modified downward closure property. Furthermore, the magnitude of C_k is far smaller than that generated by enumeration.*

4.4   Parameter Settings

The Calculation of L_f**.** To estimate L_f, an exponential mechanism based approach is proposed to select it from $R = \{1, 2, \ldots, n\}$. The scoring function used is $q(D, i) = -|\zeta_i(D) - \sigma|$, where $\zeta_i(D)$ is the maximal support of T_i-sequences in D. Since adding or removing a sequence affect the support of $\zeta_i(D)$ at most by one, thus the sensitivity $\Delta q = 1$. Then, a sample r on R is drawn with probability proportion to $exp(\epsilon_{13}q(D, r))$. The main drawback for this approach is the existence of long sequence, it causes a large R from which it has to select, making the selections inaccurate.

In real world applications, the majority of the sequences in the dataset are short and only few of them are very long [4]. Hence, the support of frequent T-sequences are captured by most short sequences. Toward this end, we utilize a heuristic way to determine the upper of n, n_p. Concretely, let α_i be the number of input sequences with length i. Starting from $j = 1$, we incrementally compute the percentage $p = 1 - (\sum\limits_{j=1}^{n_p} \alpha_j)/|D|$ until p is less than $\sigma/|D|$. To guarantee differential privacy, the noise $Lap(1/\epsilon_{11})$ and $Lap(1/\epsilon_{12})$ are injected into D and each α_j, respectively. After that, R becomes $R_h = \{1, 2, \ldots, n_p\}$, which is significantly smaller than itself.

The Estimation of c_k**.** Given the candidate T_i-sequences, we need to estimate the number of frequent T_i-sequences. To guarantee the result with high accuracy, we utilize the binary estimation method [26, 29]. It mainly leverages the idea of binary search to reduce the amount of injected noise. It starts with obtaining the noisy support of T_i-sequence with the $\frac{|C_k|}{2}$-th largest support by injecting noise $Lap(\lceil log_2|C_k|\rceil/(\epsilon_2/2))$. If it is larger than σ, this means the candidate T_i-sequences in the upper half are all above σ, so we only need to consider the candidate T_i-sequences in the lower half in the next iteration, and vice versa. This process continues until the number of candidate T_i-sequences with supports larger than σ, c_k, is determined.

The Allocation of ϵ_t **and** ϵ_p**.** In standard SVT, it uses half privacy budget to derive the noisy threshold, and uses the remaining half to calculate the noisy count of each query. However, it is observed that the ratio of these two parts can be optimized to improve the accuracy of SVT. In Sample-based Filter, SVT is used to determine potentially frequent T-sequences, i.e., we check if $q_i(D_k) + Lap(\frac{c_k}{\epsilon_p})$ is larger than $\sigma_k + Lap(\frac{1}{\epsilon_t})$. Then, the error of SVT can be expressed as $\text{Error(SVT)} = Lap(\frac{c_k}{\epsilon_p}) - Lap(\frac{1}{\epsilon_t})$. To make this comparison as accurate as possible, we want to minimize the variance of error, which is $2(\frac{c_k^2}{\epsilon_p^2}) + 2(\frac{1}{\epsilon_t^2})$. When $\epsilon_p + \epsilon_t$ is fixed, it is minimized when $\epsilon_t : \epsilon_p = \frac{1}{c_k^{2/3}}$.

4.5   Privacy Analysis of Sample-Based Filtering Approach

In this subsection, we give the privacy analysis of our sample-based filter algorithm. Since the key step is our improved SVT, we first prove it satisfies ϵ_2-differential privacy.

Theorem 3. *Our proposed improved SVT is ϵ_2-DP, where $\epsilon_t + \epsilon_p = \epsilon/2$.*

Proof. Given any two neighboring datasets D and D', such that D' is obtained by inserting a sequence into D, thus, $q_i(D) \leq q_i(D')$ and $q_i(D') - 1 \leq q_i(D) \leq q_i(D') + 1$. For any output vector $\boldsymbol{a} \in \{a_1, a_2, \ldots, a_l\}^l$, let $\boldsymbol{a}_\top = \{i : a_i = \top\}$, $\boldsymbol{a}_\perp = \{i : a_i = \perp\}$, and ρ donate the binary estimation function. As shown in [26], $Pr(\rho(D) = c) \leq e^{\frac{\epsilon_2}{2}} Pr(\rho(D') = c))$. Then, for improved SVT, we have

$Pr[\mathcal{A}(D) = a]$

$= \int_{-\infty}^{\infty} \int_{-\infty}^{\infty} Pr(\rho(D) = c) Pr(\widetilde{\sigma_k} = z) \prod_{\boldsymbol{a}_\perp} Pr(q_i(D) + v_i < z) \prod_{\boldsymbol{a}_\top} Pr(q_i(D) + v_i \geq z) \, dz \, dc$

$= \int_{-\infty}^{\infty} Pr(\rho(D) = c) [\int_{-\infty}^{\infty} Pr(\widetilde{\sigma_k} = z - 1) \prod_{\boldsymbol{a}_\perp} Pr(q_i(D) + v_i < z - 1) \prod_{\boldsymbol{a}_\top} Pr(q_i(D) + v_i \geq z - 1) \, dz] \, dc$

$\leq \int_{-\infty}^{\infty} Pr(\rho(D) = c) [\int_{-\infty}^{\infty} e^{\epsilon_t} Pr(\widetilde{\sigma_k} = z) \prod_{\boldsymbol{a}_\perp} Pr(q_i(D') - 1 + v_i < z - 1) \prod_{\boldsymbol{a}_\top} Pr(q_i(D) + v_i \geq z - 1) \, dz] \, dc$

$\leq \int_{-\infty}^{\infty} Pr(\rho(D) = c) [\int_{-\infty}^{\infty} e^{\epsilon_t} Pr(\widetilde{\sigma_k} = z) \prod_{\boldsymbol{a}_\perp} Pr(q_i(D') + v_i < z) \prod_{\boldsymbol{a}_\top} Pr(q_i(D') + v_i \geq z - 1) \, dz] \, dc$

$\leq \int_{-\infty}^{\infty} Pr(\rho(D) = c) [\int_{-\infty}^{\infty} e^{\epsilon_t} Pr(\widetilde{\sigma_k} = z) \prod_{\boldsymbol{a}_\perp} Pr(q_i(D') + v_i < z) \prod_{\boldsymbol{a}_\top} e^{\frac{\epsilon_p}{c}} Pr(q_i(D') + v_i \geq z) \, dz] \, dc$

$\leq \int_{-\infty}^{\infty} e^{\frac{\epsilon_2}{2}} Pr(\rho(D') = c) [e^{\epsilon_t + c \times \frac{\epsilon_p}{c}} \int_{-\infty}^{\infty} Pr(\widetilde{\sigma_k} = z) \prod_{\boldsymbol{a}_\perp} Pr(q_i(D') + v_i < z) \prod_{\boldsymbol{a}_\top} Pr(q_i(D') + v_i \geq z) \, dz] \, dc$

$= (e^{\frac{\epsilon_2}{2} + \epsilon_t + c \times \frac{\epsilon_p}{c}}) \int_{-\infty}^{\infty} Pr(\rho(D') = c) [\int_{-\infty}^{\infty} Pr(\widetilde{\sigma_k} = z) \prod_{\boldsymbol{a}_\perp} Pr(q_i(D') + v_i < z) \prod_{\boldsymbol{a}_\top} Pr(q_i(D') + v_i \geq z) \, dz] \, dc$

$= e^{\epsilon_2} Pr[\mathcal{A}(D') = a]$

Theorem 4. *Our proposed sample-based filtering is ϵ_s-DP, where $\epsilon_s = \epsilon_1 + \epsilon_2$.*

Proof. In our pre-processing phase, L_f is computed. Since adding (removing) a sequence only affects $|D|$ by 1, the sensitivity of computing $|D|$ is 1. Thus, injecting Laplace noise $Lap(1/\epsilon_{11})$ in this computation satisfies ϵ_{11}-DP. Similarly, the sensitivity of computing α_i (i.e., the number of i-sequences) is 1, then adding Laplace noise $Lap(1/\epsilon_{12})$ satisfies ϵ_{12}-DP. Besides, since $q(D, i)$ is one-directional and adding (removing) a sequence only affect Δq at most by one, selecting r with probability proportion to $exp(\epsilon_{13}q(D, r))$ satisfies ϵ_{13}-DP according to the exponential mechanism. Based on sequential composition, computing L_f satisfies $(\epsilon_{11} + \epsilon_{12} + \epsilon_{13} = \epsilon_1)$-DP.

Since all sample datasets are disjoint, due to the parallel composition property, the privacy budget used in computing potentially frequent T_i-sequences do not need to accumulate. In summary, based on the sequential composition property, we can conclude that our sample-based filtering algorithm satisfies $\epsilon_1 + \epsilon_2 = \epsilon_s$-DP.

5   Phase 2: Count Refining

After privately identifying potentially frequent T-sequences, we now discuss how to compute their noisy supports. A simple method is to inject independent Laplace noise in each potentially frequent T-sequence support before releasing it. However, the noise scale is proportionate to the number of identified potentially frequent T-sequences. If this value is large, the support of each frequent T-sequence is perturbed by a large amount of noise, which reduces the accuracy of the noisy supports.

One of popular schemes on static histogram publication is *grouping scheme* [25]. The benefit of this scheme is that grouping the bins of a histogram with similar counts can enjoy Laplace noise reduction proportional to the sizes of the groups. Specifically, it decomposes the histogram into disjoint groups, and averages the count in each group. Subsequently, it injects noise to each group average, and sets the result as the new count of every bin in the group. Thus, the Laplace noise injected to each unit-length range inside a group covering b such ranges has a scale of $1/b \cdot \epsilon$, compares to $1/\epsilon$ scale.

Motivated by the shortcoming of simple method and the power of grouping, Fig. 3 sketches our refine counting approach for computing the noisy supports for all T-sequences in TS ($TS = \{TS_1, TS_2, \ldots, TS_{lmax}\}$). It consists of two novel modules: a sampling rate calculation module and a counting publication module. The former calculate the sampling rate to learn the underly dataset's feature, based on which takes proper publishing strategies to improve the utility.

In more detail, the sampling rate γ_i of each TS_i is calculated as $\gamma_i = (e^{\frac{\epsilon_n}{2}} - 1)/(e^{\frac{\epsilon_n}{2} \times |TS_i|} - 1)$, where $\epsilon_n = \epsilon_r/l_{max}$. If $\gamma_i \geq \gamma_{min}$, the sampled dataset is considered representative enough to predict the group strategy of the original dataset, thus enabling to enjoy noise reduction by group-based counting; In contrast, if $\gamma_i < \gamma_{min}$, the sampled dataset may not be representative. This fact may considerably balance off the benefits from Laplace noise reduction, sometimes even completely negating them. To remedy this deficiency, we propose a greedy-based counting method. In the sequel, we introduce the details of these methods.

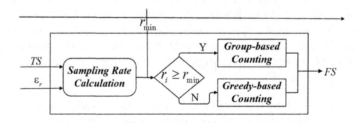

Fig. 3. Internal mechanics of count refining

5.1   Grouping-Based Counting

As discussed above, grouping-based counting could necessitate low Laplace noise for guaranteeing differential privacy. However, averaging introduces a new source of error, *approximate error* (i.e., the error that caused by replacing the true supports by the average support of each group). If the grouping strategy is not designed carefully, T-sequences with considerably different supports may be grouped together. As a result, their averaged supports (prior to Laplace noise injection) will greatly deviate from their original ones. Ideally, we must group together identified T_i-sequences with similar supports, in order to reduce the approximate error. However, determining the optimal group strategy is a challenge.

To tackle this problem, we present an effective grouping technique with low privacy cost that is based on *sampling*. In the context of differential privacy, previous research [16] has indicated the amplification effect of sampling on the privacy budget, which is formalized in Lemma 2. This lemma suggests that if the sample generated is *representative* (i.e., the statistics obtained over it are approximately the same as those of the original data), it is possible to obtain more accurate group results over the sample by taking advantage of reduced Laplace noise.

Lemma 2. *Let M be a mechanism that satisfies ϵ-differential privacy. Let M_s be another algorithm that first samples each record independently in its input dataset with probability γ, and then applies M on the sample dataset. M_s satisfies $ln(1 + \gamma(e^\epsilon - 1))$-differential privacy.*

Algorithm 2. Group_Count$(D, TS_i, \triangle t, \epsilon_n, \sigma)$

1 $D_s \leftarrow$ sampling D with rate $\gamma_i = \frac{(e^{\frac{\epsilon_n}{2}} - 1)}{(e^{\frac{\epsilon_n}{2} \times |TS_i|} - 1)}$, $d = |TS_i|$;

2 $\widetilde{o_s} = Q(D_s) + <Lap(\frac{2}{\epsilon_n}) >^d$;

3 sort them based on $\widetilde{o_s}$;

4 estimate $\widetilde{o} = \frac{o_s}{\gamma_i}$, and compute the grouping g using $\widetilde{o}$;

5 $o = Q(D)$ and compute o_g using g and $\widetilde{o}$;

6 compute $\widetilde{o_g} = o_g + <Lap(\frac{2|g|}{\epsilon_n}) >^{d_g}$;

7 recompute $\widetilde{o}$ using g and $\widetilde{o_g}$;

8 select FS_i using $\widetilde{o}$ and σ;

Algorithm 2 describes the main process of our group-based counting method. For each TS_i, the first step is to obtain the sample dataset D_s with sampling rate $\gamma_i = \frac{(e^{\frac{\epsilon_n}{2}} - 1)}{(e^{\frac{\epsilon_n}{2} \times |TS_i|} - 1)}$ (line 1). Then, the algorithm derives the local support of each T_i-sequence in TS_i, and injects noise $Lap(2/\epsilon_n)$ to produce local noisy support (line 2). Next, it sorts these T_i-sequences according to their local noisy values, and compute the grouping strategy g (lines 3–4). Specifically, g is computed as follows: it estimates the true supports of these T_i-sequences as $\tilde{o} = \frac{O_s}{\gamma_i}$, after which it employs the V-Optimal algorithm [14] to derive the group strategy g. Subsequently, the algorithm adds reduced Laplace noise $Lap(\frac{2|g|}{\epsilon_n})$ to each group average and sets the noisy average result as the final support of each identified T_i-sequence in the group (lines 5–7). At last, the algorithm selects the T_i-sequence with noisy support larger than σ as frequent (line 8).

5.2   Greedy-Based Counting

In our group-based counting, the sampling rate γ_i decreases exponentially with the sensitivity $|TS_i|$, potentially yielding a very small sample for larger $|TS_i|$. Consequently, it inflicts an amount of approximate error, which greatly destroys the utility of results. The greedy-based counting method presented below tackles this problem. Our solution is based on the following observation: for T_i-sequences with high sensitivity, merging them with others can contribute to incur smaller noise; In contrast, computing T_i-sequences with low sensitivity separately introduces less noise. To demonstrate this, we see the following example. Suppose $TS = \{TS_1, TS_2, TS_3, TS_4\}$ where $|TS_1| = 15$, $|TS_2| = 28$, $|TS_3| = 9$, and $|TS_4| = 8$. Due to the existence of long sequence, the sensitivity of computing the support of T_i-sequences ($1 \leq i \leq 4$) is $|TS_i|$. If we compute these T_i-sequences together, then the error variance of adding Laplace noise is $2 \times (60/4\epsilon_n)^2$. While if we compute each TS_i by separately, the error variances of adding Laplace noise are $2 \times (15/\epsilon_n)^2$, $2 \times (28/\epsilon_n)^2$, $2 \times (9/\epsilon_n)^2$ and $2 \times (8/\epsilon_n)^2$. Obviously, computing TS_3 and TS_4 separately incurs less noise, compared to merging them with others. In contrast, computing TS_2 shows the opposite trend.

In real world, we observed that the counting of short T-sequences are more larger, which is more resistant to noise than long T-sequences. Furthermore, the number of frequent short T-sequences is huge, dominating the whole performance. Unfortunately, computing noisy support of these short T-sequences usually incurs high sensitivity. Motivated by these facts, the goal of our greedy counting method is two-fold: (1) merging the support computation of short T_i-sequences with different length to improve the performance. (2) keeping the support computation of long T_i-sequences individually to reduce the injected noise, further making more accurate estimation.

The detailed description of our greedy counting method is shown in Algorithm 3. Firstly, the algorithm distinguish the T-sequences that required to compute separately with those that need to merge (lines 1–2). In the following, it runs a series of iterations. At each iteration, it selects two T-sequence set with the largest error reduction to merge until no error reduction by merging (lines 3–10). Finally, it perturb each T-sequence set by injecting Laplace noise(lines 11–13).

5.3   Privacy Analysis of Count Refining Approach

In what follows, we establish the privacy guarantees of our group-based counting and refine counting approach.

Algorithm 3. Greedy_Count($D, TS_d, \Delta t, \epsilon_n, \sigma$)

---

1   $n_d = |TS_d|$ and $\Delta_{total} = \sum\limits_{TS_i \in TS_d} |TS_i|$;

2   $TS_{sep} = \{TS_i | \frac{2\Delta_{total}^2}{n_d^2 \times \epsilon_n^2} \geq \frac{2|TS_i|^2}{\epsilon_n^2}\}$, $TS_{total} = TS_d - TS_{sep}$;

3   **while** ($|TS_{total}| \geq 2$) **do**

4   **for** $TS_I, TS_J (I \neq J)$ in TS_{total} **do**

5         $TS_m = TS_I \cup TS_J$, $n_m = |\{TS_i | TS_i \in TS_m\}|$, $\Delta_m = \sum\limits_{TS_i \in TS_m} |TS_i|$;

6   estimate inc $= |TS_I| \times (\frac{2|TS_I|^2}{n_I^2 \epsilon_n^2} - \frac{2\Delta_m^2}{n_m^2 \epsilon_n^2}) + |TS_J| \times (\frac{2|TS_J|^2}{\epsilon_n^2} - \frac{2\Delta_m^2}{n_m^2 \epsilon_n^2})$;

7   **if** *maximum inc* ≤ 0 **then**

8   break;

9   **else**

10   merge TS_I, TS_J with the largest inc, and update TS_{total};

11 **for** *each* TS_I *in* $TS_{seq} \cup TS_{total}$ **do**

12      $n_m = |\{TS_i | TS_i \in TS_I\}|$, $\Delta_m = \sum\limits_{TS_i \in TS_I} |TS_i|$, $\epsilon_m = \Delta_m \times \epsilon_n$;

13   perturb each T-sequence in TS_I by injecting Laplace noise $Lap(\Delta_m/\epsilon_m)$;

Theorem 5. *Our group-based counting method is ϵ_n-DP.*

Proof. Let $M : D \to \mathbf{O}$ donate the group-based counting mechanism, $\mathscr{G}$ is the module that calculates the grouping strategy g, and M_g be the mechanism that return $\widetilde{o_g}$, which is the result of $Q_g(D)$ perturbed with noise $Lap(2|g|/\epsilon_n)$. Then, in the view of an adversary, it holds that $Pr(M(D) = \widetilde{o}) = Pr(M_g(D) = \widetilde{o_g}) \cdot Pr(\mathscr{G}(D) = g)$.

Let m_{g_i} be the number that a sequence can affect the supports of the group g_i, and recall that a sequence can affect the supports of T_i-sequences by at most $|TS_i|$, thus, the sensitivity $\triangle Q_g = \frac{m_{g_1}}{|g_1|} + \frac{m_{g_2}}{|g_2|} + \ldots + \frac{m_{g_{|g|}}}{|g_{|g|}|} \leq |g|$. Then, M_g injects noise $Lap(\frac{2|g|}{\epsilon_n})$ satisfies $\frac{\epsilon_n}{2}$-differential privacy according to the Laplace mechanism, i.e., $Pr(M_g(D) = \widetilde{o_g}) \leq \frac{\epsilon_n}{2} Pr(M_g(D') = \widetilde{o_g})$. In addition, g is derived from the sample dataset. If injecting noise $Lap(2/\epsilon_n)$ on D, it satisfies $(|T_i|/(2/\epsilon_n))$-differential privacy. Based on Lemma 2, injecting this noise on D_s, it achieves $(ln(1 + \frac{(e^{\frac{\epsilon_n}{2}} - 1)}{(e^{\frac{\epsilon_n}{2} \times |TS_i|} - 1)}(e^{\frac{|T_i|}{2/\epsilon_n}} - 1)) = \frac{\epsilon_n}{2})$-differential privacy, i.e., $Pr(\mathscr{G}(D) = g) \leq \frac{\epsilon_n}{2} Pr(\mathscr{G}(D') = g)$.

To sum up, $Pr(M(D) = \widetilde{o}) = Pr(M_g(D) = \widetilde{o_g}) \cdot Pr(\mathscr{G}(D) = g) \leq \frac{\epsilon_n}{2} \cdot Pr(M_g(D') = \widetilde{o_g}) \cdot \frac{\epsilon_n}{2} Pr(\mathscr{G}(D') = g) = e^{\epsilon_n} Pr(M_g(D') = \widetilde{o_g}) \cdot Pr(\mathscr{G}(D') = g) = e^{\epsilon_n} \cdot Pr(M(D') = \widetilde{o})$, which completes the proof.

In the following, we prove that the refine counting algorithm overall guarantees ϵ_r-differential privacy.

Theorem 6. *Our proposed count refining satisfies ϵ_r-DP.*

Proof. In greedy-based counting method, the only step involves D is deriving support count (lines 11–13). Since a sequence can affect the support of TS_I by at most Δ_m, adding noise $Lap(\Delta_m/\epsilon_m)$ in this step satisfies ϵ_m-differential privacy. Besides, our refine counting approach consists of group-based counting and greedy-based counting two parts. Therefore, due to the sequential composition theorem, it gives $\epsilon_n \times |TS_s| + \sum_{TS_I \in TS_{seq} \cup TS_{total}} \epsilon_m = (|TS_s| + |TS_d|) \times \epsilon_n = lmax \times \epsilon_n = \epsilon_r$-DP.

6   Experiments

In this section, we evaluate the performance of our proposed *PrivTS* algorithm through extensive experiments. All programs are implemented on a machine with CPU Inter(R) Core(TM)i7–2600, memory 8.00 GB, frequency 3.40 GHz, hard disk 500 GB, using C++ language.

6.1   Experimental Setup

(1) **Datasets.** Our experiments are based on two publicly available real datasets[1]. Since the original data in dataset MSNBC appears as non-time-series, we attached the timestamps to each sequence by randomly. A summary of these two datasets is reported in Table 2.

Table 2. Real Dataset Parameters

Name of Dataset	Sequences	Items	Max length	Average length
MSNBC	989818	17	14975	4.7
HOUSE_POWER	40691	21	50	50

(2) **Algorithms.** We evaluate the following algorithms. (a) *PrivTS*: The algorithm is our proposed solution for mining frequent T-sequences with ϵ-differential privacy guarantee. (b) *Prefix*: It is a differentially private sequence dataset publishing algorithm proposed. More concretely, *prefix* uses noisy prefix tree (Sect. 4.4 in [6]) to publish sequence data for mining T-sequences. In *PrivTS*, the privacy budget ϵ is allocated as follows: $\epsilon_{11} = 0.025\epsilon$, $\epsilon_{12} = 0.025\epsilon$, $\epsilon_{13} = 0.05\epsilon$, $\epsilon_2 = 0.25\epsilon$, $\epsilon_r = 0.65\epsilon$, and γ_{min} is determined by the sample size formula of simple random sampling. Besides, in these experiments, the *relative threshold* is used. As both algorithms involve randomization, we run each algorithm several times and report its average results.

(3)**Metrics.** The utility metrics in these experiments are *F-score* [29] and *Relative Error (RE)* [17]. *F-score* is combination of precision and recall for the mined T-sequences, and *RE* is used to measure the error with respect to the actual supports.

---

[1] http://archive.ics.uci.edu/ml/datasets.

6.2   Experimental Results

In this subsection, we first compare the effectiveness of *PrivTS* and *Prefix*. Then, we evaluate the accuracy of our proposed refine counting approach.

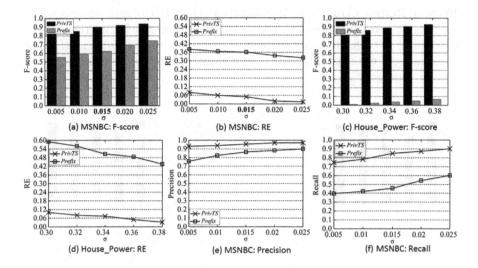

Fig. 4. Effectiveness comparison of *PrivTS* and *Prefix* under σ

Varying σ. Figure 4 shows the effect of σ on the performance of two algorithms. It can be observed that *PrivTS* substantially outperforms *Prefix*. This can be explained by the fact: *Prefix* directly deletes items exceeding the limit, which lose much information. In contrast, *PrivTS* does not limit the length of input sequence, which effectively preserves the support information. Besides, *Prefix* obtains good performance in MSNBC while not producing reasonable results in House_Power. One main reason for this phenomenon is *prefix* makes use of the prefixes of input sequences, usually less than 20 items, to construct a prefix tree. For House_Power, the average length of input sequences is long, the prefix tree cannot preserve enough frequency information, which inevitably leads to poor performance. Since *Prefix* cannot produce reasonable results in House_Power, we do not compare them on this dataset in the following experiments.

Varying Δt. Figure 5(a) and (b) illustrate the performance of the two algorithms with different Δt. From the results, it is clearly that the performance of these two algorithms is improved in terms of F-Score and RE. This is intuitive as a larger Δt imposes a weaker constraint on the maximum transition time between any two consecutive items, thus T-sequences tend to gain larger support. Further, a larger support can resist the effect of added noise, which can improve the utility of the results.

Varying ϵ. Figure 5(c) and (d) show the performance of the two algorithms as ϵ varies. Generally, *PrivTS* consistently gains better performance at the same level of privacy. In addition, larger ϵ lead to higher F-score and lower RE. This conforms to the theoretical analysis that a larger ϵ results in less noise and therefore a more accurate result.

Effect of Refine Counting. We also study the effect of our refine counting approach. We compare *PrivTS* to a *naive* algorithm which directly perturbs the support of each identified frequent T-sequence. Figure 5(e) and (f) show the accuracy of the noisy supports is significantly improved by using our refine counting approach.

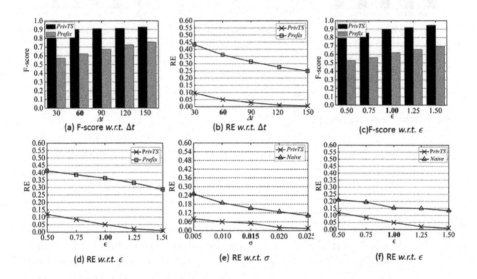

Fig. 5. Effect of different parameters on *PrivTS* and *Prefix*

7 Related Work

Nowadays, privacy-preserving has been an active research in many fields, such as spatial crowdsourcing [9,10,23,24], regression analysis [30], and frequent pattern mining. More broadly, previous differentially private frequent pattern mining studies can be divided into three groups according to the type of pattern being mined. The topic most related to ours is differentially private frequent sequence mining.

Sequence Mining. Bonomi and Xiong [4] propose a two-phase differentially private algorithm for mining consecutive item sequences. Subsequently, Xu et al. [27,28] propose the PFS² algorithm to mine general sequences and general gap-constrained sequences, regardless of whether they are consecutively. It mainly leverages a sampling based pruning technique to effectively prune the unpromising candidate sequences, further reducing the amount of added noise to improve the accuracy of the mining results. Recently, Cheng et al. [8] propose DP-MFSM for finding maximal frequent sequences under differential privacy. However, due to the time constraint of the T-sequences, these approaches are not suitable in our setting.

Several studies have been proposed to tackle the issue of publishing sequence dataset under differential privacy. Chen et al. propose two algorithms to release a sanitized dataset from which frequent sequential patterns can be mined. They are based on prefix tree [6] and variable length n-gram model [5]. These two studies differ from ours lies in that they focus on the publication of sequence dataset for mining frequent sequences, while our work aims at the release of frequent sequences.

Itemset Mining and Graph Mining. Bhaskar et al. [3] firstly develop two differentially private FIM algorithms, which are based on Laplace mechanism and exponential mechanism respectively. Subsequently, Li et al. [17] propose PrivBasis which projects the high dimensional dataset onto lower dimensions to meet the challenge of high dimensionality. Zeng et al. [29] propose SmartTrunc to deal with large noise caused by the existence of long transactions. Besides, Lee and Clifton [15] propose utilizing a generalized SVT to identify frequent itemsets. Unfortunately, it fails to satisfy differential privacy [7]. After that, a differentially private FIM algorithm based on FP-growth algorithm is proposed [22]. However, due to inherent difference between the item and the sequence, these algorithms are not applicable for our problem.

Mining frequent graph patterns under differential privacy was first proposed in [21]. In this work, both frequent graph pattern mining and the privacy guarantee are unified into a Markov Chain Monte Carlo sampling framework. However, it only achieves a relaxed differential privacy. In contrast, Xu et al. [26] propose a novel algorithm to address the issue of frequent graph mining under the rigorous differential privacy.

8   Conclusions

In this paper, we presented an efficient differentially private algorithm *PrivTS* for frequent time-constrained sequential pattern mining. It is a two-phase algorithm, which consists of *sample-based filtering* and *count refining* modules. At the first stage, the former takes advantage of an improved sparse vector technique to retrieve a set of potentially frequent sequential patterns. Utilizing this information, at the second stage, the latter computes their noisy supports and detects the final frequent patterns. Extensive experiment evaluations show the effectiveness of our proposed algorithms on large-scale real datasets.

Acknowledgments. This research was partially supported by the National Natural Science Foundation of China under Grant Nos. 61572119, 61622202, 61732003, 61729201 and U1401256, and the Fundamental Research Funds for the Central Universities under No. N150402005.

References

1. Agrawal, R., Srikant, R., et al.: Fast algorithms for mining association rules. In: Proceedings of VLDB, pp. 487–499 (1994)
2. Aoga, J.O., Guns, T., Schaus, P.: Mining time-constrained sequential patterns with constraint programming. Constraints **22**, 548–570 (2017)
3. Bhaskar, R., Laxman, S., Smith, A., Thakurta, A.: Discovering frequent patterns in sensitive data. In: Proceedings of KDD, pp. 503–512 (2010)
4. Bonomi, L., Xiong, L.: A two-phase algorithm for mining sequential patterns with differential privacy. In: Proceedings of CIKM, pp. 269–278 (2013)
5. Chen, R., Acs, G., Castelluccia, C.: Differentially private sequential data publication via variable-length n-grams. In: Computer Communication Security, pp. 638–649 (2012)
6. Chen, R., Fung, B., Desai, B.: Differentially private transit data publication: a case study on the Montreal transportation system. In: Proceedings of KDD, pp. 213–221 (2012)
7. Chen, Y., Machanavajjhala, A.: On the privacy properties of variants on the sparse vector technique. CoRR, arXiv:1508.07306 (2015)
8. Cheng, X., Su, S., Xu, S.: Differentially private maximal frequent sequence mining. Comput. Secur. **55**(C), 175–192 (2015)
9. Cheng, Y., Yuan, Y., Chen, L., Giraud-Carrier, C., Wang, G.: Complex event-participant planning and its incremental variant. In: ICDE, pp. 859–870 (2017)
10. Cheng, Y., Yuan, Y., Chen, L., Wang, G., Giraud-Carrier, C., Sun, Y.: DistR: a distributed method for the reachability query over large uncertain graphs. TPDS **27**(11), 3172–3185 (2016)
11. Dwork, C.: Differential privacy. In: ICALP, pp. 1–12 (2006)
12. Dwork, C., McSherry, F., Nissim, K., Smith, A.: Calibrating noise to sensitivity in private data analysis. In: Halevi, S., Rabin, T. (eds.) TCC 2006. LNCS, vol. 3876, pp. 265–284. Springer, Heidelberg (2006). https://doi.org/10.1007/11681878_14
13. Dwork, C., Naor, M., Reingold, O.: On the complexity of differentially private data release: efficient algorithms and hardness results. In: Proceedings of STOC, pp. 381–390 (2009)
14. Jagadish, H.V., Koudas, N., Muthukrishnan, S.: Optimal histograms with quality guarantees. VLDB **98**, 24–27 (1998)
15. Lee, J., Clifton, C.W.: Top-k frequent itemsets via differentially private fp-trees. In: Proceedings of SIGKDD, pp. 931–940. ACM (2014)
16. Li, N., Qardaji, W., Su, D.: On sampling, anonymization, and differential privacy or, k-anonymization meets differential privacy. In: Proceedings of ICCS, pp. 32–33 (2012)
17. Li, N., Qardaji, W., Su, D., Cao, J.: PrivBasis: frequent itemset mining with differential privacy. VLDB J. **5**(11), 1340–1351 (2012)
18. Mcsherry, F., Talwar, K.: Mechanism design via differential privacy. In: Proceedings of FOCS, pp. 94–103 (2007)
19. Mcsherry, F., Mironov, I.: Differentially private recommender systems: building privacy into the net. In: Proceedings of KDD, pp. 627–636 (2009)
20. Pei, J., Han, J., Wang, W.: Constraint-based sequential pattern mining: the pattern-growth methods. J. Intell. Inf. Syst. **28**(2), 133–160 (2007)
21. Shen, E., Yu, T.: Mining frequent graph patterns with differential privacy, pp. 545–553 (2013)

22. Su, S., Xu, S., Cheng, X., Li, Z., Yang, F.: Differentially private frequent itemset mining via transaction splitting. TKDE **27**(7), 1875–1891 (2015)
23. Tong, Y., Chen, L., Zhou, Z., Jagadish, H.V., Shou, L., Lv, W.: SLADE: a smart large-scale task decomposer in crowdsourcing. TKDE (2018)
24. Tong, Y., She, J., Ding, B., Wang, L., Chen, L.: Online mobile micro-task allocation in spatial crowdsourcing. In: ICDE, pp. 49–60 (2016)
25. Xu, J., Zhang, Z., Xiao, X., Yang, Y., Yu, G.: Differentially private histogram publication. VLDB J. **22**(6), 797–822 (2013)
26. Xu, S., Su, S., Xiong, L., Cheng, X., Xiao, K.: Differentially private frequent subgraph mining. In: Proceedings of ICDE, pp. 229–240 (2016)
27. Xu, S., Cheng, X., Su, S., Xiao, K., Xiong, L.: Differentially private frequent sequence mining. TKDE **28**(11), 2910–2926 (2016)
28. Xu, S., Su, S., Xiang, C., Li, Z.: Differentially private frequent sequence mining via sampling-based candidate pruning. In: Proceedings of ICDE, pp. 1035–1046 (2015)
29. Zeng, C., Naughton, J., Cai, J.: On differentially private frequent itemset mining. VLDB J. **6**(1), 25–36 (2012)
30. Zhang, J., Zhang, Z., Xiao, X., Yang, Y., Winslett, M.: Functional mechanism: regression analysis under differential privacy. VLDB J. **5**(11), 1364–1375 (2012)

Secure Range Query over Encrypted Data in Outsourced Environments

Ningning Cui[1](✉), Xiaochun Yang[1], Leixia Wang[1], Bin Wang[1], and Jianxin Li[2]

[1] School of Computer Science and Engineering,
Northeastern University, Shenyang 110819, China
willber1988@163.com, leixiawang@foxmail.com,
{yangxc,binwang}@mail.neu.edu.cn
[2] Department of Computer Science and Software Engineering,
University of Western Australia, Perth 6009, Australia
jianxin.li@uwa.edu.au

Abstract. With the rapid development of cloud computing paradigm, data owners have the opportunity to outsource their databases and management tasks to the cloud. Due to the privacy concerns, it is required for them to encrypt the databases before outsourcing. However, there is no existing techniques handling range queries in a fully secure way. Therefore, in this paper we focus exactly on secure processing of range queries over outsourced encrypted databases. To efficiently process secure range queries, the extraordinarily challenging task is how to perform fully secure range queries over encrypted data without the cloud ever decrypting the data. To address the challenge, we first propose a basic secure range queries algorithm which is not absolutely secure (i.e., leaking the privacy of access patterns and path patterns). To meet a better security, we present a fully secure algorithm that preserves the privacy of the data, query, result, *access patterns* and *path patterns*. At last, we empirically analyze and conduct a comprehensive performance evaluation using real dataset to validate our ideas and the proposed secure algorithms.

Keywords: Database outsourcing · Encrypted index
Secure range query

1 Introduction

With the rapid development of cloud computing paradigm, such as Amazon EC2, Google AppEngine, and Microsoft Azure, it is a promising choice for data owners to outsource their databases and management tasks to the cloud due to their inefficiently calculative capacity. Moreover, the databases outsourced to the cloud can provide high availability and flexibility at a relatively low cost. However, the cloud can access the original data directly without supervision. It easily leads to the leakage of sensitive data, e.g., user's location or financial records are exposed to the cloud completely. Such data leakage may cause that the data owner and client could not trust the cloud completely.

© Springer International Publishing AG, part of Springer Nature 2018
J. Pei et al. (Eds.): DASFAA 2018, LNCS 10828, pp. 112–129, 2018.
https://doi.org/10.1007/978-3-319-91458-9_7

A straightforward approach to address the above secure issue is to encrypt the databases before outsourcing to the cloud, which can guarantee the data confidentiality [1]. In addition, in order to preserve the user's query privacy, it is essential to encrypt the query with the same cryptosystem before sending to the cloud [1]. Moreover, during the query processing period, through analyzing the *access patterns* [2] and *path patterns* [12], the cloud can infer some useful information about the real data though the data and query are encrypted. Therefore, to perform fully secure queries, we aim to guarantee the five key factors: (1) the confidentiality of the real data in database; (2) the confidentiality of the query; (3) the confidentiality of the result; (4) the confidentiality of the access patterns corresponding to the query; (5) the confidentiality of the path patterns.

In real life, range query is one of the most common query types which is widely used in many areas. However, the above privacy issue is inevitable in range queries. A typical example is finding all the banks around financial center of New York within a rectangle range. In this example, the information corresponding to the rectangle range and banks are both involved in privacy. Over the years, a lot of research works have been studied to address the privacy issue for range queries. According to the literatures on how to tackle the privacy, these works can be summarized as Table 1. However, these works have weak privacy protection, especially leaking the *access patterns* privacy or *path patterns* privacy to the cloud. In addition, the approaches in [5,6,9,10,13] have the false positive rate problem (i.e., negative records that are returned as positive).

Table 1. Comparison of four privacy-preserving types with competing methods

Type	Hore [5,6]	Shi [8]	Wang [9]	Chi [7]	Chi [10]	Wang [11]	Wang [12]	Li [13]	Kim [14]
Data	√	√	√	√	√	√	√	√	√
Query	√	×	√	√	√	√	×	√	√
Result	×	×	×	×	×	×	√	×	√
Access patterns	×	×	×	×	×	×	×	×	√
Path patterns	-	×	×	×	×	×	×	×	×

Along this direction, a major challenge faced in secure range queries is how to design a fully secure and efficient algorithm over the encrypted data. In order to achieve this, we propose the secure range query algorithms on an encrypted R-tree by utilizing a *paillier cryptosystem* [15]. Firstly, we design a set of basic secure operations to develop a basic secure range queries algorithm, while ignoring the privacy of access patterns and path patterns. To address the omissive privacy, we further propose a set of fully secure operations and two obfuscations for oblivious traversal, which guarantee the confidentiality of the access patterns and path patterns scenarios, and can also be used as stand-alone building blocks for other applications.

Our main contributions are summarized as follows:

- This is the first work to address the problem of fully secure range query processing with regards to the confidentiality of the *data, query, result, access patterns* and *path patterns*.
- This is the first effort of applying encrypted R-tree by using paillier cryptosystem for fully secure range queries.
- We propose a basic secure range query algorithm which can support range queries over encrypted R-tree in a not absolutely secure but efficient way.
- To provide a better security, we propose a fully secure range queries algorithm that does not reveal any privacy to the cloud by employing the secure oblivious traversal.
- We perform extensive experiments on both real and synthetic datasets, which demonstrate efficiency and scalability of our proposed solutions.

The rest of the paper is organized as follows. We introduce the preliminaries and problem definition in Sect. 2. Sections 3 and 4 show the process of our proposed algorithms. The experimental evaluation results are shown in Sect. 5. We give an overview of related work in Sect. 6. Section 7 concludes our paper.

2 Preliminaries and Problem Definition

In this section, we firstly define the *system framework* and *security model* which are used for implementing and evaluating the secure algorithms, and then introduce the cryptosystem used in our study. Finally, we formalize our privacy preserving problem of secure range queries.

2.1 System Framework

Figure 1 shows the general system framework. It mainly consists of four components: *Data Owner, Service Cloud, Client* and *Certificate Authority*. Specifically, the framework works as follows.

- *Step 1.* Certificate authority first allocates the public key pk and secret key sk of the public-key cryptosystem to different parties.
- *Step 2.* Before outsourcing the database to the cloud, the data owner uses pk to encrypt the database and sends the encrypted index $E_{pk}(I)$ (e.g., encrypted R-tree) to cloud C_1, where E_{pk} denotes the encryption function.
- *Step 3.* Considering an authority client, in order to protect query privacy, the data owner utilizes the public key pk to encrypt the query Q (i.e., $E_{pk}(Q)$), and sends the request to cloud C_1.
- *Step 4.* Service cloud is composed of two different cloud servers, denoted by C_1 and C_2. Between them, the proposed algorithms are performed in a not-colluding and semi-honest way which is widely used in related work [1,3,4]. After the execution of the searching algorithms, both of them return partial query results which need to be integrated further to form the final results on the client. It is worthy noting that C_1 only possesses the public key pk, and C_2 possesses both public key pk and secret key sk.

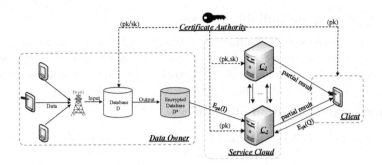

Fig. 1. System framework

2.2 Security Model

Adversary Model: There mainly exist two types of adversaries: *semi-honest* and *malicious*[4]. In semi-honest model, each participant explicitly and correctly implements the secure protocol specification, but intends to obtain the additional information of intermediate results between the corrupted participants, and uses them to analyze the transcript of messages. In the malicious model, the adversary can violate the protocol specification arbitrarily, but in practice, it is inefficient to be employed. In this paper, we adopt the semi-honest adversary model because it is a widely used model [1,3,4] and it can guarantee strong security and high efficiency for designing secure operations.

Privacy Specification: In our study, we propose a novel secure range queries over encrypted data, which attempts to protect the privacy of data and query as well as the result. Specifically, during the entire process, the proposed algorithms should achieve the following requirements.

- **Data Privacy.** The cloud servers just have the encrypted data or do not know the exact data.
- **Query Privacy.** The cloud servers can know nothing about user's query Q, and can infer nothing about user's query Q by intermediate results.
- **Result Privacy.** From the user side, no information other than the query result should be revealed to user; From the side of cloud servers, the exact query result should not be revealed to them.
- **Access Patterns Privacy.** Access patterns refer to the data corresponding to the user's query Q. For the cloud servers, they know nothing about which exact data matches the user's query Q.
- **Path Patterns Privacy.** The original traversal path should not be exposed to the cloud servers while searching over the index.

2.3 Cryptographic Building Blocks

Paillier Cryptosystem. Paillier cryptosystem [15] is a probabilistic asymmetric algorithm for public key cryptography. A notable feature of the paillier cryptosystem is its homomorphic properties along with its non-deterministic encryption. The properties of paillier cryptosystem are described as follows.

- **Homomorphic addition of plaintexts.**

$$D_{sk}(E_{pk}(a) \times E_{pk}(b)) \bmod N^2 = a + b \bmod N^2$$
$$D_{sk}(E_{pk}(a))^b \bmod N^2 = a \times b \bmod N^2$$

  where a and $b \in \mathbb{Z}_N$ are plaintexts for encryption, respectively, and N is a product of two large prime numbers. Also, let E_{pk} be the encryption function with public key pk, and D_{sk} be decryption function with secret key sk.
- **Semantic security.** Given a set of ciphertexts, the adversary can deduce nothing about the plaintexts [1].

Paillier-Based Encrypted R-Tree. In this paper, we construct an encrypted R-tree by employing the paillier cryptosystem. In the encrypted R-tree, for the non-leaf node, we only encrypt the node's minimum boundary rectangle (MBR); while for the leaf node, we encrypt the node's MBR and all points located in this leaf node. As shown in Fig. 2, the encrypted MBRs or points are dark. By padding with dummy entries, both non-leaf and leaf nodes have the same number of entries. In particular, each point in leaf nodes owns an encrypted sign bit $E_{pk}(f)$. If f is 0, it represents the point is a real data; otherwise, the point is a dummy data.

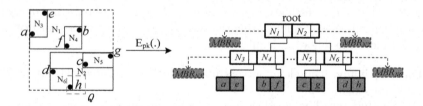

Fig. 2. Paillier-based Encrypted R-tree

2.4 Problem Definition

Definition 1 (Secure Range Query). *Consider an encrypted dataset* $D^{*(d)} = \{E_{pk}(p_1^d), \dots, E_{pk}(p_n^d)\}$ *where n represents the dataset size and d represents the dimension size, and an encrypted query range* $E_{pk}(R^d) = \{E_{pk}(R_{Min}^1), \dots, E_{pk}(R_{Min}^d)\} \times \{E_{pk}(R_{Max}^1), \dots, E_{pk}(R_{Max}^d)\}$ *where* R_{Min}^k *and* R_{Max}^k *represent the minimum and maximum of* R^d *in* k^{th} *dimension, respectively. The secure range query retrieves all points covered by query range* $E_{pk}(R^d)$ *from* $D^{*(d)}$ *without revealing the privacy of data, query, result, access patterns as well as path patterns.*

In order to address this problem, our goal is to design a secure algorithm which can make the two cloud servers cooperate mutually so that the secure range query processing can be implemented.

3 Basic Secure Range Queries Algorithm

In this section, we propose a basic secure range queries ($\mathbf{SRQ}_b$) algorithm, including basic secure node intersection operation and basic secure intermediate value operation, to support secure range queries.

3.1 Basic Secure Node Intersection Operation

As we know, since the node intersection operation is an inevitable step for the traversal over the tree-based structure, we devise a basic secure node intersection ($\mathbf{SNI}_b$) operation. It needs to check whether the query region $E_{pk}(R^d)$ intersects with the node $E_{pk}(N^d)$ (i.e., MBR in R-tree), where d denotes the domain size of dimension. Assume C_1 with $E_{pk}(R^d)$ and $E_{pk}(N^d)$.

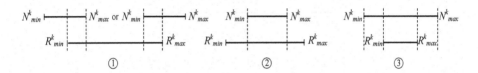

Fig. 3. Example of the node intersection in one-dimension

As shown in Fig. 3, the node intersection consists of the following three situations in each dimension: ① the segment $[E_{pk}(R_{Min}^k), E_{pk}(R_{Max}^k)]$ partially intersects with the segment range $[E_{pk}(N_{Min}^k), E_{pk}(N_{Max}^k)]$; ② the segment range $[E_{pk}(N_{Min}^k), E_{pk}(N_{Max}^k)]$ is surrounded by $[E_{pk}(R_{Min}^k), E_{pk}(R_{Max}^k)]$; ③ the segment range $[E_{pk}(R_{Min}^k), E_{pk}(R_{Max}^k)]$ is surrounded by $[E_{pk}(N_{Min}^k), E_{pk}(N_{Max}^k)]$. In order to sufficiently check whether the node intersection is true, we only discuss the SNI_b operation according to the former two situations because if the situation ① is not satisfied, the situation ③ is not satisfied surely; and if the situation ① is satisfied, no matter whether the situation ③ is satisfied, the node intersection is true. In detail, for situation ①, it needs to check whether at least one endpoint of $[E_{pk}(R_{Min}^k), E_{pk}(R_{Max}^k)]$ lies in the segment range $[E_{pk}(N_{Min}^k), E_{pk}(N_{Max}^k)]$. If it does, SNI_b assigns 0 to Γ^k; otherwise, it continues to perform the next stage. For situation ②, it only needs to check whether $E_{pk}(N_{Min}^k)$ or $E_{pk}(N_{Max}^k)$ lies in the segment range $[E_{pk}(R_{Min}^k), E_{pk}(R_{Max}^k)]$. If it does, SNI_b assigns 0 to Γ^k; otherwise, SNI_b assigns 1 to Γ^k. At last, the SNI_b returns $\Gamma = \Gamma \bigvee \Gamma^k$, for $1 \leq k \leq d$, where the initial value of Γ is 0. Similarly, we can define the basic secure point intersection operation as the way of SNI_b, denoted as $\mathbf{SPI}_b$. C_1 only needs to check whether this point $E_{pk}(p^d)$ is covered by the query region $E_{pk}(R^d)$ in each dimension.

In the process of SNI_b or SPI_b, the most important part is how to check whether $E_{pk}(R_{Min}^k)/E_{pk}(R_{Max}^k)$ lies in the range $[E_{pk}(N_{Min}^k), E_{pk}(N_{Max}^k)]$ or $E_{pk}(N_{Min})^k/E_{pk}(N_{Max}^k)$ lies in the range $[E_{pk}(R_{Min}^k), E_{pk}(R_{Max}^k)]$. To do this, we propose a basic secure intermediate value operation in Sect. 3.2.

3.2   Basic Secure Intermediate Value Operation

We devise the basic secure intermediate value operation, called $\mathbf{SIV}_b$, which is used to estimate whether the value of $E_{pk}(b)$ is between $E_{pk}(a)$ and $E_{pk}(c)$ (a, b, $c \in \mathbb{Z}_N$) in a loosely secure way, but more efficient. Consider C_1 with $E_{pk}(a)$, $E_{pk}(b)$, and $E_{pk}(c)$, and C_2 with sk. Before elaborating SIV_b, we first show an observation below.

Observation 1. Given any three numbers a, b, $c \in \mathbb{Z}_N$, let d_1, d_2, d_3 be the distances between a and b, a and c, and b and c, respectively. We have $b \in [a, c]$ iff $d_2 = d_1 + d_3$, as shown in Fig. 4. Similarly, we have $b \in [a, c]$ iff $|a - c| = |a - b| + |b - c|$.

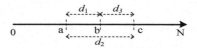

Fig. 4. Example of the SIV_b operation

Based on observation 1, initially, C_1 and C_2 need to compute the difference value of $(c - a)$, $(c - b)$, and $(b - a)$ in the ciphertext domain. Here C_1 assigns $E_{pk}(|c - a|)$ and $E_{pk}(|c - b| + |b - a|)$ to $E_{pk}(z)$ and $E_{pk}(z')$, respectively. Our SIV_b operation is equivalent to checking the equality of $[z]^1$ and $[z']$ in a secure way. Then, we employ the SBD protocol [16] to decompose the value of $E_{pk}(z)$ and $E_{pk}(z')$. Next, for $1 \le j \le m$, C_1 computes the encrypted bit-wise XOR G_j between $E_{pk}(z_j)$ and $E_{pk}(z'_j)$ by utilizing the protocol secure XOR [1]. Later, C_1 computes the encrypted bit-wise OR [1] between G_j and H_{j-1}, and assigns the result to H_j, where $H_0 = E_{pk}(0)$. Notice that $[z] = [z']$ iff $H_m = 0$, otherwise $[z] \ne [z']$ iff $H_m = 1$. After choosing a random number $r \in \mathbb{Z}_N$ and computing $H_m \times E_{pk}(r)$, C_1 sends the result to C_2. Upon receiving, C_2 decrypts it and assigns it to Ψ. Furthermore, if Ψ is even, C_2 sends $\sigma = 0$ to C_1; otherwise, C_2 sends $\sigma = 1$ to C_1. Finally, in C_1, if r is even, C_1 returns σ; otherwise, it returns $1 - \sigma$.

Example 1. Consider that $a = 4$, $b = 6$, $c = 8$ and $m = 4$. C_1 gets $E_{pk}([z]) = E_{pk}([|8 - 4|]) = \langle E_{pk}(0), E_{pk}(1), E_{pk}(0), E_{pk}(0)\rangle$ and $E_{pk}([z']) = E_{pk}([|8 - 6| + |6 - 4|]) = \langle E_{pk}(0), E_{pk}(1), E_{pk}(0), E_{pk}(0)\rangle$. Then, C_1 gets $G = \langle E_{pk}(0), E_{pk}(0), E_{pk}(0), E_{pk}(0)\rangle$ and $H = \langle E_{pk}(0), E_{pk}(0), E_{pk}(0), E_{pk}(0), E_{pk}(0)\rangle$. By generating

---

[1] $[z]$ denotes the standard binary conversion (e.g., for m = 8, [6] = '00000110', where m represents the domain size in bits).

Algorithm 1. BASIC SECURE RANGE QUERY ALGORITHM

Input: C_1: $E_{pk}(R^d)$, $E_{pk}(I)$; C_2: sk
Output: Query Result Set S

1 C_1 and C_2 :
2 Initializes the *root* node of $E_{pk}(I)$ and a stack s;
3 s.push($root$);
4 **while** $s \neq null$ **do**
5   Node $o \leftarrow s$.pop();
6   **if** o *is a non-leaf node* **then**
7   **if** $(\Gamma = SNI_b(E_{pk}(o^d), E_{pk}(R^d)))==0$ **then**
8   s.push(o_j^d), for each $o_j^d \in o^d$;

9   **else**
10   **if** $(\Gamma = SNI_b(E_{pk}(o^d), E_{pk}(R^d)))==0$ **then**
11   **for** *each point* p_j^d *in node* o^d **do**
12   **if** $(\tau = SPI_b(E_{pk}(p_j^d), E_{pk}(R^d)))==0$ **then**
13   chooses a random number $r_j \in \mathbb{Z}_N$;
14                      $E_{pk}(P_j^k) \leftarrow E_{pk}(p_j^k) \times E_{pk}(r_j)$, for $k = 1$ to d;
15                      $E_{pk}(F_j) \leftarrow E_{pk}(f_j) \times E_{pk}(r_j)$;
16   sends $E_{pk}(P_j^d)$, $E_{pk}(F_j)$ to C_2, and r_j to *Client*;

17 C_2 :
18 $P_j^d \leftarrow D_{sk}(E_{pk}(P_j^d))$, $F_j \leftarrow D_{sk}(E_{pk}(F_j))$;
19 sends P_j^d and F_j to *Client*;
20 *Client*:
21 **if** $F_j - r_j == 0$ **then**
22      $p_j^k \leftarrow P_j^k - r_j$, for $k = 1$ to d;
23      $S \leftarrow p_j^d$;

---

a random number $r = 2$, C_1 sends $H_4 \times E_{pk}(r)$ to C_2. Subsequently, C_2 decrypts it and gets $\Psi = 2$, and sends $\sigma = 0$ to C_1 since Ψ is even. Finally, since r is even, C_1 returns $\sigma = 0$.

3.3 Basic Secure Range Query Algorithm

Based on the above proposed operations, we develop our basic secure range query algorithm (**SRQ$_b$**), which is easy to understand but not fully secure. The overall step involved in the protocol SRQ$_b$ is given in Algorithm 1.

Assume that C_1 with $E_{pk}(R^d)$ and encrypted index $E_{pk}(I)$, and C_2 with sk. The SRQ$_b$ computes the range query by traversing the encrypted R-tree in a depth-first way. Initially, C_1 is assigned with the *root* node of $E_{pk}(I)$ and a stack s, and pushes the *root* into s. The SRQ$_b$ starts from the *root* and recursively visits all entries which are intersected with $E_{pk}(R^d)$. By popping up the top element o from s, the algorithm checks whether it is a non-leaf node. If it is, by exploiting the protocol SNI_b, the algorithm put all children nodes of o into s iff the returned value of the SNI_b is equal to 0. Otherwise, the algorithm first checks whether the leaf node o intersects with $E_{pk}(R^d)$. If it does, by exploiting the protocol SPI_b, the algorithm checks whether each point in o is covered by $E_{pk}(R^d)$. If it is, the point is obfuscated by computing $E_{pk}(P_j^k) = E_{pk}(p_j^k) \times E_{pk}(r_j)$ $(1 \leq j \leq |v|$, assuming the capacity of the node is $|v|)$, for $1 \leq k \leq d$, and computing $E_{pk}(F_j) = E_{pk}(f_j) \times E_{pk}(r_j)$. Here $r_j \in \mathbb{Z}_N$ is a random number. Then, C_1 sends $E_{pk}(P_j^d)$ and $E_{pk}(F_j)$ to C_2 and sends r_j to $Client$. Upon receiving, C_2 decrypts the obfuscated point $E_{pk}(P_j^k)$ and obfuscated sign bit $E_{pk}(F_j)$, and sends P_j^k and F_j to the $Client$. After receiving r_j from C_1 and P_j^k and F_j from C_2, $Client$ checks if $F_j - r_j$ is 0, $Client$ computes $p_j^k = P_j^k - r_j$, for $1 \leq k \leq d$, and puts the point p_j^d into result set S. Otherwise, the point is filtered out since the point is not covered by R^d or is the dummy data.

Table 2. Example of basic secure range query algorithm

s	o	Γ	s	o	Γ	s	o	Γ	s	o	Γ	p_1	τ	p_2	τ	r_2	P_2	F_2	$F_2 - r_2$	S
N_1	N_1	1	N_2	N_2	0	N_5	N_5	1	N_6	N_6	0	d	1	h	0	2	$h^{:\hat{h}+2}$	2	0	h
N_2						N_6														

As shown in Table 2, we illustrate the entire algorithm through the running example of Fig. 2. Please note that all column values are in encrypted form except the last two columns. In addition, in Table 2, $p^{:\hat{p}+r}$ represents that the value of point p is $\hat{p} + r$ in each dimension, where $\hat{p}$ is the original value and p is identical with $p^{:\hat{p}}$.

Correctness Analysis. According to the above process, the node intersection operation puts all points intersecting with the query range R^d into candidate set. Moreover, if the point is a dummy data, it will be filtered out by checking the sign bit. Hence, the points in final result set S are correct.

Security Analysis. Based on literature [3], the protocol (operation) is secure if the sub-protocols are secure and the intermediate results are random or pseudo-random. Above all, it is clear that the privacy of data and query is achieved by using the semantic security of the paillier cryptosystem. For the result privacy, $Client$ only knows the query result, and C_1 and C_2 do not know the exact data since the data is obfuscated by random number. Hence, the result privacy is achieved. In addition, for the access patterns and path patterns, the SBD protocol is secure based on literatur [16]. However, since the operations, SIV_b and SNI_b, both reveal some intermediate results to C_1 (e.g., the returned results σ and Γ of the operations SIV_b and SNI_b), they are secure except that the access patterns and path patterns can be revealed to the cloud. More detail, in the algorithm SRQ_b, due to the intermediate results (e.g., the node intersecting results and the traversal path on the original index) of the operation SIV_b or

SNI$_b$ are revealed to C_1, the path patterns are also revealed to C_1 and C_2. Thus, the algorithm is secure except that the access patterns and path patterns can be revealed to the cloud.

4   Fully Secure Range Query Algorithm

As discussed in Sect. 3, the algorithm SRQ$_b$ is not secure absolutely since it reveals access patterns and path patterns to the clouds. Accordingly, we first propose two fully secure operations, namely SIV$_f$ and SNI$_f$, for addressing the issues of SIV$_b$ and SNI$_b$ operations, respectively. Then, we propose two obfuscation operations that can ensure clouds to traverse the index obliviously. Finally, we show how to perform range queries in a fully secure way.

4.1   Fully Secure Operations

Intuitively, the operations SIV$_b$ and SNI$_b$ return unencrypted values which can make the cloud servers C_1 and C_2 to check whether the node intersects with the query region directly. Along this direction, we propose two fully secure operations: **SIV$_f$** and **SNI$_f$**.

- **SIV$_f$.** Unlike the SIV$_b$, SIV$_f$ returns an encrypted value. Specifically, in C_2, if Ψ mod $2 == 0$, it returns $E_{pk}(\sigma) \leftarrow E_{pk}(0)$; otherwise, $E_{pk}(\sigma) \leftarrow E_{pk}(1)$. And in the end, if r mod $2 == 0$, C_1 returns $E_{pk}(\sigma)$; otherwise, $E_{pk}(\sigma)^{N-1} \times E_{pk}(1)$.
- **SNI$_f$.** Similar to SIV$_f$, SNI$_f$ also returns an encrypted value. Specifically, based on the two conditions discussed in node intersection, for $1 \leq k \leq d$, since SIV$_f$ returns $E_{pk}(0)$ or $E_{pk}(1)$ (i.e., Γ^k is $E_{pk}(0)$ or $E_{pk}(1)$), SNI$_f$ computes $\Gamma = \Gamma \vee \Gamma^k$ (for $1 \leq k \leq d$) by utilizing SOR [1], where the initial value of Γ is $E_{pk}(0)$. At last, SNI$_f$ returns Γ. Similarly, **SPI$_f$** only implements SOR to check all dimensions of the point, and returns the result of SOR.

4.2   Obfuscation for Oblivious Traversal

To avoid revealing the traversal path, we propose two obfuscations for traversing the encrypted R-tree obliviously.

Traversal Path Obfuscation. First, we access the nodes according to the level iteratively. For each level l ($1 \leq l \leq L$, where L represents the height of encrypted R-tree), C_1 generates a perturbation function Π_l. Before sending all nodes (O_l) of level l to C_2, C_1 adopts the perturbation function Π_l to obfuscate the traversal path (e.g., $\Pi_2([N_3, N_4, N_5, N_6]) = ([N_6, N_3, N_4, N_5])$).

Node Obfuscation. Intuitively, for the perturbed nodes ($\mathbb{O}_l$) at each level l, cloud C_1 first chooses a random number $r_l \in \mathbb{Z}_N$, and computes $E_{pk}(M)^{Max2}_{Min} \times E_{pk}(r_l) \to E_{pk}(M')^{Max}_{Min}$ and $E_{pk}(Q) \times E_{pk}(r_l) \to E_{pk}(Q'_l)$. It is worth emphasizing that the operation on the point or MBR (e.g., $E_{pk}(p)$ or $E_{pk}(M)$) represents the corresponding operation in each dimension. Then, cloud C_1 sends $E_{pk}(M')^{Max}_{Min}$ of this level's nodes to C_2. C_2 processes them as the way of cloud C_1 by choosing a random number $r'_l \in \mathbb{Z}_N$, and only sends the obfuscated nodes ($\mathbb{O}'_l$) who are intersected with the query region (i.e., the initial intersecting result is $E_{pk}(0)$.) and $E_{pk}(r'_l)$ to server C_1. Notice that the node obfuscation is based on the fact that the equality condition is still satisfied by adding the identical parameter at both sides. Thus, C_1 needs to compute $E_{pk}(Q'_l) \times E_{pk}(r'_l) \to E_{pk}(Q''_l)$ for each level l. The above process is denoted as $\mathbf{N}_{obf}(.)$. Similarly, the points (P) in the leaf nodes also need to be obfuscated, denoted as $\mathbf{P}_{obf}(.)$. C_1 and C_2 choose $|v|$ random numbers γ and γ' ($|\gamma| = |\gamma'| = |v|$) $\in \mathbb{Z}_N$ to obfuscate the points in each leaf node, respectively. Also, C_2 only sends the obfuscated points ($\mathbb{P}$) who are located in the leaf node intersecting with the query region and its corresponding encrypted random numbers $E_{pk}(\gamma')$ to C_1. At last, C_1 computes the obfuscated queries $E_{pk}(\mathbb{Q}''_l)$ ($|\mathbb{Q}''_l| = |v|$) by using encrypted random numbers $E_{pk}(\gamma)$ and $E_{pk}(\gamma')$.

4.3   Fully Secure Range Query Algorithm

In this section, we propose a fully secure range query algorithm ($\mathbf{SRQ}_f$), which combines the two full operations and obfuscations for preserving the access patterns and path patterns. The steps involved in SRQ_f are shown in Algorithm 2.

Initially, C_1 first gets the node set O_l of level l ($1 \leq l \leq L$), and obtains the corresponding initial intersecting results T_l. Note that initial intersecting result of the first level $T_1 = \{E_{pk}(0)\}$. Then, by exploiting the perturbation function $\Pi_l(.)$ and obfuscation function $\mathbf{N}_{obf}(.)$, C_1 obtains the obfuscated intersecting node set $\mathbb{O}'_l$ and the obfuscated query $E_{pk}(Q''_l)$. Next, by employing $SNI_f(.)$, C_1 gets the intersecting result set W_l of $\mathbb{O}'_l$, and sends W_l to C_2. Upon receiving W_l, C_2 updates T_l by using W_l, and sends the updated intersecting result set of level l to C_1. Upon receiving, C_1 exploits the inverse perturbation function $\Pi_l^{-1}(.)$ to recover the order of the nodes at level l, and assigns recovered T'_l to their children. Repeat the above steps until the leaf level.

Next, C_1 gets the point set P_i ($1 \leq i \leq |v|^L$) in each leaf node, and perturbes the points in P_i by using $\Pi(.)$. Then C_1 obtains the obfuscated intersecting point set $\mathbb{P}$ and the obfuscated query $E_{pk}(\mathbb{Q}''_l)$ by exploiting obfuscation function $\mathbf{P}_{obf}(.)$. As for each point p in $\mathbb{P}$, C_1 obtains the intersecting result $E_{pk}(\alpha)$ of point p by using $SPI(p, E_{pk}(\mathbb{Q}''_{l_j}))$, where $E_{pk}(\mathbb{Q}''_{l_j})$ is the obfuscated query corresponding to point p, and computes $E_{pk}(F_j) = E_{pk}(f_j) \times E_{pk}(\gamma_j)$. Then, C_1 computes $E_{pk}(\beta) = E_{pk}(p \times (1 - \alpha)) \times E_{pk}(\lambda_j)$ and $\Delta = \gamma_j + \lambda_j$. Here λ_j is

---

[2] $(M)^{Max}_{Min}$ denotes the point with the minimum and the point with the maximum coordinate values in each dimension, respectively.

Algorithm 2. FULLY SECURE RANGE QUERY ALGORITHM

Input: C_1: $E_{pk}(R^d)$, $E_{pk}(I)$; C_2: sk
Output: Query Result Set S

1   C_1 and C_2 :
2   **for** $1 \leq l \leq L$ **do**
3   gets the nodes (O_l) of level l and their initial intersecting results (T_l). // for level $l = 1$, $T_l = \{E_{pk}(0)\}$.
4     $\mathbb{O}_l = \Pi_l(O_l)$;
5     $< E_{pk}(Q_l''),\mathbb{O}_l' > = \mathbf{N}_{obf}(\mathbb{O}_l)$;
6     $W_l = SNI_f(\mathbb{O}_l', E_{pk}(Q_l''))$, sends W_l to C_2;
7   C_2:
8   updates T_l by using W_l and sends the updated results to C_1;
9   C_1:
10   computes $T_l' = \Pi_l^{-1}(T_l)$ and set them as the initial intersecting results of their children;
11   gets the point set P_i in each leaf node, $1 \leq i \leq |v|^L$;
12   computes $P_i' = \Pi(P_i)$ and $< E_{pk}(Q_i''),\mathbb{P} > = \mathbf{P}_{obf}(P')$;
13   **for** *each point p in* $\mathbb{P}$ **do**
14    $E_{pk}(\alpha) = SPI_f(p, E_{pk}(Q_{l_j}''))$, $1 \leq j \leq |v|$
15    $E_{pk}(F_j) = E_{pk}(f_j) \times E_{pk}(\gamma_j)$;
16   sends $E_{pk}(\beta) = E_{pk}(p \times (1 - \alpha)) \times E_{pk}(\lambda_j)$ to C_2, and sends $\Delta = \gamma_j + \lambda_j$ and γ_j to *Client*, λ_j is a random number $\in \mathbb{Z}_N$;
17   C_2:
18    $\beta = D_{sk}(E_{pk}(\beta))$, $F_j = D_{sk}(E_{pk}(F_j))$;
19   sends $\Delta_p = \beta - \gamma_j'$ and F_j to *Client*;
20   *Client*:
21   **if** $\Delta_p - \Delta > 0 \bigwedge F_j - \gamma_j == 0$ **then**
22      $p = \Delta_p - \Delta$ mod N;
23      $S.put(p)$;

a random number $\in \mathbb{Z}_N$. After that, C_1 sends $E_{pk}(\beta)$ and $E_{pk}(F_j)$ to C_2, and sends Δ and γ_j to *Client*. Upon receiving, C_2 gets β and F_j by using decryption function $D_{sk}(.)$. On this basis, C_2 computes $\Delta_p = \beta - \gamma_j'$, and sends Δ_p and F_j to *Client*. Upon receiving, the *Client* computes the difference between Δ_p and Δ in each dimension. If $\Delta_p - \Delta > 0$ in each dimension and $F_j - \gamma_j == 0$, the *Client* inserts the point $p = \Delta_p - \Delta$ mod N into the result set S.

As shown in Table 3, we illustrate the entire algorithm through the running example of Fig. 2. Please note that all column values are in encrypted form except the last five columns. In the period of obfuscation, C_1 holds the perturbation function $\Pi_1 = \{2, 1\}$, $\Pi_2 = \{2, 4, 1, 3\}$ and $\Pi = \{2, 1\}$, respectively. In addition, in Table 3, (N) (or (p)) represents the node (or point) obfuscated by the random number of its own, and $p^{:\widehat{p}+r}$ represents that the value of point p is $\widehat{p} + r$ in each dimension, where $\widehat{p}$ is the original value and p is identical with $p^{:\widehat{p}}$.

Table 3. Example of fully secure range query algorithm

O_1 T_1 O_1' W_1T_1'	O_2 T_2 O_2' W_2T_2'	P P' T_3 γ γ'	$\mathbb{P}$ α F λ	β	Δ	Δ_p	Δ_p-Δ F-γ S
N_1 0 (N_2)0 1	N_3 1 (N_5)1 1	a e 1 2 2					
		e a 1 1 2					
	N_4 1 (N_3)1 1	b f 1 2 2					
		f b 1 1 2					
N_2 0 (N_1)1 0	N_5 0 (N_6)0 1	c g 1 2 2					
		g c 1 1 2					
	N_6 0 (N_4)1 0	d h 0 2 2	(h) 0 2 3	$h^{:\hat{h}+7}$	5	$h^{:\hat{h}+5}$	$h^{:\hat{h}}$ 0 h
		h d 0 1 2	(d) 1 1 2	$d^{:2}$	3	$d^{:0}$	$d^{:-3}$ 0

Correctness Analysis. According to the above process, the fully secure intersection operations can also guarantee that the returned nodes or points intersect with the query region surely. Notice that the node obfuscation is based on the fact that the equality condition is still satisfied by adding the identical parameters at both sides. Meanwhile, *Client* subtracts the identical parameters which can make sure the point value is unchanged. Moreover, if the point is a dummy data, it will be filtered out by checking the sign bit. Hence, the points in final result set S are correct.

Security Analysis. The security of the algorithm SRQ_f is analyzed as follows. Since the privacy data, query and result can be achieved through security analysis in Sect. 3.3, here we only analyze the privacy of access patterns and path patterns. From the view of C_1, at each level, since C_2 only sends obfuscated intersecting nodes $\mathbb{O}_l'$ to C_1, and the operations SNI_f and SPI_f both return an encrypted value W_l or $E_{pk}(\alpha)$ to C_1. Therefore, C_1 cannot infer any information from the intermediate results (i.e., the intermediate results are encrypted data). In addition, C_1 know nothing about the exact locations of obfuscated intersecting nodes, so C_1 cannot infer the traversal path over the encrypted R-tree. From the view of C_2, though C_2 can obtain the permuted traversal path, it cannot trace back to the corresponding original path. In addition, even if C_2 receives $E_{pk}(\beta)$ from C_1, and decrypts $E_{pk}(\beta)$, due to the permutation by C_1, it cannot infer whether the corresponding data point is covered by $E_{pk}(R)$. Therefore, the access patterns and path patterns are preserved from both C_1 and C_2. Moreover, from the view of *Client*, he/she can only obtain the exact points located in $E_{pk}(R^d)$ but nothing else. Hence, our algorithm SRQ_f is secure.

5   Experiments

The experiments are conducted on a PC machine with Intel Core2 Duo @ 2.93 GHz CPU and 8 GB RAM. The program is mainly coded by utilizing Java.

DataSet. We evaluate the performance of our algorithms using two real datasets, **CAR**[3] and **US**[4] [12], and a synthetic dataset **SYN**. CAR represents

---

[3] http://snap.stanford.edu/data/roadNet-CA.html.
[4] http://archive.ics.uci.edu/ml/datasets.html.

the road segments in California which includes 2,096,702 two-dimensional points, US represents part of the 1990 census in US which has 2,458,285 records and 68 categorical attributes, and SYN is randomly generated following uniform distribution and includes 100,000 points with 10 attributes which are normalized to [0, 1000] along every dimension.

Parameter Settings. To verify the effect of parameters in the experiments, we vary the dataset size **n** from 2,000 to 10,000, the query range **R** from 1% to 5% of the whole space, the key size of paillier cryptosystem **K** from 256 to 1024 bits, and the domain size of bit-decomposition **m** from 6 to 12 bits. In addition, the data dimension is varying from 2 to 6. Table 4 summarizes the parameter settings used in the experiments.

For comparison, we evaluate four algorithms, including: **SRange$_I$**, **Basic**, **SRQ$_b$** and **SRQ$_f$**, where SRange$_I$ is proposed in literature [14] which is similar to our work and Basic only performs the operation SPI$_b$ to scan and check all the points one by one. In addition, we evaluate the performance corresponding to five different parameters through fixing the other four parameters.

Table 4. Parameter settings

Parameter	Symbol	Default	Range
Dataset size	n	2,000	2,000–10,000
Query range size	R	1%	1%–5%
Encryption key size	K	512	256–1024
Domain range size	m	6	6–12
Data dimension	d	2	2–6

Evaluation of Varying n. Figure 5 shows the time performance over the datasets SYN and CAR, respectively. It is obvious that the time cost of four algorithms increases linearly. The SRQ$_b$ is the most efficient because it does not need extra encryption or decryption operations for high efficiency. The time cost of SRange$_I$ is higher than that of SRQ$_f$ because it needs to implement extra secure multiplication protocol [1] for each point. In addition, Basic has the lowest efficiency. For different datasets SYN and CAR, their time costs show similar trend when the number of data records n varies.

Evaluation of Varying R. Figure 6 compares the time performance by varying different query ranges over SYN and CAR, respectively. We observe that the time cost of SRQ$_f$ increases almost by a factor of 5 comparing with SRQ$_b$, SRange$_I$ increases almost by a factor of 6 comparing with SRQ$_b$, and Basic increases almost by a factor of 9 comparing with SRQ$_b$. In particular, When the query range varies from 1% to 5%, SRQ$_f$ increases from 258.74 to 351.38 s, while

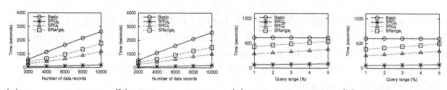

(a) Time cost of SYN. (b) Time cost of CAR.   (a) Time cost of SYN. (b) Time cost of CAR.

Fig. 5. The impact of dataset size n ($R = 1\%$, $K = 512$, $m = 6$, $d = 2$)

Fig. 6. The impact of query range R ($n = 2000$, $K = 512$, $m = 6$, $d = 2$)

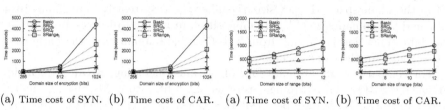

(a) Time cost of SYN. (b) Time cost of CAR.   (a) Time cost of SYN. (b) Time cost of CAR.

Fig. 7. The impact of encryption key size K ($n = 2000$, $R = 1\%$, $m = 6$, $d = 2$)

Fig. 8. The impact of bit-decomposition size m ($n = 2000$, $R = 1\%$, $K = 512$, $d = 2$)

SRange$_I$ increases from 385.53 to 506.45 s. But Basic is almost unchanged. This is because Basic needs to check all points no matter how large the query range is. For different datasets SYN and CAR, their time costs show similar trend when the number of data records R varies.

Evaluation of Varying K. In Fig. 7, we observe that the time cost of four algorithms increases exponentially when the value K increases. As shown in Fig. 7(b), the time cost of SRQ$_b$ and SRQ$_f$ increases from 10.15 to 415.36 s and from 47.43 to 1457.05 s when K varies from 256 to 1024 bits. However, the time cost of SRange$_I$ and Basic increases from 64.25 to 2158.97 s and from 100.18 to 4336.97 s, and is almost by a factor of 1.5 and 3 comparing with SRQ$_f$ when K is 1024. For different datasets, the time cost of such algorithms shows similar trend under different encryption sizes of K.

Evaluation of varying m. In Fig. 8, the time cost of all algorithms increases linearly when m increases. But SRQ$_b$ is more efficient than SRQ$_f$ and SRange$_I$. Basic is still the most expensive algorithm. As shown in Fig. 8(b), the time cost of SRQ$_f$ increases from 263.88 to 353.18 s and SRQ$_b$ increases from 59.48 to 78.53 s when m varies from 6 to 12 bits. But SRange$_I$ and Basic take the time by a factor of 1.6 and 2 comparing with SRQ$_f$ when m is 6. For different datasets, the time cost of such algorithms shows similar trend under different encryption sizes of m.

Evaluation of Varying d. Figure 9 illustrates the time performance over the datasets SYN and US, respectively. The time cost of four algorithms increases linearly when d increases. However, our approachs, SRQ_b and SRQ_f, have a better performance than $SRange_I$ by varying d. Basic is still the most time-consuming algorithm. Specifically, as shown in Fig. 9(b), the time cost of SRQ_f increases from 258.74 to 689.13 s and $SRangeI$ increases from 385.53 to 1040.59 s when d varies from 2 to 6. But Basic has to take the time from 601.83 to 2868.01 s.

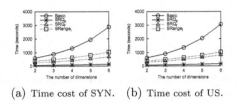

(a) Time cost of SYN.   (b) Time cost of US.

Fig. 9. The impact of dimension size d ($n = 2000$, $R = 1\%$, $K = 512$, $m = 6$)

6 Related Work

There exist a lot of related work concerning privacy-preserving techniques for range queries.

Li et al. [13] proposed a novel PBtree which can satisfy index indistinguishability and resist chosen keyword attacks. Unfortunately, the traversal path is revealed to the cloud when the range queries are performed over the PBtree, and it only support one-dimensional range queries. Chi et al. [7] proposed a scheme to build index and trapdoor by using canonical ranges and polynomials. But it is still not applicable to multiple numeric attributes.

To achieve multi-dimensional range queries, Wang et al. [12] proposed a scheme for multi-dimensional range queries by leveraging Point Predicate Encryption. But the query privacy is not protected from the cloud. Moreover, it still exposes the identifiers of the data to the cloud although the encrypted query results contain redundant data. Besides, Shi et al. [8] and Wang et al. [11] exploited R-tree to index objects and predicate encryption to encrypt nodes. However, the predicate encryption suffers from the efficiency and scalability.

In order to accelerate the efficiency of range queries over encrypted data, Wang and Ravishankar [9] and Chi et al. [10] exploited asymmetric scalar-product preserving (ASPE) and enhanced-ASPE scheme to encrypt objects. However, This scheme is not fully secure because it reveals the access patterns and traversal path to the cloud. In addition, it assumes that the client has secret key to decrypt the query result which will cause other security problems.

Moreover, Hore et al. [5,6] proposed a bucketing scheme that evaluates range queries with minimal information leakage by building privacy-preserving indices. Different from our work, this scheme requires the data owner to store and search the indices locally, rather than at the cloud.

Recently, Kim et al. [14] proposed a way to hidding data access patterns while searching over kd-tree. However, since the structure of kd-tree exposes the ordering information of points, the cloud server can infer the relative ordering information without decrypting nodes. In addition, the path patterns are also revealed to cloud server.

Meanwhile, many techniques on secure query processing are studied. Specifically, kNN [1], skyline [3] and trajectory similarity [4]. However, these techniques cannot apply to range queries directly or are infeasible to efficiently process range queries.

7   Conclusion

In this paper, we investigated and studied the problem of secure range queries in outsourced environments. To do this, we first developed a basic secure range query processing algorithm with the incomplete secure situations. To further satisfy the requirements of fully secure range queries, we developed a fully secure range query processing algorithm by designing a set of fully secure operations and two obfuscations for oblivious traversal. In addition, we analyzed the operations of the algorithms regarding to the security issue. Experimental results demonstrated that our proposed solutions can achieve an effective and efficient performance to deal with fully secure range queries. In the future, we will focus on compacting the secure operations and accelerating the query efficiency further with the strong security.

Acknowledgments. The work is partially supported by the National Natural Science Foundation of China (Nos. 61532021, 61572122, U1736104), the Project is sponsored by Liaoning BaiQianWan Talents Program, and the Fundamental Research Funds for the Central Universities (N161606002).

References

1. Elmehdwi, Y., Samanthula, B.K., Jiang, W.: Secure k-nearest neighbor query over encrypted data in outsourced environments. In: ICDE, pp. 664–675. IEEE, Chicago (2014)
2. Li, S.P., Wong, M.H.: Privacy-preserving queries over outsourced data with access pattern protection. In: IEEE International Conference on Data Mining Workshop, pp. 581–588. IEEE, Shenzhen (2014)
3. Liu, J., Yang, J., Xiong, L., et al.: Secure skyline queries on cloud platform. In: International Conference on Data Engineering, pp. 633–645. IEEE, San Diego (2017)
4. Liu, A., Kai, Z., Lu, L., et al.: Efficient secure similarity computation on encrypted trajectory data. In: International Conference on Data Engineering, pp. 66–77. IEEE, Seoul (2015)
5. Hore, B., Mehrotra, S., Tsudik, G.: A privacy-preserving index for range queries. In: Thirtieth International Conference on Very Large Data Bases VLDB Endowment, pp. 720–731. ACM, Toronto (2004)

6. Hore, B., Mehrotra, S., Canim, M., Kantarcioglu, M.: Secure multidimensional range queries over outsourced data. VLDB J. **21**(3), 333–358 (2012)
7. Chi, J., Hong, C., Zhang, M., Zhang, Z.: Privacy-enhancing range query processing over encrypted cloud databases. In: WISE, pp. 63–77 (2015)
8. Shi, E., Bethencourt, J., Chan, T.H., et al.: Multi-dimensional range query over encrypted data. In: Security and Privacy, pp. 350–364. IEEE (2007)
9. Wang, P., Ravishankar, C.V.: Secure and efficient range queries on outsourced databases using $\widehat{R}$-trees. In: IEEE International Conference on Data Engineering, pp. 314–325. IEEE, Brisbane (2013)
10. Chi, J., Hong, C., Zhang, M., Zhang, Z.: Fast multi-dimensional range queries on encrypted cloud databases. In: Candan, S., Chen, L., Pedersen, T.B., Chang, L., Hua, W. (eds.) DASFAA 2017. LNCS, vol. 10177, pp. 559–575. Springer, Cham (2017). https://doi.org/10.1007/978-3-319-55753-3_35
11. Wang, B., Li, M., Wang, H.: Geometric range search on encrypted spatial data. IEEE Trans. Inf. Forensics Secur. **11**(4), 704–719 (2016)
12. Wang, B., Hou, Y., Li, M., Wang, H., Li, H.: Maple: scalable multi-dimensional range search over encrypted cloud data with tree-based index. In: ASIA CCS, pp. 111–122. ACM, Kyoto (2014)
13. Li, R., Liu, A.X., Wang, A.L., et al.: Fast and scalable range query processing with strong privacy protection for cloud computing. IEEE/ACM Trans. Netw. **24**(4), 2305–2318 (2016)
14. Kim, H.I., Kim, H.J., Chang, J.W.: A range query processing algorithm hiding data access patterns in outsourced database environment. In: Tan, Y., Shi, Y. (eds.) DMBD 2016. LNCS, vol. 9714, pp. 434–446. Springer, Cham (2016). https://doi.org/10.1007/978-3-319-40973-3_44
15. Paillier, P.: Public-key cryptosystems based on composite degree residuosity classes. In: Stern, J. (ed.) EUROCRYPT 1999. LNCS, vol. 1592, pp. 223–238. Springer, Heidelberg (1999). https://doi.org/10.1007/3-540-48910-X_16
16. Samanthula, B.K.K., Hu, C., Jiang, W.: An efficient and probabilistic secure bit-decomposition. In: ASIA CCS, pp. 541–546. ACM, Hangzhou (2013)

TRQED: Secure and Fast Tree-Based Private Range Queries over Encrypted Cloud

Wei Yang[✉], Yang Xu, Yiwen Nie, Yao Shen, and Liusheng Huang

School of Computer Science and Technology,
University of Science and Technology of China, Hefei, China
qubit@ustc.edu.cn

Abstract. With the prevalence of cloud computing, data owners are motivated to outsource their databases to the cloud. However, sensitive information has to be encrypted to preserve the privacy, which inevitably makes effective data utilization a challenging task. Existing work either focuses on keyword searches, or suffers from inadequate security guarantees or inefficiency. In this paper, we focus on the problem of multi-dimensional range queries over dynamic encrypted cloud data. We propose a Tree-based private Range Query scheme over dynamic Encrypted cloud Data (TRQED), which enables faster-than-linear range queries and supports data dynamics while preserving the query privacy and single-dimensional privacy simultaneously. TRQED achieves provable security against semi-honest adversaries under known background model. Extensive experiments on real-world datasets show that the overhead of TRQED is desirable, and TRQED is more efficient compared with existing work.

Keywords: Range query · Cloud data · Data dynamics
Query privacy

1 Introduction

Database outsourcing has become a popular service in cloud computing. In cloud computing paradigm, a data owner prefers to remotely store their data into the cloud so as to benefit from outsourcing heavy storage and management tasks to the cloud server. To protect the privacy, sensitive data such as emails and medical records have to be encrypted before outsourcing [1]. This introduces new challenges to data utilization. How to enable effective queries over encrypted cloud data while preserving the privacy is still one of the most significant issues to be resolved. Traditional searchable encryption (SE) schemes [1–3] have been proposed to support specific type of queries, such as keyword searches [3,4]. However, considering the common SQL query operations that can be transformed into multi-dimensional range queries, deploying the above SE schemes

© Springer International Publishing AG, part of Springer Nature 2018
J. Pei et al. (Eds.): DASFAA 2018, LNCS 10828, pp. 130–146, 2018.
https://doi.org/10.1007/978-3-319-91458-9_8

directly would not be adequate, which cannot be applied to the complex multi-dimensional range queries over encrypted cloud data.

To enable multi-dimensional range queries, Boneh and Waters designed a predicate encryption scheme [5] (henceforth referred as BonehW) by using Hidden Vector Encryption (HVE) [6]. The scheme, however, incurs high computation and cannot prevent the cloud server from identifying whether two encrypted queries are from the same query, which we refer to as *query privacy*. To improve the efficiency, researchers proposed a symmetric scheme LSED$^+$ [7] and an $\widehat{R}$-tree scheme [8] (henceforth referred as WangR) by decomposing a multi-dimensional range query into single-dimensional queries. However, both schemes suffer from *single-dimensional privacy* leakage, through which the server would learn the relationship between every single-dimensional query and its corresponding results. To solve this problem, Wang *et al.* [9] proposed a scheme Maple based on HVE. As a public-key scheme, Maple causes heavy computation cost, and fails to provide query privacy. Moreover, Maple cannot support data dynamics (or data update), i.e., *insertion, deletion* and *modification*. In summary, the aforementioned multi-dimensional range query solutions either suffer from inadequate security guarantees, or fail in efficiency. Moreover, most works which achieve faster-than-linear search do not support data dynamics. Thus, the issue of private range query over dynamic encrypted cloud data remains open to date. This motivates us to design an efficient and privacy-preserving multi-dimensional range query scheme while supporting data update.

In this paper, to achieve both security and efficiency, we propose a Tree-based private Range Query scheme over dynamic Encrypted cloud Data (TRQED), which enables efficient range queries without revealing sensitive information to the cloud server. Specifically, our design goals of TRQED are as follows:

(1) *Multi-dimensional range query.* The design should enable multi-dimensional range queries over encrypted data, and return the correct results.
(2) *Privacy-preservation.* The design should protect data and index privacy, query privacy, and single-dimensional privacy from learning anything by the cloud server.
(3) *Data dynamics support.* The design supports data update including insertion, deletion and modification.
(4) *High Efficiency.* The above goals should be achieved with low system overhead and high efficiency.

To achieve the above goals, we address two technical challenges. The first challenge is how to preserve query privacy and single-dimensional privacy. We address this challenge by first proposing the perturbation-based *inner product comparison* (IPC) and extended dimensions to convert the same query to varied query vectors and obfuscate the cloud server's view, and then checking the security of our system to verify its effectiveness of privacy protection. The second technical challenges is how to reduce the system overhead. We address this challenge by designing the lightweight *point intersection predicate encryption* (PIPE) and the *range intersection predicate encryption* (RIPE), thus transforming the query processing into an efficient traversal of a tree by checking the relations between the given range query and tree nodes.

Our contributions are summarized as follows:

- *Design of Novel Range Query Scheme.* We present TRQED, the first private range query scheme which achieves faster-than-linear search and supports *data dynamics* while preserving the *query privacy* and *single-dimensional privacy* simultaneously. As a contrast, almost all the related works cannot support the above all characteristics at the same time.
- *Secure and Efficient Building Blocks.* For the objective of privacy-preservation and high efficiency, we propose two building blocks PIPE and RIPE, thus transforming private range query processing into an efficient traversal of R-tree in the ciphertext domain, and use the perturbation-based inner product comparison and extended dimensions to preserve the query privacy and single-dimensional privacy.
- *Formal Security Proof and Dataset-driven Evaluation.* We theoretically prove that TRQED is secure against *semi-honest* adversaries under *known background model*, which is the strongest threat model in the range query area to date. Furthermore, extensive experiments using real-world dataset show that the overhead of TRQED is desirable, and TRQED is more efficient compared with existing works.

2   Related Work

Our work is related to prior art in two aspects: searchable encryption and range query.

Searchable Encryption. The issue of secure query processing on encrypted cloud data has been studied in recent years. An ideal approach is using ORAM [15]. However, the efficiency of ORAM is a huge concern for real applications. Traditional Searchable Encryption (SE) schemes [1,18] have been put forward to support simple types of queries, but directly deploying them would not be adequate for practical use. Wong *et al.* proposed a SCONEDB model [19] which achieved secure kNN query over encrypted vector databases. Li *et al.* [2] proposed an authorized private keyword search framework. Subsequently, researchers studied the privacy-preserving multi-keyword ranked search [3], fuzzy search [4], and verifiable search [20]. However, the researches cloud only support keyword searches or single-dimensional range queries, which cannot be applied to the complex multi-dimensional range queries.

Range Query. BonehW [5] designed a public-key query system by using HVE, and [21] studied range queries in authenticated data structures where the privacy requirements are not considered. [22] proposed a private range query approach at the price of much more memorizing space. Moreover, all these schemes incur high computation overhead. To improve the efficiency, [7,8] are designed to improve the efficiency. As mentioned before, the schemes would cause single-dimensional privacy leakage. To solve the problem, [9] proposed a scheme Maple based on HVE and R-trees. As a public-key scheme, it suffers from the heavy computation and fails to provide query privacy. Afterwards, the authors proposed a symmetric-key MDRSE [23]. However, it should know all the range domain sizes of the dimensions in advance, and thus cannot support data update effectively. Furthermore, the client of the schemes who outsources database should be the one that searches, which does not fit our private range query model. Our work distinguishes itself from others by addressing the challenges that the above solutions either suffer from inadequate privacy guarantees, or fails in efficiency and supporting data update, which have not been addressed in research to date. Compared with Wang's work [9,23], we study the private range queries in the owner-server-user cloud computing model. TRQED achieves faster-than-linear search and supports effective data update while preserving query and single-dimensional privacy simultaneously.

3 Problem Formulation

3.1 System Model

The system model in this paper consists of three entities: a *data owner*, a *cloud server* and multiple authorized *users*, as depicted in Fig. 1. The data owner wants to outsource its database to the cloud to reduce the management cost. To protect the privacy, the sensitive data are encrypted before outsourcing. To enable effective data utilization over encrypted database, the data owner first builds a searchable index with a multi-dimensional tree, and encrypts the index. Afterwards, the owner outsources the encrypted database and index to the cloud server. The users have mutual authentication capability with the data owner, and can search the encrypted cloud data by submitting an encrypted range query to the server. Once receiving the search token, the cloud server searches the tree-based index and returns the set of matched data records.

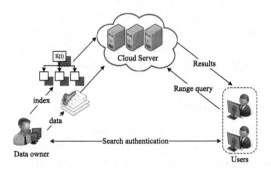

Fig. 1. Architecture of private range query over encrypted cloud data

In this paper, we employ R-tree [10] to build the index. R-tree is a type of multi-dimensional data structure which represents each leaf node with a point and each non-leaf node with a Minimum Bounding Rectangle (MBR), and grouping nearby objects (e.g., points or MBRs) in the same layer. In our work, each data record is essentially a leaf node, and each node of R-tree has the form (R, p), where R represents a MBR or point, and p is the pointer.

3.2   Threat Model

We consider a *semi-honest* (also known as *honest-but-curious*) cloud server in the threat model, which is the foundation of designing secure protocols against malicious adversaries [11]. That is, the server follows the protocol honestly, but it is curious to record all intermediate results and try to deduce useful information about the data and queries. We assume the authorized users are trusted by the data owner and do not collude with the server.

In terms of the level of privacy protection, there are three kinds of threat models, namely *Known ciphertext model*, *Chosen Plaintext model* and *Known background model* with respect to the assumption of the cloud server's ability. The threat strength increases from low to high accordingly. For example, [12–14] adopt the former two models, and thus only provide moderate security for range queries over encrypted cloud data. In this paper, we adopt the strongest threat model, namely *Known background model*, which is also considered in some existing work [3,4]. In this model, the cloud server is supposed to not only know the encrypted dataset and index as those in the former two models, but also possess more knowledge such as the distribution of queries. As an instance of possible attacks, the server could launch frequency analysis attacks combined with the background information to deduce useful information.

4   Intersection Predicate Encryption

Query processing is essentially a traversal of R-tree index, by checking the relations that whether a point (i.e., a data record C_i) is inside a MBR (i.e., a range query Q) and whether two MBRs (i.e., non-leaf node E_i and query Q) intersect. In this section, we first present a method called *inner product comparison* (IPC) to convert range query into inner product computation in the ciphertext domain, and design two predicate encryptions to determine the geometric relations.

4.1   Inner Product Comparison

Comparison Predicate. In predicate query model [7], the users send predicate functions p as queries to the cloud server. The server searches the tuples of which the attribute values d satisfies $p(d) = true$ and then returns the query results. In this, for a one-dimensional range query $[x_l, x_r]$ and a point d, we propose a function $p : \mathbb{R}^+ \to \mathbb{R}$, where $\mathbb{R}$ is the set of real numbers as: $p(d) = (d - x_l)(d - x_r)$.

The predicate function regards $p(x) \leq 0$ as true iff x is in the closed range $[x_l, x_r]$ and false otherwise.

To preserve privacy, we add perturbations δ to the predicate $p(d)$ as:

$$\widetilde{p}(d) = (d - x_l)(d - x_r)(d + \delta), \tag{1}$$

where δ is a κ bit positive number (κ is a given security parameter). Therefore, $\widetilde{p}(d) \leq 0$ iff $x_l \leq d \leq x_r$ as $d + \delta$ is positive. As such, the statistical information of transformed queries are different from that of plain ones, and the perturbation-based predicate can find the correct tuples of which $d \in [x_l, x_r]$.

Vector Extraction. We now transform the predicate $\widetilde{p}(d)$ into query and value vectors. Define the query vector $\boldsymbol{p}$ and value vector $\boldsymbol{v}$ as:

$$\boldsymbol{p} = (1, -x_l - x_r + \delta, x_l x_r - x_l \delta - x_r \delta, x_l x_r \delta)^T, \boldsymbol{v} = (d^3, d^2, d, 1)^T.$$

Thus the inner product is $\langle \boldsymbol{p}, \boldsymbol{v} \rangle = \widetilde{p}(d)$, and $\langle \boldsymbol{p}, \boldsymbol{v} \rangle \leq 0$ is true iff $d \in [x_l, x_r]$.

Vector Encryption. We employ the matrix based encryption to encrypt the vectors. The secret key for vector encryption is a four-dimensional invertible matrix. Alternatively, we can also adopt a ϕ-dimensional matrix where $\phi > 4$ as the key, and add dummy values into additional $(\phi - 4)$ dimensions which satisfy that $\sum_{i=5}^{\phi} p_i v_i = 0$. For simplicity, we adopt a four-dimensional invertible matrix M to encrypt $\boldsymbol{p}$ in this paper. Correspondingly, the encryption key for $\boldsymbol{v}$ is $\widetilde{M} = |det(M)|M^{-1}$, where $det(M)$ is the determinant of M as follows:

$$E_M(\boldsymbol{p}) = r_p M^T \boldsymbol{p} \tag{2}$$

$$E_{\widetilde{M}}(\boldsymbol{v}) = r_v |det(M)|M^{-1}\boldsymbol{v} \tag{3}$$

where r_p and r_v are randomly generated positive numbers.

In the ciphertext domain, the query processing is to find the tuples of which the encrypted value vector $E_{\widetilde{M}}(\boldsymbol{v})$ satisfies $\langle E_M(\boldsymbol{p}), E_{\widetilde{M}}(\boldsymbol{v}) \rangle \leq 0$, which is computed as

$$\langle E_M(\boldsymbol{p}), E_{\widetilde{M}}(\boldsymbol{v}) \rangle = r_p r_v |det(M)|\langle \boldsymbol{p}, \boldsymbol{v} \rangle. \tag{4}$$

As $r_p r_v |det(M)| > 0$, $\langle E_M(\boldsymbol{p}), E_{\widetilde{M}}(\boldsymbol{v}) \rangle$ depends on the $\langle \boldsymbol{p}, \boldsymbol{v} \rangle$. Consequently, $\langle E_M(\boldsymbol{p}), E_{\widetilde{M}}(\boldsymbol{v}) \rangle \leq 0$ is true iff $d \in [x_l, x_r]$.

4.2 Point Intersection Predicate Encryption (PIPE)

Based on IPC, we introduce the building block PIPE to check whether a value d is inside a range R. Inspired by the predicate encryptions in [7,9], the 1-dimensional PIPE (1dPIPE) is detailed as follows:

- Setup(1^κ). On input a security parameter κ, output a secret key SK.
- Encrypt(SK, d). On input SK and a value d, output ciphertext $C = r_v \widetilde{M} \boldsymbol{v}$, where $\boldsymbol{v} = (d^3, d^2, d, 1)^T$ and r_v is a generated random positive number.

- GenToken(SK, R). On input SK and a range $R = [x_l, x_r]$, output search token $TK = r_p M^T p$, where $p = (1, -x_l - x_r + \delta, x_l x_r - x_l \delta - x_r \delta, x_l x_r \delta)^T$, and r_p is a random positive number.
- Query(TK, C). On input TK and C, output 1 iff $\langle TK, C \rangle \leq 0$ and 0 otherwise.

We now extend the 1dPIPE to the multi-dimensional case and construct the w-dimensional PIPE (wdPIPE) to check whether a multi-dimensional point D is inside a query range. It is based on the geometric relation that if a point is inside a query range, the value of this point in every dimension is inside the corresponding single-dimension range of the MBR, and vice versa: $D \in \mathsf{MBR} \Leftrightarrow \{d_i \in R_i\}, \forall i \in [1, w]$, where $D = (d_1, \cdots, d_w)$ and $\mathsf{MBR} = (R_1, \cdots, R_w)$. The details of wdPIPE are presented in Algorithm 1. Note that the dimension w is extended to $2w$ dimensions in the inner product queries to obfuscate the cloud server's view and preserve the single-dimensional privacy.

Algorithm 1. wdPIPE

- Setup(1^κ). On input a security parameter κ, output a secret key $SK = \{M, \widetilde{M}\}$.
- Encrypt(SK, D). On input SK and a point $D = (d_1, \cdots, d_w)$, output ciphertext $C = r_v(\widetilde{M} v_1, \cdots, \widetilde{M} v_{2w})$, where r_v is a random positive number, and for $k \in [1, w]$,

$$
v_i = \begin{cases} (d_k^3, \ d_k^2, \ d_k, \ 1)^T, & i = 2k - 1 \\ (r_{k,1}, \ r_{k,2}, \ r_{k,3}, \ r_{k,4})^T, & i = 2k \end{cases}
$$

where $r_{k,j}$ is a random positive number for $1 \leq j \leq 4$.
- GenToken(SK, MBR). On input SK and a range query $\mathsf{MBR} = (R_1, \cdots, R_w)$, where $R_k = [x_{k,l}, x_{k,r}]$, output search token $TK = r_p(M^T p_1, \cdots, M^T p_{2w})$, where r_p is a random positive number, and for $k \in [1, w]$,

$$
p_i = \begin{cases} [1, \ -x_{k,l} - x_{k,r} + \delta, \ x_{k,l} x_{k,r} - (x_{k,l} + x_{k,r})\delta, \ x_{k,l} x_{k,r} \delta]^T, & i = 2k - 1 \\ (r'_{k,1}, \ r'_{k,2}, \ r'_{k,3}, \ r'_{k,4})^T, & i = 2k \end{cases}
$$

where $r'_{k,j}$ is a random negative number for $1 \leq j \leq 4$.
- Query(TK, C). On input TK and C, output 1 iff $\langle TK_i, C_i \rangle \leq 0, \forall i \in [1, 2w]$, or output 0 otherwise.

Correctness. The inner product for $i = 2k - 1, k \in [1, w]$ is $\langle TK_i, C_i \rangle = r_p r_v |det(M)| \langle p_i, v_i \rangle$, and he inner product for dimension $i = 2k, k \in [1, w]$ is $\langle TK_i, C_i \rangle = r_p r_v |det(M)| \sum r_{i,j} r'_{i,j}$, where $r_{i,j} > 0, r'_{i,j} < 0$ for $1 \leq j \leq 4$, which guarantees the inner product below 0. Thus, $\langle TK_i, C_i \rangle \leq 0$ is true if $d_k \in [x_{2k,l}, x_{2k,r}], \forall i \in [1, 2w], k \in [1, w]$. The probability of returning the point is negligible if D is not inside the range.

4.3   Range Intersection Predicate Encryption (RIPE)

Based on IPC, we propose the RIPE to check whether two MBRs intersect. We begin with the 1-dimensional case. For two ranges $R = [x_l, x_r]$ and $R' = [x'_l, x'_r]$, we find that there exists a true proposition: $R \cap R' \neq \varnothing \Leftrightarrow p = (x_l - x'_r)(x_r - x'_l) \leq 0$. Similarly, add the perturbation to the predicate, and we can obtain $\hat{p} = (x_l - x'_r)(x_r - x'_l)(x_r + \delta)$, where δ is also a positive random number. In the same way, we express the predicate as query and value vectors as follows:

$$p = [x_r + \delta, -x_r^2 - x_r\delta, -x_l x_r - x_l\delta, x_l x_r(x_r + \delta)]^T, v = (x'_l x'_r, x'_r, x'_l, 1)^T.$$

Similarly, we design the 1dRIPE, and further extend it to the multi-dimensional case, namely wdRIPE, as shown in Algorithm 2.

Algorithm 2. wdRIPE

- Setup(1^κ). On input a security parameter κ, output a secret key $SK = \{M, \widetilde{M}\}$.
- Encrypt(SK, MBR$'$). On input SK and a MBR$' = (R'_1, \cdots, R'_w)$, where $R'_k = [x'_{k,l}, x'_{k,r}]$, output ciphertext $C = r_v(\widetilde{M}v_1, \cdots, \widetilde{M}v_{2w})$, where r_v is a generated random positive number, and for $k \in [1, w]$,

$$v_i = \begin{cases} (x'_{k,l}x'_{k,r}, \ x'_{k,r}, \ x'_{k,l}, \ 1)^T, & i = 2k - 1 \\ (r_{k,1}, \ r_{k,2}, \ r_{k,3}, \ r_{k,4})^T, & i = 2k \end{cases}$$

where $r_{k,j}$ is a random positive number for $1 \leq j \leq 4$.
- GenToken(SK, MBR). On input SK and a range query MBR $= (R_1, \cdots, R_w)$, where $R_k = [x_{k,l}, x_{k,r}]$, output search token $TK = r_p(M^T p_1, \cdots, M^T p_{2w})$, where r_p is a random positive number, and for $k \in [1, w]$,

$$p_i = \begin{cases} [x_{k,r} + \delta, \ -x_{k,r}^2 - x_{k,r}\delta, \ -x_{k,l}x_{j,r} - x_{k,l}\delta, \ x_{k,l}x_{k,r}(x_{k,r} + \delta)]^T, & i = 2k - 1 \\ (r'_{k,1}, \ r'_{k,2}, \ r'_{k,3}, \ r'_{k,4})^T, & i = 2k \end{cases}$$

where $r'_{k,j}$ is a random negative number for $1 \leq j \leq 4$.
- Query(TK, C). On input TK and C, output 1 iff $\langle TK_i, C_i \rangle \leq 0, \forall i \in [1, 2w]$, or output 0 otherwise.

Using IPC, the correctness proof of wdRIPE is similar to that of wdPIPE.

5   TRQED Scheme

Now, we can present our TRQED formally.

5.1   TRQED Construction

To perform private range query, TRQED searches the R-tree index through the five polynomial algorithms (Setup, EncIndex, GenToken, Query) as:

- Setup(1^κ). On input a security parameter κ, the data owner generates a secret key $\boldsymbol{SK} = \{SK_{leaf}, SK_{nonleaf}, SK_{data}\}$, where

$$SK_{leaf} \leftarrow wdPIPE.Setup(1^\kappa),$$
$$SK_{nonleaf} \leftarrow wdRIPE.Setup(1^\kappa),$$
$$SK_{data} \leftarrow AES(1^\kappa).$$

- EncIndex($\boldsymbol{SK}$, $\mathcal{D}$). On input the $\boldsymbol{SK}$ and dataset $\mathcal{D}$, the data owner builds an R-tree $\mathcal{I} = \{D_1, \cdots, D_n, N_1, \cdots, N_t, \boldsymbol{P}\}$, where $\{D_i\}_{i=1}^n$ is a leaf node, $\{N_j\}_{j=1}^t$ is a non-leaf node, and $\boldsymbol{P}$ is the set of pointers to children nodes. Then owner encrypts every leaf and non-leaf node separately as follows:

$$C_i \leftarrow wdPIPE.Encrypt(SK_{leaf}, D_i),$$
$$E_j \leftarrow wdRIPE.Encrypt(SK_{nonleaf}, N_j).$$

where $\{C_i\}_{i=1}^n$ is an encrypted leaf node, and $\{E_j\}_{j=1}^t$ is an encrypted non-leaf node. The data records are encrypted as follows:

$$D_i^* \leftarrow AES.encrypt(SK_{data}, D_i), \text{for } 1 \leq i \leq n.$$

Afterwards, the data owner outsources the encrypted index $\mathcal{I}^* = \{C_1, \cdots, C_n, E_1, \cdots, E_t, \boldsymbol{P}\}$ and dataset $\mathcal{D}^*$ to the cloud server.

- GenToken($\boldsymbol{SK}$, Q). On input the $\boldsymbol{SK}$ and a range query Q, the owner generates a search token $\boldsymbol{TK} = \{TK_{leaf}, TK_{nonleaf}\}$, where

$$TK_{leaf} \leftarrow wdPIPE.GenToken(SK_{leaf}, Q),$$
$$TK_{nonleaf} \leftarrow wdRIPE.GenToken(SK_{nonleaf}, Q).$$

Then $\boldsymbol{TK}$ is distributed to authorized users and submitted to the cloud.

- Query($\boldsymbol{TK}$, $\mathcal{I}^*$). On input the $\boldsymbol{TK}$ and $\mathcal{I}^*$, the cloud server searches tree $\mathcal{I}^*$ level-by-level as follows:
  - For a non-leaf node E_j, if $wdRIPE.Query(TK_{nonleaf}, E_j)$ is true, it continues to search the child nodes of E_j; otherwise it stops searching this branch.
  - For a leaf node C_i, if $wdPIPE.Query(TK_{leaf}, C_i)$ is true, that indicates $D_i \in Q$, and it puts the identifier I_i of this node into the list ID_Q.

  Finally, the server returns a set ID_Q of identifiers of matched data records, and the user obtains the results.

Correctness. The correctness of TRQED is based on the correctness of wdPIPE and wdRIPE. Formally, for all $\kappa \in \mathbb{N}$, let $SK \xleftarrow{R} \text{Setup}(1^\kappa)$, $\mathcal{I}^* \xleftarrow{R}$ EncIndex($\boldsymbol{SK}, \mathcal{D}$), $TK \xleftarrow{R}$ GenToken($\boldsymbol{SK}, Q$). If $D_i \in Q$,

$$Pr[\text{Query}(\boldsymbol{TK}, \mathcal{I}^*) = ID_Q, I_i \in ID_Q] = 1.$$

If $D_i \notin Q$,

$$Pr[\mathsf{Query}(\boldsymbol{TK}, \mathcal{I}^*) = ID_Q, I_i \in ID_Q] \leq negl(\kappa),$$

where $negl(\kappa)$ is a negligible function in κ. Thus, $\mathsf{Query}(\boldsymbol{TK}, \mathcal{I}^*)$ returns I_i if and only if $D_i \in Q$.

5.2 Data Update

Insertion. To insert a new record D', we should find the lowest-layer non-leaf node of R-tree that contains this tuple. That is, the algorithm needs to traverse the index by testing whether a point (i.e., D') is inside a MBR of non-leaf node, which is based on equality query. Roughly speaking, equality query is a specific type of range query. For a new data record $D' = (d'_1, \cdots, d'_w)$, it can be expressed as a range $\mathsf{MBR}' = (R'_1, \cdots, R'_w)$, where $R'_j = [d'_j, d'_j]$, for $1 \leq j \leq w$. Then the data owner computes an update token $\boldsymbol{TK}_\pi$ with $\mathsf{GenToken}(SK, \mathsf{MBR}')$ of $wd\mathsf{PIPE}$ and $wd\mathsf{RIPE}$. After receiving the token and C' generated by $wd\mathsf{PIPE}.\mathsf{Encrypt}(SK_{leaf}, D')$, the cloud server searches the R-tree until it finds the lowest-layer non-leaf node E' that contains D'. Afterwards, the server just inserts this tuple into the non-leaf node, and puts the encrypted record into $\mathcal{D}^*$ and new pointers into $\boldsymbol{P}$.

Deletion. To delete a data record, the algorithm should search for its location in the R-tree index. Specifically, it first finds the non-leaf node in the lowest-layer that contains the record, and then checks whether a leaf-node is equal to the record. Given the record D', the owner submits the deletion token $\boldsymbol{TK}_\pi$ similarly with the above MBR'. The token consists of $(TK'_{leaf}, TK'_{nonleaf})$, which enables the server to locate the record D'. Then the cloud server deletes D', and deletes the corresponding pointers and encrypted record.

Modification. Data modification is considered as a combination of deletion of old value and insertion of new value [7]. The server just needs to perform one deletion and insertion to modify a record, and thus we omit the details here for brevity.

The $\mathsf{UpdToken}$ and Update are defined as follow:

- $\mathsf{UpdToken}(\boldsymbol{SK}, \pi, \mathsf{MBR}')$. On input $\boldsymbol{SK}$, update option $\pi \in \{insertion, deletion, modification\}$, and a target record $\mathsf{MBR}' = (R'_1, \cdots, R'_w)$, where $R'_j = [d'_j, d'_j]$, the owner outputs a token $\boldsymbol{TK}_\pi = (TK'_{leaf}, TK'_{nonleaf})$, where

$$TK'_{leaf} \leftarrow wd\mathsf{PIPE}.\mathsf{GenToken}(SK_{leaf}, \mathsf{MBR}'),$$
$$TK'_{nonleaf} \leftarrow wd\mathsf{RIPE}.\mathsf{GenToken}(SK_{nonleaf}, \mathsf{MBR}').$$

- Update($\boldsymbol{TK}_\pi$, $\mathcal{I}^*$, C'). On input the token $\boldsymbol{TK}_\pi$, the index $\mathcal{I}^*$, and ciphertext of the record C', where $C' = wd\mathsf{PIPE.Encrypt}(SK_{leaf}, D')$, the cloud server outputs a new $\mathcal{I}^*$ and updates the $\mathcal{D}^*$.
  - If $\pi = insertion$, call $\mathsf{Query}(TK'_{nonleaf}, \mathcal{I}^*)$ and find the lowest-layer non-leaf node E' where $C' \in E'$. Insert the leaf node C' in the R-tree, and add the pointers into $\boldsymbol{P}$ and encrypted record into $\mathcal{D}^*$.
  - If $\pi = deletion$, call $\mathsf{Query}(\boldsymbol{TK}', \mathcal{I}^*)$ to find the leaf node C'. If yes, delete C' from the R-tree, and delete the pointers from $\boldsymbol{P}$ and encrypted record from $\mathcal{D}^*$.
  - If $\pi = modification$, first perform the *deletion* operation, and then perform *insertion* operation.

6   Security Analysis

In this section, we prove the security of our TRQED scheme.

Lemma 1. *The intersection predicate encryption schemes are semantically secure if the encrypted messages are indistinguishable under Chosen Plaintext Attack (IND-CPA).*

Proof. We just have to prove that the probability for the probabilistic polynomial-time (PPT) adversary to break the encrypted messages of IPE schemes (i.e., PIPE and RIPE) is negligible. Suppose the challenger runs $\mathsf{Setup}(1^\kappa)$ to generate an intersection encryption system $SK = \{SK_1, SK_2\}$, and a PPT adversary $\mathcal{A}$ submits two messages m_0 and m_1, which knows SK_2 from the challenger. The challenger randomly chooses $b \in \{0, 1\}$, encrypts m_b with SK_1, and sends the ciphertext to $\mathcal{A}$. Then the adversary $\mathcal{A}$ takes a guess b' of b. Recall with the perturbation-based matrix encryption, a random number is used each time, thus transforming a plaintext into varied ciphertext with the same key. It is easy to conclude that $\mathcal{A}$ cannot guess b correctly with a probability higher than $1/2$. Thus the advantage in the security game is $\boldsymbol{Adv}_{IPE,\mathcal{A}}^{CPA}(\kappa) = |Pr[b = b'] - \frac{1}{2}| < negl(\kappa)$. Moreover, considering the invertible matrix is random, it is difficult to be cracked, as there are an infinite number of key pairs. Thus we say that the IPE schemes are semantically secure.

The indistinguishability of encrypted vectors and tokens is based on the indistinguishability of the pseudo-random number and perturbation-based matrix encryption, so that the index privacy are preserved. Based on Lemma 1, we analyze the security of TRQED.

Theorem 1. *TRQED is secure against semi-honest adversaries under the known background model.*

Before proving the Theorem 1, we introduce some notions used in [16] and adapt them for our proof.

- *History:* an interaction between the user and cloud server, determined by a dataset $\mathcal{D}$, a searchable index $\mathcal{I}$ and a set of queries $Q = (q_1, \cdots, q_\tau)$ submitted by users, denoted as the knowledge $H = (\mathcal{D}, \mathcal{I}, Q)$.
- *View:* the encrypted form of H under the secret key SK, denoted as $V(H)$, i.e., the encrypted dataset $\mathcal{D}^*$, the secure index $\mathcal{I}^*$, and the search tokens $TK(Q)$. Note that the cloud server can only see the views.
- *Trace:* given a history H, the trace $Tr(H)$ is the information which can be learned by the cloud server. It contains the access pattern $\alpha(H)$, search pattern $\sigma(H)$ and the returned identifiers $ID(Q)$. Let $ID(q)$ be the set of identifiers of the matched data records, and $\alpha(H) = \{ID(q_1), \cdots, ID(q_\tau)\}$. The search pattern $\sigma(H)$ is a $n \times \tau$ binary matrix where $\sigma(H)_{i,j}$ is 1 if I_i is returned by a query q_j, and 0 otherwise. Then we have $Tr(H) = \{ID(Q), \alpha(H), \sigma(H)\}$.

Under known background model, we assume the server obtains the $Tr(H)$, and a certain number of query and its probability pairs (q_i, p_i). Informally, given two histories with the same trace, if the server with the distribution of queries cannot distinguish which view of them is generated by the simulator, he cannot learn additional knowledge beyond what we are willing to leak (i.e., the trace), and thus our solution is secure.

Proof. Let $\mathcal{S}$ be a simulator that can simulate a view V' indistinguishable from the cloud server's view $V(H) = \{\mathcal{D}^*, \mathcal{I}^*, TK(Q)\}$. To achieve this, the $\mathcal{S}$ does the following:

- To generate $\mathcal{D}'$, $\mathcal{S}$ selects a random $D_i' \in \{0, 1\}^{|D_i^*|}$, $D_i^* \in \mathcal{D}^*$, $1 \leq i \leq |\mathcal{D}^*|$, and outputs $\mathcal{D}' = \{D_i', 1 \leq i \leq |\mathcal{D}^*|\}$.
- $\mathcal{S}$ randomly picks an invertible matrix $M_1', M_2' \in \mathbb{R}^{4 \times 4}$. Set $SK' = \{M_1', M_2'\}$.
- To generate $I'(\mathcal{D}')$, $\mathcal{S}$ first generates a vector of $2w$ elements for each $v_i \in I'(\mathcal{D}')$, $1 \leq i \leq |\mathcal{I}^*|$ as the index, then does the following:
  (1) For each element of v_i, $\mathcal{S}$ replaces it with a four-dimensional vector $(r_{i,1}, \cdots, r_{i,4})^T$, where $r_{i,k}$ is a random number for $1 \leq i \leq n, 1 \leq k \leq 4$.
  (2) $\mathcal{S}$ encrypts each v_i' with the M_2', and obtains $I'(\mathcal{D}') = \{Enc_{SK'}(v_i'), 1 \leq i \leq |\mathcal{I}^*|\}$.
- $\mathcal{S}$ constructs the query Q' and the search tokens $TK(Q')$ as follows. For each $q_i' \in Q'$, $1 \leq i \leq \tau$,
  (1) Generates a vector of $2w$ elements, denoted as u_i', and replace each element with a four-dimensional vector $(r_{i,1}', \cdots, r_{i,4}')^T$, where $r_{i,k}'$ is a random number for $1 \leq i \leq n, 1 \leq k \leq 4$. Output $Q' = \{q_i', 1 \leq i \leq \tau\}$.
  (2) Generate the search token for each q_i' by encrypting u_i' with M_1' from SK' for $1 \leq i \leq \tau$. Then $\mathcal{S}$ obtains $Enc_{SK'}(Q') = \{Enc_{SK'}(q_1'), \cdots, Enc_{SK'}(q_\tau')\}$.
- $\mathcal{S}$ outputs the view $V' = (\mathcal{D}', I'(\mathcal{D}'), TK(Q'))$.

The correctness of such construction is easy to demonstrate by querying $TK(Q')$ over $I'(\mathcal{D}')$. The index $I'(\mathcal{D}')$ and the token $TK(Q')$ generate the same trace as the one that the cloud server has. We claim that no PPT adversary can distinguish the view V' from $V(H)$. Specifically, due to the semantic security of IPE and AES, no PPT adversary can distinguish the $\mathcal{D}^*$ from $\mathcal{D}'$. Moreover, the PPT adversary with the query and probability pairs cannot distinguish which tokens are generated from the same query because of the indistinguishability of random perturbation-based comparison predicate, so that it cannot exploit the distributions of plain and encrypted queries to deduce information.   □

7   Evaluation

We evaluate the performance of TRQED on a real-world (REAL) dataset from the U.S. census bureau dataset [17]. The REAL dataset consists of 299,285 records with more than 20 attributes. We chose a number of records (from 20,000 to 100,000) to construct our datasets with needed dimensionality, and built R-tree indexes on each REAL dataset. The data owner, user and cloud server are all set on desktop computers with 3.20 GHz CPU and 4 GB RAM. We adopt GMP library and AES CTR mode for data encryption, and set the security parameter κ at 128, bit length l of each attribute at 32. For performance evaluation, we focus on overhead and efficiency as the scheme is proved to have a high search accuracy, which we omit for space reasons. 100 queries are executed and their average costs are reported.

7.1   Overhead and Efficiency

Setup. To setup the system, the data owner first encrypts all nodes of the R-tree index, of which each node requires about $\mathcal{O}(w)$ matrix-vector multiplications and bits. Here we are interested in the encryption overhead ignoring tree construction cost. As shown in Fig. 2(a) and (b), the costs of index encryption is increasing with number of dimensions w and data records n, as the TRQED should encrypt each non-leaf node of R-tree with RIPE and each leaf node with PIPE. That is, TRQED trades off the setup time and storage cost for improvement of the search efficiency.

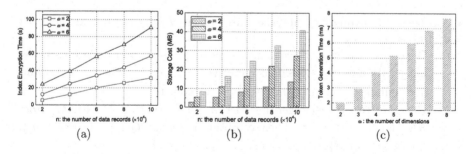

Fig. 2. Setup and token generation cost: (a) R-tree index encryption time versus n with dimensions $w = 2, 4, 6$; (b) total storage cost versus n with $w = 2, 4, 6$; (c) token generation time versus w with $n = 100,000$.

Token Generation. We evaluate the experimental results of generating search tokens in Fig. 2(c). The time of token generation per query is linearly increasing with the number of dimensions w, as TRQED needs to spend additional time to generate tokens in the added dimensions.

Search. We can see from Fig. 3(a) the impact of w and n on the search time per query. The search time shows a logarithmic increase trend as n increases, as the cloud server searches the R-tree level-by-level honestly to check whether the node satisfies $\mathsf{Query}(TK, \mathcal{I}^*) = true$. If fixing n, the cost shows a linear trend with the dimensions w, as the amount of inner product computations at each node increases with w. This verifies search time is linear with the number of dimensions and the height of R-tree. In Fig. 3(b), it can be seen that the average communication cost is linear with w, n has little effect, as the communication cost is composed of the generated search tokens and matched records.

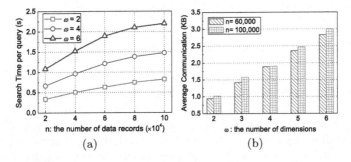

(a)   (b)

Fig. 3. Search efficiency: (a) search time versus n with $w = 2, 4, 6$; (b) communication cost versus w with $n = 60,000, 100,000$.

The efficiency of data update is similar to that of search phase, as the update operation is essentially fulfilled through search. The detailed results are omitted due to space limitation. Roughly speaking, the *insertion* and *deletion* operations require about the same computation, while the *modification* needs about the double computation overhead.

7.2   TRQED v.s. Prior Art

We compare our TRQED with the existing protocols, i.e., BonehW [5], LSED$^+$ [7], WangR [8], and Maple [9] in terms of privacy and data dynamics. As for efficiency, BonehW [5] fails to achieve faster-than-linear search and LSED$^+$ [7] achieves efficiency at the price of revealing ordering information and compromising privacy which makes it to be impractical. Therefore, we focus our efficiency comparison on WangR [8] and Maple [9], which are practical and represent the state-of-the-art multi-dimensional range query schemes.

Privacy and Data Dynamics. We compare the privacy guarantees and data dynamics support of the four related schemes and our TRQED, as shown in Table 1. As for the query and single-dimensional privacy, LSED[+] and WangR preserve the query privacy and do not provide single-dimensional privacy. BonehW and Maple preserve the single-dimensional privacy and fails in query privacy. In contrast, our TRQED preserves both the query privacy and single-dimensional privacy. As to data dynamics, only LSED[+] and our TRQED can provide data update while the other three scheme do not support.

Table 1. Comparison among different solutions

	Faster-than-linear search	Query privacy	Single-dimensional privacy	Data dynamics support
BonehW [5]	×	×	√	×
LSED[+] [7]	√	√	×	√
WangR [8]	√	√	×	×
Maple [9]	√	×	√	×
TRQED	√	√	√	√

Efficiency. The experimental results of search efficiency are depicted in Fig. 4(a) varying n from 20,000 to 100,000. They all show an increasing trend as n increases. Figure 4(b) illustrates the search time of three schemes are all linearly increasing with w. By contrast, the time cost of TRQED is slightly smaller than that of WangR because of smaller coefficient of the time complexity. Compared with Maple, TRQED is about 240 to 1,000 times faster than Maple, as Maple relies on public-key cryptography and the impact of domain limit T cannot be ignored. Thus, the proposed TRQED is more efficient.

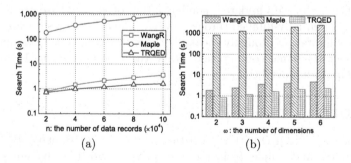

Fig. 4. Performance comparison: (a) average search time of the schemes versus n with $w = 2$; (b) average search time versus w with $n = 100,000$.

From the above comparison, we can conclude TRQED has a better performance than prior art. It not only preserves query and single-dimensional privacy, but also has a comparatively desirable overhead. On the whole, combining both security and efficiency, we think TRQED may be promising.

8 Conclusion

We have studied the problem of multi-dimensional private range queries over dynamic encrypted cloud data. Our proposed TRQED is the first private range query scheme that achieves faster-than-linear search and supports data dynamics while preserving the query privacy and single-dimensional privacy simultaneously. We have proven TRQED is secure against semi-honest adversaries under known background model. Experiments on real-world dataset have shown TRQED is more practical compared with prior art.

Acknowledgments. This work was supported by the National Natural Science Foundation of China (No. 61572456), the Anhui Province Guidance Funds for Quantum Communication and Quantum Computers and the Natural Science Foundation of Jiangsu Province of China (No. BK20151241).

References

1. Song, D.X., Wagner, D., Perrig, A.: Practical techniques for searches on encrypted data. In: Proceedings of IEEE S&P, pp. 44–55 (2000)
2. Li, M., et al.: Authorized private keyword search over encrypted data in cloud computing. In: Proceedings of IEEE ICDCS, pp. 383–392 (2011)
3. Cao, N., et al.: Privacy-preserving multi-keyword ranked search over encrypted cloud data. In: Proceedings of IEEE INFOCOM, pp. 829–837 (2011)
4. Wang, B., et al.: Privacy-preserving multi-keyword fuzzy search over encrypted data in the cloud. In: Proceedings of IEEE INFOCOM, pp. 2112–2120 (2014)
5. Boneh, D., Waters, B.: Conjunctive, subset, and range queries on encrypted data. In: Vadhan, S.P. (ed.) TCC 2007. LNCS, vol. 4392, pp. 535–554. Springer, Heidelberg (2007). https://doi.org/10.1007/978-3-540-70936-7_29
6. Okamoto, T., Takashima, K.: Hierarchical predicate encryption for inner-products. In: Matsui, M. (ed.) ASIACRYPT 2009. LNCS, vol. 5912, pp. 214–231. Springer, Heidelberg (2009). https://doi.org/10.1007/978-3-642-10366-7_13
7. Lu, Y.: Privacy-preserving Logarithmic-time search on encrypted data in cloud. In: Proceedings of NDSS (2012)
8. Wang, P., Ravishankar, C.V.: Secure and efficient range queries on outsourced databases using $\widehat{R}$-trees. In: Proceedings of IEEE ICDE, pp. 314–325 (2013)
9. Wang, B., et al.: Maple: scalable multi-dimensional range search over encrypted cloud data with tree-based index. In: Proceedings of ACM ASIACCS, pp. 111–122 (2014)
10. Guttman, A.: R-trees: a dynamic index structure for spatial searching. ACM (1984)
11. Goldreich, O.: Foundations of Cryptography: Basic Applications, vol. 2. Cambridge University Press, Cambridge (2009)
12. Wang, C., et al.: Secure ranked keyword search over encrypted cloud data. In: Proceedings of ICDCS, pp. 253–262 (2011)
13. Li, R., et al.: Fast and scalable range query processing with strong privacy protection for cloud computing. IEEE/ACM Trans. Netw. **24**(4), 2305–2318 (2016)

14. Chi, J., et al.: Fast multi-dimensional range queries on encrypted cloud databases. In: Proceedings of DASFAA, pp. 559–575 (2017)
15. Stefanov, E., et al.: Path ORAM: an extremely simple oblivious RAM protocol. In: Proceedings of ACM CCS, pp. 299–310 (2013)
16. Curtmola, R., et al.: Searchable symmetric encryption: improved definitions and efficient constructions. J. Comput. Secur. **19**(5), 895–934 (2011)
17. Lichman, M.: (1999). http://archive.ics.uci.edu/ml
18. Chang, Y.-C., Mitzenmacher, M.: Privacy preserving keyword searches on remote encrypted data. In: Ioannidis, J., Keromytis, A., Yung, M. (eds.) ACNS 2005. LNCS, vol. 3531, pp. 442–455. Springer, Heidelberg (2005). https://doi.org/10.1007/11496137_30
19. Wong, W.K., et al.: Secure kNN computation on encrypted databases. In: Proceedings of ACM SIGMOD, pp. 139–152 (2009)
20. Sun, W., et al.: Catch you if you lie to me: efficient verifiable conjunctive keyword search over large dynamic encrypted cloud data. In: Proceedings of IEEE INFOCOM, pp. 2110–2118 (2015)
21. Papadopoulos, D., Papadopoulos, S., Triandopoulos, N.: Taking authenticated range queries to arbitrary dimensions. In: Proceedings of ACM CCS, pp. 819–830 (2014)
22. Kawamoto, J., Yoshikawa, M.: Private range query by perturbation and matrix based encryption. In: Proceedings of ICDIM, pp. 211–216 (2011)
23. Wang, B., Hou, Y., Li, M., Wang, H., Li, H., Li, F.: Tree-based multi-dimensional range search on encrypted data with enhanced privacy. In: Tian, J., Jing, J., Srivatsa, M. (eds.) SecureComm 2014. LNICST, vol. 152, pp. 374–394. Springer, Cham (2015). https://doi.org/10.1007/978-3-319-23829-6_26

Search and Information Retrieval

iExplore: Accelerating Exploratory Data Analysis by Predicting User Intention

Zhihui Yang[1,2], Jiyang Gong[1,2], Chaoying Liu[1,2], Yinan Jing[1,2(✉)], Zhenying He[1,2(✉)], Kai Zhang[1,2], and X. Sean Wang[1,2,3(✉)]

[1] School of Computer Science, Fudan University, Shanghai, China
{zhyang14,jygong14,chaoyingliu14,jingyn,zhenying,zhangk,
xywangCS}@fudan.edu.cn
[2] Shanghai Key Laboratory of Data Science, Shanghai, China
[3] Shanghai Institute of Intelligent Electronics & Systems, Shanghai, China

Abstract. Exploratory data analysis over large datasets has become an increasingly prevalent use case. However, users are easily overwhelmed by the data and might take a long time to find interesting facts. In this paper, we design a system called iExplore to assist users in doing this time-consuming data exploration task through predicting user intention. Moreover, we propose an intention model to help the iExplore system have a comprehensive understanding of user's intention. Thus, the exploratory process can be accelerated by the intention-driven recommendation and prefetching mechanisms. Extensive experiments demonstrate that the intention-driven iExplore system can significantly lighten the burden of users and facilitate the exploratory process.

Keywords: User intention · Data exploration · Query log

1 Introduction

Exploratory data analysis is an effective way for users to obtain insights from large datasets, especially for non-expert users who are unfamiliar with the underlying data. In the exploratory analysis scenario, users usually pose several trial queries sequentially to explore the dataset. Such a *"human-in-the-loop"* analysis is essentially a multi-steps and time-consuming process. There are two reasons making the exploratory process time-consuming and tedious for users. (i) Users have to continuously pose a series of queries to try to figure out the underlying data space and discover interesting facts as much as possible. (ii) When users face massive amount of data, they tend to be overwhelmed by the data and fall in random walk in the data space, even without any useful information found for a long time. Basically, in data exploration, most of users are faced with such a dilemma: *How could users find what they want to know efficiently?*

The work is supported by the NSFC (No. 61732004) and the Shanghai Innovation Action Project (Grant No. 16DZ1100200).

In fact, in the exploratory process, there is hidden intention which leads users to approach their goals step-by-step, even though they do not have a clear end-goal at first. It is because users always pose suitable queries based on their previous queries and results. Therefore, if we could automatically extract this hidden intention to offer some advice and prefetch some results for users, it can greatly lightening the burden of users and accelerating the exploratory process. Additionally, it is worth noting that the intention is hidden in their previous queries and query logs from other people with similar backgrounds, reflecting in both query logs and the similarity of data. Actually, the hidden intention is the intrinsic connection between previous queries posed by the user and the next query to be posed, as users leverage their observations or experiences.

Let us consider a specific example from SDSS [1] DR7 database[1], which stores digital astronomy data. We retrieve a segment of a session, which is a series of queries posed by users from the SDSS DR7 log dataset[2], as shown in Fig. 1. The log dataset records users' successive exploratory queries. In Fig. 1, the DBObjects table stores the descriptions of all database objects, and IndexMap contains all index information. The example segment includes three queries. The first one (q_1) is that users want to see the description information of the 'HoleObj' table, and q_2 indicates that users want to figure out the detail description of the 'HoleObj' table. These two queries are similar as both of them show the description information of the same table at different level. q_3 shows that users want to make out the index information built on table 'HoleObj'.

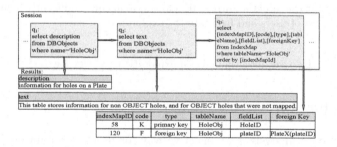

Fig. 1. Example of a session with queries and their corresponding results.

As shown above, this exploratory segment illustrates that the user wants to figure out the underlying data space, and it also implies a very strong inherent intention as follows. When a user wants to explore a table, she/he often starts with looking at the description of the table. If the user is interested in the table, she/he will further look at the index built on the table. Two points are worth highlighting: First, many users all have this common query behavior according to the statistic information of the query log dataset. This reveals that users have tremendous possibility to turn to the index information of the table, when they

---

[1] http://cas.sdss.org/dr7/en.
[2] http://skyserver.sdss.org/log/en/traffic/sql.asp.

glimpse at the description information of the table. Second, when users pose a query, results similar to this query are most likely to be selected by users next, such as q_1 and q_2. Therefore, we could extract the hidden intention from both query log and the similarity of data. The extracted intention can be used to navigate users to their end-goals and accelerate the exploratory process.

In this paper, we study the problem on how to navigate users to explore the data space and to accelerate the exploratory process through predicting users' hidden intention. However, to solve this problem, we face the following challenges: (i) The hidden intention essentially embodies a degree of uncertainty and vagueness. It is because users leverage their uncertain observations or experiences to decide their next-step query. This greatly complicates the modelling and computation of the hidden intention. (ii) We should ensure that the intention model makes the exploratory process convergent, which means the user's exploration would stop at a stable point finally under the guidance of the system. To the best of our knowledge, the investigation of the convergence of the exploratory analysis process has not been reported in the literatures.

To address these challenges, we propose an intention model to model the exploratory analysis process. A key component of this model is an intention function. This function measures the intrinsic connection between the next query and previous queries to describe the uncertainty and vagueness of the intention. Furthermore, the convergence of the intention model has been studied to figure out the characteristic of exploratory process that users have end-goals. Thus we can know whether users randomly walk or not. Finally, we propose an intention-driven exploratory (iExplore) system to assist users' exploration and accelerate the exploratory process. The main contributions are summarized as follows:

i A generalized intention model modelling the exploratory analysis process has been proposed in this paper. The intention model defines the fundamental components for modelling user's hidden intention.

ii The convergence of the intention model has been studied to figure out the characteristic of exploratory process that users have clear end-goals.

iii An intention-driven exploration system called iExplore is proposed in this paper, which incorporates an instantial intention model into the system. The iExplore offers recommendations to assist users and prefetches data to accelerate the exploratory process.

iv We experimentally demonstrate that iExplore can successfully help users to specify their querie, significantly facilitating the exploratory process.

Paper outline: In Sect. 2, we formulate our user intention problem and study the conditions of convergence of the intention model. Section 3 describes the iExplore system. Comprehensive experiments are presented in Sect. 4. We review the related work in Sect. 5 and conclude this paper in Sect. 6.

2   Problem Formulation

2.1   Intention Model

In the exploratory process, the user's hidden intention which describes the intrinsic connection between two queries determines the possibility that the user moves from one query to another query. We introduce an intention function to describe this hidden intention. Given the current query and its result, the intention function gives the next query a score according to the intensity of user's intention. The definition of this intention function is as follows and one type of detailed method of the intention function is given in Sect. 3.3.

Definition 1 (Intention Function). *For given two pairs of query and result* (q_i, r_i) *and* (q_j, r_j) *where* q_i *and* q_j *represent different queries and* r_i *and* r_j *represent corresponding result, the intention function* f *gives a score to describe the degree of user's intention that he/she moves from the current query and result* (q_i, r_i) *to the next query and result* (q_j, r_j). *The higher score, the stronger intention.* f *is* $(q_i, r_i), (q_j, r_j) \xrightarrow{f}$ *a score.*

The intention function is the critical component in our intention model. Additionally, the intention function operates according to some basic premises considering the notion of query similarity. The fundamental premises and the intention function determine the intention model, and it is defined as follows.

Definition 2 (Generalized Intention Model). *The intention model is a quadruple* $[S, D, \mathcal{F}, f((q_i, r_i), (q_{i+1}, r_{i+1}))]$ *where*

 i *S is a sequence of pairs of query q_i and result r_i posed by the current user.*
   $S = \{(q_1, r_1), (q_2, r_2), \ldots, (q_s, r_s)\}$ where (q_i, r_i), $i \in \{1, \ldots, s\}$ is the pair of query and result that the user posts at time i.
 ii *D is a set composed of sessions in the query log dataset.*
iii *$\mathcal{F}$ is a framework for modelling the representation of pairs of query and result and the relationship between different pairs, such as vectors and probability distributions.*
iv *$f((q_i, r_i), (q_{i+1}, r_{i+1}))$ is the intention function. Given that the user posts query q_i and sees its result r_i at time i, the intention function gives query q_{i+1} a score for modelling the possibility that user converts to this query at time $i + 1$, or the intention function gives a score to result r_{i+1} for modelling the possibility that this result to be selected at time $i + 1$. This intention function defines an order for the next pair of query and result (q_{i+1}, r_{i+1}).*

The intention model consists of two main tasks: (i) the conception of a logical framework for representing pairs of query and result, $\mathcal{F}$. This framework provides a method to map pairs of query and result into the measurable space, such as vector space and probability distribution. (ii) The detailed definition of the intention function $f((q_i, r_i), (q_{i+1}, r_{i+1}))$. This intention function models user's intention during data exploration and can be used to navigate users in the data space. Additionally, q_i is used to denote both the query and its representation, and r_i is used to denote both the corresponding result and its representation.

The iExplore system presented in Sect. 3 provides a prototype of the intention model especially these two main tasks. On one hand, we introduce a vector model to map pairs of query and result into the vector space and standard linear operations are allowed on this model given in Sect. 3.2. And other methods for $\mathcal{F}$ are in our future work. On the other hand, a detailed method for the intention function f which measures the hidden intention based on both query log and the similarity of data is given in Sect. 3.3 although this may be not the only method. Therefore, this intention model is a general definition of the exploratory analysis process and its components can be instanced more than the methods mentioned in this paper. Additionally, the intention model can be used not only to navigate the user in the data space, but also to automatically prefetch data during data exploration.

2.2 Convergence

In terms of the characteristics of the exploratory process, it is worth noting that the users' intention is basically consistent and does not change with the process of exploratory analysis. Therefore, we assume that the intention function does not change over time, this is called the homogeneity of the intention model, otherwise inhomogeneous. The formal definition of the homogeneity is as follows.

Definition 3 (Homogeneity). *For the given intention model* $[S, D, \mathcal{F}, f((q_i, r_i), (q_{i+1}, r_{i+1}))]$, *if the intention function* $f((q_i, r_i), (q_{i+1}, r_{i+1}))$ *is invariant during an exploratory process, we say that the intention model is homogeneous.*

For simplicity, we study the homogeneous intention model in the sequel of this paper. Additionally, we discover that if users find what they want or their end-goals in the process of exploration, users' query would stop at a stable query point. This analysis shows that such a type of exploratory analysis that users have end-goals is convergent. The convergence of the intention model is defined as follows.

Definition 4 (Convergence). *For the given intention model* $[S, D, \mathcal{F}, f((q_i, r_i), (q_{i+1}, r_{i+1}))]$, *if the user ends with a unique stationary pair of query and result* (q_s, r_s) *through multi-steps queries, that is* $(q_{s+1}, r_{s+1}) = (q_s, r_s)$, *we say that the intention model is convergent.*

During the process of exploration, if the intention model is convergent, the user's query would stably stop at the end-goal of this user. Then the exploratory analysis process is convergent and valid for the user. On the contrary, if the intention model is not convergent, the user's query randomly walks in the data space without useful information found. This kind of exploratory analysis process is useless and invalid for the user. Additionally, we also study the conditions that make the intention model convergent with the following theorem introduced.

Theorem 1. *Given the intention model* $[S, \ D, \ \mathcal{F}, \ f((q_i, r_i), (q_{i+1}, r_{i+1}))]$, *if the intention function satisfies the following two conditions: (i)the intention function is aperiodic, and (ii) no matter how many steps, every two pairs of query and result,* (q_i, r_i) *and* (q_j, r_j), *is reachable, we say that the intention function is convergent.*

Proof. Assume that the query and result space is $\{(q_1, r_1), (q_2, r_2), \cdots, (q_n, r_n)\}$, we can construct an intention table which has n rows and n columns. The value in the ith row and j^{th} column of the intention table is the value of the intention function $f((q_i, r_i), (q_j, r_j))$ which associates a score with the intensity of user's intention. Then perform the row normalization on the intention table and we can obtain a $n \times n$ stochastic matrix [9], denoted P. A sequence of queries posed by the user according to the intention model during the exploratory analysis process can form a Markov Chain [9] with transition matrix P. If every two pairs of query and result is reachable, the transition matrix P is irreducible. Therefore, these two conditions indicate that P is aperiodic and irreducible. Then the Markov Chain is convergent according to the Markov Chains Convergence Theorems [4]. Thus the intention model is convergent.

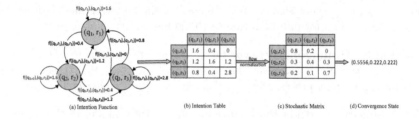

Fig. 2. Example of intention function with $n = 3$.

To illustrate the homogeneity and convergence of the intention model, consider a simple example of the intention function with $n = 3$ as shown in Fig. 2. Figure 2(a) shows the intention function in the query and result space $\{(q_1, r_1), (q_2, r_2), (q_3, r_3)\}$. The intention table constructed from these intention functions is shown in Fig. 2(b). In this example, we assume that the intention model is homogeneous, that is the intention table will not change during the exploratory analysis process. Then as shown in Fig. 2(c), we can obtain a 3×3 stochastic matrix through the row normalization on the intention table. As this matrix satisfies Theorem 1, the user's intention has the convergent state shown in Fig. 2(d). This indicates that user's query would finally stop at (q_1, r_1), and this reflects the intention of a general exploratory process of a user.

3   iExplore: Intention-Driven Exploration

3.1   iExplore Architecture

In this paper, we propose an intention-driven exploratory (iExplore) system, which uses users' intention to assist and accelerate the exploratory process for users. The overview of the iExplore system is presented in Fig. 3. The user interacts with the system by posting queries through the user interface. The query posted by the user goes to the query engine to fetch result from the database or the cache. Then the vector model map the query and its result to the vector space. The intention function computes user's next query based on vectors of pairs of query and result. Then the recommendation engine select top-k queries q_{i+1} according to the intention function, and these queries are recommended to the user. q_{i+1} is a set of k recommended queries. Finally, the prefetching engine prefetches results according to the query recommendation to accelerate the execution of user's next query.

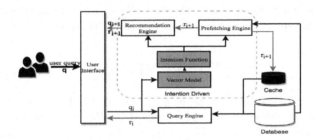

Fig. 3. The iExplore system architecture.

The iExplore system is actually an instance of the generalized intention model presented in the previous section. It is composed of the following four components: (i) **Vector Model:** it maps pairs of query and result into the vector space, and standard linear operations are allowed. (ii) **Intention Function:** this component scores user's next query from two perspectives: query logs and the data similarity. (iii) **Recommendation Engine:** this engine recommends top-k queries according to scores computed by the intention function. (iv) **Prefetching Engine:** it prefetches the result data of queries recommended to users to accelerate the exploratory process. The first two components form the intention model and the last two components show how the intention model works in the exploratory process.

3.2   Vector Model

Before measuring the hidden intention, we need to map pairs of query and result into the measurable space. On one hand, as the query is a declarative language, different users may write syntactically different queries by leveraging their custom and experiences to reflect the same information needed. This complicates greatly the measurement of different queries. On the other hand, as the query and its result are equivalent, we can use the result set to represent the query, avoiding the aforementioned problem. Therefore, we can express queries based on their result sets.

For a given database $\mathcal{D}$, all distinct tuples can construct a tuple vocabulary V_T. A query on this database can be expressed as a vector v, depending on whether the tuple appears in the result set. If the tuple t appears in the result set of query q, $q[t] = 1$. Otherwise, $q[t] = 0$. As the length of the vocabulary is too long and the vector of query has mostly zero elements, we use the sparse vector to express the query. The sparse vector records the indices of all result tuples in the tuple vocabulary. The tuple vocabulary V_T and sparse vectors of three queries are shown in Fig. 4(a) and (b). But this vector model is suitable for Select-Project-Join (SPJ) queries.

We use the Jaccard Similarity [13] to measure the similarity between two queries. The Jaccard Similarity between two queries is the ratio of the size of intersection of these two queries' vectors to the size of union vectors. And the similarity of two query q_i and q_j can be described as follows:

$$sim(q_i, q_j) = \frac{|v_i \cap v_j|}{|v_i \cup v_j|} \tag{1}$$

where v_i, v_j are the vectors of query q_i, q_j, respectively, and $|v_i \cap v_j|$ is the size of intersection of vectors v_i and v_j. The Jaccard Similarity of the three queries in Fig. 4(b) are shown in Fig. 4(c). Therefore, we can map pairs of query and result into the measurable vector space and standard linear operations are allowed. Then we can measure the hidden intention.

3.3 Intention Function

We consider the hidden intention from two aspects, which are query logs and data similarity. Therefore, the intention function consists of two parts: (i) query logs from other people with similar backgrounds, and (ii) the similarity between two adjacent queries. We denote these information as f_{ql} and f_{sim}, respectively. A parameter α can be introduced to leverage these two parts. Then the total intention function can be represented as follows:

$$f((q_i, r_i), (q_{i+1}, r_{i+1})) = \alpha f_{ql}((q_i, r_i), (q_{i+1}, r_{i+1})) + (1 - \alpha) f_{sim}((q_i, r_i), (q_{i+1}, r_{i+1})) \tag{2}$$

First, let us consider f_{ql}. If the sub-sequence $\{q_i, q_{i+1}\}$ appears frequently in the log dataset, the next query is likely to be selected by the user at time $i + 1$ when he/she has posted query q_i at time i. According to this intuition and the query log dataset D, the intention can be estimated by the conditional probability of the next query based on previous query. Therefore, the probability that user posed query q_{i+1} at time $i + 1$ can be estimated as follows:

$$f_{ql}((q_i, r_i), (q_{i+1}, r_{i+1})) = p((q_{i+1}, r_{i+1})|(q_i, r_i)) = C((q_i, r_i), (q_{i+1}, r_{i+1}))\ C(q_i, r_i) \quad (3)$$

where $C(q_i, r_i)$ is the count of query q_i appearing in D.

Then, we now model f_{sim}. If a query is similar to the previous query, it is likely to be selected. Based on this intuition, we use the similarity between two queries introduced in Sect. 3.2 to model the hidden intention f_{sim} and it is as follows:

$$f_{sim}((q_i, r_i), (q_{i+1}, r_{i+1})) = sim(q_i, q_{i+1}) \quad (4)$$

Therefore, according to Eqs. 2, 3 and 4, the intention function is as follows:

$$f((q_i, r_i), (q_{i+1}, r_{i+1})) = \alpha p((q_{i+1}, r_{i+1})|(q_i, r_i)) + (1 - \alpha)sim(q_i, q_{i+1}) \quad (5)$$

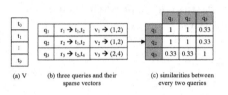

| (a) V | (b) three queries and their sparse vectors | (c) similarities between every two queries |

Fig. 4. An example of a database with ten tuples and three queries on this database.

Fig. 5. The distribution of the length of sessions.

Algorithm 1. The Intention-Driven Exploration Algorithm

Input: q_i, D, V_Q, α, k
Output: $\mathbf{q_{i+1}}$, $\mathbf{r_{i+1}}$
1: $c_i \leftarrow Count(D, q_i)$, pro
2: **for all** $q' \in V_Q$ **do**
3:    $c_{i,i+1} \leftarrow Count(D, q_i, q')$
4:    $pro[q'] \leftarrow \alpha \times c_i/c_{i,i+1} + (1 - \alpha) \times Sim(q_i, q')$
5: **end for**
6: $pro \leftarrow Sort(pro)$
7: **while** $k \geq 0$ **do**
8:    $q' \leftarrow pop(pro)$, $k = k - 1$
9:    $insert(\mathbf{q_{i+1}}, q')$, $insert(\mathbf{r_{i+1}}, fetch(q'))$
10: **end while**

3.4   Recommendation and Prefetching

In this section, we will describe how the intention model works in the exploratory analysis process. The existing users' hidden intention decides their suitable next queries in the exploratory process. This fact indicates that the hidden intention steers users to explore in the data space. Therefore, the extracted hidden intention can be used to navigate users to their end-goals and to accelerate the exploratory process. In terms of navigation, the iExplore system recommends top-k queries from the query vocabulary V_Q to the user according to the intention function. V_Q is a set of all distinct queries in the query log dataset D. In terms of acceleration, the iExplore system prefetches the result data of these top-k queries. When the user adopts the recommended queries, the result can be obtained from cache without accessing the database.

The intention-driven recommendation and prefetching algorithm is presented as pseudo-code shown in Algorithm 1. It computes the value of intention function for every query in the query vocabulary V_Q (line 1–5), where $Count(D, q_i)$ is the number of times that q_i appears in D, and $Sim(q_i, q')$ is the Jaccard Similarity between q_i and q'. Line 6 sorts according to the value of intention function. Finally, line 7–10 select top-k queries ($insert(\mathbf{q_{i+1}}, q')$), prefetch their results ($insert(\mathbf{r_{i+1}}, fetch(q'))$) and return back to users.

4   Experimental Evaluation

4.1   Experiments Setup

Experimental Environment. All experiments presented in this section are implemented in Java (JDK 1.8). They are conducted on a Ubuntu Linux 16.04 LTS machine with an Intel E5-2650v3 CPU of 40 cores. This machine is equipped with 128 GB DDR4 RAM and Seagate 2TB SATA HDD.

Dataset. Our experiments are performed on the SDSS DR7 [1] real database which can be accessed online. It is the largest astronomical data collection to date, and covers half of the Northern sky characterizing about 200M objects in 5 optical bands. The SDSS DR7 database is about 20TB containing 95 tables, 51 views, 224 functions and 90 indices.

Workloads. For the experimental evaluation, we have gathered 47,214 query logs issued by users in 2011 from the SDSS DR7 log dataset. These query logs can be separated into 12,378 sessions according to the method mentioned in [17]. We have conducted statistical analysis of these sessions. The distribution of the length of sessions is shown in Fig. 5. The length of a session is the number of queries in this session. As shown in Fig. 5, the distribution has a long tail indicating that long sessions are very few. To study the effect of different length sessions, we divide the session set into three sets according to the length of sessions, denoted D_1, D_2, D_3. The characteristics of them are summarized in Table 1. Each dataset is divided into a train set and a test set with 9:1.

Performance Metrics. The exploratory analysis system can be divided into three categories, and the time cost for two adjacent queries on three categories is shown in Fig. 6. The symbols used in Fig. 6 are described in Table 2. As the time cost of posting the first query on three category systems is the same, we

Table 1. The characteristics of D_1, D_2, D_3.

Workloads	Length of sessions	Number of sessions	Number of queries	The size of V_T
D_1	$[2, 4]$	9,680	25,494	1,044,375
D_2	$[5, 16]$	2,548	17,062	616,463
D_3	$[17, 64]$	150	4,658	2,831

Table 2. The symbols used to describe exploratory time in Fig. 6.

$T_{r\&n}$	The time of reading the results of previous query and thinking the next query
T_{wnq}	The time of writing the next query
$T_{w\hat{n}q}$	The time of modifying the next query based on the recommended queries
T_{gr}	The time of getting results of query from database
$T_{\hat{g}r}$	The time of getting results of query from cache
T_r	The time of computing the recommended queries
a_i	Adoption flag

ignore it in the following analysis. We assume that users posed N queries in one exploratory process. The time cost for three category exploratory processes are as follows.

1. **Exploration without Assistance:** The pure database system just performs queries posed by users without providing any support to help users carry out their exploratory analysis, such as SkyServer. The SkyServer is a search tool allowing users to explore the SDSS database by submitting queries. The time cost for two adjacent queries is shown in Fig. 6(a). The response time for the ith query is: $T_{r\&n,i} + T_{wnq,i} + T_{gr,i}$. And the exploratory time for users on this system is: $\sum_{i=1}^{N}(T_{r\&n,i} + T_{wnq,i} + T_{gr,i})$.

2. **Exploration with Recommendation:** It provides recommended queries to help users explore. For instance, QueRIE [8] recommends similar queries by collaborative filtering based on query logs, and AIDE [6] provides suggestions according to users' feedback. The time cost for two adjacent queries is illustrated in Fig. 6(b). The response time for the ith query is: $T_{r\&n,i} + T_{r,i} + (1 - a_i) \times T_{wnq,i} + a_i \times T_{w\hat{n}q,i} + T_{gr,i}$. And the exploratory time for users is: $\sum_{i=1}^{N}(T_{r\&n,i} + T_{r,i} + (1 - a_i) \times T_{wnq,i} + a_i \times T_{w\hat{n}q,i} + T_{gr,i})$.

3. **Exploration with Recommendation and Prefetching:** Such a kind of system offers some advice to navigate users' exploration and prefetches results to accelerate the exploratory process, such as the iExplore system. The time cost for two adjacent queries is shown in Fig. 6(c). The response time for the ith query is: $T_{r\&n,i} + T_{r,i} + (1 - a_i) \times (T_{wnq,i} + T_{gr,i}) + a_i \times (T_{w\hat{n}q,i} + T_{\hat{g}r,i})$. And the exploratory time for users on such a type of system is: $\sum_{i=1}^{N}(T_{r\&n,i} + T_{r,i} + (1 - a_i) \times (T_{wnq,i} + T_{gr,i}) + a_i \times (T_{w\hat{n}q,i} + T_{\hat{g}r,i}))$.

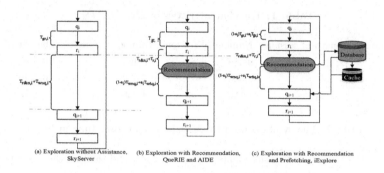

Fig. 6. Several exploratory systems.

where a_i is the adoption flag. If the user adopts the recommended queries, $a_i = 1$. Otherwise, $a_i = 0$. As the average time for two adjacent queries in sessions in D_1, D_2 and D_3 is 100 s, we set $T_{r\&n,i} = 50$ s, $T_{wnq,i} = 50$ s and $T_{w\hat{n}q,i} = 10$ s. In the following experiments, we compare iExplore with QueRIE and SkyServer, as both QueRIE and iExplore do not need users' feedback. SkyServer serves as the baseline.

4.2 Response Time Comparison

For the first group experiments, we compare iExplore, QueRIE and SkyServer through measuring the average response time in the exploratory process. We use sessions in the test set to simulate the exploratory processes and calculate the response time for every query posed by users. The average response time on three datasets are shown in Fig. 7.

As shown in Fig. 7, the average response time of iExplore is less than that of QueRIE and SkyServer, indicating that iExplore has accelerated the response time. Additionally, the fluctuation of the average response time on iExplore is because that users may adopt the recommended queries and do not need to write the next query and access the underlying database. Specifically, as the adoption flag a_i is different in different exploratory processes for the ith query, the average response time varies greatly with i. For completeness, we evaluate a_i in Sect. 4.3.

The slight fluctuation of the average response time on QueRIE is because users may not adopt the recommended queries. QueRIE is concerned with the effect of the last prediction, and does not focus on the effect of the whole exploratory process. Additionally, as we measure the average response time to offset the deviation, the response time on SkyServer has changed only a little.

4.3 Impact of Parameter

In this experiment, we now turn our attention to the problem that how the adoption flag a_i is affected by the adjustable parameter α. From an overall view, we evaluate the adoption rate p, where $p = \sum_{i=1}^{N} a_i/N$ and N is the number of

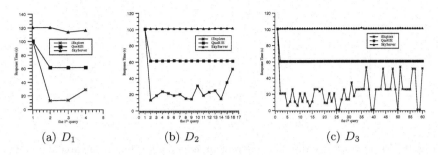

(a) D_1 (b) D_2 (c) D_3

Fig. 7. The average response time of these three systems (iExplore, QueRIE, Sky-Server) on three datasets (D_1, D_2, D_3).

queries in the exploratory process as we mentioned in Sect. 4.1. We employ our three datasets and measure p varying α. Figure 8 reports the results.

In Fig. 8, we can see that p can achieve 0.8 indicating that the iExplore can help users reduce about 80% query burden in the exploratory process. We observe that $\alpha \geq 0.5$ performs better. It is coincident with the fact that the similarity of queries may be contained in the query log dataset. Additionally, the adoption rate on D_1 is higher than that on D_2 and D_3. It is because the dataset D_1 is larger than D_2 and D_3, and there is sufficient statistics in D_1.

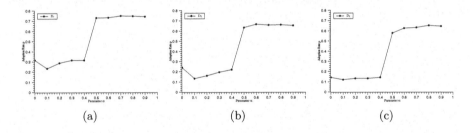

(a)   (b)   (c)

Fig. 8. The change of adoption rate with α on three datasets.

4.4 Prefetching Evaluation

We conduct an experiment to evaluate our intention-driven prefetching (IDP) strategy by comparing it with random prefetching (RP) strategy and no-prefetching (NP) strategy. RP strategy prefetches data randomly. Specifically, RP randomly selects k queries, and then prefetches their results. NP does not prefetch any data, and it serves as the baseline. We employ our three datasets and use sessions in the test set to simulate the exploratory process. The average exploratory time for users are measured and shown in Fig. 9.

In Fig. 9, the exploratory time of IDP is significantly less than that of RP and NP with 86%, 81% and 82% improvements respectively in D_1, D_2 and D_3. It is because IDP could predict user's next queries and prefetch their results.

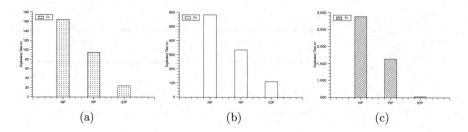

Fig. 9. The exploratory time for different prefetching strategies on three datasets.

The exploratory time of RP is less than NP (76%, 67% and 68% improvements respectively in D_1, D_2 and D_3). As RP randomly selects queries and prefetches their results, this may hit the query the user wants next and accelerate the exploratory process. It also demonstrates that prefetching is an effective strategy. And the exploratory time on D_1 is less than that on D_2 and D_3. This is due to the length of session.

4.5   Case Study

As a final step, we invite 20 users to explore the SDSS DR7 database by using the iExplore system, QueRIE and SkyServer, respectively. Every user carries out three exploratory processes, and we record their average exploratory time for three datasets and three systems in Fig. 10.

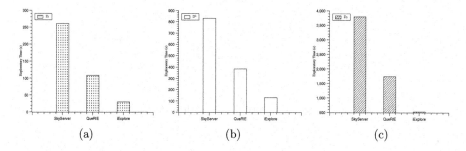

Fig. 10. The users exploratory time on three systems and three datasets.

According to Fig. 10, the iExplore system significantly accelerates the exploratory process comparing to QueRIE and SkyServer with 89%, 84.6% and 86% improvements in D_1, D_2 and D_3. It demonstrates that the iExplore system can successfully assist users to specify their queries and accelerate the exploratory process for users. Additionally, the QueRIE accelerates the exploratory process comparing to the no assistance system (SkyServer) with 59%, 53% and 54% improvements in D_1, D_2 and D_3. This is due to the recommended queries by QuerRIE accelerating the process for writing the next query.

5   Related Work

Data Exploration. During the last decade, a variety of recent research efforts focusing on data exploration [10] have been proposed. A comprehensive overview of this field is proposed in [10]. And it reviews recent researches in the emerging systems tailored for data exploration and discusses new challenges in data exploration. Furthermore, there are some efforts focus on the prototype systems from different aspects. Query Steering [6] relies on the user's feedback to provide suggestions. QueRIE [8] recommends the most likely queries, and YMALDB [7] recommends additional results based on the most interesting sets of attribute values. Additionally, DICE [11] supports efficient faceted explorations of data cubes. Finally, work in [16] aim at interactively and efficiently navigating explorers through large data space based on cluster information of the data. However, none of these systems pays attention to the characteristics of the exploratory process to the best of our knowledge.

Prefetching. The idea of speculative execution of queries and prefetching results has been extensively used in the database area to improve performance of query processing [18]. PROMISE [15] investigates the likehood of future queries based on Markov Models while it targets very specific OLAP workloads. Work in [3] predicts upcoming requests based on past observations. And Ramachandran et al. [14] focus on the speculation of exact, non-approximate drill-down queries. Ideas in these papers can be used to further prioritize our intent queries.

Query Log Mining. Several authors have proposed query log mining algorithms, either to improve the performance of database or to help users write queries. Aouiche et al. [2] mine the column names in the log dataset to select candidate materialized views and indexes. [12] recommends possible additions to query clauses by collecting relevant snippets from log dataset. [5] presents with an ambiguous query to better results by diversification. But [5] focuses on are keyworks search queries in a search engine not SQL queries. Additionally, SDSSLogViewer [19] visualizes and analyzes the SDSS log data to understand users' behaviors. But these works do not accelerate the data exploration.

6   Conclusion and Future Work

In this paper, we study the exploratory data analysis problem from the perspective of user intention. As we propose a general definition of the intention model, there are also several interesting directions for future researches. Other methods implementing the intention model can be proposed. Specially, we can also map them into the probability distribution space to support complex queries. In fact, the work reported in this paper is part of our ongoing research towards a general interactive data exploration analysis framework.

References

1. Abazajian, K.N., Adelman-McCarthy, J.K., Agüeros, M.A., Allam, S.S., Prieto, C.A., An, D., Anderson, K.S., Anderson, S.F., Annis, J., Bahcall, N.A., et al.: The seventh data release of the sloan digital sky survey. Astrophys. J. Suppl. Ser. **182**(2), 543 (2009)
2. Aouiche, K., Darmont, J.: Data mining-based materialized view and index selection in data warehouses. J. Intell. Inf. Syst. **33**(1), 65–93 (2009)
3. Bowman, I.T., Salem, K.: Semantic prefetching of correlated query sequences. In: 2007 IEEE 23rd International Conference on Data Engineering, ICDE 2007, pp. 1284–1288. IEEE (2007)
4. Brémaud, P.: Markov Chains: Gibbs Fields, Monte Carlo Simulation, and Queues, vol. 31. Springer Science & Business Media, Heidelberg (2013). https://doi.org/10.1007/978-1-4757-3124-8
5. Crane, M.: Diversified relevance feedback. In: Proceedings of the 36th International ACM SIGIR Conference on Research and Development in Information Retrieval, p. 1142. ACM (2013)
6. Dimitriadou, K., Papaemmanouil, O., Diao, Y.: AIDE: an active learning-based approach for interactive data exploration. IEEE Trans. Knowl. Data Eng. **28**(11), 2842–2856 (2016)
7. Drosou, M., Pitoura, E.: Ymaldb: exploring relational databases via result-driven recommendations. VLDB J. **22**(6), 849–874 (2013)
8. Eirinaki, M., Abraham, S., Polyzotis, N., Shaikh, N.: Querie: collaborative database exploration. IEEE Trans. Knowl. Data Eng. **26**(7), 1778–1790 (2014)
9. Gagniuc, P.A.: Markov Chains: From Theory to Implementation and Experimentation. Wiley, Hoboken (2017)
10. Idreos, S., Papaemmanouil, O., Chaudhuri, S.: Overview of data exploration techniques. In: Proceedings of the 2015 ACM SIGMOD International Conference on Management of Data, pp. 277–281. ACM (2015)
11. Kamat, N., Jayachandran, P., Tunga, K., Nandi, A.: Distributed and interactive cube exploration. In: 2014 IEEE 30th International Conference on Data Engineering (ICDE), pp. 472–483. IEEE (2014)
12. Khoussainova, N., Kwon, Y., Balazinska, M., Suciu, D.: SnipSuggest: context-aware autocompletion for SQL. Proc. VLDB Endow. **4**(1), 22–33 (2010)
13. Kosub, S.: A note on the triangle inequality for the jaccard distance (2016). arXiv preprint arXiv:1612.02696
14. Ramachandran, K., Shah, B., Raghavan, V.V.: Dynamic pre-fetching of views based on user-access patterns in an OLAP system. In: ICEIS, vol. 1, pp. 60–67 (2005)
15. Sapia, C.: PROMISE: predicting query behavior to enable predictive caching strategies for OLAP systems. In: Kambayashi, Y., Mohania, M., Tjoa, A.M. (eds.) DaWaK 2000. LNCS, vol. 1874, pp. 224–233. Springer, Heidelberg (2000). https://doi.org/10.1007/3-540-44466-1_22
16. Sellam, T., Kersten, M.: Cluster-driven navigation of the query space. IEEE Trans. Knowl. Data Eng. **28**(5), 1118–1131 (2016)
17. Singh, V., Gray, J., Thakar, A., Szalay, A.S., Raddick, J., Boroski, B., Lebedeva, S., Yanny, B.: Skyserver traffic report-the first five years. arXiv preprint cs/0701173 (2007)

18. Tauheed, F., Heinis, T., Schürmann, F., Markram, H., Ailamaki, A.: SCOUT: prefetching for latent structure following queries. Proc. VLDB Endow. **5**(11), 1531–1542 (2012)
19. Zhang, J., Chen, C., Vogeley, M.S., Pan, D., Thakar, A., Raddick, J.: SDSS log viewer: visual exploratory analysis of large-volume SQL log data. In: Visualization and Data Analysis 2012, vol. 8294, pp. 82940D. International Society for Optics and Photonics (2012)

Coverage-Oriented Diversification of Keyword Search Results on Graphs

Ming Zhong[(✉)], Ying Wang, and Yuanyuan Zhu

School of Computer, Wuhan University, Wuhan 430072, China
{clock,wysklse,yyzhu}@whu.edu.cn

Abstract. Query result diversification has drawn great research interests in recent years. Most previous work focuses on finding a locally diverse subset of a given finite result set, in which the results are as dissimilar to each other as possible. However, such a setup may not always hold. Firstly, we may need the result set to be globally diverse with respect to all possible demands behind a given query. Secondly, the result set may not be given before diversification. In this paper, we address these two problems in the scenario of keyword search on graphs. We first reasonably formalize a problem of coverage-oriented diversified keyword search on graphs. It aims to find both locally and globally diverse and also relevant results simultaneously while searching on graphs. The global diversity is defined as a query-dependent metric called coverage, which dynamically assigns weights to potential query demands with respect to their topological distances to the given keywords. Then, we present a search algorithm to solve our problem. It guarantees to return the optimal diverse result set, and can eliminate unnecessary and redundant diversity computation. Lastly, we perform both effectiveness and efficiency evaluation of our approach on DBPedia. Compared with the local diversification approach, our approach can improve the coverage and reduce the redundancy of search results remarkably.

1 Introduction

Query result diversification has drawn great research interests in recent years. In a nutshell, it is to find a set of query results, which have as high scores as possible, and meanwhile, are as dissimilar as possible to the others. Since it can improve user satisfaction by expanding the query demands covered by the returned results, it is widely used in many applications, such as Web search [1, 3,6,14,19], structured data querying/search [7–10,26], online shopping [17,21], recommender system [25], location-based service [10], etc.

In the scenario of Keyword Search on Graphs (KSoG) [5,11,13,15,16], the redundancy of results could be serious. Generally, a result to a keyword query is defined as a set of joined paths (formulated as subtrees or subgraphs) on the graph from each keyword. Thus, by combining a few of small groups of paths with the highest scores, a large number of top-ranked results can be generated, so

---

The original version of this chapter was revised: the acknowledgement section was updated. The correction to this chapter is available at https://doi.org/10.1007/978-3-319-91458-9_58

J. Pei et al. (Eds.): DASFAA 2018, LNCS 10828, pp. 166–183, 2018.
https://doi.org/10.1007/978-3-319-91458-9_10

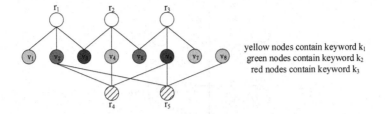

Fig. 1. A comparison of local and global diversity.

that the top-k results could contain very redundant information. This is similar to the *over-specialization* in recommender systems [25]. As a result, query result diversification is very important to KSoG applications.

However, diversifying the results of KSoG brings forth two challenges. Firstly, we hope the diversified results are not only dissimilar to each other, but also can cover more possible demands behind the given query. For convenience, these two properties of a result set are referred to as *local diversity* and *global diversity*, respectively. Let us consider the following example.

Example 1. As shown in Fig. 1, there are a set of nodes $v_1, \cdots, v_8$ (in three colors) matched by the three query keywords k_1, k_2, and k_3, respectively. On these keyword nodes, five result trees $r_1, \cdots, r_5$ can be built, which are ordered by score. Let a result be dissimilar to another if they contain at most one identical node. Thus, the top-4 results are dissimilar to each other. If we need a locally diverse top-4 result set, r_1, r_2, r_3 and r_4 will be chosen, because they have the highest scores. However, if we need the top-4 result set to cover all keyword nodes that represent different query demands, namely, be more globally diverse, r_1, r_2, r_3 and r_5 would be a better choice.

Secondly, we have to implement diversification simultaneously while generating the results, different from most previous work that only deals with a given result set. Because KSoG normally can generate enormous results for a given query due to combinatorial explosion of edges. It would be too time-consuming to enumerate all results and then diversify them separately.

Although there have been researches on either challenge, there still lacks of a practical solution to address both of them in one task like KSoG. As shown in Fig. 2, we classify the query result diversification problems into four categories, with respect to whether the diversity needs to be global and whether the query results have been given already. The problems addressed by existing approaches fall into three of them (see Sect. 2). To the best knowledge we have, the global (and local) diversification of unknown results (see the upper right quadrant) has not been well studied yet.

In order to address the problem, this paper presents an approach that searches for a both locally and globally diverse set of most relevant results for a given keyword query on graphs. Our contributions are as follows.

- We propose a novel metric of global diversity called *coverage*. Different from the previous definition of coverage [23], our definition is query-dependent.

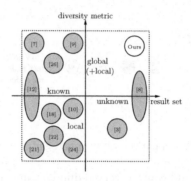

Fig. 2. A classification of diversification problems.

Specifically, given a keyword query, we will evaluate the weights of the query demands with respect to their topological distance on graph, and the more weighted demands contribute more to the overall coverage. Thus, our measurement of coverage is more precise for specific queries. Moreover, although we only give a solution to diversified KSoG problem, our idea can be applied on similar problems in many other applications, such as natural language question answering [28].

– Based on the new metric, we formalize a Coverage-Oriented Diversified KSoG (COD-KSoG) problem. It aims to find an optimal set of search results with respect to the common objectives in KSoG. Moreover, different from the traditional diversification problem [8], it does not fix the number of returned results but requires their coverage to satisfy a certain condition. For the users, coverage is a more intuitive and explicit parameter than result number to constrain the global diversity of returned results. Like the previous work, COD-KSoG also requires the returned results to be locally diverse, namely, the results are dissimilar to each other.

– To solve the COD-KSoG problem, we present a search algorithm that combines diversification into query evaluation. Our algorithm applies a pop-and-diversify framework that can avoid unnecessary diversity computation. The framework repeats two steps iteratively. Firstly, leverage the upper bound estimation method of KSoG to enumerate the top-scored results. Secondly, find diverse sets from the top results. More specifically, our algorithm treats the problem of finding the diverse sets as finding maximal independent sets on an evolving similarity graph, and uses a dynamic programming based approach to address the problem for reducing the redundant computation. Lastly, our algorithm can stop as soon as a diverse set of results have been found. The optimality of returned results is guaranteed.

– To evaluate our approach, we perform comprehensive experiments on a real world RDF graph called DBPedia [4]. Compared with the local diversification approach, our approach can improve the coverage and reduce the redundancy of search results remarkably.

The rest of this paper is organized as follows. Section 2 introduces the related work. Section 3 gives a formal definition of the problem to be addressed. Section 4 presents the measurement of diversity. Section 5 presents a solution to the problem. Section 6 introduces the experiments. Lastly, Sect. 7 concludes our work and briefly plans the future work.

2   Related Work

Query Result Diversification. Figure 2 shows a number of related work in recent years. We can see that most work [7,9,10,12,18,21,22,25,26] assumes a finite result set that has been already known. Zheng et al. [27] give a survey of such kind of query result diversification. Both local (content-based) and global (intent-based) diversification problems are discussed and a list of typical solutions are introduced in the survey.

Angel and Koudas [3] propose a diversity-aware search model on a (possibly infinite) unordered list of results. It dynamically adjusts the scores of results with respect to their "novelty" to the already selected results, thereby still falling to the category of local diversification.

Deng and Fan [8] formulate a general theoretical framework of query result diversification problems. In particular, they analyze the problems with the assumption that the result set is not available. Moreover, one of the discussed objective functions considers the global diversity as the sum of dissimilarity to all other results. However, since the result set is unknown, their algorithm has to enumerate all possible results in the search space. Obviously, it is too expensive for a large search space.

Different from the framework, our approach uses the coverage of a predefined set of query demands as the metric of global diversity, to address the problem of global diversification on unknown results. Moreover, we propose a novel and query-dependent definition of coverage in the scenario of KSoG.

In addition, most existing diversification approaches fix the number of returned results. However, estimating an appropriate number of results that are enough to cover a number of query demands is intractable in the scenario of KSoG. Thus, we use coverage to replace number of returned results as a constraint and study the COD-KSoG problem.

Keyword Search and Result Ranking on Graphs. A variety of KSoG approaches have been proposed in recent years, such as [5,11,13,15,16,20]. Yu et al. [24] review the approaches and summarize the proposed graph search algorithms. A number of state-of-the-art metrics are also proposed to rank the search results, such as size, TF/IDF, PageRank, etc.

Compared to these work, our approach focuses on combining result diversification into the search algorithm, so that the algorithm can achieve early termination. Thus, the overhead of generating the universal result set, which is typically expensive, can be reduced.

3 Problem Formulation

In the scenario of KSoG, we can generally represent the relational, XML, RDF or other data as the following data graph.

Definition 1 (Data Graph). *A data graph $G = (V, E, T, K)$ is a labeled undirected graph, where (1) $V(G)$ is a set of nodes that represent tuples, XML elements, or RDF entities; (2) $E(G) = V(G) \times V(G)$ is a set of edges that represent primary-foreign key references, parent-child relationships, or predicates between nodes; (3) $T(G)$ is a set of node types inferred from the schema, DTD, or OWL ontology; and (4) $K(G)$ is a set of terms contained by the attributes of nodes.*

Given a data graph G and a keyword query $Q \subseteq K(G)$, the results of KSoG are subtrees or subgraphs on graph G that contain all keywords in query Q, which are denoted by $Q(G)$.

Normally, the KSoG problem tries to find the top-k query results with respect to an objective function sco : $Q(G) \mapsto \mathbb{R}$ that scores them in IR style [20]. To diversify the query results, given a dissimilarity function dis : $Q(G) \times Q(G) \mapsto [0, 1]$, the dissimilarity between query results is considered as a constraint or another objective [7]. The COD-KSoG problem further takes the global diversity into consideration, which is evaluated by a coverage function cov : $2^{Q(G)} \mapsto [0, 1]$. The coverage of a result set $R \subseteq Q(G)$ measures how much semantics (namely, potential query demands) of a query Q is covered by R. Formally, the COD-KSoG problem to be addressed in this paper is as follows.

Definition 2 (COD-KSoG Problem). *Given a data graph G, a keyword query Q, two positive real numbers $\alpha, \beta \in [0, 1]$, find a set of query results $R \subseteq Q(G)$ such that (1) $\mathrm{cov}(R) \geqslant \alpha$, (2) $\min_{r, r' \in R} \mathrm{dis}(r, r') \geqslant \beta$, and (3) $\min_{r \in R} \mathrm{sco}(r)$ is maximized.*

As the major difference from the previous work, we use the coverage but not the number of results as a constraint, because the coverage is a more intuitive measurement of users' interest ranges. For a diversification problem, the users require more results means they want more semantics of the query to be demonstrated. However, more results may not contain new semantics. It is difficult to give a precise number of results that can guarantee to contain a specific proportion of query semantics.

4 Diversity Measurement

In the absence of external resources like query logs or user preferences, we treat the node types, namely, $T(G)$ as query demands. Each node type represents a set of data objects which the users may intent to see. For a query result r, we denote by $T(r) \subseteq T(G)$ the set of types of all nodes in r. Intuitively, $T(r)$ represents the query demands satisfied by r.

Dissimilarity. Firstly, we define the dissimilarity between two query results. For a pair of results, we assume the less common query demands they satisfy,

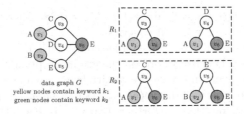

Fig. 3. An example of query-dependent coverage measurement.

the more dissimilar they are. Formally, given two query results $r, r' \in Q(G)$, we use the Jaccard distance between $T(r)$ and $T(r')$ as their dissimilarity.

$$\text{dis}(r, r') = 1 - \frac{T(r) \cap T(r')}{T(r) \cup T(r')} \tag{1}$$

Coverage. We propose an effective query-dependent metric of global diversity for KSoG. For a set of query results, we assume the more query demands they satisfy, the more globally diverse they are. In other words, we prefer the result set that has more types of nodes. While, different from the previous work [23], we assume each query represents a particular set of query demands, and thus do not treat each node type equally to a specific query. Instead, we define a weighting function $\omega : T(G) \times 2^{K(G)} \mapsto [0, 1]$ that assigns a weight to each node type with respect to a given keyword as follows

$$\omega(t, Q) = \frac{\delta_{max}(Q) - \min_{k \in Q} \delta(t, k) + 1}{\sum_{t' \in T(G)} (\delta_{max}(Q) - \min_{k \in Q} \delta(t', k) + 1)} \tag{2}$$

where $\delta(t, k)$ is the shortest distance between the nodes of type t and the nodes that contain keyword k, and $\delta_{max}(Q) = \max_{t \in T(G)} \min_{k \in Q} \delta(t, k)$. According to this weighting function, the types of node that are closer to the nodes containing keywords have higher relevances to a given keyword query. Therefore, the results that cover more relevant node types to the given query should be preferred.

Thus, we evaluate the coverage of a set of results $R \subseteq Q(G)$ for a given query Q as follows

$$\text{cov}(R, Q) = \sum_{t \in T(R)} \omega(t, Q)/|Q| \tag{3}$$

where $T(R) = \cup_{r \in R} T(r)$ is a set of all types contained by R.

Let us consider the example in Fig. 3. Given a keyword query $Q = \{k_1, k_2\}$, we have two sets of result $R_1, R_2 \subset Q(G)$. They both cover four node types, $T(R_1) = \{A, C, D, E\}$ and $T(R_2) = \{A, B, C, E\}$, so that their coverage are equal if the weights of node types are not taken into consideration. However, the type B is directly related to a keyword node, and thus should be preferred than the types of nodes that are farther from keywords (i.e., C and D). Thus, R_2 should be better than R_1 in terms of coverage. By using our coverage function, we have $\omega(A, Q) = 0.25$, $\omega(B, Q) = 0.25$, $\omega(C, Q) = 0.13$, $\omega(D, Q) = 0.13$,

$\omega(E, Q) = 0.25$, $cov(R_1, Q) = 0.76$, $cov(R_2, Q) = 0.88$. Therefore, our coverage function is more precise for specific queries.

To reduce the overhead of online coverage computation, we precompute $\delta(t, k)$ for each node type and each keyword on the graph. In practice, there is no need to consider the query demands that are too irrelevant to a given query. Thus, we fix all the maximum shortest distances $\delta_{max}(k)$ to be δ_{max}. For each keyword $k \in K(G)$, the node types that are farther than δ_{max} to k will be ignored, namely, $\omega(t, k) = 0$ if $\delta(t, k) > \delta_{max}$, in computation of coverage. Also, the overheads of indexing can be reduced considerably.

5   Diversified Search

In this section, we address the COD-KSoG problem defined in Definition 2. In order to improve search efficiency, we propose an algorithm that executes two tasks iteratively and can stop early when the optimal diverse result set has been found. The first task is to find a new candidate result by traversing the graph. We narrow the range of candidate results as much as possible to avoid the unnecessary computation in this task. The second task tries to find diverse result sets from the current candidate results, while reducing the redundant computation. We propose two functions nextTop() and searchSG() to implement these two tasks respectively. It is guaranteed that the first diverse result set found by the algorithm is optimal.

5.1   Result Generation

As the state-of-the-art KSoG algorithms (e.g. [13]), our search algorithm decomposes the problem of finding results to a keyword query Q into $|Q|$ independent subproblems, each of which is to traverse the graph and enumerate the search path from a keyword in Q with respect to specific heuristics. Meanwhile, the algorithm will check whether the enumerated search paths can be joined to generate new results constantly. The algorithm can stop early when there is a set of results that (1) satisfy all constraints and (2) have the best objective value.

Through analyzing the search procedure, we have such an important observation: since the results are not generated in a strict descending order of score, it is not necessary to diversify all results generated during the search. The computational complexity of diversification is inherently very high, so that we should avoid to diversify more results unless that is proved to be necessary.

Based on the above observation, we propose a *pop-and-diversify* framework to reduce the overheads of diversification. The KSoG algorithms normally offer effective upper bound estimation for scores of unknown results to support efficient top-k search. We refer *top results* to the results with scores higher than the current upper bound. The framework uses a function called nextTop() to enumerate the top results one by one during the search. It returns a result with the highest score among all current non-top results (including unknown results). Once a result is returned by nextTop(), our algorithm tries to find a diverse set

---

Algorithm 1. nextTop($temp, \overline{unknown}$)

1: **while** $temp$.peek() $< \overline{unknown}$ **do**
2:     $path \leftarrow$ enumeratePath();
3:     $new \leftarrow$ generateResults($path, temp$);
4:     $temp \leftarrow temp \cup new$;
5:     update $\overline{unknown}$;
6: **end while**
7: **return** $temp$.pop();

---

of top results that satisfy the constraints of diversity. Whenever a diverse set of top results have been found, it is certainly the optimal diverse result set. See the following theorem.

Theorem 1. *The pop-and-diverse framework guarantees to return the optimal diverse result set.*

Proof. Let R be the returned result set and R' be another different result set. (1) Assume R' contain a result r' that is not in the current set of top results. It is certain that $\min_{r \in R} \text{sco}(r) > \text{sco}(r')$. Thus, we have $\text{sco}(R) > \text{sco}(R')$. (2) Assume R' be composed of the current top results other than $\arg\min_{r \in R} \text{sco}(r)$. In other words, the results in R' are top results popped earlier, so that $\text{sco}(R) \leqslant \text{sco}(R')$. However, R' is certainly not a diverse set, or it has been returned before. Therefore, R must be the optimal diverse result set.

Algorithm 1 presents the pseudo codes of function nextTop(). Let $temp$ be a priority queue of currently generated results in descending order of their scores, and $\overline{unknown}$ be the upper bound of scores of results that have not been generated yet. The function will run iteratively until there is at least one result in $temp$ with a score higher than $\overline{unknown}$. At each iteration, a search path is enumerated by traversing the graph. Then, the path will be joined with the other paths from different keywords in $temp$, and new results could be found. See details of enumeratePath() and generateResults() in [5]. The new results will be added into $temp$. Then, the upper bound $\overline{unknown}$ will be updated with respect to specific scoring function. Once the loop is over, the first result in $temp$, namely, the next top result will be returned.

5.2   MIS-Based Diversification

In order to find a diverse set of top results, we introduce the concept of *similarity graph*.

Definition 3 (Similarity Graph). *Given a list of top results $S = \{r_1, r_2, ..., r_n\}$, the similarity graph of S, denoted as $G^S = (V, E)$, is an undirected graph such that (1) for each result $r \in S$, there is a corresponding node $v \in V(G^S)$, and (2) for any two results $r, r' \in S$, there is an edge $(v, v') \in E(G^S)$ if $dis(r, r') \geqslant \beta$, namely, the result r is similar to the result r'.*

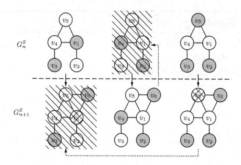

Fig. 4. The example similarity graphs and maximal independent sets.

A set of top results that are dissimilar to each other is equal to a set of nodes on similarity graph that are not adjacent to each other. Thus, searching for the sets of locally diverse results is equal to finding the *maximal independent sets* (MIS) of nodes on similarity graph. The maximal independent set is defined as follows.

Definition 4 (Maximal Independent Set). *Given a similarity graph G^S, a set of nodes $I \subseteq V(G^S)$ is an independent set, if each node in I is not adjacent to the others. Let $S_I(G^S)$ be the set of all independent sets on G^S. If there is no other independent set $I' \in S_I(G^S)$ such that $I \subset I'$, I is the maximal independent set.*

For example, Fig. 4 shows a similarity graph G_n^S. There are three maximal independent sets on G_n^S, such as $\{v_1, v_3\}$, $\{v_2, v_4\}$, and $\{v_2, v_3, v_5\}$. If the sets also meet the coverage constraint, they are diverse sets of top results.

Thus, our problem can be equivalently defined as finding a set of nodes I on similarity graph G^S, such that

1. I is a maximal independent set;
2. $\text{cov}(I) \geqslant \alpha$;
3. $\min_{v \in I} \text{sco}(v)$ is maximum.

Next, we discuss how to find the maximal independent sets on an evolving similarity graph.

5.3   A DP-Based Approach

Finding the maximal independent sets on a graph is an important NP-complete problem in graph theory. There are many studies about traversing the maximal independent sets on a certain graph. However the similarity graph we dealt with is dynamically growing. Obviously, repeating the traditional approach in each time when the similarity graph has been updated is too expensive.

We present a function searchSG(), a dynamic programming based approach, to incrementally find the new maximal independent sets on a similarity graph that is built by adding a new node into the previous similarity graph. Our approach can avoid the redundant computation that occurs while finding the maximal independent sets on a constantly evolving graph, thereby reducing the overall overheads.

Let C be the set of all the maximal independent sets on the current similarity graph G_n^S, and C_{new} be the new maximal independent sets containing the new node on the next similarity graph G_{n+1}^S to which a new node has been added. To find C_{new}, the function searchSG(C, v) has two steps. Firstly, for each maximal independent set $I_m \in C$, create a new node set $I_m' = I_m \cup \{v\}$. If there are nodes in I_m' adjacent to v, we remove them from I_m'. Otherwise, we remove I_m from C, and put I_m' into C_{new}. Secondly, remove the sets in the C_{new} that are the subsets of some other sets. Then, we can find all maximal independent sets on the new similarity graph by getting the union of C and C_{new}.

Lemma 1. *We assume G_{n+1}^S be the similarity graph by adding new node v to the G_n^S. For each maximal independent set on G_{n+1}^S without node v, it is also the maximal independent set on G_n^S.*

Proof. Assume I_m is maximal independent set without node v on G_{n+1}^S. Firstly, it is also an independent set on G_n^S. Now we prove that there is no other independent set I_m' on G_n^S such that $I_m \subset I_m'$. Now assume that the set I_m' exists, and v' is the node in I_m' but not in I_m. Because v' is also a node in G_{n+1}^S, we put the node v' into the I_m. However I_m is the maximal independent set without node v of G_{n+1}^S, there must be a node in I_m which is also in I_m' adjacent to v'. So the independent set I_m' dose not exist.

Lemma 2. *C_{new} contains all the maximal independent sets that contain node v on G_{n+1}^S.*

Proof. Assume I_m is the maximal independent set containing node v but not in C_{new}. We create a new set I by removing node v from I_m. Now, there must be a maximal independent set I_m' on G_n^S that $I \subset I_m'$. We remove all the nodes in I_m' that are adjacent to node v to get a new set I'. We still have $I \subset I'$, because there is not any node in I that is adjacent to node v. Because I_m is a maximal independent set on G_{n+1}^S, we can know $I' \cup \{v\}$ is actually the same as I_m. So there is not any maximal independent set containing node v but not in C_{new}.

Theorem 2. *The union $C \cup C_{new}$ contains all the maximal independent sets of G_{n+1}^S*

Proof. With Lemma 1, we can know all the maximal independent sets without node v are in the set C. With Lemma 2, we can know all the maximal independent sets containing node v are in the set C_{new}. So $C \cup C_{new}$ contains all the maximal independent sets of G_{n+1}^S.

With the above theorem, the correctness of searchSG(C, v) holds. Like dynamic programming (DP), it finds the best solutions to G_{n+1}^S based on the solutions to a simpler subproblem G_n^S. So we call it DP-based approach. There

Algorithm 2. searchSG(C, v)

1: $C_{new} \leftarrow \emptyset$;
2: **for all** $I_m \in C$ **do**
3: 　　$I'_m \leftarrow I_m \cup \{v\}$;
4: 　　**if** there are nodes in I_m adjacent to v **then**
5: 　　　　remove the nodes from I'_m;
6: 　　**else**
7: 　　　　remove I_m from C;
8: 　　**end if**
9: 　　$C_{new} = C_{new} \cup \{I'_m\}$;
10: **end for**
11: maximize(C_{new});
12: **return** C_{new};

are some supplementary explanations. Firstly, we remove some sets from C because they are the subsets of I'_m (Line 7). So there is no non-maximal independent set in C. Secondly, since there still may be some non-maximal independent sets in C_{new}, we use a procedure maximize(C_{new}) to remove them (Line 11). Let us consider the following example.

Example 2. As shown in Fig. 4, the similarity graph G^S_{n+1} is built by adding a new node v_6 to the previous similarity graph G^S_n. The maximal independent sets on G^S_n are $C = \{\{v_1, v_3\}, \{v_2, v_4\}, \{v_2, v_3, v_5\}\}$. By adding v_6 to these sets respectively, we have the new node sets $\{v_1, v_3, v_6\}$, $\{v_2, v_4, v_6\}$, and $\{v_2, v_3, v_5, v_6\}$. By removing the adjacent nodes of v_6 from these sets, we have new independent sets $C_{new} = \{\{v_3, v_6\}, \{v_2, v_4, v_6\}, \{v_2, v_3, v_6\}\}$ that contains v_6, and meanwhile, remove $\{v_2, v_4\}$ from C because it is a subset of elements in C_{new}. Then, we have $C_{new} = \{\{v_2, v_4, v_6\}, \{v_2, v_3, v_6\}\}$ by removing the non-maximal set. Lastly, the new maximal independent sets on G^S_{n+1} are $C \cup C_{new}$, including $\{v_1, v_3\}$, $\{v_2, v_3, v_5\}$, $\{v_2, v_4, v_6\}$, and $\{v_2, v_3, v_6\}$.

Complexity Analysis. Alekseev [2] have discussed the upper bound of the number of maximal independent sets in a general graph. We assume $\overline{bound}$ is this upper bound. We assume $\overline{bound_k}$ is the upper bound of the number of maximal independent sets of the *kth* similarity graph $G(S_k)$. The largest size of a maximal independent set is k. So the worst space complexity is o$(k * \overline{bound_k})$. Each time the algorithm $SearchGS()$ is called, we traverse the C one time to find the C_{new}. So the time complexity is o$(\overline{bound_1} + \overline{bound_2} + \cdots + \overline{bound_k})$ if we end our search at the similarity graph $G(S_k)$.

5.4　Search Algorithm

Lastly, we present a search algorithm divSearch() to address COD-KSoG based on above studies. It applies the pop-and-diversify framework and incorporates the two functions nextTop() and searchSG(). Algorithm 3 presents the pseudo codes of divSearch(). At each iteration, it calls nextTop() to get a top result. The result will be added to the similarity graph. Then, it calls searchSG() to get the new maximal independent sets that contain the result, and checks whether they satisfy the coverage constraint. Whenever a new maximal independent set does, it will be returned. At the end of iteration, current maximal independent sets will be expanded by importing the new ones. If no result set can be returned, the algorithm will stop when no other result can be found.

6 Experiments

In this section we evaluate our proposed algorithms. The experiments are performed on a Windows 2012 server with 3.3 GHz CPU and 128 GB memory. Our algorithms are implemented in Java 1.7.

6.1 Setup

Data Graph. The data graph is derived from DBPedia, a popular real-world RDF dataset which contains over two million entities and nearly ten million relationships. Moreover, the entities have 272 types, such as "Aircraft", "BaseballPlayer", "ChemicalCompound", etc.

Algorithm. The tests are executed by using three algorithms. (1) Our COD algorithm (see Algorithm 3). (2) A locally diversified search algorithm denoted by Div. Compared with COD, it replaces the requirement of coverage to the size of result set and returns a result set $R' \in Q(G)$ such that $|R| = k$, $\min_{r,r' \in R'} dis(r, r') \geq \beta$, and $\min_{r \in R'} sco(r)$ is maximized. The implement of Div is similar to COD except the termination condition. (3) A typical top-k search algorithm denoted by Topk which returns the most relevant k results without diversification.

Keyword Query. Each test query is generated by randomly choosing 2–5 keywords. Moreover, we define a metric ψ called *ambiguity* for keyword queries. It measures how diverse the results of a query could be.

$$\psi = \lambda H_0 + (1 - \lambda)H_1 \tag{4}$$

where $\lambda \in [0, 1]$, H_0 is the number of types of nodes that contain the keywords in query, and H_1 is the number of types of nodes that are one hop away from the nodes matched by keywords. The greater value of H_0 and H_1 means the keyword query is more ambiguous. Moreover, the ambiguity of keyword queries has a great influence on the process time of diversification. Therefore, we classify the test keyword queries with respect to their value of ψ into the five levels shown in Table 1. The experimental results demonstrated in the followings are actually the average of 50 keyword queries in each ambiguity level.

Algorithm 3. divSearch(G, Q)

1: $temp, C, C_{new} \leftarrow \emptyset$, initialize $\overline{unknown}$
2: **while** $v \leftarrow$ nextTop($temp, \overline{unknown}$) $\neq$ null **do**
3:    update the similarity graph;
4:    $C \leftarrow C \cup C_{new}$;
5:    **for all** $I_m \in C_{new}$ **do**
6:       **if** cov(I_m) $\geqslant \alpha$ **then**
7:          **return** I_m;
8:       **end if**
9:    **end for**
10:    $C_{new} \leftarrow SearchGS(C, v)$;
11: **end while**

Table 1. The ambiguity levels of keyword queries.

	Am_1	Am_2	Am_3	Am_4	Am_5
ψ	1–9	10–19	20–29	30–39	40–49

6.2 Effectiveness

To evaluate the effectiveness of our approach, we compare the coverage of search results of the three algorithms with queries in different levels of ambiguity. In order to ensure the comparability of different algorithms, for each query, we firstly run COD with the given value of α, and then run Div and Topk to return the top-k results, where k is the number of results returned by COD.

The experimental results are shown in Fig. 5. In general, we can see that the coverage of search results of COD is always the highest in different settings. Specifically, we have the following observations. (1) Since the coverage of COD is constrained by the input value of α, the output coverage is certainly higher than α denoted by the dashed line. While, the coverage of Div and Topk is lower than COD when they return the same number of results as COD. It indicates that, our algorithm can indeed improve the global diversity of a specific number of search results, compared with the algorithms that do not consider diversity or only consider the local diversity. (2) As shown in Fig. 5(a), given $\alpha = 0.25$, the average coverage difference of all query groups is 0.047 between COD and Div, and is 0.152 between COD and Topk. As shown in Fig. 5(b), given $\alpha = 0.5$, the corresponding differences are 0.162 and 0.323 respectively, though the coverage of Div and Topk is also increased with the increase of α. It means, the redundancy of results of Div and Topk increases quickly while more results are returned. In contrast, COD can still capture novel query demands. Thus, the coverage constraint is more effective than the number of results for declaring the interest range of users. (3) With the growth of the ambiguity of queries, the coverage differences between COD and Div increase from 0.031 to 0.059 when $\alpha = 0.25$ and from 0.121 to 0.189 when $\alpha = 0.5$. Thus, COD beats Div by a larger margin for more ambiguous queries.

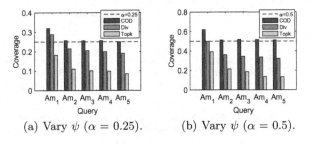

(a) Vary ψ ($\alpha = 0.25$).   (b) Vary ψ ($\alpha = 0.5$).

Fig. 5. The coverage of search results.

Moreover, we conduct a simple case study. Given a query "Beckham, Ronaldo", Table 2 shows a part of top results returned by Topk, Div and COD, respectively. For conciseness, we only give a brief summary of the search results: the types of root node and two leaf nodes on each result tree. We can see that, the top-1 results of them are the same, and are indeed the most relevant pattern to these two keywords: the teammate of both famous soccer players. However, the other results of Topk are still the same pattern, and just show the different teammates. In contrast, the other two algorithms reveal the different patterns related to the two keywords. For example, if some users are interested in "Beckham" as an album and "Ronaldo" as a musical artist, they will find that the album and the artist share some sort of musical genre. Although both algorithms can improve the diversity of search results, we can see COD is obviously more effective, by comparing their search results. Also, we offer the statistics of the top-12 results. Except the coverage, we denote by redundancy rate the percentage of redundant results. The redundant results are the results whose node types are all included in other returned results, like the top-5 result of Div, which are not helpful for global diversity. The comparison of redundancy rate further verifies the effectiveness of COD.

6.3   Efficiency

For evaluating the efficiency of our approach, we test the average response time and answer size of specified approaches. The followings are our observations.

1. Figure 6(a) and (b) illustrates the average response time of Topk, Div and COD with $\beta = 0.7$. In the experiments, for each query, we firstly run COD with the given value of α, and then run Div and Topk to return the top-k results, where k is the number of results returned by COD. With the increase of ψ, the average response time of Topk is almost stable, because the computational overhead of it will not be much increased by the growth of redundant search results. With the growth of the ambiguity of queries, the average response time of COD increase from 1.8 s to 25.1 s when $\alpha = 0.25$ and from 20.5 s to 100.2 s when $\alpha = 0.5$. The average growth rate of response time is 89.56% and 51.25% respectively. As for the Div, the average growth rate of

Table 2. Results of keyword search.

Approach	Topk	Div	COD
Top-1	Soccer Player SoccerPlayer & SoccerPlayer	SoccerPlayer SoccerPlayer & SoccerPlayer	SoccerPlayer SoccerPlayer & SoccerPlayer
Top-2	Soccer Player SoccerPlayer & SoccerPlayer	MusicGenre Album & MusicalArtist	BroadcastNetWork SettleMent & School
Top-3	Soccer Player SoccerPlayer & SoccerPlayer	Book Book & School	MusicGenre Album & MusicalArtist
Top-4	Soccer Player SoccerPlayer & SoccerPlayer	SoccerManager Settlement & School	Book Book & School
Top-5	Soccer Player SoccerPlayer & SoccerPlayer	SoccerManager SoccerPlayer & Album	Village Settlement & Person
. . .	. . .	. . .	. . .
Coverage of top-12	0.12	0.31	0.50
Redundancy rate of top-12	83%	33%	0%

response time is 86% and 54% respectively. The average growth rate is almost same for Div and COD which means their response time will not differ too much, however COD is much effective for the diversity of results.

2. Figure 6(c) illustrates the average response time of COD with $\beta = 0.7$ and Fig. 6(d) illustrates the average response time of COD with $\alpha = 0.25$. The average growth rate of response time with varying α is $29.35\%(Am_2)$, $59.9\%(Am_3)$ and $77\%(Am_4)$ and $11.17\%(Am_2)$, $10.77\%(Am_3)$ and $17.16\%(Am_4)$ for varying β. The result shows that α has a greater impact on the response time. For the Fig. 6(c), the main increasing in response time is from the growth of the result set with the growth of coverage, especially when the query ambiguity is relatively high. For Fig. 6(d), on the one hand, as β increases, it becomes more difficult to search for results which are dissimilar with each other. On the other hand, because the results are dissimilar with each other, we need fewer results in the answer set to reach α. Thus the growth of response time is relatively low.

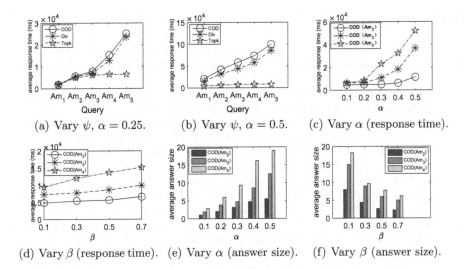

Fig. 6. Efficiency experiment.

3. Figure 6(e) and (f) illustrates the average answer size of COD with varying α and β. For Fig. 6(e), α equals 0.25. For Fig. 6(f), β equals 0.7. The answer size is positively related to the α and negatively related to β. With the growth of the ambiguity of queries, the answer size grows both of them.

7 Conclusions and Future Work

In this paper, we study a novel query result diversification problem, namely, given a keyword query, finding a both locally and globally diverse set of most relevant results while searching on the graph. Since the query results are generated continuously in realtime, there rise two challenges. Firstly, how to measure the global diversity. Secondly, how to extract an optimal diverse set from the enormous streaming results efficiently. We address them respectively. The experimental results demonstrate the effectiveness and efficiency of our solutions.

In the future, we plan to consider the coverage as an additional objective but not a constraint, and study the multi-objective optimization problem. Naturally, more efficient algorithms that do not have to yield the optimal solutions are needed due to the increased complexity.

Acknowledgement. This paper was supported by National Natural Science Foundation of China under Grant No. 61202036, 61502349 and 61572376 and Natural Science Foundation of Hubei Province under Grant No. 2018CFB616.

References

1. Agrawal, R., Gollapudi, S., Halverson, A., et al.: Diversifying search results. In: ACM International Conference on Web Search and Data Mining, pp. 5–14. ACM (2009)
2. Alekseev, V.E.: An upper bound for the number of maximal independent sets in a graph. Discrete Math. Appl. **17**(4), 355–359 (2007)
3. Angel, A., Koudas, N.: Efficient diversity-aware search. In: ACM SIGMOD International Conference on Management of Data, pp. 781–792. ACM (2011)
4. Auer, S., Bizer, C., Kobilarov, G., Lehmann, J., Cyganiak, R., Ives, Z.: DBpedia: a nucleus for a web of open data. In: Aberer, K., Choi, K.-S., Noy, N., Allemang, D., Lee, K.-I., Nixon, L., Golbeck, J., Mika, P., Maynard, D., Mizoguchi, R., Schreiber, G., Cudré-Mauroux, P. (eds.) ASWC/ISWC -2007. LNCS, vol. 4825, pp. 722–735. Springer, Heidelberg (2007). https://doi.org/10.1007/978-3-540-76298-0_52
5. Hulgeri, A., Nakhe, C.: Keyword searching and browsing in databases using BANKS. In: Proceedings of International Conference on Data Engineering, pp. 431–440. IEEE (2002)
6. Capannini, G., Nardini, F.M., Perego, R., et al.: Efficient diversification of web search results. Proc. VLDB Endow. **4**(7), 451–459 (2011)
7. Demidova, E., Fankhauser, P., Zhou, X., et al.: DivQ: diversification for keyword search over structured databases, pp. 331–338. ACM (2010)
8. Deng, T., Fan, W.: On the complexity of query result diversification. ACM (2014)
9. Drosou, M., Pitoura, E.: DisC diversity: result diversification based on dissimilarity and coverage. Proc. VLDB Endow. **6**(1), 13–24 (2012)
10. Fraternali, P., Martinenghi, D., Tagliasacchi, M.: Top-k bounded diversification. In: ACM SIGMOD International Conference on Management of Data, pp. 421–432. ACM (2012)
11. Golenberg, K., Kimelfeld, B., Sagiv, Y.: Keyword proximity search in complex data graphs. In: ACM SIGMOD International Conference on Management of Data, pp. 927–940. ACM (2008)
12. Gollapudi, S., Sharma, A.: An axiomatic approach for result diversification, pp. 381–390 (2009)
13. He, H., Wang, H., Yang, J., et al.: BLINKS: ranked keyword searches on graphs. In: ACM SIGMOD International Conference on Management of Data, pp. 305–316. ACM (2007)
14. Hu, S., Dou, Z., Wang, X., et al.: Search result diversification based on hierarchical intents, pp. 63–72 (2015)
15. Kacholia, V., Pandit, S., Chakrabarti, S., et al.: Bidirectional expansion for keyword search on graph databases. In: International Conference on Very Large Data Bases, Trondheim, Norway, 30 August - September, pp. 505–516. DBLP (2005)
16. Li, G., Ooi, B.C., Feng, J., et al.: EASE: an effective 3-in-1 keyword search method for unstructured, semi-structured and structured data. In: ACM SIGMOD International Conference on Management of Data, pp. 903–914. ACM (2008)
17. Liu, Z., Sun, P., Chen, Y.: Structured search result differentiation. VLDB Endow. 313–324 (2009)
18. Qin, L., Yu, J.X., Chang, L.: Diversifying top-k results. Proc. VLDB Endow. **5**(11), 1124–1135 (2012)
19. Rafiei, D., Bharat, K., Shukla, A.: Diversifying web search results. In: International Conference on World Wide Web, WWW 2010, Raleigh, North Carolina, USA, April 2010, pp. 781–790. DBLP (2010)

20. Tran, T., Wang, H., Rudolph, S., et al.: Top-k exploration of query candidates for efficient keyword search on graph-shaped (RDF) data. In: IEEE International Conference on Data Engineering, pp. 405–416. IEEE (2009)

21. Vee, E., Srivastava, U., Shanmugasundaram, J., et al.: Efficient computation of diverse query results. In: IEEE International Conference on Data Engineering, pp. 228–236. IEEE (2008)

22. Vieira, M., Razente, H., et al.: On query result diversification. In: ICDE Proceedings, pp. 1163–1174 (2011)

23. Wu, Y., Yang, S., Srivatsa, M., et al.: Summarizing answer graphs induced by keyword queries. Proc. VLDB Endow. **6**(14), 1774–1785 (2013)

24. Qin, L., Yu, J.X., Chang, L.: Keyword search in databases: the power of RDBMS. In: ACM SIGMOD International Conference on Management of Data, SIGMOD, Providence, Rhode Island, USA, 29 June - July, pp. 681–694. DBLP (2009)

25. Cong, Y., Lakshmanan, L., et al.: It takes variety to make a world: diversification in recommender systems. In: EDBT Proceedings, pp. 368–378 (2009)

26. Zhao, F., Zhang, X., Tung, A.K.H., et al.: BROAD: diversified keyword search in databases. Proc. VLDB Endow. **4**(12), 1355–1358 (2012)

27. Zheng, K., Wang, H., Qi, Z., et al.: A survey of query result diversification. Knowl. Inf. Syst. **51**(1), 1–36 (2017)

28. Zou, L., Huang, R., Wang, H., et al.: Natural language question answering over RDF: a graph data driven approach. ACM (2014)

Novel Approaches to Accelerating the Convergence Rate of Markov Decision Process for Search Result Diversification

Feng Liu[1], Ruiming Tang[2], Xutao Li[1], Yunming Ye[1(✉)], Huifeng Guo[1], and Xiuqiang He[2]

[1] Shenzhen Key Laboratory of Internet Information Collaboration, Shenzhen Graduate School, Harbin Institute of Technology, Shenzhen 518055, China
liufeng@stmail.hitsz.edu.cn, lixutao@hitsz.edu.cn,
yeyunming@hit.edu.cn, huifengguo@yeah.net
[2] Noah's Ark Lab, Huawei, China
{tangruiming,hexiuqiang}@huawei.com

Abstract. Recently, some studies have utilized the Markov Decision Process for diversifying (MDP-DIV) the search results in information retrieval. Though promising performances can be delivered, MDP-DIV suffers from a very slow convergence, which hinders its usability in real applications. In this paper, we aim to promote the performance of MDP-DIV by speeding up the convergence rate without much accuracy sacrifice. The slow convergence is incurred by two main reasons: the large action space and data scarcity. On the one hand, the sequential decision making at each position needs to evaluate the query-document relevance for all the candidate set, which results in a huge searching space for MDP; on the other hand, due to the data scarcity, the agent has to proceed more "trial and error" interactions with the environment. To tackle this problem, we propose MDP-DIV-kNN and MDP-DIV-NTN methods. The MDP-DIV-kNN method adopts a k nearest neighbor strategy, i.e., discarding the k nearest neighbors of the recently-selected action (document), to reduce the diversification searching space. The MDP-DIV-NTN employs a pre-trained diversification neural tensor network (NTN-DIV) as the evaluation model, and combines the results with MDP to produce the final ranking solution. The experiment results demonstrate that the two proposed methods indeed accelerate the convergence rate of the MDP-DIV, which is 3x faster, while the accuracies produced barely degrade, or even are better.

Keywords: Search result diversification · Markov decision process
Convergence rate

1 Introduction

In real web search scenarios, a large number of queries are ambiguous or multi-faceted. For instance, the query "apple" can be a kind of delicious fruit or the

---

The work is done when Feng Liu works as an intern in Noah's Ark Lab, Huawei.

© Springer International Publishing AG, part of Springer Nature 2018
J. Pei et al. (Eds.): DASFAA 2018, LNCS 10828, pp. 184–200, 2018.
https://doi.org/10.1007/978-3-319-91458-9_11

great IT company; the huge vehicle "rocket" can also be mentioned as the Houston Rocket basketball team. In order to satisfy the users with different information needs, search result diversification approaches, which provide the search results that covered with a wide range of subtopics for a query, have been widely studied. The approaches work by ranking documents or webpages take both relevance and information novelty (diversification) into considerations.

A majority of traditional methods for search result diversification are heuristic methods with manually defined functions [2,5,13–16]. Their key rationale is that the subsequent document should be "different" from the ones already ranked. As a representative work, the maximal marginal relevance (MMR) [2] is proposed to formulate the construction of a diverse ranking as a process of sequential document selection. In MMR, the marginal relevance is defined as a sum of query-document relevance and the maximal document distance as novelty by a predefined document distance function.

Recently, in order to avoid heuristic methods with manually defined evaluation functions, machine learning methods have been proposed and applied to search result diversification [21,22,24,25,28]. The basic idea is to automatically learn a diverse ranking model from the labeled training data. Typical approaches include the relational learning to rank (R-LTR) [28] and its variations [21,22,24]. In [21,28], the novelty of a document with respect to the previously selected documents is encoded as a set of handcrafted novelty features. In [22], the neural tensor networks are extended to model the novelty among them.

However, all these methods model utility of a candidate document either based on carefully designed heuristics or handcrafted relevance features and novelty features. The utility perceived from the preceding documents is not fully utilized. To avoid this, the latest work for search result diversification, Markov decision process diversification model (MDP-DIV) [23] is proposed, which formalizes the construction of a diverse ranking as a sequential decision making process and models the process with Markov decision process (MDP). Reinforcement learning technique, the policy gradient algorithm of REINFORCE [18], is adopted to adjust the model parameters. MDP-DIV outperforms the state-of-the-art baselines on the TREC benchmark datasets. However, its low convergence rate, often requiring tens of thousands iterations to converge, is unacceptable, especially for industrial applications.

In this paper, we aim to promote the performance of MDP-DIV by speeding up the convergence rate and maintaining the accuracy. The primary reasons for low convergence rate are the large action space and data scarcity. On the one hand, the sequential decision making at each position needs to evaluate all the remaining documents of relevance, which forms a huge search space; on the other hand, the data scarcity compels the agent to proceed more "trial and error" interactions with the environment. To address the problem, we propose MDP-DIV-kNN and MDP-DIV-NTN methods. The MDP-DIV-kNN method adopts a k nearest neighbor strategy to linearly reduce the action space at each position. Specifically, it removes the k nearest neighbors of the recent selected action (document). Different from the MDP-DIV-kNN, the MDP-DIV-NTN employs a pre-trained diversification neural tensor network (NTN-DIV) as the evaluation model, and combines the results with MDP to produce the final ranking list. There are two instantiations of MDP-DIV-NTN. Specifically, the MDP-DIV-NTN(D) method directly filters the pre-ranked list; while the MDP-DIV-NTN(E) method sequentially models the novelty of candidate document with respect to previously selected documents. The main contributions of this paper can be summarized as follows:

- We analyze the reasons for the slow convergence of MDP-DIV, and find that it is mainly due to the large action space and the data scarcity.
- We propose the MDP-DIV-kNN and MDP-DIV-NTN methods, which can promote the convergence rate while maintaining the accuracy of MDP-DIV for search result diversification.
- Extensive experiments are carried out on 09-12 TREC benchmark datasets, and the results demonstrate the proposed methods indeed fasten MDP-DIV and outperforms the state-of-the-art competitors.

The remainder of the paper is structured as follows. In Sect. 2, we briefly review the related works. In Sect. 3, the Markov decision process, MDP-DIV, and NTN-DIV are introduced as preliminaries. The proposed methods are presented in Sect. 4. Experimental results are provided in Sect. 5 to demonstrate the effectiveness of the proposed methods.

2   Related Work

2.1   Search Result Diversification

One of the key problems in search result diversification is the diverse ranking. Formalizing the construction of diverse ranking as a process of sequential document selection is a common practice. This ranking strategy provides us a more rational way to model the utility of a candidate document which not only depends on the document itself but also the preceding documents. Existing approaches can be classified into two categories, namely heuristic methods [2,5–7,16] and machine learning methods [21–24,28].

The representative work in the first kind is the maximal marginal relevance (MMR) [2] criterion to guide the design of diverse ranking models. In MMR, the sequential document selection is based on the marginal relevance score, which is a linear combination of query-document relevance score and document novelty score. A variation of MMR is the probabilistic latent MMR model proposed by Guo and Scanner [6]. PM-2 [5] tackles the problem from the perspective of proportionality. xQuAD [16] explicitly models the relationships between the documents retrieved for the query and the possible sub-queries coverage. The authors in [7] propose to combine the implicit and explicit topic representations for constructing diverse ranking. All these methods model the utility of candidate document based on carefully designed heuristics with manually defined evaluation functions. However, it is hard to design an unified similarity function for different tasks.

Recently, machine learning approaches have been proposed for search result diversification issue. The ranking score for diverse ranking is based on a linear combination of relevance features and novelty features, and the parameters can be automatically adjusted from the training data. Zhu et al. [28] propose the relational learning to rank (R-LTR) framework by optimizing the objective function to construct the diverse ranking model. With different definitions of the objective functions and optimization techniques, different diverse ranking algorithms have been proposed [21,22,24]. Xia et al. [21] learn a maximal marginal relevance model via directly optimizing diversity evaluation measures. The authors in [22] utilize the neural tensor network to model the novelty relations. To avoid the handcrafted features and fully utilize the utility in preceding documents, Xia et al. [23] propose to adapt reinforcement learning techniques to formalize the diverse ranking as a process of sequential decision making which can be modeled with MDP, where the parameters can be trained by policy gradient algorithm of REINFORCE [18].

2.2   Reinforcement Learning for Information Retrieval

Reinforcement learning (RL) techniques are widely used in information retrieval (IR) applications. The aforementioned MDP for diverse ranking in [23] is a representative work in this kind. What's more, MDP also can be extended to learning to rank problems [20], in which the proposed MDPRank model utilizes the MDP to directly optimize the NDCG at all ranking positions. Wang et al. [19] propose a game theoretical minimax game to iteratively optimize the generative retrieval and discriminative retrieval models, in which the generative retrieval model is optimized by the policy gradient algorithm of REINFORCE. In session search, Luo et al. [10] propose to utilize the partially observed Markov decision process (POMDP) to model session search as a dual-agent stochastic game for constructing a win-win search framework. The authors in [27] propose to utilize the log-based document re-ranking, which is modeled as a POMDP to improve the ranking performance. Moreover, RL techniques are also utilized in recommender systems. For instance, Guy et al. [17] designed a MDP based recommender system which employs a strong initial model to converge quickly. The multi-armed bandits technique is also utilized for diverse ranking [12]. Lu and Yang [9] propose a neural-optimized POMDP model for building a collaborative filtering recommender system.

Recent advances in reinforcement learning techniques make the research in IR one step further, and promising performances are delivered, such as MDP-DIV, MDPRank, etc. However, MDP-DIV suffers from a very slow convergence, which hinders the usability in real applications. In this paper, we aim to promote the performance of MDP-DIV by speeding up the convergence rate without much accuracy sacrifice.

3   Preliminaries

3.1   Markov Decision Process

The search result diversification issue considered in this paper could be formulated with a continuous state Markov decision process (MDP) [11,18] which is usually utilized for sequential decision making. An MDP is comprised of states, actions, rewards, policy, and transition, and can be represented by a tuple $\langle S, A, T, R, \pi \rangle$.

States S is a set of states. In [23], states can be defined as tuples consisting of preceding ranked documents, candidate documents, and the utility that the agent perceives from the preceding documents as well as the query.

Actions A is a discrete set of actions that an agent can take. The possible actions at each time step depend on the current state s, denoted as $A(s)$.

Transition T is the state transition function $s_{t+1} = T(s_t, a_t)$ which maps a state s_t into a new state s_{t+1} in response to the selected action a_t.

Reward $r = R(s, a)$ is the immediate reward, also known as reinforcement. It gives the agent an immediate reward when taking action a under state s.

Policy $\pi(a|s)$ describes the behaviors of an agent which is a sequence mapping from states to actions. Generally speaking, π is optimized to decide how to move around in the state space to achieve the optimal long-term discounted reward $\sum_{t=1}^{\infty} \gamma^t r_t$.

The agent interacts with the environment at each time step. For instance, at time step t, the agent receives the environment's state $s_t \in S$, and then selects an action $a_t \in A(s_t)$ based on the current state s_t, where $A(s_t)$ is the set of actions available under state s_t. As a consequence of the action taken, the agent receives a numerical reward $r_{t+1} \in \mathbb{R}$ and the state changes to $s_{t+1} = T(s_t, a_t)$ simultaneously in the next time step.

3.2   MDP-DIV

MDP-DIV is proposed by Xia et al. [23], which is the latest and the first approach that utilizes the reinforcement learning techniques for search result diversification. The construction of diverse ranking is formalized as a process of sequential decision making, which is modeled with a continuous state Markov decision process (MDP). The user's perceived utility can be treated as a part of its MDP state.

More specifically, at time step t, the agent receives the environment's state s_t which models the user's dynamic state on the perceived utility, starting from the first ranking position. Based on the received state, the agent chooses an action $a_t \in A(s_t)$ depending on the policy that the agent has learned recently. The policy in MDP-DIV is formulated as a *softmax* type of function that maps from the current state to a probability distribution of selecting each possible actions. According to the selected action (document), the user perceives some additional utility, also known as the immediate reward, from the recently-selected document. Here the reward is defined as the quality improvement of the selected documents in terms of α-DCG or Subtopic recall (S-recall), which are two widely used metrics in search result diversification. Then the system transits to a new state. The transition function, which maps old state and the selected document to a new state, is implemented in a recurrent manner. Reinforcement Learning techniques, the policy gradient algorithm of REINFORCE [18], is adopted to coordinate the model parameters for the sake of maximizing the expected long-term discounted rewards.

The end-to-end MDP-DIV model unifies the relevance and novelty as the criterion for selecting documents which directly optimizes a diversity evaluation measure, and outperforms the state-of-the-art baselines on the TREC benchmark datasets. However, the low convergence rate of needing tens of thousands iterations in the training phase is indeed unacceptable, especially for industrial applications. The reasons are two fold: (i) In the training stage, for decision making at each ranking position, the agent has to go through the whole remaining candidate set which introduces high computational complexity. Suppose we are given N training queries, and each query is associated with a set of M retrieved documents[1]. The diverse ranking process will cost $N(\frac{1}{2}M(M+1))$ times of query-document relevance evaluations for just one iteration. Moreover, the reinforcement learning process often needs large numbers of iterations to converge. Therefore, it is really a catastrophe if we are unfortunately facing to a large discrete action space, i.e. M is large; (ii) The retrieved documents are too scarce to train, which means that the agent has to proceed more "trial and error" interactions with the environment. For instance, more than 70% of data utilized in MDP-DIV are not labeled (i.e., no subtopics is contained). Worse still, some queries are associated with completely irrelevant (unlabeled) documents.

3.3 NTN-DIV

The NTN-DIV model is proposed by Xia et al. [22] that models document novelty with neural tensor networks. Intuitively, the neural tensor networks model the relationships between two entities with a bilinear tensor product. This idea could be naturally extended to model the novelty relation of a document with respect to the other documents for search result diversification. Suppose we are given a set of M candidate documents $X = \{d_j\}_{j=1}^{M}$, where each document is characterized with its preliminary representation with embedding models, such as the doc2vec model. The novelty score of a candidate document $d \in X$ with its preliminary representation v, and a set of ranked documents $S \in X \backslash \{d\}$ with their representations $\{v_1, ..., v_{|S|}\}$ can be defined as a neural tensor network with z hidden slices. The ranking function can be defined in Eq. (1):

$$f_n(v, S) = \omega^T v + \mu^T \max\{\tanh(v^T W^{[1:z]}[v_1, ..., v_{|S|}])\} \tag{1}$$

---

[1] For the ease of explaination, we suppose each query is associated with the same number of documents.

where the first term is the relevance score[2], and ω weights the embedding feature v. The second term is the novelty score computed by neural tensor network. Specially, $W^{[1:z]}$, a z dimensional three-way tensor, represents the relationship of the documents, where W_{ijk} stands for the k-th feature of relationship between documents d_i and d_j. And μ weights the importance of the slices of the tensor. The primary merit of using neural tensor network to model the document novelty is that the tensor can relate the candidate document and the selected documents multiplicatively, instead of only going through a predefined similarity function or through a linear combination of novelty features. To the best of our knowledge, the NTN-DIV model is the latest and the best approach for search result diversification except for MDP-DIV.

4   Methodology

As aforementioned that large action space and data scarcity will lead to low convergence rate, in this paper, we propose two kinds of strategies to deal with this issue. The first one is the k nearest neighbor strategy, which discards the k nearest neighbors of the recently-selected action (document); The second strategy relies on the pre-trained NTN-DIV [22] model, which employs a pre-trained NTN-DIV as the evaluation model, and combines the results with MDP to produce the final ranking solution. The two strategies are, respectively, realized by the proposed MDP-DIV-kNN and MDP-DIV-NTN methods in this paper. Both methods are based on the original MDP-DIV, and they differ from each other in the sampling procedure of the episode. Suppose we are given N labeled training data $D = \{(q^{(n)}, X^{(n)}, J^{(n)})\}_{n=1}^{N}$, where each query $q^{(n)}$ is associated with a set of retrieved documents $X^{(n)} = \{x_1^{(n)}, ..., x_M^{(n)}\}$, and $J^{(n)}$ denotes the labels on the documents, in the form of a binary matrix. $J^{(n)}(i, j) = 1$ if document $x_i^{(n)}$ contains the j-th subtopics of $q^{(n)}$ and 0 otherwise. The reward function $R(s_t, a_t) = \alpha\text{-}DCG[t+1] - \alpha\text{-}DCG[t]$ is based on α-DCG. As an overview of our approaches, we first summarize main procedure in Algorithm 1. Clearly, similar to the MDP-DIV model, our approaches also work in an iterative manner. The main improvements come from the step 4, where two different sampling methods are developed to efficiently search the action space. Next, we will elaborate the two methods.

4.1   K Nearest Neighbors Strategy

The action evaluation is always a parameterized function that takes both state and action as input. Hence, each time to select an action, $|\mathcal{A}|$ evaluations have to be performed first, where $|\mathcal{A}|$ is the size of action space. However, this quickly becomes intractable, especially if the parameterized function is costly to evaluate.

---

[2] In order to learn end-to-end, we use the embedding features instead of handcrafted relevance features.

Algorithm 1. MDP-DIV-kNN and MDP-DIV-NTN

---

input : Labeled training set $D = \{(q^{(n)}, X^{(n)}, J^{(n)})\}_{n=1}^{N}$, learning rate η, discount factor
γ, reward function R, and the size of returned list m
output: All the parameters Θ
1 Randomly initialize Θ in $[-1, 1]$
2 **while** *not converge* **do**
3   **for** $(q, X, J) \in D$ **do**
4        $(s_0, a_0, r_1, ..., s_{M-1}, a_{M-1}, r_M) \leftarrow$ SampleEpisode(Θ, q, X, J, R) with kNN strategy
      for MDP-DIV-kNN or pre-trained NTN-DIV strategy for MDP-DIV-NTN
5   **for** $t = 0$ to $m - 1$ **do**
6           $G_t \leftarrow \sum_{k=0}^{M-1-t} \gamma^k r_{t+k+1}$
7           $\Theta \leftarrow \Theta + \eta\gamma^t G_t \nabla_\Theta \log \pi(a_t | s_t; \Theta)$

8 **return** Θ

In MDP-DIV, the policy $\pi(a|s)$ is defined as a normalized softmax function whose input is the bilinear product of the utility and the selected document in Eq. (2):

$$\pi(a_t | [Z_t, X_t, h_t]) = \frac{\exp\{x_{m(a_t)}^T U h_t\}}{Z}$$

$$Z = \sum_{a \in A(s_t)} \exp\{x_{m(a)}^T U h_t\} \tag{2}$$

where U is the parameter in the bilinear product and Z is the normalization factor. The perceived utility of information h_t could be computed in a recurrent manner in Eq. (3):

$$h_t = \sigma(V x_{m(a_t)} + W h_{t-1}) \tag{3}$$

where V is the document-state transformation matrix that adds the newly perceived utility from the recently-selected document. W is the state-state transformation matrix which determines the utility remained across time step. Generally speaking, at each time step, the utility perceived by users for fulfilling the information needs has to take all the previously selected documents into account, i.e., the later, the more complicated. Unfortunately, the execution complexity grows quadratically with $|\mathcal{A}|$ which makes this approach inefficient. This motivate us to reduce the computational complexity.

Since the complexity of MDP-DIV closely relates to $|\mathcal{A}|$, it is natural to find a way to "shrink" the action space, i.e. reduce the complexity. To maintain the accuracy not degrading, the "shrink" strategy guarantees such foundations that: (i) It has the ability to smartly prune part of the redundant (highly similar) actions; (ii) The shrunken action evaluation can nearly generalize over actions. For search result diversification, our goal is to return the most relevant documents to the queries and ensure the diversity of the documents simultaneously. Therefore, consider such a situation: a_i and a_j are highly alike and both are closely relevant to the queries, can we just return a_i (or a_j)? The answer is positive, because learning about a_i also inform us about a_j. Moreover, in order to guarantee the diversity of the selected documents, returning them both is not a reasonable choice. Therefore, we propose a k nearest neighbor based strategy

(MDP-DIV-kNN) to reduce the complexity of MDP-DIV. The basic idea of the MDP-DIV-kNN is to discard the k nearest neighbors of the recently-selected action (document) at each time step. In particular, the strategy is instantiated in Algorithm 2. Each time we adopt an action $a_t \in A(s_t)$, at the same time, we remove the k nearest neighbors of a_t from the action space, where the neighbors are computed by using the document embeddings[3] with Euclidean distance as:

$$f_k(a_t) = \underset{a \in A(s_t)}{\arg\min}^{k} \|a_t - a\|_2 \tag{4}$$

The kNN lookup is a lightweight operation than the action evaluation execution although they are of the same complexity of the action space. Therefore, the kNN based strategy offers us three merits here: (i) It provides sub-quadratic complexity with respect to the action space; (ii) It avoids heavy cost of evaluating all actions while retraining generalization over actions; (iii) It directly optimizes the diversity of the selected documents.

Algorithm 2. SampleEpisode with kNN strategy

---
    input : Θ, q, X, J, R, and m
    output: An episode
1 Initialize s_0 and E=()
2 **for** $t = 0$ *to* $m - 1$ **do**
3   sample $a_t \in A(s_t)$ according to $\pi(a_t|s_t; \Theta)$
4       $r_{t+1} = R(s_t, a_t)$
5   change s_t to s_{t+1} according to the transition function
6   discard k nearest neighbor of a_t in X_t according to Eq.(4)
7   append (s_t, a_t, r_{t+1}) to the tail of E
8 **return** E;

4.2 Pre-trained NTN-DIV Strategy

The other method we propose to speed up the convergence rate of MDP-DIV is to use a pre-trained diversity ranking model. As aforementioned that the large action space and the data scarcity will lead to low convergence rate. The proposed k nearest neighbors strategy in turn reduces the action space at each position by filtering out the k nearest neighbors of the recently-selected action (document). It is apparent that this strategy will efficiently shrink the action space to speed up the convergence. However, it cannot deal with the data scarcity. Because, in the incipient phase, once the document is selected, we will delete the k nearest neighbors of the selected document, but we cannot make sure that it is relevant to the query or is the right one to rank at the current position. To deal with this problem, we propose the MDP-DIV-NTN method, which has two instantiations, i.e., MDP-DIV-NTN(D) and MDP-DIV-NTN(E), to promote the performance of MDP-DIV.

---

[3] All the queries and documents are embedded with doc2vec [8] embedding model.

The first instantiation adopts the pre-trained NTN-DIV model to rank the candidate set first and then takes actions in part of the pre-ranked list by applying the MDP-DIV. As a result, the action space is reduced as the NTN-DIV model can provide accurate candidates with good diversity. The MDP-DIV-NTN(D) offers us two merits: (i) It directly shrinks the candidate set, i.e., the action space in MDP-DIV; (ii) It straightforwardly takes out part of the irrelevant documents (the documents with none subtopics). Although the MDP-DIV-NTN(D) methods is effective, it may loss a bit of information because the NTN-DIV model is indeed not perfectly accurate.

The second variant is more precise. We utilize the pre-trained NTN-DIV model **at each position**, i.e., each time to adopt an action. Similar to kNN strategy, we summarize its sampling strategy in Algorithm 3. It can be seen that, at each step time of the training, once the agent chooses an document, we utilize the pre-trained NTN-DIV model to find the documents which are novelty to the previously selected documents and relevant to the query simultaneously. For the next time step, the agent only needs to learn on the filtered the candidate set. Moreover, this approach also provides more considerable advantages: (i) It precisely shrinks the action space; (ii) It accurately takes out the irrelevant documents.

Algorithm 3. SampleEpisode with pre-trained NTN-DIV strategy

> **input** : Θ, q, X, J, R, m, K, and pre-trained NTN-DIV model
> **output**: An episode
> 1   Initialize s_0, $D = ()${empty set of selected docs}, and E=(){empty episode}
> 2   **for** $t = 0$ *to* $m - 1$ **do**
> 3   sample $a_t \in A(s_t)$ according to $\pi(a_t|s_t; \Theta)$ and add a_t to D
> 4 $r_{t+1} = R(s_t, a_t)$
> 5   change s_t to s_{t+1} according to the transition function
> 6   rank the documents in X_t with D and the pre-trained NTN-DIV model
> 7   choose the first K documents of X_t as X_{t+1}
> 8   append (s_t, a_t, r_{t+1}) to the tail of E
> 9   **return** E;

However, the training of the NTN-DIV model using the original implementation is time consuming[4], because it is executed sequentially on CPU. In order to accelerate the training, we re-implement this model with Tensorflow [1] on a NVIDIA® Tesla® K80 GPU because all the tensor product can be computed parallelly. Finally, we obtain a slightly better performance with less than 30 min to train instead of more than 5 h training of the original CPU version. We also note that the NTN-DIV is trained off-line and its GPU implementation brings no improvement on the convergence for the MDP-DIV-NTN.

---

[4] https://github.com/sweetalyssum/DiverseNTN.

5 Experimental Study

5.1 Datasets and Evaluation Metrics

The dataset is provided by the authors[5] which is a combination of four TREC benchmark datasets: TREC 2009–2012 Web Track. The retrieved documents are carried out on the ClueWeb09 Category B data collection[6], which is comprised of 50 million English web documents. We note that the large number of parameters in MDP-DIV needs lots of labeled data to train, which is the reason why the four benchmark datasets are merged together. In total, there are 200 queries. Each query includes several subtopics identified by the TREC assessors. Moreover, the documents' relevance labels are made at the subtopic level, which are binary with 0 denoting irrelevant and 1 denoting relevant.

We employ three widely-used evaluation metrics to assess the diverse ranking models. They are α-NDCG [4], subtopic recall [26] (denoted as "S-recall"), and ERR-IA [3]. The α-NDCG and ERR-IA adopt the default settings in official TREC evaluation program[7], which measure relevance and diversity of the ranking list by explicitly rewarding diversity and penalizing redundancy observed at each rank. The parameter α in these two evaluation metrics are set to 0.5. The traditional diversity metric S-recall measures the coverage rate of the retrieved subtopics for each query. All of the measures are computed over the top-k search results ($k = 5$ and $k = 10$).

5.2 Experimental Setup

All the experiments are conducted with 5-fold cross-validation. We randomly re-split the queries into five even subsets[8]. For each fold, three subsets are utilized for training, one is for validation, and the rest one for testing. Moreover, for fair comparison, we run each fold five times, and the results reported are presented with average and standard deviation values over the total 25 trials. All the experiments are performed on an intel® Xeon® Processor E5 V4 server with NVIDIA® Tesla® K80 GPU and over 256 GB memory.

We compare the proposed methods with the latest state-of-the-art baselines in search result diversification, including the NTN-DIV [22] and MDP-DIV [23]. We do not compare conventional models because previous studies have shown that their performances are inferior [22,23].

NTN-DIV: As mentioned in Sect. 3.3, as a state-of-the-art method, the model computes a ranking score by taking both relevance and novelty into account with a neural tensor network. To speed up the training, we implement this method with Tensorflow on GPU which is extremely much faster than the original CPU version. The tensor slices is 100.

---

[5] The datasets and source code are available at https://github.com/sweetalyssum/ RL4SRD.

[6] http://lemurproject.org/clueweb09/.

[7] http://trec.nist.gov/data/web/12/ndeval.c.

[8] The authors does not provide the split results, therefore we re-split the queries.

MDP-DIV: As introduced in Sect. 3.2, this is the latest and state-of-the-art method based on the MDP. We set parameters following [23], because the datasets utilized are exactly the same as in [23]. As our methods employ α-DCG as reward function, the α-DCG version MDP-DIV is thus adopted for a fair comparison.

MDP-DIV-kNN: The parameter k is set to be $10\% \times |\mathcal{A}|$, $20\% \times |\mathcal{A}|$, and $30\% \times |\mathcal{A}|$, denoted as MDP-DIV-kNN(10), MDP-DIV-kNN(20), and MDP-DIV-kNN(30), respectively. The other parameters follow the settings in MDP-DIV.

MDP-DIV-NTN: The tensor slices of the pre-trained NTN-DIV model is 100, and the learning rate is 0.009. The size of both pre-ranked list in MDP-DIV-NTN(D) and MDP-DIV-NTN(E) is set to $50\% \times |\mathcal{A}|$. Again, the other parameters follow the setting in MDP-DIV.

In the experiments, the query vector and document vector are represented as the embeddings generated by the Doc2vec model, which is trained on all the documents in Web Track datasets. When training of the Doc2vec model, the number of dimension is set to 100, the learning rate is set to 0.025 and 8 is utilized as the window size.

5.3   Results and Analysis

Performance Comparison for Search Result Diversification. Table 1 shows the performance of all the methods on TREC web track datasets. From the table, we can see that the re-implemented GPU version of NTN-DIV needs half an hour to train which is extremely faster than all the other methods. However, its performance (accuracy) is significantly inferior to the other approaches.

Table 1. Performance comparison of all methods on TREC web track dataset. (The best results are marked in bold format)

Method	α-NDCG@10	α-NDCG@5	S-recall@10	S-recall@5	ERR-IA@10	ERR-IA@5	time (:h)
NTN-DIV(GPU)	0.4617	0.4124	0.6205	0.5140	0.3446	0.3186	**0.5**
MDP-DIV	0.4874	0.4480	0.6639	0.5599	0.3697	0.3477	65
MDP-DIV-kNN(10)	0.4915	0.4462	0.6731	0.5435	0.3725	**0.3539**	43
MDP-DIV-kNN(20)	0.4869	0.4461	0.6582	0.5463	0.3723	0.3506	25
MDP-DIV-kNN(30)	0.4844	0.4464	0.6489	0.5467	0.3721	0.3517	16
MDP-DIV-NTN(D)	0.4912	0.4470	0.6738	0.5464	0.3727	0.3493	26
MDP-DIV-NTN(E)	**0.4937**	**0.4485**	**0.6795**	**0.5627**	**0.3735**	0.3497	53

Compared to the original MDP-DIV, the proposed MDP-DIV-kNN methods and MDP-DIV-NTN methods are all faster, with a barely degraded or even slightly better accuracy. Among the MDP-DIV-kNN methods, the fastest one is the MDP-DIV-kNN(30) which discards 30% of the current actions by the nearest neighbor strategy. It takes 16 h to train which is 3x faster than the MDP-DIV

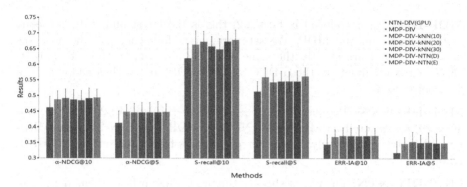

Fig. 1. Performance of stability comparison of all the methods on TREC web track dataset.

(taking 65 h). Moreover, the MDP-DIV-kNN(10) shows best accuracy among the three. We observe that it is slightly better than the original MDP-DIV, while the other two (i.e., MDP-DIV-kNN(20) and MDP-DIV-kNN(30)) are slightly worse. The reasons are two fold: (i) The k nearest neighbors strategy can help produce a more diverse ranking list; (ii) Filtering nearest neighbors may also result in a information loss. The lager the k, the more the information loss is. Therefore, the performance is a trade-off between the complexity and the accuracy.

As to the MDP-DIV-NTN methods, the performance is better compared to MDP-DIV. For MDP-DIV-NTN(D), the pre-trained NTN-DIV model offers us a pre-ranked list which helps to shrink the action space and filters part of the irrelevant document; For MDP-DIV-NTN(E), at each time step, we model the novelty of the candidate document based on both the query and preceding selected documents which provides us a more accurate pre-ranked list. Hence, its performance (accuracy) is not only better than MDP-DIV, but also better than MDP-DIV-NTN(D). However, the computing on GPU at each time step will cost some time. This is the reason that MDP-DIV-NTN(E) (taking 53 h) does not run as fast as MDP-DIV-NTN(D) (taking 26 h).

In Fig. 1, we report the error-bar of the comparison methods. From the figure, we can see that all the approaches show relatively consistent standard deviation, which indicates the proposed methods achieve stably better or comparable performance than NTN-DIV and MDP-DIV.

Next, we present some results to analyze the efficiency and effectiveness of the proposed methods in details.

Efficiency Analysis. To analyze the efficiency, We draw a shaded-line figure in Fig. 2 to show the time cost for α-NDCG@10 performance of the models based on 5-fold cross validation. The horizon axis is the α-NDCG@10 performance, and the vertical axis is the time cost to achieve the α-NDCG@10 performance. The curve in the figure means the average time cost for α-NDCG@10 performance, and the shade is the standard deviation. From the figure we can see that the proposed MDP-DIV-kNN and MDP-DIV-NTN methods are all trained faster

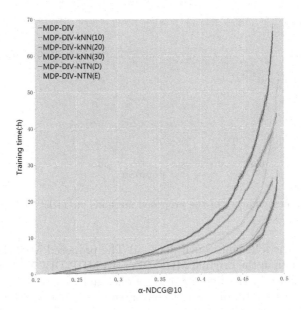

Fig. 2. Efficiency analysis of the proposed methods on TREC web track dataset.

than the original MDP-DIV. Specially, with the increase of the k value, the MDP-DIV-kNN models converge faster. Although the accuracy of the final convergence will sacrifice, it is still relatively acceptable. The MDP-DIV-NTN(D) trained faster than other models before the α-NDCG@10 performance reaches 0.48. However, the promotion of α-NDCG@10 after 0.48 becomes very time-consuming. In terms of α-NDCG@10, after convergence, MDP-DIV-NTN(D) performs worse than MDP-DIV-NTN(E) which achieves the best accuracy.

Compared to the original MDP-DIV, for instance, to achieve the α-NDCG@10 performance at 0.48, MDP-DIV-kNN(30) and MDP-DIV-NTN(D) are almost 3 times faster, MDP-DIV-kNN(20) is 1.4 times faster, MDP-DIV-kNN(10) is 0.4 times faster, and MDP-DIV-NTN(E) is 0.54 times faster than MDP-DIV. According to the observations, we draw the following conclusions: (i) The proposed MDP-DIV-kNN and MDP-DIV-NTN models fulfill the target of accelerate the convergence of MDP-DIV without much accuracy sacrifice; (ii) The MDP-DIV-kNN methods converge fast with a relatively acceptable accuracy, and the MDP-DIV-NTN methods converge fast and show better accuracy than MDP-DIV.

Effectiveness Analysis. Another promotion comes from the accuracy performance. Here we draw a shaded-line figure in Fig. 3 to show the α-NDCG@10 performance against the number of iterations. From this figure, we observe that during the first 2000 iterations, MDP-DIV-NTN(E) shows a significant improvement of α-NDCG@10 up to 0.05 over the MDP-DIV. As the training phase goes on, the improvement becomes gentle. Finally, when both the methods converge, MDP-DIV-NTN(E) still delivers better performance than MDP-DIV. In sum-

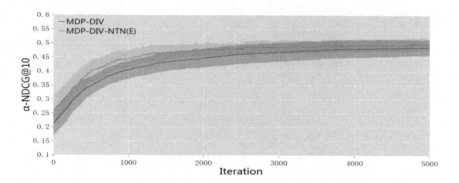

Fig. 3. Effectiveness analysis of the proposed methods on TREC web track dataset.

mary, we draw the following conclusions: (i) The proposed MDP-DIV-NTN(E) converges faster than the original MDP-DIV; (ii) MDP-DIV-NTN(E) can reach a high performance in the first 2000 iterations, and the converge performance is also better. The reason of the fast convergence rate is that we utilize an offline NTN-DIV model to shrink the search space and filter part of the irrelevant documents.

6   Conclusion

In this paper, we aim to promote the performance of MDP-DIV by speeding up its convergence rate without much accuracy sacrifice. After analysis, we find the slow convergence of MDP-DIV is mainly due to the two reasons: the large action space and data scarcity. On the one hand, the sequential decision making at each position needs evaluate the query-document relevance for all the candidate set, which results in a huge searching space for MDP; on the other hand, due to the data scarcity, the agent has to proceed more "trial and error" interactions with the environment. To tackle this problem, we propose MDP-DIV-kNN and MDP-DIV-NTN methods. The experiment results demonstrate that the two proposed methods indeed accelerate the convergence rate of the MDP-DIV, while the accuracies produced barely degrade, or even become better.

Acknowledgement. This research was supported in part by Shenzhen Science and Technology Program under Grant No. JCYJ20160330163900579, and NSFC under Grant Nos. 61572158 and 61602132.

References

1. Abadi, M., Agarwal, A., Barham, P., Brevdo, E., Chen, Z., Citro, C., Corrado, G.S., Davis, A., Dean, J., Devin, M., et al.: Tensorflow: large-scale machine learning on heterogeneous distributed systems. arXiv preprint arXiv:1603.04467 (2016)
2. Carbonell, J., Goldstein, J.: The use of MMR, diversity-based reranking for reordering documents and producing summaries. In: SIGIR 1998, pp. 335–336. ACM (1998)
3. Chapelle, O., Ji, S., Liao, C., Velipasaoglu, E., Lai, L., Wu, S.L.: Intent-based diversification of web search results: metrics and algorithms. Inf. Retr. **14**(6), 572–592 (2011)
4. Clarke, C.L., Kolla, M., Cormack, G.V., Vechtomova, O., Ashkan, A., Büttcher, S., MacKinnon, I.: Novelty and diversity in information retrieval evaluation. In: SIGIR 2008, pp. 659–666. ACM (2008)
5. Dang, V., Croft, W.B.: Diversity by proportionality: an election-based approach to search result diversification. In: SIGIR 2012, pp. 65–74. ACM (2012)
6. Guo, S., Sanner, S.: Probabilistic latent maximal marginal relevance. In: SIGIR 2010, pp. 833–834. ACM (2010)
7. He, J., Hollink, V., de Vries, A.: Combining implicit and explicit topic representations for result diversification. In: SIGIR 2012, pp. 851–860. ACM (2012)
8. Le, Q., Mikolov, T.: Distributed representations of sentences and documents. In: ICML 2014, vol. 32, pp. 1188–1196. PMLR, Beijing, 22–24 June 2014
9. Lu, Z., Yang, Q.: Partially observable Markov decision process for recommender systems. arXiv preprint arXiv:1608.07793 (2016)
10. Luo, J., Zhang, S., Yang, H.: Win-win search: dual-agent stochastic game in session search. In: SIGIR 2014, pp. 587–596. ACM (2014)
11. Puterman, M.L.: Markov Decision Processes: Discrete Stochastic Dynamic Programming. Wiley, Hoboken (2014)
12. Radlinski, F., Kleinberg, R., Joachims, T.: Learning diverse rankings with multi-armed bandits. In: ICML 2008, pp. 784–791. ACM (2008)
13. Rafiei, D., Bharat, K., Shukla, A.: Diversifying web search results. In: WWW 2010, pp. 781–790. ACM (2010)
14. Raman, K., Shivaswamy, P., Joachims, T.: Online learning to diversify from implicit feedback. In: SIGKDD 2012, pp. 705–713. ACM (2012)
15. Santos, R.L., Macdonald, C., Ounis, I.: Exploiting query reformulations for web search result diversification. In: WWW 2010, pp. 881–890. ACM (2010)
16. Santos, R.L.T., Peng, J., Macdonald, C., Ounis, I.: Explicit search result diversification through sub-queries. In: Gurrin, C., He, Y., Kazai, G., Kruschwitz, U., Little, S., Roelleke, T., Rüger, S., van Rijsbergen, K. (eds.) ECIR 2010. LNCS, vol. 5993, pp. 87–99. Springer, Heidelberg (2010). https://doi.org/10.1007/978-3-642-12275-0_11
17. Shani, G., Heckerman, D., Brafman, R.I.: An MDP-based recommender system. J. Mach. Learn. Res. **6**(Sep), 1265–1295 (2005)
18. Sutton, R.S., Barto, A.G.: Reinforcement Learning: An Introduction, vol. 1. MIT Press, Cambridge (1998)
19. Wang, J., Yu, L., Zhang, W., Gong, Y., Xu, Y., Wang, B., Zhang, P., Zhang, D.: IRGAN: a minimax game for unifying generative and discriminative information retrieval models. In: SIGIR 2017, pp. 515–524. ACM, New York (2017)
20. Wei, Z., Xu, J., Lan, Y., Guo, J., Cheng, X.: Reinforcement learning to rank with Markov decision process. In: SIGIR 2017, pp. 945–948. ACM, New York (2017)

21. Xia, L., Xu, J., Lan, Y., Guo, J., Cheng, X.: Learning maximal marginal relevance model via directly optimizing diversity evaluation measures. In: SIGIR 2015, pp. 113–122. ACM (2015)

22. Xia, L., Xu, J., Lan, Y., Guo, J., Cheng, X.: Modeling document novelty with neural tensor network for search result diversification. In: SIGIR 2016, pp. 395–404. ACM (2016)

23. Xia, L., Xu, J., Lan, Y., Guo, J., Zeng, W., Cheng, X.: Adapting Markov decision process for search result diversification. In: SIGIR 2017, pp. 535–544. ACM, New York (2017)

24. Xu, J., Xia, L., Lan, Y., Guo, J., Cheng, X.: Directly optimize diversity evaluation measures: a new approach to search result diversification. ACM Trans. Intell. Syst. Technol. (TIST) 8(3), 41 (2017)

25. Yu, H.T., Jatowt, A., Blanco, R., Joho, H., Jose, J., Chen, L., Yuan, F.: A concise integer linear programming formulation for implicit search result diversification. In: Proceedings of the Tenth ACM International Conference on Web Search and Data Mining, pp. 191–200. ACM (2017)

26. Zhai, C.X., Cohen, W.W., Lafferty, J.: Beyond independent relevance: methods and evaluation metrics for subtopic retrieval. In: SIGIR 2003, pp. 10–17. ACM (2003)

27. Zhang, S., Luo, J., Yang, H.: A POMDP model for content-free document re-ranking. In: SIGIR 2014, pp. 1139–1142. ACM (2014)

28. Zhu, Y., Lan, Y., Guo, J., Cheng, X., Niu, S.: Learning for search result diversification. In: SIGIR 2014, pp. 293–302. ACM (2014)

Structures or Texts? A Dynamic Gating Method for Expert Finding in CQA Services

Zhiqiang Liu[1,2(✉)] and Yan Zhang[1]

[1] School of Electronics Engineering and Computer Science, Peking University,
Beijing, China
lucien@pku.edu.cn, zhy@cis.pku.edu.cn
[2] Academy for Advanced Interdisciplinary Studies, Peking University, Beijing, China

Abstract. Expert finding plays an important role in community question answering websites. Previously, most works focused on assessing the user expertise scores mainly from their past question-answering semantic features. In this work, we propose a gating mechanism to dynamically combine structural and textual representations based on past question-answering behaviors. We also use some user activities including temporal behaviors as the features, which determine the gate values. We evaluate the performance of our method on the well-known question answering sites Stackexchange and Quora. Experiments show that our approach can improve the performance on expert finding tasks.

Keywords: Expert finding · Representation learning
Gating mechanism · Neural Tensor Network

1 Introduction

Community-based question answering (CQA) is an Internet-based web service that enables users to post their questions on a CQA website, which might be answered by other users later [7]. Expert finding is an essential problem in CQA sites [9], which arises in many real applications such as question routing and the identification of best answers [2]. Most of the existing works consider the expert finding problem as a text-based expert recommendation task, which learns the user representations via deep semantic models [1] and then predicts users' performance for answering the questions. Although these expert finding methods have achieved promising performance, most of them still suffer from the insufficiency of discriminative feature representations for question contents [7] and the sparsity of CQA data. To solve the issue of data sparsity, some methods have been proposed to learn answerer embeddings by utilizing related network structure information. However, the optimal combination of the structure-aspect and text-aspect representations is not well studied in these methods.

© Springer International Publishing AG, part of Springer Nature 2018
J. Pei et al. (Eds.): DASFAA 2018, LNCS 10828, pp. 201–208, 2018.
https://doi.org/10.1007/978-3-319-91458-9_12

In this paper, we propose a novel vector gate architecture to learn the expert representation by utilizing structural information of the CQA heterogeneous networks and textual information of questions. Specifically, we learn a joint representation of each user from two information sources: one is structural information, and another is textual information. The joint representation is the combination of the structures and texts with gating mechanism. Each dimension of the vector gate controls how much information is flowed from the structural and textual representations respectively and then the gate can generate the final joint representation. The major contribution of this paper is listed below:

1. Unlike previous methods, we propose a gating mechanism to dynamically combine structure-aspect and text-aspect representations based on question-answering behaviors.
2. To take full advantage of users' limited activities, we employ the features including users' temporal behaviors to guide the combination between structures and texts more accurately.

2   Related Work

The existing work for the task of expert finding can be mainly categorized as authority-based approaches and topic-based approaches called as TSPM [2,4]. In recent years, with the prevalence of online social networks in CQA sites, some researchers adopt the Network Representation Learning to exploit the rich social information from heterogeneous networks to solve the sparsity problem in CQA tasks. Authors of [11] tackle the expert finding problem via graph regularized matrix completion and metric network learning named as RMNL. Based on previous work [5], we integrate the structural and textual information of experts into joint deep representations for expert finding tasks. Recently, it has gained lots of interests to jointly learn the embeddings of structure and text information [6,10]. The most commonly used method is to concatenate these two representations.

3   Expert Finding via Jointly Embedding Texts and Structures

In this section, we will describe our dynamic gating method for expert finding via jointly embedding structures and texts. We will first introduce the settings of expert finding tasks and then describe our structure-text gate, definition of user temporal behaviors and neural tensor layer approaches for Q-U matching respectively.

3.1   Problem Description and Formulation

We denote the set of user representations by $U = \{u_1, u_2, \ldots, u_m\}$ where u_i is the embedding vector for the latent expertise of the i-th user and the semantic representations of the questions by $Q = \{q_1, q_2, \ldots, q_n\}$ that are learnt via the BiLSTM model. We first construct the heterogeneous CQA network based on User nodes and Question nodes to obtain user structure-aspect representations, which is a undirected graph. We denote the proposed heterogeneous CQA network by $G = (V, E)$ where the set of nodes V is composed of question contents Q and users U. The set of edges E consists of question-user relations and user-user relations. Secondly, the users' text-aspect representations need to be obtained using Bi-directional LSTM. We define the text-aspect representations as $\mathbf{U}_t$. The major descriptions of edged E is listed below:

(a) **user-user relations.** We define user-user connections with users' common keywords features in Stackexchange Datasets and define user-user connections with users' friend relations in Quora Datasets. Thus, we denote the set of edges between users by $E^{(1)} \in \mathcal{R}^{m \times m}$. The entry $e_{ij}^{(1)} = 1$ if the i-th user and the j-th user are friends or there are more than λ common badge keywords between them, otherwise $e_{ij}^{(1)} = 0$. The parameter λ is defined that there is an edge between two users in G if there are more than λ common badges between them in Stackexchange Datasets.

(b) **question-user relations.** We denote the set of edges between questions and users by $E^{(2)} \in \mathcal{R}^{n \times m}$. The entry $e_{ij}^{(2)} = 1$ if the i-th question is answered by the j-th user, otherwise $e_{ij}^{(2)} = 0$.

An overview of the heterogeneous CQA network is illustrated in Fig. 1. We learn users' structure-aspect representations $\mathbf{U}_s$ in the heterogeneous network by *Node2vec* [3] method.

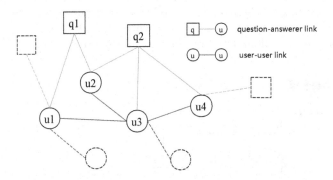

Fig. 1. The heterogeneous CQA network. The network contains two types of nodes: user, question. and the edges include question-answerer and user-user.

3.2   Gating Mechanism

Since both the structures and texts provide valuable information for a user, we wish to integrate these information into a joint representation. In this Section, we propose a united model to learn a joint representation of both structural and textual information. The whole model can be end-to-end trained. For a user $\mathbf{u}$, we denote $\mathbf{u}_s$ to be its structure-aspect representation, $\mathbf{u}_t$ to be its text-aspect representation. Then, the main concern is how to combine $\mathbf{u}_s$ and $\mathbf{u}_t$. To integrate two kinds of representations of users, we use gating mechanism to decide how much the joint representation depends on structures or texts. The joint representation $\mathbf{u}$ is a linear fusion between the $\mathbf{u}_s$ and $\mathbf{u}_t$.

Given a text encoding $\mathbf{u}_t$ and a structure encoding $\mathbf{u}_s$, we now describe how to compute the vector representation $\mathbf{u} = f(\mathbf{u}_s, \mathbf{u}_t)$ for the user. Previous methods defined f using the textual representation $\mathbf{u}_t$, the structural representation $\mathbf{u}_s$, or the concatenation $[\mathbf{u}_t; \mathbf{u}_s]$. Unlike these methods, we propose to use a vector gate to dynamically choose between the textual and structural representations based on the features of the users' activities. Let $\mathbf{v}$ denote the feature vector that encodes the features. In this work, we use the concatenation of structure features, structure embeddings and user temporal behaviors to form the feature vector $\mathbf{v}$. Then the gate is computed as follows:

$$\mathbf{g} = \sigma(\mathbf{W}_g \mathbf{v} + \mathbf{b}_g) \tag{1}$$

where $\mathbf{W}_g$ and $\mathbf{b}_g$ are the model parameters with shapes $d_u \times d_v$ and d_u, and σ denotes an element-wise sigmoid function. The final representation is computed using a gating mechanism,

$$\mathbf{u} = f(\mathbf{u}_s, \mathbf{u}_t) = \mathbf{g} \odot \mathbf{u}_s + (1 - \mathbf{g}) \odot \mathbf{u}_t \tag{2}$$

where $\odot$ denotes element-wise product between two vectors.

All illustration of our gating mechanism is shown in Fig. 2. Intuitively speaking, when the gate $\mathbf{g}$ has high values, more information flows from the structural representation to the final representation; when the gate $\mathbf{g}$ has low values, the final representation is dominated by the textual representation.

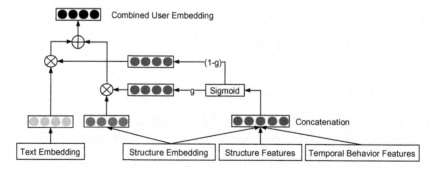

Fig. 2. The dynamic gating framework for user representations

3.3   Matching U-Q with Neural Tensor Network

To model the interactions between new questions and the user, we need utilize some metrics to measure their relevance. In this paper, we model the matching degree of two vectors with a non-linear tensor layer, which has been successfully applied to explicitly model multiple interactions of relational data [8]. Given a new question representation q and some users' representations. Following the Neural Tensor Network (NTN) [8], we place a tensor layer to model the relations of the questions and each user. The final output is the matching score of Q-U pairs. Figure 3 shows a visualization of our general architecture.

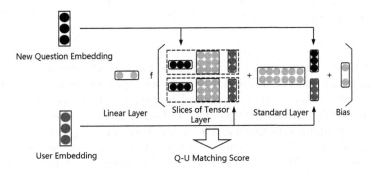

Fig. 3. Visualization of the Neural Tensor Network. Each dashed box represents one slice of the tensor, in this case there are k = 2 slices.

The tensor layer calculates the matching degree of a question-user pair by the following score function:

$$s(\mathbf{q}, \mathbf{u}) = \mu^T f(\mathbf{q}^T \mathbf{W}^{[1:z]} \mathbf{u} + \mathbf{V} \begin{bmatrix} \mathbf{q} \\ \mathbf{u} \end{bmatrix} + \mathbf{b}) \qquad (3)$$

where f is a standard nonlinearity applied element-wise, $W^{[1:z]} \in \mathcal{R}^{d_q \times d_u \times z}$ is a tensor and we conduct the bilinear tensor product $q^T W^{[1:z]} u$ results in a vector $h \in \mathcal{R}^d$, where each entry is computed by one slice $i = 1, \ldots, z$ of the tensor: $h_i = q^T W^{[1:z]}$. The other parameters are the standard form of a neural network: $V \in \mathcal{R}^{z \times 2d_u}$, $b \in \mathcal{R}^z$ and $\mu \in \mathcal{R}^z$.

4   Experiments

We evaluate the performance of our method on the well-known question answering sites Stackexchange[1] and Quora[2]. The themes of these two forums are "Academia" and "Askubuntu" in Stackexchange datasets. Our two datasets of

---

[1] https://www.stackexchange.com.
[2] https://www.quora.com.

Stackexchange cover 7 years from Sep 2010 to Aug 2017 divided into training data (Sep 2010 to Aug 2016) and testing data (Sep 2016 to Aug 2017) as shown in Table 1. In addition, We also evaluate the performance of our method using the Quora dataset, which is obtained from a popular question answering site Quora. The detail is also shown in Table 1.

Table 1. User set, training set and testing set numbers

Forums	# of users U	# of training questions Q_{train}	# of testing questions Q_{test}
Quora	3,352	11,599	3,464
Academia	15,047	75,901	12,954
Askubuntu	34,924	187,241	33,154

We evaluate the performance of our proposed method based on three widely-used ranking evaluation criteria for the problem of expert finding in CQA site: Mean Reciprocal Rank, Mean Average Precision and Precision@1. For ground truth, we consider all the corresponding answerers as the candidate user set and their received thumbs-up/down as the ground truth ranking scores. The experts for the questions tend to receive more scores. Given the testing question set Q, we denote the predicted ranking of all the users for question q and the ranked user on i-th position. We compare our proposed method with other state-of-the-art methods for the problem of expert finding in CQA sites as follows:

(1) **AuthorityRank** method [2] computes the user authority based on the number of best answers provided, which is an in-degree method.
(2) **TSPM** method [4] is a topic-sensitive probabilistic model for expert finding in CQA sites, which learns question representations via LDA-based model.
(3) **Node2vec** method [3] only learns the structure embeddings of users based on the CQA network as users' representations.
(4) **BiLSTM** method only learns the text embedding of users based on the history questions answered by users as users' representations.
(5) **QR-DSSM** [1] method uses deep semantic similarity model (DSSM) to extract semantic similarity features using deep neural networks.
(6) **RMNL** [11] method is a ranking metric network learning framework for expert finding by exploiting both users' relative quality rank to given questions and their social relations.

Among them, the AuthorityRank and Node2vec methods learn the user model for expert finding only based on network structures while the TSPM, BiLSTM and QR-DSSM methods learn the user model based on both the question contents. The RMNL method learns the user model from the proposed CQA network. Unlike previous works, we learn user representation via jointly embedding structures and texts.

Table 2 shows the evaluation results on MRR, MAP and Precision@1 of three forums. With these experimental results, we observe that our framework can

Table 2. Experiment results on three forums.

Method	Quora			Academia			Askubuntu		
	MRR	MAP	P@1	MRR	MAP	P@1	MRR	MAP	P@1
AuthorityRank	0.5711	0.5595	0.4217	0.6955	0.6765	0.5248	0.6653	0.6486	0.5255
TSPM	0.5495	0.528	0.3743	0.6551	0.6345	0.5122	0.6374	0.6156	0.4959
Node2vec	0.5873	0.5624	0.4455	0.7167	0.6988	0.5852	0.6967	0.6789	0.5788
BiLSTM	0.6179	0.5952	0.4671	0.749	0.7215	0.6087	0.7245	0.711	0.6134
QR-DSSM	0.6608	**0.6456**	**0.5108**	0.7755	0.7442	0.6519	0.7632	0.7401	0.6538
RMNL	0.6542	0.6355	0.4955	0.7656	0.7347	0.6533	0.7467	0.7255	0.6683
Our model	**0.6629**	0.6424	0.5104	**0.8177**	**0.7895**	**0.685**	**0.7904**	**0.7767**	**0.6872**

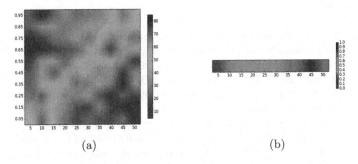

(a)   (b)

Fig. 4. Visualization of gate values in the Askubuntu CQA forum. Red means high number of users and blue means low in (a), while red means high gate values and blue means low in (b). (Color figure onlin)

outperform other state-of-the-art solutions to the problem, which suggests that our model is appropriate and effective for expert finding tasks. In addition, we visualize the model parameter **g** as described in Fig. 4. The horizontal axis is user degrees, while the vertical axis is mean values of the gates in 4(a). We notice rare degrees in CQA Networks that means there are less edges linked to the users tend to use text-aspect representations, while others tend to use structure-aspect representations. To further justify the argument, we also compute the mean values of gate in each degree-level that is defined a level per 5 degrees. The result is shown in Fig. 4(b). The same conclusion can be obtained based on Fig. 4(b). We conclude this method can make use of structure information and text information of CQA data to detect experts or route questions. In addition, in this way, we can dynamically obtain more valuable information of data and it is convenient for the end-to-end applications.

5   Conclusions

In this paper, we present a gating method that dynamically combines structure-aspect and text-aspect representations for expert finding tasks. Experiments on the Stackexchange and Quora datasets show that the gating model outperforms other state-of-the-art methods. This work opens to several interesting directions for future work. Firstly, it is of interest for us to explore other question-answering features to enhance the performance in this way. Secondly, We plan to apply the dynamic gating mechanism for combining other aspects of representations and apply it to other domains, such as commodity recommendation and link prediction in heterogeneous networks with rich texts.

Acknowledgements. This work is supported by NSFC under Grant No.61532001, and MOE-ChinaMobile under Grant No.MCM20170503.

References

1. Azzam, A., Tazi, N., Hossny, A.: A question routing technique using deep neural network for communities of question answering. In: Candan, S., Chen, L., Pedersen, T.B., Chang, L., Hua, W. (eds.) DASFAA 2017. LNCS, vol. 10177, pp. 35–49. Springer, Cham (2017). https://doi.org/10.1007/978-3-319-55753-3_3
2. Bouguessa, M., Wang, S.: Identifying authoritative actors in question-answering forums: the case of yahoo! answers. In: ACM SIGKDD International Conference on Knowledge Discovery and Data Mining, pp. 866–874 (2008)
3. Grover, A., Leskovec, J.: Node2vec: scalable feature learning for networks. In: Proceedings of the 22nd ACM SIGKDD international conference on Knowledge discovery and data mining, pp. 855–864. ACM (2016)
4. Guo, J., Xu, S., Bao, S., Yu, Y.: Tapping on the potential of q & a community by recommending answer providers. In: Proceedings of the 17th ACM conference on Information and Knowledge Management, pp. 921–930. ACM (2008)
5. Li, H., Jin, S., Shudong, L.: A hybrid model for experts finding in community question answering. In: 2015 International Conference on Cyber-Enabled Distributed Computing and Knowledge Discovery (CyberC), pp. 176–185. IEEE (2015)
6. Lin, Y., Liu, Z., Sun, M., Liu, Y., Zhu, X.: Learning entity and relation embeddings for knowledge graph completion. In: AAAI, pp. 2181–2187 (2015)
7. Qiu, X., Huang, X.: Convolutional neural tensor network architecture for community-based question answering. In: IJCAI, pp. 1305–1311 (2015)
8. Socher, R., Chen, D., Manning, C.D., Ng, A.: Reasoning with neural tensor networks for knowledge base completion. In: Advances In Neural Information Processing Systems, pp. 926–934 (2013)
9. Wang, G., Gill, K., Mohanlal, M., Zheng, H., Zhao, B.Y.: Wisdom in the social crowd: an analysis of quora. In: Proceedings of the 22nd International Conference on World Wide Web, pp. 1341–1352. ACM (2013)
10. Yang, Z., Dhingra, B., Yuan, Y., Hu, J., Cohen, W.W., Salakhutdinov, R.: Words or characters? fine-grained gating for reading comprehension. arXiv preprint arXiv:1611.01724 (2016)
11. Zhao, Z., Yang, Q., Cai, D., He, X., Zhuang, Y.: Expert finding for community-based question answering via ranking metric network learning. In: IJCAI, pp. 3000–3006 (2016)

Query Processing and Optimizations

Collusion-Resistant Processing of SQL Range Predicates

Manish Kesarwani[1], Akshar Kaul[1], Gagandeep Singh[1], Prasad M. Deshpande[2], and Jayant R. Haritsa[3(✉)]

[1] IBM India Research Lab, Bangalore, India
{manishkesarwani,akshar.kaul,gagandeep_singh}@in.ibm.com
[2] KENA Labs, New Delhi, India
prasadmd@acm.org
[3] Indian Institute of Science, Bangalore, India
haritsa@iisc.ac.in

Abstract. Prior solutions for securely handling SQL range predicates in outsourced cloud-resident databases have primarily focused on *passive* attacks in the Honest-but-Curious adversarial model, where the server is only permitted to *observe* the encrypted query processing. We consider here a significantly more powerful adversary, wherein the server can launch an *active* attack by clandestinely issuing specific range queries via *collusion* with a few compromised clients. The security requirement in this environment is that data values from a plaintext domain of size N should not be leaked to within an interval of size H. Unfortunately, all prior encryption schemes for range predicate evaluation are easily breached with only $O(log_2\psi)$ range queries, where $\psi = N/H$. To address this lacuna, we present SPLIT, a new encryption scheme where the adversary requires *exponentially more* – $\mathbf{O}(\psi)$ – range queries to breach the interval constraint, and can therefore be easily detected by standard auditing mechanisms.

The novel aspect of SPLIT is that each value appearing in a range-sensitive column is first segmented into two parts. These segmented parts are then independently encrypted using a *layered composition* of a Secure Block Cipher with the Order-Preserving Encryption and Prefix-Preserving Encryption schemes, and the resulting ciphertexts are stored in separate tables. At query processing time, range predicates are rewritten into an equivalent set of table-specific sub-range predicates, and the disjoint union of their results forms the query answer. A detailed evaluation of SPLIT on benchmark database queries indicates that its execution times are well within a factor of *two* of the corresponding plaintext times, testifying to its efficiency in resisting active adversaries.

1 Introduction

Cloud computing has led to the emergence of the "Database-as-a-Service" (DBaaS) model for outsourcing databases to third-party service providers (e.g., Amazon RDS, IBM Cloudant). Accordingly, considerable efforts have been made

© Springer International Publishing AG, part of Springer Nature 2018
J. Pei et al. (Eds.): DASFAA 2018, LNCS 10828, pp. 211–227, 2018.
https://doi.org/10.1007/978-3-319-91458-9_13

over the last decade to devise encryption mechanisms that organically support query processing without materially compromising on data security. Here, we investigate this issue specifically with regard to *range predicates*, the core building blocks of decision-support (OLAP) queries on data warehouses.

Security Architecture. A typical DBaaS setup consists of the entities shown in Fig. 1, including: (i) a Service Provider (SP), who maintains the cloud infrastructure; (ii) a Data Owner (DO), who is the data source; (iii) a set of Query Clients (QC), who are authorized to issue queries over the data stored by DO on SP's platform, and (iv) a Security Agent (SA), who acts as the bridge connecting the DO and QC with the SP.

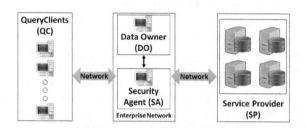

Fig. 1. System entities in DBaaS model

The SA is a *trusted* entity, and could be a simple proxy in the DO's enterprise network. Alternatively, it could be located at the SP, implemented using secure threads or secure co-processors. Although all queries pass through the SA, it is a light-weight component since it is responsible only for query rewriting and decryption of the final results.

Adversary Model. The SP, on the other hand, is always untrusted and treated as the primary adversary. We assume that the SP is only interested in deciphering the encrypted data, and not in affecting the functionality of the database system. That is, the query processing engine is in pristine condition, and all client queries are answered correctly and completely. Further, the SP maintains compliance with the standard access control and auditing mechanisms.

The Query Clients (QC) can either be trusted or untrusted, giving rise to the following alternative adversarial models:

(a) **Honest-but-Curious (HBC)**, in which the clients are trusted. Here, only *passive* attacks by the SP are possible – that is, the SP can try to breach the plaintext values solely by *observing* the encrypted data, and the computations executed by the database engine on this data. This model has been widely considered in the literature (e.g. [1,5,12–14,17,18]).

(b) **Honest-but-Curious with Collusion (HCC)**, in which the SP can unleash *active* attacks through *collusion* with a few compromised clients – specifically, the SP can *inject* range queries of its choice through the compromised QC, and then observe how these queries are processed by the database engine hosted at its site. Further, these injected queries can be constructed *adaptively*, using the results of previous queries. This powerful attack model was also recently considered in [8], as an *adaptive semi-honest* adversary.

1.1 Example Security Breach Under HCC

Consider a bank that has outsourced its relational database to the Cloud. Let the schema include a table LOAN *(CustName, LoanAmt, Collateral)* capturing the loans taken by customers, and the collaterals furnished to obtain these loans, as shown in Fig. 2a. In order to simultaneously maintain security on the Cloud and support range query processing, the current practice is to employ one of the contemporary range encryption schemes – e.g. OPE [5] – on the sensitive *LoanAmt* and *Collateral* data columns, as shown in Fig. 2b[1].

CustName	LoanAmt	Collateral
Alice	50000	40000
Bob	24576	25000
Charlie	32000	28000
Dave	10000	8000

CustName (AES)	LoanAmt (OPE)	Collateral (OPE)
Wsg^5j	5000340	173364
Sg*2js	1634009	35463
Uywhs@	4237461	65463
h7F&a1	738263	12073

(a) Plaintext LOAN Table   (b) Encrypted LOAN_OPE Table

Fig. 2. Plaintext and OPE banking database

Assume that the bank provides a form-based interface to third-parties, such as auditors, analysts, etc. to query the encrypted data. For instance, a form to generate a report that lists all the loans of a customer (say *Alice*) in a given range – say [15000 : 40000], and the associated collaterals in another range – say [13000 : 33000]. The corresponding plaintext SQL query that is internally generated from the Web form is shown in Fig. 3.

SELECT * **FROM** *Loan* **WHERE**
LoanAmt BETWEEN 15000 *AND* 40000 **AND**
Collateral BETWEEN 13000 *AND* 33000 **AND** *CustName* = 'Alice';

Fig. 3. Form-based SQL query with range predicates

---

[1] The CustName column is encrypted with AES for additional security.

Now suppose the HCC adversary comprises of the SP and the authorized auditors of customer *Alice*. In this setting, the security goal is to protect the adversary from learning the plaintext values of *LoanAmt* (and *Collateral*) for an *unrelated* customer from the encrypted LOAN_OPE table. However, the OPE-based encryption scheme can be easily breached for any target cell with just a few injected queries by *Alice's* auditors on LOAN_OPE. For instance, say the adversary selects the shaded tuple in LOAN_OPE as the target cell – corresponding to customer *Bob*. Then the attack proceeds as follows:

- The adversary first injects a query Q_1, similar to that of Fig. 3, with the *LoanAmt* range set to [OPE(32768):OPE(65535)], *Collateral* range set to [OPE(40000):OPE(40000)]^2 and CustName set to [AES('*Alice*')]. When Q_1 is processed by the database engine, the SP observes whether or not *Bob's* encrypted LoanAmt lies in this range (note that the SP has unrestricted read access over encrypted data).
- Since it happens to lie outside the range, the adversary injects Q_2, which is identical to Q_1 except that the *LoanAmt* range is now set to [OPE(16384):OPE(32767)]. When Q_2 is executed, the SP finds that Bob's encrypted *LoanAmt* lies in the target range.
- The adversary then injects another similar query, Q_3, with *LoanAmt* now set to [OPE(24576):OPE(32767)].
- Since OPE(24576) is equal to Bob's encrypted *LoanAmt* value in LOAN_OPE, the HCC adversary learns that Bob's loan amount is 24576.

The above process is representative of an injection-based *binary search attack* (BSA) that becomes feasible via collusion. As explained in [20], it is also the *strongest* feasible attack in the HCC environment, and applicable to all security systems that store the encryption of a plaintext table in a single ciphertext table.

1.2   Range Predicate Security (RPS)

Before we address the above weakness, it is necessary to formalize the security definition in the HCC model. In this scenario, a plausible security formulation for SQL range predicates is that data values from a plaintext domain of size N should not be leaked to within an interval of size H on this domain. For instance, the bank may require that no loan amount should be leaked to within an interval of size 15000 from its actual value. Note that setting H to 1 corresponds to the special case where a security breach occurs only if a plaintext is *fully* leaked – this typically applies to identificatory attributes such as Social Security numbers.

Unfortunately, as highlighted in the BSA attack example, all previous schemes for range security can be breached under HCC with a sequence of only $O(\log_2 \psi)$ range queries, where $\psi = N/H$. To address this lacuna, we present here a new encryption scheme, called **SPLIT**, in which the HCC adversary requires *exponentially more* – i.e. $O(\psi)$ – range queries to breach the interval

---

^{2} The *Collateral* range is fixed to a single value since the objective is to breach *LoanAmt*. A similar exercise can be carried out to break the *Collateral* column.

constraint. Such extended query patterns can be easily detected by standard auditing mechanisms, or incur impractically long durations to achieve covertly, thereby effectively satisfying the interval security requirement.

We present a detailed evaluation of SPLIT on benchmark databases, and demonstrate that its execution times are always within *twice* the corresponding plaintext times, thus providing an attractive security-performance tradeoff against an extremely strong adversary. Further, while SPLIT does incur large storage overheads, the extremely low resource costs on the Cloud allow it to retain viability. Finally, SPLIT is attractive from a deployment perspective also since it can be implemented as a security layer over existing database engines, without necessitating internal changes.

Organization. The rest of the paper is organized as follows: We begin with the formal problem framework in Sect. 2. The new SPLIT encryption scheme, and its associated range query processing technique, are described in Sects. 3 and 4, respectively. The security of SPLIT is analysed in Sect. 5, and the experimental results are presented in Sect. 6. Related work is reviewed in Sect. 7, and our conclusions are summarized in Sect. 8.

2   Problem Framework

As mentioned previously, the OPE and PPE schemes are currently in vogue for the secure handling of range queries, and are defined as follows:

Order-Preserving Encryption [5]: An order-preserving encryption function E_o is a one-to-one function from $A \subseteq \mathbb{N}$ to $B \subseteq \mathbb{N}$ with $|A| \leq |B|$, such that, for any two plaintext numbers $i, j \in A$, $E_o(i) > E_o(j)$ iff $i > j$.

Prefix-Preserving Encryption [18]: A prefix-preserving encryption function E_p is a one-to-one function from $\{0,1\}^n$ to $\{0,1\}^n$ such that, given two plaintext numbers a and b sharing a k-bit prefix, their corresponding ciphertexts $E_p(a)$ and $E_p(b)$ also share a k-bit prefix.

2.1   Adversary Objective

In accordance with the DBaaS model, the DO provides authorized access to portions of the data stored on the Cloud to individual QCs, using an access control mechanism and fixed query form templates. Further, the DO also defines the interval constraint size H. Given this environment, the adversary (i.e. SP + colluding QC) chooses to attack a *target cell* from an encrypted tuple which is *outside* of its authorized access, with the objective of breaching the Range Predicate Security (RPS) interval constraint H on this target cell.

Formally, the adversary $\mathcal{A}$ is given a set M^* consisting of m ciphertexts, and the interval constraint size H. $\mathcal{A}$ selects a challenge ciphertext $x^* \in M^*$ and its objective is to identify a plaintext interval (a, b) containing x^* such that $|b - a| < H$. In its attack, $\mathcal{A}$ is allowed to issue a polynomial(λ) number of range queries and observe their computations and results – here λ is the security parameter, corresponding to the bit-lengths of the plaintext values.

In the full version of this paper [20], the above attack model is formalized in the form of a *game* between the challenger $\mathcal{C}$ and the adversary $\mathcal{A}$ for a deterministic encryption scheme $\mathcal{SE}$ that supports range query execution. We hereafter refer to this game as **Chosen Range Attack (CRA)**.

2.2   Notations

The following notations are used in the remainder of this paper:

- $x_p x_{p+1} \cdots x_q$ denotes extraction of bits p through q from the (big-endian) binary representation of x.
- $x_1 \| \cdots \| x_k$ denotes the concatenation of bits $x_1, \cdots, x_k$, from which each x_i is uniquely recoverable.
- $\mathcal{P}$ denotes the plaintext domain. Further, given a plaintext value x, its encrypted version is denoted by x^*.
- N denotes the size of the plaintext domain, and H represents the size of the RPS interval constraint specified by the Data Owner. The *normalized* plaintext domain size is denoted by $\psi = \frac{N}{H}$.

3   Database Encryption with SPLIT

In this section, we present the design of the SPLIT encryption scheme, which is conceptually based on two main ideas of *splitting* and *layered encryption*. Subsequently, we describe how a plaintext database is converted to an encrypted database, followed by a rationale for the design choices.

3.1   Splitting of Data

If we consider plaintexts sourced from an n-bit integer domain, the entire set of these plaintexts can be represented by a complete binary tree of height n, referred to as the **Plaintext Tree (PT)**. The leaf level containing 2^n nodes is denoted as L_0, the level above it is denoted as L_1, and so on. For example, consider the plaintext tree for *4-bit* integers shown in Fig. 4(a). In this case, n is 4 and PT contains nodes at 5 different levels, L_0 through L_4. Every node at the leaf level of PT is associated with *n-bits* of information characterizing its path from the root to level L_0.

SPLIT partitions the levels of the *PT* into two contiguous groups, referred to as **Range Safe (RS)** and **Brute-force Safe (BS)**, respectively, and associated encrypted tables RS and BS are created based on this partitioning. The RS partition consists of the *top* levels of *PT*. For example, in Fig. 4(a), levels L_2 through L_4 belong to the RS partition, and the bits corresponding to these levels are encrypted for range query processing (this procedure is explained later in Sect. 3.2). Thus, in the encrypted RS table, for each plaintext value, the upper

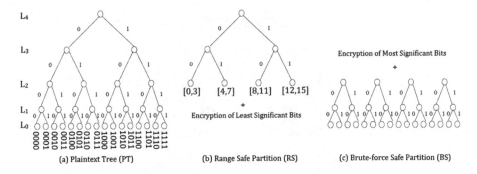

Fig. 4. Basic SPLIT scheme

bits are encrypted for range query processing and the remaining bits are blinded using a Secure Block Cipher (SBC). Hence, in this example, nodes at level L_2 effectively serve as leaf nodes and the associated range for every such node is of granularity 2^2 integers, as shown in Fig. 4(b).

The BS partition is comprised of the remaining levels of PT from level L_0 up to the level where the RS partition ends. In the current example, levels L_0 through L_2 are assigned to the BS partition, and the bits corresponding to these levels are encrypted for range query processing. Thus, in the encrypted BS table, the lower bits are encrypted for range query processing while the upper bits are blinded using SBC. This represents a set of trees, with the prefixes blinded, as shown in Fig. 4(c).

3.2 Layered Encryption

SPLIT uses three encryption schemes as black boxes, namely, Secure Block Cipher ($\mathcal{E}_{SBC}$), Order Preserving Encryption ($\mathcal{E}_{OPE}$) and Prefix Preserving Encryption ($\mathcal{E}_{PPE}$). The SPLIT encryption scheme for plaintext domain $\mathcal{P}$ is constructed as a tuple of polynomial-time algorithms SPLIT = $(KeyGen, \mathcal{E}_{BS}, \mathcal{E}_{RS}, \mathcal{E}_{SBC}, \mathcal{D}_{BS}, \mathcal{D}_{RS}, \mathcal{D}_{SBC})$, where $KeyGen$ is probabilistic and the rest are deterministic.

Key Generation [$sk \leftarrow KeyGen(\lambda, w, d)$]. $KeyGen$ is a probabilistic algorithm that takes the following as input: The security parameter λ, the total number of table columns w, and the number of columns on which range predicates can be simultaneously applied d. It then outputs the secret key sk, which consists of $d * 2^d$ equi-length secret keys $(K_O^1, K_O^2, ..., K_O^{d*2^d})$ of the OPE encryption algorithm ($\mathcal{E}_{OPE}$), $d * 2^d$ equi-length secret keys $(K_P^1, K_P^2, ..., K_P^{d*2^d})$ of the PPE encryption algorithm ($\mathcal{E}_{PPE}$) and $w * 2^d$ equi-length secret keys $(K_S^1, K_S^2, ..., K_S^{w*2^d})$ of a Secure Block Cipher ($\mathcal{E}_{SBC}$).

Encryption Algorithms. SPLIT incorporates two encryption algorithms $\mathcal{E}_{BS}$ and $\mathcal{E}_{RS}$. Both the algorithms are deterministic and take the following as input:

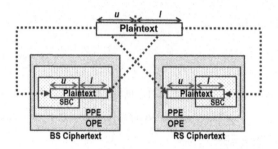

Fig. 5. SPLIT ciphertext construction

the plaintext data item m, key for OPE encryption K_O, key for PPE encryption K_P, key for SBC K_S and number of bits u in the RS partition. The $\mathcal{E}_{BS}$ algorithm outputs the BS ciphertext (c^*_{BS}) while $\mathcal{E}_{RS}$ outputs the RS ciphertext (c^*_{RS}) corresponding to message m encrypted under the given keys. Let $l = n - u$, $m' = m_{n-1}m_{n-2}\cdots m_l$ and $m'' = m_{l-1}m_{l-2}\cdots m_0$, thus, $m = m'||m''$. Then,

– *Encryption for* BS $[\mathcal{E}_{\mathbf{BS}}(\mathbf{m}, \mathbf{K_O}, \mathbf{K_P}, \mathbf{K_S}, \mathbf{u})]$

$$c^*_{BS} \leftarrow \mathcal{E}^{K_O}_{OPE}(\mathcal{E}^{K_P}_{PPE}(\mathcal{E}^{K_S}_{SBC}(m')||m'')) \tag{1}$$

– *Encryption for* RS $[\mathcal{E}_{\mathbf{RS}}(\mathbf{m}, \mathbf{K_O}, \mathbf{K_P}, \mathbf{K_S}, \mathbf{u})]$

$$c^*_{RS} \leftarrow \mathcal{E}^{K_O}_{OPE}(\mathcal{E}^{K_P}_{PPE}(m'||\mathcal{E}^{K_S}_{SBC}(m''))) \tag{2}$$

The entire set of data encryption steps for a given plaintext value, as described above, is pictorially shown in Fig. 5. The corresponding decryption algorithm is comprised of similar equations and is presented in [20].

3.3 Data Transformation

Consider a plaintext table with w columns, from which we wish to support range predicates on d columns. The plaintext values for each of the d columns are independently encrypted 2^{d-1} times using $\mathcal{E}_{BS}$ and $\mathcal{E}_{RS}$ each, thus creating 2^d ciphertext columns. Further, 2^d encrypted tables are created by capturing all BS and RS combinations of these columns. The remaining columns in the plaintext table – on which range queries will not be issued, are simply encrypted using an SBC.

We illustrate this data transformation process with the help of an example. Say our plaintext table is *Loan* with schema as enumerated in Fig. 2a – then, $w = 3$. Assume that range predicates can only be asked on *LoanAmt* and *Collateral* columns, i.e. $d = 2$. First, we call $KeyGen(\lambda, 3, 2)$, which returns secret keys consisting of eight $(2 * 2^2)$ OPE keys $(K^1_O, K^2_O, \ldots, K^8_O)$, eight $(2 * 2^2)$ PPE keys $(K^1_P, K^2_P, \ldots, K^8_P)$, and twelve $(3 * 2^2)$ SBC keys $(K^1_S, K^2_S, \ldots, K^{12}_S)$. Next, we create four encrypted tables, as shown in Fig. 6, which contain all combinations of the BS and RS partitions of *LoanAmt* and *Collateral*. Further, the physical row orderings of the tables are *randomized* to prevent *position-based* linkages across their tuples.

CustName	LoanAmt	Collateral	CustName	LoanAmt	Collateral	CustName	LoanAmt	Collateral	CustName	LoanAmt	Collateral
$E_{SBC}(K^1_l)$	$E_{BS}(K^2_o, K^1_p, K^2_l)$	$E_{BS}(K^2_o, K^1_p, K^1_l)$	$E_{SBC}(K^1_l)$	$E_{BS}(K^1_o, K^1_p, K^2_l)$	$E_{BS}(K^2_o, K^1_p, K^1_l)$	$E_{SBC}(K^1_l)$	$E_{RS}(K^1_o, K^1_p, K^2_l)$	$E_{RS}(K^2_o, K^1_p, K^1_l)$	$E_{SBC}(K^1_l)$	$E_{RS}(K^1_o, K^1_p, K^2_l)$	$E_{RS}(K^2_o, K^1_p, K^1_l)$
Kjhd*&	7981328	18718	&3y9W2	81927347	82723	&weu7w	7643	92837	Uye7^y	736473	83827
Rhwe#5	8374237	43628	ye@3^5	173687111	276372	Hwe^2h	2387	73648	82&^ey	546378	74812
Ywtw^2	237282	876213	tsfgU7	23193821	72376	7Shsu%	38272	27381	usgE6&	738272	328363
iduhu7	918237	63782	lwUn7e	5362819	92836	Sah#8s	2938	46372	hs&6Hj	283749	83636

(a) Loan_BS_BS   (b) Loan_BS_RS   (c) Loan_RS_BS   (d) Loan_RS_RS

Fig. 6. SPLIT banking database

3.4   Design Rationale

The motivation for row randomization and layered encryption in SPLIT is to *prevent linkages* of tuples across the various encrypted tables. For example, there should be no linkage between tuples in LOAN_RS_RS and LOAN_BS_RS, both of which correspond to the RS partition of *Collateral*. If such a linkage exists, it can be used to connect the tuples on the *Collateral* column in the two tables, thereby enabling a binary search attack by keeping this column fixed, and searching on the other *LoanAmt* column.

Further, the *Collateral* values are encoded using the same RS *Encrypt* function, but with different keys in LOAN_RS_RS and LOAN_BS_RS. This is where the layered encryption, using OPE and PPE, plays a role. In both these columns, the lower l bits are blinded using an SBC with different keys, so it is not possible to link tuples based on the lower bits. However, if no further encryption is used, i.e. the upper u bits are kept as plaintext, it would be possible to link the tuples based on the upper bits. So, further encryption that enables range queries based on the upper u bits is necessary. Clearly, OPE and PPE are possible schemes that can be used. However, OPE by itself is not sufficient. Consider a set of values $\mathcal{V}$ encrypted using OPE with two different keys giving sets $\mathcal{V}_1$ and $\mathcal{V}_2$. Since OPE preserves order, the order of encrypted values in $\mathcal{V}_1$ and $\mathcal{V}_2$ is identical. Thus, by sorting these sets, one could link their values.

Similarly, PPE by itself is not secure since it preserves the structure of the tree corresponding to the binary representation. In some cases, it may be possible to map nodes across two PPE trees by using the structure. For example, if in the plaintext domain, there is a single value with bit $n-1$ as 1 and all others have bit $n-1$ as 0, then this value can be linked across different PPE trees, irrespective of whether bit $n-1$ gets flipped or not.

In a nutshell, the advantage of OPE is that it destroys the structure of the tree and the advantage of PPE is that it destroys the order information. Thus, by combining OPE with PPE, we remove both order and structure-based linkages.

4   Range Query Processing

In this section, we explain how a range query is executed over a SPLIT-encrypted database. The main idea is to transform the query range into a disjoint set of prefix ranges of the form $b_{n-1}b_{n-2}\cdots b_j*$, where each b_i is a bit taking value 0 or 1, and $*$ can match any value. Smaller ranges, corresponding to $j < l$, are answered from the BS tables and the larger ranges from the RS tables. Formally Range Query Processing consists of two main steps – Range Query Mapping and Range Query Execution, as described below.

4.1 Range Query Mapping

The steps to map range predicates from the plaintext domain to the RS and BS partitions are shown in RQM Algorithm 1. The mapping process starts by converting the input range r into a set of ranges $\mathcal{R}$ represented by prefixes (Line 1). The maximum number of such ranges is $2 * (n - 1)$, where n is the number of bits used for representing the attribute values [18]. For each prefix in $\mathcal{R}$, a value with that prefix is chosen – the remaining unspecified bits are set to 0 (Line 4). Then, depending on the size of the range represented by the prefix, it is mapped to either the RS or the BS partition. For a BS range, the higher order bits are encrypted with the SBC (Line 7). Then the value is encrypted with PPE encryption (Lines 8, 10). The lower and upper bounds of the range in the PPE encrypted domain are computed by replacing the remaining lower j bits by all 0 and by all 1 (Lines 12–13). Finally, these lower and upper bits are further encrypted using OPE encryption with the appropriate keys and the range is added to R_{BS} or R_{RS}, depending on the size of the range (Lines 14–20). It can be seen that due to the prefix-preserving property of PPE and the order preserving property of OPE, this mapping produces the correct range on the encrypted domain. The ranges in R_{RS} are answered from the RS partition, and those from R_{BS} are answered from the BS partition.

The above walkthrough shows the range mapping for a single column. If there are ranges on multiple columns, each range is split into prefixes and the set of all combinations of prefixes together represents the full range of the original query. Each combination is answered from the table corresponding to the range types. For example, a BS range on the *LoanAmt* column combined with BS range on the *Collateral* column is answered from the LOAN_BS_BS table.

4.2 Range Query Execution

The next step is to execute the ciphertext queries at SP. We illustrate this process through the example plaintext query specified in Fig. 3. The following steps are performed to evaluate this query in SPLIT:

1. QC sends the plaintext query to the SA.
2. SA calls RQM Algorithm 1 and identifies sub-ranges over ciphertext tables.
3. Using output of Step 2, SA creates ciphertext sub-queries and sends them to SP.
4. SP executes the sub-queries and sends (encrypted) result tuples to the SA.
5. SA computes the union of the tuples returned from each sub-query, and then decrypts the result tuples. (The union is efficiently computable because it is apriori known that the sub-queries access *disjoint* sets of tuples.)

Algorithm 1. *Range Query Mapping (RQM)*

Input: Range r on plaintext attribute. OPE keys K_O^1 and K_O^2, PPE keys K_P^1 and K_P^2, SBC keys K_S^1 and K_S^2 for RS and BS partition respectively. The number of bits in RS partition 'u'

Output: Set of ranges on RS partition R_{RS}, set of ranges on BS partition R_{BS}

1: Convert r into a set of ranges $\mathcal{R}$ of form $b_{n-1}b_{n-2}\cdots b_j*$ {using technique in [18]}
2: Let $l = n - u$
3: **for all** $(r_i = b_{n-1}b_{n-2}\cdots b_j*)$ in $\mathcal{R}$ **do**
4:     $v \leftarrow b_{n-1}b_{n-2}\cdots b_j 0 \cdots 0$ {set lower bits to 0}
5:     $v_U \leftarrow v_{n-1}v_{n-2}\cdots v_l;\ v_L \leftarrow v_{l-1}v_{l-2}\cdots v_0$
6:     **if** $(j < l)$ **then** {BS range}
7:         $v^* \leftarrow \mathcal{E}_{K_S^2}(v_U)\|v_L$
8:         $e_v^* \leftarrow \mathcal{E}_{K_P^2}(v^*)$
9:     **else** {RS range}
10:         $e_v^* \leftarrow \mathcal{E}_{K_P^1}(v)$
11:     **end if**
12:     Let $c_n c_{n-1} \cdots c_0$ be the bit representation of e_v^*
13:     $r_L \leftarrow c_{n-1}c_{n-2}\cdots c_j 0\cdots 0;\ r_U \leftarrow c_n c_{n-1}\cdots c_j 1\cdots 1$
14:     **if** $(j < l)$ **then**
15:         $r_L^* \leftarrow \mathcal{E}_{K_O^2}(r_L)\ ;\ r_U^* \leftarrow \mathcal{E}_{K_O^2}(r_U)$
16:         Add (r_L^*, r_U^*) to R_{BS}
17:     **else**
18:         $r_L^* \leftarrow \mathcal{E}_{K_O^1}(r_L)\ ;\ r_U^* \leftarrow \mathcal{E}_{K_O^1}(r_U)$
19:         Add (r_L^*, r_U^*) to R_{RS}
20:     **end if**
21: **end for**
22: **return** R_{RS}, R_{BS}

---

5   Security Analysis of SPLIT

In this section, we evaluate the Range Predicate Security offered by the SPLIT scheme against a Honest-but-Curious with Collusion adversary mounting a Chosen Range Attack. Specifically, in a binary search attack as the range is refined, the table from which the query is answered is switched from RS to BS according to the RQM Algorithm 1. So, a target RS cell cannot be guessed to a range of size less than 2^l. And there is no way to reach the corresponding target cell in BS table in $log(\psi)$ steps unless the rows in the tables can be linked. Without linkage, binary searches over all the ψ sub-trees in the BS partition will be needed. We prove that the table rows cannot be correlated in the following discussion.

For ease of understanding, a diagrammatic view of the layered SPLIT encryption scheme is shown in Fig. 7.[3] The various ways in which RPS for the *LoanAmt* column can be breached are highlighted through the numbered dotted lines, which are explained below – a similar reasoning holds for the *Collateral* column.

---

[3] For visual clarity, *CustName* is not shown in the figure, but its encrypted form, *CustName_Enc*, is present in all four tables.

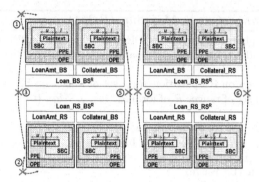

Fig. 7. Ensuring security of *LoanAmt* values

The SPLIT scheme protects against all these breaches, as explained in the remainder of this section.

To begin with, the HCC adversary is unable to independently break the BS and RS ciphertexts (dotted lines 1 and 2, respectively) because these were generated by SBC-encrypting the upper and lower half bits of the plaintext value, respectively. Secondly, the BS and RS ciphertexts (dotted lines 3 and 4) corresponding to a given *LoanAmt* plaintext value, cannot be associated, because there is no value linkage between these ciphertexts – again due to the blinding of the lower half bits in the RS table and the upper half bits in the BS table using a SBC. Further, the linkages of row locations between these tables have been removed due to the randomization (denoted by R in the figure) of the physical row orderings of the tables. Preventing this association ensures a break in the chain of attack queries.

Apart from these direct attacks on *LoanAmt*, there could also be *indirect* attacks launched on it via the sibling *Collateral* attribute. Specifically, the linkage between a pair of BS ciphertexts corresponding to a *Collateral* plaintext value (dotted line 5), or a pair of RS ciphertexts corresponding to a *Collateral* plaintext value (dotted line 6), could be used to launch a BSA on *LoanAmt*. This is prevented because physical randomization ensures the absence of row linkages between the encrypted *Collateral* columns, while value linkages are eliminated by the three-layered SBC-PPE-OPE encryption, using different keys for each table, as described in Sect. 3.

In a nutshell, the security of the SPLIT encryption scheme is established based on the following points (the complete set of formal claims and proofs are available in [20]):

1. The BS and RS encryptions are independently secure (dotted lines 1 and 2 in Fig. 7).
2. For any plaintext table, there is no linkage between the corresponding BS and RS ciphertext tables (dotted lines 3 and 4 in Fig. 7).
3. For any plaintext table, there is no linkage between a pair of corresponding BS (or two RS) ciphertext tables (dotted lines 5 and 6 in Fig. 7).

6   Experimental Evaluation

The importance of range predicates in OLAP environments can be gauged from the fact that more than half the queries in the TPC-H and TPC-DS decision support benchmarks feature such predicates. In this section, we move on to empirically evaluating SPLIT's efficiency with regard to handling range predicates in the encrypted domain.

Our experimental setup consisted of two identical server machines, with one representing the SP hosting the DO's encrypted data, and the other representing the SA interfacing with the QCs. PostgreSQL 9.4 was used as the database engine on the SP server, and all queries were issued through a Java program, which converted the plaintext queries to their SPLIT ciphertext equivalents.

The experiments were carried out on 10 GB versions of the TPC-H and TPC-DS benchmark databases. For TPC-H, the queries having range predicates on 4 attributes were constructed, with a range of selectivities on LINEITEM, the largest table in the TPC-H schema with 60 million rows. For TPC-DS, the standard benchmark tables sizes [21] were used and three benchmark queries (Query 82, Query 87 and Query 96) were executed to evaluate the performance.

6.1   Query Execution Time

The execution times taken for range query processing by the SPLIT and plaintext algorithms on the TPC-H and TPC-DS databases, as per the above experimental framework, are captured in Figs. 8a and b, respectively, The results in these figures consistently show that the performance of SPLIT is within a factor of *two* of the plaintext query execution. For instance, in Fig. 8a at 50% selectivity, the plaintext query takes around 30 s while SPLIT completes in 52 s. Similarly, in Fig. 8b, Query 82 takes 32 s in the plaintext environment, and is computed in 45 s with SPLIT encryption.

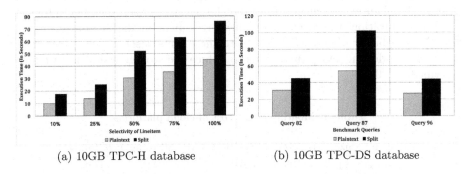

(a) 10GB TPC-H database   (b) 10GB TPC-DS database

Fig. 8. Query execution time on benchmark databases

At first look it may seem that SPLIT will incur a performance slowdown equal to the storage blowup. However such worst case scenario will require a query containing multi dimensional range predicate where each predicate has high selectivity requiring a full table scan. In general cases, if indexes are present and are chosen by the optimizer, then the number of tuples fetched from the disk will be equal to the size of the final result set. In these cases the performance overhead will be within two times since the ciphertext size is twice the size of the plaintext. Further note that since the query rewriting leads to multiple queries, each with predicates having lesser selectivity, the probability that the optimizer decides to use indexes is higher.

Note that the good performance of SPLIT is *inspite* of the large number of sub-queries in the transformed query. This is because each sub-query accesses a *disjoint* set of tuples, meaning that the total work done is almost equivalent to that of the single query in the plaintext domain, particularly if indexes are used in the query plan. This points to the practicality of the SPLIT scheme.

An important observation here is that the SPLIT implementation in these experiments lacked any *parallelization*. However, the many sub-queries (one per encrypted table) in the transformed query over the encrypted database can, in principle, all be executed in parallel. If this optimization were to be implemented, the time overheads will be further reduced.

6.2　Storage Cost

The size of the plaintext TPC-H database with indexes is 21 GB, whereas the corresponding SPLIT encrypted database is 335 GB. This is because we are handling 4D range predicates, resulting in the encrypted database being roughly 16 times the size of the plaintext database. Though this blowup is certainly large, the overall impact on the system *dollar cost* is substantively lower, since storage is relatively cheap. For instance, Table 1 shows the monthly costs for attaining same throughput with both the plaintext and SPLIT schemes, estimated using the rates charged by Amazon's AWS service [19] for machines similar to our experimental configuration. Since the execution time of SPLIT is within twice of the plaintext execution time, and the resource cost is dominated by the VM rental duration, the overall monetary investment in the SPLIT scheme is also within a factor of *two* with respect to the plaintext scheme. Further, various workload-dependent optimizations to reduce the storage overheads are also described in [20].

Table 1. Monthly dollar cost of cloud platforms

Scheme	Size (GB)	$/VM	$/GB	$(VM)	$(Storage)	$(Total)
Plaintext	21	288	0.045	288	0.945	288.945
SPLIT	335	288	0.045	576	15.075	591.075

7   Related Work

Several schemes have been proposed over the last decade for securely processing range predicates over outsourced encrypted databases. The most prominent among them have been **OPE** [1,5,6,11,14] and **PPE** [12,18], which inevitably leak order-based and structure-based characteristics respectively, of plaintext data. In **PBtree** [13] the authors have proposed an encrypted tree-based index structure, but this scheme requires significant changes to the underlying database engine which may hinder its adoption by industry.

Subsequently, alternative tree-based encryption schemes have been proposed in [7,8] and **Bucketing Schemes** are proposed in [9,10]. These schemes provide stronger security guarantees than **OPE** schemes in Honest-but-Curious model. However the fundamental problem is that, these schemes return *false positives* in the query results.

Another line of research [2,3,15–17] has focused on building complete systems which support secure execution of entire SQL queries over encrypted databases. In **CryptDB** [15], multiple encryption schemes are used to encrypt the data in an "onion"-style layering. At query processing time, the outer layers of the appropriate onions are removed as dictated by the query predicates. **MONOMI** [16] also uses multiple encryption schemes, albeit without the onion-based layering. It assumes instead that the clients also have a local database engine, and each query is split into two parts – the first part is executed on the encrypted data at the Cloud server, and its result is transferred to the client, decrypted and loaded into the local database. The second part of the query is then run on this local plaintext database.

Systems such as **TrustedDB** [3] and **Cipherbase** [2] assume the availability of trusted hardware at the server, which can be used to decrypt and process the data in a secure manner. In TrustedDB, the whole database engine runs inside the trusted hardware, whereas in Cipherbase, the database engine is aware of the encryption requirements and integrates tightly with trusted hardware.

The common limitation of all the above systems is that they are susceptible to a CRA attack in the HCC model, as described in detail in [20].

8   Conclusions

In this paper we considered a Honest-but-Curious with Collusion adversary on Cloud-resident databases. This model represents a significantly more powerful attack than the traditional HBC adversary, and is capable of easily launching Chosen Range Attack to breach the encrypted data. We proposed the SPLIT encryption scheme to securely process range predicates in the presence of such adversaries, with the key features being splitting of data values and layered encryption. With this scheme the adversary requires exponentially more queries to breach the data, making the attack unviable in practice. SPLIT was implemented and evaluated on benchmark environments, and the experimental results demonstrate that its strong security guarantees can be supported without incurring more than a doubling of the plaintext response time, even under sequential execution. When parallel execution is implemented, these performance overheads will be much smaller.

In the full version of this paper [20], we have shown how SPLIT can be extended to handle updates and other database operators, as well as serve as a potent and efficient replacement for OPE in complete systems such as Cipherbase. Therefore, in an overall sense, SPLIT promises to be a viable and desirable component for securely handling OLAP queries.

In our future work, we plan to compare the efficiency of our work with other solutions in the HBC model (ex. **PBtree** [13]), and to design encryption schemes to securely handle additional SQL operators (ex. θ join) against HCC adversaries.

References

1. Agrawal, R., Kiernan, J., Srikant, R., Xu, Y.: Order-preserving encryption for numeric data. In: Proceedings of ACM SIGMOD Conference (2004)
2. Arasu, A., Blanas, S., Eguro, K., Kaushik, R., Kossmann, D., Ramamurthy, R., Venkatesan, R.: Orthogonal security with cipherbase. In: Proceedings of CIDR Conference (2013)
3. Bajaj, S., Sion, R.: TrustedDB: a trusted hardware based outsourced database engine. PVLDB **4**(12), 1359–1362 (2011)
4. Bellare, M., Ristenpart, T., Rogaway, P., Stegers, T.: Format-preserving encryption. In: Jacobson, M.J., Rijmen, V., Safavi-Naini, R. (eds.) SAC 2009. LNCS, vol. 5867, pp. 295–312. Springer, Heidelberg (2009). https://doi.org/10.1007/978-3-642-05445-7_19
5. Boldyreva, A., Chenette, N., Lee, Y., O'Neill, A.: Order-preserving symmetric encryption. In: Joux, A. (ed.) EUROCRYPT 2009. LNCS, vol. 5479, pp. 224–241. Springer, Heidelberg (2009). https://doi.org/10.1007/978-3-642-01001-9_13
6. Boldyreva, A., Chenette, N., O'Neill, A.: Order-preserving encryption revisited: improved security analysis and alternative solutions. In: Rogaway, P. (ed.) CRYPTO 2011. LNCS, vol. 6841, pp. 578–595. Springer, Heidelberg (2011). https://doi.org/10.1007/978-3-642-22792-9_33
7. Chi, J., Hong, C., Zhang, M., Zhang, Z.: Fast multi-dimensional range queries on encrypted cloud databases. In: Candan, S., Chen, L., Pedersen, T.B., Chang, L., Hua, W. (eds.) DASFAA 2017. LNCS, vol. 10177, pp. 559–575. Springer, Cham (2017). https://doi.org/10.1007/978-3-319-55753-3_35
8. Demertzis, I., Papadopoulos, S., Papapetrou, O., Deligiannakis, A., Garofalakis, M.: Practical private range search revisited. In: Proceedings of ACM SIGMOD Conference (2016)
9. Hacigümüs, H., Iyer, B.R., Li, C., Mehrotra, S.: Executing SQL over encrypted data in the database-service-provider model. In: Proceedings of ACM SIGMOD Conference (2002)
10. Hore, B., Mehrotra, S., Tsudik, G.: A privacy-preserving index for range queries. In: Proceedings of VLDB Conference (2004)
11. Kerschbaum, F.: Frequency-hiding order-preserving encryption. In: Proceedings of CCS Conference (2015)
12. Li, J., Omiecinski, E.R.: Efficiency and security trade-off in supporting range queries on encrypted databases. In: Jajodia, S., Wijesekera, D. (eds.) DBSec 2005. LNCS, vol. 3654, pp. 69–83. Springer, Heidelberg (2005). https://doi.org/10.1007/11535706_6

13. Li, R., Liu, A.X., Wang, A.L., Bruhadeshwar, B.: Fast range query processing with strong privacy protection for cloud computing. PVLDB **7**(14), 1953–1964 (2014)
14. Popa, R.A., Li, F.H., Zeldovich, N.: An ideal-security protocol for order-preserving encoding. In: Proceedings of IEEE Symposium on Security and Privacy (2013)
15. Popa, R.A., Redfield, C.M.S., Zeldovich, N., Balakrishnan, H.: CryptDB processing queries on an encrypted database. Commun. ACM **55**(9), 103–111 (2012)
16. Tu, S., Kaashoek, M.F., Madden, S., Zeldovich, N.: Processing analytical queries over encrypted data. PVLDB **6**(5), 289–300 (2013)
17. Wong, W.K., Kao, B., Cheung, D.W., Li, R., Yiu, S.: Secure query processing with data interoperability in a cloud database environment. In: Proceedings of ACM SIGMOD Conference (2014)
18. Xu, J., Fan, J., Ammar, M.H., Moon, A.B.: Prefix-preserving IP address anonymization: measurement-based security evaluation and a new cryptography-based scheme. In: Proceedings of ICNP Conference (2002)
19. https://aws.amazon.com/ec2/pricing/
20. http://dsl.cds.iisc.ac.in/publications/report/TR/TR-2016-01.pdf
21. http://www.tpc.org/tpcds/

Interactive Transaction Processing
for In-Memory Database System

Tao Zhu, Donghui Wang, Huiqi Hu$^{(\boxtimes)}$, Weining Qian, Xiaoling Wang,
and Aoying Zhou

East China Normal University, Shanghai, China
{tzhu,donghuiwang}@stu.ecnu.edu.cn,
{hqhu,wnqian,ayzhou}@dase.ecnu.edu.cn, xlwang@sei.ecnu.edu.cn

Abstract. In-memory transaction processing has gained fast development in recent years. Previous works usually assume the one-shot transaction model, where transactions are run as stored procedures. Though many systems have shown impressive throughputs in handling one-shot transactions, it is hard for developers to debug and maintain stored procedures. According to a recent survey, most applications still prefer to operate the database using the JDBC/ODBC interface. Upon realizing this, the work targets on the problem of interactive transaction processing for in-memory database system. Our key contributions are: (1) we address several important design considerations for supporting interaction transaction processing; (2) a coroutine-based execution engine is proposed to handle different kinds of blocking efficiently and improve the CPU usage; (3) a lightweight and latch-free lock manager is designed to schedule transaction conflicts without introducing many overhead; (4) experiments on both the TPC-C and a micro benchmark show that our method achieves better performance than existing solutions.

Keywords: Transaction · Concurrency control · Network interaction

1 Introduction

In-memory database systems have gained a rapid development in recent years. These systems store the entire database in the main memory and totally removes the performance bottleneck resulted from slow disk I/O. And the major design consideration has evolved into better utilization of the multi-core and multi-socket CPUs [3]. To achieve that, in-memory databases usually **assume** the one-shot transaction model (see Fig. 1), where transactions are run as stored procedures and no client-server interaction was allowed once a transaction got started. One-shot transactions do not worry about I/O-related stall any more, and can keep being processed to the completion if no conflict access is witnessed. Based on that, many lightweight concurrency control schemas are proposed [8,15,16]. All introduce little maintaining overhead in identifying access conflicts. On the other hand, conflicts are resolved by simply aborting and retrying transactions,

© Springer International Publishing AG, part of Springer Nature 2018
J. Pei et al. (Eds.): DASFAA 2018, LNCS 10828, pp. 228–246, 2018.
https://doi.org/10.1007/978-3-319-91458-9_14

Fig. 1. The one-shot transaction   **Fig. 2.** The interactive transaction

which is actually quite time-consuming [17,18]. Thus, the efficiency of these methods is promised based on the fact that most one-shot transactions finish execution in a short time and conflict with each other seldomly.

Although, one-shot transaction processing achieves impressive throughput, another very important kind of workload is rarely studied, namely the interactive transaction processing (see Fig. 2). In this case, applications operate database system with SQLs and JDBC/ODBC interface. Transactions are invoked by sending SQLs one-by-one to the server. A recent survey [10] shows transaction processing in interactive way is much more common than that in one-shot way. It is reported that 54% responders never or seldom use stored procedure in their DBMSs. Only 16% responders have more than half of transactions ran as stored procedures. Several reasons for the situation are: (1) stored procedures are difficult to maintain and debug; (2) they lack portability, making it hard to deploy applications on different platforms and databases.

Such fact has made us think about how to design in-memory database system for interactive transaction processing. We conclude the major differences between the interactive model and the one-shot model are: (1) a transaction can be stalled by network I/O, which requires the execution engine to handle the network I/O blocking efficiently; (2) a transaction lasts much longer since network latencies are included, which results in more access conflicts. The concurrency control is expected to be efficient in both identifying and resolving conflicts. Our contributions can be summarized as follows: (1) we examine the design space, and conclude several design considerations in implementing interactive transaction processing. (2) A new execution model is designed to interleave transaction execution efficiently. It can fully utilize the CPUs and does not waste much time on handling different kinds of blocking. (3) A lightweight and latch-free lock manager, named as *iLock*, is proposed to schedule conflict operations efficiently. It introduces little overhead regardless of the workload containing lots of contention or not. (4) Experiments on well-known benchmarks show our method achieves the better throughput than existing techniques.

The paper is organized as follows: Sect. 2 analyzes the properties of interactive transaction processing and gives the design consideration. Section 3 presents a new execution engine which efficiently running concurrent transactions on multi-core hardwares. Section 4 designs a new lock manager and analyzes its correctness. Section 5 shows the experiment results. Related works are discussed in Sect. 6 and the work is concluded in Sect. 7.

2   Design Consideration

The section firstly emphasizes two important facts about the interactive transaction workload, and then gives our design considerations.

Frequent Network I/O Blocking. As a transaction is executed by interactively sending SQLs to the server, its processing is frequently blocked by network I/O. As a result, a transaction has itself suspended when a SQL request is finished, and get resumed after the next request arrives. The database system is responsible for handling transaction's suspending and resuming efficiently.

Long Transaction Duration. Transaction duration is the time from the beginning of a transaction to the end. Given a one-shot transaction accessing tens of record, its processing phase only contains memory access and CPU calculation (the commit phase is not included, which requires one disk write). Therefore, it has a very short duration and can usually be finished in less than $100\,\mu s$. On the other hand, an interactive transaction requires multiple network communications between the client and the database system. In a cluster, it takes nearly $100\,\mu s$ to do a simple message round-trip between two servers through Ethernet. If there is 10 client-server interactions, a transaction will last longer than $1000\,\mu s$.

Based on these observations, we consider that the in-memory transaction engine should have the following design principles.

CPU-Efficient Execution Model. Consider that a transaction t_x is running on a CPU core c_i. If t_x is blocked by network I/O, c_i becomes idle. In order to make full use of CPU, c_i ought to process another transaction t_y before the next SQL of t_x is received. Hence, a large number of transaction should run concurrently on a relatively small number of CPU cores. To achieve that, existing systems use the Transaction-To-Thread execution model, where a transaction binds its execution with the same thread during its lifetime. If a thread is blocked and a core becomes idle, another thread would be switched in and run on the same core. However, the limitation is that each time a thread is blocked by network I/O, it results in one *context switch*. For instance, on an *Intel E5-2620 CPU*, it takes typically 8–$13\,\mu s$ to finish one context switch. Such overhead makes the execution model less appealing. Hence, we target on designing a new execution model, which handles the network-related blocking more efficiently.

Lightweight Lock-Based Concurrency Control. Since interactive transactions have the long duration, the risk that two transactions access the same data item is greatly increased. As a result, transaction conflicts happen more frequently. (*i*) Many in-memory systems [8,15] favor the optimistic concurrency control [7]. But it is not suitable for scheduling interactive transactions. Since the protocol identifies conflicts at the end of a transaction with a validation phase, a client has to process a transaction from the scratch once its validation fails. As mentioned in the above, conflicts happen more frequently under the interactive workload, OCC would waste a plenty of CPU time and network resources on retrying failed transactions. (*ii*) With that in mind, we consider the lock-based concurrency control (i.e. 2PL) as a better choice. Under the lock-based protocol, access conflicts are identified by asking each transaction to acquire locks. Conflict can be resolved by blocking a transaction until the required lock is released. The protocol has the balanced performance in scheduling workloads exhibiting either high lock contention or low. In further, in order to work efficiently in the main memory environment, the lock-based protocol requires a lightweight implementation to reduce the CPU overhead. However, We observe existing solutions do not satisfy the requirement. (*a*) Disk-based systems implement the protocol with a centralized lock table. The structure is proven to have considerable maintaining overhead [5,11]. (*b*) Some in-memory systems optimize the protocol by simplifying lock manager (i.e. row locking [8,11]). It is only efficient in identifying access conflicts. When a conflict is witnessed, it forces a transaction to abort and get retried, which consumes much CPU time. Hence, it is not suitable for the case where the workload contains much contention.

In all, the interactive transaction processing calls for (1) an execution engine, which is efficient in handling blocking caused by network I/O or access conflicts; and (2) a lightweight lock manager, which introduces little overhead in identifying and resolving conflicts.

3   Execution Model

This section presents an execution model, which is efficient in handling blocking coming from network I/O and lock conflicts. Overall, our goal is to minimize the number of context switches resulted from blocking. In the following, we start with a simple execution model which is able handle network-related blocking efficiently. Based on that, we take the conflict-related blocking into consideration. And then, a new execution model is given with its features clearly discussed.

3.1   SQL-To-Thread

Here we firstly assume there is no conflict-related blocking. As discuss in the previous section, transaction-to-thread model is not a good design because each time a thread waits for the next SQL request to arrive, it is blocked and generates one context switch. A possible design is to bind a SQL request with a thread. As a client interacts with the server in degree of SQL request, a thread can execute a request to the completion without being blocked. Once a request is processed, the thread continues to process a SQL of another transaction.

In the SQL-To-Thread model, a batch of I/O threads are responsible for communicating with clients. Several working threads are created to handle the transaction execution. The number of working threads **is equivalent to** the number of available CPU cores. Once a request is received, an I/O thread would push it into a task pool. Each working thread keeps pulling a request from the

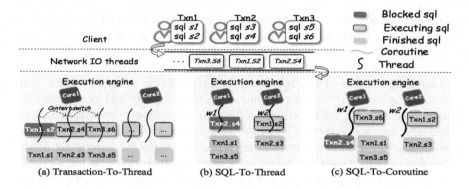

Fig. 3. Different kinds of execution models.

		caller	callee (function instance)	caller	callee (coroutine instance)
co-init	allocate resources required by a coroutine instance				
co-resume	resume the execution of a coroutine instance				
co-yield	exit the current coroutine instance temporarily				

Fig. 4. Interface for coroutine usage   **Fig. 5.** Coroutine mechanism

task pool for processing. After a request is processed, a working thread informs an I/O thread to send results back to the client, and continues to pull a request from the task pool. Figure 3(b) gives an example. Two working threads w_1, w_2 are started on two cores. Here, w_1 processes SQL s1 of transaction Txn1 in the first, and then Txn2.s4. When the next request Txn1.s2 is arrived, w_1 is occupied by Txn2.s4 but w_2 is available. Hence, w_2 begins to process Txn1.s2. In the model, each thread keeps itself busy by servicing requests of different transactions. It eliminates the context switches generated by network I/O.

3.2 SQL-To-Coroutine

Conflict-Related Blocking. In addition to network-related blocking, a transaction can also be blocked by access conflicts. The SQL-To-Thread model cannot suffi-ciently handle this kind of blocking. Considering a SQL s_1 attempts to acquire a lock l_i and the l_i is already held by another transaction t_y, s_1 can not pro-ceed any more before t_y releases l_i. Facing this situation, s_1 has two choices: (1) abort and retry itself later (only abort s_1, not the whole transaction); (2) wait until the lock is released. The first choice is not acceptable. As analyzed in the previous section, an interactive transaction tends to have a long duration. It is very likely that s_1 will be retried for many times before s_1 acquires the lock and all those failed attempts only waste CPU time without any other return. Hence, we prefer the second choice, which let s_1 wait for t_y to release l_i.

SQL-Wait. The next problem is how to make a SQL wait in a thread. Two common solutions are possible: (1) busy-waiting and (2) condition-waiting. In busy-waiting, the thread keeps checking whether the lock is released. Obviously, it is even worse than the abort-retry strategy, as it not only wastes lots of CPU time on checking the lock condition, but also prevents other requests from being processed. In condition-waiting, the thread gives up the CPU and is awakened later when t_y releases l_i. Before the thread is waked up, its CPU core becomes idle (recall that exact one thread is created for each core). As a result, using condition-waiting will under utilize the CPU resource.

Ideally, we expect the execution engine to process SQLs in the following way. If s_1 is blocked due to a lock conflict, the thread temporarily leaves its execution, and starts to process the next request. After l_i is released, the thread resumes the execution of s_1 again. To achieve that, we propose a SQL-To-Coroutine model. Figures 4 and 5 illustrates the mechanism of coroutine [6] briefly. As we can see in Fig. 5(b), a caller A uses *co-init* to allocate memory space and do some initializations for a coroutine instance B. The caller invokes *co-resume* to begin the execution of B. In the middle of the execution, B can invoke *co-yield* to exit its execution temporarily. In the back-end, *co-yield* saves B's stack into the memory region allocated by *co-init*. Later, when A calls *co-resume*, B would have its stack copied back into the thread's execution context again. This time, B continues its execution from where the last *co-yield* is called. Essentially, the coroutine is similar with the function if *co-yield* is never used. The difference is that a coroutine instance holds all states and variables before it is ended. It can exit in the middle of its execution and later return to the point where it leaves. More information about coroutine can be found in [6].

The SQL-To-Coroutine model processes each SQL request as a coroutine instance. During processing, if there is no lock conflict, the coroutine instance executes the same as a normal function call. Otherwise, if lock conflict does exist, *co-yield* can be called to exit the current execution temporarily. Once the coroutine instance yields, the thread begins to process a new request. In later, if the thread is *informed* that the lock has been released, it gets back the pointer of the coroutine instance, and invoke *co-resume* to resume the execution. The informing mechanism would be left to the next section.

Figure 3(c) gives an example for executing SQL requests as coroutines. In the example, Txn2.s4 is blocked due to conflict data access. The thread w_1 calls *co-yield* to exit the execution of Txn2.s4 temporarily. Then w_1 begins to process the next request Txn3.s6 in the task pool. And Txn2.s4 is resumed when another transaction releases the lock that it is waiting on. As we can see, in the SQL-To-Coroutine model, a thread will not be blocked by transaction conflicts.

3.3   Discussion and Refinement

Running SQL requests as coroutine instances also introduces some inherent overheads, which come from two sources: coroutine initialization and switching.

Initialization. co-init allocates about 1 M bytes for a coroutine instance to save
its execution context. If it is always called for each request, lots of CPU time
are wasted on memory allocation and deallocation. A simple refinement is to
build a resource pool on each thread. To process a request, a thread tries to
reuse a coroutine instance in its pool. If the pool is empty, *co-init* is called to
create a new instance. After a request is finished, its coroutine instance is put
back into the pool. Another concern is whether using coroutines would consume
lots of memory resource. Obviously, each thread requires an instance for its in-
processing request. In addition, an instance is also required for each blocked
transaction that is waiting for a lock. Actually, the number of working threads
and blocked requests would not get very big. For example, give a system with
30 CPU cores and 70 transactions waiting for locks, it takes about 100 M bytes
memory space, which is relatively small as an in-memory database is deployed
with more than hundreds of gigabytes memory.

Coroutine Switching. As discussed in the above, a coroutine instance gives up
the thread by using *co-yield*, which saves its stack into the preallocated memory
region. The thread uses *co-resume* to copy its stack back and continues its execution. A problem is whether such memory copying takes lots of CPU time. On
a *Intel E5-2620 CPU*, a micro-benchmark showed that it usually takes about
less than 50 ns to finish one switch. In a comparison, it takes about 10 μs to
done one context switch. And it takes about several tens of microseconds (μs)
to handle a point-get request in a prototype main memory system. Hence, the
overhead introduced by coroutine switching is negligible. In addition, switching
only happens when a transaction is blocked. If all requests are processed without
any conflicts, coroutine switching never happens.

The section discusses how to handle different kinds of blocking efficiently
with a SQL-To-Coroutine execution engine. Based on that, we are required to
discuss how to schedule concurrent transactions correctly and efficiently.

4   Lock Manager

This section proposes a lightweight, latch-free lock manager, named as *iLock*.
It produces little CPU overhead in both identifying and resolving transaction
conflicts. In the following, we firstly present related data structures, and then
design its acquisition/releasing algorithms, as well as the deadlock avoidance
mechanism. In the last, the correctness of iLock is analyzed.

A **record** stores the following information in its header:

- *lock-state*, a 64-bit variable, representing the state of the lock;
- *write-waiter*, a 64-bit variable, encoding who waits for writing the record;

– *read-waiter*, a 64-bit variable, encoding who waits for reading the record;

The first bit of *lock-state* tells the lock mode of the record, using 1 for write-mode and 0 for read-mode. In write mode, the last 63 bits of *lock-state* stores the transaction identifier of the owner. In read mode, the rest bits tells how many transactions have acquired the read lock. Clearly, the record is not locked if *lock-state* = 0, i.e. all bits of the lock state are zeros.

The *i-th* bit of *write-waiter* represents whether any transaction in the *i-th* working thread is trying to acquire the write lock of the record. Similarly, *read-waiter* tells which thread is waiting for the read lock. Concretely, *write-waiter* uses N bits if there are N working threads. Here N is determined by the number of CPU cores used by transaction processing. Since we leverage the SQL-To-Coroutine model and modern servers are usually equipped with no more than 64 cores, 64 bits are enough for most cases. Besides there is no limitation to use a large-sized *write-waiter* to work with an advanced platform equipped with more CPU cores.

A **transaction** has the following fields in iLock.

– *tid*, a unique identifier for a transaction;
– *co-pointer*, the pointer to the current coroutine instance.

Here, *tid* is allocated as a 63-bit positive integer. Note that 0 is not used as *tid* because the number is used by a dummy write lock, which will be explained in the later. *co-pointer* is only used when a transaction has its SQL execution blocked due to lock conflicting.

A **working thread** maintains some thread-local structures:

– *thd*, a unique index for a thread, numbered starting from 0.
– *wait-map*, a hash map, organizing all suspended transactions.
– *lock-queue* buffers all lock entries received from ended transactions.

The *wait-map* is a hash map permitting multiple entries with the same key. Its key field is a locking request, and the value field is a pointer of a transaction. When a transaction t_x has its lock request l_i blocked due to conflict, it adds an entry $<l_i, t_x>$ into the *wait-map*.

The *lock-queue* buffers all lock entries received from finished transactions. When a transaction releases its lock, it checks the *write-waiter* and *read-waiter* fields of the record, and sends the lock to a proper thread by adding a lock entry into the *lock-queue*. Later, a thread uses the received lock entry to wake up a proper transaction in the *wait-map*.

4.1   Lock Acquisition

Here we discuss how to identify conflict locking requests and to resolve conflicts by suspending a transaction. In the following, we consider the scenario where a transaction t_x is trying to acquire the write/read lock of the record r.

```
Input: Record r, Transaction t_x
1  nstate ← t_x.tid | mask;
2  if 0 ≠ atomic-cas(r.lock-state, 0, nstate) then
3  |  changed ← atomic-set(r.write-waiter, thd) ;
4  |  atomic-synchronize();
5  |  if 0 = atomic-cas(r.lock-state, 0, t_x.tid) then
6  |  |  if changed then
7  |  |  └ atomic-unset(r.write-waiter, thd) ;
8  |  └ return locked ;
9  |  t_x.co-pointer ← co-self() ;
10 |  wait-map.put(r.rid, write-mode, t_x) ;
11 |  co-yield(t_x.co-pointer) ;
12 |  if timed-out(t_x) then
13 |  └ return aborted ;
14 └ atomic-cas(r.lock-state, mask, nstate) ;
15 return locked ;
```

Fig. 6. acquire-write-lock

```
Input: Record r, Transaction t_x
1  do
2  |  ostate ← r.lock-state;
3  |  if write-locked(ostate) then
4  |  |  changed ← atomic-set(r.read-waiter, thd) ;
5  |  |  atomic-synchronize();
6  |  |  if write-locked(r.lock-state) then
7  |  |  |  t_x.co-pointer ← co-self() ;
8  |  |  |  wait-map.put(r.rid, read-mode, t_x) ;
9  |  |  |  co-yield(t_x.co-pointer) ;
10 |  |  |  if timed-out(t_x) then
11 |  |  |  └ return aborted ;
12 |  |  else if changed then
13 |  |  |  atomic-unset(r.read-waiter, thd) ;
14 |  |  ostate ← r.lock-state;
15 |  nstate ← ostate + 1 ;
16 while ostate ≠ atomic-cas(r.lock-state, ostate, nstate);
17 return locked ;
```

Fig. 7. acquire-read-lock

Identifying Conflicts. To lock a record r down, (1) t_x checks whether its lock request is compatible with the current lock state $r.lock\text{-}state$. If the result is yes, (2) t_x updates the field with a new state (the field is updated in different ways for the read lock and the write lock as introduced later). The two steps are done atomically using the compare-and-swap instruction (atomic-cas in figures), which is a hardware-assisted synchronization primitive (See Wikipedia).

Figure 6 gives the pseudo-code for the write lock acquisition. Here, Line 1 creates a new lock state, whose first bit is set as 1 and the rest bits are set as tid of t_x, representing a write lock held by t_x. Then Line 2 checks whether r.lock-state $= 0$, i.e. r is not locked by anyone, and try to atomically update $r.lock\text{-}state$ with the new lock state calculated in Line 1. Figure 7 gives the pseudo-code for the read lock acquisition. (1) Line 2–3 fetches a snapshot of $r.lock\text{-}state$ and checks whether the record is locked in write mode (i.e. the first bit of $r.lock\text{-}state$ is 1). (2) If the result is yes, a conflict is identified, otherwise, Line 15 creates a new lock state by increasing the number of readers by 1. (3) Line 16 checks that the latest state $r.lock\text{-}state$ is equal with the snapshot, and update $r.lock\text{-}state$ with the new value computed in Line 15. (4) Line 16 may find that $r.lock\text{-}state$ has been changed as other threads can modify the variable in parallel. In this case, the above steps have to be done again.

Resolving Conflicts. If $r.lock\text{-}state$ is in an incompatible state, t_x is suspended with the following steps. (1) *Turn the wait bit on.* First of all, the thd-th bit of $r.write\text{-}waiter$ (or $r.read\text{-}waiter$) is turned on. It is used to inform the lock holder that a transaction in the thd-th thread is waiting for the lock. Then a memory fence (i.e. atomic-synchronize in figures) is added to ensure the wait bit is modified before the following steps are executed. (2) *Try locking again.* t_x must check the locking state again in case that the record is just unlocked after the previous attempt. If t_x can secure the lock, it continues its processing; otherwise, t_x will still be suspended. (3) *Suspend the execution.* t_x saves the pointer of its coroutine instance into $r.co\text{-}pointer$, and then add an entry into the *wait-map*, which maps the blocked lock request to t_x itself. After that, *co-yield* is invoked to exit its execution temporarily.

```
Input: Record r, Transaction t_x
1  ostate ← r.lock-state;
2  nstate ← 0 ;
3  if r.write-waiter ≠ 0 ∨ r.read-waiter ≠ 0 then
4  └ nstate ← mask;

5  atomic-cas(r.lock-state, ostate, nstate) ;
6  atomic-synchronize();
7  if nstate = 0 then
8  │  if r.write-waiter ≠ 0 ∨ r.read-waiter ≠ 0 then
9  │  │  if 0 = atomic-cas(r.lock-state, 0, mask) then
10 │  │  │  nstate ← mask;
11 │  │  └  atomic-synchronize();

12 if nstate = mask then
13 │  if r.read-waiter ≠ 0 then
14 │  │  send-read-lock(r) ;
15 │  else if r.write-waiter ≠ 0 then
16 └  │  send-write-lock(r) ;
```

Fig. 8. release-write-lock

```
Input: Record r, Transaction t_x
1  do
2  │  ostate ← r.lock-state;
3  │  nstate ← ostate − 1 ;
4  │  if nstate = 0 then
5  │  │  if r.write-waiter ≠ 0 ∨ r.read-waiter ≠ 0 then
6  │  └  └  nstate = mask;

7  while ostate ≠ atomic-cas(r.lock-state, ostate, nstate);
8  atomic-synchronize();
9  if nstate = 0 then
10 │  if r.write-waiter ≠ 0 ∨ r.read-waiter ≠ 0 then
11 │  │  if 0 = atomic-cas(r.lock-state, 0, mask) then
12 │  │  │  nstate ← mask;
13 │  │  └  atomic-synchronize();

14 if nstate = mask then
15 │  if r.write-waiter ≠ 0 then
16 │  │  send-write-lock(r) ;
17 │  else if r.read-waiter ≠ 0 then
18 └  │  send-read-lock(r) ;
```

Fig. 9. release-read-lock

In Fig. 6, Line 3–11 are detail steps used to suspend a write lock request. Line 3–4 turn the thd-th bit of $write\text{-}waiter$ on. Line 5–6 try locking again. If Line 3 does flip the wait bit, Line 7 turns that off if the lock is acquired successfully. It is because no transaction in the thread is waiting for the lock once t_x is granted. If the second lock attempt is failed, Line 9–11 suspend the execution of t_x. In Fig. 7, Line 4–14 suspends a read lock request. Line 4–5 turn the thd-th bit of $read\text{-}waiter$ on. Line 6 checks whether the read lock is acquirable again. If not, Line 7–9 suspend the execution of t_x. Otherwise, Line 12–13 clear the wait bit if it is flipped in the previous. And Line 14–17 try locking again. Here, a write lock request may be starved when new readers are always allowed to acquire the read lock. To fix the issue, we can refine the algorithm in Fig. 7 by blocking any new reader if there is a writer waiting for the lock (i.e. $r.write\text{-}waiter \neq 0$).

4.2   Lock Releasing

Releasing a lock on a record is decomposed into two stages. The first stage reverts the $lock\text{-}state$ field of the record. The second stage informs some particular threads to wake up their blocked transactions if no one is holding the lock any more. In the following, we consider the scenario where a transaction t_x is releasing its lock on a record r.

Reverting Lock State. If multiple transactions are holding read locks on r, t_x simply decreases $r.lock\text{-}state$ by 1. Otherwise, t_x becomes the last lock holder. In this case, we need check whether any transaction is waiting for accessing r. (A.) *Have a waiter.* If there is a blocked request, $r.lock\text{-}state$ is reverted to a special state $mask = 0x80000000$, whose first bit is 1 and the rest are zeros. It can be viewed as a **dummy write lock** held by a dummy transaction with $tid = 0$. Recall that $tid = 0$ is never used by real transactions. It prevents ongoing transactions from *stealing* the lock, and ensure the lock would be granted to a blocked transaction. (B1.) *Have no waiter.* If no one is blocked, the record is

Input: Record r
1 state ← $r.write\text{-}waiter$;
2 **for** i ← 1 ; i ≤ $thd\text{-}num$; ++i **do**
3   j ← (i + thd) mod $thd\text{-}num$;
4   **if** 0 ≠ (state & (1 ≪j)) **then**
5          $lock\text{-}queue[j]$.push(rid, write-mode) ;
6   break ;

Fig. 10. send-write-lock

Input: Record r
1 state ← $r.read\text{-}waiter$;
2 reader-number ← $popcount$(state) ;
3 atomic-cas($r.lock\text{-}state$, $mask$, read-number) ;
4 atomic-synchronize();
5 **for** i ← 0 ; i < $thd\text{-}num$; ++i **do**
6   **if** 0 ≠ (state & (1 ≪i)) **then**
7          $lock\text{-}queue[i]$.push(rid, read-mode) ;

Fig. 11. send-read-lock

truly unlocked by set $r.lock\text{-}state$ ← 0. (B2.) *Have a new waiter.* If the record is unlocked, $r.write\text{-}waiter$ and $r.read\text{-}waiter$ should be checked again since a transaction t_y may be just suspended after t_x does the first check. If a new waiter is found, we try to acquire the dummy write lock again and resume t_y.

Figures 8 and 9 give procedures used to release write locks and read ones respectively. In Fig. 8, Line 1–6 reverts $r.lock\text{-}state$ to $mask$ or 0 based on whether a waiter exists or not. Line 7–11 check waiter information again, and try to acquire the dummy write lock if a new waiter is found. In Fig. 9, Line 1–8 reverts the lock state based on the waiter information, while Line 9–13 tries to acquire the dummy write lock if a new waiter arrives. In both algorithms, a memory fence is added after modifying the lock state in order to avoid the execution order is changed by the compiler or CPU.

Resuming Transactions. If r is protected by a dummy write lock, we try to wake up blocked transactions (by Line 12–16 in Fig. 8 and by Line 14–18 in Fig. 9).

Figure 10 is used to wake up a suspended write lock request. Here, (1) *Sending lock entry.* t_x sends a write lock entry to a thread who is waiting. Line 2–4 start from the $(thd + 1)$-th bit, and tries to find a non-zero bit in $write\text{-}waiter$. Let j-th bit be the non-zero bit. Then, Line 5 pushes a write lock entry into the $lock\text{-}queue$ of the j-th thread. (2) *Resuming transaction.* When j-th thread pops the lock entry from its $lock\text{-}queue$, it gets a transaction t_y from $wait\text{-}map$, who is waiting for the lock entry. Then co-resume is invoked on $t_y.co\text{-}pointer$, and t_y resumes its execution from Line 12 in Fig. 6. This moment, t_y must succeed in acquiring the lock in Line 14. (3) *Clearing wait bit.* the j-th thread calls atomic-unset($r.write\text{-}waiter$, j) to clear its wait bit, if no transaction is waiting for the write lock on the j-thread any more.

Figure 11 is used to wake up read lock requests, which is a little different from the write one. Before sending read lock entries to any thread, Line 2 calculates the number of threads those require the read lock[1]. Line 3–4 acquires **a dummy read lock** for each of them. It is used to ensure that r is always locked in read mode before each thread has waked all its blocked transactions up. The dummy read lock is released when a thread has finished resuming transactions. After that, Line 5–7 distribute read lock entries to $lock\text{-}queue$. Then each thread resumes all transactions blocked by the read lock. And a transac-

---

[1] popcount is an efficient algorithm for calculating the hamming weight of a bit array.

Schedule	Result	
(1) R5 < A2	A2 acquire r.lock	correct
(2) A2 < R5 < A5	(i) if A3 < R3, R4-5 acquire the dummy write lock	
	(ii) if R3 < A3 < R8, R8-9 acquire the dummy write lock or A5 acquire r.lock correct	
	(iii) if R8 < A3, A5 acquire r.lock for Txn B	
(3) A5 < R5 (imply A3 < R8)	(i) if A3 < R3, R4-5 acquire the dummy write lock	
	(ii) if R3 < A3, R8-9 acquire the dummy write lock correct	

Lock Release(t$_x$)	Lock Acquire(t$_y$)
R3: if r.write-wait ≠ 0	A2: if(0 ≠ cas(r.lock-state,
R4: nstate <- mask	0, ..)
R5: cas(r.lock-state, ..,	A3: set(r.write-waiter, ..)
nstate)	A5: if(0 = cas(r.lock-state,
R8: if r.write-wait ≠ 0	0, ..)
R9: cas(r.lock-state, ..,	A7: unset(r.writer-watier, ..)
mask)	

Fig. 12. Critical path of acquiring and releasing a lock

Fig. 13. All possible schedules

tion resumes its execution from Line 10 in Fig. 7. In the last, each thread calls atomic-unset(r.*read-waiter, thd*) to clear its wait bit and release the dummy read lock.

4.3   Deadlock

Deadlock happens when multiple transactions are waiting for the others to release a lock. Here we adopt a time-out strategy to prevent deadlocks from happening. It allows each transaction to wait for a lock within a limited time. When time runs out, a transaction is aborted. It introduces little maintaining overhead and does not generate many unnecessary aborts. The drawback is that a deadlock can not be identified and solved very fast. We consider that a well-designed application would not generate many deadlocks. Hence, the time-out strategy is chosen by iLock to handle deadlocks. In implementation, each thread periodically checks whether any local blocked transaction is timed-out. A transaction is resumed if it is timed-out. In Figs. 6 and 7, a timed-out transaction aborts itself after being resumed. There are also some other solutions, such as using a wait-graph or a wait-die policy [2]. But, maintaining the wait-graph introduces lots of CPU overhead. It is worthwhile when deadlock happen frequently. On the other hand, the wait-die policy would result in many unnecessary aborts.

4.4   Correctness

The security property requires that no conflict lock requests will be granted at the same time. Detailed speaking, two requests are considered to be conflict if they are for the same record and one is a write lock request. In Fig. 6, a write lock is acquirable only when *lock-state* $= 0$ (Line 1) or *lock-state* $= mask$ (Line 14). If r.*lock-state* $= 0$, the record is not locked by any others. If r.*lock-state* $= mask$, the record is just unlocked. In both cases, no one is holding the lock of the record. Hence, iLock does not break the security property.

Another correctness concern is that concurrent acquisition and releasing operations execute correctly. Figure 12 illustrates the scenario where a transaction t_x is about to release its write lock on r. Concurrently, transaction t_y tries to acquire the write lock. We must guarantee that the end of t_x will wake up t_y. Figure 13 gives all schedules resulting from the interleaving of the acquisition and releasing operations.

- *Case 1*, t_x releases its lock **(R5)** before t_y does its first locking attempt **(A2)**, then t_y must succeed in lock acquisition.
- *Case 2*, t_x releases its lock **(R5)** between the two locking attempts of t_y **(A2, A5)**. Here, either t_y directly acquire the lock in the second attempt; or t_x finds that t_y is waiting, acquires the dummy write lock and wake up t_y **(R8–9)**.
- *Case 3*, the lock is released **(R5)** after the second locking attempt from t_y **(A5)**, t_x must find that t_y is in-waiting, and then acquire the dummy write lock for t_y.
  In all, t_y is always able to acquire the write lock after t_x releases that. It is similar to verify the correctness of read lock acquisition/releasing.

5   Experiment

We implement an in-memory database prototype with 12,850 lines of C++ codes to verify the efficiency of our method. It is equipped with an in-memory B+-Tree, a query engine and a concurrency control component. All experiments are conducted on a cluster with 5 servers. Each has *two 2.00 GHz 6-Core E5-2620 processors, 192 GB* DRAM, connected by a *10 GB switch*. Four servers are used to simulate clients, and one is used to deploy the database.

Three methods are compared in the following experiments.

- *iLock* uses the SQL-To-Coroutine model and the new lock manager proposed.
- *Row-Locking* uses the SQL-To-Thread model and the simplified lock manager (row locking). If a lock conflict happens, the transaction gets its SQL request aborted and retried. The row locking mechanism is proposed and used by VLL [11] and pessimistic transactions in Hekaton [1].
- *Lock-Table* uses the Transaction-To-Thread model and a centralized lock table [5]. Such design is adopted by many disk-based database systems.

Each method uses a time-out mechanism to avoid deadlocks. The TPC-C and a micro benchmark are used to evaluate the performance of different methods. For the TPC-C, we ran a standard transaction mix. On the other hand, the micro benchmark has a table with two columns: key, 64-bit integer and value, 64-bit integer. A transaction would read N records and write M records ($N + M$ SQLs in total). Records are selected with a Zipfian distribution. It is used to generate workloads with varied read/write mixes and access distributions.

5.1   Varying Number of Clients

Figure 14 uses TPC-C to evaluate performances by varying the number of clients. 200 warehouses are populated in the database. Overall, *iLock* has the best performance. Its performance increases with more clients are used, and reaches the

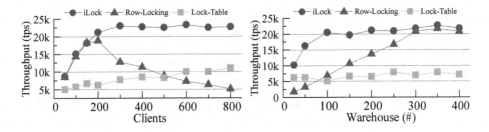

Fig. 14. TPC-C: varying clients   **Fig. 15.** TPC-C: varying warehouses

peak throughput (about 23k tps) when 300 clients are used. The performance stabilizes because the server has reached the maximal CPU usage. The performance of *Row-Locking* increases when a few clients are used. Its throughput is almost the same as that of *iLock*, because both methods have little CPU overhead if conflicts happen rarely. With the number of clients increased, the performance of *Row-Locking* sees a rapid decrease. It is because access conflicts happen more frequently now, and *Row-Locking* would waste a plenty of CPU time on retrying many failed SQL requests. In contrast, *iLock* is much more efficient. It would suspend a transaction and resume that until the lock is available. In the last, The performance of *Lock-Table* increases slowly. Its peak throughput is about 10k tps. The poor performance is mainly because much CPU time is consumed by context switching and accessing the heavy-weighted lock table. As a result, *Lock-Table* exhausts the CPU easily.

5.2   Varying Number of Warehouses

Next we increase the number of warehouses populated to decrease the workload contention. Figure 15 shows the results. 300 clients are running in default. Overall, *iLock* has the best performance. Its performance increases at the beginning, because when more warehouses are populated, fewer transactions would be blocked by lock conflicts. After the number of warehouses reaches 100, the performance of *iLock* converges as the CPU is fully utilized. The throughput of *Row-Locking* increases almost linearly against the number of warehouses. Under a small number of warehouses, *Row-Locking* is worse than *iLock* because the workload contains many conflicts and the former is less efficient than the latter in handling them. Using more warehouses reduces the probability of lock conflicts, and also reduces the total number of SQL requests retried by *Row-Locking*. As a result, CPU time is saved for processing real work. When 300 warehouses are used, *Row-Locking* reaches its peak throughput, which is similar to that of *iLock*. In the last, *Lock-Table* has a stable but low performance under each case. Decreasing the workload contention has little effect because the performance is always dominated by CPU usage.

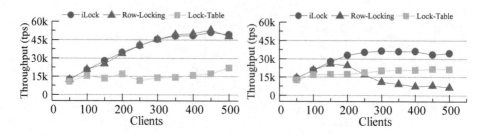

Fig. 16. Micro: 10 reads   **Fig. 17.** Micro: 5 reads + 5 writes

5.3 Varying Workload Characteristics

Figures 16 and 17 give micro benchmark performances using different mixes of read/write requests. The Zipfian distribution uses $\theta = 0.6$ to generate requests. Figure 16 uses a read-only workload and each transaction issues 10 read requests. Figure 17 uses a write-intensive workload and each transaction issues 5 read requests and 5 write ones. (1) Under the read-only workload, *iLock* and *Row-Locking* almost have the same performance. It is because there is no conflict in the workload. And both are very efficient in scheduling a conflict-free workload. Their peak throughputs reach 45k tps when the CPU is fully utilized. On the other hand, the throughput of *Lock-Table* is always about 15k tps under each case, which is only one-third the performance of *iLock*. Its poor performance results from spending much valuable CPU time on context switches and maintaining the lock table. (2) Under the write-intensive workload, when 500 clients are used, the performance of *iLock* is about 5.3x that of *Row-Locking*, and 1.6x that of *Lock-Table*. By increasing the number of clients, its performance gets stabilized because more transactions are blocked by lock conflicts. The performance of *Row-Locking* increases at the beginning and get decreased as adding clients results in more lock conflicts. *Lock-Table* shows the similar performance in both the read-only and the write-intensive workloads. The performance gap between *iLock* and *Lock-Table* is smaller than that in Fig. 16. It is because the transaction conflict is the dominating factor for the performance now.

Figure 18 gives throughputs by varying the skewness θ of the Zipfian distribution. Here, we simulate the same write-intensive workload used by Fig. 17. And 200 clients are running concurrently. Figure 19 displays the normalized performances of different methods against that of *iLock*. As we can see, throughputs of all methods are decreasing when the access distribution becomes more skewed. It is because some records become "hot" and being frequently accessed. Transactions are more likely blocked when accessing hot records. When $\theta < 0.6$, there is little contention in the workload, and performances of different methods are limited by the CPU usage. When $\theta > 0.6$, workload contention becomes non-negligible. Here, the throughput of *Row-Locking* drops faster than that of *iLock*. It is because *iLock* is more efficient in scheduling lock conflicts while *Row-Locking* wastes lots of CPU time on retrying failed lock requests. When

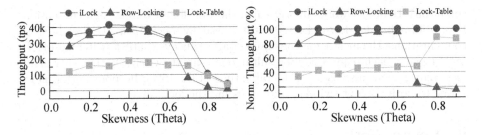

Fig. 18. Micro: skewness (raw)   **Fig. 19.** Micro: skewness (normalized)

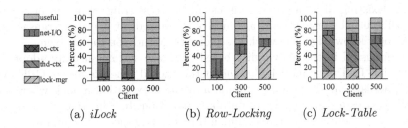

(a) *iLock*   (b) *Row-Locking*   (c) *Lock-Table*

Fig. 20. CPU time breakdown

$\theta \geq 0.8$, *iLock* and *Lock-Table* show very similar performance. At the moment, throughputs are largely limited by the workload contention. Figure 19 shows the performance gaps among different methods clearly. Given a workload containing low contention, a lightweight lock manager gains the better performance. Given a workload containing high contention, the lock manager, which is more efficient in resolving conflicts, works better. Overall, *iLock* shows nice performance under different kinds of workloads.

5.4 Breaking CPU Time down

In Fig. 20, we use VTune, a CPU profiling tool, to track where a working thread has its time gone. Here, we run the TPC-C workload with 200 warehouses. The number of clients is varied to change transaction conflicts in the workload. In Fig. 20(a), *iLock* spends little time on lock management (lock-mgr), context switching (thd-ctx) or coroutine switching (co-ctx). It means that *iLock* handles both network-related and conflict-related blocking efficiently. The network I/O (net-I/O) uses a considerable percent of CPU time. It mainly includes serializing results into a packet and informing I/O threads to respond clients. Most time (about 70%) are spent on real work (e.g. accessing the index, writing records). Figure 20(b) shows the CPU time used by *Row-Locking*. When 100 clients are running concurrently, the workload contains little contention, and *Row-Locking* spends most time on useful work. By increasing the number of clients, there are more conflicts in the workload. *Row-Locking* wastes a large percent of time (41.3%–53.2%) on locking (e.g. retrying failed lock requests, checking the lock condition). Hence, it works poorly when the workload exhibits high contention. Figure 20(c) breaks down the CPU time used by *Lock-Table*. Under different configurations, the most time-consuming part is always context switching (41%–58%) because each request could result in one context switch. On the other hand, lock management also uses considerable time (13%–16%). Its poor performance is due to both context switches and accessing the lock table.

6   Relate Work

Interactive transaction processing is rarely studied under in-memory database system. Existing disk-based systems usually process transactions in interactive way, while in-memory systems are mostly optimized for one-shot transactions. As our best knowledge, the work is a pioneering study on the problem.

As discussed in the above, disk-based database systems usually adopts the Transaction-To-Thread model. But the method could generate lots of CPU overhead. Pandis et al. [9] proposes the data-oriented execution model, where each thread services a disjoint part of database partition. It helps reduce contention because less resources are shared among threads. Transactions should be decomposed into small actions in advance based on the partition, so that each action runs on the proper thread. However, it does not work for interactive transaction since the pre-process is unfeasible.

Disk-based database systems schedule transactions with a centralized lock manager. It is proven to generate much contention on multi-core hardwares. Jung et al. [5] proposes a multi-core scalable lock manager by reducing latch usages. However, acquiring/releasing locks still consumes considerable CPU instructions. Johnson et al. [4] designs a lock inheritance mechanism to reduce the number of lock table access. A transaction can inherit locks directly from an ending one. The mechanism does not work if two transactions have no locks in common. The iLock we proposed here is a latch-free and lightweight structure. It has multi-core scalability and does not generate too much maintaining overhead.

In-memory databases are mostly designed for one-shot transactions. Some new execution models are proposed. Stonebraker [12] proposes a serial execution model, which queues and executes transactions one-by-one. Similarly, Thomson [13,14] proposes the deterministic transaction execution. Concurrency control is done before the real transaction execution. They take full advantage of CPU resources by eliminating all blocking during transaction processing. But, they are limited to handle only one-shot requests.

Many concurrency control schemas are designed for one-shot transactions processing. Ren et al. [11] and Larson et al. [8] design lightweight lock managers, which put the lock information in the header of a record. Tu et al. [15] improves the OCC by eliminating its centralized contention point. Wu et al. [18] adds a healing phase for OCC, which helps reduce operations retried when a transaction has its validation failed. Wang et al. [17] chops a transaction into multiple pieces in an offline phase, and isolates transactions in the degree of pieces. Efficiencies of these methods rely on the one-shot property. Some should analyze transaction logics in advance. And others would like transactions to be short and do not conflict with each other frequently. As a result, they are not suitable for scheduling interactive transactions.

7 Conclusion

This work studies the interactive transaction processing for the in-memory database system. We propose a new execution model and a new lock manager to efficiently process and schedule interactive transactions. Experiments on well-known benchmarks show our method achieves good performance in processing interactive workloads. A future direction is to design a transaction engine for processing the hybrid workload, containing both one-shot requests and interactive ones.

Acknowledgement. This is work is partially supported by National High-tech R&D Program(863 Program) under grant number 2015AA015307, National Science Foundation of China under grant numbers 61702189, 61432006 and 61672232, and Youth Science and Technology - "Yang Fan" Program of Shanghai under grant number 17YF1427800.

References

1. Diaconu, C., Freedman, C., Ismert, E., Larson, P.A., et al.: Hekaton: SQL server's memory-optimized OLTP engine. In: SIGMOD, pp. 1243–1254 (2013)
2. Gray, J.: Transaction Processing: Concepts and Techniques. Elsevier, Amsterdam (1992)
3. Harizopoulos, S., Abadi, D.J., Madden, S., Stonebraker, M.: OLTP through the looking glass, and what we found there. In: SIGMOD, pp. 981–992. ACM (2008)
4. Johnson, R., Pandis, I., Ailamaki, A.: Improving OLTP scalability using speculative lock inheritance. VLDB **2**(1), 479–489 (2009)
5. Jung, H., Han, H., Fekete, A., Heiser, G., Yeom, H.Y.: A scalable lock manager for multicores. TODS **39**(4), 29 (2014)
6. Knuth, D.E.: The Art of Computer Programming: Fundamental Algorithms, vol. 1. Addison Wesley Longman Publishing Co., Inc., Boston (1997)
7. Kung, H.T., Robinson, J.T.: On optimistic methods for concurrency control. TODS **6**(2), 213–226 (1981)
8. Larson, P.Å., Blanas, S., Diaconu, C., Freedman, C., Patel, J.M., Zwilling, M.: High-performance concurrency control mechanisms for main-memory databases. VLDB **5**(4), 298–309 (2011)
9. Pandis, I., Johnson, R., Hardavellas, N., Ailamaki, A.: Data-oriented transaction execution. VLDB **3**(1–2), 928–939 (2010)
10. Pavlo, A.: What are we doing with our lives?: nobody cares about our concurrency control research. In: SIGMOD, p. 3. ACM (2017)
11. Ren, K., Thomson, A., Abadi, D.J.: Lightweight locking for main memory database systems. VLDB **6**, 145–156 (2012)
12. Stonebraker, M., Madden, S., Abadi, D.J., et al.: The end of an architectural era: (it's time for a complete rewrite). In: VLDB, pp. 1150–1160 (2007)

13. Thomson, A., Abadi, D.J.: The case for determinism in database systems. VLDB **3**(1–2), 70–80 (2010)
14. Thomson, A., Diamond, T., Weng, S.C., Ren, K., et al.: Calvin: fast distributed transactions for partitioned database systems. In: SIGMOD, pp. 1–12 (2012)
15. Tu, S., Zheng, W., Kohler, E., Liskov, B., Madden, S.: Speedy transactions in multicore in-memory databases. In: SOSP, pp. 18–32 (2013)
16. Wang, T., Kimura, H.: Mostly-optimistic concurrency control for highly contended dynamic workloads on a thousand cores. VLDB **10**(2), 49–60 (2016)
17. Wang, Z., Mu, S., Cui, Y., Yi, H., Chen, H., Li, J.: Scaling multicore databases via constrained parallel execution. In: SIGMOD, pp. 1643–1658. ACM (2016)
18. Wu, Y., Chan, C.Y., Tan, K.L.: Transaction healing: scaling optimistic concurrency control on multicores. In: SIGMOD, pp. 1689–1704. ACM (2016)

An Adaptive Eviction Framework for Anti-caching Based In-Memory Databases

Kaixin Huang, Shengan Zheng, Yanyan Shen$^{(\boxtimes)}$, Yanmin Zhu, and Linpeng Huang

Shanghai Jiao Tong University, Shanghai, China
{kaixinhuang,venero1209,shenyy,yzhu,lphuang}@sjtu.edu.cn

Abstract. Current in-memory DBMSs suffer from the performance bottleneck when data cannot fit in memory. To solve such a problem, anti-caching system is proposed and with proper configuration, it can achieve better performance than state-of-the-art counterpart. However, in current anti-caching eviction procedure, all the eviction parameters are fixed while real workloads keep changing from time to time. Therefore, the performance of anti-caching system can hardly stay in the best state. We propose an adaptive eviction framework for anti-caching system and implement four tuning techniques to automatically tune the eviction parameters. In particular, we design a novel tuning technique called window-size adaption specialized for anti-caching system and embed it into the adaptive eviction framework. The experimental results show that with adaptive eviction, anti-caching based database system can outperform the traditional prototype by 1.2x–1.8x and 1.7x–4.5x under TPC-C benchmark and YCSB benchmark, respectively.

Keywords: In-memory database · Anti-caching · Database tuning

1 Introduction

In-memory DBMSs remove heavy components such as data buffers and locks, thus providing higher OLTP throughput than traditional disk-oriented DBMSs [1,24]. However, a fundamental problem of in-memory DBMSs is that the improved performance is only achievable when database is smaller than the amount of available physical memory. If database grows larger than main memory while executing transactions, operating system will start to page virtual memory, and accesses to main memory will cause page faults; the performance of in-memory DBMSs may suffer a rapid decrease. One widely adopted method to enhance performance is to apply a main memory distributed cache, such as Memcached [23], in front of a disk-based DBMS. Such implementations with a two-tier model, however, come with a problem of double data buffering, causing a serious waste of memory resources.

As a better solution, anti-caching system [4] is proposed. In an anti-caching based in-memory database, when memory is exhausted, the DBMS gathers the

© Springer International Publishing AG, part of Springer Nature 2018
J. Pei et al. (Eds.): DASFAA 2018, LNCS 10828, pp. 247–263, 2018.
https://doi.org/10.1007/978-3-319-91458-9_15

"coldest" tuples and writes them to disk with minimal translation from their main memory format, thereby freeing up space for more recently accessed tuples. As such, the "hotter" data resides in main memory, while the colder data resides on disk in the anti-cache portion. Unlike a traditional DBMS architecture, each tuple is in either memory or a disk block, but never in both places at the same moment. Anti-caching system is not bound with any specified database systems, it is a design architecture which can be applied to any in-memory DBMSs aimed at dealing with OLTPs [3].

However, the configuration for anti-caching system prototype is fixed. Eviction parameters such as eviction size, eviction threshold and eviction check interval, are all preset by DBMS using a configuration file. For different types of workloads, an in-memory DBMS with anti-caching cannot run at its best performance with these fixed eviction parameters. Furthermore, anti-caching system with a fixed eviction configuration cannot always work at a high performance level when transaction workloads change from time to time.

One natural method is to manually modify the anti-caching configuration file each time a new workload comes. For example, DBMSs such as Oracle and MySQL are equipped with tuning manuals for DBAs and users. But for performance tuning, manual method comes with obstacles such as inflexibility, inefficiency and time consumption.

The drawbacks of both fixed anti-caching configuration and labored manual tuning motivate us to develop an adaptive tuning design, which is able to support evicting data dynamically with respect to temporary workload. Instead of tuning components like buffer pool in traditional disk-oriented DBMSs, we tune eviction-related parameters for anti-caching in a main memory DBMS. The whole procedure is automatic and we name it adaptive eviction framework in this paper.

We modify a few existing tuning techniques in order to embed them into the adaptive eviction framework. However, it is hard for these methods to exert their full potential because none of them is anti-caching oriented.

To overcome such a challenge, we propose a novel tuning technique called window-adaption specialized for anti-caching system. By shrinking and extending the window size related to anti-caching eviction parameters according to the workloads and system information, unsuitable eviction parameters will be adaptively tuned to a more proper set.

Our contributions can be concluded as follows.

- We make an observation into the relationship between different workloads and anti-caching parameter configurations; we find that fixed eviction parameter configuration limits the potential of anti-caching system.
- To the best of our knowledge, we are the first to propose an adaptive eviction framework for anti-caching based in-memory databases.
- We implement a variety of tuning methods aimed for in-memory anti-caching system based on previous work and propose a novel tuning technique called window-size adaption with high efficiency.

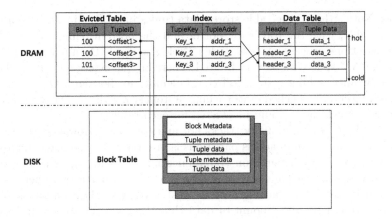

Fig. 1. Anti-caching architecture

- We conduct extensive experiments with standard TPC-C and YCSB benchmarks. The experimental results show that with adaptive eviction framework, anti-caching based database system can outperform the previous one by 1.2x–1.8x and 1.7x–4.5x under TPC-C benchmark and YCSB benchmark, respectively.

The remainder of this paper is organized as follows. Section 2 introduces the background of anti-caching system and the motivation of our proposed adaptive eviction framework. Section 3 provides the design of adaptive eviction framework and introduces four tuning techniques for adaptive eviction. The experimental results are presented in Sect. 4 and we review the related work in Sect. 5. The paper is concluded in Sect. 6.

2   Background and Motivation

2.1   Anti-caching Background

Anti-caching system aims at alleviating the pressure of in-memory DBMS while the data grows larger than memory space with transaction processing. The basic idea is to manage data in tuple granularity instead of page granularity. Since a data tuple is always much smaller than a page size (usually 4 KB), tuple-grained eviction can avoid evicting hot tuples to disk when a page includes both hot and cold tuples but considered to be a cold page by OS.

Figure 1 shows the storage architecture of anti-caching system, which includes four main components: Data Table, Index Table, Evicted Table and Block Table. Among them, Data Table, Index Table and Evicted Table reside in DRAM while Block Table is in disk.

Similar to other in-memory DBMSs, in an anti-caching based database system, the whole database stays in the memory when transactions begin. All the data tuples reside in Data Table. The Index Table is built by sampling the tuple accesses. Notice that Data Table stores data tuples in a LRU linked list for each database table, with the top being hotter and the bottom being colder. Along with transactions processing, both Index Table and Tuple Data may grow larger and the DRAM space can be very limited. Therefore, an eviction decision should be made by anti-caching system to free up memory for hot data. Once the eviction check interval is reached, anti-caching system scans all the Data Tables and decides how much data should be evicted for each table.

Anti-caching system globally tracks hot and cold data. However, the cost of maintaining a single chain across partitions is prohibitively expensive due to the added overhead of inter-partition communication. Instead, anti-caching maintains a separate LRU Chain per table that is local to a partition. Therefore, in order to evict data, anti-caching system must determine (1) from which tables to evict data and (2) the amount of data that should be evicted from a given table. In its current implementation, the DBMS answers these questions by the relative skew of accesses to tables. The amount of data accessed at each table is monitored, and the amount of data evicted from each table is inversely proportional to the amount of data accessed in the table since last eviction. Therefore, the hotter a table is, the less data will be evicted from it.

Block Table is a disk-resident hash table that stores evicted tuples in block format. Since Block Table stores relatively colder data, it's less accessed than the memory-resident Data Table. Data tuples stay in either DRAM or disk, thus the memory space occupation and coherence maintaining overhead is efficiently reduced. To track the tuples in Block Table, Evicted Table is designed to map evicted tuples to block ids. When a transaction requires data not in Data Table, the database system then looks up the Evicted Table, obtains its corresponding block id, and finally locates the binding block address for data access.

Main memory DBMSs, like H-Store [2], owe their performance advantage to processing algorithms which assume that data is in main memory. But any system will slow down if a disk read must be processed in the middle of a transaction. Anti-caching avoids stalling transaction execution at a partition whenever a transaction accesses an evicted tuple by applying a pre-pass process [4].

2.2 Motivation

With proper configuration, anti-caching system can behave much better than traditional disk-oriented DBMSs with large-than-memory database, due to its fine-grained eviction data control. However, the eviction parameters of anti-caching system prototype is fixed, thus unsuitable for multiple workloads or changing workload. The key parameters of anti-caching system fall on three: eviction threshold, eviction size and eviction check interval. Their functions are: deciding when to evict, how much to evict and how often should the system check if eviction is needed, respectively. In this section we give a few observations of the fixed configuration performance for anti-caching system and then analyze some typical cases to give insight about the drawbacks of fixed eviction parameter configuration.

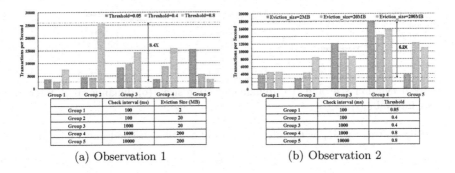

| (a) Observation 1 | (b) Observation 2 |

Fig. 2. Performance range with different eviction parameter groups

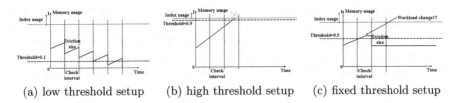

(a) low threshold setup   (b) high threshold setup   (c) fixed threshold setup

Fig. 3. Anti-caching with different types of threshold setup

Observations: We conduct a variety of experiments using the same type of YCSB benchmark to observe the limitations of fixed eviction configuration. Figure 2 shows that under different anti-caching configuration for three eviction parameters listed above, the performance of anti-caching system distinguishes from each other a lot. We can see that the best versus the worst ratio on transaction performance is 8.4x, 6.2x for each. We notice that the best performance group in Fig. 2(a) is <Check_interval = 100 ms, Eviction_size = 20 MB, Threshold = 0.8>. However, for other groups with Check_interval = 100 ms, Eviction_size = 20 MB, their performance seems to be very poor. It might be considered that the threshold makes sense. The <Check_interval = 10000 ms, Eviction_size = 200 MB, Threshold = 0.8> group, however, just gives the extreme reverse result. Therefore, it's hard to find the optimized eviction parameter group pattern for a certain workload. The deeper reason is that the performance of anti-caching system is not bound with a certain parameter group, but tightly related to workload state (e.g., skew, read/write ratio) and system state (e.g., memory usage). **In conclusion, we argue that fixed eviction parameter configuration is improper for anti-caching system.**

Case Analysis: In this part we offer some cases to further explain why fixed eviction parameter configuration is improper for anti-caching system. To make the explanation simpler, we assume that the index size is stable (even though it can grow larger and larger under real workloads).

Figure 3 shows the case how different types of eviction threshold configuration can affect the anti-caching processing. Figure 3(a) illustrates that when low

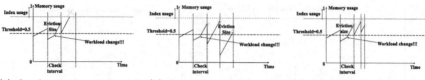

(a) fixed eviction size and   (b) dynamic eviction size   (c) dynamic check interval
check interval

Fig. 4. Anti-caching with different types of eviction size and interval setup

threshold is set up by a DBA or system maintainer, the memory utilization rate would be very low. Even if the user workload is slow and steady without too much data expansion in memory, eviction process will still be invoked frequently, naturally deriving the conclusion that such a fixed low-threshold configuration is not optimized. Figure 3(b) shows the case when an improper high threshold is set up. It might be considered that high threshold can better utilize the available memory. However, once the total database memory is not enough, page swap will happen automatically. Since page swap is transparent to user transactions, when the anti-caching engine checks whether all the data tuples needed for a specific transaction, it may consider those tuples in pages which are swapped before still reside in memory. However, such accesses need disk I/O indeed, thus resulting in much longer latency. If a high threshold is set up, then fewer evictions will occur with the price of page swap during a transaction execution. We argue that such a fixed high-threshold configuration is also not proper. Figure 3(c) shows how a fixed threshold fails to utilize available memory when meeting with workload changes, say, from write-heavy to read-heavy. The main difference between these two workloads in anti-caching environment is that the latter workload results in much fewer data appended into memory, thus only a little more memory is used in addition. In such cases, the available memory can be abundant. However, the anti-caching system just performs conservatively with regular evictions. We argue that the fixed eviction threshold configuration is not optimized.

Figure 4 shows how eviction size and check interval can affect the transaction performance. Figure 4 shows the case when fixed eviction size and check interval are set up. For example, if a normal read-heavy workload turns into a write-heavy workload, then the data in memory will increase very quickly. Page swap might happen during a transaction processing and the database performance can suffer a rapid descending. To better illustrate the inefficiency of fixed configuration for eviction size and check interval, we compare it with the cases that either eviction size or check interval changes along with the workload change. These two cases are shown in Fig. 4(b) and (c), respectively. With workload-aware adjustment for eviction size and check interval, page swap can be avoided to some extent, thus maintaining a high transaction performance. Therefore, we argue that both fixed eviction size and check interval are not optimized.

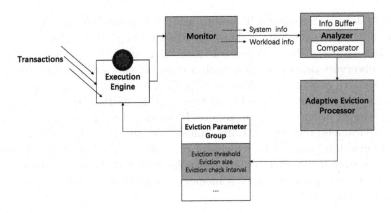

Fig. 5. Adaptive eviction framework

3   Adaptive Eviction

3.1   Overview

Figure 5 presents the framework of adaptive eviction. The three important components are Monitor, Analyzer and Adaptive Eviction Processor (AEP). The Monitor is responsible for gathering information about transaction workload and system state. Workload information includes read/write times, query numbers in a certain time period, accessed tuple numbers; system information includes real-time transaction performance and memory usage. The raw information collected by Monitor is conveyed to Analyzer for further computation. For example, the read/write ratio of the workload is computed by $rw\_ratio = read\_number/total\_access\_number$; the access skew is computed by $access\_skew = total\_access\_number/access\_tuple\_number$. The Info Buffer of Analyzer is used to store the historic workload and system information while the Comparator is used to compare its current results with the history, which implies whether there is a change of workload characteristics or system state. The result computed by Analyzer is then transferred to AEP. AEP is the core tuning model of adaptive eviction framework. We can implement different tuning techniques in AEP to modify the eviction parameter group for anti-caching based database systems.

Algorithm 1 further introduces the work flow of adaptive eviction framework. First, necessary components go through either startup or initiation. During transactions processing, the Monitor keep tracking the information of both system and workload, and then transfer them to Analyzer. Once the eviction check interval is reached, Analyzer deals with the workload/system information data received during the check interval period. It counts the access skew, read/write ratio and transaction rate. Then it makes a comparison of these data with corresponding ones computed last time to tell whether eviction parameters should be reconfigured or not. In our current implemented version of adaptive eviction framework, when either one of the following two conditions are satisfied, a reconfiguration action should be performed: (1) available memory space change is over 10%; (2) workload state change is over 10% (e.g., 10% write-heavier, 10% more skew access). The choice of our threshold 10% is based on a few experimental observations, which show such a configuration outperforms than most other ones. We expect to explore more critical knowledge about the choice of threshold representing changes in our future work. AEP is the core component which performs actual tuning procedure. It decides the concrete method of how eviction parameters change. We implement four tuning techniques as AEP to tune the eviction parameters, which are presented in next part of this section. Notice that even if the eviction parameters have been changed, the data eviction do not need to happen if only the memory occupation of data is smaller than the eviction threshold.

Algorithm 1. *AdaptiveAntiCaching*

---

1 $P \leftarrow default\_eviction\_parameters$
2 **while** *transaction_state is ON* **do**
3 | Analyzer $\leftarrow$ Monitor.getInfo()
4 | **if** *Monitor.timer.last() == P.check_interval* **then**
5 | | Analyzer.compute()
6 | | Analyzer.compare()
7 | | AEP $\leftarrow$ Analyzer.changeInfo()
8 | | **if** *AEP.shouldChangeConfig() is True* **then**
9 | | |_ AEP.tune(P)
10 | | **if** *data_memory > P.threshold* **then**
11 | | |_ Evict(database, P);
12 |_   Monitor.timer.continue()

3.2   Tuning Techniques

We design our adaptive eviction framework as a pluggable platform for equipping various tuning techniques. Four tuning methods are implemented in our study. They are simple-rule based tuning (SRB), experiment-reflected tuning (ER), candidate block replacement tuning (CBR) and window-size adaption tuning (WSA). Among them, the first three refer to previous research [12–14] and WSA is an efficient tuning technique we propose for better adapting to anti-caching system. Next we will introduce each of the tuning technique with more details.

SRB: Simple Rule-Based. Pavlo et al. [14] introduce machine learning into in-memory database to tune the indexes, views, storage layout and etc. We adopt the simpler off-line machine learning version to obtain the patterns with certain <workload info, system info, eviction parameter set> format. It is fast to make a tuning decision and can often work better than the default configuration of anti-caching. However, it is not accurate indeed and can suffer serious performance degrading in a few cases.

Table 1. Design issues for different tuning techniques

	SRB	ER	CBR	WSA
In-memory DBMS oriented	Yes	No	No	Yes
Anti-caching oriented	No	No	No	Yes
Tuning time	Short	Long	Middle	Short
Memory overhead	Small	Neutral	Large	Small
Performance improvement	Low	Middle	Middle	High

ER: Experiment-Reflected. Duan et al. [13] proposes this feedback-driven tuning method based on adaptive sampling. Adaptive sampling analyzes the samples collected so far to understand how the surface looks like (approximately), and where the good settings are likely to be. Based on this analysis, more experiments are done to collect new samples that add maximum utility to the current samples.Although this technique is able to be quite adjacent to the optimized eviction parameter values for single-type workload, it is time-consuming and suffer from serious performance bottleneck under changing workload.

CBR: Candidate Block Replacement. Storm et al. [12] describes such a self-tuning approach based on cost-benefit analysis. This method previously attempts to solve the problem of developing a database-wide, memory-tuning algorithm by considering each of the memory consumers such as compiled cache-pool, buffer-pool, sort-buffer, etc., wherein each has a different usage. It accumulates the cost savings in processing time for each component with which a database process is interacting with the memory subcomponent. This technique is also called shared memory management technique (SMMT) and has been incorporated in IBM's DB2. It behaves in a block-grained memory replacement, thus losing the flexibility of tuple-grained data control.

WSA: Window-Size Adaption. The idea of window-size adaption comes from the TCP flow control mechanism in network communication. Figure 6 shows the implementation of WSA. The aim of WSA is to balance the trade-off between utilizing memory space and avoiding page swap. If more data can reside in the memory, access to disk will become less, thus promoting the overall transaction performance and decreasing the average delay. However, if the system greedily holds too much data tuples in DRAM, the sum of space used by indexes, data tuples, buffers and evicted tables may exceed the available memory space, thus causing heavy OS page swap. To explore the full potential of anti-caching, we design two phases for WSA. One is Relaxation Phase, during which transaction burden is not heavy and memory space is sufficient, anti-caching system tries to hold more data tuples in the memory and reduce the tuning overhead (i.e., decrease eviction size, increase eviction threshold and check interval); the increasing/decreasing ratio is chosen to be 0.1 in our current implementation version. The other is Shrinking Phase, where transaction burden is detected to be heavy or memory resource is inadequate, anti-caching system makes a radical change

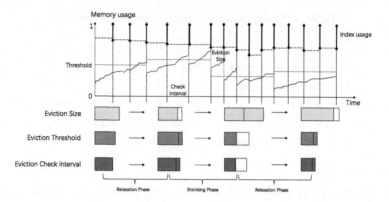

Fig. 6. Window-size adaption

for its eviction parameters (i.e., double eviction size, halve eviction threshold and check interval). Notice that for most workloads the Relaxation Phase can improve transaction behavior with optimizing the eviction parameters gradually and this process is slow and steady. As for Shrinking Phase, it changes rapidly to avoid the danger of OS page swap by sacrificing the memory usage and adding more frequent checking overhead. However, compared to access tuples in the method of OS page swap, it is more reasonable to evict data in advance to save memory space. This is because anti-caching itself allow asynchronously fetching disk-resident tuples while executing next transactions which only concern data in memory. By taking advantage of this intrinsic feature of anti-caching system, WSA becomes a well-suited tuning method in our study.

Table 1 makes a conclusion of four tuning techniques used in our adaptive eviction framework. Compared the other three tuning techniques, WSA is anti-caching oriented, thus it can obtain considerable performance improvement with small overhead and short tuning time. SRB is also a fast-tuning technique with small memory occupation. However, it is unaware of the dynamic work-load/system changing, which limits its ability. As for ER and CBR, they are previously designed for disk-oriented DBMSs to optimize buffer resources, thus cannot fully exert their potential for anti-caching based database systems.

4   Experiments

We implement our adaptive eviction framework in H-Store and compare its performance against traditional anti-caching system. Four kinds of adaptive eviction techniques introduced in Sect. 3.2 are tested in our experiments. We first describe the two benchmarks and the DBMS configurations used in our analysis.

4.1   Benchmarks

TPC-C: This benchmark is the current industry standard for evaluating the performance of OLTP systems. It consists of nine tables and five procedures that simulate a warehouse-centric order processing application. Only two of these procedures modify or insert tuples in the database, but they make up 88% of the benchmark's workload. For our experiments, we use a 10 GB TPC-C database containing 100 warehouses and 100,000 items. For this benchmark, we set the available memory to the system to 12 GB. As the benchmark progresses and more orders accumulate, the data size will continue to grow, eventually exhausting available memory, at which point the anti-caching system will begin evicting cold data from the data tables to disk.

YCSB: The Yahoo! Cloud Serving Benchmark is a collection of workloads that are representative of large-scale services created by Internet-based companies. For all of the YCSB experiments in this paper, we use a 20 GB YCSB database containing a single table with 20 million records. Each YCSB tuple has 10 columns, each with 100 bytes of randomly generated string data. The workload consists of two types of transactions; one that reads a single record and one that updates a single record. We use the write-heavy transaction workload mixtures (i.e., 50% reads/50% updates). We also vary the amount of skew in workloads to control how often a tuple is accessed by transactions. In our experiments, we use a Zipfian skew with values of s between 0.5 and 1.5.

4.2   System Setup

We deploy latest H-Store with our adaptive eviction framework on a single node with a dual socket Intel Xeon E5-2620 CPU (12 cores per socket, 15M Cache, 2.00 GHz) processor running 64-bit Ubuntu Linux 14.04. All transactions are executed with a serializable isolation level. The benchmark clients in each experiment are deployed on a separate node in the same cluster. In each trial except one that tests the connection between performance change and running time, H-Store is allowed to "warm-up" for two minutes. During the warm-up phase, transactions are executed as normal but throughput is not recorded in the final benchmark results. For H-Store, cold data is evicted to the anti-cache and hot data is brought into memory. After the warm-up, each benchmark is run for a duration of ten minutes, during which average throughput is recorded. The final throughput is the number of transactions completed in a trial run divided by the total time (excluding the warm-up period). Each benchmark is run five times and the throughputs from these runs are averaged for a final result. To properly control the data size for experimental presentation, we write extra codes for benchmark testing. Once the goal (e.g., 2x memory size) is achieved for one experiment, new data will not be generated throughout this experiment. For each trial we test the performance of six database configurations: one is pure DBMS without anti-caching system (i.e., No AC), another is anti-caching system without eviction parameter configuration as baseline (i.e., Default), the other four are anti-caching systems equipped with our proposed adaptive eviction framework, which includes four different tuning techniques (i.e., SRB, ER, CBR, WSA). We don't perform manually-tuned experiments because such a method fails to access the optimal performance, due to the random changing patterns and slow human reactions to the OLTP workloads.

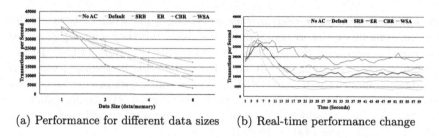

(a) Performance for different data sizes   (b) Real-time performance change

Fig. 7. Adaptive eviction performance under TPC-C benchmark

4.3   Results and Analysis

We now discuss the results of executing two benchmarks with our adaptive eviction framework across different size configurations and workload skews.

TPC-C: The results for running the TPC-C benchmark are shown in Fig. 7. We first examine the performance of different tuning methods with the change of data size, whose result is depicted in Fig. 7(a). It can be simply observed that with data size being larger and larger, the performance of DBMS degrades again and again. This is because more data tuples are evicted to disk thus more disk I/Os should be taken. When the data size can just fit into memory (data/memory = 1), DBMS with no anti-caching behaves the best because the overhead of eviction process is removed. However, when data size become larger, the performance of DBMS without anti-caching decrease rapidly. It can be concluded that using tuning methods with our adaptive eviction framework can obtain higher average throughputs than anti-caching system with default eviction parameter configuration. Among the four tuning techniques, window-size adaption is the best and it beats default anti-caching system in transaction throughput with 21%, 52% and 118% under 2x, 4x and 8x memory data size, respectively. Compared with other tuning techniques, WSA wins 43% to 75% when the data size is 8x memory. WSA can efficiently balance the trade-off between utilizing memory space and avoiding page swap. Therefore, it is more proper than the other implemented tuning methods for anti-caching system.

Figure 7(b) shows the performance change of each tuning technique along with time. Notice that we just choose the first 60 s to give explanations for better visual effect. The time-sequential throughput change can tell more detailed information about what happens for each tuning technique while transactions are being executed. When transactions just begin, the memory space is sufficient and DBMS without anti-caching works at its best state with fetching all the target tuples from memory. Since no eviction threshold is limited for it, the performance will not be worse until OS page swap happens. However, once the data size start to exceed the available memory, page swap can occur frequently for DBMS without anti-caching and the performance goes down rapidly and finally reaches a stable state. Other anti-caching choices, start with relatively lower transaction throughputs, grow steadily until the same bottleneck occur

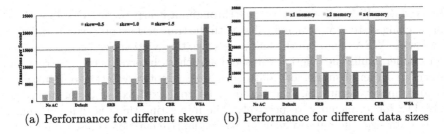

(a) Performance for different skews   (b) Performance for different data sizes

Fig. 8. Adaptive eviction performance under YCSB benchmark

for them: the data size become too large and only a small fraction of them can reside in memory. The interesting thing is that compared to the baseline, all tuning techniques have a jump-after-fall phenomenon. For example, WSA experiences a performance fall from 10 s to 12 s and just after this, it has an obvious performance jump to 17 s. Another instance is for the CBR tuning method. Its performance falls from 8 s to 15 s and jumps from 15 s to 22 s. The reason for the fall-and-jump style is simple: when performance degrading is captured by Analyzer in our adaptive eviction framework, a tuning decision should be made by AEP, configuring the eviction parameters to a relatively proper set. After such a procedure, anti-caching system fits more to the current workload than before and obtains a performance growth, which we regard as a jump action. Among the four tuning techniques, WSA's stable state is the highest and it gains 1.25x better performance than the baseline.

YCSB: The results in Fig. 8 are for running the YCSB benchmark with write-heavy workload across a range of workload skews and data sizes. Figure 8(a) shows the performance difference of each tuning technique with different workload skews. We can observe that under the workload of same skew, anti-caching system can obtain higher average transaction throughput by taking advantage of our proposed adaptive eviction framework. With the skew becoming higher, more transaction accesses reach the same group of hot data in memory, DBMS can obtain a natural throughput improvement. However, in the case of low skew, anti-caching system with default eviction parameter configuration behaves poorly and is only 52% better than no anti-caching choice. While using adaptive eviction, the performance can obtain considerable improvement, up to 1.7x–4.5x compared with the baseline. In particular, under low-skew workload, WSA also beats other tuning techniques by 2.6x, 2.2x and 2.3x compared with SRB, ER and CBR, respectively. The reason is that when transaction accesses randomly fall into data tuples, WSA gently changes its window size to fit more cold data in memory while promising that the memory resource is not over-used. It keeps more access in memory rather than evicts a large amount to disk. In this way WSA fits anti-caching system better than other tuning techniques.

Figure 8(b) presents the performance behaviors of different tuning techniques for three data size setups. The performance difference among all the anti-caching choices in the 1x memory data size scenario is slight, but it can still be clearly observed that WSA obtains an average throughput nearly to pure DBMS without anti-caching system. We can infer that with the transactions proceeding, different windows of WSA behave in the following style: the eviction size becomes smaller and smaller, while the eviction threshold and check interval becomes larger and larger. In larger-than-memory data scenarios, adaptive eviction obtains more obvious improvement than the baseline, up to 1.2x–1.8x for 2x memory and 2.3x–4.6x for 4x memory, respectively. The interesting thing is that with much larger data size, the improvement of using adaptive eviction is also larger. This is reasonable because with adaptive eviction framework, anti-caching system is able to leave more hot data in memory with more proper eviction parameters.

5   Related Work

Anti-caching Data Management. [3] concludes different kinds of "anti-caching" data management mechanisms and divides them into three categories: user-space, kernel space, hybrid of user- and kernel-space. H-Store anti-caching [4] falls into the user-space approach. Project Siberia [6–8] also adopts a user-space "anti-caching" approach for Hekaton [5]. Instead of maintaining an LRU like H-Store anti-caching, Siberia performs offline classification of hot and cold data by logging tuple accesses first, and data in Hekaton is evicted to or fetched from a cold store in disk. Kernel-space approaches mainly refers to OS paging, which is an important part of virtual memory management in most contemporary general purpose operating systems. OS paging tracks data access in the granularity of pages and it allows a program to process more data than the physically available memory [9]. As for hybrid of user- and kernel-space approach, efficient OS paging [10] and access observer method [11] are proposed to better assist the hot/cold data classification. Our work is based on H-Store Anti-caching as it is the state-of-the-art in-memory DBMS data management architecture and deals with workload online to serve multiple application scenarios.

Performance Tuning. Performance tuning of database systems has been an interesting and active area of research in the last three decades and recent trend has been in developing self-tuning database systems with little or no human intervention. Several methods have been proposed in the literature [12,15,16] to implement self-tuning techniques ranging from use of histograms, gradient descent technique, creation of index, use of materialized views, etc. There have been several attempts to self-manage the DBMS memory [13,17–19] for improved performance. Oracle 10 g uses automated shared memory management (ASMM) to resize the subcomponents of the shared memory pool based on current workload. When switched on, the ASMM controls the sizes of certain components in the SGA by making sure that the workload gets the memory it needs. It does that by shrinking the components which are not using all of the memory allocated to them, and growing the ones which need more than the allocated memory. Microsoft SQL Server also has an automatic memory tuning manager.

However, none of these techniques can fully exert their potential for anti-caching architecture because of either disk-oriented design or heavy and redundant buffer components.

Workload Characterization. The key to successful implementation of a self-tuning database system is the knowledge base [21,22] on two aspects. One is helping the system to identify important performance bottlenecks. The other is collecting information about tuning impact of each tuning parameter on the system performance under different workloads and user load conditions. There have been attempts to identify key tuning parameters that have significant tuning impact on performance. [20] has presented the impact of various tuning parameters on the performance and the parameters are ranked using statistical approach. In our study we consider that the workload characteristics are bound with anti-caching memory environment, since the workload itself does not really affect the system performance.

6   Conclusion

In this paper, we propose an adaptive eviction framework for anti-caching based in-memory databases to figure out the problems caused by fixed eviction parameter configuration. With adaptive eviction for anti-caching system, in-memory DBMS is able to collect workload and system information to adaptively adjust the eviction parameters, taking more advantage of anti-caching system. We propose a novel window-size adaption strategy based on our designed general adaptive eviction framework; by extending and shrinking the eviction parameters along with the workload characteristics and system information change, an in-memory DBMS can smartly avoid weak memory utilization or slow OS page swap. The experimental results show that with adaptive eviction, an anti-caching based database system can obtain higher transaction performance under both TPC-C and YCSB benchmarks. In particular, window-size adaption tuning technique can outperform the base line up to 2.2x and 4.5x under TPC-C and YCSB benchmark, respectively. We conclude that for OLTP workloads, the results of this study demonstrate that adaptive eviction can efficiently improve the transaction performance of in-memory DBMS with anti-caching system.

Acknowledgment. This research is supported in part by 863 Program (no. 2015AA015303), NSFC (no. 61772341, 61472254, 61170238, 61602297 and 61472241), Singapore NRF (CREATE E2S2), and 973 Program (no. 2014CB340303). This work is also supported by the Program for Changjiang Young Scholars in University of China, and the Program for Shanghai Top Young Talents.

References

1. Harizopoulos, S., et al.: OLTP through the looking glass, and what we found there. In: Proceedings of the 2008 ACM SIGMOD International Conference on Management of Data. ACM (2008)
2. Kallman, R., et al.: H-store: a high-performance, distributed main memory transaction processing system. Proc. VLDB Endow. **1**(2), 1496–1499 (2008)
3. Zhang, H., et al.: Anti-caching based elastic memory management for big data. In: 2015 IEEE 31st International Conference on Data Engineering (ICDE). IEEE (2015)
4. DeBrabant, J., et al.: Anti-caching: a new approach to database management system architecture. Proc. VLDB Endow. **6**(14), 1942–1953 (2013)
5. Diaconu, C., et al.: Hekaton: SQL server's memory-optimized OLTP engine. In: Proceedings of the 2013 ACM SIGMOD International Conference on Management of Data. ACM (2013)
6. Eldawy, A., Levandoski, J., Larson, P.-Å.: Trekking through Siberia: managing cold data in a memory-optimized database. Proc. VLDB Endow. **7**(11), 931–942 (2014)
7. Levandoski, J.J., Larson, P.-Å., Stoica, R.: Identifying hot and cold data in main-memory databases. In: 2013 IEEE 29th International Conference on Data Engineering (ICDE). IEEE (2013)
8. Alexiou, K., Kossmann, D., Larson, P.-Å.: Adaptive range filters for cold data: avoiding trips to Siberia. Proc. VLDB Endow. **6**(14), 1714–1725 (2013)
9. Tanenbaum, A.S.: Modern Operating System. Pearson Education Inc., Upper Saddle River (2009)
10. Stoica, R., Ailamaki, A.: Enabling efficient OS paging for main-memory OLTP databases. In: Proceedings of the Ninth International Workshop on Data Management on New Hardware. ACM (2013)
11. Funke, F., Kemper, A., Neumann, T.: Compacting transactional data in hybrid OLTP&OLAP databases. Proc. VLDB Endow. **5**(11), 1424–1435 (2012)
12. Storm, A.J., et al.: Adaptive self-tuning memory in DB2. In: Proceedings of the 32nd International Conference on Very Large Data Bases. VLDB Endowment (2006)
13. Duan, S., Thummala, V., Babu, S.: Tuning database configuration parameters with iTuned. Proc. VLDB Endow. **2**(1), 1246–1257 (2009)
14. Pavlo, A., et al.: Self-driving database management systems. In: CIDR (2017)
15. Benoit, D.G.: Automatic diagnosis of performance problems in database management systems. In: Proceedings of the Second International Conference on Autonomic Computing, ICAC 2005. IEEE (2005)
16. Tran, D.N., et al.: A new approach to dynamic self-tuning of database buffers. ACM Trans. Storage (TOS) **4**(1), 3 (2008)
17. Chen, A.N.K.: Robust optimization for performance tuning of modern database systems. Eur. J. Oper. Res. **171**(2), 412–429 (2006)
18. Xu, J.: Rule-based automatic software performance diagnosis and improvement. Perform. Eval. **69**(11), 525–550 (2012)
19. Jeong, J., Dubois, M.: Cache replacement algorithms with nonuniform miss costs. IEEE Trans. Comput. **55**(4), 353–365 (2006)
20. Debnath, B.K., Lilja, D.J., Mokbel, M.F.: SARD: a statistical approach for ranking database tuning parameters. In: IEEE 24th International Conference on Data Engineering Workshop, ICDEW 2008. IEEE (2008)

21. Melcher, B., Mitchell, B.: Towards an autonomic framework: self-configuring network services and developing autonomic applications. Intel Technol. J. **8**(4), 279–290 (2004)
22. Wiese, D., Rabinovitch, G.: Knowledge management in autonomic database performance tuning. In: Fifth International Conference on Autonomic and Autonomous Systems, ICAS 2009. IEEE (2009)
23. Fitzpatrick, B.: Distributed caching with memcached. Linux J. **2004**(124), 5 (2004)
24. DeWitt, D.J., et al.: Implementation techniques for main memory database systems. **14**(2) (1984)

Efficient Complex Social Event-Participant Planning Based on Heuristic Dynamic Programming

Junchang Xin[1(✉)], Mo Li[1], Wangzihao Xu[2], Yizhu Cai[1], Minhua Lu[3], and Zhiqiong Wang[2]

[1] School of Computer Science and Engineering, Northeastern University, Shenyang, China
xinjunchang@mail.neu.edu.cn
[2] Sino-Dutch Biomedical and Information Engineering School, Northeastern University, Shenyang, China
[3] College of Biomedical Engineering, Shenzhen University, Shenzhen, China

Abstract. To manage the Event Based Social Networks (EBSNs), an important task is to solve the Global Event Planning with Constraints (GEPC) problem, which arranges suitable social events to target users. Existing studies are not efficient enough because of the two-step framework. In this paper, we propose a more efficient method, called Heuristic-DP, which asynchronously considers all the constraints together. Using this method, we improve the computational complexity from $O(|E|^2 + |U||E|^2)$ to $O(|U||E|)$, where $|U|$ is the number of users and $|E|$ is the number of events in an EBSN platform. We also propose an improved heuristic strategy in one function of the heuristic-DP algorithm, which slightly increases the time cost, but can obtain a more accurate result. Finally, we verify the effectiveness and efficiency of our proposed algorithms through extensive experiments over real and synthetic datasets.

1 Introduction

In recent years, Event Based Social Network (EBSN) platforms, such as Meetup[1] and Plancast[2] increasingly show its importance in citizens' daily life, specially with the popularity of *Online to Offline* (O2O) services [18,22]. These platforms help users to online create and manage social events, and make personalized plans for offline joining, which is increasingly attracting attention from both industry and academia [12]. Meetup, as the largest current EBSN platform, for example, has 16 million registered users and involved in aggregate 300,000 events held each month. Thus, how to efficiently make suitable plans for such a large number of users and events becomes an urgent problem.

---

[1] http://www.meetup.com/.
[2] http://plancast.com/.

© Springer International Publishing AG, part of Springer Nature 2018
J. Pei et al. (Eds.): DASFAA 2018, LNCS 10828, pp. 264–279, 2018.
https://doi.org/10.1007/978-3-319-91458-9_16

Planning events for users' participation over EBSNs is first proposed by She et al. in [16]. Aiming at the limitation that they failed to consider the lower bound of participants of each event, Cheng et al. in [2] proposed the GEPC problem. We use the following example to describe the GEPC problem in detail.

Example 1 (The GEPC Problem). Figure 1 is a 2-D grid, which shows the locations of users and events. For each user, a utility score is assigned to each event, which shows his/er interest to each event. These utility scores are shown in Table 1. Each user provides a travel budget, which is attached in the parenthesis after each user in Row 1 of Table 1. Events together with their respective participation lower and upper bounds are shown in column 1, and their start and end times are shown in column 10.

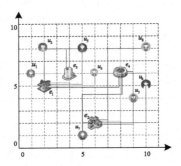

Fig. 1. Location of events and users

Table 1. Utility between events and users and time of events and utility scores

$e_j(\xi_j, \eta_j)$	$u_1(10)$	$u_2(12)$	$u_3(13)$	$u_4(15)$	$u_5(17)$	$u_6(19)$	$u_7(20)$	$u_8(23)$	Time
$e_1(1,5)$	**0.3**	**0.4**	0.8	0.9	**0.8**	**0.4**	0.5	0.3	1:00–2:00 p.m.
$e_2(1,5)$	0.2	0.3	**0.8**	**0.6**	0.7	0.7	**0.9**	**0.7**	1:00–4:00 p.m.
$e_3(3,6)$	**0.6**	**0.7**	0.4	0.3	**0.9**	0.5	0.5	0.8	3:00–7:00 p.m.
$e_4(2,6)$	0.8	0.2	**0.5**	**0.6**	0.6	**0.4**	**0.6**	**0.7**	5:00–9:00 p.m

A global plan, P, is a set of feasible plans that users are assigned to events, such that the sum of utility scores of all users is the largest under the constraints as follows. (1) Users' plans are designed with no time conflicts. That is, the holding time of the events planned for the same user should not have overlaps. Specifically, in Example 1, events e_1 and e_2 have time conflicts, so they cannot appear in the plan of one user at the same time. (2) The number of participants of every event should be larger than the lower bound and smaller than the upper bound. In Example 1, the number of participants of event e_1 should be larger than 1 and smaller than 5. (3) The travel cost of each user is not larger than his/er travel budget, which is attached after each user in parenthesis in

Row 1 of Table 1. Here, we simply use Manhattan distance to calculate the travel cost. In Example 1, if u_1 attends to e_1 and e_3, his/er travel cost is $D_1 = d(u_1, e_1) + d(e_1, e_2) + d(e_2, u_1) = 2 + 3 + 3 = 8$, which should be smaller than 10. Under the above constraints, a feasible plan of Example 1 is shown by the bold cells in Table 1, and its total utility score is 9.9. The GEPC problem aims at finding such a feasible plan whose total utility score is the largest.

Although the authors of [2] provide two approximate algorithms to solve the GEPC problem, these algorithms are not efficient enough. They both follow a two-step framework, in which each step satisfies parts of the constrains. Even the one with a smaller computational complexity is still $O(|E|^2 + |U||E|^2)$, where $|U|$ is the number of users, $|E|$ is the number of events in the EBSN platform. To overcome this shortcoming, we propose a more efficient algorithm, called *Heuristic-DP algorithm*. The main idea is using dynamic programming to reduce the time cost, asynchronously considering all the constraints together instead of a two-step method. The complexity of our algorithm is only $O(|U||E|)$. Furthermore, we propose an improved heuristic strategy of the Heuristic-DP algorithm. This strategy spends a little more time, but can obtain a more accurate result. Our experiments show that the total utilities of our approximate algorithms are larger than those of [2], which indicate that the approximation of our proposed algorithms not only improve the efficiency, but also ensure the accuracy.

To summarize, the contribution of our paper is,

- We propose a more efficient method, called Heuristic-DP algorithm, to solve the GEPC problem than that proposed by [2], which improves the computational complexity from $O(|E|^2 + |U||E|^2)$ to $O(|U||E|)$.
- We also propose an improved heuristic strategy, which spends a little more time, but can obtain a more accurate result.
- Extensive experiments show that our proposed algorithms are more efficient and accurate than those of [2].

The rest of our paper is organized as follows. In Sect. 2, we formally define the GEPC problem. In Sect. 3, we summarize the related works. Then we describe our Heuristic-DP method in Sect. 4, and improved strategy in Sect. 5. We report our experiment results and corresponding analysis in Sect. 6, and conclude our work in Sect. 7.

2   Problem Definition

In this section, we introduce a formal mathematical definition of the GEPC problem. We assume that there is a set $U = \{u_i\}$ of n users and a set $E = \{e_j\}$ of m events in EBSN problem.

A tuple (l_{u_i}, B_i) consisting of the location and travel budget of u_i is used to describe each user $u_i \in U$. A 5-tuple $(l_{e_j}, \xi_j, \eta_j, t_j^s, t_j^t)$ denotes each event $e_j \in E$, consisting of a location, participation lower bound, participation upper bound, start time, and end time. Each user participates each event, forming a utility score, $\mu(u_i, e_j) \geq 0$, signifying u_i's level of interest in e_j. If $\mu(u_i, e_j) = 0$, user u_i will not attend event e_j. The notations of symbols are summarized in Table 2.

Table 2. Summary of symbol notations

Notation	Description
$E = \{e_1, \ldots, e_{|E|}\}$	Set of events
$U = \{u_1, \ldots, u_{|U|}\}$	Set of users
$\mu(e, u)$	Utility scores when u_i participates e_j
$cost(u, e)$	Travel costs between u and e
$cost(e_i, e_j)$	Travel costs from e_i to e_j
$\{\xi^e, \eta^e\}$	Lower bound and upper bound of e
$\{t_1^e, t_2^e\}$	Start time and end time of e
b_u	Travel budgets of u

2.1   Complex Event Planning: GEPC Problem

The EBSN's global utility score for a plan P, denoted $\mathcal{U}_P$, is the sum of the every user's utility scores in P_u.

Definition 1 (GEPC problem [2]). *Given an EBSN, the GEPC problem is to find a feasible global plan $P^\star$, such that $\mathcal{U}_{P^\star} = \max_P \mathcal{U}_P$, subject to the following constraints:*

1. *Users' plans have no time conflicts, i.e., $\forall i \; \forall e_k \neq e_h \in P_i \quad t_{e_k}^s < t_{e_h}^s \Rightarrow t_{e_k}^t < t_{e_h}^s$.*
2. *Users' travel costs are within budget, i.e., $\forall i \; D_i \leq B_i$.*
3. *Events' participation upper bounds are satisfied, i.e., $\forall j \; |\{P_i : e_j \in P_i\}| \leq \eta_j$*
4. *Lower bounds are satisfied, i.e., $\forall j \; |\{P_i : e_j \in P_i\}| \geq \xi_j$*

Example 2. The cells with bolded entries in Table 1 correspond to a global plan. It is easy to verify that all constraints are satisfied. There are no time conflicts among assigned events. All travel costs are within budget. All events' participation upper bounds are met. Finally, all events' participation lower bounds are also met. The EBSN's global utility score under the given plan is $\mu(u_1, e_1) + \mu(u_1, e_3) + \cdots + \mu(u_8, e_4) = 9.9$.

According to [2], the GEPC problem is NP-hard, and we propose a more efficient heuristic algorithm than those of [2] to solve this problem.

3   Related Work

Studies on EBSNs. Different with Location-Based Social Networks (LBSNs) [1,8,11,13,14,19] focusing on maximizing users' individual utilities, EBSNs [12] focus on maximizing system total utilities. Feng et al. [4] formulate a problem of mining influential cover set (ICS), combining influence maximization problem [7] and the team formation problem [9], to discover influential event organizers who are essential to the overall success of social events.

Liu et al. [12] first investigated EBSNs properties, then studied problems of community detection and information flow considering both online and offline social interactions. Zhang et al. [24] focused on event-based group recommendation, proposed a method considering location features, social features to model interactions between users and events to provide better solution. Du et al. [3] aimed to exploit individual behaviors in EBSNs, and proposed a novel SVD-MFN algorithm to predict activity attendance by integrating heterogeneous factors into a single framework. Pham et al. [15] modeled the rich information with a heterogeneous graph, and considers the three kinds of recommendation problems (i.e. recommending groups to users, recommending tags to groups, and recommending events to users) as a query-dependent node proximity problem. These works all focused on individual user recommendations, while ours focus on a global satisfiable planning.

Social Event Organization and its variants were investigated by [10,16,17, 20,21,23], assigning a group of users to attend a set of events by maximizing overall satisfaction. Cheng et al. [2] proposed approximate solutions to solve the GEPC and its incremental variant problems, considering the aspects of the events' participation lower bound. However, this algorithm is not efficient enough since it satisfy all the constraints in order rather than in parallel. The main contribution of our paper is synchronously considering all the constraints to accelerate this process and gain a litter bigger overall satisfaction.

Studies on Heuristic DP. Many optimization problems in various fields have been solved by diverse algorithms, such as greedy, dynamic programming (DP). Although these algorithms can guarantee global optimization in simple and ideal models, there also exists some drawbacks. For instance, in DP the increase in the number of variables would exponentially increase the number of the recursive functions. Geem et al. in [5] proposed a heuristic optimization techniques based on simulation to overcome the shortage above. Researchers in [6] presented an algorithm for planning with time and resources, based on heuristic search. In this paper, we combine the heuristic strategy and dynamic programming process.

4   Heuristic-DP Algorithm

The main idea of heuristic-DP algorithm is to transfer the EBSN problem into $|U|$ sub-problems. In each sub-problem, we satisfy one user's constraint until all users are dispatched. During this process, all the constraints are taken into account asynchronously, overcoming the shortage of the algorithm in [2] which synchronously use the greedy-based algorithm and then the algorithm in [16].

When designing plans for users, it must predefine a time horizon, $\mathcal{H}$, which the EBSN operated on. For simplicity, we assume a time horizon is one day, so that every user are provided with their individualized "Plan for today".

In Heuristic-DP algorithm, we scan each user in order until all the users have been dispatched with the feasible plan. During the process of scanning, each user has two states. One is called "incomplete", in which not all the events satisfy their the lower bound of participants. The other is called "complete", in which all events satisfy their the lower bound of participants. In the state of uncomplete, we firstly assign users to the event which is chosen by the heuristic strategy, and then consider the utility score and travel budget of users. While in the state of complete, we only take the utility scores and travel costs of users into account, meeting the needs of users. The pseudo-code is shown in Algorithm 1.

Algorithm 1. Heuristic-DP Algorithm

---

input : E, U, $\mu(e,u)$, $\{cost\}$, $\{\xi^e, \eta^e\}$, $\{t_1^e, t_2^e\}$, b_u
output: A feasible schema P

1   sort all the events in E in non-descending order by t_2^e
2   sort all the users in U in non-descending order by b_u
3   $C_1^r := 0$
4   **for** *each user* u_r **do**
5   **if** *not all events meet their lower bound* **then**
6         $P := P \cup iTDP(u_r)$
7   update C_1^r according to the selected events
8   **else**
9         $P := P \cup TDP(u_r)$
10   update C_1^r according to the selected events

11 **return** P;

We firstly sort the events in E in non-descending order according to the end time t_2^e of each event (Line 1). Secondly, we sort the users in U by non-descending order according to the travel budgets b_u (Line 2). Initialize the number of selected participants for each event C_1^e before the first iteration (Line 3). We scan all the users in order (Line 4–10), and check whether the lower bounds are been satisfied in each iterator. If not, it falls into the "incomplete" state. We conduct iTDP algorithm to obtain a feasible plan of user u_r with the assignment satisfying both the user's need and the lower bound of events, and update the number of selected users of every event (Line 5–7). If so, it falls into the "complete" state, and we call the TDP Algorithm to obtain a feasible plan of user u_r considering his/er travel budgets and the whole time slice. Then we update the number of selected users of every event (Line 8–10). Finally, we combine the sequence of events for all users after finishing the process of scanning, and every user's feasible schemes is obtained (Line 11).

4.1 iTDP Algorithm

When not all events satisfying the lower bound on number of participants, i.e., the "incomplete" state, the iTDP algorithm is operated. The main idea of iTDP algorithm is to distribute user u_r to the event selected by the heuristic strategy priority, and then meet user's need to gain a better utility scores. Specifically, the heuristic strategy here is to select the event whose difference between the selected number of participants and the lower bound is the largest. Then, we record the start time and end time of the selected event as the end time and start time of the two new time scales, whose original start time is 0 o'clock and end time is 24 o'clock. At last, the TDP algorithm is invoked to obtain the feasible scheme that has the biggest total utility scores in the two new time scales. The approach is summarized in Algorithm 2.

Algorithm 2. iTDP Algorithm

---

input : E, u_r, $\mu(e, u)$, $\{cost\}$, $\{\xi^e, \eta^e\}$, $\{t_1^e, t_2^e\}$, b_u
output: P_{u_r}: a feasible schema of user u_r

1 $P_{u_r} := \emptyset$
2 Find the event e_t by heuristic strategy
3 $end = t_1^{e_t} - 1$, $start = t_2^{e_t} + 1$
4 $B = b_{u_r} - cost(u_r, e_t)$
5 $P_{u_r} = P_{u_r} \cup TDP(0, end, B)$
6 Update B according the selected events
7 $P_{u_r} = P_{u_r} \cup TDP(start, 24, B)$
8 **return** P_{u_r};

We first initialize the set of plan for user u_r, P_{u_r}, as empty set (Line 1). Then, heuristic strategy is used to select one event who has the biggest difference from the already selected number of participants and the lower bound among all the unsatisfied events (Line 2). In addition, the travel costs between the location of selected event and u_r cannot exceed u_r's travel budgets. After that, we get the end time end and start time $start$ of the new time scale according to the start time $t_1^{e_t}$ and end time $t_2^{e_t}$ of event e_t (Line 3). Meantime, we get the residue travel budgets B of user u_r according to the travel costs $cost(u_r, e_t)$ from the location of user u_r to the event e_t (Line 4). Particularly, the distance will be infinity when the number of participants of event e_t reaches its upper bound, so that no user will be dispatched to event e_t. At that time, we invoke TDP algorithm to calculate the list of events adding to the set of user u_r's plan, which maximizes the utility scores of user u_r during 0 o'clock and end with the travel budgets B (Line 5). Then update the residue travel budgets B of user u_r by his/er selected events (Line 6). After that, the event lists are obtained by invoking TDP algorithm, which maximizes the utility scores of user u_r during $start$ and 24 o'clock with the updated travel budget B (Line 7). Finally, a feasible schema of user u_r is gained (Line 8).

Example 3. Consider the problem defined in Example 1. When we dispatch event lists for user u_1, we should judge whether all events have been satisfied their lower bound. Obviously it is not, so iTDP algorithm is used to design schedule for u_1. We first choose the event by heuristic strategy, so we add e_3 into P_{u_1}. Then, a "before" time scale is designed with its start time been 0 o'clock and end time been the start time of e_3 e.g. 3:00 p.m. Meantime, the start time of "after" time scale is the end time of e_3, 7:00 p.m. and end time is 24 o'clock. The TDP algorithm is invoked to calculate the event lists making the utility scores biggest with u_1's residue travel budgets in the "before" and "after" time scales. The blue font in Table 3 is the feasible schema of users in the "incomplete" state.

Table 3. Plans obtained by iTDP algorithm

$e_j(\xi_j, \eta_j)$	$u_1(10)$	$u_2(12)$	$u_3(13)$	$u_4(15)$	$u_5(17)$	$u_6(19)$	$u_7(20)$	$u_8(23)$	Time
$e_1(1,5)$	**0.3**	**0.4**	0.8	**0.9**	0.8	0.4	0.5	0.3	1:00–2:00 p.m.
$e_2(1,5)$	0.2	0.3	**0.8**	0.6	**0.7**	0.7	0.9	0.7	1:00–4:00 p.m.
$e_3(3,6)$	**0.6**	**0.7**	0.4	**0.3**	0.9	0.5	0.5	0.8	3:00–7:00 p.m.
$e_4(2,6)$	0.8	0.2	**0.5**	0.6	**0.6**	0.4	0.6	0.7	5:00–9:00 p.m

Theorem 1. *All the events will achieve their lower bound on number of partic-ipants if $\sum_{i=1}^{M} \xi_i \leq N$.*

Proof. All the events will achieve their lower bound on number of participants if the sum of the lower bound of every event is less than the number of users. In our algorithm, if the lower bound of each event has not been satisfied, we invoke the iTDP algorithm to dispatch users to events, it is clear that at least one of the unsatisfied events will be dispatched users at one iteration, so all the events' lower bounds will be meet at most during $\sum_{i=1}^{M} \xi_i$.

4.2 TDP Algorithm

The aim of this algorithm is to design a feasible schema P_{u_r} for user u_r in the r-th iteration, where $1 \leq r \leq |U|$. The main idea of this algorithm is to use two-dimensional dynamic programming algorithm, considering both the time conflicts and travel costs. Particularly, the key of the TDP algorithm is to divide the time horizon, $\mathcal{H}$, into several time slices, recording the biggest total utility scores in every time slice of user u_r. For simplicity, we use $\Omega^r(i,j)$ to donate the biggest utility scores obtained in the r-th iterator by the i time with j travel budgets of user u_r.

In TDP algorithm, we scan all events and find the biggest total utility scores qualifying the constraints, and denote $\hat{e}_{i,k}$ as the last event in the schema at different time slices. Firstly, the travel costs to the location of event e_k not exceed the user u_r's travel budgets. That is, when $cost(u_r, e_1) + \sum_{i=2}^{|P_u|} cost(e_{i-1}, e_i) + cost(\hat{e}_{i,k}, e_k) > b_{u_r}$, the event e_k could not be included in the schema of u_r. The second constraint is that for the event chosen by the user u_r, its start time cannot be earlier than S time and end time cannot be later than T time. Meanwhile, the start time and end time between the chosen events cannot overlap.

TDP algorithm takes both time conflicts and the travel costs between the locations of users and the events into consideration. Two-dimensional dynamic programming algorithm scans all the events in every time slices to judge whether event e_k satisfies the constraints, and calculate the biggest total utility scores $\Omega^r(i,j)$ in current moment with current travel budgets. The specific computational formula of $\Omega^r(i,j)$ is defined as follows:

$$\Omega^r(i,j) = \begin{cases} \Omega^r(i-1,j) & j < cost(\hat{e}_i, e_k) \\ \max(\Omega^r(i-1,j), \Omega^r(\xi^{e_j}, Y) + \mu^r(u_r, e_k)) & \xi^{e_k} \geq S, \eta^{e_k} \leq T, \\ Y = j - cost(\hat{e}_{\xi^{e_j},k}, e_k) & j > cost(\hat{e}_i, e_k) \end{cases}$$

The equation means whether user u_r chooses the current event e_k. If there is a time conflict or exceeding travel budgets, u_r do not select the current event e_k, so the $\Omega^r(i,j)$ is equal to $\Omega^r(i-1,j)$ at the last moment with the same travel budgets. If it will get bigger utility scores with no time conflicts and no exceeding travel budgets, u_r would select e_k and add it to the schema. So $\Omega^r(i,j)$ is equal to the current utility scores add to $\Omega^r(\xi^{e_k}, j - cost(\hat{e}_{i,k}, e_k))$, whose start time is as same as the start time of e_k and the travel budgets is current residue travel budgets. The entire dynamic programming process obtains the biggest utility scores of every moments and every travel costs, and get a feasible schema S_{u_r} of the user u_r by recalling the dynamic programming process by $\Omega^r(i,j)$. The specific description of the TDP algorithm is illustrated as Algorithm 3.

Algorithm 3. two-dimensional DP Algorithm

---

input : E_r, u_r, $\hat{e}_i$,$\mu(e,u)$, $cost(e,u)$, $\{\xi^e, \eta^e\}$, $\{t_1^e, t_2^e\}$, b_{u_r}, S, T
output: the best feasible schema of u_r, S_{u_r}.

1 $j = b_{u_r}$, Initialize the $\hat{e}_0$ as the location of u_r
2 **for** *each moment between S and T* **do**
3   **for** *each event e_k* **do**
4   **if** $j > cost(\hat{e}_{i,k}, e_k)$ **then**
5   **if** $\Omega(\xi^{e_k} - 1, j - cost(\hat{e}_{i,k}, e_k)) + \mu^r(u_r, e_k) > \Omega(i-1,j)$ **then**
6                 $\Omega^r(i,j) = \Omega^r(\xi^{e_k}, j - cost(\hat{e}_{i,k}, e_k)) + \mu^r(u_r, e_k)$
7                 $j = j - cost(\hat{e}_{i,k}, e_k)$
8   add e_k into $path(i,j)$
9   **else**
10                 $\Omega^r(i,j) = \Omega^r(i-1,j)$
11   **else**
12   **if** $j < cost(\hat{e}_{i,k}, e_k)$ **then**
13                 $\Omega^r(i,j) = \Omega^r(i-1,j)$

14 find the largest $\Omega(i,j)$
15 construct S_{u_r} according to $path(i,j)$
16 **return** S_{u_r};

In Algorithm 3, we first set a variable j to record the residue travel budgets of u_r, and initialize the $\hat{e}_0$ as the location of u_r (Line 1). Then we scan all the moments and events in order to obtain the list of events who has the biggest utility scores with the travel budgets j according to the Eq. (1) (Line 2–13). In the process of scanning, if the travel costs between the location of last selected event $\hat{e}_{i,k}$ and event e_k is no more than the residue travel budgets of u_r, and the start time ξ^{e_j} and end time η^{e_k} are both in the period between S time and T time (Line 4). Then we continue to judge whether the utility scores will be bigger by adding event e_k to the event lists (Line 5). If so, we update the $\Omega^r(i,j)$ and the residue travel budgets, meantime, record the selected event e_k (Line 6–8). If not, $\Omega^r(i,j)$ is equal to $\Omega^r(i-1,j)$ at the last moment with the same travel budgets (Line 9–10). If the travel costs between the location of last selected event $\hat{e}_{i,k}$ and event e_k is more than the residue travel budgets of user u_r, we update $\Omega^r(i,j)$ by the $\Omega^r(i-1,j)$ at the last moment with the same travel budgets (Line 12–13). After finishing the scanning, we find the biggest utility scores (Line 14). Then rollback the process according to the biggest utility scores (Line 15). Finally, the algorithm returns the best feasible schema S_{u_r} of user u_r (Line 16).

Example 4. Since the 5-th iteration, we have already satisfied every events' lower bound, we use the TDP algorithm to continue obtain feasible plans for users. In the 6-th iteration, we could obtain the biggest utility scores from 0:00–24:00 with the travel budgets b_{u_r}, then rollback the process getting the feasible schemas of every users. So u_6 chooses e_1 and e_4 as his/er schedule. We can easily find that there is no time conflicts or no exceeding the travel budgets of u_6, meanwhile, every events' lower bound and upper bound are also satisfied. By the same way, we get every users' feasible plans with red colored fond in the Table 4.

Table 4. Plans obtained by TDP algorithm

$e_j(\xi_j,\eta_j)$	$u_1(10)$	$u_2(12)$	$u_3(13)$	$u_4(15)$	$u_5(17)$	$u_6(19)$	$u_7(20)$	$u_8(23)$	Time
$e_1(1,5)$	0.3	0.4	0.8	0.9	0.8	0.4	0.5	0.3	1:00–2:00 p.m.
$e_2(1,5)$	0.2	0.3	0.8	0.6	0.7	0.7	0.9	0.7	1:00–4:00 p.m.
$e_3(3,6)$	0.6	0.7	0.4	0.3	0.9	0.5	0.5	0.8	3:00–7:00 p.m.
$e_4(2,6)$	0.8	0.2	0.5	0.6	0.6	0.4	0.6	0.7	5:00–9:00 p.m

4.3   Complexity Analysis

Algorithm 1 scans every users, so the number of iterations is $|U|$. When user in the stage that some events not been satisfied the lower bound on number of participants, it need to invoke Algorithm 2 to obtain the feasible schema with the start time and after the end time of the event, which is chosen by the heuristic strategy, the complexity of this process is $O(|E| + |E|\,|T|)$, where $|T|$ is a constant indicting the number of time slices. After all events satisfy the lower bound, we invoke the Algorithm 3 to calculate the feasible schedule during the whole day with user's travel budgets, which complexity is $O(|E|\,|T|)$. Thus, the computational complexity of our heuristic-DP algorithm is $O(|U|\,|E| + |U|\,|E|\,|T|)$.

5   Improving Heuristic Strategy

Although Heuristic-DP algorithm speeds up the process of GEPC, and gets a larger total utility scores, it still can be further improved. This is because the heuristic strategy in Heuristic-DP algorithm only considers the lower bound of the events, not takes the interests of users into account. It may appear the phenomenon that user u_1 has no interests in event e_1, but since the difference between lower bound of e_1 and the already selected number of participants are the biggest, the event e_1 are still included in the feasible schema of user u_1. To overcome this shortage, we propose an improved heuristic strategy that takes the demand of user and the constraints of events into account at the same time. Specifically, the event with the biggest utility score for user u_r among all the unsatisfied events will be selected. This method may spends a little more time, but obtains a larger total utility scores.

Example 5. Consider the same problem defined in Example 3. If we use the user-oriented heuristic strategy, we first choose the event with biggest utility scores $\mu(u_1, e_k)$ in all the unsatisfied events, whose travel distance to u_1 also not exceed the travel budgets of u_1, so we add e_3 into P_{u_1}. Then, a "before" time scale and a "after" time scale is defined as the same way in Example 3. The TDP algorithm is invoked to calculate the event lists making the utility scores biggest with u_1's residue travel budgets in the "before" and "after" time scale. The blue fonts in Table 5 are the feasible schemas of users in the state of not all event been satisfied their lower bounds.

Table 5. Plans obtained by iTDP algorithm with improved heuristic strategy

$e_j(\xi_j, \eta_j)$	$u_1(10)$	$u_2(12)$	$u_3(13)$	$u_4(15)$	$u_5(17)$	$u_6(19)$	$u_7(20)$	$u_8(23)$	Time
$e_1(1,5)$	**0.3**	**0.4**	0.8	0.9	**0.8**	0.4	0.5	0.3	1:00–2:00 p.m.
$e_2(1,5)$	0.2	0.3	**0.8**	**0.6**	0.7	0.7	0.9	0.7	1:00–4:00 p.m.
$e_3(3,6)$	**0.6**	**0.7**	0.4	0.3	**0.9**	0.5	0.5	0.8	3:00–7:00 p.m.
$e_4(2,6)$	0.8	0.2	**0.5**	**0.6**	0.6	0.4	0.6	0.7	5:00–9:00 p.m

6   Experimental Evaluation

6.1   Experimental Environment and Datasets

The algorithm are implemented in C++, and the experiments are performed on a Windows 10 machine with Inter i7-6700 3.40 GHZ CPU and 8 GB memory. The time costs reported here are calculated by the system clock. It is unlikely for a user to attend the events in different cities, so we focus on a single situation that the users and events are located in the same city. We use the Meetup and Plancast datasets as real database, and we choose four different cities as our

Table 6. Real datasets

City	$|U|$	$|E|$	Mean of ξ	Mean of η
Beijing	113	16	10	100
Singapore	1500	87	10	100
Vancouver	2012	225	10	100
Auckland	569	37	10	100

research object, covering Beijing, Singapore, Vancouver and Auckland. Table 6 presents the parameters of the data.

To further evaluate the stable of our algorithm, we also use synthetic datasets which are extracted by the Meetup dataset. Various settings are shown in Table 7, where we mark our default settings in bold font.

Table 7. Synthetic datasets

Factor	Setting
$|E|$	20, **50**,100,200,500
$|U|$	200, 500, 1000, **5000**

6.2 Results

In this section, we will compare the efficiency of the GAP-based, greedy-based, our heuristic-DP algorithm, and heuristic-DP algorithm with an improved heuristic stategy by solving the GEPC problem. Since the heuristic-DP algorithm is an extension of the GAP-Based and greedy-based algorithms, we use the GAP-based and greedy-based algorithms as a baseline to evaluate the results and effects of our proposed heuristic-DP algorithm.

Tables 8, 9, 10 and 11 depict the results and effects on the real datasets. By comparing Tables 8 and 10, we can easily find that the total utility scores obtained by the GAP-based algorithm is a little larger than the total utility obtained by the heuristic-DP algorithm. However, the time costs and the memory costs of the GAP-based algorithm are much larger than that of our heuristic-DP algorithm. This suggests that the heuristic-DP algorithm is much more efficient than the GAP-based algorithm. Furthermore, it is clear that the total utility scores gained by the heuristic-DP algorithm are larger than that of greedy-based algorithm, meantime, the time costs and the memory costs of the greedy-based algorithm are also much larger than the time costs of the heuristic-DP algorithm by comparing Tables 9 and 10. From Tables 10 and 11, we can see that the total utility scores of heuristic-DP algorithm with an improved strategy are a little larger, while the time costs are a little larger than that of heuristic-DP algorithm, and the memory costs of both two methods are the same. In conclusion, the heuristic-DP algorithm may be as effective and more efficient than the GAP-based and greedy-based algorithms. By contrasting our heuristic-DP algorithm, and the heuristic-DP algorithm with an improved strategy may get a little larger total utility scores, while the heuristic-DP algorithm may costs a little smaller time.

Table 8. GAP algorithm for GEPC on real datasets

Datasets	Greedy-based		
	Total utility	Time costs (s)	Memory costs (MB)
Beijing	223.19	1.34	3.9
Singapore	5753.68	137.83	56.7
Vancouver	6821.83	13539	545.6
Auckland	1290.13	7.95	44.3

Table 9. Greedy-based algorithm for GEPC on real datasets

Datasets	Greedy-based		
	Total utility	Time costs(s)	Memory costs (MB)
Beijing	209.65	0.063	1.7
Singapore	4702.00	232.14	52.5
Vancouver	5522.34	13.56	341.2
Auckland	1158.98	1.75	19.3

Table 10. Heuristic-DP algorithm on real datasets

Datasets	Heuristic-DP algorithm		
	Total utility	Time costs (s)	Memory costs (MB)
Beijing	239.48	0.041	1.2
Singapore	5593.17	186.74	47.7
Vancouver	6641.85	10.17	305.2
Auckland	1245.63	1.06	13.1

Figures 2, 3 and 4 depict the performance of the GAP-based, greedy-based and our heuristic-DP algorithm (Event-DP) and our improved method (User-DP) on different datasets. We first study the effective and efficient of different number of users $|U|$. Figures 2a, 3a and 4a show the results when the number of events $|E| = 50$ and varying the number of users $|U|$ from 100 to 5000. We can learn that the total utility scores obtained by the GAP-based algorithm are

Table 11. Heuristic-DP algorithm with improved strategy on real datasets

Datasets	Heuristic-DP algorithm with improved strategy		
	Total utility	Time costs (s)	Memory costs (MB)
Beijing	246.27	0.047	1.2
Singapore	5607.21	198.4	47.7
Vancouver	6657.48	11.28	305.2
Auckland	1261.45	1.23	13.1

the biggest, and that obtained by the greedy-based algorithm are the smallest. However, the time costs and memory costs of heuristic-DP algorithm are the smallest, and the costs of GAP-based algorithm are much larger than that of heuristic-DP and greedy-based algorithms. Hence, the heuristic-DP algorithm is more practical.

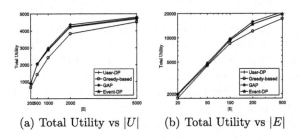

(a) Total Utility vs $|U|$ (b) Total Utility vs $|E|$

Fig. 2. Total utility of algorithm for GEPC

We then study the effective and efficient of different number of events $|E|$. Figures 2b, 3b and 4b show the results when the number of users $|U| = 5000$ and varying the number of events $|E|$ from 20 to 500. By comparison, we can see that the total utility scores obtained by the GAP algorithm are larger, but the time costs and the memory costs of this algorithm are also the largest. The total utility scores obtained by heuristic-DP algorithm are the second largest,

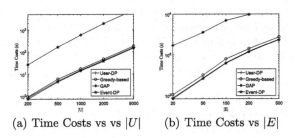

(a) Time Costs vs vs $|U|$ (b) Time Costs vs $|E|$

Fig. 3. Time costs of algorithm for GEPC

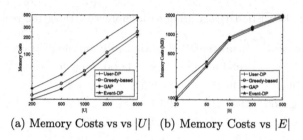

(a) Memory Costs vs vs $|U|$ (b) Memory Costs vs $|E|$

Fig. 4. Memory costs of algorithm for GEPC

and the costs are the smallest. Therefore, these results further suggests that the heuristic-DP algorithm is more effective and efficient than the GAP-based and greedy-based algorithms.

7   Conclusion

In this paper, we propose Heuristic-DP algorithm to solve the complex social event-participant planning efficiently. To solving the GEPC problem, the Heuristic-DP algorithm asynchronously considers all the constraints together with the computational complexity been only $O(|U||E|)$. To obtain a more accurate result, we also have proposed a heuristic strategy in the first stage of Heuristic-DP algorithm. At last, Experiments over real and synthetic datasets demonstrate the efficiency of our proposed algorithm.

Acknowledgment. The work has been supported by the National Natural Science Foundation of China (NSFC) under Grant Nos. 61472069, 61402089, 61332006 and U1401256; and the Fundamental Research Funds for the Central Universities under Grant Nos. N161602003 and N171607010.

References

1. Chen, C., Zhang, D., Guo, B., Ma, X., Pan, G., Wu, Z.: TripPlanner: personalized trip planning leveraging heterogeneous crowdsourced digital footprints. T-ITS **16**, 1259–1273 (2014)
2. Cheng, Y., Yuan, Y., Chen, L., Giraud-Carrier, C., Wang, G.: Complex event-participant planning and its incremental variant. In: 2017 IEEE 33rd International Conference on Data Engineering (ICDE), pp. 859–870. IEEE (2017)
3. Du, R., Yu, Z., Mei, T., Wang, Z., Wang, Z., Guo, B.: Predicting activity attendance in event-based social networks: content, context and social influence. In: UbiComp (2014)
4. Feng, K., Cong, G., Bhowmick, S.S., Ma, S.: In search of influential event organizers in online social networks. In: SIGMOD (2014)
5. Geem, Z.W., Kim, J.H., Loganathan, G.V.: A new heuristic optimization algorithm: harmony search. Simulation **76**(2), 60–68 (2001)

6. Haslum, P., Geffner, H.: Heuristic planning with time and resources. In: Sixth European Conference on Planning (2014)
7. Kempe, D., Kleinberg, J., Tardos, É.: Maximizing the spread of influence through a social network. In: SIGKDD (2003)
8. Khrouf, H., Troncy, R.: Hybrid event recommendation using linked data and user diversity. In: RecSys (2013)
9. Lappas, T., Liu, K., Terzi, E.: Finding a team of experts in social networks. In: SIGKDD (2009)
10. Li, K., Lu, W., Bhagat, S., Lakshmanan, L.V., Yu, C.: On social event organization. In: SIGKDD (2014)
11. Liao, G., Zhao, Y., Xie, S., Yu, P.S.: An effective latent networks fusion based model for event recommendation in offline ephemeral social networks. In: CIKM (2013)
12. Liu, X., He, Q., Tian, Y., Lee, W.C., McPherson, J., Han, J.: Event-based social networks: linking the online and offline social worlds. In: SIGKDD, pp. 1032–1040 (2012)
13. Lu, E.H.C., Chen, C.Y., Tseng, V.S.: Personalized trip recommendation with multiple constraints by mining user check-in behaviors. In: GIS (2012)
14. Minkov, E., Charrow, B., Ledlie, J., Teller, S., Jaakkola, T.: Collaborative future event recommendation. In: CIKM (2010)
15. Pham, T.A.N., Li, X., Cong, G., Zhang, Z.: A general graph-based model for recommendation in event-based social networks. In: ICDE (2015)
16. She, J., Tong, Y., Chen, L.: Utility-aware social event-participant planning. In: SIGMOD (2015)
17. She, J., Tong, Y., Chen, L., Cao, C.C.: Conflict-aware event-participant arrangement. In: ICDE (2015)
18. She, J., Tong, Y., Chen, L., Cao, C.C.: Conflict-aware event-participant arrangement and its variant for online setting. IEEE Trans. Knowl. Data Eng. 28(9), 2281–2295 (2016)
19. Sun, Y.-C., Chen, C.C.: A novel social event recommendation method based on social and collaborative friendships. In: Jatowt, A., et al. (eds.) SocInfo 2013. LNCS, vol. 8238, pp. 109–118. Springer, Cham (2013). https://doi.org/10.1007/978-3-319-03260-3_10
20. Tong, Y., Cao, C.C., Chen, L.: TCS: efficient topic discovery over crowd-oriented service data, pp. 861–870 (2014)
21. Tong, Y., She, J., Ding, B., Chen, L., Wo, T., Xu, K.: Online minimum matching in real-time spatial data: experiments and analysis. Proc. VLDB Endow. 9(12), 1053–1064 (2016)
22. Tong, Y., She, J., Meng, R.: Bottleneck-aware arrangement over event-based social networks: the max-min approach. World Wide Web 19(6), 1151–1177 (2016)
23. Tong, Y., Wang, L., Zhou, Z., Ding, B., Chen, L., Ye, J., Xu, K.: Flexible online task assignment in real-time spatial data. Proc. VLDB Endow. 10(11), 1334–1345 (2017)
24. Zhang, W., Wang, J., Feng, W.: Combining latent factor model with location features for event-based group recommendation. In: SIGKDD (2013)

Data Quality and Crowdsourcing

Repairing Data Violations with Order Dependencies

Yu Qiu[1,2], Zijing Tan[1,2(✉)], Kejia Yang[3], Weidong Yang[1,2],
Xiangdong Zhou[1,2], and Naiwang Guo[4]

[1] School of Computer Science, Fudan University, Shanghai, China
zjtan@fudan.edu.cn
[2] Shanghai Key Laboratory of Data Science, Shanghai, China
[3] Computer Science and Mathematical Science, University of Michigan,
Ann Arbor, USA
[4] State Grid Shanghai Municipal Electric Power Company, Shanghai, China

Abstract. Lexicographical order dependencies (ODs) are proposed to describe the relationships between two lexicographical ordering specifications with respect to lists of attributes, and are proved to be useful in query optimizations concerning ordered attributes. To take full advantage of ODs, the data instance is supposed to satisfy OD specifications. In practice, data are often found to violate given ODs, as demonstrated in recent studies on discovery of ODs. This highlights the quest for data repairing techniques for ODs, to restore consistency of the data with respect to ODs. New challenges arise since ODs convey order semantics beyond functional dependencies, and are specified on lists of attributes. In this paper, we make a first effort to develop techniques for repairing data violations with ODs. (1) We formalize the data repairing problem for ODs, and prove that it is NP-hard in the size of the data. (2) Despite the intractability, we develop effective heuristic algorithms to address the problem. (3) We experimentally evaluate the effectiveness and efficiency of our algorithms, using both real-life and synthetic data.

1 Introduction

Data consistency is one of the central aspects of data quality, where inconsistencies in the data are generally identified as violations of data dependencies. In light of this, various dependency proposals are presented to express application semantics that data are required to satisfy, and fundamental theoretical problems, dependency discoveries and data repairing techniques related to these dependencies are also studied in literature, among other things.

Lexicographical order dependencies (ODs) [13,15] are recently introduced to state the relationship between two lexicographical ordering specifications on lists of attributes. Ordered attributes, *e.g.,* date, time and price, are prevalent in data values and are well employed in SQL queries. As stated in [13], 85 out of the 99 queries in the TPC-DS benchmark involve date. Sorting is one of the most basic database operations, and ODs are shown to play important roles in

© Springer International Publishing AG, part of Springer Nature 2018
J. Pei et al. (Eds.): DASFAA 2018, LNCS 10828, pp. 283–300, 2018.
https://doi.org/10.1007/978-3-319-91458-9_17

	no	year	month	day	time	country	city	accum_expenses
t_1:	00000001	2017	12	12	20:00	AUS	BNE	$3800
t_2:	00000002	2017	12	26	09:00	CHN	SHH	$5200
r t_3:	00000003	2018	01	22	16:00	UK	LON	$6300
t_4:	00000004	2018	02	08	07:00	FRA	LON	$7980
t_5:	00000005	2018	01	15	08:00	FRA	PAR	$8680

Fig. 1. A relational instance r.

query optimizations concerning sorting [13,15]. In addition to the theoretical foundations of ODs, discoveries of ODs [10,14] are also recently studied. The researches reveal that inconsistencies of ODs exist in real data sets, and have to be treated as noises in the discovery process. To improve data quality and further to facilitate query optimization, the data consistency should be restored with respect to ODs. With this comes the need for repairing techniques for OD violations. To our best knowledge, no such algorithms are in place yet.

Example 1: Figure 1 presents a relation r about a round-the-world tour. Each tuple specifies the time (year, month, day, time), the place (country, city) and the accumulated expenses (accum_expenses) to that time, and each tuple carries an auto-increment number in its no attribute.

Formal definitions of ODs will be reviewed in Sect. 2. Intuitively, An OD " X *orders* Y" (written as X $\mapsto$ Y) states that if we sort tuples by X, they are also sorted by Y. Here X (resp. Y) is a *list* of attributes, and sorting by X = [A, B, ...] means sorting by attribute A first, and then breaking ties by attribute B, etc. This specification is in accordance with the SQL order by clauses.

By analyzing the semantics, the following ODs can be defined.

φ_1 : [no] $\mapsto$ [year, month, day, time]
φ_2 : [year, month, day, time] $\mapsto$ [accum_expenses]
φ_3 : [city] $\mapsto$ [city, country]

(1) φ_1 states that when tuples are sorted by no, they are also sorted by [year, month, day, time] in lexicographical order (new records with larger no are added as time goes by); similarly for φ_2. As stated in [13,15], ODs can be employed in query optimization. On a relation that satisfies φ_1, a query in the form of "order by year, month, day, time" can be readily rewritten to "order by no". This is beneficial when index is only built on no and helps reduce indexing space [13].

(2) φ_3 is a "re-interpretation" of the functional dependency (FD) city $\rightarrow$ country. Any FD can be mapped to an equivalent OD by prefixing the left-hand side (LHS) attribute onto the right-hand side (RHS) in the list [13]. Therefore, techniques on ODs can be readily applied when both FDs and ODs are taken into account.

Note that relation r violates given ODs, and is hence inconsistent. As an example, values on [year, month, day, time] do not comply with the ordering imposed by no, which violates φ_1. As another example, tuples t_3, t_4 agree on their city values, but have different values for country; this violates φ_3.

The goal of this paper is to develop repairing techniques for OD violations. We highlight features of ODs that complicate this issue.

(1) OD X ↦ Y states that values on Y are monotonically non-decreasing with respect to values on X. Specifically, (a) Each OD X ↦ Y implies a FD $\mathcal{X} \rightarrow \mathcal{Y}$ [13]. Here set $\mathcal{X}$ (resp. $\mathcal{Y}$) denotes the set of elements in list X (resp. Y). Recall that FDs are defined on *sets* of attributes. When tuples agree on X's values, they must agree on Y's values; otherwise when tuples are sorted by X, it cannot be guaranteed that they are also sorted by Y. (b) ODs convey additional order semantics, since ordering is imposed on tuples with different X' values. The richer expressiveness of ODs necessarily comes at a cost. [13] proves that a sound and complete axiomatization for ODs consists of 6 inference rules, while a well-known axiomatization for FDs consists of only 3 rules.

(2) In contrast to traditional dependencies, ODs are specified on *lists* of attributes, and the order of attributes on the LHS and RHS matters. As an example, neither [no] ↦ [year, day, month, time] nor [time, year, month, day] ↦ [accum_expenses] holds, since any given year (resp. month; day) corresponds to several different months (resp. days; times). Indeed, the order of attributes in the list implies a hierarchy of attributes: the value of month (resp. day; time) is relative to that of year (resp. month; day). We will formalize this observation in Sect. 3, as a guide for data repairing.

(3) FDs are typically given in a *minimal* form with a single RHS attribute; other constraints, *e.g.,* conditional FDs [3], denial constraints (DCs) [4], differential dependencies (DDs) [12], can be easily converted to similar forms. However, RHS attributes in an OD are typically considered as a whole and *may not* be splitted. For example, neither [no] ↦ [month] nor [no] ↦ [day] holds, since any year (resp. month) corresponds to several different months (resp. days).   □

Contributions. We make a first effort to investigate data repairing with ODs.

(1) We formalize the data repairing problem for OD violations (Sect. 3). In addition to a cost model based on the number of modified attribute values, we take into account the *hierarchy* of ordered attributes. We show that it is NP-hard to repair OD violations with our framework.

(2) Despite the intractability, we develop efficient heuristic algorithms for data repairing with ODs (Sect. 4). We tackle this problem by unifying the approach enforcing *equivalence* and the technique enforcing *ordering* among tuples. Since any FD can be mapped to an equivalent OD, our techniques can be readily applied to the setting when both FDs and ODs are considered.

(3) Using both real-life and synthetic data, We conduct an extensive experimental study to verify the effectiveness and efficiency of our algorithms (Sect. 5).

Related Work. Lexicographical order dependencies (ODs) [11,13,15] are proposed to describe the relationships between two lexicographical ordering specifications with respect to *lists* of attributes. As stated in [15], Lexicographical ODs properly subsume functional dependencies (FDs), which are specified on *sets* of attributes. As for query optimization, OD is to *order-by* what FD is to *group-by* in SQL statements. After the theoretical foundations of lexicographical ODs are discussed in [13,15], the automatic discoveries of ODs are studied in [10,14], aiming to find ODs in a given data set. OD validations are also discussed in [10,14] as basic steps of OD discoveries, which are to verify whether given ODs hold on a data set. OD validation is polynomial in the data size, while repairing with ODs is proved to be intractable for a simple cost model and update operation (Sect. 3). We argue that violations of ODs should be cleaned in advance to fully take advantage of ODs in query optimization, which motivates this research.

In literature a different OD, referred to as *pointwise* order dependency, is proposed in [8]. Unlike lexicographical OD on lists of attributes, pointwise OD is defined on sets of attributes. The pointwise OD $\mathcal{X} \hookrightarrow \mathcal{Y}$ holds if for any two tuples s and t, for every attribute A in $\mathcal{X}$, $val(s[A])\ op\ val(t[A])$ implies $val(s[B])$ $op\ val(t[B])$ for every attribute B in $\mathcal{Y}$, where $op \in \{<, \leq, >, \geq, =\}$. Here $\mathcal{X}$ and $\mathcal{Y}$ are sets of attributes, and $val(t[A])$ denotes the value of A in tuple t.

Data repairing is one of issues central to data quality, and is well studied (see *e.g.,* [1–4,6,7,9,16,17]). Specifically, data repairing is discussed for different constraint models, such as FDs [1,9], conditional FDs [2,3], editing rules [7], DCs [4] and fixing rules [16], among others. To our best knowledge, neither of former works discusses the problem of data repairing with lexicographical ODs. As opposed to above-mentioned dependencies, lexicographical ODs specify *ordering* semantics on *lists* of attributes. This necessarily introduces new challenges, as demonstrated in existing researches with ODs.

[15] states that pointwise OD subsumes lexicographical OD, and denial constraint (DC) subsumes pointwise OD. Therefore, one naive solution for our problem is to map lexicographical ODs to pointwise ODs, then to map pointwise ODs to DCs, and to employ existing repairing algorithms for DCs, *e.g.,* [4]. However, a single lexicographical OD of the form X ↦ Y needs to be expressed in a set of DCs with size $|Y| + |X| \cdot |Y|$, where $|X|$ (resp. $|Y|$) is the number of attributes in list X (resp. Y). As remarked earlier, lexicographical ODs may not be converted to the form with a single RHS attribute; that is, $|Y|$ is typically larger than 1. As an example, 8 DCs are required to encode the OD φ_1 : [no] ↦ [year, month, day, time] in Example 1. This large number of DCs necessarily has negative effects on the efficiency and on the repair quality, and makes the naive approach impractical. Moreover, as shown in Example 1, lists of attributes in ODs essentially imply a hierarchy of ordered attributes, and this hierarchical structure should be reflected in the repair framework. Obviously, this important factor is not taken into account in existing repairing techniques. A good solution for OD repairing should exploit the nature of ODs itself.

2   Preliminaries

In this section, we review basic notations and definitions of ODs [10,13–15].

Relation. For a relation schema $R(A_1, \ldots, A_m)$, each A_j denotes a single attribute of R. Given an instance r of R, t^r, s^r denotes tuples in r. Each tuple is associated with a distinct identifier (id), which is not subject to updates. We use t^r and t_i^r interchangeably when i is the identifier of t^r. $t^r[A]$ denotes attribute A of tuple t in r, called a *cell*, and $val(t^r[A])$ denotes the value of this cell.

We abbreviate t_i^r as t_i if r is clear from the context, and also use t_i to denote the id of tuple t_i when it is clear from the context.

One subtle issue is that ODs are specified on *lists* of attributes, while traditional dependencies, *e.g.*, FDs, are typically specified on *sets* of attributes.

Sets and Lists.

(1) $\mathcal{X}$ and $\mathcal{Y}$ denote sets of attributes of schema R, while X and Y denote lists of attributes of R. Specifically, {} (resp. [ ]) denotes the empty set (resp. empty list). XY is a shorthand for the concatenation of X and Y.
(2) By convention, a non-empty list X can be expressed as $[A|Y]$, where *head A* is a single attribute, and *tail* Y is the remaining list by removing A from X.
(3) To simplify notation, for a list X, set $\mathcal{X}$ denotes the set of elements in X, and $t[\mathcal{X}]$ denotes the projection of tuple t on $\mathcal{X}$.

ODs define lexicographic orders commonly found in the SQL order-by clause.

Order Operator $\preceq_X$ **on Lists.** For tuples t, s and an attribute list X, $t \preceq_X s$ if

(1) $X = [\,]$; or
(2) $X = [A|Y]$ and (a) $val(t[A]) <_A val(s[A])$; or (b) $val(t[A]) = val(s[A])$, and $t \preceq_Y s$.

Here $<_A$ is an order operator defined on the domain of attribute A. $<_A$ can be naturally defined for numbers, strings and dates, among other things.

Note that $t \preceq_X s$ when $\forall A \in \mathcal{X}$, $val(t[A]) = val(s[A])$. Let $t \prec_X s$ iff $t \preceq_X s$ but $s \not\preceq_X t$.

Example 2: Recall Fig. 1. $t_5 \prec_{[year,month,day,time]} t_4$, while $t_4 \prec_{[no]} t_5$.   □

Order Dependency [10,13–15]. For two lists of attributes X, Y on R,
(1) X $\mapsto$ Y denotes an *order dependency*, read as X orders Y. An instance r of R satisfies OD $\varphi = X \mapsto Y$, denoted $r \models \varphi$, if for any two tuples $t, s \in r$, when $t \preceq_X s$, $t \preceq_Y s$. (2) We write $r \models \Sigma$ for a set Σ of ODs, when $\forall \varphi \in \Sigma, r \models \varphi$.

Theorem 1: *[13, 15]*.

(1) if an OD X $\mapsto$ Y holds, then the FD $\mathcal{X} \rightarrow \mathcal{Y}$ holds.
(2) $\mathcal{X} \rightarrow \mathcal{Y}$ holds, iff X $\mapsto$ XY holds, for any list X (resp. Y) over the attributes of $\mathcal{X}$ (resp. $\mathcal{Y}$).   □

As stated in [15], the dependency class of ODs *generalizes* that of FDs, in the sense that there is a semantically preserving mapping of any FD γ into a set Σ of ODs, and γ holds iff Σ holds (Theorem 1 (2)).

Violations of Order Dependency [13,15]. For an OD $\varphi = \mathsf{X} \mapsto \mathsf{Y}$, two sources of OD violations are discussed in [13,15]:

(1) A split *w.r.t.* φ is a pair of tuples s and t such that $val(t[\mathcal{X}]) = val(s[\mathcal{X}])$, but $val(t[\mathcal{Y}]) \neq val(s[\mathcal{Y}])$.
(2) A swap *w.r.t.* φ is a pair of tuples s and t such that $t \prec_\mathsf{X} s$ but $s \prec_\mathsf{Y} t$.

Example 3: (1) A split is actually a violation of FD, since each $\mathsf{X} \mapsto \mathsf{Y}$ implies $\mathcal{X} \rightarrow \mathcal{Y}$. In Fig. 1, t_3 and t_4 incur a split *w.r.t.* [city] $\mapsto$ [city, country]: they agree on city values, but not on country values.
(2) t_4 and t_5 cause a swap: $t_5 \prec_{[year,month,day,time]} t_4$, while $t_4 \prec_{[no]} t_5$.   □

3   Framework of Repairing Order Dependency Violations

In this section, we first present the definition of *repair* for OD violations, followed by two repair quality metrics based on the distance of instances and the concept of *attribute hierarchy*. We then formalize the problem of data repairing with order dependencies, and finally prove the complexity of this problem.

Repair *w.r.t.* OD. Given an instance r of schema R and a set Σ of ODs on R, r is inconsistent when there exist OD violations in r *w.r.t.* Σ. A repair of r is an instance r' of R such that (1) r' has the same set of tuples (with identifiers) as r, possibly with modified values; and (2) $r' \models \Sigma$.

Example 4: We get a repair of r in Fig. 1, when modifying the value of $t_5[month]$ to be 02, and $t_4[country]$ to be "UK".   □

Attribute value modification is employed as the only repair operation in our definition, similar to repairing techniques for FDs and DCs, *e.g.*, [2–4,9]. Obviously, value modifications suffice to restore the consistency of any instance *w.r.t.* ODs. There may be a large or even infinite number of repairs. To this end, some metrics are required to evaluate the quality of a repair.

Distance and Cost. A *distance* function d is employed to measure the dissimilarity between two instances carrying the same set of tuples (identifiers), with modified attribute values. Specifically, d is defined as follows [1,3,4,9]:

$$d(r,r') = \sum_{i \in [1,n], j \in [1,m]} \triangle(t_i{}^r[A_j], t_i{}^{r'}[A_j])$$

r, r' are instances of schema $R(A_1, \ldots, A_m)$, both carrying a set of tuple identifiers $\{t_1, \ldots, t_n\}$. For an inconsistent instance r and its repair r', the cost of r' is typically defined as the distance between r and r': $\mathbf{cost}(r,r') = d(r,r')$.

$\triangle(t_i{}^r[A_j], t_i{}^{r'}[A_j])$ is a distance between values of cells $t_i{}^r[A_j]$ and $t_i{}^{r'}[A_j]$. There are varying measurements for different types of values, *e.g.*, string, number, date, time. To avoid depending on a particular approach, we consider a binary distance between values: 0 for equal values, and 1 otherwise. This is a typical setting adopted in data repairing (cardinality-based repairs), *e.g.*, [4,9], where the cost of a repair is measured as the number of changed cells. Note that to find a repair with the smallest cost is NP-hard in this setting [4,9].

Example 5: The repair in Example 4 has a cost of 2.   □

In Example 1, year, month, day and time form a hierarchy of attributes: the value of month is relative to that of year, and the value of day is relative to that of month, etc. This suggests that it essentially incurs more changes to the instance when modifying year attribute, compared to modifying month. We introduce the following concept to encode this observation in the framework.

Attribute Hierarchy. Given a relation schema $R(A_1, \ldots, A_m)$, an attribute hierarchy on R is a strict partial order $<_R$ defined on certain pairs of attributes of R, written as $A_k <_R A_j$. $A_k <_R A_j$ indicates that A_j precedes A_k in the order: A_k is directly or indirectly relative to A_j.

Example 6: In Fig. 1, we define *time* $<_R$ *day* $<_R$ *month* $<_R$ *year*.   □

Based on attribute hierarchy, we present an additional metric to judge the "goodness" of repairs with a same cost. As an auxiliary notion, we denote by $mod(r, r')$ the set of cells modified in repair r' of r.

Metric on Attribute Hierarchy. Given two repairs r', r'' of r with schema R, we say r'' is preferable to r' in terms of attribute hierarchy $<_R$ on R, written as $r'' \rightsquigarrow r'$, if (1) there is a one-to-one mapping from each cell $c' \in mod(r, r')$ to cell $c'' \in mod(r, r'')$ via function ρ or τ, where (a) ρ is an *identity* function: $\rho(c') = c' = c''$, *i.e.*, c', c'' are the same cell, and (b) τ is a *hierarchy reduction* function: $\tau(c') = c''$, such that $c' = t_i[A_j]$, $c'' = t_i[A_k]$, and $A_k <_R A_j$; and (2) at least one cell c' is mapped to c'' via τ.

Example 7: We get a repair r' by setting $val(t_5[year]) = 2019, val(t_4[country]) =$ "UK", and another repair r'' by setting $val(t_5[month]) = 02, val(t_4[country]) =$ "UK". $r'' \rightsquigarrow r'$.   □

Remark. (1) $r'' \rightsquigarrow r'$ implies that $cost(r, r') = cost(r, r'')$, since there is a one-to-one mapping between modified cells in r' and r''. (2) In accordance with the binary distance adopted in the distance function, identity function ρ concerns only the cell (position), not the cell value.

We are now ready to formalize the data repairing problem with ODs.

Repairing OD Violations. Given a relation r of schema $R(A_1, \ldots, A_m)$ with attribute hierarchy $<_R$, when r is inconsistent *w.r.t.* a set Σ of ODs, the problem of *repairing* r *w.r.t.* Σ is to find a repair r' of r, such that there does not exist a repair r'' of r, where (1) $cost(r, r'') < cost(r, r')$, or (2) $r'' \rightsquigarrow r'$.

Intuitively, repairing OD violations aims to find a repair that either has the minimum cost among all repairs, or incurs minimum changes in terms of attribute hierarchy among repairs with the same cost. This problem is necessarily hard.

Theorem 2: *The problem of data repairing with ODs is NP-hard.*   □

4   Data Repairing with Order Dependencies

Despite the intractability of data repairing with ODs, we present efficient heuristic algorithms to address the problem. Given an OD $\varphi = X \mapsto Y$, an order o_φ is specified by its LHS attribute list X. o_φ is then imposed on the RHS attribute list Y, where possible violations are to be repaired. We formalize orders specified by LHS attributes, then present techniques to repair violations on RHS attributes, and finally tackle the repairing problem when multiple ODs are involved.

4.1   Order Specified by LHS Attributes

We first introduce one auxiliary notation.

Equivalence Class. Equivalence class (EC) [1–3,6] is a common technique in data repairing with FDs, for keeping track of equivalence relationships between cell values. We use the following notations: (1) an EC e is a set of tuple *ids*; (2) any tuple *id* t_i belongs to one EC, denoted by $ec(t_i)$; (3) e^A is the projection of EC $e = \{t_i, \ldots, t_k\}$ on attribute A, *i.e.*, a set of cells $\{t_i[A], \ldots, t_k[A]\}$; and (4) all cells in e^A are assigned a same value, referred to as the target value of e^A.

Order Specified by an OD. On a relation r with n tuples $\{t_1, \ldots, t_n\}$, OD $\varphi = X \mapsto [A]$ specifies an order o_φ. Specifically, (1) By lexicographically sorting tuples on X, we get a list of tuple *ids* $[t_{i_1}, \ldots, t_{i_n}]$, where $i_j \in [1, n]$ ($j \in [1, n]$), such that tuples $t_{i_1} \preceq_X \cdots \preceq_X t_{i_n}$. (2) We denote order o_φ as a list of equivalence classes (ECs) $[e_1, \ldots, e_k]$. o_φ is constructed in a linear scan of $[t_{i_1}, \ldots, t_{i_n}]$, by collecting successive tuple *ids* with equivalent X's values in an EC.

Example 8: Consider the instance r given in Fig. 2(a). By sorting on $[A, B]$, we get a list of tuple *ids* $[t_1, t_2, t_3, t_4, t_5]$. Then, the order o_φ specified by $\varphi = [A, B] \mapsto [E]$ is $[\{t_1\}, \{t_2\}, \{t_3, t_4\}, \{t_5\}]$, in a list of ECs.   □

Merge Multiple Orders. Several ODs may share (part of) their RHS attributes; in this case multiple orders are imposed on the same RHS attributes, which may cause violations. We generally *cannot* resolve violations for these orders one by one, since fix violations for one order may break another. One better approach is to merge these orders, say $o_{\varphi_1}, \ldots, o_{\varphi_m}$, into a new order o_φ, such that if o_φ is satisfied, then $o_{\varphi_1}, \ldots, o_{\varphi_m}$ are all satisfied.

Algorithm. Algorithm Merge takes as input orders $o_{\varphi_1}, \ldots, o_{\varphi_m}$, and produces a single order o_φ, such that if o_φ is satisfied, then any o_{φ_i} ($i \in [1, m]$) is satisfied.

(a) instance r (b) tuple relationships

Fig. 2. Instance r for Example 8 to Example 11

Algorithm 1. Merge

---

input : *orders* $o_{\varphi_1}, \ldots, o_{\varphi_m}$

    o_{φ_i} ($i \in [1, m]$) is a list of ECs with size i': $o_{\varphi_i} = [e_{i1}, \ldots, e_{ii'}]$;

    e_{ij} ($j \in [1, i']$) is a set of tuple *ids* with size j': $e_{ij} = \{t_{ij1}, \ldots, t_{ijj'}\}$.

output: *An order* o_φ

1 **for** each tuple t_l in relation r **do** add a vertex v_l to graph G ;

2 **for** $i := 1; i \leq m; i{+}{+}$ **do**

3   **for** $j := 1; j \leq i'; j{+}{+}$ **do**

4   **for** $k := 1; k < j'; k{+}{+}$ **do**

5   add an edge from v_{ijk} to $v_{ij(k+1)}$;

6   add an edge from $v_{ijj'}$ to v_{ij1};

7   **if** $j \neq i'$ **then** add an edge from v_{ij1} to $v_{i(j+1)1}$;

8 find strongly connected components (SCCs) in G; convert G into a directed acyclic graph (DAG), by treating each SCC as a single vertex; relabel vertices in the DAG as V_h (if V_h is a simple vertex in G, $V_h = v_h$);

9 **for** *each vertex* V_h *in the DAG* **do**

10   construct an EC e_h: (a) if V_h is a SCC in G, then $t_l \in e_h$ for all v_l in V_h.

11   (b) otherwise, $e_h = \{t_h\}$, *i.e.*, a single tuple *id* corresponding to vertex v_h.

12 find a topological order o on the vertices in the DAG; put EC e_h into o_φ according to V_h in o ;

---

(1) A graph G is used to represent relationships between tuples (*ids*), and each tuple t_l is denoted as a vertex v_l in G (line 1);

(2) (a) Merge adds edges between tuple *ids* (vertices in G) in a same EC e_{ij} in an order o_{φ_i}, such that these vertices form a circle (lines 4–6). (b) Merge then adds an edge from v_{ij1} in e_{ij} to $v_{i(j+1)1}$ in $e_{i(j+1)}$ for $j \in [1, i'-1]$; this follows the order in list o_{φ_i} (line 7). Merge does (a) and (b) for all orders $o_{\varphi_1}, \ldots, o_{\varphi_m}$.

(3) Merge converts G into a DAG by treating each SCC as a single vertex. Merge then transforms vertices back to ECs in o_φ. Specifically, vertices in a SCC are converted into tuple *ids* in a same EC, and orders of ECs in o_φ follows the topological order in the DAG (lines 8–11).

Example 9: Recall instance r in Fig. 2(a). The order specified by $[A] \mapsto [C]$ is $[\{t_1\}, \{t_2, t_3, t_4\}, \{t_5\}]$ (solid lines) and the order specified by $[D] \mapsto [C]$ is $[\{t_1\},$

$\{t_5\}, \{t_3, t_4\}, \{t_2\}]$ (dotted lines) in Fig. 2(b). By merging the two orders, we get the order as $[\{t_1\}, \{t_2, t_3, t_4, t_5\}]$.   □

Complexity. The time complexity of Merge is $O(m \cdot n)$, where m is the number of orders as input, and n is the number of tuples. This is because the number of edges in G is in $O(m \cdot n)$, and finding SCCs takes linear time in the number of vertices and edges with well-known algorithm, *e.g.*, Tarjan's algorithm [5].

4.2   Fix Violations on RHS Attributes

We start from the basic case, to fix violations for one RHS attribute.

Violations of OD. Suppose order o_φ is imposed on a single RHS attribute A. Recall that OD violations are categorized into split and swap (Sect. 2). To repair OD violations, we need to enforce equivalence between cells in the same EC for resolving split, and simultaneously, enforce ordering between cells in different ECs for resolving swap. This is further complicated since each EC on A can have arbitrary number of different values, and worse, we aim to make the fewest modifications.

Algorithm. We present algorithm Fix to resolve violations for a single RHS attribute. Fix is optimal in that it incurs the minimal number of modifications.

Algorithm 2. Fix

---

    input : *order $o_\varphi = [e_1, \ldots, e_k]$ and attribute A on which o_φ is to be imposed*
1 **for** $j:=1; j \le k; j++$ **do**
2   initialize an empty list L' of dpUnit;
3   **for** each t_i in e_j **do**
4   **if** L' has a dpUnit dp such that $dp.value = val(t_i[A])$ **then**
5           $dp.weight := dp.weight+1$
6   **else** add a new dpUnit $dp'(value=val(t_i[A]), ecno=j, weight=1)$ to L';
7   add all dpUnits in list L' to a list L;
8 sort dpUnits dps in list L by $dp.value$, breaking ties by $dp.ecno$;
9 initialize two lists DP and π with size l, where l is the size of L ;
10 $DP[1] := L[1].weight; \pi[1] := $ -1;
11 **for** $j:=2; j \le l; j++$ **do**
12   **if** $\exists\, i < j$ such that $L[i].ecno < L[j].ecno$ **then**
13        $DP[j]:=L[j].weight+\max_i(DP[i])$;
14        $\pi[j] := argmax_i(DP[i]); (i \in [1, j-1], L[i].ecno < L[j].ecno)$
15   **else** $DP[j] := L[j].weight; \pi[j] := $ -1 ;
16 $index := argmax_{i \in [1,l]} DP[i]$;
17 **while** $index! = -1$ **do**
18   assigns $L[index].value$ as the target value of e_m^A when $m = L[index].ecno$;
19     $index := \pi[index]$;
20 determines target values for *unresolved* ECs by adjacent *resolved* ECs;

Fix takes as inputs the order $o_\varphi = [e_1, \ldots, e_k]$ and the attribute A on which o_φ is to be imposed. Fix modifies some $t_i[A]$ values such that o_φ is satisfied on A.

(1) On attribute A, the projection of an EC e_j in o_φ, i.e., e_j^A, results in a set of cells. For each distinct value v of cells in this set, Fix creates a tuple dp of the structure $(value = v,\ ecno = j,\ weight = w)$, referred to as a dpUnit, when there are w cells in e_j^A having v as their values (lines 3–6). Intuitively, when assigning a target value to e_j^A, all of the cells in a same dpUnit should be modified or not be modified, which incurs a repair cost of w or 0. All dpUnits are collected in a list L for all ECs of o_φ (line 7).

(2) Fix sorts dpUnits in list L by $value$ first, then breaks ties by $ecno$ (line 8).

(3) Fix finds the maximum number of cells on A that can remain unchanged without violating o_φ; this implies finding a *sublist* of L in strictly ascending order of $ecno$, and with the maximal summed $weights$. Fix employs dynamic programming with two lists DP and π (line 9); each list is of size l (l is the size of list L). Fix iteratively computes DP, π for dpUnits in L (lines 11–15). (a) If there is some $L[i]$ such that $L[i].ecno < L[j].ecno$, $L[j]$ can be attached to the existing sublist ended with $L[i]$ having the maximal $DP[i]$. DP and π are maintained accordingly: $DP[j]$ is to save the maximal summed weights of any sublist ended with $L[j]$, and $\pi[j]$ is to save the index of the $(j$-1)th elements in the sublist. (lines 12–14). (b) Otherwise a new sublist starts from $L[j]$ (line 15).

(4) Fix finds the *index* that maximizes $DP[index]$ (line 16), and backtracks all indices of list L that forms the optimal sublist (line 19); corresponding dpUnits in L are *not* modified by Fix. Fix finds the EC e_m^A (the projection of e_m in o_φ on A) such that $m = L[index].ecno$, and assigns $L[index].value$ as the target value of e_m^A. That is, Fix sets all values of cells in e_m^A to be $L[index].value$ when they have different values (line 18). We call e_m^A *resolved*.

(5) For each *unresolved* EC e_h^A, Fix finds its most adjacent *resolved* ECs: $i_1 = min(i)$ (e_{h-i}^A is resolved), $i_2 = min(i)$ (e_{h+i}^A is resolved). Note that (a) at least one of i_1, i_2 exists, and that (b) there may be other *unresolved* ECs between e_h^A and its most adjacent *resolved* ECs. Fix determines target values for e_h^A, following the imposed order restrictions by $e_{h-i_1}^A$, $e_{h+i_2}^A$.

Example 10: Recall $[A, B] \mapsto [E]$ specifies order $o_\varphi = [\{t_1\}, \{t_2\}, \{t_3, t_4\}, \{t_5\}]$ in Fig. 2(a). When enforced on E, we get $[\{val(t_1[E]) = 5\}, \{3\}, \{4, 1\}, \{5\}]$. Fix then computes and sorts dpUnits in list L as $[(value = 1, ecno = 3, weight = 1)$, **(3,2,1)**, **(4,3,1)**, $(5,1,1)$, **(5,4,1)**$]$. By computing DP and π, Fix finds the optimal sublist (dpUnits in bold); they are in strictly ascending order of $ecno$ and have the maximal summed $weights$ of 3. Those dpUnits are not modified. Fix determines target values of ECs $\{t_2\}$ ($ecno = 2$), $\{t_3, t_4\}$ ($ecno = 3$) and $\{t_5\}$ ($ecno = 4$) as 3, 4 and 5 respectively. $t_4[E]$ is in EC $\{t_3, t_4\}$, and hence Fix sets $val(t_4[E]) = 4$. For the *unresolved* EC $\{t_1\}$, Fix sets its value v according to the order restrictions $val(t_1[E]) \leq 3$, since 3 is the target value of EC $\{t_2\}$. Suppose we want to minimize the quadratic objective function $(v - 5)^2$ under this restriction, where 5 is the original value of $t_1[E]$. We can set $v = 3$ in this case. Existing techniques, *e.g.*, quadratic programming, can be employed in this process, complementary to our method. Similar techniques exist for string values as well, see *e.g.*, [4]. □

Theorem 3: *Algorithm* Fix *is optimal, with the fewest value modifications.*   □

Complexity. The worst-case time complexity of Fix is $O(n^2)$, where n is the number of tuples. Specifically, (1) lines 1–7 takes $O(n)$ by hashing on cell values, since each tuple *id* belongs to one EC; (2) line 8 takes $O(l \cdot log(l))$, l is the size of list L, and $l \leq n$. There are typically many duplicate values in each EC; l is much smaller than n; (3) lines 11–15 takes $O(l^2)$; and (4) lines 17–20 takes $O(n)$.

We further present techniques to repair violations on multiple RHS attributes.

Algorithm. We present algorithm FixM, which takes as inputs order o_φ and RHS attribute list $Y = [A|T]$. FixM modifies some RHS attribute values such that o_φ is satisfied. FixM is recursively defined, and employs algorithm Fix to resolve violations for each attribute in Y. Specifically, FixM works as follows.

(1) FixM calls Fix with inputs o_φ and A, to resolve violations on A (line 1).

(2) Based on the results of repairing cells on A, FixM divides o_φ into several partitions (orders), to be further enforced on the remaining list T; each e_j in o_φ belongs to one of these partitions. FixM generates these partitions one by one in o'_φ, and initially puts e_1 into o'_φ (line 3). Recall that in Fix, to satisfy the ordering between two successive ECs e_{j-1} and e_j on attribute A, the target value v_{j-1} of e_{j-1}^A must be less than or equal to the target value v_j of e_j^A. (a) If v_{j-1} $= v_j$, the ordering between e_{j-1} and e_j is copied in o'_φ (line 5). (b) Otherwise, the ordering between e_{j-1} and e_j is already guaranteed to be satisfied on Y in lexicographical order. In this case, FixM discards the order restriction between e_{j-1} and e_j in o'_φ, and recursively calls FixM with o'_φ on T (line 7). After that, FixM starts a new partition of o_φ in o'_φ, with e_j as the first EC (line 8).
(3) A final call of FixM is required for remaining ECs in o'_φ (line 9).

Example 11: We repair violations of $[A, B] \mapsto [E, F]$ in Fig. 2(a). Suppose following Example 10, FixM sets $val(t_4[E]) = 4$ and $val(t_1[E]) = 3$ in repairing E.

Algorithm 3. FixM

---

    input : order $o_\varphi=[e_1,\dots, e_k]$ and RHS attribute list $Y = [A|T]$
1 Call algorithm Fix with o_φ and A as inputs;
2 **if** $T = [\ ]$ **then return**;
3 $o'_\varphi := [e_1]$;
4 **for** $j:=2; j \leq k; j++$ **do**
5   **if** e_{j-1}^A *and* e_j^A *have a same target value* **then** add e_j to the tail of o'_φ ;
6   **else**
7   call algorithm FixM with o'_φ and T as inputs;
8         $o'_\varphi := [e_j]$;
9 call algorithm FixM with o'_φ and T as inputs ;

Then the order $o_\varphi = [\{t_1\}, \{t_2\}, \{t_3, t_4\}, \{t_5\}]$ specified by $[A, B]$ is divided into three orders to be imposed on F: $[\{t_1\}, \{t_2\}], [\{t_3, t_4\}], [\{t_5\}]$. This is because $val(t_2[E]) = 3$ and $val(t_3[E]) = 4$, this already guarantees $t_2 \preceq_{[E,F]} t_3$; similarly for t_4 and t_5. Note that the three orders are disjoint in terms of tuples ids, and each of them is imposed on a horizontal partition of r on attribute F.

Consider $[A] \mapsto [A, B]$, which is essentially a FD $A \to B$. The order o specified by $[A]$ is $[\{t_1\}, \{t_2, t_3, t_4\}, \{t_5\}]$, when applying o on A in FixM, no value modification occurs. However, this order is divided into three orders $[\{t_1\}], [\{t_2, t_3, t_4\}], [\{t_5\}]$, enforcing equivalence in each EC on B with further call of FixM. □

Complexity. The worst-case time complexity of FixM is $O(|Y| * n^2)$, where n is the number of tuples and $|Y|$ is the number of RHS attributes. Note that FixM partitions the input order into several orders enforced on disjoint set of cells when possible, to reduce the computational complexity in practice.

Remark. FixM considers the preference of attribute hierarchy: it delays value modifications in RHS attributes whenever possible.

4.3 Repairing Violations for Multiple ODs

We put our techniques together to repair violations for multiple ODs.

Algorithm. RepairOD is to fix OD violations for a set Σ of ODs on relation r. (1) To facilitate repair computations, we encode the relationships between attributes in an *attribute graph* G, for identifying the sequence of attributes in the repairing process. Specifically, for each OD φ in Σ, where $\varphi = X \mapsto Y$, $X = [A_1, \ldots, A_k]$ and $Y = [B_1, \ldots, B_l]$, we add in G (a) vertices for $A_1, \ldots, A_k$, $B_1, \ldots, B_l$, respectively; (b) a *composite* vertex $[A_1, \ldots, A_k]$; (c) an edge from each vertex A_i ($i \in [1, k]$) to vertex $[A_1, \ldots, A_k]$; (d) an edge from vertex $[A_1, \ldots, A_k]$ to B_1; and (e) an edge from B_i to B_{i+1} for each $i \in [1, l-1]$.

Example: we show in Fig. 3(b) the attribute graph for ODs in Fig. 3(a).
(2) We find SCCs in G, convert G into a DAG by treating each SCC as a single vertex, and find a topological order on the vertices in the DAG.

Example: The topological order for Fig. 3(b) can be A, SCC (B, $[A, B]$, C), E, D.

Algorithm 4. RepairOD

---

    input : a set Σ of ODs, relation r
1 construct *attribute graph* G ;
2 find SCCs in G; convert G into a DAG by treating each SCC as a single vertex; find a topological order on the vertices in the DAG ;
3 **foreach** *vertex v in the DAG, by following the topological order* **do**
4   **if** *v is not a SCC* **then** repair the attribute corresponding to v;
5   **else** make a traversal of all vertices in the SCC, and repair corresponding attribute one by one, possibly with iterative computations;

$$[AB] \mapsto [CD] \quad [C] \mapsto [B]$$

$$[E] \mapsto [D] \quad [C] \mapsto [E]$$

(a) A set Σ of ODs　　　　　　(b) attribute graph

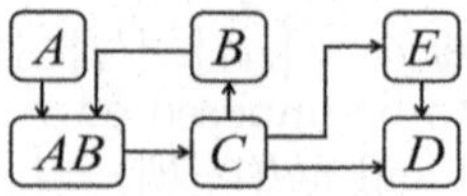

Fig. 3. Example of RepairOD

(3) We deal with vertices in the DAG one by one in the topological order. If v is not a SCC, we repair the corresponding attribute, say A. Specifically, we collect all ODs with A as RHS attributes, and employ algorithm Merge to find a single order o to be imposed on A. When A is not the left-most attribute in the RHS attribute list, we may employ FixM to compute *partitioned* orders on A, to be merged with orders specified by other ODs. Finally, we employ algorithm Fix or FixM to repair A with the order o.

Example: Consider attribute D. There are two ODs with D in RHS: $[A, B] \mapsto [C, D]$ and $[E] \mapsto [D]$. When calling FixM $[A, B] \mapsto [C, D]$, after repairing attribute C (C is in a SCC, see (4) below), we get the order (or several partitioned orders) to be imposed on D. We use Merge to merge this order (these orders) with the order specified by E together, and employ Fix on D with this merged order. This improves efficiency and repair quality, as will be seen in Sect. 5.

(4) For a SCC in the DAG, we deal with vertices in that SCC one by one in a traversal of all vertices. We repair corresponding attributes just like (3). However, iterative computations may be required when repairing attributes involved in a SCC, due to interactions between ODs. More specifically, we have to repair an attribute A_i when repairing attribute A_j incurs changes and in the SCC (a) there is an edge from A_j to A_i, or (b) there is an edge from A_j to a composite vertex v' and an edge from v' to A_i. To this end, we label attributes once they are repaired. When an attribute A is repaired again, we apply a slightly modified version of Fix in its final step. For each *unresolved* EC e_h^A, Fix finds its most adjacent *resolved* EC, w.l.o.g., say e_{h-i}^A, and sets the target value of e_h^A the same as that of e_{h-i}^A; this guarantees the reduction of number of ECs on A.

Example: Consider the SCC $(B, [A, B], C)$ in Fig. 3(b). We start from a vertex, say B, and repair it according to $[C] \mapsto [B]$. No computation is required for the composite vertex $[A, B]$. We then repair C with $[A, B] \mapsto [C, D]$. Note that D is after the SCC in the topological order, we hence do not repair D at this time. When any changes occur on C, we repair B again; similarly for C.

Remark. RepairOD always terminates and generates a repair. For those attributes not involved in a SCC, repairing them according to the topological order in G suffices. While for those attributes A in a SCC, RepairOD guarantees termination because at each step the total number of ECs on A is reduced.

5   Experimental Study

Experimental Setting. We use one machine with 24 core Intel Xeon CPU and 64GB RAM, run each experiment 5 times and report the average here.

Data. (1) Real-life data (FLI) is about US flights. We obtain more than 100K tuples with 12 attributes (https://www.transtats.bts.gov/). This dataset is used in OD discoveries [10,14]; we choose 7 ODs. (2) Synthetic data SYN is to evaluate our approach with intricate ODs. We generate up to 100K tuples with 10 attributes, and design 8 ODs on it (extension of ODs in Fig. 3).

Algorithms. We implement the following algorithms in Java: (1) RepairOD (with Fix, FixM and Merge) and (2) algorithm VC for comparison. VC follows the same framework as RepairOD, and only differs in repairing a single RHS attribute, *i.e.,* algorithm Fix. VC handles OD violations on a RHS attribute with *conflict graph*, a technique in repairing violations for FDs [9] and DCs [4]. VC builds a conflict graph with an edge between each pair of cells that causes a violation, employs a 2-approximation algorithm for a minimum vertex cover (MVC) of the graph, and modifies cells in the MVC similar to the last step of Fix.

Metrics. Cardinality-based repairs aim to correctly identify error cells (positions), so as to minimize number of changes. We measure the number of changes ($\#C$) in the repair. Based on the known ground truth, we evaluate precision (P, the fraction of correct changes), recall (R, correct changes over the total number of errors), and F-measure (F, $2 \times (P \times R)/(P+R)$), in terms of correct positions.

All experiments are controlled by 3 parameters: (1) n: the number of tuples; (2) $|\Sigma|$: the number of ODs; and (3) θ: the ratio of dirty data, the number of introduced dirty cells to the number of cells involved in ODs in the dataset.

Exp-1. Using FLI data, we compare RepairOD against VC. We set $n = 30K$, $|\Sigma| = 4$, $\theta = 5\%$ by default, and vary one parameter in each of the experiments.

Varying $|n|$. We first evaluate all algorithms by varying n from 10K to 90K. Figure 4(a) shows results of running time; we omit results over 30 min. We see RepairOD significantly outperforms VC and scales well with n: it takes less than 73 s when $n = 90k$. We reduce the number of value comparisons in VC to $O(n \cdot log(n))$ with sorted partitions [10], and the most expensive part of VC is to maintain its large conflict graph. In RepairOD, finding the optimal sublist governs the overall time, with a complexity of $O(l^2)$. l is the number of dpUnits in the list L (Algorithm 2: Fix), which is much smaller than n on FLI.

Figure 4(b) shows results of the number of changes ($\#C$). We find RepairOD significantly outperforms VC; RepairOD finds repairs with much smaller costs. Recall that Fix is proved to be optimal in terms of changes for a given OD with one RHS attribute. Figure 4(c) shows results of F-measure (F). We see that F values of RepairOD are high, in the range of [0.85, 0.88]; RepairOD can correctly identify error cells in most cases. VC exhibits lower F values compared to RepairOD, mainly due to lower precision values (not shown in figures); the 2-approximation vertex cover algorithm causes some unnecessary changes.

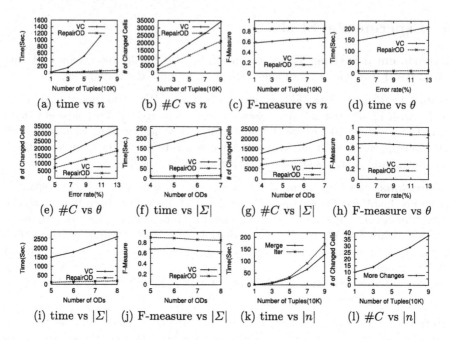

Fig. 4. Experimental results

In practice, we may also want to find correct values (last step of Fix). When F-measures are considered in terms of both positions and values, they reduce to the range of [0.45, 0.49] on FLI (not shown), as expected. This is because (1) randomly introduced errors are not helpful in deducing original values; and (2) many values are identified based on order restrictions, say $v_1 \leq v \leq v_2$, when the difference between v_1, v_2 is large, it is very difficult to find exactly the correct value of v. When data follow some mathematical distributions, statistical methods [17] can be combined with our approach, as a future work.

Varying θ. We then evaluate all algorithms by varying θ from 5% to 13%. Figure 4(d) shows results of running times. The number of edges in the conflict graph of VC almost grows linearly with θ, so are the running times. As θ increases, we see times of RepairOD increase from 11 s to 14 s, due to more dpUnits in list L (Algorithm 2: Fix). Figure 4(e) shows results of $\#C$. Since the number of errors increases when θ increases, more changes are required for both algorithms. The $\#C$ values of RepairOD are almost linear with θ, as expected, and the F-measure values are stable as θ increases (not shown).

Varying $|\Sigma|$. We then vary the number of ODs ($|\Sigma|$) from 4 to 7. Figure 4(f) shows that times of the two algorithms increase almost linearly with $|\Sigma|$. This is because (1) errors are evenly distributed in cells involved in ODs; (2) ODs found in FLI are relatively simple such that the attribute graph has no circle (Algorithm 4: RepairOD); and (3) VC follows the same framework as RepairOD, which repairs attributes one by one in the topological order. In Fig. 4(g), we can see more changes are required to fix violations for both algorithms, as expected. When $|\Sigma|$ increases, the number of introduced errors increases with a fixed θ.

Exp-2. Using SYN data, we further verify our approach with more intricate ODs. We set $n = 90K$, $|\Sigma| = 5$, $\theta = 5\%$ by default in this set of experiments.

Varying θ. By varying θ from 5% to 13%, Fig. 4(h) shows F-measure values. The results confirm our observations on FLI: RepairOD has high F values and outperforms VC. We find F values slightly decrease (about 5%) for both algorithms as θ increases. More ODs with multiple RHS attributes are used in Exp-2, and some introduced dirty cells are at the tail of RHS attribute list, which do not cause violations and are hence not repaired. This negatively affects values of recall.

Varying $|\Sigma|$. By varying $|\Sigma|$ from 5 to 8, Fig. 4(i) and (j) show values of time and F-measure, respectively. RepairOD scales well: the time increases from 111 s to 202 s as $|\Sigma|$ increases. F-measure values slightly decrease when $|\Sigma|$ increases, due to more interactions between ODs and some error values at the tail of RHS attribute list do not cause violations and are not repaired.

Exp-3. We demonstrate the benefits of Merge, to merge orders into one before fixing attributes. For comparison, we implement another approach, denoted by Iter, which repairs RHS attributes with several orders one by one and guarantees termination by merging ECs in the repairing. We set $|\Sigma| = 4$, $\theta = 5\%$, vary n from 10K to 90K on SYN data. Figure 4(k) shows running times of both algorithms. We find RepairOD (with Merge) outperforms Iter, and this becomes more evident as n increases. The number of repairing attributes with Fix is reduced by capitalizing on Merge, and the time complexity of Merge is only $O(n)$. Figure 4(l) shows that Iter incurs more number of changes ($\#C$) compared to Merge; Merge also helps improve repair quality by reducing the number of changes in the repair.

6   Conclusions

We have formalized the problem of repairing OD violations, studied its computational complexity, presented algorithms, and experimentally verified our techniques. We are currently developing distributed repairing techniques for OD violations to enhance the scalability, so as to leverage more resources with distributed computations. Another topic for future work is to combine our techniques with statistical methods, in helping identify correct values.

Acknowledgements. This work is supported by NSFC 61572135, NSFC 61370157, National High Technology Research and Development Program (863 Program) of China (2015AA050203), State Grid Rsearch Project No. 52094016000A, Shanghai Science and Technology Project (No. 16DZ1100200, 16DZ1110102), Aircraft Risk Management Database Project, National Nonprofit Ocean Research Project (No. 201405031-04).

References

1. Bohannon, P., Fan, W., Flaster, M., Rastogi, R.: A cost based model and effective heuristic for repairing constraints by value modification. In: SIGMOD (2005)
2. Beskales, G., Ilyas, I., Golab, L., Galiullin, A.: Sampling from repairs of conditional functional dependency violations. VLDB J. **23**(1), 103–128 (2014)
3. Cong, G., Fan, W., Geerts, F., Jia, X., Ma, S.: Improving data quality: consistency and accuracy. In: VLDB (2007)
4. Chu, X., Ilyas, I., Papotti, P.: Holistic data cleaning: putting violations into context. In: ICDE (2013)
5. Cormen, T., Leiserson, C., Rivest, R., Stein, C.: Introduction to Algorithms. MIT Press, Cambridge (2009)
6. Dallachiesa, M., Ebaid, A., Eldawy, A. Elmagarmid, A., Ilyas, I., Ouzzani, M., Tang, N.: NADEEF: a commodity data cleaning system. In: SIGMOD (2013)
7. Fan, W., Li, J., Ma, S., Tang, N., Yu, W.: Towards certain fixes with editing rules and master data. VLDB J. **21**(2), 213–238 (2012)
8. Ginsburg, S., Hull, R.: Order dependency in the relational model. TCS **26**(1), 149–195 (1983)
9. Kolahi, S., Lakshmanan, L.: On approximating optimum repairs for functional dependency violations. In: ICDT (2009)
10. Langer, P., Naumann, F.: Efficient order dependency detection. VLDB J. **25**(2), 223–241 (2016)
11. Ng, W.: An extension of the relational data model to incorporate ordered domains. TODS **26**(3), 344–383 (2001)
12. Song, S., Chen, L.: Differential dependencies: reasoning and discovery. TODS **36**(3), 16:1–16:41 (2011)
13. Szlichta, J., Godfrey, P., Gryz, J.: Fundamentals of order dependencies. PVLDB **5**(11), 1220–1231 (2012)
14. Szlichta, J., Godfrey, P., Golab, L., Kargar, M., Srivastava, D.: Effective and complete discovery of order dependencies via set-based axiomatization. PVLDB **10**(7), 721–732 (2017)
15. Szlichta, J., Godfrey, P., Gryz, J., Zuzarte, C.: Expressiveness and complexity of order dependencies. PVLDB **6**(14), 1858–1869 (2013)
16. Wang, J., Tang, N.: Towards dependable data repairing with fixing rules. In: SIGMOD (2014)
17. Zhang, A., Song, S., Wang, J.: Sequential data cleaning: a statistical approach. In: SIGMOD (2016)

Multi-Worker-Aware Task Planning in Real-Time Spatial Crowdsourcing

Qian Tao[1], Yuxiang Zeng[2], Zimu Zhou[3], Yongxin Tong[1(✉)], Lei Chen[2], and Ke Xu[1]

[1] SKLSDE Lab and BDBC, Beihang University, Beijing, China
{qiantao,yxtong,kexu}@buaa.edu.cn
[2] The Hong Kong University of Science and Technology, Hong Kong SAR, China
{yzengal,leichen}@cse.ust.hk
[3] Laboratory TIK, ETH Zurich, Zurich, Switzerland
zzhou@tik.ee.ethz.ch

Abstract. Spatial crowdsourcing emerges as a new computing paradigm with the development of mobile Internet and the ubiquity of mobile devices. The core of many real-world spatial crowdsourcing applications is to assign suitable tasks to proper workers in real time. Many works only assign a set of tasks to each worker without making the plan how to perform the assigned tasks. Others either make task plans only for a single worker or are unable to operate in real time. In this paper, we propose a new problem called the _Multi-Worker-Aware Task Planning (MWATP)_ problem in the online scenario, in which we not only assign tasks to workers but also make plans for them, such that the total utility (revenue) is maximized. We prove that the offline version of MWATP problem is NP-hard, and no online algorithm has a constant competitive ratio on the MWATP problem. Two heuristic algorithms, called Delay-Planning and Fast-Planning, are proposed to solve the problem. Extensive experiments on synthetic and real datasets verify the effectiveness and efficiency of the two proposed algorithms.

Keywords: Spatial crowdsourcing · Task assignment · Task planning

1 Introduction

The development of mobile devices has triggered the fast growing of spatial crowdsourcing. Unlike traditional crowdsourcing where workers perform tasks via webs [1], workers in spatial crowdsourcing need to physically go to the location of a task to perform it [2]. Spatial crowdsourcing extends traditional crowdsourcing to the physical world and has seen many applications in daily life [3,4]. For example, Waze[1] provides a dynamic traffic navigation by collecting the GPS information; Uber[2] offers an efficient real-time taxi-calling service; Gigwalk[3] performs location-based micro tasks via crowds, _etc._

---

[1] http://www.waze.com.
[2] http://www.uber.com.
[3] http://www.gigwalk.com.

© Springer International Publishing AG, part of Springer Nature 2018
J. Pei et al. (Eds.): DASFAA 2018, LNCS 10828, pp. 301–317, 2018.
https://doi.org/10.1007/978-3-319-91458-9_18

Table 1. Release time, expiration time and utility

Task/worker	t_1	w_1	t_2	t_3	w_2	t_4	t_5
Release time	1	1	1.5	2	3	5	5.5
Expiration time	4	6.8	3	5	8.2	6.2	7
Utility (revenue)	5	/	2	3	/	2	1

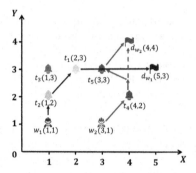

Fig. 1. Initial locations of workers and tasks (Color figure online)

One of the most important issues in spatial crowdsourcing research is how to assign tasks to proper workers [5–8]. Imagine the following scenario. Suppose Alice is off duty at 5:00 p.m. from her office, and she wants to perform some tasks passingly from Gigwalk on her way home. However, she has to reach home before 6:00 p.m. to have dinner with her family. Thus Alice wants to receive not only the guidance of which tasks to perform, but also a plan (order) to perform them. Every performed task contributes a revenue to the platform. When *multiple workers* raise such demands *in real time*, the platform faces a new problem in spatial crowdsourcing: *how to make plans of tasks for multiple workers in online scenario, such that the total revenue of the platform is maximized?*

We further illustrate our motivation using the following example.

Example 1. Suppose there are two workers $w_1 - w_2$ and five tasks $t_1 - t_5$ appearing on the platform, whose release and expiration times (in minutes) are shown in Table 1, and the locations are shown in Fig. 1. The "utility" of each task is also shown in Table 1, representing the revenue contributed to the platform when the task is performed. For ease of presentation, the coordinates in Fig. 1 have been transformed to the corresponding time. Workers and tasks can be observed after their release times, and tasks cannot be performed after their expiration times. At time 1, only the task t_1 and the worker w_1 are observed. Along a route $\langle (1,1), (1,2), (2,3), (5,3) \rangle$ (blue arrows in Fig. 1), w_1 performs $\{t_1, t_2\}$, and reaches his/her destination d_{w_1} at time 6.41 (we calculate the time accurately to 2 decimal places), which is earlier than w_1's expiration time. Since t_5 appears at time 5.5, w_1 cannot reach his/her destination d_{w_1} earlier than time $r_{t_5} + dis(l_{t_5}, d_{w_1}) = 7.5$ if he/she performs t_5. w_2 appears at time 3, but no task can be accomplished by w_2 right away. Suppose we let w_2 move to location $(4, 2)$ and stay. At time 5, t_4 appears, and it can be accomplished by w_2. We then alter w_2's plan to be "accomplishing t_4 and moving to his/her destination" (gray arrow in Fig. 1). However, at time 5.5, observing that t_5 appears and it can be accomplished by w_2 (w_2 is at location $(4, 2.5)$ right now), we redirect w_2 to perform t_5 before he/she moves to his/her final destination. w_2 can finally accomplish t_4 and t_5. Based on the above plans for w_1 and w_2, we obtain a total utility (revenue) of 10, which is the optimal planning in this instance.

Many works model the task assignment problem as an online bipartite graph matching and only assign a set of tasks to the workers without indicating an order to perform them [2, 9, 10]. Some pioneer works have explored task planning [11–16]. However, they either are designed for a single worker [11, 13], or cannot handle the real-time (*i.e.*, online) scenario [12, 14–16].

In this paper, we propose a new task assignment problem for real-time spatial crowdsourcing, called the *Multi-Worker-Aware Task Planning (MWATP)* problem. We attempt to not only assign a set of tasks to multiple workers, but also make plans for them, to maximize the total utility (revenue) contributed to the platform, in the *two-sided online* scenario (*i.e.* both workers and tasks appear on the platform dynamically). In summary, we make the following contributions.

- We formulate the Multi-Worker-Aware Task Planning *(MWATP)* problem, which assigns tasks and makes plans for multiple workers in online scenario, such that the total utility (revenue) is maximized. We prove that the offline MWATP problem is NP-hard, and any online algorithm for the online MWATP problem has no constant competitive ratio.
- We propose two heuristic algorithms, called Delay-Planning and Fast-Planning to solve the online MWATP problem.
- We conduct extensive experiments on both synthetic and real datasets. Evaluations verify the effectiveness and efficiency of our proposed algorithms.

The rest of the paper is organized as follows. We formally define the MWATP problem and prove its hardness in Sect. 2. Two heuristic algorithms are proposed in Sect. 3. We present the experimental evaluations in Sect. 4, review related work in Sect. 5, and finally conclude this work in Sect. 6.

2 The MWATP Problem

In this section, we first formally define the *Multi-Worker-Aware Task Planning (MWATP)* problem, and then prove its hardness.

2.1 Problem Definitions

This subsection presents the formal definition of the *Multi-Worker-Aware Task Planning* problem.

Definition 1 (Task). *A task t is a tuple $\langle l_t, r_t, e_t, u_t \rangle$, where l_t is the location which requires the worker to reach, r_t and e_t are the release time and expiration time of task t, and u_t is the utility (revenue) contributed to the platform if the task is accomplished. The task t can be observed only after its release time r_t, and it cannot be performed after its expiration time e_t. The time interval $[r_t, e_t]$ is called the valid interval of t.*

We assume the release time of a task is always no greater than its expiration time, *i.e.,* $r_t \leq e_t$ for $\forall t \in T$, because otherwise the task will never be finished. We also assume a task can only be performed by one worker. Besides, we use "utility" to represent the revenue of a task hereafter.

Definition 2 (Worker). *A worker w is a tuple $w = \langle s_w, d_w, c_w, r_w, e_w \rangle$, where w appears at his/her release time r_w with the initial location s_w, and needs to reach his/her destination d_w before the expiration time e_w. We use c_w to denote the current location of w at a certain time T^*. Specifically, c_w equals the initial location s_w when w appears on the platform. A worker can accomplish a task t if he/she can reach the location of t within the valid interval $[r_t, e_t]$, which will add a utility value of u_t for the platform.*

We model the locations on a metric space (M, dis), where M is a set of locations, and dis is a function, $dis : M \times M \rightarrow R$, and assume each worker can reach his/her destination before his/her expiration time, *i.e.,* $e_w \geq r_w + dis(s_w, d_w)$ for $\forall w \in W$. For simplicity, each worker is assumed to travel at the same constant speed. Consequently, a distance can be represented by a time period, and we will follow this rule whenever there is no ambiguity.

Definition 3 (Guidance). *A guidance for a worker w is a tuple $g = \langle t_g, DIR(l) \rangle$, which means at time t_g, worker w needs to head to the location l from his/her current location. Note that l can be the same as the worker's current location. In this case, we denote $DIR(l)$ as STAY.*

Definition 4 (Plan). *A plan for a worker w is a vector of guidance $p_w = \langle g_1, g_2, \ldots, g_{|p_w|} \rangle$, where $t_{g_i} < t_{g_{i+1}}$ for $i = 1, 2, \ldots, |p_w - 1|$. A plan is valid if w can reach his/her destination d_w before his/her expiration time e_w following the plan p_w. Given a set of tasks T, we further denote $AT(T, p_w)$ as the set of tasks that can be accomplished by w following p_w.*

Note that the plan p_w for a worker w can be updated by the platform when new tasks appear on the platform. However, the updated plan should always be valid. With a plan p_w, we can generate a route of the worker. Based on this route, we can check whether a task can be accomplished, *i.e.* a task $t \in T$ belongs to $AT(T, p_w)$, by checking whether w can reach l_t within t's valid interval $[r_t, e_t]$. Given a set of plans P for multiple workers on a set of tasks T, we define the set of accomplished tasks $AT(T, P) = \cup_{p \in P} AT(T, p)$.

Definition 5 (Online Multi-Worker-Aware Task Planning Problem).
Given a set of workers W and a set of tasks T, where workers and tasks arrive

one by one according to their release times, the problem is to find a valid plan set P for W, such that the total utility of accomplished tasks, i.e.,

$$U(T, P) = \sum_{t \in AT(T,P)} u_t \qquad (1)$$

is maximized.

In this paper we mainly study the online MWATP problem. If not explicitly specified, we will use "MWATP" to refer to the online MWATP problem.

Example 2. Assume the same settings as in Example 1. Then the plan for w_1 is $p_{w_1} = \langle \langle 1, DIR((1,2)) \rangle, \langle 2, DIR((2,3)) \rangle, \langle 3.41, DIR((5,3)) \rangle \rangle$, which generates a route $\langle (1,1), (1,2), (2,3), (5,3) \rangle$. Similarly, the plan for w_2 is $p_{w_2} = \langle \langle 3, DIR ((4,2)) \rangle, \langle 4.41, STAY \rangle, \langle 5, DIR((4,4)) \rangle, \langle 5.5, DIR((3,3)) \rangle, \langle 6.62, DIR((4,4)) \rangle \rangle$, which also generates a route $\langle (3,1), (4,2), (4,2.5), (3,3), (4,4) \rangle$. The set of tasks accomplished from $P = \{p_{w_1}, p_{w_2}\}$ is $AT(T, P) = AT(T, p_{w_1}) \cup AT(T, p_{w_2}) = \{t_1, t_2, t_4, t_5\}$.

2.2 Hardness of MWATP Problem

In this subsection, we first show that the offline MWATP problem is NP-hard, and then prove that no algorithm can achieve a constant competitive ratio on the MWATP problem.

Definition 6 (Offline Multi-Worker-Aware Task Planning Problem). *Given a set of workers W and a set of tasks T, where the release times of workers and tasks are known a priori, the problem is to decide a valid plan set P for W, such that the total utility of the accomplished tasks is maximized.*

Theorem 1. *The offline MWATP problem is NP-hard.*

Proof. We prove the NP-hardness of offline MWATP problem by reducing the orienteering problem [17] to it. The decision version of MWATP problem is to decide if there is a valid plan set P, such that total utility is no less than U. The decision version of orienteering problem is defined as follows. Given n nodes, where one is the start node s_1, one is the end node s_n, and each of the other $n - 2$ nodes is associated with a score, the objective is to find a route of nodes starting from s_1 and ending at s_n, such that the total score is no less than S, with a time constraint T_{MAX}. For an instance I of the orienteering problem, we map the start node s_1 to the worker's start location, the end node s_n to the worker's destination, the other $n - 2$ nodes to $n - 2$ tasks, and the decision threshold S to U. Let the release time and expiration time of each task and the worker be 0 and T_{MAX}, respectively. Now we get an instance I' of the offline MWATP problem. In I', a task can be performed at any time, as long as the worker can reach his/her destination on time. This means that as long as there is a route for the worker in I' achieving utility U, then there must be a route in the orienteering problem gaining the same scores, and vice versa. Since the decision version of orienteering problem is NP-complete, the optimization version of offline MWATP problem is NP-hard.   □

Next we prove that for the MWATP problem, neither deterministic nor randomized online algorithm can yield a constant competitive ratio. Although a similar claim of a special case of the MWATP problem has been considered in [13], it neglects the proof on randomized algorithms.

Lemma 1. *No deterministic algorithm for the MWATP problem has a constant competitive ratio.*

Proof. The problem in [13] is a special case of the MWATP problem with a single worker. Since the problem in [13] does not have a deterministic algorithm with constant competitive ratio, the MWATP problem does not have a deterministic algorithm with constant competitive ratio either.   □

Lemma 2. *No randomized algorithm for the MWATP problem has a constant competitive ratio.*

Fig. 2. An instance that randomized algorithms perform bad

Proof. We prove the lemma by showing that the MWATP problem with exactly one worker does not have a constant competitive ratio. Consider an instance shown in Fig. 2. We omit the Y axis since tasks and workers appear on the X axis. l_0 is the origin with coordinate $(0,0)$. Let m be an arbitrary positive integer and $\epsilon = \frac{1}{m}$. At time 1, with probability $\frac{1}{m}$, n tasks appear at location l_i with expiration time $1 + \frac{\epsilon}{2}$. All of the tasks have a utility value of 1. This yields a probability distribution $\mathcal{X}$ over the input of the tasks. At time 0, a worker w appear at l_0, with the destination $l_m = (1,0)$ and expiration time 2. No matter where the n tasks appear, in the optimal solution w can wait at the location until tasks appear, and then go to the destination before his/her expiration time. Therefore the optimal result on $\mathcal{X}$ is $\mathbb{E}_{\mathcal{X}}[OPT] = n$. Now consider a generic deterministic online algorithm ALG. The worker at most reach one location of $l_1, l_2, \ldots, l_m$ before the tasks' expiration time, no matter where he/she is located at time 1. This means that the expectation of the utility value under the input distribution $\mathcal{X}$ is at most $\mathbb{E}_{\mathcal{X}}[ALG] \leq \frac{1}{m} \cdot n = n\epsilon$. This yields

$$\frac{\mathbb{E}_{\mathcal{X}}[ALG]}{\mathbb{E}_{\mathcal{X}}[OPT]} \leq \frac{n\epsilon}{n} = \epsilon \tag{2}$$

The ratio for any deterministic online algorithm becomes unbounded when ϵ is small enough. From Yao's Principle [18], no randomized algorithm for the MWATP problem can achieve a constant competitive ratio.   □

Theorem 2. *No online algorithm, neither deterministic nor randomized, can achieve a constant competitive ratio on MWATP problem.*

Proof. The theorem is a direct result from Lemmas 1 and 2.   □

Algorithm 1. BenefitGreedy

---

input : a worker w and a set of tasks T
output: worker w with new plan

1 Sort t in T according to $\text{BEN}(w, t)$ in descending order;
2 $end\_loc \leftarrow c_w$;
3 $end\_time \leftarrow$ the current time;
4 **foreach** $task\ t \in T$ **do**
5   **if** $end\_time + \text{dis}(end\_loc, l_t) \le e_t$ **and**
        $end\_time + \text{dis}(end\_loc, l_t) + \text{dis}(l_t, d_w) \le e_w$ **then**
6   Append t to S_w;
7          $end\_time \leftarrow end\_time + \text{dis}(end\_loc, l_t)$;
8          $end\_loc \leftarrow l_t$;
9          $T \leftarrow T - t$;

10 **if** *no task is assigned to w* **then**
11   Let w move toward to d_w;

---

3 Solutions to MWATP Problem

Although no deterministic or randomized algorithms can achieve a constant competitive ratio, we propose two efficient heuristic algorithms, Delay-Planning and Fast-Planning, to solve the MWATP problem.

3.1 The Delay-Planning Algorithm

Main Idea. In the Delay-Planning algorithm, a worker neglects the new tasks while he/she is executing his/her current plan. Once the current plan is finished, the worker is assigned a new plan with the delayed (previously neglected) tasks.

Algorithm Details. We use a task pool to store the tasks that have not been assigned to workers. Whenever a new worker arrives or a worker finishes his/her last plan, the algorithm finds a new plan for the worker from the task pool.

We apply a succinct greedy function to make new plans for a worker by considering both the utility and the distance from the worker's current location to the task. *(i)* A task with a higher utility is preferred. *(ii)* A larger distance between the task and the worker leads to a higher risk of the expiration of the task. Combining these two considerations, we use the ratio between the utility and the distance from the worker, denoted by $BEN(w, t) = \frac{u_t}{dis(c_w, l_t)}$, to measure the *benefit* of a task. The function greedily chooses the next task with the largest benefit that can be accomplished on time by the worker.

Algorithm 1 illustrates the procedure of the greedy function. In line 1, the tasks in the task pool are sorted according to their benefits from w. In lines 2–3, two variables end_loc and end_time are defined to represent the location and time when the worker finishes his/her current plan. For each task t, we judge if the worker can accomplish it and reach the destination on time if t is appended to the tail of S_w in line 5. Note that S_w represents the task sequence of w, as is shown in Algorithm 2. If "yes", then the algorithm assigns t to w, updates *end_time* and *end_loc*, and removes t from the set T in lines 6–9.

The Delay-Planning algorithm is built upon the *BenefitGreedy* function (see Algorithm 2). In lines 1–2, we initialize a task pool *taskPool*, and a free worker set *freeWorkerSet*. In line 3, each worker $w \in W$ is associated with a task sequence S_w, *i.e.*, a plan. Whenever a task arrives ("true" judgement in line 5), we first attempt to assign it to the workers in *freeWorkerSet* in lines 6–8. If fail, we add the task to *taskPool* in lines 9–10. When a worker arrives, we assign tasks and update his/her plan from the task pool *taskPool* for him/her in line 13. A worker who has just finished his/her current plan is regarded as a new worker in Delay-Planning (see lines 12–15).

Example 3. Back to our running example in Example 1. When w_1 appears, there is one task, t_1, in the task pool. We then invoke $BenefitGreedy(.,.)$ to make a plan for w_1. Since w_1 can reach l_{t_1} before e_{t_1}, and reach his/her destination on time, w_1's new plan is to accomplish $\{t_1\}$. Then t_2 and t_3 appear at time 1.5 and 2 but there is no worker in *freeWorkerSet*. So they are added to *taskPool*. At time 3, w_2 appears. Now *taskPool* is $\{t_3\}$, because t_2 has expired. However, w_2 cannot accomplish t_3 before the expiration time of t_3. Thus w_2 directly moves to d_{w_2}. At time 3.24, w_1 finishes the last task sequence $\langle t_1 \rangle$ and now *taskPool* $= \{t_3\}$. w_1 cannot accomplish t_3 before the expiration time of w_1. Hence w_1 directly moves to d_{w_1}. At time 5, t_4 appears. Currently *freeWorkerSet* $= \{w_1, w_2\}$. The locations of w_1 and w_2 are $(3.76, 3)$ and $(3.63, 2.90)$, respectively. t_4 is assigned to w_2. We cannot choose w_1 because of w_1's expiration time. At time 5.5, t_5 appears. Neither w_1 nor w_2 can accomplish it. Note that at this time, w_2 is still accomplishing his/her current task sequence $\langle t_4 \rangle$. Finally, $S_{w_1} = \langle t_1 \rangle$ and $S_{w_2} = \langle t_4 \rangle$. The total utility is $u_{t_1} + u_{t_4} = 7$.

Time Complexity. We apply the amortized analysis to analyze the complexity of Algorithm 2. Assume n and m are the number of workers and tasks, respectively. First, the time complexity for calling Algorithm 1 is $O(m \log m)$. In Algorithm 2, the time complexity of lines 6–8 is $O(n)$, and they are executed at most m times. The time complexity of lines 5–10 is $O(mn)$. In lines 11–15, a worker may become a request more than one time. However, this happens only when he/she appears for the first time or just accomplishes a plan, which means that lines 11–15 are executed at most $O(m + n)$ times (n for appearing and m for accomplishing a plan). The total complexity of lines 11–15 is $O((m + n)m \log m)$. Combing these two parts, the time complexity of Delay-Planning is $O((m + n)m \log m)$.

3.2   The Fast-Planning Algorithm

The Delay-Planning algorithm defers the processing of tasks for a certain time, which potentially leads to the expiration of some tasks. Thus we further propose the Fast-Planning algorithm to fasten the process of making new plans, and therefore, potentially increase the total utility.

Algorithm 2. Delay-Planning

---

    input : A set of workers W, a set of tasks T
    output: Plans for $w \in W$

1   $taskPool \leftarrow \emptyset$;
2   $freeWorkerSet \leftarrow \emptyset$;
3   Set S_w an empty task sequence for each $w \in W$;
4   **for** *each new arrival request* **do**
5   **if** *the request is a task t* **then**
6   **if** *there exists a worker $w' \in freeWorkerSet$ can accomplish t with largest BEN* **then**
7   Append t to $S_{w'}$;
8               $freeWorkerSet \leftarrow freeWorkerSet - \{w'\}$;
9   **else**
10              $taskPool \leftarrow taskPool \cup \{t\}$;
11   **else**
12   // Denote the arrival worker by w.
13   BenefitGreedy($w, taskPool$);
14   **if** *there is no task appended to S_w* **then**
15              $freeWorkerSet \leftarrow freeWorkerSet \cup \{w\}$;

Main Idea. Whenever a task appears, the Fast-Planning algorithm immediately assigns the task to a worker and makes a new plan for the worker. To make the new plan efficiently, the algorithm only attempts to combine the new task with the current plan, rather than going through all possible permutations.

Algorithm Details. Algorithm 3 illustrates the procedure of the Fast-Planning algorithm. In line 1, we initialize two sets, $aWorkerSet$ and $freeTaskSet$, representing the available worker set and the unassigned task set, respectively. Whenever a worker w arrives ("true" judgement in line 3), we make a new plan for w from the $freeTaskSet$, as shown in lines 4–5. Otherwise if a task t appears, we try to combine t with the task sequence (plan) of a worker in $aWorkerSet$, with minimized increased travel distance (lines 8–15). If such combination does not exist, t is added to $freeTaskSet$ and waits to be assigned to prospective workers, as shown in lines 16–17.

Example 4. We use the settings in Example 1 to run the Fast-Planning algorithm. At time 1, w_1 arrives and moves to t_1, which is the same as in the Delay-Planning algorithm. At time 1.5, t_2 appears and we try to combine it with w_1's task sequence S_{w_1}. At this time, w_1 is at $(1.22, 1.45)$. With simple calculation, it results in a smaller increased travel distance by performing t_2 first than performing t_1 first. Therefore $S_{w_1} = \langle t_2, t_1 \rangle$. At time 3, w_2 arrives. Now $freeTaskSet = \{t_3\}$, but w_2 cannot accomplish t_3 on time. Hence w_2 directly moves to his/her destination d_{w_2}. t_4 and t_5 appear at time 5 and 5.5, respectively, and only t_4 can be accomplished by w_2. This process is similar to that in the example of the Delay-Planning algorithm and we omit the details. Finally, t_1, t_2 and t_4 are accomplished, and we obtain a utility value of 9.

Algorithm 3. Fast-Planning

---

    input : A set of workers W, a set of tasks T

    output: Plans for $w \in W$

1   $aWorkerSet \leftarrow \emptyset, freeTaskSet \leftarrow \emptyset$;

2 **for** *each new arrival request* **do**

3   **if** *the request is a worker w* **then**

4   BenefitGreedy$(w, freeTaskSet)$;

5          $aWorkerSet \leftarrow aWorkerSet \cup \{w\}$;

6   **else**

7   // Denote the arrival task by t.

8          $w_{best} \leftarrow None, bestComPos \leftarrow -1, minCost \leftarrow \infty$;

9   **foreach** $w_a \in aWorkerSet$ **do**

10   **foreach** *combination position ComPos in S_{w_a}* **do**

11                $tmpCost \leftarrow$ extra distance if combine t to $ComPos$;

12   **if** $tmpCost < minCost$ **then**

13                    $minCost \leftarrow tmpCost, bestComPos \leftarrow ComPos, w_{best} \leftarrow w_a$;

14   **if** $minCost < \infty$ **then**

15   Combine t with $S_{w_{best}}$ according to $bestComPos$.

16   **else**

17             $freeTaskSet \leftarrow freeTaskSet \cup \{t\}$;

Table 2. Experiments settings

$\lvert T \rvert$	$100, 200, \mathbf{300}, 400, 500$
$\lvert W \rvert$	$1000, 2000, \mathbf{3000}, 4000, 5000$
ts_t	$\sigma = \mathbf{10}, \mu = 30, 60, \mathbf{90}, 120, 150$
ts_w	$\sigma = \mathbf{10}, \mu = 60, 120, \mathbf{180}, 240, 300$
U_{max}	$2, 4, \mathbf{6}, 8, 10$
Scalability$(\lvert T \rvert \times \lvert W \rvert)$	$10k \times 1k, 20k \times 2k, 30k \times 3k, 40k \times 4k,$ $50k \times 5k, 60k \times 6k, 70k \times 7k, 80k \times 8k,$ $90k \times 9k, 100k \times 10k, 200k \times 20k, 300k \times 30k,$ $400k \times 40k, 500k \times 50k$

Time Complexity. We still use n and m to denote the number of workers and tasks, respectively. Lines 3–5 are executed at most $O(n)$ times, and the time complexity per execution is $O(m \log m)$. Thus the total time complexity of lines 3–5 is $O(mn \log m)$. When the request is a task, lines 11–13 are executed $O(nm^2)$ times ($O(n)$ for line 9, and $O(m)$ for line 6 and line 10). Line 8 and lines 14–17 can be executed in $O(1)$ time, and they are iterated at most $O(m)$ times. The total time complexity of lines 6–17 is $O(nm^2)$. Combing these two parts, the time complexity of the Fast-Planning algorithm is $O(nm^2)$.

4   Experimental Study

4.1   Experimental Setup

Datasets. We evaluate the performance of the proposed algorithms on both synthetic and real datasets. Table 2 shows the settings of the synthetic dataset, where the default settings are marked in bold. Tasks and workers are randomly sampled on a 600 × 600 metric space, with different values of $|T|$ and $|W|$. We also change the extra expiration time span of tasks and workers (ts_t and ts_w). Motivated by [19], the waiting time of a task (worker) follows a Gaussian distribution with the settings as in Table 2. The utilities of tasks are randomly sampled between $[1, U_{max}]$. Similarly to [20], we generate the release time of tasks and workers by the Poisson distribution, with a parameter $\lambda = 2/min$ for workers, and $\lambda = 20/min$ for tasks. We also generate datasets with large scales to test the scalability of the algorithms. For real data, we use the taxi order data, collected from a real taxi-calling service platform, to generate the locations of workers and tasks. Specifically, the location of a task is generated from an order's starting location. The initial location and the destination of a worker are generated from an order's starting location and destination, respectively. Other settings are the same as in the synthetic dataset.

Baselines. In addition to the two proposed algorithms, we also evaluate the performance of two baseline algorithms. The first is the NNH algorithm in [9], and the second is the GMCS algorithm in [13]. Both of them solve the single-worker task planning problem, and perform best in [9,13] respectively. To extend them to the MWATP problem, whenever a task appears, we find a candidate worker set (satisfying the expiration constraint) and randomly assign the task to a worker in the set. Each worker runs the corresponding single-worker algorithm to accomplish the tasks. The two baselines are denoted by Baseline-NNH and Baseline-GMCS, respectively.

Implementation. All the algorithms are implemented in C++, and the experiments were performed on a machine with 40 Intel(R) Xeon(R) E5 2.30 GHz CPUs and 512 GB memory.

4.2   Experiment Results

Effect of $|T|$. Figure 3a–c show the results of varying $|T|$. Delay-Planning and Fast-Planning outperform the two baselines in terms of the total utility value while Fast-Planning performs the best. The utility obtained by Delay-Planning is stable, while that of the other three increases with $|T|$. This might be because in each batch of Delay-Planning, tasks have been overflowed, and more tasks do not increase the utility. All the algorithms consume more time when $|T|$ increases, because more tasks lead to a larger searching space. Delay-Planning is the most time-efficient, while Fast-Planning consumes more time, because a combination

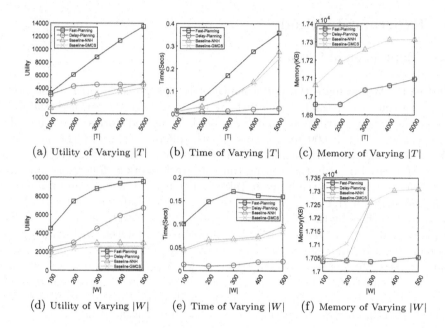

Fig. 3. Results on varying $|T|$ and $|W|$

inspection for all workers is called whenever a task appears. For memory, Delay-Planning and Fast-Planning consume more space when $|T|$ increases, but are still more efficient than baselines.

Effect of $|W|$. Figure 3d–f show the results of varying $|W|$. For the total utility value, Delay-Planning and Fast-Planning perform better than the baselines. The total utility of Delay-Planning and Fast-Planning increases with $|W|$, yet that of the baselines remain almost constant. The running time of all the algorithms are stable as the increase of $|W|$. This might be because the tasks are overflowed, and more tasks do not lead to more efficient plans. Delay-Planning is still the most time-efficient. For memory, Delay-Planning and Fast-Planning consume stable space as $|W|$ increases, which is better than the baselines.

Effect of ts_t. Figure 4a–c show the results of varying ts_t. The total utility of Delay-Planning and Fast-Planning increases as ts_t increases, but that of the baselines decrease as ts_t increases. This is probably because when μ of ts_t increases, a worker can be assigned to a task far away from him/her, which wastes too much time in one task, which leads to a decrease in the total utility. Delay-Planning and Fast-Planning still get larger total utility than the baselines (except for Delay-Planning when $\mu = 60$), and Fast-Planning performs the best. As μ of ts_t increases, the running time of Delay-Planning and Fast-Planning increases, but that of the baselines tends to be stable. This is because the size of the candidate task set for a worker is restricted by the spare time of workers, even though tasks have more waiting time. All the four algorithms consume stable memory, but the two proposed algorithms require less memory.

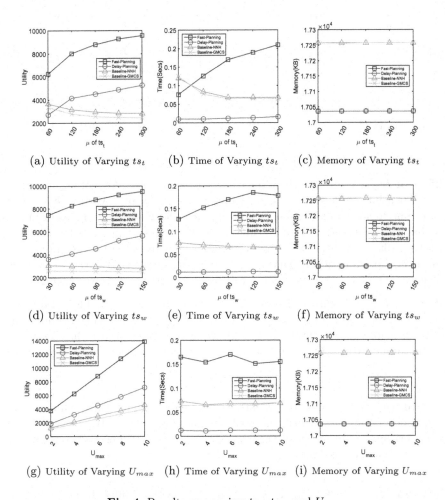

(a) Utility of Varying ts_t (b) Time of Varying ts_t (c) Memory of Varying ts_t

(d) Utility of Varying ts_w (e) Time of Varying ts_w (f) Memory of Varying ts_w

(g) Utility of Varying U_{max} (h) Time of Varying U_{max} (i) Memory of Varying U_{max}

Fig. 4. Results on varying ts_t, ts_w and U_{max}

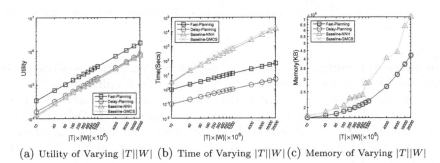

(a) Utility of Varying $|T||W|$ (b) Time of Varying $|T||W|$ (c) Memory of Varying $|T||W|$

Fig. 5. Results on scalability

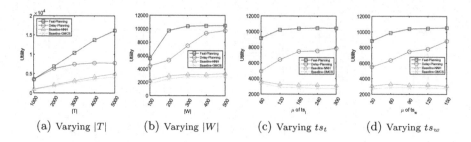

(a) Varying $|T|$ (b) Varying $|W|$ (c) Varying ts_t (d) Varying ts_w

Fig. 6. Utility results on real datasets

Effect of ts_w. Figure 4d–f show the results of varying ts_w. As μ of ts_w increases, the total utility values of our algorithms increase, because workers have more spare time to accomplish tasks. Delay-Planning and Fast-Planning perform better than the baselines. As μ of ts_w increases, the running time of Delay-Planning and Fast-Planning increases, while that of the baselines tend to be stable. The reason is similar as when varying ts_t. Delay-Planning is still the most time-efficient. The memory of the four algorithms are similar as when varying ts_t.

Effect of U_{max}. Figure 4g–i show the results of varying U_{max}. The total utility value of all the algorithms increases linearly as U_{max} increases. The running time of all the algorithms remains stable, indicating the utility of tasks has no impact on the running time. The trend of the memory consumption is similar as when ts_t varies.

Scalability. The scalable results are shown in Fig. 5. The utility and the running time of the four algorithms increase linearly as $|T| * |W|$ increases, and our algorithms perform better than baselines. Our algorithms also consume less memory than the baselines.

Performance on Real Datasets. Figure 6 shows the results of the total utility value on real datasets. The results are similar to those on the synthetic datasets. The results for the running time and memory are also similar to those on the synthetic datasets. Thus we omit the figures of memory and time due to the limited space.

Summary of Results. The Delay-Planning algorithm, though performs worse than the Fast-Planning algorithm in terms of the total utility value, has the most efficient running time. The Fast-Planning algorithm obtains the largest total utility at the cost of a slightly longer running time than the Delay-Planning algorithm. Both the two proposed algorithms can fit the scalable environment in terms of total utility and running time. The results are also similar on the real-world datasets.

5   Related Work

Our work is related to the domains of **Spatial Crowdsourcing** and **Orienteering Problem**.

5.1   Spatial Crowdsourcing

Spatial crowdsourcing has attracted extensive research interest since [2].

Task Assignment and Planning. Task assignment, task planning in particular, is one of the most important issues in spatial crowdsourcing [2,9,21–23]. In [11], the authors make plans for a single worker in the offline scenario, with the objective to maximize the number of accomplished tasks. The model is generalized to multiple workers in [12,15], but still only for the offline scenario. Both models [11,12] try to find approximate plans for workers. In [22], a protocol is proposed for protecting the privacy while task assignment. One recent work [16] makes one step further to find the exact plans of maximizing accomplished tasks for the offline scenario by using dynamic programming and graph partition.

Online Models. Since many real-world spatial crowdsourcing applications are real-time, recent studies have proposed various online models. In [24] and [25], the authors study the maximizing weighted bipartite matching in the one-sided online scenario, where only nodes on one side appear dynamically. The two-sided online scenario is explored in [10], and a solution with a competitive ratio of $\frac{1}{4}$ is proposed. [23] further studies the online trichromatic matching problem. However, these works focus on task assignment as a bipartite matching problem, which is invalid for task planning in our work. The closest related work is [13], which studies the route planning problem for a single worker in the one-sided online scenario (*i.e.*, only tasks appear dynamically).

5.2   Orienteering Problem

Given a worker with a starting location, an end location, and a time budget, and a set of n nodes in the plane, each of which is associated with a score, the orienteering problem aims to make a scheduling for the worker to gain maximal scores, with the constraint of costing less time than the time budget [17]. Many variants of the orienteering problem have been proposed [26]. Among them, the Team Orienteering Problem with Time Windows (TOPTW) [27] is the closest to our work. In this problem, each node is associated with a valid time window and we need to find a proper scheduling for a team of workers. However, these time windows can be observed at the beginning of the system, which means that the TOPTW is still an offline scenario. Furthermore, in TOPTW the workers are foreknown, while in our MWATP problem, the arrivals of both workers and tasks are unknown beforehand.

6   Conclusion

In this paper, we propose a new online task planning problem, called *Multi-Worker-Aware Task Planning* (MWATP) problem. We prove that the offline MWATP problem is NP-hard and no online algorithm has a constant competitive ratio. We then propose two heuristic algorithms, called Delay-Planning and Fast-Planning to solve the MWATP problem. We finally evaluate the effectiveness and the efficiency of the proposed algorithms on both synthetic and real datasets.

Acknowledgment. Qian Tao, Yongxin Tong and Ke Xu's works are partially supported by the National Science Foundation of China (NSFC) under Grant No. 61502021 and 71531001, National Grand Fundamental Research 973 Program of China under Grant 2014CB340300, the Base construction and Training Programme Foundation for the Talents of Beijing under Grant No. Z171100003217092, and the Science and Technology Major Project of Beijing under Grant No. Z171100005117001. Yuxiang Zeng and Lei Chen's works are partially supported by the Hong Kong RGC GRF Project 16207617, the National Science Foundation of China (NSFC) under Grant No. 61729201, Science and Technology Planning Project of Guangdong Province, China, No. 2015B010110006, Webank Collaboration Research Project, and Microsoft Research Asia Collaborative Research Grant.

References

1. Tong, Y., Chen, L., Zhou, Z., Jagadish, H.V., Shou, L., Lv, W.: SLADE: a smart large-scale task decomposer in crowdsourcing. IEEE Trans. Knowl. Data Eng. (2018)
2. Kazemi, L., Shahabi, C.: GeoCrowd: enabling query answering with spatial crowdsourcing. In: GIS, pp. 189–198 (2012)
3. Zeng, Y., Tong, Y., Chen, L., Zhou, Z.: Latency-oriented task completion via spatial crowdsourcing. In: ICDE (2018)
4. Tong, Y., Chen, Y., Zhou, Z., Chen, L., Wang, J., Yang, Q., Ye, J., Lv, W.: The simpler the better: a unified approach to predicting original taxi demands on large-scale online platforms. In: SIGKDD, pp. 1653–1662 (2017)
5. Chen, L., Shahabi, C.: Spatial crowdsourcing: challenges and opportunities. IEEE Data Eng. Bull. **39**(4), 14–25 (2016)
6. Tong, Y., Chen, L., Shahabi, C.: Spatial crowdsourcing: challenges, techniques, and applications. PVLDB **10**(12), 1988–1991 (2017)
7. Tong, Y., She, J., Ding, B., Chen, L., Wo, T., Xu, K.: Online minimum matching in real-time spatial data: experiments and analysis. PVLDB **9**(12), 1053–1064 (2016)
8. She, J., Tong, Y., Chen, L., Cao, C.C.: Conflict-aware event-participant arrangement and its variant for online setting. IEEE Trans. Knowl. Data Eng. **28**(9), 2281–2295 (2016)
9. Kazemi, L., Shahabi, C., Chen, L.: GeoTruCrowd: trustworthy query answering with spatial crowdsourcing. In: GIS, pp. 304–313 (2013)
10. Tong, Y., She, J., Ding, B., Wang, L., Chen, L.: Online mobile micro-task allocation in spatial crowdsourcing. In: ICDE, pp. 49–60 (2016)
11. Deng, D., Shahabi, C., Demiryurek, U.: Maximizing the number of worker's self-selected tasks in spatial crowdsourcing. In: GIS, pp. 314–323 (2013)

12. Deng, D., Shahabi, C., Zhu, L.: Task matching and scheduling for multiple workers in spatial crowdsourcing. In: GIS. pp. 21:1–21:10 (2015)
13. Li, Y., Yiu, M.L., Xu, W.: Oriented online route recommendation for spatial crowd-sourcing task workers. In: Claramunt, C., Schneider, M., Wong, R.C.-W., Xiong, L., Loh, W.-K., Shahabi, C., Li, K.-J. (eds.) SSTD 2015. LNCS, vol. 9239, pp. 137–156. Springer, Cham (2015). https://doi.org/10.1007/978-3-319-22363-6_8
14. She, J., Tong, Y., Chen, L.: Utility-aware social event-participant planning. In: SIGMOD, pp. 1629–1643 (2015)
15. Deng, D., Shahabi, C., Demiryurek, U., Zhu, L.: Task selection in spatial crowd-sourcing from worker's perspective. GeoInformatica 20(3), 529–568 (2016)
16. Zhao, Y., Li, Y., Wang, Y., Su, H., Zheng, K.: Destination-aware task assignment in spatial crowdsourcing. In: CIKM, pp. 297–306 (2017)
17. Golden, B.L., Levy, L., Vohra, R.: The orienteering problem. Nav. Res. Logist. 34(3), 307–318 (1987)
18. Yao, A.C.C.: Probabilistic computations: toward a unified measure of complexity. In: FOCS, pp. 222–227 (1977)
19. Tong, Y., Wang, L., Zhou, Z., Ding, B., Chen, L., Ye, J., Xu, K.: Flexible online task assignment in real-time spatial data. PVLDB 10(11), 1334–1345 (2017)
20. Roy, S.B., Lykourentzou, I., Thirumuruganathan, S., Amer-Yahia, S., Das, G.: Task assignment optimization in knowledge-intensive crowdsourcing. VLDB J. 24(4), 467–491 (2015)
21. To, H., Shahabi, C., Kazemi, L.: A server-assigned spatial crowdsourcing frame-work. ACM Trans. Spat. Algorithms Syst. 1(1), 2:1–2:28 (2015)
22. Liu, A., Wang, W., Shang, S., Li, Q., Zhang, X.: Efficient task assignment in spatial crowdsourcing with worker and task privacy protection. Geoinformatica 3, 1–28 (2017)
23. Song, T., Tong, Y., Wang, L., She, J., Yao, B., Chen, L., Xu, K.: Trichromatic online matching in real-time spatial crowdsourcing. In: ICDE, pp. 1009–1020 (2017)
24. Mehta, A.: Online matching and ad allocation. Found. Trends Theor. Comput. Sci. 8(4), 265–368 (2013)
25. Ting, H., Xiang, X.: Near optimal algorithms for online maximum edge-weighted b-matching and two-sided vertex-weighted b-matching. Theor. Comput. Sci. 607, 247–256 (2015)
26. Gunawan, A., Lau, H.C., Vansteenwegen, P.: Orienteering problem: a survey of recent variants, solution approaches and applications. Eur. J. Oper. Res. 255(2), 315–332 (2016)
27. Vansteenwegen, P., Souffriau, W., Berghe, G.V., Oudheusden, D.V.: Iterated local search for the team orienteering problem with time windows. Comput. OR 36(12), 3281–3290 (2009)

MT-MCD: A Multi-task Cognitive Diagnosis Framework for Student Assessment

Tianyu Zhu[1], Qi Liu[1], Zhenya Huang[1], Enhong Chen[1(✉)], Defu Lian[2], Yu Su[3], and Guoping Hu[4]

[1] Anhui Province Key Laboratory of Big Data Analysis and Application,
University of Science and Technology of China, Hefei 230026, China
{zhtianyu,huangzhy}@mail.ustc.edu.cn, {qiliuql,cheneh}@ustc.edu.cn
[2] University of Electronic Science and Technology of China, Chengdu 610054, China
dove.ustc@gmail.com
[3] Anhui University, Hefei 230039, China
firesysysy@163.com
[4] Anhui USTC IFLYTEK Co., Ltd., Hefei, China
gphu@iflytek.com

Abstract. Student assessment aims to diagnose student latent attributes (e.g., skill proficiency), which is a crucial issue for many educational applications. Existing studies, such as cognitive diagnosis, mainly focus on exploiting students' scores on questions to mine their attributes from an independent exam. However, in many real-world scenarios, different students usually participate in different exams, where the results obtained from different exams by traditional methods are not comparable to each other. Therefore, the problem of conducting assessments from different exams to obtain precise and comparable results is still underexplored. To this end, in this paper, we propose a Multi Task - Multidimensional Cognitive Diagnosis framework (MT-MCD) for student assessment from different exams simultaneously. In the framework, we first apply a multidimensional cognitive diagnosis model for each independent assessment task. Then, we extract features from the question texts to bridge the connections with each task. After that, we employ a multi-task optimization method for the framework learning. MT-MCD is a general framework where we develop two effective implementations based on two representative cognitive diagnosis models. We conduct extensive experiments on real-world datasets where the experimental results demonstrate that MT-MCD can obtain more precise and comparable assessment results.

Keywords: Student assessment · Cognitive diagonosis
Item Response Theory · Multi-task learning

1 Introduction

Educational Data Mining (EDM) is an emerging research field which seeks to develop methods for exploring data from educational settings (e.g., schools

J. Pei et al. (Eds.): DASFAA 2018, LNCS 10828, pp. 318–335, 2018.
https://doi.org/10.1007/978-3-319-91458-9_19

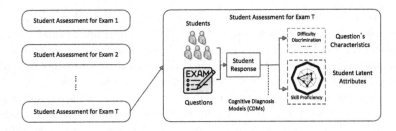

Fig. 1. Student assessment for exams

or learning systems). It contributes to learning theories, especially extracting instructive patterns from student learning, which helps understand students better and improve their learning [19,21].

One of the most important research issues in EDM is student assessment [4,17], where the goal is to discover student latent attributes (e.g., skill proficiency) based on their learning activities, such as exam scores [20] and feedback records in systems [26]. For better illustration, Fig. 1 shows a toy example of the general process of student assessment. From the figure, after collecting the responses of the students for each exam, the general goal of student assessment is to develop effective models to evaluate and diagnose student skills with the corresponding question characteristics (e.g., difficulty, discrimination). As the assessment results could be a fundamental task for various educational applications [22], such as targeted knowledge training and question recommendation, this issue has caused a great attention from both researchers and general publics [1].

In the literature, researchers have proposed many cognitive diagnosis models (CDMs) for the assessment along this line [12]. Existing CDMs have achieved a great success for student assessment in an independent exam, in which we argue that student A is more capable than student B if A gets a higher score than B. However, in most real-world scenarios, such as Graduate Record Examinations (GRE), students are allowed to take part in different exams [14]. If A gets a higher score than B when they participate in different exams, can we believe that A has a higher ability than B? In fact, educational psychologists claim that scores for students who participated in different exams could not compare directly [14]. Thus, in Fig. 1, it is not satisfied if we directly apply traditional CDM to conduct student assessment for all T exams. To this end, there is a urgent problem of conducting assessments from different exams simultaneously and it is necessary to propose an unified solution in such situation.

However, there are many challenges along this line. First, it is challenging to design a general unified framework to connect different exams for student assessment. Second, how to bridge connection with independent exams is a nontrivial problem. At last, in order to obtain comparable results for students, it is also difficult to find an appropriate way to estimate student latent attributes from independent exams simultaneously.

In this paper, inspired by the idea multi-task methods that can associate similar tasks together, we propose a Multi Task - Multidimensional Cognitive Diagnosis (MT-MCD) framework to conduct several independent student assessment tasks simultaneously. In this framework, given a set of exams containing response records of students and corresponding text information of questions, we first view the assessment in each exam as a single task and apply an existing CDM for each independent task. Then, we extract features from the question texts and develop an mapping matrix to bridge the connection with different tasks, which helps make tasks comparable. After that, we present a multi-task optimization method for the framework learning. Specifically, MT-MCD is a general framework and we propose two implementations based on two cognitive diagnosis models. i.e., M2PL model and M2PNO model. Finally, we conduct extensive experiments on real-world datasets, in which the experimental results demonstrate that MT-MCD can obtain more precise and comparable assessment results. The main contributions of this paper are summarized as follows:

- By conducting several independent student assessment tasks simultaneously, MT-MCD framework can estimate comparable student latent attributes. To the best of our knowledge, this is the first attempt to conduct several independent student assessment tasks at the same time.
- MT-MCD framework utilizes question's text information as supplemental material to bridge the connections among all assessment task, which ensures the comparability of student cognitive results.
- MT-MCD is a general framework which can apply many cognitive diagnosis models. Meanwhile, several student assessment tasks could be conducted simultaneously.

2   Related Work

In this section, we will introduce two aspects of related work: student assessment and multi-task learning.

2.1   Student Assessment

Student assessment is designed to measure specific knowledge structures and skills of students, which aims to find student latent attributes and provide information about their cognitive strength and weakness [5,12,16]. Educational psychologists have proposed a number of CDMs for student assessment [11].

Different CDMs are applied in specific occasions which can generate different types of student latent attributes (e.g. skill proficiency, guessing and slip factors) [25]. According to the assessment result, CDMs could be classified into two main categories: unidimensional CDM and multidimensional CDM. Unidimensional CDMs represent student latent attribute by a single dimensional variable [8]. For example, Item Response Theory (IRT) applies a mathematical expression that shows the relation between characteristics of a student (e.g., a latent trait) and the characteristics of the questions [13]. IRT provides a collection of models such as Two-Parameter Normal Ogive (2PNO) model and Two-Parameter Logistic (2PL) model [18]. One of the violates assumptions is the uni-dimensionality in the latent trait structure [15]. When the single dimensional variable is insufficient to indicate the complex and diverse student latent attributes, multidimensional CDMs would be necessary. Multidimensional Item Response Theory (MIRT) is a nature extension of IRT [19], and also contains a collection of models such as multidimensional extension of the 2PNO model (M2PNO [24]) and multidimensional extension of 2PL model (M2PL [18]). These MIRT models represent student latent attributes by a vector [18]. Multidimensional CDMs can assess a more complex student latent attributes.

However, most traditional CDMs aimed to do student assessment for an individual exam. In many real-word situations, student in different schools usually participate in different exams. So, it is eager to considered a framework which can conduct several independent student assessment simultaneously.

2.2   Multi-task Learning

Multi-task Learning (MTL) is a subfield of machine learning, in which several learning tasks are solved simultaneously by exploiting commonalities and differences across tasks [29].

MTL aims to improve the performance of each task by learning them jointly, which is different from single task learning. When adopting multi-task learning methods, independent tasks are learned simultaneously by utilizing shared information through tasks [28]. Multi-task learning has been applied in many different research fields, which utilizes the similarity information to conduct several tasks simultaneously to get higher performance [3,28], especially for those research problems where the amount of data per task is small. For example, Bansal et al. used multi-task method in text recommendations which a combination of content recommendation is trained by the text encoder network [2]. Yu et al. conducted image privacy protection by a deep multi-task learning algorithm to jointly learn more representative deep convolutional neural networks and more discriminative tree classifier [27].

In the research field of student assessment, it suffered from the problem that records available for each exam are limited. Therefore, applying MTL in student assessment may expand the sample size and generate more accuracy estimation. Therefore, it is necessary to consider a multi-task framework to optimize several independent student assessment tasks together based on the shared information.

3   Multi Task - Multidimensional Cognitive Diagnosis

In this paper, we propose a Multi Task - Multidimensional Cognitive Diagnosis (MT-MCD) framework which can implement several independent student assessment tasks simultaneously to generate more comparable and accurate student latent attributes than traditional CDMs. First, we formulate our problems in Sect. 3.1. Then we describe our MT-MCD framework in Sect. 3.2. At last, we illustrate wo implementations on the basis of MT-MCD in Sect. 3.3.

3.1 Problem Formulation

Given a set of exams $E = \{E_1, E_2, \cdots, E_T\}$, and student set $U_t = \{U_{t1}, U_{t2}, \cdots, U_{tU}\}$, question set $V_t = \{V_{t1}, V_{t2}, \cdots, V_{tV}\}$ for each exam $E_t (t = 1, 2, \cdots, T)$, we consider each student assessment on exam E_t as an independent task T_t ($t = 1, 2, \cdots, T$). Note that, none of these students or questions sets overlaps among different tasks. In this paper, independent tasks are implemented simultaneously to generate comparable results.

Students' responses to questions are represented by matrix Y_t for task t, where Y_{tuv} is the student U_{tu}'s response on question V_{tv}. Usually, in traditional CDMs, Y_{tuv} equals 1 when U_{tu} answered V_{tv} correctly, and equals 0 otherwise. Therefore, each student response matrix Y_t is a binary matrix composed of 0 and 1. In addition, we also collect corresponding question's text information as a supplement to connect independent assessment tasks. For each task t, we have questions' text feature F_t which is generated from text information. Specifically, $F_t = (F_{t1}, F_{t2}, \cdots, F_{tV})$ is composed of row vector F_{tv} which represent the text feature for question V_{tv}. Therefore, our problem could be defined as:

Problem Definition: *Given a set of exams $E = \{E_1, E_2, \cdots, E_T\}$, student set U_T and question set V_t for each exam E_t, student response matrix Y_t and question information matrix F_t for each exam E_t, the main propose of our MT-MCD framework is: (1) Implement T independent student assessment tasks for each exam simultaneously to obtain comparable and accurate student latent attributes and question's characteristics (e.g., discrimination, difficult); (2) Predict student's performance on questions based on the student latent attributes and question's characteristics assessed by MT-MCD.*

For better illustration, Table 1 shows some important math notations.

Table 1. Some important notations

Notation	Description
T	Task number
U_t, V_t	Students and questions in task t
Y_t	Students' response matrix for task t
F_t	Questions' text feature matrix for task t
Ξ_t	Questions' parameter matrix for task t
ξ_{tv}	Parameters for vth question in task t
Θ_t	Student latent attributes for task t
θ_{tu}	Latent attributes for uth student in task t
W_t	Mapping matrix for questions in task t
M	Dimension of student latent attributes
D	Dimension of question's text feature

3.2   Framework

We propose the MT-MCD framework to conduct several independent student assessment tasks simultaneously. Figure 2 illustrate MT-MCD framework.

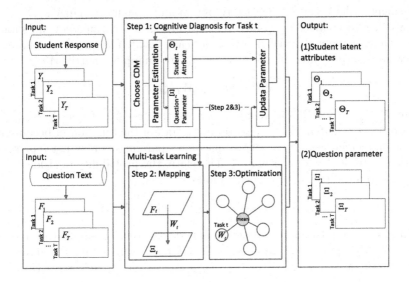

Fig. 2. MT-MCD framework

3.2.1   Step 1: CDM for Single Assessment Task

Step 1 of our proposed MT-MCD framework is to apply an existing multidimensional CDM to each individual exam. Therefore, we need to select a basic multidimensional CDM. The CDM which could be applied in MT-MCD framework can be constructed in the following way:

$$P(Y_{uv} = 1|\boldsymbol{\theta}_u, \boldsymbol{\xi}_v) \equiv f(\boldsymbol{\theta}_u, \boldsymbol{\xi}_v), \tag{1}$$

where $\boldsymbol{\theta}_u$ is a M-dimensional column vector which represents the latent attributes of student u (we will discuss the effectiveness of hyperparameter M experimentally in Sect. 4.4), and $\boldsymbol{\xi}_v$ is a row vector which represents the parameters of question v. When there are several independent tasks for different exams to be assessed simultaneously, function of CDM that t assessment tasks givens as (2):

$$f(\boldsymbol{\theta}_u, \boldsymbol{\xi}_v) = \prod_{t=1}^{T} f(\boldsymbol{\theta}_{tu}, \boldsymbol{\xi}_{tv}). \tag{2}$$

There are many existing CDMs which could be applied for each task separately to assess student latent attributes, however, the student latent attributes estimated from different individual assessment task are not comparable to each other. To solve this problem, step 2 and step 3 of MT-MCD framework connect independent tasks, and conduct these assessment tasks simultaneously to estimate accurate and comparable student latent attributes.

3.2.2 Step 2: Connecting Questions in Different Tasks

Questions play a significant role in student assessment, and it can be a great help to bridge the difference among diverse tasks. In our proposed MT-MCD framework, question's text feature are utilized as a supplement to those separate assessment tasks because it is easy to obtain and remain unchanged.

In order to connect question's text feature to its' parameters, we suppose there is a mapping matrix $W_t \in \mathbb{R}^{D \times M}$ for each task t. The question's parameter ξ_{tv} could be represented by it's feature F_{tv} and mapping matrix W_t:

$$\xi_{tv} = m(F_{tv}, W_t, \xi_{tv}), \tag{3}$$

where F_{tv} is a $1 \times D$ row vector represent the feature of question V_{tv}. The questions's parameters ξ_{tv}, appear on both sides of the function m because in a specific implementation, part of the question's parameters may not represented by it's feature and mapping matrix. Therefore, the probability of student U_{tu}'s response to question V_{tv} is defined as follow:

$$P(Y_{tuv} = 1 | \theta_{tu}, \xi_{tv}) \equiv f(\theta_{tu}, m(F_{tv}, W_t, \xi_{tv})). \tag{4}$$

In step 2, we introduce the question's text feature as a supplement and connect it to the question's parameter. Therefor, we can obtain the interaction between students and question's text feature based on the selected CDM.

3.2.3 Step 3: Multi-task Learning Optimization

After we applied question's text feature matrix F_t into each assessment task t in step 2, we need to connect these individual tasks. There are two basic assumptions in our framework:

Assumption 1. Similar questions are similar in text feature.

Assumption 2. Similar questions should have similar parameters.

Based on these two assumptions, questions which have similar text feature should have similar parameters even in different tasks. Therefore, we assume that mapping matrix W_t for all tasks are close to each other.

Evgeniou and Pontil presented a multi-task learning method based on the minimization of regularization functionals, which is a natural extension to existing methods for single task learning [9]. Inspired by this, we define the optimization function of the mapping matrix W_t for each task who's regularization function penalizes the deviation from the mean:

$$\min_W \frac{1}{2} \sum_{t=1}^{T} \parallel \hat{Y}_t - Y_t \parallel_F^2 + \lambda \sum_{t=1}^{T} \parallel W_t - \frac{1}{T} \sum_{s=1}^{T} W_s \parallel_F^2$$

$$= \min_W \frac{1}{2} \sum_{t=1}^{T} (\sum_{u,v} (f(\theta_{tu}, m(F_{tv}, W_t, \xi_{tv})) - Y_{tuv})^2) + \lambda \sum_{t=1}^{T} \parallel W_t - \frac{1}{T} \sum_{s=1}^{T} W_s \parallel_F^2 .$$

$$\tag{5}$$

The first part of Eq. (5) is the loss function which ensures the accuracy of estimation. The second part is the regularization which tries to make $\boldsymbol{W}_t$ closer to each other. If question's text feature mapping matrix $\boldsymbol{W}_t$ from different tasks are made close, then question's parameters for similar questions in different tasks will be closer.

We apply gradient descent (GD) method to optimize the Eq. (5). The gradient $\boldsymbol{W}_t$ is as follow, where f' is the first derivative of f:

$$\nabla \boldsymbol{W}_t = \sum_{u,v}(f(\boldsymbol{\theta}_{tu}, m(\boldsymbol{F}_{tv}, \boldsymbol{W}_t, \boldsymbol{\xi}_{tv})) - Y_{tuv})f'(\boldsymbol{\theta}_{tu}, m(\boldsymbol{F}_{tv}, \boldsymbol{W}_t, \boldsymbol{\xi}_{tv}))(\boldsymbol{\theta}_{tu}\boldsymbol{F}_{tv})^T$$

$$+ \lambda((-\frac{1}{T})\sum_{s=1}^{T}\boldsymbol{W}_s + \frac{2T-1}{T}\boldsymbol{W}_t). \tag{6}$$

By optimizing the mapping matrix $\boldsymbol{W}_t$, we can update the question parameter $\boldsymbol{\xi}_t$ for each question V_{tv} according to Eq. (3), and new question parameters are used in next estimation epoch in Step 1.

3.2.4 Output and Predicting

After processing by the 3-steps framework, MT-MCD, we could generate latent attributes for each student and parameters for every single question.

The outputs of MT-MCD framework are student latent attribute matrix $\boldsymbol{\Theta}_t$ and question parameters $\boldsymbol{\Xi}_t$ for each task t. For the students, we get latent attribute matrix $\boldsymbol{\Theta}_t = (\boldsymbol{\theta}_{t1}, \boldsymbol{\theta}_{t1}, \cdots, \boldsymbol{\theta}_{tU})$ for task t, which is composed of column vector $\boldsymbol{\theta}_{tu}$ represent the student U_{tu}'s latent attribute. For questions, we estimate its' parameter matrix $\boldsymbol{\Xi}_t = (\boldsymbol{\xi}_{t1}, \boldsymbol{\xi}_{t2}, \cdots, \boldsymbol{\xi}_{tV})^T$ for task t where $\boldsymbol{\xi}_{tv}$ is a row vector represented the question V_{tv}'s parameter.

Since similar questions gain close parameters, student latent attributes assessment corresponding to the questions they have answered would be more comparable. Thus, MT-MCD framework not only guarantee the accuracy of student latent attributes and question parameters, but also make questions and students from independent task more comparable.

The second purpose of MT-MCD is to predict student's performance on questions. MT-MCD helps us to assess comparable student latent attributes and corresponding question's parameters. Since the basic CDM we choose in step 1 describes the interaction between students and questions. We could easily adopt the output of MT-MCD to predict student performance by utilizing the probabilistic function for the selected basic CDM.

3.3 MT-MCD Implementation

As we mentioned before, many existing CDMs could be applied in MT-MCD framework to generate comparable student assessment result.

There are many existing CDMs proposed for student assessment, among which, Multidimensional Item Response Theory (MIRT) provides a collection of models that describe how questions and students interact to produce probabilistic response of correct or incorrect [8,23]. MIRT model is assumed to be a continuous probability function relating the student latent attribute $\boldsymbol{\theta}$ to the probability of correct response to a question with specified structural parameters. In this section, we illustrate MT-MCD framework with two MIRT models.

3.3.1   MT-MCD with M2PL Model

Multidimensional extension of the two-parameter logistic (M2PL) model [7] is a widely used MIRT model. First, we use M2PL model to illustrate how MT-MCD work.

When we select M2PL model as basic CDM, the cognitive diagnosis function Eq. (1) would be replaced by Eq. (7), which defines the probability that student U_{tu} answered question V_{tv} correctly by the changing shape of the standard logistic function [18] as:

$$f(\boldsymbol{\theta}_{tu}, \boldsymbol{\xi}_{tv}) = f(\boldsymbol{\theta}_{tu}, (\boldsymbol{\alpha}_{tv}, \beta_{tv})) = \frac{e^{(\alpha_{tv}\theta_{tu}+\beta_{tv})}}{1 + e^{(\alpha_{tv}\theta_{tu}+\beta_{tv})}}, \tag{7}$$

where the question's parameter $\boldsymbol{\xi}_{tv} = (\boldsymbol{\alpha}_{tv}, \beta_{tv})$ is composed of discrimination parameters $\boldsymbol{\alpha}_{tv} = (\alpha_{tv1}, \alpha_{tv2}, \cdots, \alpha_{tvM})$ and difficulty parameter β_{tv} [7]. We suppose that the mapping matrix $\boldsymbol{W}_t$ is connecting question's text feature $\boldsymbol{F}_{tv}$ and discrimination parameters $\boldsymbol{\xi}_{tv}$ as:

$$m(\boldsymbol{F}_{tv}, \boldsymbol{W}_t, \boldsymbol{\xi}_{tv}) = (\boldsymbol{F}_{tv}\boldsymbol{W}_t, \beta_{tv}), \tag{8}$$

Thus, the probability of student U_{tu}'s response to question V_{tv} correctly (Eq. (4)) could be replaced by Eq. (9) when selecting M2PL model as basic CDM:

$$f(\boldsymbol{\theta}_{tu}, (\boldsymbol{F}_{tv}\boldsymbol{W}_t, \beta_{tv})) = \frac{e^{(\boldsymbol{F}_{tv}\boldsymbol{W}_t\theta_{tu}+\beta_{tv})}}{1 + e^{(\boldsymbol{F}_{tv}\boldsymbol{W}_t\theta_{tu}+\beta_{tv})}}. \tag{9}$$

Then, for the multi-task learning optimization in step 3, the first derivative f' in gradient descent (Eq. (6)) could be replace by Eq. (10):

$$f'(\boldsymbol{\theta}_{tu}, (\boldsymbol{F}_{tv}\boldsymbol{W}_t, \beta_{tv})) = \frac{e^{(\boldsymbol{F}_{tv}\boldsymbol{W}_t\theta_{tu}-\beta_{tv})}}{(1 + e^{(\boldsymbol{F}_{tv}\boldsymbol{W}_t\theta_{tu}-\beta_{tv})})^2}. \tag{10}$$

3.3.2   MT-MCD with M2PNO Model

Besides M2PL model, there are many other forms of MIRT model. Multidimensional extension of the two-parameter normal ogive (M2PNO) model [24] derives from the assumption of normally distributed measurement error an is theoretically appealing on that bia s [6]. M2PNO is another widely used MIRT model, and When M2PNO is selected as basic CDM, the cognitive diagnosis function Eq. (1) would be replace by Eq. (11):

$$f(\boldsymbol{\theta}_{tu}, \boldsymbol{\xi}_{tv}) = f(\boldsymbol{\theta}_{tu}, (\boldsymbol{\alpha}_{tv}, \beta_{tv})) = \frac{1}{\sqrt{2\pi}} \int\limits_{-\infty}^{\alpha_{tv}\theta_{tu}-\beta tv} e^{-\frac{t^2}{2}} dt = \Phi(\alpha_{tv}\theta_{tu} - \beta tv),$$

$$\tag{11}$$

where the question's parameter $\boldsymbol{\xi}_{tv} = (\boldsymbol{\alpha}_{tv}, \beta_{tv})$ is also composed of discrimination parameter $\boldsymbol{\alpha}_{tv}$ and difficulty parameter β_{tv} [24], and $\Phi(z) = \frac{1}{\sqrt{2\pi}} \int\limits_{-\infty}^{z} e^{-\frac{x^2}{2}} dx$ is the normal cumulative density function (normal CDF). After the question's parameter $\boldsymbol{\xi}_{tv}$ is replaced by Eq. (8), the response probability for student U_{tu} on question V_{tv} (Eq. (4)) could be replaced by Eq. (12):

$$f(\boldsymbol{\theta}_{tu}, (\boldsymbol{F}_{tv}\boldsymbol{W}_t, \beta_{tv})) = \Phi(\boldsymbol{F}_{tv}\boldsymbol{W}_t\boldsymbol{\theta}_{tu} - \beta tv). \tag{12}$$

Correspondingly, the first derivative f' in gradient descent (Eq. (6)) is replaced by the following equation:

$$f'(\boldsymbol{\theta}_{tu}, (\boldsymbol{F}_{tv}\boldsymbol{W}_t, \beta_{tv})) = \varphi(\boldsymbol{F}_{tv}\boldsymbol{W}_t\boldsymbol{\theta}_{tu} - \beta tv) = \frac{1}{\sqrt{2\pi}} e^{\frac{(\boldsymbol{F}_{tv}\boldsymbol{W}_t\boldsymbol{\theta}_{tu} - \beta tv)^2}{2}}. \tag{13}$$

3.3.3   Conclusion

As we can see, many different existing CDMs could be applied in MT-MCD framework. Apart from this, there are many existing CDM could by applied in MT-MCD framework such as multidimensional partial credit model and multidimensional extension of Rasch model [18]. Therefore, MT-MCD framework could implement several independent student assessment tasks simultaneously and improve the accuracy and comparability of traditional CDMs. The effectiveness of MT-MCD would be proved in Sect. 4.

4   Experiment

In this section, we conduct extensive experiments to demonstrate the effectiveness of MT-MCD framework. Specifically, we use two implementations, which denoted as MT-MCD(M2PL) and MT-MCD(M2PNO), introduced in Sect. 3.3.

In the following section, we first introduce our experimental datasets and setups in Sect. 4.1. Then, we report experimental results of MT-MCD framework from the following four aspects:

- **Student Score Prediction**: Evaluate the accuracy of student assessment for each task in Sect. 4.2.
- **Student Attribute Evaluation**: Comparability evaluation of student attributes in Sect. 4.3.
- **Dimension Sensitivity of Student Attributes**: Evaluate the accuracy of MT-MCD with different dimensions of student attribute M in Sect. 4.4.
- **Question Parameter Evaluation**: Question analysis via the learned parameters in Sect. 4.5.

4.1   Dataset and Setups

4.1.1   Experimental Dataset

In the experiments, we use two real-world datasets supplied by iFLYTEK Co., Ltd., i.e., $MATH1$ and $MATH2$, to evaluate the effectiveness of MT-MCD

Table 2. Task statistics

(a) MATH1

T	E	School	U_t	V_t	Records
T_1	E_1	S_1	910	40	36,400
T_2	E_2	S_2	892	40	35,680
T_3	E_3	S_3	885	20	17,700
T_4	E_4	S_4	628	20	12,560
T_5	E_5		$3,315^1$	40	132,600
Total:			3,315	160	243,940

(b) MATHA2

T	E	School	U_t	V_t	Records
T_6	E_6	S_5	698	22	15,356
T_7	E_7	S_6	437	51	22,287
T_8	E_8	S_7	711	42	29,862
T_9	E_9	S_8	842	19	15,998
T_{10}	E_{10}	S_9	849	50	42,450
T_{11}	E_{11}	S_{10}	1,575	46	72,450
T_{12}	E_{12}	S_{11}	726	21	15,246
T_{13}	E_{13}	S_{12}	523	21	10,983
T_{14}	E_{14}		$2,839^2$	20	56,780
Total:			6,361	292	281,412

[1] All students in School S_1 to S_4
[2] Part of students in School S_5 to S_{12}

framework. Both datasets are about mathematics exam records for high school students collected from different schools in China.

In both datasets, students of the same school take the same exam, and each exam is taken by at least one school's students. Specifically, in $MATH1$, there are 4 senior high school students (S_1, S_2, S_3, S_4) participating in 5 different exams $(E_1, E_2, \cdots, E_5)$. In $MATH2$, 8 senior high schools $(S_5, S_6, \cdots, S_{12})$ participates in 9 different exams $(E_6, E_7, \cdots, E_{14})$. For task partition, we take each exam as a student assessment task in our MT-MCD framework. Therefore, there are 5 (9) tasks in $MATH1$ and $MATH2$, respectively. Table 2 shows the statistics of both datasets. In the following experiments, we take the first 4 (8) tasks for training, and the remaining one for testing.

We collect student records and the original texts of questions in all exams. For preprocessing, we first utilize the open source software $Jieba^1$ tool to segment each question's original text into a word sequence. Then, we extract question features by averaging the word embedding vector in the dimensions of $D = 60$.

4.1.2   Setups

We select the M2PL model and M2PNO model to illustrate MT-MCD framework, which have been introduced in Sect. 3.3.

When selecting M2PL as basic model, we apply a Maximum Likelihood Estimation (MLE) method in step 1 of MT-MCD framework [15]. In the following experiments, we set the numbers of MLE iterations to 1,500 for each task. When applying M2PNO model in MT-MCD, we apply a 5-step Gibbs Sampler [24] in step 1. In the following experiments, we set the number of iterations of gibbs sampler to 1,500 and estimate the parameter based on the last 1,000 samples to guarantee the convergency of the Markov Chain. Besides, we set regularization parameter λ in Eq. (5) to 0.001 in all of the following experiments.

---

[1] https://github.com/fxsjy/jieba.

4.1.3 Baseline Approaches

To demonstrate the effectiveness of MT-MCD framework, we compare two implementations i.e., MT-MCD(M2PL) and MT-MCD(M2PNO), with many models from various perspectives. First, we consider the traditional CDMs without MT-MCD framework on multiple tasks to evaluate whether MT-MCD improve the performance, we introduce M2PL_m and M2PNO_m method. Then, to evaluate the effectiveness of MT-MCD framework by applying a multi-task learning method in multiple tasks, we introduce M2PL_s and M2PNO_s method. At last, introduce a traditional multi-task learning (MTL) method from data mining area as the baseline. The details of them are as follows:

(1) M2PL_m [15]: Use M2PL model (Eq. (7)) on each task independently to generate parameters of students and questions.
(2) M2PNO_m [18,24]: Conduct the M2PNO model (Eq. (11)) on each task independently to generate parameters of students and questions.
(3) M2PL_s: Consider all tasks as a whole and applied M2PL model to do student assessment.
(4) M2PNO_s: Consider all tasks as a whole and apply M2PNO model to do student assessment.
(5) MTL [9,28]: A multi-task learning method to optimize several related classification task simultaneously. In this baseline approach, we use $(\bar{Y_t}u, \bar{Y_t}v, \boldsymbol{F_t}\boldsymbol{v})$ as a feature vector or each response record for student u on question v in task t.

4.2 Student Score Prediction

One of the problems to be solved by MT-MCD is to obtain accurate student latent attributes and corresponding question's parameters. In this section, we evaluate the accuracy of the results assessed by MT-MCD. We compare the performance on predicting student's score against the baseline approaches. In other words, we evaluate the precision of predicting the students response to prove the accuracy of parameter estimation [25].

In this experiment, we evaluate the performance of MT-MCD from both regression and classification perspectives. For regression, we adopt *root mean square error* (RMSE) and *mean absolute error* (MAE) to quantify the distance between predicted scores and the actual ones. The smaller these values are, the better the results have. For classification, we consider the predicted scores which bigger than 0.5 as 1 and those less than 0.5 as 0, to compute *precision, recall* and $F1$, and the larger, the better.

Figure 3 shows the predicting results of our MT-MCD framework and baseline approaches on dataset $MATH1$ and $MATH2$. First, we construct different size of training sets with 90%, 80%, 70% and 60% of records for each student to observe how MT-MCD behave at different sparsity levels. Then, we set the dimensions of student latent attributes $M = 3$ to observe the effectiveness of MT-MCD framework. From this figure, we observe that, MT-MCD framework could improve the accuracy of the basic CDM which demonstrates that improve the accuracy of estimation for students and questions of basic CDM. This is because MT-MCD framework introduces the question's text feature as a supplement to do a multi-task optimization on several independent student assessment tasks. Second, the performance of MT-MCD frame work beats MTL method, this is because MT-MCD framework applied student assessment method to observe student latent attributes and question's parameters.

In many real-world occasions, students usually participate in different test, thus, MT-MCD helps to improve the student assessment accuracy.

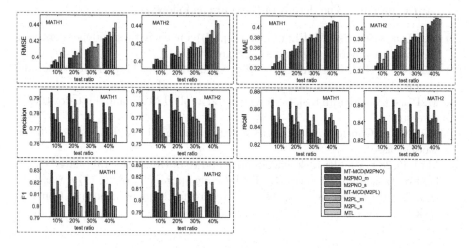

Fig. 3. Predicting student performance

4.3   Student Attribute Evaluation

In this subsection, we evaluate the comparability of student latent attributes assessed by MT-MCD framework. Intuitively, if student a masters better than student b on a specific dimension of latent attributes, a will have a higher probability to get larger score than student b when they participated in the same exam. We adopt $Degree of Agreement(DOA)$ [10] metric for a specific dimension m, which is defined as:

$$DOA(m) = \sum_{a=1}^{U} \sum_{b=1}^{U} \frac{\delta(\theta_{am}, \theta_{bm}) \cap \delta(Sum_a, Sum_b)}{\delta(\theta_{am}, \theta_{bm})}, \tag{14}$$

where m refers to the ability dimensions, θ_{im} represent student ith ability on dimension m which assessed from task T_1 to T_4 in dataset $MATH1$ or T_6 to T_9 in dataset $MATH2$. Besides, Sum_i is the total score for student i in task T_5 or T_{14}. The higher the DOA value, the stronger comparability of student latent attributes.

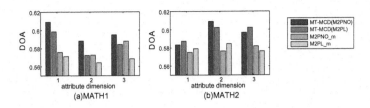

Fig. 4. DOA

Figure 4 shows the result of *DOA* for MT-MCD(M2PNO), MT-MCD (M2PL), and comparative approach M2PNO_m, M2PL_m when the dimension of students is set to 3. As can be seen from the figure, the comparability of student latent attributes assessed by MT-MCD framework is higher than the basic CDM.

4.4 Dimension Sensitivity of Student Attributes

In this subsection, we apply MT-MCD(M2PNO) and MT-MCD(M2PL) to evaluate when the dimension of student latent attributes M is set to different values.

We set the dimension of student latent attribute M equals 2 to 5. Then, construct the size of training sets with 90% of records in dataset $MATH1$ and $MATH2$ in this experiment.

Figure 5 shows the results of MT-MCD framework whit different dimensions M. As we can see from this figure, as dimensions of student latent attributes increases, the performance of MT-MCD framework firstly increases but decreases when dimensions surpasses 3 with both MT-MCD(M2PNO) and MT-MCD (M2PL) in both datasets $MATH1$ and $MATH2$. Therefore, we can summarize that performance of $M = 3$ is better and more stable, and set $M = 3$ in Sects. 4.2 and 4.3 to obtaining the best results.

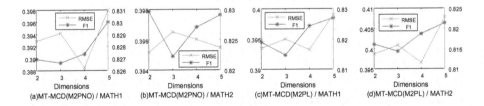

Fig. 5. Results of different dimensions

4.5 Question Parameter Evaluation

We emphasize that MT-MCD framework can make parameters of similar questions in different tasks closer, therefore, we evaluate the question parameters (discrimination, difficulty) estimated by MT-MCD framework with M2PNO basic CDM to prove the effectiveness in this section.

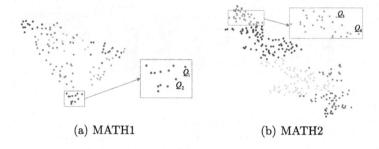

(a) MATH1	(b) MATH2

Fig. 6. Clustering result

For this experiment, we cluster the question's parameters estimated by MT-MCD to illustrate that similar question's parameters are closer. Specifically, we set the dimension parameter $M = 3$. Then we use the K-means clustering method to cluster the result of question's parameters into 7 categories. Finally, we adopt *t-SNE*[2] program to visualize these questions. Figure 6(a) and (b) shows the clustering result of question's parameter for dataset $MATH1$ and $MATH2$. Each dot in Fig. 6(a) represent a question in task T_1 to task T_4, and each dot in Fig. 6(b) represent a question in task T_6 to task T_{13}. The dots of same color belong to the same class clustered by K-means.

We check all the categories clustered by K-means, and questions in same category are similar to each other. For example, we find that all questions corresponding to these blue dots in Fig. 6(a) are about 'function' knowledge point, and all these green dots in Fig. 6(b) are about 'triangle' knowledge point. Further, a case study of several questions in these two categories are listed in Table 3. This experiment proves that MT-MCD framework makes parameters of similar questions in different student assessment tasks closer.

Table 3. Case study

	Task	Question description	Parameters (α,β)
Q_1	T_1	The number of zero points for function $f(x) = 3x^2 + 2x - 4$ is?	$((0.25, 0.66, 0.89), -1.77)$
Q_2	T_2	The range of function $f(x) = x + \sqrt{1 - 2x}$ is?	$((0.23, 0.50, 0.94), -1.21)$
Q_3	T_8	A, B, C is the inner corner of the triangle, therefore, $sin(A + B) = sinC$?	$((1.56, 1.69, 2.54), 0.99)$
Q_4	T_{11}	$A = (5.1)$, $B = (1,1)$, $C = (2.3)$, The shape of triangle $\triangle ABC$ is?	$((1.45, 1.38, 2.07), 1.05)$

---

[2] https://lvdmaaten.github.io/tsne/.

5   Conclusion and Future Work

In this paper, we proposed a MT-MCD framework to conduct several independent student assessment task simultaneously to generate accurate and comparable student latent attributes for students who participated in different exams. Specifically, we first applied an existing multidimensional cognitive diagnosis model to each independent student assessment task to estimate student latent attributes and corresponding question's parameters (e.g., discrimination, difficulty). Second, we introduced question's text information as a bridge to connect each independent assessment tasks. Then, we and employed a multi-task optimization method to make parameters of similar questions closer. New question's parameters updated by multi-task learning method will be adopted in cognitive diagnosis model for each student assessment task to obtain comparable student latent attributes. Extensive experiments on the real-world datasets clearly demonstrated the effectiveness of our propose framework MT-MCD which can assess accurate and comparable student latent attributes and question's parameters from independent student assessment tasks.

In the future, there are some directions for further studies. First, we will consider to find more relatedness between independent student assessment tasks. For example, student is an important aspect in assessment, the characteristics of students may connect individual student assessment together. Second, many natural language processing (NLP) method could be used for the pre-processing of question's text information.

Acknowledgments. This research was partially supported by grants from the National Natural Science Foundation of China (Grants No. 61672483, U1605251 and 91546103), and the Youth Innovation Promotion Association of CAS (No. 2014299).

References

1. Baker, R.S.J.D., Yacef, K.: The state of educational data mining in 2009: a review and future visions. JEDM-J. Educ. Data Min. **1**(1), 3–17 (2009)
2. Bansal, T., Belanger, D., McCallum, A.: Ask the GRU: multi-task learning for deep text recommendations. In: Proceedings of the 10th ACM Conference on Recommender Systems, pp. 107–114. ACM (2016)
3. Bickel, S., Bogojeska, J., Lengauer, T., Scheffer, T.: Multi-task learning for HIV therapy screening. In: Proceedings of the 25th International Conference on Machine Learning, pp. 56–63. ACM (2008)
4. Cox, K., Imrie, B.W., Miller, A.: Student Assessment in Higher Education: A Handbook for Assessing Performance. Routledge, London (2014)
5. Cui, Y., Li, J.: Evaluating person fit for cognitive diagnostic assessment. Appl. Psychol. Meas. **39**(3), 223–238 (2015)
6. De La Torre, J., Minchen, N.: Cognitively diagnostic assessments and the cognitive diagnosis model framework. Psicología Educativa **20**(2), 89–97 (2014)
7. DiBello, L.V., Roussos, L.A., Stout, W.: 31A review of cognitively diagnostic assessment and a summary of psychometric models. Handb. Stat. **26**, 979–1030 (2006)

8. DiBello, L.V., Stout, W.: Guest editors' introduction and overview: IRT-based cognitive diagnostic models and related methods. J. Educ. Meas. **44**(4), 285–291 (2007)
9. Evgeniou, T., Pontil, M.: Regularized multi-task learning. In: Proceedings of the Tenth ACM SIGKDD International Conference on Knowledge Discovery and Data Mining, pp. 109–117. ACM (2004)
10. Fouss, F., Pirotte, A., Renders, J.-M., Saerens, M.: Random-walk computation of similarities between nodes of a graph with application to collaborative recommendation. IEEE Trans. Knowl. Data Eng. **19**(3), 355–369 (2007)
11. Huebner, A.: An overview of recent developments in cognitive diagnostic computer adaptive assessments. Pract. Assess. Res. Eval. **15**(3), n3 (2010)
12. Huff, K., Goodman, D.P.: The demand for cognitive diagnostic assessment (2007)
13. Klaus, D., Kubinger, K.D.: On the revival of the rasch model-based LLTM: from constructing tests using item generating rules to measuring item administration effects. Psychol. Sci. **50**(3), 311 (2008)
14. Kuncel, N.R., Hezlett, S.A., Ones, D.S.: A comprehensive meta-analysis of the predictive validity of the graduate record examinations: implications for graduate student selection and performance. Psychol. Bull. **127**(1), 162 (2001)
15. Lee, J.: Multidimensional Item Response Theory: An Investigation of Interaction Effects Between Factors on Item Parameter Recovery Using Markov Chain Monte Carlo. Michigan State University, Measurement and Quantitative Methods (2012)
16. Leighton, J., Gierl, M.: Cognitive Diagnostic Assessment for Education: Theory and Applications. Cambridge University Press, Cambridge (2007)
17. Liu, Q., Runze, W., Chen, E., Guandong, X., Yu, S., Chen, Z., Guoping, H.: Fuzzy cognitive diagnosis for modelling examinee performance. ACM Trans. Intell. Syst. Technol. (TIST) **9**(4), 48 (2018)
18. Reckase, M.: Multidimensional Item Response Theory, vol. 150. Springer, New York (2009). https://doi.org/10.1007/978-0-387-89976-3
19. Romero, C., Ventura, S., Pechenizkiy, M., d Baker, R.S.J.: Handbook of Educational Data Mining. CRC Press, Boca Raton (2010)
20. Saxon, P.D., Morante, E.A.: Effective student assessment and placement: challenges and recommendations. J. Dev. Educ. **37**(3), 24 (2014)
21. Scheuer, O., McLaren, B.M.: Educational data mining. In: Seel, N.M. (ed.) Encyclopedia of the Sciences of Learning, pp. 1075–1079. Springer, Boston (2012). https://doi.org/10.1007/978-1-4419-1428-6
22. Serrano-Laguna, Á., Torrente, J., Moreno-Ger, P., Fernández-Manjón, B.: Tracing a little for big improvements: application of learning analytics and videogames for student assessment. Procedia Comput. Sci. **15**, 203–209 (2012)
23. Sheng, Y.: Markov chain Monte Carlo estimation of normal ogive IRT models in MATLAB. J. Stat. Softw. **25**(8), 1–15 (2008)
24. Sheng, Y., Headrick, T.C.: A gibbs sampler for the multidimensional item response model. ISRN Appl. Math. **2012**, 14 (2012)
25. Wu, R., Liu, Q., Liu, Y., Chen, E., Su, Y., Chen, Z., Hu, G.: Cognitive modelling for predicting examinee performance. In: IJCAI, pp. 1017–1024 (2015)
26. Wu, R., Xu, G., Chen, E., Liu, Q., Ng, W.: Knowledge or gaming?: cognitive modelling based on multiple-attempt response. In: Proceedings of the 26th International Conference on World Wide Web Companion, pp. 321–329. International World Wide Web Conferences Steering Committee (2017)

27. Jun, Y., Zhang, B., Kuang, Z., Lin, D., Fan, J.: iPrivacy: image privacy protection by identifying sensitive objects via deep multi-task learning. IEEE Trans. Inf. Forensics Secur. **12**(5), 1005–1016 (2017)
28. Zhou, J., Chen, J., Ye, J.: Malsar: multi-task learning via structural regularization. Arizona State University, vol. 21 (2011)
29. Zhou, J., Chen, J., Ye, J.: Multi-task learning: theory, algorithms, and applications. In: https://www.siam.org/meetings/sdm12/zhou_chen_ye.pdf. Citeseer (2012)

Towards Adaptive Sensory Data Fusion for Detecting Highway Traffic Conditions in Real Time

Yanling Cui[1,2], Beihong Jin[1,2(✉)], Fusang Zhang[1,2], and Tingjian Ge[3]

[1] State Key Laboratory of Computer Sciences, Institute of Software,
Chinese Academy of Sciences, Beijing, China
jbh@otcaix.iscas.ac.cn
[2] University of Chinese Academy of Sciences, Beijing, China
[3] University of Massachusetts, Lowell, USA

Abstract. The key challenge of detecting highway traffic conditions is to achieve it in a fully-covered, high-accuracy, low-cost and real-time manner. We present an approach named Megrez on the basis of treating mobile phones and probe vehicles as roving sensors, loop detectors as static sensors. Megrez can admit one or multiple types of data, including signaling data in a mobile communication network, data from loop detectors, and GPS data from probe vehicles, to carry out the traffic estimation and monitoring. In order to accurately reconstruct traffic conditions with full road segment coverage, Megrez provides a practical way to overcome the sparsity and incoherence of sensory data and recover the missing data in light of recent progresses in compressive sensing. Moreover, Megrez incorporates the characteristics of traffic flows to rectify the estimates. Using large-scale real-world data as input, we conduct extensive experiments to evaluate Megrez. The experimental results show that, in contrast to three other fusion methods, the results from our approach have high precisions and recalls. In addition, Megrez keeps the errors of estimates low even when not all three types of data are available.

Keywords: Data fusion · Traffic condition detection · Mobile signaling
Compressive sensing · Adaptation

1 Introduction

In recent years, various types of heterogeneous data have been providing great opportunities for the improvement and reconstruction of information systems in all areas [1]. In particular, with the aid of smart sensors and pervasive networks, large amounts of transportation data which relate to people, vehicles, roads and environments can be continuously recorded and collected. Taking full advantage of these data makes it possible to provide high-quality services, such as personalized travel services, real-time monitoring services, and intelligent decision support services.

In this paper, we focus on detecting traffic conditions for highways. It is well known that real-time highway traffic conditions can be used for vehicle navigation.

© Springer International Publishing AG, part of Springer Nature 2018
J. Pei et al. (Eds.): DASFAA 2018, LNCS 10828, pp. 336–352, 2018.
https://doi.org/10.1007/978-3-319-91458-9_20

More importantly, such information can assist traffic administrative departments in traffic monitoring, emergency management, and real-time dispatching.

For road traffic condition detection, what is highly desired is to utilize the sensory data for high-accuracy real-time detection with full road coverage but at a low cost. Unfortunately, existing methods cannot achieve the requirements satisfactorily. For example, some methods require deploying expensive devices, which are not scalable to large areas. Some methods cannot handle low-quality sensory data well (e.g., sparse, inconsistent, and inaccurate data). Therefore, in this paper, we present a novel approach called *Megrez* (which is the name of a star—with the hope that our approach will shed some light on traffic control). Megrez fuses three types of data: signaling data, loop detector data, and GPS data from probe vehicles.

Signaling data [2] come from signaling monitoring systems of mobile operators and play a crucial role in controlling and recording activities in mobile communication networks. When users with phones move with the vehicles, the signaling data are generated at the cell towers along the travel trajectories. According to these signaling data, the speeds of corresponding vehicles can be estimated, and then the traffic conditions of the corresponding road segments can be speculated. On the other hand, *probe vehicles* refer to vehicles equipped with the satellite positioning devices and are mostly taxis and buses in the cities, which appear *infrequently* on highways. We notice that, in China, two kinds of long-distance buses (i.e., intercity buses and tour buses) and the vehicles specially used for transporting hazardous goods (hereafter, referred to as 2K1W vehicles, an abbreviation in Chinese pinyin) can be viewed as the probe vehicles, since they have to comply with the provisions of administrative regulations to install satellite positioning devices. Furthermore, loop detectors are fixed, static devices installed on highways, from which the speeds of passing vehicles can be obtained.

We collect and analyze some real-world data, including the signaling data in a mobile communication network, the data from loop detectors, and the GPS data from 2K1W vehicles. We find that all these data are indeed poor in quality, and they are highly correlated in both temporal and spatial dimensions.

Inspired by the data characteristics observed from our analyses, we design the Megrez approach which consists of the following three steps: (1) a concrete function is proposed to get the first-cut estimates of vehicle speeds associated with road segments; (2) missing vehicle speeds at certain road segments are completed using compressive sensing; and (3) vehicle speeds are finally rectified by incorporating the characteristics of traffic flows. These three steps provide high-accuracy guarantees over all road segments in the traffic condition detection. Besides, a parallel linear algebra library is employed to speed up and provide real-time road traffic conditions.

Megrez is adaptive to both temporal and spatial variations of data. It is originally designed for three types of data sources. However, it can work well using only two types of data sources. Furthermore, our approach can deal with changes in the data distribution of incoming data streams. In reality, at some time slices, a particular type of data may be obtained only at certain regions. The coverage of different types of data may be either overlapping or disjoint. For example, signaling data and GPS data can both be captured when vehicles are driving at open areas, and only the signaling data can be captured when the vehicles drive in tunnels. In either case, our approach can detect the whole road traffic conditions seamlessly.

The remainder of the paper is organized as follows. Section 2 introduces the related work, and Sect. 3 gives some analyses of real-world data, including signaling data, loop detector data and GPS data. In Sect. 4, we describe the Megrez approach in detail, and discuss our experimental evaluations in Sect. 5. Finally, we conclude in Sect. 6.

2 Related Work

So far, there have been a lot of research efforts on traffic condition detection. The data which have been used for detecting road traffic conditions include loop detector data [3], images [4], GPS data from probe vehicles [5–9], signaling data from mobile communication networks [10, 11] and CDRs (Call Detail Records) [12, 13] which can be viewed as a subset of signaling data.

Acquiring data from loop detectors and cameras requires deploying the devices in advance, but leveraging these data can only obtain the traffic conditions on the specific cross-sectional points. Besides, these devices cannot be deployed to cover all the road segments of highways due to device investment costs and their limited working lifetime. These innate weaknesses make fixed device-based solutions alone difficult to acquire satisfactory traffic conditions.

Signaling data have the advantage of their wide coverage, but they suffer from the poor data quality. They not only have the low positioning accuracy, but also have irregular frequencies, depending on the mobile users' behaviors. On the whole, the temporal and spatial distributions of the signaling data are uneven. Moreover, a number of factors, including the radio propagation characteristics (signal fading, multipath effect, etc.), the locations and radiuses of cell towers as well as their realistic loads, will affect the handovers, increasing the uncertainty of handovers. In general, before signaling data are used for highway monitoring, the data cleaning is necessary, e.g., the ping-pong handovers and cell oscillation need to be identified and removed. However, the signaling based approaches cannot avoid the dilemma that the accuracy is not high enough.

GPS data have received much attention in detecting road traffic conditions due to their high location accuracy. In [6], the authors provide a multi-channel singular spectrum analysis (MSSA) to iteratively estimate urban traffic conditions. But the estimated results have large errors. What is worse, the MSSA-based method is so time consuming that it cannot be used for real-time monitoring. Further, in [7], the authors give a compressive sensing based method to recover the vehicle speeds at the positions where no nearby GPS data can be obtained. Their work shows that compressive sensing can be applied to GPS data from taxis. However, there have been no in-depth studies reported on applying compressive sensing to data fusion tasks.

Adopting multi-source data provides a promising way to improve the accuracy in detecting traffic conditions [14–17]. For example, different data streams are used in DynaMIT2.0 [17], including data from inductive loops, cameras and probe cars, incident information feed, as well as data from the Internet (e.g., special events websites, weather forecasts, and social networks). DynaMIT2.0 uses a SP-EKF method (simultaneous perturbation extended Kalman filter) to calibrate the traffic parameters, and uses microscopic traffic simulator MITSIMLab to conduct closed loop experiments on Singapore Expressway to verify the method.

In contrast to previous work, we combine the signaling data from mobile operators, the data from loop detectors, and the GPS records from probe vehicles for the first time, and estimate the fused vehicle speeds in three steps, i.e., estimating the initial vehicle speeds by a concrete function, filling in the missing speeds via compressive sensing, and rectifying the speeds using traffic flow features. Our comprehensive and carefully designed approach ensures that our traffic condition estimates are closer to the ground truth, as demonstrated in our experiments.

3   Real-World Data Analyses

In order to understand the characteristics of signaling data, loop detector data and GPS data on highways, we collect real data generated along the nine highways (i.e., G3, G15, G25, S35, G70, G72, G76, G1501 and G1514) in Fujian Province, China. The data include the signaling records from CMCC (China Mobile Communications Corporation) Fujian branch, the loop detector data, and the GPS records of 2K1W vehicles running on the highways in Fujian.

3.1   Spatial-Temporal Distribution Analyses

On the nine highways whose total length is about 2,620 km, there are 431 loop detectors with 1,289 coil sensors. We count the numbers of coil sensors from which no items are reported (referred to as *silent coil sensors*) on each day of October, November, and December 2015, which are shown in Fig. 1. From Fig. 1, we can see that in the last three months of 2015, the number of daily silent coil sensors is at most 470, at least 340, and with an average of 406. Moreover, we find 302 coil sensors (accounting for 23.43%) and the corresponding 126 detectors (accounting for 29.23%) do not report any data in the three months. These data indicate that there is a high probability that the loop detectors do not work normally and the data stream from the loop detectors is unstable. The sparse geographic distribution of loop detectors plus their abnormal states lead to the result that loop detector data are sparse in space and time.

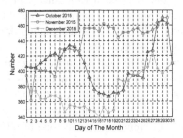

Fig. 1. Daily variation on silent coil sensors

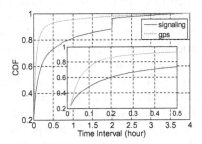

Fig. 2. CDFs of time intervals of signaling data and GPS records

Next, we calculate the time intervals between two consecutive signaling records from the same mobile phone and time intervals between two consecutive GPS records from the same vehicle. Figure 2 shows the CDFs (Cumulative Density Functions) of time intervals of signaling records and GPS records generated on November 1, 2015. As shown in Fig. 2, over 70% of time intervals between two consecutive GPS records are shorter than 5 min. Whereas only 45% of the time intervals between two consecutive signaling records are shorter than 5 min, and about 57% are shorter than 10 min. There is a sudden change in the CDF curve of time intervals of signaling data when time interval reaches 2 h. It is because the mobile operator will proactively contact a mobile phone via signaling data if the phone has not been used in the past 2 h.

We also plot the spatial-temporal graphs for signaling records and GPS records to observe their spatial-temporal coverage. A typical spatial-temporal graph is shown in Fig. 3, where the X axis denotes the time of day and the Y axis denotes the distance to G76's start location Xiamen and white space denotes no data at that time and location. The more blank space in Fig. 3(b) illustrates that GPS data are sparse and signaling data are not as sparse in terms of sheer numbers. Meanwhile, the GPS data and the signaling data are non-uniform in spatial-temporal distributions.

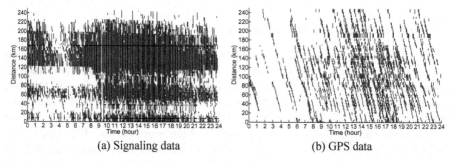

(a) Signaling data   (b) GPS data

Fig. 3. Spatial-temporal graph of Fujian section of G76 (from Chengdu to Xiamen) on November 12, 2015

3.2 Data Inherent Structure Discovery

For each road segment on highways, the vehicle speeds at each time slice can be measured by three types of data (i.e., signaling data, GPS data, and loop detector data). We call the vehicle speeds obtained by loop detectors, GPS records and signaling data the detector speeds, GPS speeds, and signaling speeds, respectively.

Then, we employ a matrix to record vehicle speeds, where each column denotes the time series of vehicle speeds over a specific road segment and each row denotes the vehicle speeds on every road segment at a specific time slice (5 min by default).

Taking the three types of data during November 1–7, 2015 as input, we construct three matrices, one for a single type (i.e., one matrix of signaling speeds, one of GPS speeds, and one of detector speeds). We find that many elements in the three matrices are missing. We search for the corresponding *maximum dense square submatrices* in each of these three matrices and find their sizes are 273, 90, and 4, respectively. Then, for each matrix, we extract multiple dense square submatrices with different sizes, compute the singular values of each extracted submatrix and then normalize them by setting the largest singular value to 1.

Figures 4(a)–(c) show the magnitude (ratio to the maximum) of singular values of different-sized dense square submatrices in the three matrices, respectively. In Fig. 4, the X axis denotes the singular values sorted from large to small, and the legend shows the size of the square submatrix as well as the number of non-overlapping same-sized submatrices whose singular values are averaged into the curve shown. We find that three kinds of submatrices show similar singular value distributions.

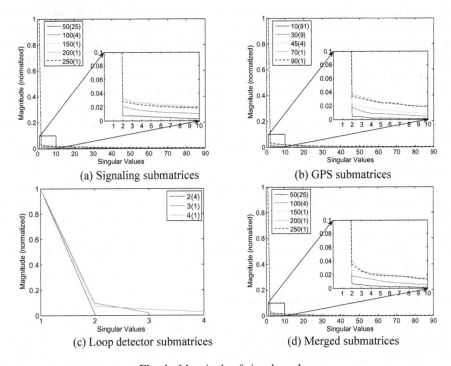

(a) Signaling submatrices

(b) GPS submatrices

(c) Loop detector submatrices

(d) Merged submatrices

Fig. 4. Magnitude of singular values

Sharp knees can be observed in Figs. 4(a)–(c), which illustrates that the ranks of different-sized dense square submatrices are low and indicates that vehicle speeds are tightly correlated in both temporal and spatial dimensions.

4   Megrez Approach

In this paper, detecting traffic conditions refers to estimating vehicle speeds on highways and then mapping them into road traffic states including *congested*, *slow* and *free* states. For highways with n road segments in total, we use an $m \times n$ nonnegative matrix to describe highway traffic conditions within m time slices. Let $X_{m \times n}$ and $\widehat{X}_{m \times n}$ be the real and estimated vehicle speeds matrices, respectively. Given a signaling stream, a loop detector data stream, and a GPS data stream of probe vehicles, the problem of detecting traffic conditions is to find the $\widehat{X}_{m \times n}$ with the minimum $\|X - \widehat{X}\|_F$. Here, $\|\cdot\|_F$ is the Frobenius norm of a matrix, i.e., $\|X\|_F := \sqrt{\sum_{i,j} (X_{ij})^2}$.

To obtain an accurate $\widehat{X}_{m \times n}$, we present our Megrez approach, which consists of three steps, i.e., multi-source speed merging, compressive sensing based interpolation, and feature-driven filtering.

In the first step, we obtain the vehicle speeds from loop detectors, calculate the vehicle speeds via GPS records, and estimate the vehicle speeds from signaling data. These vehicle speeds are referred to as *detector speeds* (hereafter, DSs for short), *GPS speeds* (GSs for short), and *signaling speeds* (SSs for short), respectively. Then, we merge three speeds (if available) into the *merged speeds* (MSs for short) by a concrete function and then form the measurement matrix of traffic conditions, denoted as $M_{m \times n}$.

Due to the sparsity of original data, the matrix $M_{m \times n}$ is not complete where values of many elements are missing. However, in light of features of traffic flows, many rows should be linearly dependent on each other, and so are many columns. That is, the rank of the matrix $M_{m \times n}$ should be relatively low. Based on these observations, in the second step, we employ the *compressive sensing* technique to achieve the interpolation of matrix $M_{m \times n}$ and fill in missing values. The speeds in the completed matrix are called *interpolated MSs*, or IMSs for short.

After the above step, we filter the elements in the interpolated matrix to restrain the interference of random factors. What we design is in fact a type of exponential smoothing algorithm, but we integrate the spatial-temporal features of traffic flows into the filter so as to conform to the real road traffic conditions. After filtering, we get the final vehicle speed estimates which are called *filtered IMSs*, or FIMSs for short. In the following, we describe the Megrez approach in detail.

4.1   Multi-source Vehicle Speed Merging

As initial processing, we divide the roads into segments by the locations of cell towers along the roads. Next, we clean the signaling records so as to get valid ones. In detail, we glean the signaling records whose cell towers are along the roads. On receiving these signaling records, we filter the signaling data with abnormal movement features (e.g., teleportation), and further identify and get rid of the ping-pong handovers and cell oscillation. Then, we discern whether a user is in a vehicle and moving along the road. If so, the user is marked as being in-motion state. The signaling data are thus mapped into mobile users' trajectories. In order to obtain the vehicle speeds, the trajectories of users with in-motion states within several hours are cached in memory. So we can calculate the moving speed of the vehicle where the user is in, and then obtain the average vehicle speed for every road segment.

At the same time, we can obtain the vehicle speeds by GPS data of 2K1W vehicles, i.e., GSs, and the ones from loop detectors, i.e., DSs.

Then, we apply the following concrete function to merge SSs, GSs and DSs on the same road segment and same time slice into MSs, and construct $M_{m \times n}$, the measurement matrix of traffic conditions.

$$MS = \begin{cases} DS, & \text{if } DS \neq \perp \\ GS, & \text{if } DS = \perp \land GS \neq \perp \\ SS, & \text{if } DS = \perp \land GS = \perp \land SS \neq \perp \\ \perp, & \text{otherwise} \end{cases} \qquad (1)$$

where $\perp$ denotes that the value is missing.

4.2 Data Completion via Compressive Sensing

We notice that $M_{m \times n}$ still has missing values. Therefore, in this section, we devise a compressive sensing model, and exploit the relations hidden in the measurement matrix to complete the matrix. Our aim is to estimate the traffic condition matrix that approximates the real traffic conditions as closely as possible.

Designing the Compressive Sensing Model
Since $M_{m \times n}$ is a sparse matrix, which has many missing elements compared to $X_{m \times n}$, we introduce an indicator matrix $B_{m \times n}$ which makes the following hold:

$$M_{m \times n} = B_{m \times n} .\times \widehat{X}_{m \times n}, B_{m \times n} = [b_{ij}] = \begin{cases} 0, & \text{if no sensory data for road} \\ & \text{segment } j \text{ in time slice } i \\ 1, & \text{otherwise} \end{cases} .$$

where $.\times$ is an operator of Hadamard (element wise) product.

Our objective is to obtain $\widehat{X}_{m \times n}$ given $M_{m \times n}$.

Due to the correlation characteristics of traffic flows on roads, the measurement matrix $M_{m \times n}$ must have correlations between different rows or columns, which implies that $M_{m \times n}$ has a relatively low rank with a high probability. That is, for traffic condition scenarios, it is reasonable to approximate $M_{m \times n}$ by a matrix of lower rank.

The method in Sect. 3.2 can be used to observe the property of measurement matrix $M_{m \times n}$. For example, for the measurement matrix $M_{m \times n}$ constructed by the three types of data during November 1–7, 2015, the singular values of dense square submatrices of $M_{m \times n}$ are shown in Fig. 4(d) and these submatrices are considered to have low ranks. Therefore, $M_{m \times n}$ are regarded to have a low rank with a high probability. This feature of $M_{m \times n}$ allows us to apply compressive sensing techniques to complete the matrix $M_{m \times n}$. In other words, if $M_{m \times n}$ can be approximated by a low rank matrix, the accurate recovery of $M_{m \times n}$ from a small number of matrix elements is possible by using compressive sensing [18].

Designing the Algorithm for Data Completion
To solve the traffic detection problem, we are required to make an estimate close to the measurement matrix. What we desire is to find the low-rank estimate $\widehat{X}_{m \times n}$ as shown in Eq. (2) [18].

$$\min \ rank(\widehat{X}_{m \times n})$$
$$s.t. \ B_{m \times n} \cdot \times \ \widehat{X}_{m \times n} = M_{m \times n} \tag{2}$$

Minimizing $rank(\widehat{X}_{m \times n})$ is a non-convex optimization problem and is NP-hard. As the nuclear form $|| \cdot ||_*$ of the matrix is the tightest convex envelop for the matrix rank, it is usually adopted to replace the rank. Based on it, the non-convex problem in Eq. (2) can be converted to a convex problem as shown in Eq. (3) [19, 20]. Specifically, if the restricted isometry property holds, minimizing the nuclear form equals to the rank minimization exactly for a matrix of low rank [21].

$$\min \ ||\widehat{X}_{m \times n}||_*$$
$$s.t. \ B_{m \times n} \cdot \times \ \widehat{X}_{m \times n} = M_{m \times n} \tag{3}$$

Here $||\widehat{X}_{m \times n}||_* := \sum_{i=1}^{rank(\widehat{X}_{m \times n})} \chi_i(\widehat{X}_{m \times n})$, and $\chi_i(\widehat{X}_{m \times n})$ is the i-th largest singular value of $\widehat{X}_{m \times n}$. To obtain $\widehat{X}_{m \times n}$ satisfying Eq. (3), we make use of the SVD-like factorization, and let $\widehat{X}_{m \times n} = U\Sigma V^T = LR^T$, where $L = U\Sigma^{1/2}$ is a $m \times r$ matrix, and $R = V\Sigma^{1/2}$ is a $n \times r$ matrix. There exist many possible factorization results for $\widehat{X}_{m \times n}$; however, what we need is to find the matrices L and R that minimize the summation of their Frobenius norms, i.e., L and R should satisfy the following Eq. (4).

$$\min \ ||L||_F^2 + ||R||_F^2$$
$$s.t. \ B_{m \times n} \cdot \times \ (LR^T) = M_{m \times n} \tag{4}$$

Besides, $r \geq rank(X_0)$ should be satisfied as a constraint, where X_0 is a solution to Eq. (2). If so, then Eq. (3) is equivalent to Eq. (4) [21]. In practice, L and R that strictly satisfy Eq. (4) are likely to lead to an undesirable result due to two reasons. First, a traffic condition matrix is usually approximately low-rank but may be not really low in the rank. Second, there is noise in the measurement matrix, and strictly satisfying the constraints may lead to over-fitting. Considering these factors, according to the Lagrange multiplier method, we convert Eq. (4) into a convex optimization in Eq. (5).

$$\min \ ||B_{m \times n} \cdot \times \ (LR^T) - M_{m \times n}||_F^2 + \lambda(||L||_F^2 + ||R||_F^2) \tag{5}$$

The parameter λ controls a tunable tradeoff between rank minimization and accuracy fitness. λ is set to 50 by default.

To solve Eq. (5), we propose an algorithm to compute L and R alternatively with regularized least-square estimation. First, with a random initialization of L, the algorithm optimizes the matrix R by least-square estimation. We notice that solving each row of R has no influence on each other in the optimization; so it is efficient to quickly obtain R by solving each row of R separately. Next, R is fixed and L is computed. The above least-square estimation is executed iteratively where the iteration number is a parameter constrained by convergence condition and real-time requirements (we set it to 200 by default in experiments).

As a result, we can obtain the traffic condition matrix $\widehat{X}_{m \times n}$, in which the row values at the current time slice indicate the real-time traffic conditions. Meanwhile, we utilize a fast linear algebra library jblas [22] to accelerate matrix multiplications. As such, while in Fujian highway data set the measurement matrix has a size of 2016 × 3038 with $r = 2$ and the iteration number 200, it only takes about 164 s to execute the algorithm. This can fully satisfy the requirements of monitoring traffic conditions in real time.

4.3   Data Filtering by Traffic Flow Features

After data completion, we carry out a filtering procedure, whose main functionality is to rectify the data by integrating the characteristics of different traffic flows. The following gives the description of the filtering.

Given the existing speed sequence $\{x(t_i, p_j)\}$, where t_i denotes a time point and p_j denotes a location, what we do as follows is to estimate the speed at time point t and location p, i.e., $x(t, p)$, where $p \in \{p_j\}$ and $t \geq t_i$. Let $x_f(t, p)$ be the estimated speed in the free flow, and let $x_c(t, p)$ be the estimated speed in the congested flow, we can determine the estimated speed by:

$$x(t, p) = [1 - w(t, p)]x_f(t, p) + w(t, p)x_c(t, p) \tag{6}$$

where $w(t, p)$ is the weight function of the filtering method. A key intuition is that the current vehicle speed at a road segment is affected by the historical speeds at the road segments nearby and that the effects of historical speeds exponentially decay over both space and time distances. Therefore, we define the above two speeds via Eq. (7).

$$
\begin{aligned}
x_c(t, p) &= \frac{1}{N_c(t, p)} \sum_{p_j, dist(p, p_j) \leq \sigma} \sum_{t_i = t - \tau}^{t} x(t_i, p_j) \phi_c(t_i - t, p_j - p) \\
x_f(t, p) &= \frac{1}{N_f(t, p)} \sum_{p_j, dist(p, p_j) \leq \sigma} \sum_{t_i = t - \tau}^{t} x(t_i, p_j) \phi_f(t_i - t, p_j - p)
\end{aligned}
\tag{7}
$$

Here σ is the maximum distance to p, τ is the earliest time point prior to t, $N_c(t, p)$ and $N_f(t, p)$ are the normalization coefficients. $\phi_f(\Delta t, \Delta p)$ and $\phi_c(\Delta t, \Delta p)$ are the weight functions for free and congested flows, respectively. Further, we design $\phi(\Delta t, \Delta p)$ to be $\phi(\Delta t, \Delta p) = \exp(-\frac{|\Delta t|}{\alpha} - \frac{|\Delta p|}{\beta})$, where α and β are given as the perturbation range in time and space dimensions, respectively. And then, we design $\phi_f(\Delta t, \Delta p)$ and $\phi_c(\Delta t, \Delta p)$ as follows:

$$
\begin{aligned}
\phi_f(\Delta t, \Delta p) &= \phi(\Delta t - \Delta p / \gamma_f, \Delta p) \\
\phi_c(\Delta t, \Delta p) &= \phi(\Delta t - \Delta p / \gamma_c, \Delta p)
\end{aligned}
\tag{8}
$$

where γ_f is the free traffic propagation speed in a free flow and γ_c is the congestion traffic propagation speed in a congested flow [23].

The weight function $w(t,p)$ should be designed to have the following properties. (a) $w(t,p)$ should tend to 0 for a free flow and to 1 for a congested flow. (b) From the traffic observation, we find that if a free flow meets a congested flow at any road segment, then the latter usually overrides the former. Therefore, as for the estimated speeds, a large weight should be assigned to $x_f(t,p)$ if both $x_c(t,p)$ and $x_f(t,p)$ are greater than v_c, or to $x_c(t,p)$ if $x_c(t,p)$ or $x_f(t,p)$ is less than v_c, where v_c denotes the crossover from a free flow to a congested flow. Therefore, we employ a sigmoid function as the kernel of $w(t,p)$. As a result,

$$w(t,p) = \frac{1}{1+e^{-\frac{v_c-\min(x_f(t,p),x_c(t,p))}{\Delta v}}} \tag{9}$$

where v_c is set to 60 km/h and Δv is set to 20 km/h which is used for normalization of speed variation. As long as the parameters $\alpha, \beta, \gamma_c, \gamma_f, \sigma, \tau$ are specified (we set them to 5 min, 3 km, -15 km/h, 80 km/h, 16 km and 60 min by default), we can obtain the final vehicle speed estimates by Eqs. (6), (7), (8) and (9).

5   Evaluation

We evaluate Megrez from different spatial granularities. We start with segment-level comparisons, then we take a highway as the granularity, and finally we treat all the highways as a whole.

5.1   Case Study

First, we compare the speeds estimated by the Megrez approach (FIMSs) with the detector speeds (DSs) on specific road segments where loop detectors are installed. In experiments, we estimate FIMSs using signaling data, GPS data, and half of the loop detector data, and take the other half of detector data as ground truth.

We select the 6th road segment of Fujian section of G25 (from Changchun to Shenzhen) and the 3rd road segment of Fujian section of G15 (from Hainan to Shenyang), and then obtain the DSs and FIMSs on the road segments on November 8, 10 and 13 of 2015, respectively. Figure 5 shows the comparison between the two kinds of speeds.

As we can see from Fig. 5, the speeds are relatively low at night, which may be attributed to the poor visibility. As a result, the drivers have to slow down for safety. Meanwhile, FIMSs are smooth and can fit the profile of DSs well. Besides, in Fig. 5(c), the congestion is observed from 10 a.m. to 12 a.m. FIMSs can capture the change of speeds accurately. There is a slight delay in the FIMSs, this is due to the sampling delays of signaling data and GPS data.

Next, we plot the spatial-temporal graph of Fujian section of G72 (from Nanning to Quanzhou) on November 10, 2015 based on the DSs and FIMSs, respectively. In Fig. 6, the X axis is the time of the day, and the Y axis is the distance to Quanzhou, which is the start of Fujian section of G72. In addition, green color in Fig. 6 represents the free traffic state (vehicle speeds higher than 60 km/h), blue color represents the slow state (vehicle speeds in 40–60 km/h), and red color represents the congestion state (vehicle speeds lower than 40 km/h).

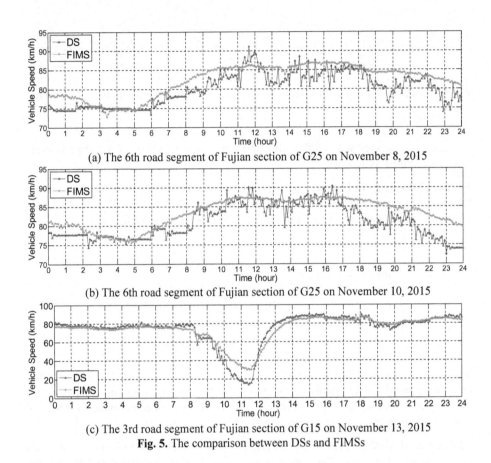

(a) The 6th road segment of Fujian section of G25 on November 8, 2015

(b) The 6th road segment of Fujian section of G25 on November 10, 2015

(c) The 3rd road segment of Fujian section of G15 on November 13, 2015

Fig. 5. The comparison between DSs and FIMSs

Fig. 5.   The comparison between DSs and FIMSs

From Fig. 6(a), we can see that the traffic conditions given by DSs are sparse, and no traffic conditions are reported on many road segments. On the other hand, the traffic conditions given by FIMSs offer full coverage in time and space, as shown in Fig. 6(b). Notice that from 0 a.m. to 7 a.m., the vehicle speeds are low at the location which is about 120 km away from Quanzhou, which is consistent with DSs.

5.2 Speed Accuracy and Traffic Condition Accuracy Evaluation

Next we give the statistical comparisons for the nine highways in Fujian, comparing Megrez with three other methods: Kalman filtering, MSSA-based method [6] and

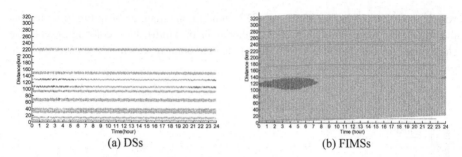

Fig. 6. Spatial-temporal graph of Fujian section of G72 on November 10, 2015 (Color figure online)

Megrez⁻ (i.e., Megrez without the third step). The methods chosen for comparison are briefly described as follows.

Kalman Filtering: In Kalman filtering [24], the merged speed at the *n-th* time slice is taken as the observed value $z(n)$, and the estimated speed of the *n-th* time slice, denoted as $\hat{x}(n)$, is calculated by performing an iteration procedure according to

$$prediction\ process\begin{cases} x'(n) = \hat{x}(n-1) \\ P'(n) = P(n-1) + Q \end{cases}$$

$$correction\ process\begin{cases} K(n) = P'(n)/(R + P'(n)) \\ \hat{x}(n) = x'(n) + K(n)[z(n) - x'(n)] \\ P(n) = [1 - K(n)]P'(n) \end{cases}$$

where $x'(n)$ denotes the prediction speed at current time slice, $P'(n)$ is a priori covariance matrix of estimation error, $K(n)$ is a Kalman gain matrix, $P(n)$ is a posteriori covariance matrix of estimation error, Q is a system noise covariance matrix, and R is a measurement noise covariance matrix.

MSSA-Based Method (MSSA for Short): MSSA is often used to solve missing data problems, e.g., for geographic data and meteorological data. It is a data adaptive and nonparametric method based on the embedded lag-covariance matrix. We adopt an iterative procedure proposed in [6] that utilizes the internal periodicity of traffic conditions. The parameter M is set to 288 (i.e., one day) as suggested by [6].

Megrez⁻: Megrez⁻ is the Megrez approach without the third step, i.e., feature-driven filtering.

For evaluating the accuracy of the speeds, we take Mean Absolute Error (MAE), Mean Absolute Percentage Error (MAPE), Root Mean Square Error (RMSE), and Normalized Mean Absolute Error (NMAE) as metrics.

Experiments are conducted with the merging of signaling data, GPS data and half of loop detector data occurred along the nine highways in Fujian as input. We treat these data as an incoming stream arrived in chronological order and execute the above four methods every 5 min. In addition, the data that span a duration of one week prior to the current time slice are adopted to form the measurement matrix. We examine the estimated speeds at locations with loop detectors as the ground truth (of course, at those locations, loop detector data are not used/merged in our speed estimations). We evaluate the speed accuracy using the four metrics mentioned above.

In particular, we take as input the data on November 8 (weekend) and 10 (weekday) of 2015, and execute the four methods, respectively. The results are shown in Fig. 7, which is the performance of the four methods in terms of estimation errors. We can see that Megrez outperforms the other three methods under all metrics. Megrez first fills the missing speeds via compressive sensing, and then filters the speeds by leveraging the characteristics of traffic flows. Since the inherent features of sensory data and characteristics of traffic flows are both utilized, the estimated results are better than other methods.

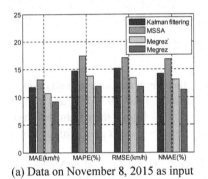

(a) Data on November 8, 2015 as input

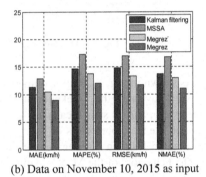

(b) Data on November 10, 2015 as input

Fig. 7. Speed errors under different methods

The performance of the Kalman filtering method is not as good as Megrez. The reason behind it is that Kalman filtering optimizes the estimates by using linear state equations, but the characteristics of realistic traffic flows are not linear. We find that MSSA takes a long time to get results. As far as Fujian highways are concerned, under the same hardware and software configurations, MSSA spends about 2.4 h to estimate the traffic conditions of one time slice while Megrez takes only about 3 min. So MSSA is not efficient in execution time and cannot be used in real-time scenarios especially when the number of road segments is large. Megrez performs better than Megrez⁻, which is attributed to the filtering. The filtering incorporates the characteristics of the traffic flows, which is consistent with real traffic patterns.

To evaluate traffic condition accuracy, we introduce state consistency and speed consistency. The former means that the estimated traffic condition (congested, slow, or free) is consistent with the traffic condition indicated by the DS, while the latter means that the difference between the estimated speed and the corresponding DS is within

20 km/h. We consider as metrics the precision and recall of the estimated speed with respect to state consistency or speed consistency. For the sake of convenience, state precision, state recall, speed precision and speed recall are used to replace the full names of the metrics. The results are shown in Fig. 8.

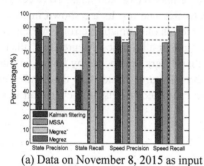
(a) Data on November 8, 2015 as input

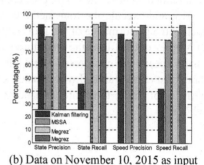
(b) Data on November 10, 2015 as input

Fig. 8. The precisions and recalls of the four methods

As shown in Fig. 8, Megrez outperforms the other three methods in both cases. For the weekday and the weekend, its state precisions reach 93.41% and 93.50%, respectively, and its state recalls are 47.82% and 37.19% higher than Kalman filtering, although the state precisions of Kalman filtering are close to Megrez. This is because the standard Kalmen filtering cannot fill in missing data.

The behaviors of the MSSA method and Megrez⁻ are not as good as Megrez. This is because the filtering process has not been imposed on MSSA and Megrez⁻; as such, the vehicle speeds show a large fluctuation.

5.3   Adaptability of Megrez

As mentioned earlier, Megrez is adaptive to the variation of data in temporal and spatial distributions. In this subsection, we evaluate the adaptability of Megrez by conducting the following experiments.

We execute Megrez with different combinations of three sets of sensory data on nine highways in Fujian on November 8 (weekend) and 10 (weekday), 2015 respectively. The first data set is the signaling data from CMCC; the second data set is the GPS data from 2K1W vehicles; and the third data set is the half of loop detector data. The results are shown in Fig. 9.

From Fig. 9, we find that Megrez can accurately estimate vehicle speeds with any two data sets, and the accuracy of using only two types of data is slightly lower than that with all three data sources. In other words, even if some types of data are not available, Megrez can still report quite accurate traffic conditions.

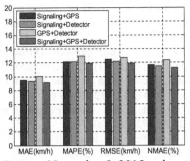
(a) Data on November 8, 2015 as input

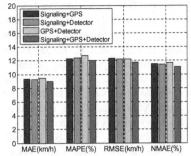
(b) Data on November 10, 2015 as input

Fig. 9. Speed errors using different data sets

6   Conclusions

The biggest hurdle to fully achieving multiple detection goals comes from the low quality of available data. The Megrez approach proposed in this paper performs adaptive data fusion to maximize the potential of available data. Especially, Megrez overcomes the sparsity and incoherence of data, and integrates the characteristics of traffic flows. Therefore, it can correct the distortions in road condition reports as much as possible. Based on a large number of real data, we conduct extensive experiments and evaluate Megrez from different perspectives. The experimental results show that the vehicle speeds estimated by Megrez have low errors, and that our approach can accurately detect the traffic conditions on highways.

Acknowledgement. This work was supported by the National Natural Science Foundation of China under Grant No. 61472408 and the Ministry of Transportation of China under Grant No. 2015315Q16080. Tingjian Ge was supported in part by the NSF grants IIS-1149417 and IIS-1633271.

References

1. Wu, X., Zhu, X., Wu, G.Q., Ding, W.: Data mining with big data. IEEE Trans. Knowl. Data Eng. **26**(1), 97–107 (2014)
2. Russell, T.: Signaling System 7, 6th edn. McGraw-Hill Education, New York (2014)
3. Deng, D., Shahabi, C., Demiryurek, U., Zhu, L., Yu, R., Liu, Y.: Latent space model for road networks to predict time-varying traffic. In: ACM International Conference on Knowledge Discovery and Data Mining, pp. 1525–1534 (2016)
4. Hajimolahoseini, H., Amirfattahi, R., Soltanian-Zadeh, H.: Robust vehicle tracking algorithm for nighttime videos captured by fixed cameras in highly reflective environments. Comput. Vis. IET **8**(6), 535–544 (2014)
5. Wang, F., Hu, L., Zhou, D., Sun, R., Hu, J., Zhao, K.: Estimating online vacancies in real-time road traffic monitoring with traffic sensor data stream. Ad Hoc Netw. **35**(C), 3–13 (2015)

6. Zhu, H., Zhu, Y., Li, M., Ni, L.M.: SEER: metropolitan-scale traffic perception based on lossy sensory data. In: IEEE International Conference on Computer Communications, Rio De Janeiro, Brazil, pp. 217–225 (2009)

7. Zhu, Y., Li, Z., Zhu, H., Li, M., Zhang, Q.: A compressive sensing approach to urban traffic estimation with probe vehicles. IEEE Trans. Mob. Comput. **12**(11), 2289–2302 (2013)

8. Liu, Z., Li, Z., Li, M., Xing, W.: Mining road network correlation for traffic estimation via compressive sensing. IEEE Trans. Intell. Transp. Syst. **17**(7), 1–12 (2016)

9. Zhan, X., Zheng, Y., Yi, X., Ukkusuri, S.V.: Citywide traffic volume estimation using trajectory data. IEEE Trans. Knowl. Data Eng. **29**(2), 272–285 (2017)

10. Huang-Fu, C.C., Lin, Y.B.: Deriving vehicle speeds from standard statistics of mobile telecom switches. IEEE Trans. Vehic. Technol. **61**(7), 3337–3341 (2012)

11. Janecek, A., Valerio, D., Hummel, K.A., Ricciato, F., Hlavacs, H.: The cellular network as a sensor: from mobile phone data to real-time road traffic monitoring. IEEE Trans. Intell. Transp. Syst. **16**(5), 2551–2572 (2015)

12. Becker, R.A., Caceres, R., Hanson, K., Ji, M.L., Urbanek, S., Varshavsky, A., Volinsky, C.: Route classification using cellular handoff patterns. In: Proceedings of the 13th International Conference on Ubiquitous Computing, pp. 123–132 (2011)

13. Caceres, N., Romero, L.M., Benitez, F.G., Del Castillo, J.M.: Traffic flow estimation models using cellular phone data. IEEE Trans. Intell. Transp. Syst. **13**(3), 1430–1441 (2012)

14. Calabrese, F., Colonna, M., Lovisolo, P., Parata, D., Ratti, C.: Real-Time urban monitoring using cell phones: a case study in Rome. IEEE Trans. Intell. Transp. Syst. **12**(1), 141–151 (2011)

15. Aslam, J., Lim, S., Pan, X., Rus, D.: City-scale traffic estimation from a roving sensor network. ACM Conference on Embedded Network Sensor Systems, pp. 141–154 (2012)

16. Bouillet, E., Chen, B., Cooper, C., Dahlem, D., Verscheure, O.: Fusing traffic sensor data for real-time road conditions. In: International Workshop on Sensing and Big Data Mining, pp. 1–6 (2013)

17. Yang, L., Pereira, F.C., Seshadri, R., O'Sullivan, A., Antoniou, C., Ben-Akiva, M.: DynaMIT2.0: architecture design and preliminary results on real-time data fusion for traffic prediction and crisis management. In: IEEE 18th International Conference on Intelligent Transportation Systems, pp. 2250–2255 (2015)

18. Candés, E.J., Tao, T.: The power of convex relaxation: near-optimal matrix completion. IEEE Trans. Inf. Theory **56**(5), 2053–2080 (2010)

19. Bach, F.R.: Consistency of trace norm minimization. J. Mach. Learn. Res. **9**(2) (2008)

20. Fazel, M.: Matrix rank minimization with applications. Ph.D. dessertation, Department of electrical engineering, Stanford University, California (2002)

21. Recht, B., Fazel, M., Parrilo, P.A.: Guaranteed minimum-rank solutions of linear matrix equations via nuclear norm minimization. Siam Rev. **52**(3), 471–501 (2007)

22. Jblas. http://jblas.org/

23. Cassidy, M.J., Windover, J.R.: Methodology for assessing dynamics of freeway traffic flow. Transp. Res. Rec. **1484**, 73–79 (1995)

24. Kalman, R.E.: A new approach to linear filtering and prediction problems. Trans. ASME-J. Basic Eng. **82**, 35–45 (1960)

On the Interaction of Functional
and Inclusion Dependencies
with Independence Atoms

Miika Hannula$^{(\boxtimes)}$ and Sebastian Link

Department of Computer Science, The University of Auckland,
Auckland, New Zealand
{m.hannula,s.link}@auckland.ac.nz

Abstract. Infamously, the finite and unrestricted implication problems
for the classes of (i) functional and inclusion dependencies together,
and (ii) embedded multivalued dependencies alone are each undecidable.
Famously, the restriction of (i) to functional and unary inclusion dependencies in combination with the restriction of (ii) to multivalued dependencies yield implication problems that are still different in the finite
and unrestricted case, but each are finitely axiomatizable and decidable
in low-degree polynomial time. An important embedded tractable fragment of embedded multivalued dependencies are independence atoms.
These stipulate independence between two attribute sets in the sense
that for every two tuples there is a third tuple that agrees with the first
tuple on the first attribute set and with the second tuple on the second attribute set. Our main results show that finite and unrestricted
implication deviate for the combined class of independence atoms, unary
functional and unary inclusion dependencies, but both are axiomatizable
and decidable in low-degree polynomial time. This combined class adds
arbitrary independence atoms to unary keys and unary foreign keys,
which frequently occur in practice as surrogate keys and references to
them.

Keywords: Functional dependency · Inclusion dependency
Independence atom · Implication problem

1 Introduction

Databases represent information about some domain of the real world. For
this purpose, data dependencies provide the main mechanism for enforcing the
semantics of the given application domain within a database system. As such,
data dependencies are essential for most data management tasks, including
database design, query and update processing, as well as data cleaning, exchange,
and integration. The usability of a class $\mathcal{C}$ of data dependencies for these tasks
depends critically on the computational properties of its associated implication

The authors were supported by the Marsden Fund grant 3711702.

© Springer International Publishing AG, part of Springer Nature 2018
J. Pei et al. (Eds.): DASFAA 2018, LNCS 10828, pp. 353–369, 2018.
https://doi.org/10.1007/978-3-319-91458-9_21

problem. The implication problem for C is to decide whether for a given finite set $\Sigma \cup \{\varphi\}$ of data dependencies from C, Σ implies φ, i.e. whether every database that satisfies all the elements of Σ also satisfies φ. If we require databases to be finite, then we speak of the finite implication problem, and otherwise of the unrestricted implication problem. While the importance of data dependencies continues to hold for new data models, the focus of this article is on the implication problems for important classes of data dependencies in the relational model of data. In this context, data dependency theory is deep and rich [37]. Our submission is from the area of database theory, on which DASFAA's call for paper has solicited original contributions.

Functional and inclusion dependencies constitute the most commonly used classes of data dependencies in practice. In particular, functional dependencies (FDs) are more expressive than keys, and inclusion dependencies (INDs) are more expressive than foreign keys, thereby capturing Codd's principles of entity and referential integrity, respectively, on the logical level. An FD $R : X \rightarrow Y$ with attribute subsets X, Y on relation schema R expresses that the values on attributes in Y are uniquely determined by the values on attributes in X. In particular, $R : X \rightarrow R$ expresses that X is a key for R. An inclusion dependency (IND) $R[A_1, \ldots, A_n] \subseteq R'[B_1, \ldots, B_n]$, with attribute sequences $A_1, \ldots, A_n$ on R and $B_1, \ldots, B_n$ on R', expresses that for each tuple t over R there is some tuple t' over R' such that for all $i = 1, \ldots, n$, $t(A_i) = t'(B_i)$ holds. If $n = 1$ we call the IND unary (UIND).

A fundamental result in dependency theory is that the unrestricted and finite implication problems for the combined class of FDs and INDs differ and each is undecidable [10,32,33]. Interestingly, for the expressive sub-class of FDs and UINDs, the unrestricted and finite implication problems still differ but each are axiomatizable and decidable in low-degree polynomial time [12].

Another important expressive class of data dependencies are *embedded multivalued dependencies* (EMVDs). An EMVD $R : X \rightarrow Y \perp Z$ with attribute subsets X, Y, Z of R expresses that the projection $r[XYZ]$ of a relation r over R on the set union XYZ is the join $r[XY] \bowtie r[XZ]$ of its projections on XY and XZ. Another fundamental result in dependency theory is that the unrestricted and finite implication problems for EMVDs differ, each is not finitely axiomatizable [36] and each is undecidable [23,24]. An important fragment of EMVDs are multivalued dependencies (MVDs), which are a class of full dependencies in which XYZ covers the full underlying set R of attributes. In fact, MVDs are the basis for Fagin's fourth normal form [13]. For the combined class of FDs, MVDs, and UINDs, finite implication is axiomatizable and decidable in cubic time, while unrestricted implication is axiomatizable and decidable in almost linear time [12,26].

Independence atoms (IAs) constitute an expressive subclass of EMVDs and FDs. An IA $X \perp Y$ with attribute subsets X, Y of R expresses that $X \cap Y$ is constant (i.e., the FD $\emptyset \rightarrow X \cap Y$ holds) and that the EMVD $\emptyset \rightarrow X \setminus Y \perp Y \setminus X$ holds. The latter expresses that the projection of a relation r on XY equals the cartesian product of its projections on X and Y, i.e., $r[XY] = r[X] \times r[Y]$.

For disjoint X and Y, the independence atoms $X \perp Y$ thus form a subclass of EMVDs. For the class of IAs, the finite and unrestricted implication problems coincide, they are finitely axiomatizable and decidable in low-degree polynomial time [27].

Given the usefulness of EMVDs, FDs, and INDs for data management, given their computational barriers, and given the attractiveness of IAs as a tractable fragment of EMVDs, it is a natural question to ask how IAs, FDs, and INDs interact. Our article helps address the current gap in the existing rich theory of relational data dependencies. Adding further to the challenge it is important to note that IAs still form an embedded fragment of EMVDs, in contrast to MVDs which are a class of full dependencies. Somewhat surprisingly, already the interaction of IAs with just keys is intricate [19,20]. For example, unrestricted implication is finitely axiomatizable but finite implication is not for keys and unary IAs (those with singleton attribute sets), while the finite and unrestricted implication problems coincide and enjoy a finite axiomatization for IAs and unary keys (those with a singleton attribute set). In contrast, the extension of INDs with IAs, although being more expressive than the class of INDs alone, does not add further complexity to the latter. For INDs and IAs taken together both implication problems still coincide and are finitely axiomatizable, PSPACE-complete, and fixed-parameter tractable in their arity [9,21].

Examples. We few examples will illustrate how knowledge about IAs advances data management. FDs and INDs do not require further motivation but the more we know about the interaction of IAs with FDs and INDs, the more we can advance data management.

Our first example is query processing. In particular, we show how the validity of independence atoms is intrinsically linked to the optimization of the famous division operator. The operator $\pi_{XY}(R) \div \pi_Y(R)$ returns all those X-values x such that for every Y-value y there is some tuple t with $t(X) = x$ and $t(Y) = y$ [11]. The ability of the division operator to express universal quantification makes it very powerful for expressing natural queries. The following result establishes the intrinsic link.

Theorem 1. *For all relations r over R, $\pi_{XY}(R)(r) \div \pi_Y(R)(r) = \pi_X(R)(r)$ if and only if r satisfies $X \perp Y$.*

Proof. The division operator is defined as follows:

$$\pi_{XY}(R)(r) \div \pi_Y(R)(r) = \pi_X(R)(r) - \pi_X((\pi_X(R)(r) \times \pi_Y(R)(r)) - \pi_{XY}(R)(r)),$$

and r satisfies $X \perp Y$ if and only if $\pi_X(R)(r) \times \pi_Y(R)(r) = \pi_{XY}(R)(r)$. The result follows directly.

In particular, the validity of an IA reduces the quadratic complexity of the division operator to a linear complexity of a simple projection [28]. The reduction in complexity also applies to the expression complexity of a query. Suppose we would like to return those entities x that occur together with all entities y (for example, suppliers that supply all products), then we need to express the division operator $\pi_{X,Y}(R) \div \pi_Y(R)$ in SQL by double-negation as in:

$$
\begin{aligned}
&\text{SELECT } R0.X \text{ FROM } R \text{ AS } R0 \\
&\text{WHERE NOT EXISTS} \\
&\quad \text{SELECT } * \text{ FROM } R \text{ AS } R1 \\
&\quad \text{WHERE NOT EXISTS} \\
&\qquad \text{SELECT } * \text{ FROM } R \text{ AS } R2 \\
&\qquad \text{WHERE } R2.X = R1.X \text{ AND} \\
&\qquad \quad R2.Y = R0.Y \ ;
\end{aligned}
$$

where $R.X$ is short for $\bigwedge_{A \in X} R.A$, and $R2.X = R1.X$ ($R2.Y = R0.Y$) for $\bigwedge_{A \in X} R2.A = R1.A$ ($\bigwedge_{B \in Y} R2.B = R1.B$). However, if a query optimizer can notice that the IA $X \perp Y$ is implied by the enforced set Σ of constraints, then the query can be rewritten into

$$
\begin{aligned}
&\text{SELECT } X \\
&\text{FROM } R;
\end{aligned}
$$

   Our second example is database security. More specifically, the aim of inference control is to protect private data under inferences that clever attacks may use to circumvent access limitations [7]. For example, the combination of a particular patient name (say Jack) together with a particular medical examination (say angiogram) may be considered a secret, while access to the patient name and access to the medical examination in isolation may not be a secret. However, in some given context such as a procedure to diagnose some condition, all patients may need to undergo all examinations. That is, the information about the patient is independent of the information about the examination. Now, if the secret (Jack, angiogram) must not be revealed to an unauthorized user that can query the data source, then this user must not learn both: that Jack is a patient undergoing the diagnosis of the condition, and that angiogram is a medical examination that is part of the process for diagnosing the condition. Being able to understand the interaction of independence atoms with other database constraints can therefore help us to protect secrets under clever inference attacks.

   Our final example is data profiling. Here we would like to demonstrate that independence atoms do occur in real-world data sets. For that purpose, we have mined some well-known publicly available data sets that have been used for the mining of other classes of data dependencies before [34]. We report the basic characteristics of these data sets in the form of their numbers of rows and columns, and list the number of maximal IAs and the maximum arity of those found. Here, an IA $X \perp Y$ is maximal in a given set of IAs if there is no other IA $V \perp W$ in the set such that $V \subseteq X$ and $W \subseteq Y$ holds. The arity of an IA is defined as the total number of attribute occurrences.

Data set	Number of columns	Number of rows	Number of IAs	Maximum arity
Bridges	13	108	4	3
Echocardiogram	13	132	5	4
Adult	14	48,842	9	3
Hepatitis	20	155	855	6
Horse	27	368	112	3

Table 1. Subclasses of FD+IND+IA. We write "ui" and "fi" for unrestricted and finite implication, respectively.

Class	ui = fi	Complexity: ui/fi	Finite axiomatization: ui/fi
FD	Yes [4]	Linear time [5]	Yes (2-ary) [4]
IND	Yes [9]	PSPACE-complete [9]	Yes (2-ary) [9]
IA	Yes [15, 27, 35]	Cubic time [15, 27]	Yes (2-ary) [15, 27, 35]
IND+IA	Yes	PSPACE-complete [21]	Yes (3-ary) [21]
FD+IA, FD+UIA	No [20]	?/?	?/no
FD+IND	No [10, 32]	Undecidable/undecidable [10, 32]	No/no [10, 32]
FD+UIND	No [12]	Cubic time/cubic time [12]	Yes/no (infinite) [12]
UFD+UIND	No [12]	Linear time/linear time [12]	Yes /no (infinite) [12]
UFD+UIND+IA	No	Cubic time/cubic time	Yes /no (infinite)

It should be stressed that the usefulness of these IAs is not restricted to those that are semantically meaningful. For example, the optimizations for the division operator also apply to IAs that "accidentally" hold on a given data set.

1.1 Contributions

In this article we make the following contributions.

(1) We illustrate the relevance of independence atoms for data management, such as their intrinsic link to the optimization of the division operator, more precise cardinality estimations for choosing better query plans, and database security. Moreover, we show that they occur in real-world data sets.

(2) For the combined class of FDs and IAs, finite and unrestricted implication differ [19, 20]. We show that finite implication is not finitely axiomatizable, already for binary FDs (those with a two-element attribute set on the left-hand side) and unary IAs. For the combined class of IAs and unary FDs, we show that finite and unrestricted implication coincide and establish a finite axiomatization. Hence, the situation for the combined class of FDs and IAs is more intricate than for the combined class of FDs and MVDs, where finite and unrestricted implication coincide, which enjoy an elegant finite axiomatization [6], and for which implication can be decided in almost linear time [14].

(3) For the combined class of IAs, unary FDs, and UINDs, we establish axiom-atizations for their finite and unrestricted implication problems, and show that both are decidable in low-degree polynomial time. This is analogous to the results for the combined class of FDs, MVDs, and UINDs. To the best of our knowledge, the class of IAs, unary FDs, and UINDs is only the second known class for which the finite and unrestricted implication differ but both are decidable in low-degree polynomial time. The class is practically relevant as it covers arbitrary IAs on top of unary keys and unary foreign keys, and already unary keys and unary foreign keys occur readily in practice [12]. The significant difference to FDs, MVDs, and UINDs is the more intricate interaction between FDs and IAs in comparison to FDs and MVDs. Note that unary FDs and INDs frequently occur in practice as surrogate keys and foreign keys that reference them. For example, 6 out of 8 keys are unary and 8 out of 9 foreign keys are unary in the TPC-H benchmark, while 20 out of 32 keys are unary and 44 and out 46 foreign keys are unary in the TPC-E benchmark[1]. The ability to reason efficiently about IAs, UFDs, and UINDs is good news for data management. Finally, trading in restrictions of the arity on INDs and FDs for restrictions on the arity of IAs cannot be successful: Finite implication for unary IAs and binary FDs is not finitely axiomatizable, see (2).

(4) For the combined class of IAs and FDs we establish tractable conditions suffi-cient for non-interaction in both the finite and unrestricted cases. Instances of the finite or unrestricted implication problems that meet the non-interaction conditions can therefore be decided efficiently by using already known algo-rithms for the sole class of IAs and the sole class of FDs. The decidability of the finite and unrestricted implication problems for IAs and FDs are both still open.

Organization. In Sect. 2 we present all the necessary definitions for the article. Section 3 addresses the combined class of FDs and IAs. In Sect. 4 we focus on the combination of UFDs, UINDs, and IAs, and establish axiomatizations for their finite and unrestricted implication problems. Section 5 identifies polynomial-time criteria for the non-interaction between INDs and IAs, and also between FDs and IAs. Finally, in Sect. 6 we discuss the computational complexity of the impli-cation problems. Due to lack of space we refer the reader to Appendix for any remaining proofs. The appendix can be found in [22].

2   Preliminaries

We denote by $A, B, C, \ldots$ attributes and by $X, Y, Z, \ldots$ either sets or sequences of attributes. For two sets (sequences) X and Y, we write XY for their union (concatenation). Similarly, we may write A instead of the single element set or sequence that consists of A. The size of a set (or length of a sequence) X is written as $|X|$.

---

[1]   http://www.tpc.org.

A *relation schema* consists of attributes A, each equipped with a set of *domain* values denoted by $\text{Dom}(A)$. By *database schema* we denote a pairwise disjoint sequence of relations schemata. Given a relation schema R, a *tuple* over R is a function that maps each attribute A from R to $\text{Dom}(A)$. A *relation* r over R is then a non-empty set of tuples over R, and a database d over $\mathcal{R} = (R_1, \ldots, R_n)$ is a sequence $(r_1, \ldots, r_n)$ where each r_i is a relation over $R_i{}^2$. We sometimes write $r[R]$ to denote that r is a relation over R, and similarly we may write $d[\mathcal{R}]$. A relation is called *finite* if the underlying set of tuples is finite, and a database is finite if it is a sequence of finite relations. For a tuple t and a relation r over R and a subset (or subsequence) X of R, $t(X)$ is the *restriction* of t to X, and $r(X)$ is the set of all restrictions $t(X)$ where $t \in r$.

Next we define the syntax and semantics of functional and inclusion dependencies and independence atoms.

Functional Dependency. Let X and Y be two sets of attributes from a relation schema R. Then $R : X \to Y$ is a *functional dependency* that is satisfied by a database $d = (r[R])$ iff for all $t, t' \in r$, $t(X) = t'(X)$ implies $t(Y) = t'(Y)$.

Inclusion Dependency. Let $A_1, \ldots, A_n$ and $B_1, \ldots, B_n$ be two sequences of distinct attributes from relation schemata R and R', respectively. Then $R[A_1 \ldots A_n] \subseteq R'[B_1 \ldots B_n]$ is an *inclusion dependency* that is satisfied by a database $d = (r[R], r'[R'])$ iff for all $t \in r_i$ there is $t' \in r_j$ such that $t(A_1) = t'(B_1), \ldots, t(A_n) = t'(B_n)$.

Independence Atom. Let X and Y be two (not necessarily disjoint) attribute sets from a shared relation schema R. Then $R : X \perp Y$ is an *independence atom* that is satisfied by a database $d = (r[R])$ iff for all tuples $t, t \in r$ there is a tuple $t'' \in r$ such that $t''(X) = t(X)$ and $t''(Y) = t'(Y)$. A *disjoint* independence atom (DIA) is an IA $X \perp Y$ where $X \cap Y$ is empty.

Regarding all the aforementioned dependencies, if the relation schema R is not needed in the context, we will drop it from the syntax. E.g., we will write $X \perp Y$ instead of $R : X \perp Y$.

We say that an IND is *k-ary* if it is of the form $A_1 \ldots A_k \subseteq B_1 \ldots B_k$. An IA $X \perp Y$ and an FD $X \to Y$ are called *k-ary* if $\max\{|X|, |Y|\} = k$. A class of dependencies is called *k-ary* if it contains at most *k-ary* dependencies. Most of the subclasses that we consider are only unary, so we add "U" to a class name to denote its unary subclass. For instance, UIND denotes the class of all unary INDs. In general, for $k \geq 2$, we add "k" to a class name to denote its *k-ary* subclass. We use "+" to denote unions of classes, e.g., IND+IA denotes the class of all inclusion dependencies and independence atoms.

---

^{2} We exclude empty relations from our definition. This is a practical assumption with no effect when single relation schemata are considered only. However, on multiple relations it has an effect, e.g., the rule $\mathcal{UI}3$ in Table 2 becomes unsound.

Notice that the semantic condition for IAs $X \perp Y$ holds only if the values of the common attributes of X and Y are constant. In other words, the following holds:

* $d \models R : X \perp X$, if for all $s, s' \in r$ it holds that $s(X) = s'(X)$.

Hence, we also call unary FDs of the form $\emptyset \to A$ and unary IAs of the form $A \perp A$ *constancy atoms* (CAs).

The *restriction* of a dependency σ to a set of attributes R, written $\sigma \upharpoonright R$, is $X \cap R \to Y \cap R$ for an FD σ of the form $X \to Y$, and $X \cap R \perp Y \cap R$ for an IA σ of the form $X \perp Y$. If σ is an IND of the form $A_1 \ldots A_n \subseteq B_1 \ldots B_n$ and $i_1, \ldots, i_k$ lists $\{i = 1, \ldots, n : A_i \in R \text{ and } B_i \in R\}$, then $\sigma \upharpoonright R = A_{i_1} \ldots A_{i_k} \subseteq B_{i_1} \ldots B_{i_k}$. For a set of dependencies Σ, the restriction of Σ to R, written $\Sigma \upharpoonright R$, is the set of all $\sigma \upharpoonright R$ for $\sigma \in \Sigma$. For attributes A and B from R, we denote by $\sigma(R : A \mapsto B)$ the dependencies obtained from σ by replacing any number of occurrences of A with B.

A set of axioms σ and rules of the form $\sigma_1, \ldots, \sigma_n \Rightarrow \sigma$ is called an *axiomatization*. A rule is called *n-ary* if its antecedent part has n conjuncts. An axiomatization consisting of at most *n*-ary rules is called *n-ary*. A *deduction* from a set of dependencies Σ by an axiomatization $\mathfrak{R}$ is a sequence of dependencies $(\sigma_1, \ldots, \sigma_n)$ where each σ_i is either an element of Σ, an axiom, or follows from $\sigma_1, \ldots, \sigma_{i-1}$ by an application of a rule in $\mathfrak{R}$. In such an occasion we write $\Sigma \vdash_{\mathfrak{R}} \sigma$, or simply $\Sigma \vdash \sigma$ if $\mathfrak{R}$ is known.

Given a finite set of database dependencies $\Sigma \cup \{\sigma\}$, the (finite) unrestricted implication problem is to decide whether all (finite) databases that satisfy Σ also satisfy σ, written $\Sigma \models \sigma$ ($\Sigma \models_{\text{fin}} \sigma$). An axiomatization $\mathfrak{R}$ is *sound* for the unrestricted implication problem of a class of dependencies $\mathcal{C}$ if for all finite sets $\Sigma \cup \{\sigma\}$ of dependencies from $\mathcal{C}$, $\Sigma \vdash_{\mathfrak{R}} \sigma \Rightarrow \Sigma \models \sigma$; it is *complete* if $\Sigma \models \sigma \Rightarrow \Sigma \vdash_{\mathfrak{R}} \sigma$. Soundness and completeness for finite implication are defined analogously.

Some of our proofs use the chase algorithm that was invented in the late 70s [3,31]. For a detailed exposition of this technique we refer the reader to [2].

Axiomatizations. Tables 2 and 3 present the axiomatizations considered in this article. In Table 2, the axiomatization $\mathfrak{I} := \{\mathcal{I}1, \ldots, \mathcal{I}5\}$ is sound and complete for independence atoms alone [20,27]. The rules $\mathcal{F}1, \mathcal{F}2, \mathcal{F}3$ form the Armstrong axiomatization for functional dependencies [4], and the rules $\mathcal{FI}1$ and $\mathcal{FI}2$ describe simple interaction between independence atoms and functional dependencies. Table 3 depicts the sound and complete axiomatization of inclusion dependencies introduced in [8,9]. Table 2 presents rules describing interaction between inclusion dependencies and independence atoms [21].

We leave it to the reader to check the soundness of the axiom systems in Tables 2 and 3. The proof does not include anything unexpected; we only note that soundness of $\mathcal{UI}3$ follows only if databases are not allowed to contain empty relations.

Theorem 2. *The axiomatization* $\mathfrak{A} \cup \mathfrak{B} \cup \mathfrak{C}$ *is sound for the unrestricted and finite implication problems of FD+IND+IA.*

Lastly, we note that, for notational clarity only, we will restrict attention to the uni-relational case in all our proofs. That is, we will consider only those cases where databases consist of a single relation.

3   IAs+FDs

First we consider the interaction between FDs and IAs. Already keys and IAs combined form a somewhat intricate class: Their finite and unrestricted implication problems differ and the former lacks a finite axiomatization [20]. In Sect. 3.1 we extend these results to the classes FD+IA and 2FD+UIA. However, the interaction between unary FDs and IAs is less involved. In Sect. 3.2 we show that for UFD+IA unrestricted and finite implication coincide and the axiomatization $\mathfrak{A}^*$ given in Table 2 forms a sound and complete axiomatization.

3.1   Implication Problem for FDs and IAs

The following theorem enables us to separate the finite and unrestricted implication problems for FD+IA as well as for FD+UIA.

Theorem 3 [19]. *The unrestricted and finite implication problems for keys and UIAs differ.*

Table 2. Axiomatizations $\mathfrak{A}$ for FDs and IAs and $\mathfrak{C}$ for IAs and INDs. We define $\mathfrak{I} := \{\mathcal{I}1, \ldots, \mathcal{I}5\}$ and $\mathfrak{A}^* := \mathfrak{A} \setminus \{\mathcal{I}5, \mathcal{F}3\}$.

$$\frac{}{\emptyset \perp X} \text{ (trivial independence, } \mathcal{I}1)$$

$$\frac{X \perp Y}{Y \perp X} \text{ (symmetry, } \mathcal{I}2)$$

$$\frac{X \perp YZ}{X \perp Y} \text{ (decomposition, } \mathcal{I}3)$$

$$\frac{X \perp Y \quad XY \perp Z}{X \perp YZ} \text{ (exchange, } \mathcal{I}4)$$

$$\frac{X \perp Y \quad Z \perp Z}{X \perp YZ} \text{ (weak composition, } \mathcal{I}5)$$

$$\frac{}{XY \rightarrow Y} \text{ (reflexivity, } \mathcal{F}1)$$

$$\frac{X \rightarrow Y \quad Y \rightarrow Z}{X \rightarrow Z} \text{ (transitivity, } \mathcal{F}2)$$

$$\frac{X \rightarrow Y}{XZ \rightarrow YZ} \text{ (augmentation, } \mathcal{F}3)$$

$$\frac{X \perp Y \quad X \rightarrow Y}{\emptyset \rightarrow Y} \text{ (constancy, } \mathcal{FI}1)$$

$$\frac{X \perp YZ \quad Z \rightarrow V}{X \perp YZV} \text{ (composition, } \mathcal{FI}2)$$

Axiomatization $\mathfrak{A}$

$$\frac{R[X] \subseteq R'[Z] \quad R[Y] \subseteq R'[W] \quad R'[Z \perp W]}{R[XY] \subseteq R'[ZW]} \text{ (concatenation, } \mathcal{UI}1)$$

$$\frac{R[XY] \subseteq R'[ZW] \quad R'[ZW] \subseteq R[XY] \quad R'[Z \perp W]}{R[X \perp Y]} \text{ (transfer, } \mathcal{UI}2)$$

$$\frac{R[X] \subseteq R'[Y] \quad R' : Y \perp Y}{R'[Y] \subseteq R[X]} \text{ (symmetry, } \mathcal{UI}3)$$

$$\frac{R[X] \subseteq R'[Y] \quad R' : Y \perp Y}{R : X \perp X} \text{ (constancy, } \mathcal{UI}4)$$

$$\frac{R[A] \subseteq R'[C] \quad R[B] \subseteq R'[C] \quad R' : C \perp C \quad \sigma}{\sigma(R : A \mapsto B)} \text{ (equality, } \mathcal{UI}5)$$

Axiomatization $\mathfrak{C}$

Table 3. Axiomatization $\mathfrak{B}$ for INDs

$R[X] \subseteq R[X]$ (reflexivity, $\mathcal{U}1$)	$\dfrac{R[X] \subseteq R'[Y] \quad R'[Y] \subseteq R''[Z]}{R[X] \subseteq R''[Z]}$ (transitivity, $\mathcal{U}2$)	$\dfrac{R[A_1 \ldots A_n] \subseteq R'[B_1 \ldots B_n]}{R[A_{i_1} \ldots A_{i_m}] \subseteq R'[B_{i_1} \ldots B_{i_m}]}^{(*)}$ (projection and permutation, $\mathcal{U}3$)
		$(^*)$ i_j are pairwise distinct and from $\{1, \ldots, n\}$

This theorem was proved by showing that $\Sigma \models_{\mathrm{fin}} \sigma$ and $\Sigma \not\models \sigma$, for $\Sigma :=$ $\{A \perp B, C \perp D, BC \rightarrow AD, AD \rightarrow BC\}$ and $\sigma := AB \rightarrow CD$. In [19] it was shown that this counterexample can be extended to a non-axiomatizability results for finite implication of keys and IAs. By an analogous line of reasoning this results carries over to the class of FDs and IAs, as well (see Appendix).

Theorem 4. *The finite implication problem for FD+IA (2FD+UIA) is not finitely axiomatizable.*

The implicit assumption in the above theorem is that an axiomatization must be *attribute-bounded*, meaning that it may not introduce new attributes [10]. It is easy to see that with this prerequisite finite axiomatization entails decidability. Contrarily, there are finite axiomatizations for undecidable implication problems that do not adhere to this assumption [16–18,33].

To the best of our knowledge, decidability is open for both FD+IA and FD+UIA with respect to their finite and unrestricted implication problems. It is worth noting here that the unrestricted (finite) implication problem for FD+UIA is as hard as that for FD+IA. For this, we demonstrate a simple reduction from the latter to the former. Let $\Sigma \cup \{\sigma\}$ be a set of FDs and IAs, and let Σ' denote the set of FDs and IAs where each IA of the form $X \perp Y$ is replaced with dependencies from $\{A \perp B, X \rightarrow A, A \rightarrow X, Y \rightarrow B, B \rightarrow Y\}$ where A and B are fresh attributes. If σ is an FD, then Σ (finitely) implies σ iff Σ' (finitely) implies σ. Also, if σ is of the form $X \perp Y$, then we have $\Sigma \models \sigma$ iff $\Sigma'' \models \sigma'$, where

$$\Sigma'' := \Sigma' \cup \{X \rightarrow A, A \rightarrow X, Y \rightarrow B, B \rightarrow Y\},$$

$\sigma' := A \perp B$, and A and B are fresh attributes.

3.2 Implication for UFDs and IAs

Next we turn to the class UFD+IA. Extending the scope and methods from [20], which presented a finite axiomatization for unary keys and IAs, we show that the axiomatization $\mathfrak{A}^*$ (see Table 2) is sound and complete for UFD+IA in both with respect to finite and unrestricted implication. Hence, compared to UIAs and FDs, the interaction between IAs and UFDs is relatively tame. Combined, however, these two may entail new restrictions to column sizes. For instance, in the finite $A \rightarrow B_1$, $A \rightarrow B_2$, and $B_1 \perp B_2$ imply $|r(B_1)| \cdot |r(B_2)| \leq |r(A)|$. The proof of the following completeness theorem is obtained by a chase-based model construction (see Appendix).

Theorem 5. *The axiomatization $\mathfrak{A}^*$ is sound and complete for the unrestricted and finite implication problems of UFD+IA.*

As the same axiomatization characterizes both finite and unrestricted implication, we obtain the following corollary.

Corollary 1. *The finite and unrestricted implication problems coincide for UFD+IA.*

4   IAs+UFDs+UINDs

Next we turn attention to the combined class of FDs, INDs, and IAs. In the previous section we noticed that the finite implication problem for binary FDs and unary IAs is not finitely axiomatizable. On the other hand, both the finite and unrestricted implication problems for unary FDs and binary INDs are undecidable [32]. Hence, in this section we restrict to unary FDs and unary INDs, a class for which the two implication problems already deviate [12]. It turns out that the combination UFD+UIND+IA can be axiomatized with respect to both problems. However, in the finite case the axiomatization is infinite as one needs to add so-called cycle rules for UFDs and UINDs.

An axiomatization for unrestricted implication follows from results in Sect. 3.2 and [12]. For the proof, see Appendix.

Theorem 6. *The axiomatization $\mathfrak{A}^* \cup \{\mathcal{U}1, \mathcal{U}2, \mathcal{U}\mathcal{I}3, \mathcal{U}\mathcal{I}4\}$ is sound and complete for the unrestricted implication problem of UFD+UIND+IA.*

For finite implication a complete axiomatization of UFD+UIND+IA is found by extending $\mathfrak{A}^* \cup \{\mathcal{U}1, \mathcal{U}2, \mathcal{U}\mathcal{I}3, \mathcal{U}\mathcal{I}4\}$ with the so-called cycle rules [12] (see Table 2) and by removing $\mathcal{U}\mathcal{I}3, \mathcal{U}\mathcal{I}4$ which become redundant. However, the completeness proof is now more involved and proved in two steps. We will combine the chase-based approach of the proof of Theorem 5 with the graph-theoretic approach from [12]. The latter method was used to prove a complete axiomatization for the finite implication problem of UIND+FD. For the graph-theoretic approach, we commence by introducing multigraphs with two sorts of edges: red ones which encode UFDs and black ones which encode UINDs.

Table 4. Cycle rules for finite implication

$A_1 \rightarrow A_2$	$A_2 \supseteq A_3$	$\cdots$	$A_{2n-1} \rightarrow A_{2n}$	$A_{2n} \supseteq A_1$
$A_1 \leftarrow A_2$	$A_2 \subseteq A_3$	$\cdots$	$A_{2n-1} \leftarrow A_{2n}$	$A_{2n} \subseteq A_1$

(cycle rule for n, $\mathcal{C}_n$)

Definition 1 [12]. *For each set Σ of UINDs and UFDs over R, let $G(\Sigma)$ be the multigraph that consists of nodes R, red directed edges (A, B), for $A \rightarrow B \in \Sigma$, and black directed edges (A, B), for $B \subseteq A \in \Sigma$. If $G(\Sigma)$ has red (black) directed edges from A to B and vice versa, then these edges are replaced with an undirected edge between A and B.*

Given a multigraph $G(\Sigma)$, we first topologically sort its strongly connected components which form a directed acyclic graph [25]. That is, each component is assigned a unique *scc-number*, greater than the scc-numbers of all its descendants. For an attribute A, denote by $scc(A)$ the scc-number of the component node A belongs to. Note that $scc(A) \geq scc(B)$ if (A, B) is an edge in $G(\Sigma)$. Denote also by scc_i the set of attributes A with $scc(A) = i$, and let $scc_{\leq i} := \bigcup_{j \leq i} scc_j$ and define $scc_{\geq i}$, $scc_{<i}$, and $scc_{>i}$ analogously. The following lemma is a simple consequence of the definition.[3]

Lemma 1 [12]. *Let Σ be a set of UFDs and UINDs, closed under $\{\mathcal{F}1, \mathcal{F}2, \mathcal{U}1, \mathcal{U}2\} \cup \{\mathcal{C}_k : k \in \mathbb{N}\}$. Then every node in $G(\Sigma)$ has a red and a black self-loop. The red (black) subgraph of $G(\Sigma)$ is transitively closed. The subgraphs induced by the strongly connected components of $G(\Sigma)$ are undirected. In each strongly connected component, the red (black) subset of undirected edges forms a collection of node-disjoint cliques. Note that the red and black partitions of nodes could be different.*

We now apply this graphical approach to earlier techniques presented in this paper. Theorem 7 shows completeness of the axiomatization $\mathfrak{A}^* \cup \{\mathcal{U}1, \mathcal{U}2\} \cup \{\mathcal{C}_n : n \in \mathbb{N}\}$ for the finite implication problem of UFD+UIND+IA by using the relation generated in Lemma 2. The proof of this lemma describes an incremental modification of the base relation, taken from the proof of Theorem 5, that is shown to reflect a growing number of inclusion dependencies in its composition. This is achieved by an inductive re-organization of the column values according to the underlying scc-numbering while at the same time maintaining the integrity of the UFD and IA dependencies in the base relation. The proof of the theorem and the lemma can be found in Appendix.

Lemma 2. *Let Σ be a set of UFDs, UINDs, and IAs over R, partitioned respectively to Σ_{UFD}, Σ_{UIND}, and Σ_{IA}. Assume that $\Sigma_{\text{UFD}} \cup \Sigma_{\text{UIND}}$ contains all UFDs and UINDs derivable from Σ by $\mathfrak{A}^* \cup \{\mathcal{U}1, \mathcal{U}2\} \cup \{\mathcal{C}_k : k \in \mathbb{N}\}$, and assume that we have assigned an scc-numbering to $G(\Sigma_{\text{UFD}} \cup \Sigma_{\text{UIND}})$. Let E be either the empty set or a single attribute, and let $R' := \{B \in R : E \to B \notin \Sigma\}$. Then there exists a finite relation r and tuples $t_0, t_1 \in r$ such that:*

(i) $\Sigma \vdash X \perp Y$ *if* $X, Y \subseteq R'$ *and for some* $t \in r$, $t(X) = t_0(X)$ *and* $t(Y) = t_1(Y)$;

(ii) $r \models \Sigma_{\text{UFD}} \cup \Sigma_{\text{IA}}$;

(iii) $r(A)$ *is (strictly) included in* $r(B)$ *if* $scc(A)$ *is (strictly) less than* $scc(B)$.

Theorem 7. *The axiomatization $\mathfrak{A}^* \cup \{\mathcal{U}1, \mathcal{U}2\} \cup \{\mathcal{C}_n : n \in \mathbb{N}\}$ is sound and complete for the finite implication problem of UFD+UIND+IA.*

---

[3] Lemma 1 is a reformulation of Lemma 4.2. in [12] where the same claim is proved for a set of FDs and UINDs that is closed under $\{\mathcal{F}1, \mathcal{F}2, \mathcal{F}3, \mathcal{U}1, \mathcal{U}2\} \cup \{\mathcal{C}_k : k \in \mathbb{N}\}$. We may omit $\mathcal{F}3$ here since, when restricting attention to UFDs, $\mathcal{F}3$ is not needed in the proof.

5   Polynomial-Time Conditions for Non-interaction

The interaction-freeness between the class FD+IND has been well-studied in the literature [29, 30]. Here, we examine the frontiers for tractable reasoning about the class FD+IA in both the finite and unrestricted cases. For IND+IA these questions have been studied in [21]. The idea is to establish sufficient criteria for the non-interaction between IAs and FDs. There is a trade-off between the simplicity and generality of such criteria. While simple criteria may be easier to apply, more general criteria allow us to establish non-interaction in more cases. Our focus here is on generality, and the criteria are driven by the corresponding inference rules. We define non-interaction between two classes as follows.

Definition 2. *Let Σ_0 and Σ_1 be two sets of dependencies from classes C_0 and C_1, respectively. We say that Σ_0, Σ_1 have no interaction with respect to unrestricted (finite) implication if*

- *for σ from C_0, σ is (finitely) implied by Σ_0 iff σ is (finitely) implied by $\Sigma_0 \cup \Sigma_1$.*
- *for σ from C_1, σ is (finitely) implied by Σ_1 iff σ is (finitely) implied by $\Sigma_0 \cup \Sigma_1$.*

Let us now define two syntactic criteria for describing non-interaction. We say that an IA $X \perp Y$ *splits* an FD $U \to V$ if both $(X \setminus Y) \cap U$ and $(Y \setminus X) \cap U$ are non-empty. An IA $X \perp Y$ splits an IND $Z \subseteq W$ if both $X \cap W$ and $Y \cap W$ are non-empty. Furthermore, $X \perp Y$ *intersects* $U \to V$ if $XY \cap U$ is non-empty. Notice that both these concepts give rise to possible interaction between two different classes. We show that lacking splits implies non-interaction for FD+IA in the unrestricted case. Non-interaction for FD+IA in the finite is guaranteed by the stronger condition defined in terms of lacking intersections.

For IND+IA lack of splits entail non-interaction [21].

Theorem 8 [21]. *Let Σ_{IND} and Σ_{IA} be respectively sets of INDs and IAs. If no IA in Σ_{IA} splits any IND in Σ_{IND}, then Σ_{IND} and Σ_{IA} have no interaction with respect to unrestricted (finite) implication.*

We proceed with the non-interaction results for FD+IA. The proofs are located in Appendix. For unrestricted implication the idea is to first apply the below polynomial-time algorithm which transforms an assumption set Σ to an equivalent set Σ^*. The set Σ^* is such that it has no interaction between FDs and IAs provided that none of its FDs split any IAs.

For a set of FDs Σ, let us denote by $\text{Cl}(\Sigma, X)$ the closure set of all attributes A for which $\Sigma \models X \to A$. This set can be computed in linear time by the Beeri-Bernstein algorithm [5]. The non-interaction condition for unrestricted implication is now formulated using $\Sigma_{\text{IA}}^* = \{X_1 \perp Y_1, \ldots, X_n \perp Y_n\}$ and $\Sigma_{\text{FD}}^* = \Sigma_{\text{FD}} \cup \{\emptyset \to Z\}$ where $Z, X_i Y_i$ are computed using the following algorithm that takes an FD set Σ_{FD} and an IA set $\Sigma_{\text{IA}} = \{U_1 \perp V_1, \ldots, U_n \perp V_n\}$ as an input.

Algorithm 1. Algorithm for computing Z, X_i, Y_i

Require: Σ_{FD} and $\Sigma_{\mathrm{IA}} = \{U_i \perp V_i \mid i = 1, \ldots, n\}$
Ensure: Z and $\Sigma_{\mathrm{IA}}^* = \{X_i \perp Y_i \mid i = 1, \ldots, n\}$
 1: **Initialize:** $V \leftarrow \emptyset, X_i \leftarrow U_i, Y_i \leftarrow V_i$
 2: **repeat**
 3:     $Z \leftarrow V$
 4:     **for** $i = 1, \ldots, n$ **do**
 5:         $X_i \leftarrow \mathrm{Cl}(\Sigma_{\mathrm{FD}}, X_i V)$
 6:         $Y_i \leftarrow \mathrm{Cl}(\Sigma_{\mathrm{FD}}, Y_i V)$
 7:         $V \leftarrow V \cup (X_i \cap Y_i)$
 8: **until** Z=V

From the construction we obtain that $\Sigma_{\mathrm{FD}}^* \cup \Sigma_{\mathrm{IA}}^*$ is equivalent to $\Sigma_{\mathrm{FD}} \cup \Sigma_{\mathrm{IA}}$ and that

(1) for $Z_1 \perp Z_2 \in \Sigma_{\mathrm{IA}}^*$ and $i = 1, 2$, $\Sigma_{\mathrm{FD}}^* \models Z_i \to X$ implies $X \subseteq Z_i$;
(2) $\Sigma_{\mathrm{FD}}^* \cup \Sigma_{\mathrm{IA}}^* \models \emptyset \to A$ iff $A \in Z$.

Recall that the closure set $\mathrm{C}(\Sigma_{\mathrm{FD}}, X)$ can be computed in linear time by the Beeri-Bernstein algorithm. Now, at stage 5 (or stage 6) the computation of the closure set is resumed whenever V introduces attributes that are new to X_i (Y_i). Since the number of the closures considered is $2|\Sigma_{\mathrm{IA}}|$, we obtain a quadratic time bound for the computation of Z, X_i, Y_i.

Theorem 9. *Let Σ_{FD} and Σ_{IA} be respectively sets of FDs and IAs, and let Σ_{FD}^* and Σ_{IA}^* be obtained from Σ_{FD} and Σ_{IA} by Algorithm 1. Then the following holds:*

– *if no IA in Σ_{IA}^* splits any FD in Σ_{FD}^*, then Σ_{FD}^* and Σ_{IA}^* have no interaction with respect to unrestricted implication;*
– *if no IA in Σ_{IA} intersects any FD in Σ_{FD}, then Σ_{FD} and Σ_{IA} have no interaction with respect to finite implication.*

To illustrate the necessity for a stronger condition in the finite case, recall from Sect. 3.1 that $AB \to CD$ is finitely implied by $\{A \perp B, C \perp D, BC \to AD, AD \to BC\}$, and notice that $AB \to CD$ is not finitely implied by $\{BC \to AD, AD \to BC\}$. However, Algorithm 1 does not produce any fresh assumptions, and neither $A \perp B$ nor $C \perp D$ splits any FD assumption. Therefore, lack of splits is not sufficient for non-interaction in the finite case.

6   Complexity Results

Next we examine the computational complexity of the discussed implication problems. We show that both implication problems for UFD+UIND+IA can be solved in low-degree polynomial time, even though the problems differ from one another. The associated decision procedures, found in Appendix, transform the implication problems first to graphs, as earlier in this paper, and subsequently modify them according to appropriate inference rules. The only difference with finite implication is that an application of the cycle rules is included in the process. The implication problem then reduces, for UFDs and UINDs, to reachability in the graph, and for IAs, to an IA-implication instance which reflects the topology of the graph. Consequently, the stated time bounds follow.

Theorem 10. *Let $\Sigma_{\mathrm{UFD}}, \Sigma_{\mathrm{UIND}}, \Sigma_{\mathrm{IA}}$ be respectively sets of UFDs, UINDs, and IAs over a relation schema R. The unrestricted and finite implication problems for σ by $\Sigma_{\mathrm{UFD}} \cup \Sigma_{\mathrm{UIND}} \cup \Sigma_{\mathrm{IA}}$ can be decided in time:*

- *$O(|\Sigma_{\mathrm{IA}}| \cdot |\Sigma_{\mathrm{UFD}}| + |\Sigma_{\mathrm{UIND}}|)$ if σ is an UFD or UIND;*
- *$O(|\Sigma_{\mathrm{IA}}| \cdot (|\Sigma_{\mathrm{UFD}}| + |R|^2) + |\Sigma_{\mathrm{UIND}}|)$ if σ is a IA.*

7   Conclusion and Outlook

In view of the infeasibility of EMVDs and of FDs and INDs combined, the class of FDs, MVDs and unary INDs is important as it is low-degree PTIME decidable in the finite and unrestricted cases. As independence atoms form an important tractable embedded sub-class of EMVDs, we have delineated axiomatizability and tractability frontiers for sub-classes of FDs, INDs, and IAs. The most interesting class is that of IAs, unary FDs and unary INDs, for which finite and unrestricted implication differ but each is axiomatisable and decidable in low-degree polynomial time. The results form a basis for the advancement of several data processing tasks, including cardinality estimation, database security, and query optimization.

Even though research on dependency theory has been rich and deep, there are many problems that warrant future research. Theoretically, the decidability remains open for both independence atoms and functional dependencies as well as unary independence atoms and functional dependencies, both in the finite and unrestricted case. This line of research should also be investigated in the probabilistic setting of conditional independencies, fundamental to multivariate statistics and machine learning. Practically, implementations and experimental evaluations of the algorithms can complement the findings in the research. Of direct practical use for data profiling would be algorithms that compute the set of IAs that hold on a given relation, as would algorithms to mine notions of approximate IAs [1].

References

1. Abedjan, Z., Golab, L., Naumann, F.: Profiling relational data: a survey. VLDB J. **24**(4), 557–581 (2015)
2. Abiteboul, S., Hull, R., Vianu, V.: Foundations of Databases. Addison-Wesley, Boston (1995)
3. Aho, A.V., Beeri, C., Ullman, J.D.: The theory of joins in relational databases. ACM Trans. Database Syst. **4**(3), 297–314 (1979)

4. Armstrong, W.W.: Dependency structures of data base relationships. In: Proceedings of IFIP World Computer Congress, pp. 580–583 (1974)
5. Beeri, C., Bernstein, P.A.: Computational problems related to the design of normal form relational schemas. ACM Trans. Database Syst. **4**(1), 30–59 (1979)
6. Beeri, C., Fagin, R., Howard, J.H.: A complete axiomatization for functional and multivalued dependenciesin database relations. In: SIGMOD, pp. 47–61 (1977)
7. Biskup, J., Bonatti, P.A.: Controlled query evaluation for enforcing confidentiality in complete information systems. Int. J. Inf. Sec. **3**(1), 14–27 (2004)
8. Casanova, M.A., Fagin, R., Papadimitriou, C.H.: Inclusion dependencies and their interaction with functional dependencies. In: PODS, pp. 171–176 (1982)
9. Casanova, M.A., Fagin, R., Papadimitriou, C.H.: Inclusion dependencies and their interaction with functional dependencies. J. Comput. Syst. Sci. **28**(1), 29–59 (1984)
10. Chandra, A.K., Vardi, M.Y.: The implication problem for functional and inclusion dependencies is undecidable. SIAM J. Comput. **14**(3), 671–677 (1985)
11. Codd, E.F.: Relational completeness of data base sublanguages. In: Rustin, R. (ed.) Database Systems, pp. 65–98. Prentice Hall and IBM Research Report RJ 987, San Jose (1972)
12. Cosmadakis, S.S., Kanellakis, P.C., Vardi, M.Y.: Polynomial-time implication problems for unary inclusion dependencies. J. ACM **37**(1), 15–46 (1990)
13. Fagin, R.: Multivalued dependencies and a new normal form for relational databases. ACM Trans. Database Syst. **2**, 262–278 (1977)
14. Galil, Z.: An almost linear-time algorithm for computing a dependency basis in a relational database. J. ACM **29**(1), 96–102 (1982)
15. Geiger, D., Paz, A., Pearl, J.: Axioms and algorithms for inferences involving probabilistic independence. Inf. Comput. **91**(1), 128–141 (1991)
16. Hannula, M.: Reasoning about embedded dependencies using inclusion dependencies. In: LPAR-20, pp. 16–30 (2015)
17. Hannula, M., Kontinen, J.: A finite axiomatization of conditional independence and inclusion dependencies. In: FoIKS, pp. 211–229 (2014)
18. Hannula, M., Kontinen, J.: A finite axiomatization of conditional independence and inclusion dependencies. Inf. Comput. **249**, 121–137 (2016)
19. Hannula, M., Kontinen, J., Link, S.: On independence atoms and keys. In: CIKM, pp. 1229–1238 (2014)
20. Hannula, M., Kontinen, J., Link, S.: On the finite and general implication problems of independence atoms and keys. J. Comput. Syst. Sci. **82**(5), 856–877 (2016)
21. Hannula, M., Kontinen, J., Link, S.: On the interaction of inclusion dependencies with independence atoms. In: LPAR-21, pp. 212–226 (2017)
22. Hannula, M., Link, S.: On the interaction of functional and inclusion dependencies with independence atoms. Report CDMTCS-518. Centre for Discrete Mathematics and Theoretical Computer Science, University of Auckland, Auckland, New Zealand, February 2018
23. Herrmann, C.: On the undecidability of implications between embedded multivalued database dependencies. Inf. Comput. **122**(2), 221–235 (1995)
24. Herrmann, C.: Corrigendum to on the undecidability of implications between embedded multivalued database dependencies. Inf. Comput. **204**(12), 1847–1851 (2006)
25. Kahn, A.B.: Topological sorting of large networks. Commun. ACM **5**(11), 558–562 (1962)
26. Kanellakis, P.C.: Elements of relational database theory. In: Handbook of Theoretical Computer Science, pp. 1073–1156 (1990)

27. Kontinen, J., Link, S., Väänänen, J.A.: Independence in database relations. In: WoLLIC, pp. 179–193 (2013)
28. Leinders, D., Van den Bussche, J.: On the complexity of division and set joins in the relational algebra. In: PODS, pp. 76–83 (2005)
29. Levene, M., Loizou, G.: How to prevent interaction of functional and inclusion dependencies. Inf. Process. Lett. **71**(3–4), 115–125 (1999)
30. Levene, M., Loizou, G.: Guaranteeing no interaction between functional dependencies and tree-like inclusion dependencies. Theor. Comput. Sci. **254**(1–2), 683–690 (2001)
31. Maier, D., Mendelzon, A.O., Sagiv, Y.: Testing implications of data dependencies. ACM Trans. Database Syst. **4**(4), 455–469 (1979)
32. Mitchell, J.C.: The implication problem for functional and inclusion dependencies. Inf. Control **56**(3), 154–173 (1983)
33. Mitchell, J.C.: Inference rules for functional and inclusion dependencies. In: PODS, pp. 58–69 (1983)
34. Papenbrock, T., Ehrlich, J., Marten, J., Neubert, T., Rudolph, J.-P., Schönberg, M., Zwiener, J., Naumann, F.: Functional dependency discovery: an experimental evaluation of seven algorithms. PVLDB **8**(10), 1082–1093 (2015)
35. Paredaens, J.: The interaction of integrity constraints in an information system. J. Comput. Syst. Sci. **20**(3), 310–329 (1980)
36. Parker Jr., D.S., Parsaye-Ghomi, K.: Inferences involving embedded multivalued dependencies and transitive dependencies. In: SIGMOD, pp. 52–57 (1980)
37. Thalheim, B.: Dependencies in Relational Databases. Teubner, Stuttgart (1991)

Source Selection for Inconsistency Detection

Lingli Li[(✉)], Xu Feng, Hongyu Shao, and Jinbao Li[(✉)]

Department of Computer Science and Technology,
Heilongjiang University, Harbin, China
{lilingli,jbli}@hlju.edu.cn, fx_hlju@163.com, shy_hlju@163.com

Abstract. Inconsistencies in a database can be detected based on violations of integrity constraints, such as functional depencies (FDs). In big data era, many related data sources give us the chance of detecting inconsistency extensively. That is, even though violations do not exist in a single data set D, we can leverage other data sources to discover potential violations. A significant challenge for violation detection based on data sources is that accessing too many data sources introduces a huge cost, while involving too few data sources may miss serious violations. Motivated by this, we investigate how to select a proper subset of sources for inconsistency detection. To address this problem, we formulate the gain model of sources and introduce the optimization problem of source selection, called SSID, in which the gain is maximized with the cost under a threshold. We show that the SSID problem is NP-hard and propose a greedy approximation approach for SSID. To avoid accessing data sources, we also present a randomized technique for gain estimation with theoretical guarantees. Experimental results on both real and synthetic data show high performance on both effectiveness and efficiency of our algorithm.

1 Introduction

Consistency is one of the central criteria for data quality. Inconsistencies in a database can be captured by the violations of integrity constraints, such as functional dependencies (FDs). Specifically, given a database D and a set of FDs Σ, tuples in D that violate rules in Σ are inconsistent and need to be repaired. However, D might still have errors when there is no violations in D. To address this problem, we can compare D with other data sets to identify more inconsistencies and capture more errors. Fortunately, with the dramatic growth of useful information nowadays, data sets can be collected from various sources, i.e., websites, data markets, and enterprises. It brings opportunities and challenges for using data sources to detect inconsistencies (or errors). We give an example below to illustrate this issue.

Example 1. Consider S_0, S_1 and S_2 in Table 1. S_0 is the data for inconsistency detection, which is called the *target data*. S_1 and S_2 are two data sources. For

S_0-S_2, the data items in *italics* are incorrect. For S_0, there are three errors: x_1[name], x_1[street] and x_2[state]. We assume the schemas of $S_0 - S_2$ have been mapped by existing schema mapping techniques. A set of FDs, $\varphi_1 - \varphi_6$ (shown in Table 1 a), are used for inconsistency detection, in which S_0 satisfies $\varphi_1 - \varphi_6$, S_1 satisfies $\varphi_5 - \varphi_6$, and S_2 satisfies $\varphi_1 - \varphi_3$. Clearly, since there is no violation of FDs in S_0, none of the above errors can be detected. However, if we compare S_0 with S_1, inconsistencies can be discovered. For example, since x_2[zip] $= y_2$[zip] $=$ 10012, x_2[state] $\neq y_2$[state], they are inconsistent according to FD φ_6. Therefore, x_2[state] might be incorrect. Similarly, according to the inconsistencies detected based on sources S_1 and S_2, we can conclude that the following data items (underlined) in S_0 might be incorrect: x_1[name], x_1[street], x_2[state], x_3[name] and x_3[state]. We call these items *candidate errors*. It can be seen that involving extra data sources can help us detect more inconsistencies in target data. The candidate errors will be cleaned by a further step, i.e., posing queries on multiple data sources for truth discovery, which has been widely studied [12]. This paper focuses on how to select sources for inconsistency detection.

Table 1. Motivating example

(a) FDs

φ_1: phone→ name
φ_2: phone→ street
φ_3: phone→ city
φ_4: phone→ state
φ_5: zip→ city
φ_6: zip→ state

(b) target data S_0 for inconsistency detection

id	phone	name	street	city	state	zip
x_1	949-1212	*Ali Smith*	*27 bridge*	Midville	AZ	05211
x_2	555-8145	Bob Jones	5 valley rd	Centre	*NJ*	10012
x_3	212-6040	Carol Black	9 mountain	Davis	CA	07912

(c) data source S_1

id	zip	city	state
y_1	05211	Midville	AZ
y_2	10012	Centre	NY
y_3	07912	Davis	AA

(d) data source S_2

id	phone	name	street	city
z_1	949-1212	Alice Smith	17 bridge	Midville
z_2	555-8195	Bob Jones	5 valley rd	Centre
z_3	212-6040	*Carike Blake*	9 mountain	Davis

However, accessing all the sources for inconsistency detection is neither worthwhile nor impractical. On one hand, most of the data in sources might be irrelevant or redundant. On the other hand, integrating data from sources requires cost. Moreover, some sources charge for their data [6], such as GeoLytics and American Business Database. Therefore, this paper studies how to select a proper subset of sources for inconsistency detection. Moreover, such selection process should not require accessing data sources. Otherwise, it is not worthwhile due to its large cost. This problem brings three major challenges: (1) How to evaluate the gain of data sources for inconsistency detection? (2) How to define the optimization problem of source selection to balance between gain and cost? (3) How to implement the source selection algorithm without accessing sources? Aiming at the aforementioned problems, this paper makes the following main contributions.

1. Based on patterns of tuples over FDs, we define the gain model of sources and formulate the problem of source selection for inconsistency detection, called SSID. As far as we know, this is the first study on how to select sources for inconsistency detection. We show that the SSID problem is NP-hard (Sect. 2).
2. A greedy approximation algorithm for SSID is presented (Sect. 3). To avoid accessing data sources, we develop a randomized approach for intersection set size estimation based on min-hashing (Sect. 4). Such estimation has a theoretical guarantee.
3. Using real-life and synthetic data, the effectiveness, efficiency and scalability of our proposed algorithm are experimentally verified (Sect. 5).

2   Problem Definition

In order to formalize the source selection problem, we first introduce the symbols and notations in this paper.

Let S_0 be the *target* data for inconsistency detection and $\mathbb{S} = \{S_i | 1 \leq i \leq m\}$ be the set of *sources*, where each S_i $(0 \leq i \leq m)$ includes a dataset D_i with schema R_i and a FD set Σ_i. We assume that the schemas of sources have been mapped by existing schema mapping techniques. Table 2 shows the symbols that will be used throughout the rest of this paper.

Table 2. Symbols

Symbols	Meaning
S_0	The target data
$\mathbb{S} = \{S_1, \cdots, S_m\}$	The set of data sources
S	A subset of sources $\mathbb{S}$
$h_1, \cdots, h_k$	k random hash functions
K	The threshold of cost
A, B	$P[S_0, \varphi]$, $P[S, \varphi]$
U	The universe of A and B, with $|U| = n$
a, b	$|A|$, $|B|$
x, y, s	$|A \cap B|$, $|A \cup B|$, $s(A, B)$

Now we introduce the notion of pattern to describe tuples over FDs. Given a tuple t in S_i and a FD φ in Σ_i, the *pattern* of t under φ, denoted as $P[t, \varphi]$, is defined as $P[t, \varphi] = (t[\mathsf{LHS}(\varphi)], \mathsf{RHS}(\varphi))$, where $\mathsf{LHS}(\varphi)$ ($\mathsf{RHS}(\varphi)$) is the left-hand side (right-hand side) of φ, and $t[\mathsf{LHS}(\varphi)]$ is the $\mathsf{LHS}(\varphi)$ value of t. For example, for tuple x_1 in Table 1(b) and FD φ_1 in Table 1(a), pattern $P[x_1, \varphi]$ is ("949-1212",name). Accordingly, the *pattern set* of S_i on FD $\varphi \in \Sigma_i$, denoted $P[S_i, \varphi]$, is defined as

$$P[S_i, \varphi] = \{(t[\mathsf{LHS}(\varphi)], \mathsf{RHS}(\varphi)) | t \in S_i\} \tag{1}$$

The *pattern set* of S_i is

$$P[S_i] = \cup_{\varphi \in \Sigma_i} P[S_i, \varphi] \tag{2}$$

Similarly, for source set $\mathcal{S}$, we have

$$P[\mathcal{S}, \varphi] = \cup_{S_i \in \mathcal{S}} P[S_i, \varphi] \tag{3}$$

$$P[\mathcal{S}] = \cup_{S_i \in \mathcal{S}} P[S_i] \tag{4}$$

Thus, using data sources to detect inconsistency is transformed into a pattern matching problem, as the property below.

Property 1. Given tuple $s \in S_0$ and FD $\varphi \in \Sigma_0 \cap (\cup_{S_i \in \mathcal{S}} \Sigma_i)$, s can be *checked* by $\mathcal{S}$ iff $P[s, \varphi] \in P[\mathcal{S}, \varphi]$. In another word, s can be *checked* by $\mathcal{S}$ iff $\exists t \in \cup_{S_i \in \mathcal{S}} S_i$ such that $P[s, \varphi] = P[t, \varphi]$. For such tuple t, we say that s and t are inconsistent on φ if $s[\mathsf{RHS}(\varphi)] \neq t[\mathsf{RHS}(\varphi)]$; otherwise we say s and t are consistent on φ.

Now we present the gain model of sources, called *coverage*. Given a source set $\mathcal{S}$, we call the number of patterns in target S_0 been covered by $\mathcal{S}$ as the *coverage* of $\mathcal{S}$, denoted as $cov(\mathcal{S})$. Formally,

$$cov(\mathcal{S}) = |P[S_0] \cap P[\mathcal{S}]| \tag{5}$$

According to Property 1, $cov(\mathcal{S})$ represents the number of patterns in target data whose consistency can be checked by $\mathcal{S}$.

Ideally, we wish to maximize the gain, i.e. $cov(\mathcal{S})$, and minimize the cost, i.e. $|\mathcal{S}|$. However, it is impractical to achieve both goals. Thus, this paper attempts to find a subset of sources that maximizes the coverage with the number of sources no more than a given threshold. Note that, we assume unit cost of each source. We call this problem as $\underline{S}$ource $\underline{S}$election for $\underline{I}$nconsistency $\underline{D}$etection (SSID). The formal definition is shown below.

Definition 1 *(Source Selection for Inconsistency Detection (SSID)).* *Given a target S_0, a source set $\mathbb{S}$ and a positive integer K, the source selection for inconsistency detection is to find a subset $\mathcal{S}$ of $\mathbb{S}$, such that $|\mathcal{S}| \leq K$, and $cov(\mathcal{S})$ is maximized.*

Discussion. In our definition of coverage, each pattern in S_0 has only two states: covered or uncovered. One question is: whether it is enough for each pattern to be covered only once? We use the following example to demonstrate that this measure is reasonable.

Example 2. Consider a pattern p in $P[S_0, \varphi]$ which is covered by source S_i, i.e., $p \in P[S_0, \varphi]$ and $p \in P[S_i, \varphi]$. Suppose $\varphi : X \to Y$. Let $s \in S_0$ and $t \in S_i$ be the two tuples s.t. $P[s, \varphi] = P[t, \varphi] = p$. Therefore, we have $s[X] = t[X]$. For $s[Y]$ and $t[Y]$, there are two cases: (a) $s[Y] \neq t[Y]$, then we don't need to access any other sources because an inconsistency has been detected; (b) otherwise, we still don't need to access any other sources because $s[Y]$ is unlikely being incorrect unless S_0 and S_i have copying relationship [13].

Theorem 1. *SSID is NP-hard.*

The proof is by reduction from the classic unweighted maximum coverage problem [11]. The detail is omitted due to space limitation.

3   Algorithm for SSID

Due to the NP-hardness of SSID, we devise a greedy approximation algorithm for SSID (shown in Algorithm 1), denoted by Greedy-SSID. At each stage, the marginal gain of coverage, $cov(\mathbb{S} \cup S_i) - cov(\mathbb{S})$, is computed for each unselected source S_i (line 3); and a source that provides the largest marginal gain is selected (line 4). Note that, in order to compute the accurate values of coverage (CompCoverage in line 3), all the data sources have to be accessed, which is unacceptable. To address this problem, we present a randomized method for coverage estimation, which will be discussed in Sect. 4.

Algorithm 1. Greedy-SSID

Input: S_0, $\mathbb{S}$, K
Output: a subset $\mathcal{S}$ of $\mathbb{S}$ with $|\mathcal{S}| \leq K$;
1: **while** $|\mathcal{S}| < K$ **do**
2:    **for** each $S_i \in \mathbb{S}$ **do**
3:        $cov(\mathcal{S} \cup S_i) \leftarrow$ CompCoverage$(\mathcal{S} \cup S_i)$;
4:    $S_{opt} \leftarrow \arg\max_{S_i \in \mathbb{S}} cov(\mathcal{S} \cup S_i) - cov(\mathcal{S})$;
5:    add S_{opt} into $\mathcal{S}$;
6:    remove S_{opt} from $\mathbb{S}$;
7: **return** $\mathcal{S}$;

The following theorems demonstrate the time complexity and the approximation ratio bound of Algorithm 1.

Theorem 2. *The expected time of Algorithm 1 is $O(K * n * m)$, where K is the number of selected sources, n is the maximum size of S_i and m is the number of sources in $\mathbb{S}$.*

Theorem 3. *Algorithm 1 is a $(1 - 1/e)$-approximation algorithm.*

Proof. Since the coverage function is submodular and monotone, this algorithm has $1 - 1/e$ approximation ratio based on the submodular theory [16].   □

4   Coverage Estimation

In this section, we will introduce a randomized approximation method for coverage estimation. The main idea is to build small sketches for each data source (including the target S_0) in an offline phase. These sketches could be used for coverage estimation without accessing data. Given target data S_0 and source set $\mathcal{S}$, we use these sketches to estimate the coverage of $\mathcal{S}$ on S_0.

According to the definition of coverage in Sect. 2, we have that,

$$cov(\mathcal{S}) = |P[S_0] \cap P[\mathcal{S}]| = \sum_{\varphi \in \Sigma_0} |P[S_0, \varphi] \cap P[\mathcal{S}, \varphi]| \tag{6}$$

As can be seen, $cov(\mathcal{S})$ is the sum of the sizes of each intersection set $P[S_0, \varphi] \cap P[\mathcal{S}, \varphi]$ for each FD φ. Thus, estimating the size of each intersection set is the key component. Unfortunately, as far as we know, there has been little work about intersection set size estimation. [21] proposed an overlap-estimation algorithm to estimate overlaps between sources under the maximum entropy principle. However, this approach requires some prior statistics overlap information of sources from third parties.

For convenience, we denote $P[S_0, \varphi]$ as set A and $P[\mathcal{S}, \varphi]$ as set B (shown in Table 2). Our goal is to estimate $|A \cap B|$. According to inclusion-exclusion principle, the intersection set size can be computed based on the Jaccard similarity $(s(A, B) = \frac{|A \cap B|}{|A \cup B|})$. We have

$$|A \cap B| = s(A, B) * (|A| + |B|)/(1 + s(A, B)) \tag{7}$$

For simplification, we also denote $|A|$, $|B|$, $|A \cap B|$ and $s(A, B)$ by a, b, x and s respectively (see Table 2). Then Eq. (7) can be written as

$$x = s * (a + b)/(1 + s) \tag{8}$$

From Eq. (8), to obtain x, values of a, b and s should be known. a can be computed directly because $a = |P[S_0, \varphi]|$, which is determined by the target data. Since $b = |P[\mathcal{S}, \varphi]|$ and $s = \frac{|P[S_0, \varphi] \cap P[\mathcal{S}, \varphi]|}{|P[S_0, \varphi] \cup P[\mathcal{S}, \varphi]|}$, which are determined by sources $\mathcal{S}$ that cannot be accessed, thus these two values need to be estimated.

To tackle this problem, we build a sketch $\tilde{P}[S_i, \varphi]$ for each $P[S_i, \varphi]$ offline and use them to estimate the coverage of sources online. The framework for coverage estimation is shown in Algorithm 2. Firstly, we estimate similarity s and set size b based on sketches (line 3–4) for each FD φ. Then the intersection set size, x, is estimated based on Eq. (8) (line 5). Finally, we sum up all the $\hat{x}$s to get $cov(\mathcal{S})$ based on Eq. (6) (line 6). Note that, this algorithm is embedded into our greedy source selection algorithm (Algorithm 1). We call the greedy source selection with Algorithm 2 embedded *MH-Greedy* algorithm.

Algorithm 2. CompCoverage

Input: $\{\tilde{P}[S_i, \varphi] | S_i \in \mathcal{S}, \varphi \in \Sigma_i\}$, S_0
Output: $\hat{cov}(\mathcal{S})$
1: $\hat{cov}(\mathcal{S}) \leftarrow 0$;
2: **for** each $\varphi \in \Sigma_0$ **do**
3:     $\hat{s} =$ Estimate_Sim($\tilde{P}[\mathcal{S}, \varphi], \tilde{P}[S_0, \varphi]$);
4:     $\hat{b} =$ Estimate_SetSize($\tilde{P}[\mathcal{S}, \varphi]$);
5:     $\hat{x} = \hat{s}(a + \hat{b})/(1 + \hat{s})$;
6:     $\hat{cov}(\mathcal{S}) \leftarrow \hat{cov}(\mathcal{S}) + \hat{x}$;
7: **return** $\hat{cov}(\mathcal{S})$;

Thus, the core of Algorithm 2 is the estimation of s, b and x. The sketch building approach and the estimation of s will be introduced in Sect. 4.1 and the sketch-based estimation of b and x will be discussed in Sect. 4.2 and Sect. 4.3.

4.1   Sketch

Min-hashing [2] is a widely-used hashing method for quickly estimating the Jaccard similarity. We also find that min-hash values can be applied to estimate set size effectively. Therefore, in order to estimate s and b effectively, we apply min-hash values to be the sketch of each $P[S_i, \varphi]$.

We first introduce the min-hashing method. To ensure the min-hashing technology can be applied, we assume that all the attributes in FDs are in finite domains. Note that, an infinite domain can be mapped to an approximate finite domain based on piecewise-linear functions. Let U denote the domain of both A and B. A random hash function h is used to imitate random permutation of all the elements in U. To achieve this goal, h randomly maps each element in U to $[1, |U|]$. For any subset X of U, the min-hash value of X, denoted by $h(X)$, is the smallest hash value whose corresponding element is in X. The most important property of min-hashing is as below.

Theorem 4. $Pr[h[A] = h[B]] = s(A, B)$.

To obtain an accurate estimation, it is necessary to determine multiple (say k) independent min-hash values. Consider k random hash functions $h_1, \cdots, h_k$, the estimated Jaccard similarity of A and B is defined as follows.

$$\hat{s} = |\{j | 1 \le j \le k \text{ and } h_j(A) = h_j(B)\}|/k \tag{9}$$

Based on the following theorem [17], we can derive a guarantee on the error bound of $\hat{s}$ (shown in Theorem 6) when k is sufficiently large.

Theorem 5. *Consider a set of r independent identically distributed random variables $\{X_1, \cdots, X_r\}$ such that $-\Delta \le X_i \le \Delta$ and $E[X_i] = 0$ for each $i \in [1, r]$. Let $M = \sum_{i=1}^{r} X_i$ (a sum of X_is). Then for any $\alpha \in (0, 1/2)$,*

$$Pr[|M| > \alpha] \le 2 \exp(-\alpha^2/2r\Delta^2)$$

We then have the following property for the estimated value $\hat{s}$.

Theorem 6. *Let ϵ be the error bound, δ be the error probability, and k be the number of random hash functions. When $k = \frac{2}{\epsilon^2} \ln 2/\delta$, we have the following property,*

$$Pr[|s - \hat{s}| < \epsilon] > 1 - \delta$$

That is, the Jaccard similarity is within ϵ error with probability at least $1-\delta$, if we use $k = \frac{2}{\epsilon^2} \ln 2/\delta$ random hash functions. Therefore, for a sketch of each $P[S_i, \varphi]$, we use k random hash functions $h_1, \cdots, h_k$ to get k min-hash values of $P[S_i, \varphi]$, denoted by $h_1(P[S_i, \varphi]), \cdots, h_k(P[S_i, \varphi])$. The sketch building algorithm for each source S_i is described in Algorithm 3.

The following theorem demonstrates the time complexity of this algorithm.

Theorem 7. *The expected time of Algorithm 3 is $O(m * t * n_{avg} * k)$, where $t = \max_{0 \le i \le m} |\Sigma_i|$, $n_{avg} = \text{avg}_{0 \le i \le m} |S_i|$.*

Although the complexity of Algorithm 3 is quite high, such sketches are computed offline. Hence it will not affect the performance of our source selection algorithm.

Algorithm 3. Sketch Building

Input: $S_0, \cdots, S_m, h_1, \cdots, h_k$
Output: $\{h_c(P[S_i, \varphi]) | 0 \leq i \leq m, \varphi \in \Sigma_i, 1 \leq c \leq k\}$
 1: **for** each S_i **do**
 2:   **for** each φ in Σ_i **do**
 3:     $h_c(P[S_i, \varphi]) \leftarrow \infty$;
 4:     **for** each $j \leftarrow 1$ to $|dom(\mathsf{LHS}(\varphi))|$ **do**
 5:       **if** j-th item is in S_i **then**
 6:         **for** $c \leftarrow 1$ to k **do**
 7:           **if** $h_c(j) < h_c(P[S_i, \varphi])$ **then**
 8:             $h_c(P[S_i, \varphi]) \leftarrow h_c(j)$;
 9: **return** $\{h_c(P[S_i, \varphi]) | 0 \leq i \leq m, \varphi \in \Sigma_i, 1 \leq c \leq k\}$;

4.2 Estimation of Set Size

Based on the sketches of sources, we first study how to compute the min-hash value $h(B)$ and then how to use $h(B)$ to estimate the size of B, denoted by b. As $B = P[\mathcal{S}, \varphi] = \cup_{S_i \in \mathcal{S}} P[S_i, \varphi]$, which is a union set, a brute-force method for computing $h(B)$ is to obtain B by merging all the sets at first. This method is impractical due to the large cost of accessing sources. Fortunately, min-hashing has a good property (shown in Theorem 8): the min-hash value of a union set can be obtained based on the min-hash value of each set.

Theorem 8. $h(A_1 \cup A_2) = \min(h(A_1), h(A_2))$.

Based on the above theorem, for each random hash function h_j, the min-hash value of union set B can be computed as below without generating set B.

$$h_j(B) = h_j(P[\mathcal{S}, \varphi]) = \min_{S_i \in \mathcal{S}} h_j(P[S_i, \varphi]) \tag{10}$$

The following equation shows how to use these min-hash values of B to estimate b. The estimator of b is denoted by $\hat{b}$.

$$\hat{b} = nk / \sum_{1 \leq j \leq k} h_j(B) \tag{11}$$

Lemma 1 shows that $\hat{b}$ is a good estimator when the number of random hash functions k is sufficiently large.

Lemma 1. *Given ϵ, δ, let $\alpha = \min\{\epsilon^2, \epsilon/n\}$. When $k \geq \frac{2}{\alpha^2} \ln \frac{2}{\delta}$, we have the following property.*

$$Pr[|\frac{\hat{b} - b}{b}| < \epsilon] > 1 - \delta$$

Proof. For each random hash function h_j, B can be modeled as a vector v_j. For each row i, the probability of $v_j[i] = 1$ is a Bernoulli distribution, where $p = b/n$. Since $h_j(B)$ is the minimum row which contains a 1, $h_j(B)$ is a geometric distribution. Specifically, for each random hash function h_j, $E(h_j(B)) = 1/p = n/b$. We then have $E(\frac{1}{b} - \frac{1}{b}) = 0$. Let $X_i = \frac{h_i(B)}{nk} - \frac{1}{bk} (1 \le i \le k)$, $M = \sum_{i=1}^{k} X_i = \frac{1}{b} - \frac{1}{b}$, $\Delta = 1$ and $r = k = \frac{2}{\alpha^2} \ln \frac{2}{\delta}$. $Pr[|\frac{1}{b} - \frac{1}{b}| < \alpha] > 1 - \delta$ can be easily proved based on Theorem 5.   □

4.3   Estimation of Intersection Set Size

Recall that the estimator $\hat{x}$ of the intersection set size x is defined as below,

$$\hat{x} = \hat{s}(a + \hat{b})/(1 + \hat{s}) \tag{12}$$

Lemma 2 shows that $\hat{x}$ is also a good estimator of x when the number of random hash functions k is sufficiently large.

Lemma 2. *Given ϵ, δ, let $\alpha = \frac{\epsilon^2}{3}$, and $\delta' = 1 - \sqrt{1 - \delta}$. When $k \ge \frac{2}{\alpha^2} \ln \frac{2}{\delta'}$, we have the following property.*

$$Pr[|\frac{\hat{x} - x}{x}| < \epsilon] > 1 - \delta$$

Proof. Let $\epsilon_0 = \epsilon^2(1 + \epsilon)/(3 - \epsilon^2)$, $\epsilon_1 = \epsilon(a + \hat{b})/(3\hat{b} + \epsilon a)$. Since $\epsilon_0 > \epsilon^2/3$ and $\epsilon_1 > \epsilon^2/3$, when $k \ge \frac{2}{\alpha^2} \ln \frac{2}{\delta'}$, we have $Pr[|\hat{s} - s| < \epsilon_0] > 1 - \delta'$ and $Pr[|\frac{\hat{b} - b}{b}| < \epsilon_1] > 1 - \delta'$. Thus, $Pr[|\hat{s} - s| < \epsilon_0$ and $|\frac{\hat{b} - b}{b}| < \epsilon_1] > (1 - \delta')^2 = 1 - \delta$. Note that, we require $\hat{s}$ to be larger than the threshold $\epsilon + \epsilon_0$. Otherwise, we do not consider this sketch $P[S_i, \varphi]$ during the coverage computation due to its small contribution to the coverage. As $|\hat{s} - s| < \epsilon_0$ and $\hat{s} > \epsilon + \epsilon_0$, we have $s \ge \epsilon$.

Supposing that $|\hat{s} - s| < \epsilon_0$ and $|\frac{\hat{b} - b}{b}| < \epsilon_1$, we prove $|\frac{\hat{x} - x}{x}| < \epsilon$. Clearly,

$$|\frac{\hat{x} - x}{x}| = |\frac{\hat{s}(a + \hat{b})(1 + s) - s(a + b)(1 + \hat{s})}{s(a + b)(1 + \hat{s})}| \tag{13}$$

Since $|\frac{\hat{x} - x}{x}|$ monotonically increases in $\hat{s}$ and $\hat{b}$, plugging the maximum of $\hat{s}$, i.e. $s + \epsilon_0$ and the maximum of $\hat{b}$, i.e. $b + \epsilon_1 b$ into $|\frac{\hat{x} - x}{x}|$ gives

$$|\frac{\hat{x} - x}{x}| \le \frac{\epsilon_0}{s(1 + s + \epsilon_0)} + \frac{\epsilon_1 b(1 + s)}{y(1 + s + \epsilon_0)} + \frac{\epsilon_0 \epsilon_1 b(1 + s)}{ys(1 + s + \epsilon_0)}$$

$$< \frac{\epsilon_0}{s(1 + s + \epsilon_0)} + \frac{\epsilon_1 b}{y} + \frac{\epsilon_0 \epsilon_1 b}{ys} \tag{14}$$

where we denote $a + b$ as y for simplicity. As $\frac{\epsilon_0}{s(1+s+\epsilon_0)} + \frac{\epsilon_1 b}{y} + \frac{\epsilon_0 \epsilon_1 b}{ys}$ monotonically decreases with s and $s \ge \epsilon$, Eq. (14) can be written as

$$|\frac{\hat{x} - x}{x}| < \frac{\epsilon_0}{s(1 + s + \epsilon_0)} + \frac{\epsilon_1 b}{y} + \frac{\epsilon_0 \epsilon_1 b}{ys} \le \frac{\epsilon_0}{\epsilon(1 + \epsilon + \epsilon_0)} + \frac{\epsilon_1 b}{y} + \frac{\epsilon_0 \epsilon_1 b}{y\epsilon} \tag{15}$$

As $\epsilon_0 = \epsilon^2(1 + \epsilon)/(3 - \epsilon^2)$, we have

$$\frac{\epsilon_0}{\epsilon(1 + \epsilon + \epsilon_0)} = \epsilon/3 \tag{16}$$

As $\epsilon_1 = \epsilon(a+\hat{b})/(3\hat{b}+\epsilon a)$, $\epsilon_1 = \frac{\epsilon a}{3\hat{b}}(1-\epsilon_1) + \frac{\epsilon}{3} < \frac{\epsilon(a+b)}{3b} = \frac{\epsilon y}{3b}$, we get

$$\frac{\epsilon_1 b}{y} < \frac{\epsilon y}{3b} * \frac{b}{y} = \epsilon/3 \tag{17}$$

$$\frac{\epsilon_0 \epsilon_1 b}{\epsilon y} < \frac{b}{\epsilon y} * \frac{\epsilon^2(1+\epsilon)}{3-\epsilon^2} * \frac{\epsilon(a+\hat{b})}{(3\hat{b}+\epsilon a)} \leq \frac{\epsilon}{3}\left(\frac{\epsilon+\epsilon^2}{3-\epsilon^2}\right) \leq \frac{\epsilon}{3} \tag{18}$$

Embedding (16)–(18) into (15), we have $|\frac{\hat{x}-x}{x}| < 3 * \frac{\epsilon}{3} = \epsilon$ if $|\hat{s}-s| < \epsilon_0$ and $|\frac{\hat{b}-b}{b}| < \epsilon_1$. Since $Pr[|\hat{s}-s| < \epsilon_0, |\frac{\hat{b}-b}{b}| < \epsilon_1] > 1-\delta$, $Pr[|\frac{\hat{x}-x}{x}| < \epsilon] > 1-\delta$. □

4.4 Properties of MH-Greedy Algorithm

In this section, we analyze the time complexity and approximation ratio bound of MH-Greedy with the properties of the estimators in Sects. 4.2 and 4.3. We firstly show the time complexity of MH-Greedy.

Lemma 3. *The expected time of CompCoverage (Algorithm 2) is $O(|\Sigma_0| * k)$, where we assume $k \geq |S|$ without loss of generality.*

Embedding Algorithm 2 into Algorithm 1, the expected time complexity of MH-Greedy (Algorithm 1) is shown as follows.

Theorem 9. *The expected time of MH-Greedy is $O(K * |\Sigma_0| * k * m)$.*

We proceed to study the approximation ratio of our algorithm.

Definition 2. *Given $0 < \varepsilon$, $\delta < 1$, an algorithm A is said to be a (ε, δ)-approximation algorithm for the SSID problem if for any SSID instance with the optimal solution S_{OPT}, the algorithm returns a solution S such that*

$$Pr[|\frac{cov(S) - cov(S_{OPT})}{cov(S_{OPT})}| < \varepsilon] > 1-\delta \tag{19}$$

According to Lemma 2, since $c\hat{o}v(S)$ is the sum of $\hat{x}$ for each FD $\varphi \in \Sigma_0$, the following property holds.

Lemma 4. *Let $0 < \epsilon$, $\delta < 1$, $\alpha = \frac{\epsilon^2}{3|\Sigma_0|}$, $\delta' = 1 - \sqrt{1-\delta}$. When $k \geq \frac{2}{\alpha^2} \ln\frac{2}{\delta'}$, $\forall S \subseteq \mathbb{S}$, we have,*

$$Pr[|\frac{c\hat{o}v(S) - cov(S)}{cov(S)}| < \epsilon] > 1-\delta.$$

We proceed to prove the approximation ratio bound of our algorithm as follows.

Theorem 10. *Given ϵ, δ, let $\alpha = \frac{\epsilon^2}{3|\Sigma_0|}$, $\delta' = 1 - \sqrt{1-\delta}$. When $k \geq \frac{2}{\alpha^2} \ln\frac{2}{\delta'}$, MH-greedy achieves a $\frac{(1-\epsilon)(1-1/e)}{(1+\epsilon)}$-approximation ratio for the SSID problem with a probability larger than $1-\delta$.*

Proof. We denote the optimum solution of SSID under the true values of coverage as α, the optimum solution under the estimated values of coverage based on min-hashing as γ, our solution is θ. Let cov ($c\hat{o}v$) denote the cost function of solution under true coverage (estimated coverage based on min-hash values).

Based on Lemma 4, for any solution x, when $k \geq \frac{2}{\alpha^2} \ln \frac{2}{\delta'}$, we have

$$Pr[|\frac{c\hat{o}v(\mathcal{S}) - cov(\mathcal{S})}{cov(\mathcal{S})}| < \epsilon] > 1 - \delta \tag{20}$$

We assume that, $\forall x \subseteq \mathbb{S}$:

$$\frac{c\hat{o}v(x) - cov(x)}{cov(x)} < \epsilon \tag{21}$$

Now we prove $cov(\theta) \geq (1 - 1/e)(1 - \epsilon)/(1 + \epsilon)cov(\alpha)$. According to (21), we get

$$(1 - \epsilon)cov(\gamma) \leq c\hat{o}v(\gamma) \leq (1 + \epsilon)cov(\gamma) \tag{22}$$

$$(1 - \epsilon)cov(\alpha) \leq c\hat{o}v(\alpha) \leq (1 + \epsilon)cov(\alpha) \tag{23}$$

$$(1 - \epsilon)cov(\theta) \leq c\hat{o}v(\theta) \leq (1 + \epsilon)cov(\theta) \tag{24}$$

As γ is the optimum solution under the estimated coverage, we have

$$c\hat{o}v(\gamma) \geq c\hat{o}v(\alpha) \tag{25}$$

According to Theorem 3, θ achieves a $(1 - 1/e)$-ratio under the estimated coverage,

$$c\hat{o}v(\theta) \geq (1 - 1/e)c\hat{o}v(\gamma) \tag{26}$$

Combining (22)–(26), we then give

$$cov(\theta) \geq \frac{c\hat{o}v(\theta)}{1 + \epsilon} \geq \frac{1 - 1/e}{1 + \epsilon}c\hat{o}v(\gamma) \geq \frac{1 - 1/e}{1 + \epsilon}c\hat{o}v(\alpha) \geq \frac{(1 - \epsilon)(1 - 1/e)}{1 + \epsilon}cov(\alpha). \tag{27}$$

According to Eqs. (20) and (27), we get $Pr[\frac{cov(\theta)}{cov(\alpha)} \geq \frac{(1-\epsilon)(1-1/e)}{1+\epsilon}] > 1 - \delta.$ $\square$

5   Experimental Results

In this section, we study the proposed algorithms experimentally. The goals of our study are to investigate (1) the quality of the results produced by MH-Greedy, and (2) how MH-Greedy performs in terms of execution efficiency.

5.1   Experiment Setup

We conducted our experiments over two real-world datasets: Book and Flight [6]. In addition, to investigate the impact of parameters and scalability of our algorithm, we evaluated the performance of MH-Greedy on synthetic datasets that yielded more sources and more tuples.

The *Book* data set contains 1263 books and 894 data sources. The *Flight* data set contains 1200 flights and 38 sources from the flight domain. The *Synthetic Data* is synthetic data sets with various data source number and data size. The relation to clean, denoted by R, contains 10 attributes $A_1 - A_{10}$. Each data source contains a subset of $A_1 - A_{10}$. The *Synthetic Data* is controlled by the following parameters: (a)perc_attr: the probability of each attribute in $A_1 - A_{10}$ been chosen in each data source. (b)(no_tuple, min_perc, max_perc): the size of each data set is a random number in the range of [min_perc*no_tuple, max_perc*no_tuple]; (c)size_domain: the size of the domain of each attribute $A_j(1 \leq j \leq 10)$; (d)#Sources: the number of data sources; and (e)#Hash: the number of random hash functions. For each attribute A_j in each tuple, the value is a random value in [1, size_domain]. Table 3 shows the parameters used for generating the data sets and the default settings for the parameters. We used following ten FDs: $A_1 \rightarrow A_6$, $A_1 \rightarrow A_7$, $A_1 \rightarrow A_8$, $A_2 \rightarrow A_9$, $A_2 \rightarrow A_{10}$, $A_5 \rightarrow A_9$, $A_5 \rightarrow A_{10}$, $[A_3, A_4] \rightarrow A_6$, $[A_3, A_4] \rightarrow A_7$ and $[A_3, A_4] \rightarrow A_8$. Each generated source should contain at least one of the FDs.

Table 3. Parameters for data generation

Param	Val. for target	Val. for source
perc_attr	100%	20%
no_tuple	20000	20000
(min_perc, max_perc)	(100%, 100%)	(0%, 100%)
size_domain	1000	1000
#Sources	1	100
#Hash	100	100

We implemented two selection algorithms: one is our approximate greedy algorithm based on estimated coverage, called MH-Greedy; the other is the greedy source selection algorithm based on exact coverage information which are obtained by accessing to all the sources, called Greedy. We also implemented the exact algorithm. However, we did not compare our approach with the exact algorithm since the exact algorithm takes prohibitive amount of time due to its expected time exponential in K. From our experiments, the exact source selection algorithm takes more than 27 h with the following setting: $K = 5$, #Sources= 100, no_tuple= 20,000. Additionally, MH-Greedy is an approximation of Greedy which is the polynomial time algorithm that achieves the best approximation ratio for SSID.

Thus, we compare MH-Greedy algorithm with Greedy algorithm to verify both the effectiveness and efficiency of our algorithm. All experiments are implemented in C and executed on a PC with Windows 10, a 8 GB of RAM and a 2.9 GHz Intel i7-7500U CPU.

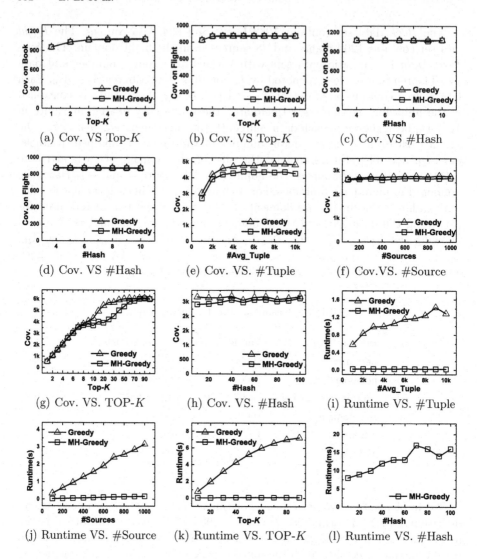

(a) Cov. VS Top-K (b) Cov. VS Top-K (c) Cov. VS #Hash

(d) Cov. VS #Hash   (e) Cov. VS. #Tuple   (f) Cov.VS. #Source

(g) Cov. VS. TOP-K (h) Cov. VS. #Hash   (i) Runtime VS. #Tuple

(j) Runtime VS. #Source   (k) Runtime VS. TOP-K (l) Runtime VS. #Hash

Fig. 1. Experimental results

5.2   Comparison

We firstly compare the effectiveness of methods with K varying from 1 to 10 and #Hash varying from 4 to 10 on real data sets. The comparison results are reported in Fig. 1(a)–(d). We have similar observations on both real-world data sets. (1) The coverage of MH-Greedy is almost as good as that of Greedy. This is because in real data sets, the differences between data sources are large, and both MH-Greedy and Greedy could easily find several proper data sources. (2) MH-Greedy performs quite well on a small number of hash functions. This is also caused by the significant differences between data sources, which could be distinguished with even a small number of hash functions.

5.3 Impact of Parameters

In this section, we evaluate the impact of average data size (#Avg_Tuple), the number of sources (#Sources), TOP-K and #Hash on the performance of our algorithm. The results are reported in Fig. 1(e)–(h). We have the following observations. (1) MH-Greedy achieves a comparable performance on effectiveness as Greedy in the synthetic data sets, where the approximation ratio is around 90%. (2) Fig. 1(e) shows that #Avg_Tuple has little impact on the effectiveness of MH-Greedy. (3) Fig. 1(f) shows that the effectiveness of MH-Greedy is not sensitive to the number of sources. (4) Fig. 1(g) reports the impact of K. It can be seen that, with the increase of K, the gap between MH-Greedy and Greedy is widened. When K is larger than 20, such gap is narrowing. This is because the gap between MH-Greedy and Greedy is added up in each iteration. The more iterations are executed, the larger the gap is. However, when K becomes sufficiently large, no matter how we select sources, most patterns have been covered, so the gap between MH-Greedy and Greedy becomes small. (5) Fig. 1(h) shows that with the increase of #Hash, the gap between two algorithms becomes smaller, the ratio increases from 0.90 to 0.98. These results on synthetic data are consistent with our expectation.

We also note that the differences between two algorithms for the synthetic data set are larger than that for real data sets. This is because there are many similar data sources for synthetic data, and more accurate coverage estimation are required. It is also the reason that the accuracy of MH-Greedy changes more significantly with the number of hash functions on synthetic data.

5.4 Results for Efficiency

In this section, we investigate the runtime performance of MH-Greedy. Figure 1 (i) reports the runtime of both algorithms with varying the data size. We observe that the runtime of Greedy grows with the data size, while the runtime of MH-Greedy is very stable, which is around 0.024 s. This result is consistent with our analysis in Sect. 4.4. In Fig. 1(j), we varied the number of sources from 100 to 1000. The runtime of both algorithms grows linearly with #Sources, and MH-Greedy outperforms Greedy significantly. In Fig. 1(k), we varied TOP-K from 10 to 100. The runtime of Greedy is still linear to TOP-K, while the runtime of MH-Greedy grows slowly with K and outperforms Greedy significantly. Such results shows the benefit of MH-Greedy on efficiency and scalability. Figure 1 (l) shows the runtime with varying #Hash from 10 to 100, the runtime of MH-Greedy grows approximately linearly as #Hash increases. These results verifies the scalability of MH-Greedy. There are also some fluctuations, it is because the runtime of coverage estimation is determined by the number of FDs in $\mathcal{S} \cup S_i$. Thus the runtime varies as we choose different $\mathcal{S}$ based on different sketches.

Summary. From the experimental results, we can draw the following conclusions. (a) The sources output by MH-Greedy is comparable to that of Greedy on both real-data sets and synthetic data sets. (b) MH-Greedy outperforms Greedy both on efficiency and scalability significantly. (c) The effectiveness of MH-Greedy is insensitive to the data size and the number of hashes. (d) MH-Greedy scales well on both the data size and the number of sources.

6   Related Work

Inconsistency detection and repairing [1,3,15] have been widely studied due to its importance (see [7] for a survey). [8] studied the problem of CFD violations detection for centralized data. [9,10] studied the problem of detecting FD and CFD violations in a distributed database.

Source selection [6,19–21] has been recently studied. Previous work of source selection focus on choosing proper sources for querying or integration while the objective for SSID is to detect FD violations. Source selection techniques can also be applied on inconsistency detection under other integrity constraints, including conditional functional dependencies (CFDs) and denial constrains (DCs), but we leave that for future work. [21] is the most relevant to our work. It proposed an overlap estimation strategy of sources. However, this method requires some prior statistics overlap information of sources from third parties. This assumption is not required in this paper.

Data fusion [4,5,12,14,18] aims at finding true values of conflicting data items. Various fusion models have been proposed. The most basic and simple fusion method is to take the value provided by the majority vote. Advanced methods assign weights to sources according to their reliability. SSID can be viewed as a preprocessing work of data fusion.

7   Conclusion

We studied the source selection problem to effectively detect inconsistencies in database in this paper. We introduced coverage model describing the gain of selected sources and defined the source selection problem for inconsistency detection, where we select the near-optimal source based on the estimated coverage values in each iteration. We provided theoretical guarantees for our algorithm.

We experimentally evaluated our algorithms on both real and synthetic data sets. The experimental results show that our algorithm finds solutions that are competitive with the near-optimal greedy algorithm and achieves a better performance on both efficiency and scalability without accessing to data sources.

In the future, we plan to extend our method to more general cases, such as selecting sources with different costs based on multiple quality measures.

Acknowledgements. This work was supported by NSFC61602159, 61370222 and Program for Group of Science Harbin technological innovation 2015RAXXJ004. The authors wish to thank Hongzhi Wang, Rong Zhu and Ran Bi for helpful discussions of this paper.

References

1. Chu, X., Ilyas, I.F., Papotti, P.: Holistic data cleaning: putting violations into context, pp. 458–469 (2013)
2. Cohen, E., Datar, M., Fujiwara, S., Gionis, A., Indyk, P., Motwani, R., Ullman, J.D., Yang, C.: Finding interesting associations without support pruning. TKDE **13**(1), 64–78 (2001)
3. Dallachiesa, M., Ebaid, A., Eldawy, A., Elmagarmid, A., Ilyas, I.F., Ouzzani, M., Tang, N.: NADEEF: a commodity data cleaning system. In: SIGMOD, pp. 541–552. ACM (2013)
4. Dong, X.L., Berti-Equille, L., Srivastava, D.: Integrating conflicting data: the role of source dependence. VLDB **2**(1), 550–561 (2009)
5. Dong, X.L., Berti-Equille, L., Srivastava, D.: Truth discovery and copying detection in a dynamic world. VLDB **2**(1), 562–573 (2009)
6. Dong, X.L., Saha, B., Srivastava, D.: Less is more: selecting sources wisely for integration. In: VLDB, vol. 6, pp. 37–48. VLDB Endowment (2012)
7. Fan, W.: Dependencies revisited for improving data quality. In: PODS
8. Fan, W., Geerts, F., Jia, X., Kementsietsidis, A.: Conditional functional dependencies for capturing data inconsistencies. TODS **33**(2), 6 (2008)
9. Fan, W., Geerts, F., Ma, S., Müller, H.: Detecting inconsistencies in distributed data. In: ICDE, pp. 64–75. IEEE (2010)
10. Fan, W., Li, J., Tang, N., et al.: Incremental detection of inconsistencies in distributed data. TKDE **26**(6), 1367–1383 (2014)
11. Hochbaum, D.S.: Approximating covering and packing problems: set cover, vertex cover, independent set, and related problems. In: Approximation Algorithms for NP-hard Problems, pp. 94–143. PWS Publishing Co. (1996)
12. Li, X., Dong, X.L., Lyons, K., Meng, W., Srivastava, D.: Truth finding on the deep web: is the problem solved? VLDB **6**(2), 97–108 (2012)
13. Li, X., Dong, X.L., Lyons, K., Meng, W., Srivastava, D.: Scaling up copy detection. In: ICDE, pp. 89–100 (2015)
14. Liu, X., Dong, X.L., Ooi, B.C., Srivastava, D.: Online data fusion. VLDB **4**(11), 932–943 (2011)
15. Mayfield, C., Neville, J., Prabhakar, S.: ERACER: a database approach for statistical inference and data cleaning. In: SIGMOD, pp. 75–86. ACM (2010)
16. Nemhauser, G.L., Wolsey, L.A., Fisher, M.L.: An analysis of approximations for maximizing submodular set functionsi. Math. Program. **14**(1), 265–294 (1978)
17. Phillips, J.M.: Chernoff-hoeffding inequality and applications. arXiv preprint arXiv:1209.6396 (2012)
18. Pochampally, R., Das Sarma, A., Dong, X.L., Meliou, A., Srivastava, D.: Fusing data with correlations. In: SIGMOD, pp. 433–444. ACM (2014)
19. Rekatsinas, T., Dong, X.L., Getoor, L., Srivastava, D.: Finding quality in quantity: the challenge of discovering valuable sources for integration. In: CIDR (2015)
20. Rekatsinas, T., Dong, X.L., Srivastava, D.: Characterizing and selecting fresh data sources. In: SIGMOD, pp. 919–930. ACM (2014)
21. Salloum, M., Dong, X.L., Srivastava, D., Tsotras, V.J.: Online ordering of overlapping data sources. VLDB **7**(3), 133–144 (2013)

Effective Solution for Labeling Candidates with a Proper Ration for Efficient Crowdsourcing

Zhao Chen[1], Peng Cheng[1], Chen Zhang[1,2(✉)], and Lei Chen[1]

[1] The Hong Kong University of Science and Technology, Kowloon, Hong Kong
{zchenah,pchengaa,czhangad,leichen}@cse.ust.hk
[2] Shandong University of Finance and Economics, Jinan, China

Abstract. One of the core problems of crowdsourcing research is how to reduce the cost, in other words, how to get better results with a limited budget. To save budget, most researchers concentrate on internal steps of crowdsourcing while in this work we focus on the pre-processing stage: how to select the input for crowds to contribute. A straightforward application of this work is to help budget-limited machine learning researchers to get better balanced training data from crowd labeling. Specifically, we formulate the prior information based input manipulating procedure as the Candidate Selection Problem (CSP) and propose an end-squeezing algorithm for it. Our results show that a considerable cost reduction can be achieved by manipulating the input to the crowd with the help of some additional prior information. We verify the effectiveness and efficiency of these algorithms through extensive experiments.

1 Introduction

With the flourishing and easily-accessible web based crowdsourcing platforms, such as Amazon Mechanical Turk (AMT), crowdsourcing becomes a popular paradigm to utilize human intelligence. Among various applications of crowdsourcing [1,3,20], labeling is arguably the most common and natural practice. Compared with hiring a number of experts, the monetary cost of crowdsourcing is fairly low [2]. Existing studies focus on reducing the cost by considering incentive design, task assignment and answer aggregation [12,16,19,21]. However, if the *raw data set* is already very biased, existing methods cannot guarantee good balanced results with limited budgets.

To address the bias issue of raw data sets, in this paper, we study an essential problem in crowdsourcing systems, namely *candidate selection problem* (CSP), which is to select a subset of candidates from the raw data set to satisfy the required distribution of the final results with the highest probability. We illustrate the CSP problem by a motivation example of image labeling as follows.

Example 1. A researcher wants to use AMT to label some portraits as male/female, then utilizes them to train a classifier. To train a good classifier,

© Springer International Publishing AG, part of Springer Nature 2018
J. Pei et al. (Eds.): DASFAA 2018, LNCS 10828, pp. 386–394, 2018.
https://doi.org/10.1007/978-3-319-91458-9_23

Table 1. Prior confidences of the raw portraits.

Portrait ID	Male confidence	Female confidence
c_1	0.95	0.05
c_2	0.95	0.05
c_3	0.94	0.06
c_4	0.92	0.08
...	...	...
c_{n-3}	0.05	0.95
c_{n-2}	0.03	0.97
c_{n-1}	0.01	0.99
c_n	0.01	0.99

it is better to input a balanced 1:1 male/female portrait dataset with enough entries (e.g., portraits) [8], as the distribution of the training data set may significantly affect the accuracy of the classifiers. She crawled a large number of raw portraits (candidates) from the Web, however, and only has $100 budget to use AMT to label the portraits. The average cost of a reliable portrait label is 10 cents, thus at most 1000 portraits can be labeled. She needs to construct a 1:1 balanced training data set with as many as possible portraits. The optimal result is that she can select 1,000 candidates and results in 500 male portraits and 500 female ones. She first uses some existing classifiers (e.g., the Viola-Jone detector [23]) to estimate the portraits into difference confidences (e.g., c_1 has 95% confidence of being male portrait) as shown in Table 1. Then, the problem she faced is how to select 1,000 candidates from the raw portraits with given confidences to be labeled by crowds such that the answers returned are balanced in the highest probability.

Motivated by the example above, in this paper, we first formalize the CSP problem, which aims to select a subset of candidates from the raw data set, given their prior confidences of being possible choices, for the crowds to label such that the probability of that the results returned by the crowds satisfy the required distribution is maximized. The CSP problem essentially concerns the quality of the final results returned by the crowds. Existing studies on quality controlling in crowdsourcing usually target on designing aggregation methods [14, 22], proposing fair mechanisms to encourage crowds to contribute reliably [4] or matching suitable crowds to particular tasks [6], which rarely consider of the preprocessing of the raw data sets. To the best of our knowledge, there is no previous studies that focus on selecting the most promising candidates for crowds to label. However, efficiently selecting a subset of candidates from the raw data set with given prior confidences and limited budgets is not easy. We propose an exact algorithm, namely *end-squeezing algorithm*, which can select candidates from the ends of a list of candidates sorted by the prior confidence values to achieve the optimal subset (proved by Theorem 1). Finally, through extensive experiments, we demonstrate the efficiency and effectiveness of our approach.

To summarize, in this paper we have made the following contributions.

1. We formulate the Candidate Selection Problem (CSP) in Sect. 2;
2. We propose one end-squeezing exact approach in Sect. 3.
3. We conduct extensive experiments on real and synthetic data sets, and show the efficiency and effectiveness of our CSP approaches in Sect. 4.

For the rest parts, Sect. 5 revises previous related works. Section 6 concludes this paper and discusses the future works.

2   Preliminary

In this section, we introduce the preliminaries used in this work.

Definition 1 (Binary Labeling Candidates). *Let $C = \{c_1, c_2, \ldots, c_n\}$ be a set of n binary labeling candidates. For each candidate c_i, it will be labeled by workers with either 0/1. Each candidate c_i has a true label φ_i and crowdsourced label ϕ_i, which is aggregated from the answers of crowds.*

Each candidate c_i can be an image of a male or female noted as it true label φ_i. To improve the accuracies of the image labels, we may utilize crowds to label the images with binary choices (male/female or 0/1). Although the crowds may be fallible, there are a lot of existing methods [5, 11] to guarantee that the qualities of the answers returned by crowds can satisfy the required accuracies. Then, the resulted label ϕ_i suggested by crowds can be used to train better detectors.

Definition 2 (Candidate Selection Problem). *Given a set, $C = \{c_1, c_2, \ldots, c_n\}$, of n binary (0/1) labeling candidates with their prior probabilities $p_i = Pr(\varphi_i = 1) = 1 - Pr(\varphi_i = 0)$ and the minimum label result requirements l_0, l_1, the candidate selection problem is to select a set C_k of $k(\geq l_0 + l_1)$ candidates that maximize the probability: $Pr(x_0 \geq l_0, x_1 \geq l_1)$, where $x_1 = \sum_{c_i \in C_k} \phi_i, x_0 = k - x_1$.*

Some existing methods (e.g., the Viola-Jone detector [23]) can estimate the prior confidence p_i of each candidate being male (or 0). Then, we need to solve the CSP problem to select a subset of candidates such that the required minimum number of male candidates and female candidates can be satisfied with the maximum probability.

3   An End-Squeezing Approach

In this section we will propose an exact algorithm, namely *end-squeezing approach*, to the CSP problem, which selects candidates from the ends of a list of raw candidates sorted by their prior probabilities to squeeze to the optimal positions (the optimal result). We first introduce some mathematical basis.

Definition 3 (Poisson Binomial Variable). *Let a sequence of* n *independent Bernoulli variables* $X_1, X_2, \ldots, X_n$ *have success probabilities* $\boldsymbol{p} = (p_1, p_2, \ldots, p_n)$, *then* $X = X_1 + X_2 + \ldots + X_n$, *i.e. the total number of successes Bernoulli trials is called a Poisson Binomial Variable with parameter* $\boldsymbol{p}$.

Definition 4 (Poisson Binomial Distribution (PBD)). *The probability of* $k(0 \le k \le n)$ *success trials is:* $Pr(X = k) = \sum\limits_{S \in S_k} \prod\limits_{i \in S} p_i \prod\limits_{0 < j \le n, j \notin S} (1 - p_j)(0 \le k \le n)$, *where* S_k *is the set of all* k *size subsets of* $\{1, 2, 3, \ldots, n\}$.

Below we show some properties of PBD and CSP.

Unimodality. From [26] we know that the p.m.f of any PBD is unimodal. The optimization goal of CSP is actually the sum of probabilities in a continuous span on such a p.m.f. Considering one of the selected $p_x \in \boldsymbol{p}$ ($\boldsymbol{p}$ is the set of probabilities of all candidates) as a variable with domain $[0, 1]$, the peak of the p.m.f moves to the right when p_x increases and to the left when p_x decreases.

Non-submodularity. Submodularity is an important property of optimization problems (A comprehensive introduction of it can be found in [7]). Maximization of submodular functions with even simple constraints is usually NP-hard. Let $f(X) = Pr(x_0 \ge l_0, x_1 \ge l_1 | X)$, $k = 2$, $l_0 = l_1 = 1$, $S = \{0.5, 0.6\}$ and $T = \{0.1, 0.2\}$, then we have: $f(S) + f(T) = 0.76 < f(S \cup T) + f(S \cap T) = 0.85$. So CSP is not submodular.

Recursiveness. Except Definition 4, PBD also has several recursive forms [26]. An intuitive one is:

Definition 5 (Recursive PBD). *Let* $Pr_{n,k}$ *be the probability of a Poisson Binomial Variable* $X = \sum\limits_{i=1..n} X_i$ *to be* k. *We have:*

$$Pr_{n,k} = p_n Pr_{n-1,k-1} + (1 - p_n) Pr_{n-1,k}, \text{ where } 0 \le k \le n, \ p_n = Pr(X_n = 1)$$

With $p \in \boldsymbol{p}$ and $\boldsymbol{p}' = \boldsymbol{p} - \{p\}$, the CSP goal function can be represented in as:

$$CSP(\boldsymbol{p}, l_0, l_1, k) = CSP(\boldsymbol{p}', l_0, l_1 - 1, k - 1)p + CSP(\boldsymbol{p}', l_0 - 1, l_1, k - 1)(1 - p)$$

Next we propose an algorithm that gives the exact solution of any CSP problem based on the properties above.

Briefly, Algorithm 1 first sorts all candidates in a row by their probabilities, then iterates over the candidate sets that only choose from the head and tail of the row, and finally returns the one with largest cumulated PBD.

If we use the fast PBD calculation method in [9], the time complexity of $CUM{-}PBD$ will be $O(k^3)$. Then the time complexity of $END{-}SQUEEZING$ is $O(k^4 + n \log n)$.

Algorithm 1 considers only candidates "squeezed" to the two ends of all candidates, thus we call it the end-squeezing algorithm. The below theorem shows its correctness.

Theorem 1. *The end-squeezing algorithm is correct.*

Algorithm 1. The End-Squeezing Algorithm

1: input: $\boldsymbol{p} = p_1, ..., p_n$, (l_0, l_1) and k
2: output: the maximum probability
3: **procedure** END-SQUEEZING($\boldsymbol{p}, l_0, l_1, k$)
4:     sort $\boldsymbol{p}$ increasingly
5:     **for** i from 0 to k **do**
6:         let $\mathbb{C}_i$ be the set of first i candidates and the last $k - i$ candidates.
7:         calculate the probability $cumPBD(\mathbb{C}_i, l_0, l_1)$
         return $\mathbb{C}_{max}$ with the largest probability.

8:
9: input: a PBD X and the range k_{min}, k_{max}
10: output: p
11: **procedure** CUM-PBD
12:     Let $p_{cum} \leftarrow 0$
13:     **for** k from k_{min} to k_{max} **do**
14:         $p_{cum} += PBD(X, k)$
         return p_{cum}

Proof. For any CSP instance $< \boldsymbol{p}, l_0, l_1, k >$, suppose that $\boldsymbol{p}'$ is the optimal candidate set and $OPT = CSP(\boldsymbol{p}, l_0, l_1, k)$ is the optimal probability.

For any $p' \in \boldsymbol{p}'$, let $Pos(p') = CSP(\boldsymbol{p}' - \{p'\}, l_0, l_1 - 1, k - 1)$ and $Neg(p') = CSP(\boldsymbol{p}' - \{p'\}, l_0 - 1, l_1, k - 1)$. Then, from the recursiveness of CSP, we know the optimal probability $OPT = p' * Pos(p') + (1 - p') * Neg(p')$.

Assume that there are two candidates p_x, p_y not selected into $\boldsymbol{p}'$ and $p_x < p' < p_y$. If $Pos(p') \geq Neg(p')$, we can replace p_y with p' to get a larger optimal probability, i.e. $OPT_x = p_y * Pos(p') + (1 - p_y) * Neg(p') > OPT$. Otherwise, if $Pos(p') < Neg(p')$, p' can be replaced by p_x.

Note that $\boldsymbol{p}'$ is the optimal candidate set, so the assumption is wrong. It means either (1) there is only one candidate not selected into $\boldsymbol{p}'$ or (2) there are at least two such candidates but they all lay on one side of p'. Formally, we have:

$$\forall p' \in \boldsymbol{p}', \nexists p_x, p_y \in \boldsymbol{p} - \boldsymbol{p}' s.t. p_x < p' < p_y$$

That is, the optimal candidate set must be composed of only the largest and smallest candidates. □

4 Experimental Study

4.1 Experimental Methodology

Data Sets. We use both real and synthetic data sets to test our CSP approaches. Specifically, for the real data set, we use the LFW [10] Face Dataset, which is a collection of 13, 233 face images of 5, 749 individuals. Note that the face images in this dataset were selected on the bias that they could be detected by the Viola-Jone detector [23]. We first choose 2000 images randomly from LFW Face Database data set as the candidates set D. Then, we use a mature CNN based

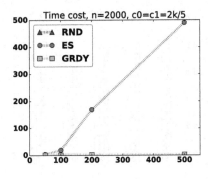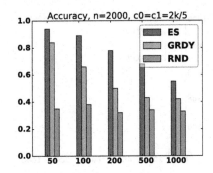

Fig. 1. Results on synthetic data. X-axis is k.

face detection/annotation model (a variant from the model in [25]) to generate a group of face image features from the rest part of data set D' (more than 8000 faces). Next, we utilize these features and D' as training data to build a SVM based male/female classifier C. For the prior confidences of images, we apply C to generate the prior probabilities of whether the input faces are male/female for candidate set D. We also test the effectiveness over a set of synthetic data in which the confidences follow a normal distribution with $\mu = 0.6, \sigma = 0.1$. For both real and synthetic datasets, we set both l_0, l_1 to $\frac{2}{5}k$.

CSP Approaches and Measures. We compare the end-squeezing approach (ES) with two baseline solutions, a random approach (RND) and a greedy approach (GRDY). Specifically, RND will return $2k$ elements from the given n probabilities randomly. GRDY tries to reduce the entropy as much as possible and greedily refrains the candidates with the highest probabilities to increase the entropy most.

For each set of experiments, we report the running times and the accuracies of the tested approaches. Particularly, accuracy is measured base on the distance between the optimal probability and the one returned by the testing approach. All codes are written in Python 2.7 and all the experiments are conducted on an Intel(R) Core(TM) i7 3.40 GHz PC with 8 GB memory.

4.2 Experiments on Synthetic Data

For synthetic data, we fix n to be 2000 and vary k in $\{50, 100, 200, 500\}$.

We first compare the efficiency of all the tested approaches (Figure 1). The time cost of ES increases when either k or n increases while both two base line methods can finish very fast for all k values. Overall, the time cost result is as expected. For accuracy, GRDY is always better than RND and its performance even approaches the optimal result from ES if there are enough "good" (high confidence) candidates available.

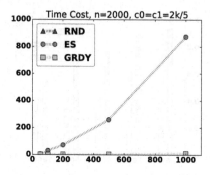 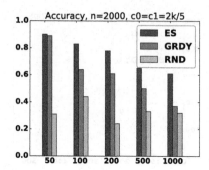

Fig. 2. Results on LFW data. X-axis is k.

4.3 Experiments on Real Data

Results of LFW Data are shown in Fig. 2. The results from LFW data are not quite different from the results of synthetic data above. A difference worth noting is: the accuracy of both GRDY here is a little bit better than synthetic data. A possible reason is the diversify of probabilities in LFW data is relatively higher. This also explains that the RND performance is not as stable as other. The LFW data experiment is an example of how pre-selection can improve the crowd-labeling result. If we select randomly, as the RND result indicates, the probability that we get the requirement ratio is only around 30% while the maximum probability could be higher than 80%.

5 Related Works

Sampling Methods. The procedure of our problem, choosing a subset from a whole data set, looks similar to the procedure of some sampling methods, e.g. stratified sampling [18] and adaptive sampling [17]. But actually they are different. Sampling aims to get the best characteristics estimation of the whole data by only checking a subset of it. While our problem is about how to get a subset with the highest probability to have a specified feature. So the goal of our problem is radically different from sampling's.

Poisson Binomial Distribution Based Prediction. Researchers in mathematics and statistics has studied some prediction methods based on the Poisson binomial distribution model. The approximation methods in [13,24] are used in these papers but it is in a different way as used in ours. The hard part of these prediction problems is the lacking of accurate prior possibilities which is the given information in our problem. They predict the best result by constructing a model while we get the best result by choosing the most suitable candidates.

Imbalanced Learning. Imbalanced learning is about how to train a better model from unbalanced input data. Researchers proposed some techniques [8,15] try to overcome the imbalance issue but still what they can do is to improve the learning process only. In this paper we also want to fix the imbalance problem but from another prospective to control the balance from the beginning.

6   Conclusion

In this paper, we show a new approach to reduce crowdsourced labeling cost and formulate the pre-selection process as the Candidates Selection Problem (CSP). We propose one exact algorithm for CSP and verify its effectiveness and efficiency on both synthetic and real data from LFW through experiments.

Acknowledgment. The work is partially supported by the Hong Kong RGC GRF Project 16207617, National Grand Fundamental Research 973 Program of China under Grant 2014CB340303, the National Science Foundation of China (NSFC) under Grant No. 61729201, Science and Technology Planning Project of Guangdong Province, China, No. 2015B010110006, Webank Collaboration Research Project, and Microsoft Research Asia Collaborative Research Grant.

References

1. Amsterdamer, Y., Grossman, Y., Milo, T., Senellart, P.: Crowdminer: mining association rules from the crowd. Proc. VLDB Endow. **6**(12), 1250–1253 (2013)
2. Buhrmester, M., Kwang, T., Gosling, S.D.: Amazon's mechanical turk a new source of inexpensive, yet high-quality, data? Perspect. Psychol. Sci. **6**(1), 3–5 (2011)
3. Cao, C.C., She, J., Tong, Y., Chen, L.: Whom to ask?: jury selection for decision making tasks on micro-blog services. Proc. VLDB Endow. **5**(11), 1495–1506 (2012)
4. Cao, C.C., Tong, Y., Chen, L., Jagadish, H.: Wisemarket: a new paradigm for managing wisdom of online social users. In: Proceedings of the 19th ACM SIGKDD International Conference on Knowledge Discovery and Data Mining, pp. 455–463. ACM (2013)
5. Dawid, A.P., Skene, A.M.: Maximum likelihood estimation of observer error-rates using the EM algorithm. Appl. Stat. 20–28 (1979)
6. Fan, J., Li, G., Ooi, B.C., Tan, K.-L., Feng, J.: icrowd: An adaptive crowdsourcing framework. In: Proceedings of the 2015 ACM SIGMOD International Conference on Management of Data, pp. 1015–1030. ACM (2015)
7. Fujishige, S.: Submodular Functions and Optimization, vol. 58. Elsevier, New York City (2005)
8. Galar, M., Fernandez, A., Barrenechea, E., Bustince, H., Herrera, F.: A review on ensembles for the class imbalance problem: bagging-, boosting-, and hybrid-based approaches. IEEE Trans. Syst. Man Cybern. Part C: Appl. Rev. **42**, 463–484 (2012)
9. Hong, Y.: On computing the distribution function for the poisson binomial distribution. Comput. Stat. Data Anal. **59**, 41–51 (2013)
10. Huang, G.B., Ramesh, M., Berg, T., Learned-Miller, E.: Labeled faces in the wild: a database for studying face recognition in unconstrained environments. Technical report, Technical Report 07–49, University of Massachusetts, Amherst (2007)
11. Ipeirotis, P.G., Provost, F., Wang, J.: Quality management on amazon mechanical turk. In: Proceedings of the ACM SIGKDD Workshop on Human Computation. ACM (2010)

12. Karger, D.R., Oh, S., Shah, D.: Budget-optimal task allocation for reliable crowd-sourcing systems. Oper. Res. **62**(1), 1–24 (2014)

13. Le Cam, L., et al.: An approximation theorem for the poisson binomial distribution. Pac. J. Math. **10**(4), 1181–1197 (1960)

14. Li, Q., Li, Y., Gao, J., Su, L., Zhao, B., Demirbas, M., Fan, W., Han, J.: A confidence-aware approach for truth discovery on long-tail data. Proc. VLDB Endow. **8**(4), 425–436 (2014)

15. Lopez, V., Fernandez, A., Garcia, S., Palade, V., Herrera, F.: An insight into classification with imbalanced data: empirical results and current trends on using data intrinsic characteristics. Inf. Sci. **250**, 113–141 (2013)

16. Mason, W., Watts, D.J.: Financial incentives and the performance of crowds. ACM SigKDD Explor. Newsl. **11**(2), 100–108 (2010)

17. Thompson, S., Seber, G.: Adaptive Sampling. Wiley series in probability and statistics. Wiley, Hoboken (1996). Show all parts in this series

18. Thompson, S.K.: Sampling. Wiley CourseSmart series, 3rd edn. Wiley, Hoboken (2012)

19. Tong, Y., Chen, L., Zhou, Z., Jagadish, H.V., Shou, L., Lv, W.: Slade: a smart large-scale task decomposer in crowdsourcing. IEEE Trans. Knowl. Data Eng. (2018)

20. Tong, Y., She, J., Ding, B., Chen, L., Wo, T., Xu, K.: Online minimum matching in real-time spatial data: experiments and analysis. PVLDB **9**, 1053–1064 (2016)

21. Tong, Y., She, J., Ding, B., Wang, L., Chen, L.: Online mobile micro-task allocation in spatial crowdsourcing. In: ICDE (2016)

22. Verroios, V., Garcia-Molina, H.: Entity resolution with crowd errors. In: 2015 IEEE 31st International Conference on Data Engineering (ICDE), pp. 219–230. IEEE (2015)

23. Viola, P., Jones, M.J.: Robust real-time face detection. Int. J. Comput. Vis. **57**(2), 137–154 (2004)

24. Volkova, A.Y.: A refinement of the central limit theorem for sums of independent random indicators. Theor. Probab. Appl. **40**(4), 791–794 (1996)

25. Wang, D., Hoi, S.C.H., He, Y.: A unified learning framework for auto face annotation by mining web facial images. In: Proceedings of the 21st ACM International Conference on Information and Knowledge Management, pp. 1392–1401. ACM (2012)

26. Wang, Y.H.: On the number of successes in independent trials. Stat. Sin. **3**, 295–312 (1993)

Handling Unreasonable Data in Negative Surveys

Jianwen Xiang[1], Shu Fang[1], Dongdong Zhao[1(✉)], Jing Tian[1],
Shengwu Xiong[1(✉)], Dong Li[2], and Chunhui Yang[1,2]

[1] School of Computer Science and Technology,
Wuhan University of Technology, Wuhan, China
{jwxiang,sfang,zdd,jtian,swxiong}@whut.edu.cn,
yangch@ceprei.com
[2] Software Quality Engineering Research Center, CEPREI, Guangzhou, China
lidong@vip.ceprei.com

Abstract. Negative survey is a method of collecting sensitive data. Compared with traditional surveys, negative survey can effectively protect the privacy of participants. Data collector usually has some background knowledge about the survey, and background knowledge could be effectively used for estimating aggregated results from the collected data. Traditional methods for estimating aggregated results would get some unreasonable data, such as negative values, and some values inconsistent with the background knowledge. Handling these unreasonable data could improve the accuracy of the estimated aggregated results. In this paper, we propose a method for handling values that are inconsistent with the background knowledge and negative values. The simulation results show that, compared with NStoPS, NStoPS-I and NStoPS-BK, more accurate aggregated results could be estimated by the proposed method.

Keywords: Negative survey · Unreasonable data · Background knowledge
Aggregated results · Data adjustment

1 Introduction

Negative survey is a promising model for privacy protection [1–4]. In traditional surveys (also called positive surveys), participants choose a category that is in line with their actual situation (this category is called the positive category), which could easily lead to the disclosure of personal privacy of the participants. Inspired by the negative selection mechanism in Biological Immune System [5], Esponda [6, 7] proposed the concept of negative survey.

Negative surveys require participants to choose a category that does not fit their own situation (this category is called the negative category). Negative survey does not require participants to give the appropriate information directly, and thus it can protect the privacy of participants when there are at least 3 categories for each question. Negative survey has been applied in several applications, e.g., anonymous data collection [8], location and trace privacy [9, 10], credits rating [11], etc.

© Springer International Publishing AG, part of Springer Nature 2018
J. Pei et al. (Eds.): DASFAA 2018, LNCS 10828, pp. 395–403, 2018.
https://doi.org/10.1007/978-3-319-91458-9_24

The data we actually need is the aggregated data about the positive survey which can be effectively reconstructed from the collected negative data. The first reconstruction algorithm NStoPS is proposed by Esponda. Bao et al. [12] showed that NStoPS algorithm might produce negative values when reconstructing aggregated results, and the negative values are unreasonable. Bao et al. [12] proposed two algorithms for handling negative values, i.e. NStoPS-I and NStoPS-II. However, NStoPS-I and NStoPS-II cannot handle some other unreasonable values in negative surveys, such as the data inconsistent with background knowledge. NStoPS-II is only suitable to the uniform negative surveys (in which participants select each negative category with the same probability). NStoPS-I is a kind of iterative method, which is less efficient [13] (i.e., need more computations) than NStoPS-II. In real world, the data collector often has some background knowledge about the negative survey or participants, for example, in a survey of a certain disease, hospitals (data collector) often know the incidence of the disease. Basically, the random selection in the negative survey, the human preferences during the process, these are likely to result in the unreasonable data in the reconstruction of the positive survey. Zhao et al. [14] proposed an algorithm called NStoPS-BK, which demonstrated that the use of background knowledge could effectively improve the accuracy of the reconstructed aggregated results.

Aiming at handling the unreasonable data, for example, negative values and values inconsistent with background knowledge, this paper proposed a new method. The proposed method is based on the idea of adjusting negative survey results according to the survey rule and background knowledge to eliminate unreasonable values. Experimental results show that the proposed method could obtain more accurate aggregated results than NStoPS, NStoPS-II. The proposed method in this paper can achieve the same effect as NStoPS-II when there is no extra background knowledge and performs better than NStoPS-BK. In summary, the contributions of this paper are listed as follows.

1. A new reconstruction algorithm (called NStoPS-UD) was proposed to handle the unreasonable data for uniform negative surveys. Then, a general expression for the algorithm was presented, which is suitable for general negative surveys.
2. Two experiments were carried out and the effectiveness of the proposed method on handling unreasonable data, i.e., negative values and values inconsistent with background knowledge was demonstrated.

This paper is organized as follows. Section 2 introduces the reconstruction algorithm proposed in this paper; Sect. 3 shows the experimental results and some discussion; Sect. 4 discusses the general expression for the proposed algorithm; we conclude this work and present some future work in Sect. 5.

2   Reconstruction Algorithm

In this section, we introduce the proposed algorithm for handling unreasonable values and reconstructing positive data, which is called NStoPS-UD. Specifically, NStoPS-UD is designed for the uniform negative survey, but we will give its general expression in Sect. 4. Assuming that the number of categories is c, the proportion of participants who belong to the $i^{th}(i = 1\ldots c)$ category in the negative survey (the data is called negative data) is r_i, and $R = \{r_1, r_2, \ldots, r_c\}, r_1 + r_2 + \ldots + r_c = 1$. The proportion of participants who belong to the $i^{th}(i = 1\ldots c)$ category in the positive survey (the data is called positive data) is t_i, and $T = \{t_1, t_2, \ldots, t_c\}, t_1 + t_2 + \ldots + t_c = 1$. p_{ij} represents the probability that a participant who chooses category i in the positive survey, but chooses category j in the negative survey. It is assumed that the positive data obtained by NStoPS is $T = \{t_1^*, t_2^*, \ldots, t_c^*\}$ and $t_1^* + t_2^* + \ldots + t_c^* = 1$. After handling the unreasonable data with the NStoPS-UD, the estimated positive data is $X = \{x_1, x_2, \ldots, x_c\}$, and x_i satisfy $x_1 + x_2 + \ldots + x_c = 1$.

First, we consider the situation in which the reconstruction result of a category is unreasonable, and without loss of generality, we assume that this category is c^{th} category. The estimated positive data of category c is t_c^*, and it should be adjusted to a reasonable value x_c. Due to the unreasonable selection of some participants, the difference Δr_c (could be negative) between t_c^* and x_c is induced. The algorithm NStoPS-UD first adjusts the unreasonable t_c^* to x_c. Because the sum of $r_1, \ldots, r_c$ should be 1, the change Δr_c in r_c will result in a change $-\Delta r_c$ over the other $c - 1$ categories. Each of the other $c - 1$ categories will have an appropriate proportion (denoted as $\Delta r_1, \Delta r_2, \ldots, \Delta r_{c-1}$) according to the survey rule. The problem can be attributed to the following equation:

$$\begin{cases} \sum_{i=1}^{c} x_i \times p_{i,1} = r_1 + \Delta r_1 \\ \sum_{i=1}^{c} x_i \times p_{i,2} = r_2 + \Delta r_2 \\ \quad\quad\quad \vdots \\ \sum_{i=1}^{c} x_i \times p_{i,c} = r_c + \Delta r_c \end{cases} \tag{1}$$

For uniform negative survey, when $i \neq j$, p_{ij} in (1) equals to $1/c - 1$, and when $i = j$, p_{ij} equals to 0. According to the rule that participants choose each negative category with the same probability, the proportions of unreasonable selection $-\Delta r_c$ from the participants that actually belong to categories $1\ldots c - 1$ are expected to be $d_{x_1}, \ldots, d_{x_{c-1}}$, and d_{x_i} $(i = 1, \ldots, c - 1)$ can be estimated as follow.

$$d_{x_i} = \frac{x_i \frac{1}{c-1}}{\sum_{j=1}^{c-1} x_j \frac{1}{c-1}} \times (-\Delta r_c) = \frac{x_i}{\sum_{j=1}^{c-1} x_j} \times (-\Delta r_c) \tag{2}$$

Because it is the uniform negative survey, d_{x_i} is re-assigned to the other $c - 2$ categories of the negative survey with an equal probability, i.e., $1/(c - 2)$. So, the Δr_i is calculated as follow:

$$\Delta r_i = \sum_{j=1, j\neq i}^{c-1} d_{x_j} \frac{1}{c - 2} \tag{3}$$

Finally, NStoPS is used again to calculate positive data using the adjusted negative data.

$$x_i = 1 - (c-1)(r_i + \Delta r_i) \tag{4}$$

Deduced from (1) (2) (3) (4), the results of estimated positive data are

$$x_i = 1 - x_c + \frac{x_c - (c-1)r_i}{1 + \frac{(1-x_c)-(c-1)r_c}{(1-x_c)(c-2)}} \tag{5}$$

Suppose there are n $(n \le c)$ categories which have unreasonable results among the positive data obtained by NStoPS, and without loss of generality, we assume that they are the $c^{th}, (c-1)^{th}, \ldots, (c-n+1)^{th}$ categories. The estimated results for these categories by NStoPS are $t_c^*, t_{c-1}^*, \ldots, t_{c-n+1}^*$, respectively, and the adjusted results for these categories are $x_c, x_{c-1}, \ldots, x_{c-n+1}$. Similar to the above derivation process, we have

$$d_{x_i} = \frac{x_i}{\sum_{j=1}^{c-n} x_j} \times (-\Delta r_c) + \ldots + \frac{x_i}{\sum_{j=1}^{c-n} x_j} \times (-\Delta r_{c-n+1}), \ (i = 1 \ldots c - n) \tag{6}$$

$$\Delta r_i = \sum_{j=1, j \ne i}^{c-n} d_{x_j} \frac{1}{c-n-1} \tag{7}$$

Finally, we can get the general formula for x_i

$$x_i = 1 - x_s + \frac{x_s - (c-1)r_i}{1 + \frac{(n-x_s)-(c-1)r_s}{(1-x_s)(c-n-1)}} \tag{8}$$

Where $x_s = x_c + x_{c-1} + \ldots + x_{c-n+1}, r_s = r_c + r_{c-1} + \ldots + r_{c-n+1}$. According to (8), we design an algorithm for handling unreasonable values and reconstructing positive data, its pseudo code is shown in Algorithm 1. The background knowledge [14] could be presented as value ranges of the positive data: $l_i \le t_i \le u_i, (i = 1 \ldots c)$. Where $0 \le l_i \le 1, 0 \le u_i \le 1$, l_i is the lower bound and u_i is the upper bound. It means that the proportion of participants who chooses category i in the positive survey is not less than l_i and not larger than u_i.

In Algorithm 1, Step 1 is to calculate the estimated value by NStoPS, Steps 3–9 are introduced to adjust the unreasonable data of NStoPS according to the background knowledge. Then the other categories' results are calculated by (8) in Step 15, and when the number of adjusted categories in Steps 3–9 is up to $c - 1$, the sum of the positive data may not equal to 1, so Steps 10–12 aim at solving the situation. If the number of adjusted categories is up to c, the results need to be scaled to keep the sum of the positive data unchanged in Steps 17–19.

3   Experiments

In this paper, we use the positive data and background knowledge in [14] to carry out the simulation experiments, which is convenient to compare the result with NStoPS-BK. For each positive data, a negative category is randomly chosen with the same probability as the corresponding negative data (in the remaining categories, one category is chosen randomly as the negative data with the same probability), and the data is shown in Table 1.

Algorithm 1: NStoPS-UD	
Input:	The negative survey results $R=\{r_1, r_2, ..., r_c\}$.
	Background knowledge $BK=([l_1, u_1], ..., [l_c, u_c])$.
Output:	Positive data $X=\{x_1, x_2, ..., x_c\}$.
1.	$x_i \leftarrow 1-(c-1)r_i \quad (i=1...c)$
2.	$x_s \leftarrow 0, \ r_s \leftarrow 0$
3.	**for** $i \leftarrow 1$ to c
4.	**if** $x_i < l_i$
5.	$x_i \leftarrow l_i$
6.	$x_s \leftarrow x_s + x_i, \ r_s \leftarrow r_s + r_i$
7.	**else if** $x_i > u_i$
8.	$x_i \leftarrow u_i$
9.	$x_s \leftarrow x_s + x_i, \ r_s \leftarrow r_s + r_i$
10.	**if** $c-1$ results have been adjusted, the last x_i :
11.	$x_i \leftarrow 1 - x_s$
12.	repeat steps 4 - 9
13.	**for** $i \leftarrow 1$ to c
14.	**if** x_i is not been adjusted
15.	calculate with formula 8
16.	repeat steps 4 - 9
17.	**if** $x_s \neq 1.0$
18.	**for** $i \leftarrow 1$ to c
19.	$x_i \leftarrow x_i/x_s$
20.	**return** $X=\{x_1, ..., x_c\}$

In the experiment, the traditional NStoPS, NStoPS-II and NStoPS-BK and the algorithm in this paper are used to reconstruct positive data from negative data and background knowledge. The accuracy of each algorithm is compared by calculating the error between the reconstructed results s_i and the original positive data. Experiments were performed 1000 times independently, and then the average error was calculated. The error is calculated as follow [9].

Table 1. Background knowledge and results of original positive surveys and negative surveys

Distribution	Background knowledge	Positive survey	Negative survey
Uniform	([0, 100], [0, 500], [0, 500], [100, 500], [0, 500])	(99, 89, 103, 109, 100)	(95, 110, 112, 98, 85)
Normal	([0, 20], [0, 500], [250, 500], [0, 500], [0, 10])	(10, 102, 290, 96, 2)	(126, 105, 55, 107, 107)
Log-normal	([0, 10], [200, 500], [200, 500], [0, 50], [0, 10])	(5, 269, 205, 19, 2)	(118, 72, 66, 125, 119)
Exponential	([250, 500], [0, 200], [0, 100], [0, 20], [0, 10])	(317, 130, 38, 10, 5)	(37, 102, 117, 113, 131)

$$error = \sqrt{\sum_{i=1}^{c} (s_i - t_i)^2} \tag{9}$$

Table 2 lists the average error of each algorithm. For three distributions (i.e., uniform distribution, normal distribution and exponential distribution), the proposed algorithm can obtain higher accuracy than other algorithms when dealing with unreasonable data that is inconsistent with the background knowledge. For the log-normal distribution, the error of our algorithm is almost the same as that of NStoPS-BK. Since three accurate background intervals [0, 10], [0, 50] and [0, 10] are used for the log-normal distribution, based on these three background intervals, the NStoPS-BK can also get good results. Because the results of the three accurate categories are similar, for the two less accurate background intervals [200, 500], the results for both NStoPS-BK and NStoPS-UD are either fixed at 200, or with a small difference. Therefore, for the log-normal distribution, the error of our algorithm is similar to that of NStoPS-BK.

Table 2. Average error of reconstructed positive data by different algorithms

Distribution	Uniform	Normal	Log-normal	Exponential
NStoPS	0.145201	0.144154	0.143554	0.146301
NStoPS-II	0.144537	0.117060	0.098597	0.107892
NStoPS-BK	0.126520	0.094933	**0.060932**	0.092924
NStoPS-UD	**0.124682**	**0.089995**	0.063179	**0.086889**

We tried to change the background knowledge of each distribution to [0, 500] (in this case, we only have the background knowledge that the reconstructed result should not be negative or larger than 500), and the average error of each algorithm is obtained by repeating the above experiment. The results are shown in Table 3. The experimental results show that the performance of NStoPS-BK on dealing with the unreasonable negative values is poor, and it results in a larger average error. However, the algorithm proposed in this paper can achieve good performance on dealing with negative values, and its performance is the same as that of NStoPS-II. The experimental results show that our algorithm can effectively deal with both the negative values and the unreasonable data which are inconsistent with the background knowledge.

Table 3. Average error of reconstructed positive data by different algorithms when handling negative values

Distribution	Uniform	Normal	Log-normal	Exponential
NStoPS	0.144997	0.144553	0.144955	0.146063
NStoPS-II	0.144394	**0.116390**	**0.097277**	**0.110176**
NStoPS-BK	**0.144350**	0.118865	0.109463	0.117183
NStoPS-UD	0.144394	**0.116390**	**0.097277**	**0.110176**

4 Extension to General Negative Surveys

In this part, the extension of our algorithm which can be applied to general negative surveys will be discussed. At present, most of the reconstruction algorithms and the simulation experiments are based on the uniform negative survey, but, in reality, there is a possibility that participants will choose one or more options in a biased way. Our algorithm can be extended to support the general negative survey in which the selection probability is not forced to follow any distribution. First, we also take the category c as an example, in the same way, we need to solve the following equations

$$
\begin{cases}
\sum_{i=1}^{c} x_i \times p_{i,1} = r_1 + \Delta r_1 \\
\sum_{i=1}^{c} x_i \times p_{i,2} = r_2 + \Delta r_2 \\
\quad \vdots \\
\sum_{i=1}^{c} x_i \times p_{i,c} = r_c + \Delta r_c \\
\sum_{i=1}^{c} x_i = 1
\end{cases}
$$

$$
\Delta r_i = \sum_{j=1}^{c-1} \frac{x_j \times p_{j,c}}{r_c + \Delta r_c} \times (-\Delta r_c) \times \frac{p_{j,i}}{1 - p_{j,c}}, \quad i = 1, \ldots, c-1
$$

Note that x_c is fixed to a known value, and $x_1, \ldots, x_{c-1}$ and Δr_c are unknown. Therefore, the above is an equation set which contains c variables and c equations.

Finally, we also need to get the general formula for the situation that the results obtained by NStoPS contain unreasonable values for n ($n \le c$) categories. The analysis is similar to Sect. 2, and finally we have:

$$
\Delta r_i = \sum_{j=1}^{c-n} \frac{x_j \times p_{j,c}}{r_c + \Delta r_c} \times (-\Delta r_c) \times \frac{p_{j,i}}{1 - (p_{j,c} + \ldots + p_{j,c-n+1})} + \ldots
$$
$$
+ \sum_{j=1}^{c-n} \frac{x_j \times p_{j,c-n+1}}{r_{c-n+1} + \Delta r_{c-n+1}} \times (-\Delta r_{c-n+1}) \times \frac{p_{j,i}}{1 - (p_{j,c} + \ldots + p_{j,c-n+1})}
$$

In specific applications, matrix P and c are known values, the equation sets will be easy to solve. For example, when the reconstruction result of a category is adjusted and c is equal to 3, it is a quadratic equation with one unknown variable finally. There are many methods for solving this kind of equation sets, such as Newton iteration method, Evolutionary Algorithm, and neural network algorithm. Therefore, the algorithm in this paper can also be used when the participants do not follow a uniform selection probability.

5   Conclusions and Future Work

In this paper, we proposed a new algorithm to handle the unreasonable data and two experiments are carried out. The experimental results demonstrated that this method could produce more reasonable aggregated results. Finally, we extend the algorithm to general negative survey, and give a general expression for the method.

In future work, we will choose an efficient method to obtain the result expression of the general equation sets. Moreover, in this paper, we only study the negative survey in which each participant should select only one negative category, and we will investigate the way to handling unreasonable values in multiple-selection negative surveys [15] in future work.

Acknowledgment. This work was partially supported by the National Natural Science Foundation of China (Grant No. 61672398), the Hubei Provincial Natural Science Foundation of China (Grant No. 2017CFA012), the Key Technical Innovation Project of Hubei (Grant No. 2017AAA122), the Applied Fundamental Research of Wuhan (Grant No. 20160101010004), and the Open Fund of Hubei Key Lab. of Transportation of IoT (Grant No. 2017III028-004).

References

1. Sun, X., Wang, H., Li, J., et al.: Publishing anonymous survey rating data. Data Min. Knowl. Discov. **23**(3), 379–406 (2011)
2. Esponda, F., Ackley, E.S., Helman, P., Jia, H., Forrest, S.: Protecting data privacy through hard-to-reverse negative databases. Int. J. Inf. Secur. **6**, 403–415 (2007)
3. Esponda, F.: Everything that is not important: negative databases. IEEE Comput. Intell. Mag. **3**, 60–63 (2008)
4. Liu, R., Luo, W., Yue, L.: Classifying and clustering in negative databases. Front. Comput. Sci. **7**(6), 864–874 (2013)
5. Esponda, F.: Negative representations of information. Ph.D. thesis, University of New Mexico (2005)
6. Esponda, F.: Negative surveys (2006). arXiv:math/0608176
7. Esponda, F., Guerrero, V.M.: Surveys with negative questions for sensitive items. Stat. Probab. Lett. **79**, 2456–2461 (2009)
8. Horey, J., Groat, M., Forrest, S., Esponda, F.: Anonymous data collection in sensor networks. In: The Fourth Annual International Conference on Mobile and Ubiquitous Systems: Computing, Networking and Services, Philadelphia, USA, pp. 1–8 (2007)

9. Horey, J., Forrest, S., Groat, M.M.: Reconstructing spatial distributions from anonymized locations. In: The 2012 IEEE 28th International Conference on Data Engineering Workshops (ICDEW), Arlington, VA, pp. 243–250 (2012)

10. Luo, W., Lu, Y., Zhao, D., et al.: On location and trace privacy of the moving object using the negative survey. IEEE Trans. Emerg. Top. Comput. Intell. **PP**(99), 1 (2017)

11. Luo, W., Jiang, H., Zhao, D.: Rating credits of online merchants using negative ranks. IEEE Trans. Emerg. Top. Comput. Intell. **1**(5), 354–365 (2017)

12. Bao, Y., Luo, W., Zhang, X.: Estimating positive surveys from negative surveys. Stat. Probab. Lett. **83**, 551–558 (2013)

13. Lu, Y., Luo, W., Zhao, D.: Fast searching optimal negative surveys. In: ICINS 2014 - 2014 International Conference on Information and Network Security, p. 27 (2014)

14. Zhao, D., Luo, W., Yue, L.: Reconstructing positive surveys from negative surveys with background knowledge. In: Tan, Y., Shi, Y. (eds.) Data Mining and Big Data. DMBD (2016). LNCS, vol. 9714, pp. 86–99. Springer, Cham. https://doi.org/10.1007/978-3-319-40973-3_9

15. Esponda, F., Huerta, K., Guerrero, V.M.: A statistical approach to provide individualized privacy for surveys. PLoS ONE **11**(1), 1–14 (2016)

Learning Models

Multi-view Proximity Learning
for Clustering

Kun-Yu Lin, Ling Huang, Chang-Dong Wang$^{(\boxtimes)}$, and Hong-Yang Chao

School of Data and Computer Science, Sun Yat-sen University, Guangzhou, China
kunyulin14@outlook.com, huanglinghl@hotmail.com, changdongwang@hotmail.com,
isschhy@mail.sysu.edu.cn

Abstract. In recent years, multi-view clustering has become a hot research topic due to the increasing amount of multi-view data. Among existing multi-view clustering methods, proximity-based method is a typical class and achieves much success. Usually, these methods need proximity matrices as inputs, which can be constructed by some nearest-neighbors-based approaches. However, in this way, neither the intra-view cluster structure nor the inter-view correlation is considered in constructing proximity matrices. To address this issue, we propose a novel method, named *multi-view proximity learning*. By introducing the idea of representative, our model can consider both the relations between data objects and the cluster structure within individual views. Besides, the spectral-embedding-based scheme is adopted for modeling the correlations across different views, i.e. the view consistency and complement properties. Extensive experiments on both synthetic and real-world datasets demonstrate the effectiveness of our method.

Keywords: Multi-view clustering · Proximity learning
Representative · Spectral embedding

1 Introduction

Recently, multi-view data, whose data features are collected from multiple heterogenous but related views, have arisen in a number of fields [1–8], such as pattern recognition, data mining, natural language processing, etc. For instance, a web page can be described in two views, one contains the words occurring in the page and the other contains the words occurring in the hyperlinks pointing to that page [4]. Another example is the multilingual document, which is available in several languages such that each language is taken as a separate view [5]. In these fields, data clustering is a basic but widely used technique [9]. Considering clustering the multi-view data, it is difficult to produce good results by using only one view of feature, since usually each view only provides partial information [10]. Therefore, it is necessary to properly combine information from all views together to improve the clustering performance. This leads to the emergence of multi-view clustering.

© Springer International Publishing AG, part of Springer Nature 2018
J. Pei et al. (Eds.): DASFAA 2018, LNCS 10828, pp. 407–423, 2018.
https://doi.org/10.1007/978-3-319-91458-9_25

Proximity-based method is a kind of typical method for multi-view clustering. These methods integrate the information from different views by making use of the predefined proximity matrices together. One of the most straightforward scheme for view integration is weighted combination, which combines the proximity matrices from all views together by weighted addition via an adaptive weighting parameter for each view [11,12]. Besides, some other useful methods are developed. In [13,14], co-training based approaches are adopted to share information among views, which improves the proximity matrices to fit multi-view data. Co-regularized approaches are also effective approaches to view integration [15,16]. Wang et al. propose a method which considers the neighborhood consistency of different views [17], while Xia et al. consider the low-rank and sparse properties of proximity matrices [18].

Despite the success of the aforementioned proximity-based methods, they suffer from some common problems. First, proximity matrices are needed as inputs for these methods, while usually data features are given rather than proximity matrices. In this case, some nearest-neighbors-based methods are applied on data features to construct proximity matrices, such as k-nearest neighbors, Gaussian proximity [19] and self-tuned Gaussian [20]. However, these proximity construction methods do not consider the underlying cluster structures, such that the constructed proximities may not exhibit good properties for clustering. Moreover, these methods only consider separately the information in individual views, leading to the loss of the inter-view correlations.

In order to address these problems, we propose a new *multi-view proximity learning* (MVPL) method for multi-view clustering. In the multi-view proximity learning, both the relations between data objects in individual views and the correlations across different views are considered. For the intra-view relations, a novel idea of data representative is adopted, such that the cluster structure is also taken into account during the learning process. Besides, spectral-embedding-based scheme is designed for modeling the inter-view correlations, such that both the view consistency and complement properties can be utilized for improving the clustering performance. Accordingly, an objective function is designed and an alternative iteration scheme is proposed to optimize the objective. Extensive experiments conducted on both synthetic and real-world datasets demonstrate the effectiveness of the proposed model.

2   The Proposed Model

In order to address the proximity learning problem for multi-view data, our model should consider two parts. One is the intra-view learning quality, which means that the learning process should consider the relations between data objects within each view. Inspired by [21], the proposed model discovers these relations based on the idea of representative. It can transform the original view feature into a more suitable representation for proximity learning, by which the cluster structures are also considered. In particular, in each view, each feature vector has a dedicated representative that is very similar to itself, and representatives of data objects with higher proximity should be similar to each other.

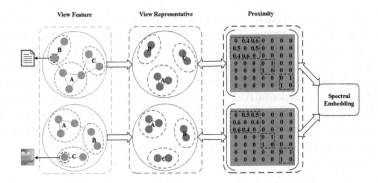

Fig. 1. Illustration of the main idea of our method. In this simple example, the dataset contains a document view and an image view.

The other part is to model the correlations across different views, such that both the view consistency and complement properties can be utilized for improving clustering performance. The view consistency property implies that the proximities learnt from different views will reach a certain degree of consistency, while the view complement property implies that one view will provide complementary information for the other views. Accordingly, a well-designed inter-view criterion function is proposed based on spectral embedding. For clarity, Fig. 1 illustrates the main idea of our method by a two-view example. From the figure, we find that the data representatives are determined by both view features and learnt proximities. Similarly, the learnt proximities are derived from intra-view data representatives and further mutually affected in a inter-view manner by spectral embedding. In what follows, we will introduce the model in detail.

2.1   The Objective Function

Given a dataset containing n objects whose features are collected from m views, the features in the v-th view are represented by matrix $X^v = [\mathbf{x}_1^v, \mathbf{x}_2^v, \ldots, \mathbf{x}_n^v] \in \mathbb{R}^{d^v \times n}$, where d^v is the dimensionality of the v-th view and $\mathbf{x}_i^v$ is the feature vector for the i-th object in the v-th view. The goal of the multi-view proximity learning is to learn a proximity set $\{S^1, S^2, \ldots, S^m\}$, where $S^v = [s_{ij}^v]_{n \times n}$ is the proximity matrix for the v-th view with s_{ij}^v representing the proximity between the i-th and j-th objects in the v-th view. According to the discussion above, the learning process should consider both intra-view and inter-view criteria.

Intra-view Criterion. In order to discover the relations between data objects in individual views, we introduce the idea of data representatives, which are better representations with clearer cluster structures for data objects. Intuitively in this process, original data point is moved to a better position for clustering according to its relations with other data points. We use $U^v = [\mathbf{u}_1^v, \mathbf{u}_2^v, \ldots, \mathbf{u}_n^v] \in \mathbb{R}^{d^v \times n}$ to denote the representative matrix where $\mathbf{u}_i^v$ is the representative for feature vector $\mathbf{x}_i^v$ in the v-th view. Treating the original feature as important basis for learning representative, $\mathbf{u}_i^v$ should not be far from $\mathbf{x}_i^v$, otherwise the topological structure will be destroyed. Besides, the learning of representative should also consider the proximities between objects. If two data objects have higher proximity in one view then their representatives should be relatively closer. Similarly, the proximity learning should consider the relations between data representatives. If two data representatives $\mathbf{u}_i^v$ and $\mathbf{u}_j^v$ are close in the v-th view, then s_{ij}^v should be relatively large. In other words, the learning processes of representatives and proximities are mutually affected by each other. According to the above discussion, the intra-view criterion is defined as follows,

$$\Phi^v(U^v, S^v) = \frac{1}{n} \sum_{i=1}^{n} \|\mathbf{x}_i^v - \mathbf{u}_i^v\|_2^2 + \frac{\alpha}{n^2} \left(\sum_{i=1}^{n} \sum_{j=1}^{n} s_{ij}^v \|\mathbf{u}_i^v - \mathbf{u}_j^v\|_2^2 + \beta \|S^v\|_F^2 \right)$$

$$\text{s.t.} \sum_{j=1}^{n} s_{ij}^v = 1, s_{ij}^v \geq 0, \forall i, j \tag{1}$$

where $\| \cdot \|_2$ is the L_2 norm of vector, $\| \cdot \|_F^2$ is the Frobenius norm of matrix and $\alpha, \beta > 0$ are trade-off parameters. In our paper, the probabilistic proximities are used. Therefore constraint $\sum_{j=1}^{n} s_{ij}^v = 1$ and $s_{ij}^v \geq 0$ should be introduced. The term $\beta \|S^v\|_F^2$ is adopted for controlling the sparsity of learnt proximity. If β is large, the learnt proximity matrix will be relatively dense, while a smaller β will make the matrix sparser.

Inter-view Criterion. The inter-view criterion considers both the view consistency and view complement properties. We design such criterion by introducing the concept of spectral embedding. Spectral embedding is a low-dimensional representation of data object, which is obtained through spectral decomposition on specific matrix. In our model, spectral embedding is the representation integrating information from all views. By denoting the embedding matrix as $F = [\mathbf{f}_1, \mathbf{f}_2, \ldots, \mathbf{f}_n] \in \mathbb{R}^{c \times n}$ with $\mathbf{f}_i$ being the c-dimensional spectral embedding of the i-th data object, the relation between F and the learnt proximity S^v can be modeled by

$$\frac{1}{2n^2} \sum_{i=1}^{n} \sum_{j=1}^{n} s_{ij}^v \|\mathbf{f}_i - \mathbf{f}_j\|_2^2 \quad \text{s.t.} \ FF^T = I \tag{2}$$

where I is the identity matrix. Here $FF^T = I$ is a widely used constraint for weakening the relations between the features of embedding, which makes F a better representation [19]. If the distance between $\mathbf{f}_i$ and $\mathbf{f}_j$ is small, it implies that i-th and j-th data objects may have higher proximity in all views. If the value of (2) is smaller, the learnt proximity of the v-th view is more consistent with the spectral embedding F. Since the spectral embedding F carries information from all views, the high consistency between F and S^v implies that information of other views is transferred to the v-th view, which reflects the view complement property. Moreover, proximities from different views can reach a certain degree of consistency through F. Here F is regarded as a medium for inter-view interactions, which reflects the view consistency property. Considering all views together, we get the inter-view criterion as follows,

$$\Psi(\{S^v\}, F) = \frac{1}{2n^2} \sum_{v=1}^{m} \sum_{i=1}^{n} \sum_{j=1}^{n} s_{ij}^v \|\mathbf{f}_i - \mathbf{f}_j\|_2^2 \quad \text{s.t. } FF^T = I \tag{3}$$

which models the inter-view correlations through the spectral embedding.

The Overall Objective Function According to the discussion above, we can use $\Phi^v(U^v, S^v)$ to measure the intra-view learning quality and $\Psi(\{S^v\}, F)$ to measure the inter-view consistency and complement properties. By integrating them together, we can get the overall objective function as follows,

$$\min_{\{U^v\},\{S^v\},F} \sum_{v=1}^{m} \Phi^v(U^v, S^v) + \gamma \Psi(\{S^v\}, F)$$

$$\text{s.t. } \sum_{j=1}^{n} s_{ij}^v = 1, s_{ij}^v \geq 0, \forall i, j, v, FF^T = I \tag{4}$$

where $\gamma > 0$ is the trade-off parameter balancing the intra-view criterion and the inter-view criterion. By minimizing the objective function (4), both the learning quality of proximities in all views and the interactions between different views are considered, such that suitable proximities for multi-view data can be obtained. Following the convention of spectral clustering, the dimensionality of spectral embedding can be set as the predefined number of clusters [19].

2.2 Determination of Parameter β

In the proposed model, three parameters are needed as inputs for proximity learning. Parameter α is adopted to control the distances between data features and data representatives, while parameter γ is adopted for controlling the view consistency. Both parameters should be determined according to the properties of datasets. In comparison, β is adopted for controlling the sparsity of learnt proximities, which has less variability. Therefore, it is necessary to design a method for determining its value more easily.

Inspired by [22], we propose a method based on k-nearest neighbors to determine β. It also induces a method for constructing single-view proximity, which will be used in our experiments. Considering data feature in certain view, whose data matrix is $X = [\mathbf{x}_1, \ldots, \mathbf{x}_n] \in \mathbb{R}^{d \times n}$ (here we ignore the superscript specifying view index for simplicity), we can learn the proximity vector $\mathbf{w}_i = [w_{i1}, w_{i2}, \ldots, w_{in}]^T$ associated with $\mathbf{x}_i$ by solving the following model

$$\min_{\mathbf{w}_i} \frac{1}{2} \left\| \mathbf{w}_i + \frac{\mathbf{d}_i^{\mathbf{x}}}{2\beta_i} \right\|_2^2 \quad \text{s.t. } \mathbf{w}_i^T \mathbf{1} = 1, \mathbf{w}_i \geq 0, \tag{5}$$

where $\beta_i > 0$ is the sparsity parameter, $\mathbf{1}$ is the all-one vector and $\mathbf{w}_i \geq 0$ means all elements of vector $\mathbf{w}_i$ are not less than 0. We assume the original distance vector as $\hat{\mathbf{d}}_i^{\mathbf{x}} = [\hat{d}_{i1}^{\mathbf{x}}, \hat{d}_{i2}^{\mathbf{x}}, \ldots, \hat{d}_{in}^{\mathbf{x}}]^T$, where $\hat{d}_{ii}^{\mathbf{x}}$ is set as a very large number (i.e. ignoring $\mathbf{x}_i$ itself) and $\forall j \neq i, \hat{d}_{ij}^{\mathbf{x}} = \|\mathbf{x}_i - \mathbf{x}_j\|_2^2$. The distance vector $\mathbf{d}_i^{\mathbf{x}}$ in (5) is defined by $\mathbf{d}_i^{\mathbf{x}} = [d_{i1}^{\mathbf{x}}, d_{i2}^{\mathbf{x}}, \ldots, d_{in}^{\mathbf{x}}]^T$, which is the sorted vector of $\hat{\mathbf{d}}_i^{\mathbf{x}}$ such that $d_{i1}^{\mathbf{x}} \leq d_{i2}^{\mathbf{x}} \leq \cdots \leq d_{in}^{\mathbf{x}}$. In the model, the parameter β_i determines the number of nonzero elements in the proximity information vector $\mathbf{w}_i$. If $\beta_i = 0$, there will be only one nonzero element in the vector, corresponding to the nearest neighbor of object $\mathbf{x}_i$. If $\beta_i \to \infty$, all elements will be nonzero except the one corresponding to $\mathbf{x}_i$. Aiming to solve problem (5), we write down its Lagrangian function as

$$\mathcal{L}(\mathbf{w}_i, \eta, \mu_i) = \frac{1}{2}\left\|\mathbf{w}_i + \frac{\mathbf{d}_i^{\mathbf{x}}}{2\beta_i}\right\|_2^2 - \eta\left({\mathbf{w}_i}^T \mathbf{1} - 1\right) - \mu_i^T \mathbf{w}_i \qquad (6)$$

where η and $\mu_i \geq 0$ are Lagrangian multipliers. According to the KKT condition, the optimal solution of $\mathbf{w}_i$ is given by

$$w_{ij} = \max\left(-\frac{d_{ij}^{\mathbf{x}}}{2\beta_i} + \eta, 0\right). \qquad (7)$$

If there are exactly k nonzero elements in the vector $\mathbf{w}_i$, we get the value of Lagrangian multiplier $\eta = \frac{1}{k} + \frac{1}{2k\beta_i}\sum_{j=1}^{k} d_{ij}^{\mathbf{x}}$ [22]. These k nonzero elements of $\mathbf{w}_i$ correspond to the k-nearest neighbors of $\mathbf{x}_i$ and the elements of $\mathbf{w}_i$ satisfy $\forall 1 \leq j \leq k, w_{ij} > 0$ and $\forall j \geq k + 1, w_{ij} = 0$. According to the constraint $\mathbf{w}_i^T \mathbf{1} = 1$, the sparsity parameter β_i can be set as

$$\beta_i = \frac{k}{2}d_{i,k+1}^{\mathbf{x}} - \frac{1}{2}\sum_{j=1}^{k} d_{ij}^{\mathbf{x}}, \qquad (8)$$

such that the resulting $\mathbf{w}_i$ will have exactly k nonzero elements. Considering all data objects, the sparsity parameter β can be set as the average of β_i, which is given by

$$\beta = \frac{1}{n}\sum_{i=1}^{n}\left(\frac{k}{2}d_{i,k+1}^{\mathbf{x}} - \frac{1}{2}\sum_{j=1}^{k} d_{ij}^{\mathbf{x}}\right). \qquad (9)$$

Using the method above, we can determine the sparsity parameter according to the number of neighbors k, which is much easier to tune. Furthermore, the single-view weighted k-nearest neighbors proximity can be constructed after k is determined. For multi-view data, since different views may have different distance distributions, it is more reasonable to use different sparsity parameters for different views. Therefore, the modified intra-view criterion function is given by

$$\widetilde{\Phi}^v(U^v, S^v) = \frac{1}{n}\sum_{i=1}^{n}\|\mathbf{x}_i^v - \mathbf{u}_i^v\|_2^2 + \frac{\alpha}{n^2}\left(\sum_{i=1}^{n}\sum_{j=1}^{n} s_{ij}^v\|\mathbf{u}_i^v - \mathbf{u}_j^v\|_2^2 + \beta^v\|S^v\|_F^2\right) \qquad (10)$$

where $\beta^v > 0$ is the sparsity parameter for the v-th view determined by the aforementioned method via the number of neighbors k. Finally, our objective is given by

$$\min_{U^v, S^v, F} \sum_{v=1}^{m} \widetilde{\Phi}^v (U^v, S^v) + \gamma \Psi(\{S^v\}, F)$$

$$\text{s.t. } \sum_{j=1}^{n} s_{ij}^v = 1, s_{ij}^v \geq 0, \forall i, j, v, FF^T = I. \tag{11}$$

Although more sparsity parameters are introduced to control the model in (11) compared with (4), they can be determined via the same number of nearest neighbors k.

2.3 Optimization

In this subsection, the alternative iteration scheme is used to solve problem (11).

Update U^v. When S^v and F are fixed, the subproblem with respect to U^v is given by

$$\min_{U^v} \frac{1}{n} \sum_{i=1}^{n} \|\mathbf{x}_i^v - \mathbf{u}_i^v\|_2^2 + \frac{\alpha}{n^2} \sum_{i=1}^{n} \sum_{j=1}^{n} s_{ij}^v \|\mathbf{u}_i^v - \mathbf{u}_j^v\|_2^2. \tag{12}$$

In order to rewrite the subproblem into matrix form, we introduce the property [19] as

$$\text{Tr}\left(H L_G H^T\right) = \frac{1}{2} \sum_{i=1}^{n} \sum_{j=1}^{n} g_{ij} \|\mathbf{h}_i - \mathbf{h}_j\|_2^2 \tag{13}$$

where $\text{Tr}(\cdot)$ is the trace operator for matrix, $G = \{g_{ij}\} \in \mathbb{R}^{a \times a}$ and $H = [\mathbf{h}_1, \mathbf{h}_2, \dots, \mathbf{h}_a] \in \mathbb{R}^{b \times a}$. L_G is the unnormalized Laplacian matrix of G defined by $L_G = D_G - G$, where D_G is the degree matrix of G. Using the property (13), the subproblem can be transformed as

$$\min_{U^v} \|X^v - U^v\|_F^2 + \frac{2\alpha}{n} \text{tr}\left(U^v L_S^v U^{vT}\right) \tag{14}$$

where L_S^v is the unnormalized Laplacian matrix of $(S^v + S^{vT})/2$. Setting the derivative with respect to U^v to zero, we find that U^v satisfies the equation as follows

$$U^v \left(I + \frac{2\alpha}{n} L_S^v\right) = X^v, \tag{15}$$

which can be solved by matrix inversion. Besides, the problem is essentially a least-square problem, which can also be solved in many efficient ways.

Update S^v. When U^v and F are fixed, the subproblem with respect to S^v is given by

$$\min_{S^v} \sum_{i=1}^{n}\sum_{j=1}^{n} s_{ij}^v \|\mathbf{u}_i^v - \mathbf{u}_j^v\|_2^2 + \beta^v \sum_{i=1}^{n}\sum_{j=1}^{n} s_{ij}^{v\,2} + \frac{\gamma}{2\alpha} \sum_{i=1}^{n}\sum_{j=1}^{n} s_{ij}^v \|\mathbf{f}_i - \mathbf{f}_j\|_2^2$$

$$\text{s.t. } \sum_{j=1}^{n} s_{ij}^v = 1, s_{ij}^v \geq 0, \forall i, j, v. \tag{16}$$

By denoting $d_{ij}^v = \|\mathbf{u}_i^v - \mathbf{u}_j^v\|_2^2 + \frac{\gamma}{2\alpha}\|\mathbf{f}_i - \mathbf{f}_j\|_2^2$, $\mathbf{d}_i^v = [d_{i1}^v, d_{i2}^v, \ldots, d_{in}^v]^T$ and $\mathbf{s}_i^v = [s_{i1}^v, s_{i2}^v, \ldots, s_{in}^v]^T$, we translate the problem into vector form as follows

$$\min_{\mathbf{s}_i^v} \left\| \mathbf{s}_i^v + \frac{\mathbf{d}_i^v}{2\beta^v} \right\|_2^2$$

$$\text{s.t. } \mathbf{s}_i^{v^T}\mathbf{1} = 1, \mathbf{s}_i^v \geq 0, \tag{17}$$

which is equivalent to computing the Euclidean projection of point $-\mathbf{d}_i^v/(2\beta^v)$ onto the probability simplex. The problem has a unique solution, which can be solved by using the method proposed in [23].

Update F. When U^v and S^v are fixed, the subproblem with respect to F is to solve a trace minimization problem as

$$\min_{FF^T=I} \text{Tr}(FL_SF^T) \tag{18}$$

where $L_S = \sum_{v=1}^{m} L_S^v$. The optimal F is a matrix formed by the c eigenvectors of L_S corresponding to the c smallest eigenvalues.

By alternatively update U^v, S^v and F, the objective value will decrease and finally converge as the iteration goes, from which the solution of problem (11) can be obtained. The optimization algorithm is summarized in Algorithm 1. After learning the proximity matrices, the spectral clustering is applied on the proximity matrices to obtain the clustering results.

3   Experiment

In this section, extensive experiments are conducted to demonstrate the effectiveness of the proposed method on one synthetic dataset and four real-world datasets. On the synthetic dataset, we will show how the proposed method works. While on the real-world datasets, parameter analysis, convergence analysis and comparison experiments will be conducted. The code of our method and the testing datasets are available on dropbox[1].

---

[1] The code is available on https://www.dropbox.com/s/tj5zc7yry0ing3l/MVPL_PCode.zip?dl=0 and the password for decompression is "DASFAA2018".

Algorithm 1. Multi-view proximity learning

Input: Data matrices of m views $\{X^1, X^2 \ldots, X^m\}$, parameters α, γ and k, number
 of clusters c.
1: Initialize representative matrix U^v as X^v.
2: Initialize S^v and determine β^v by the strategy in Section 2.2.
3: Initialize F by solving Eq. (18).
4: **repeat**
5:      Update $U^v, \forall v$ by solving Eq. (15).
6:      Update $S^v, \forall v$ by solving Eq. (17).
7:      Update F by solving Eq. (18).
8: **until** Convergence or reaching the maximum number of iterations.
Output: Proximity matrices $\{S^1, S^2, \ldots, S^m\}$.

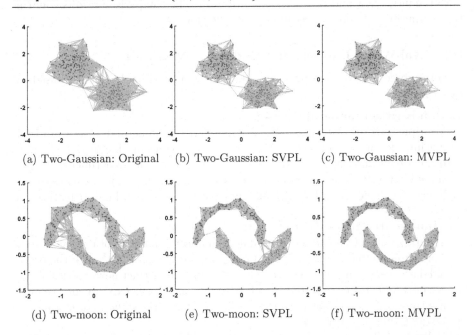

(a) Two-Gaussian: Original   (b) Two-Gaussian: SVPL   (c) Two-Gaussian: MVPL

(d) Two-moon: Original   (e) Two-moon: SVPL   (f) Two-moon: MVPL

Fig. 2. Synthetic experiment. In the figures, points in the first class are in blue while
those in the second class are in red. Green lines are edges representing the proximities
between data objects, i.e., if the proximity between two data objects in certain view is
larger than zero then there is an edge between them. (Color figure online)

3.1   Synthetic Experiment

A synthetic dataset consisting of two views, namely Two-Gaussian and Two-
moon, is used for demonstrating how the proposed method works. Figure 2(a)
and (d) plot the original data points in both views with edges representing
the initial proximities learnt by the method introduced in Sect. 2.2. In order to
show the significance of considering inter-view criterion, a variant of our method,
called SVPL, is introduced by setting $\gamma = 0$. It is a single-view proximity learning

method which considers only the intra-view criterion. Figure 2(b) and (e) show the results of SVPL, where points denote the learnt representatives. From the subfigures, we find that SVPL transforms the original data view into a more suitable state for clustering based on representatives. It is essentially equivalent to making the data points move in such a manner that the intra-class connections are stronger and the inter-class connections are weaker. However, the learnt proximity is not good enough since there are still edges between the two clusters. Therefore, we need to consider the inter-view information. Figure 2(c) and (f) show the results of MVPL, where points denote representatives learnt by MVPL. From these two subfigures, we find that there are no edges between clusters in both views. This implies that much better proximities are learnt by considering both the intra-view and the inter-view criterion. The comparison results confirm the significance of considering inter-view criterion.

3.2   Real-World Datasets and Evaluation Measures

In this subsection, we will first introduce the four real-world datasets used in experiments.

1. **Handwritten numeral dataset**

   Multiple features (Mfeat) dataset is a handwritten numeral dataset from UCI machine learning repository [24]. The dataset contains handwritten digits from 0 to 9 and each category has 200 objects. In our experiment, we use three kinds of feature to represent images, namely 216 profile correlations, 76 Fourier coefficients and 47 Zernike moments, where each kind of features is regarded as a view.

2. **Multi-source news dataset**

   3Sources dataset[2] is a multi-source news dataset consisting of news collected from three sources, namely BBC, Guardian and Reuters. Although the original dataset contains 984 news articles covering 416 distinct news stories, there are only 169 stories reported by all three medias. In our experiment, we only use these 169 news objects so that each object has three views of features.

3. **Object image datasets**

   Caltech101 [25] is an image dataset consisting of 101 categories of images for object recognition problem. Following the previous work [26], two subsets are selected to generate two datasets for experimental purpose. The first subset is called Caltech101-7, containing 1474 images from 7 widely used categories. The second one is a larger subset called Caltech101-20, which contains 2386 images of 20 categories. Three kinds of features are extracted from the images to generate three views, namely 1984-dimensional HOG feature, 512-dimensional GIST feature and 928-dimensional LBP feature.

   The statistic of the four real-world datasets is shown in Table 1.

   In order to evaluate the clustering performance of the proposed method and the compared methods, three widely used measures are adopted in our experiments, namely accuracy (ACC), normalized mutual information (NMI) and

---

[2] http://mlg.ucd.ie/datasets/3sources.html.

Table 1. Statistic of the four real-world datasets.

	Mfeat	3Sources	Caltech101-7	Caltech101-20
View1	fac(216)	BBC(3560)	hog(1984)	hog(1984)
View2	fou(76)	Guardian(3631)	gist(512)	gist(512)
View3	zer(47)	Reuters(3068)	lbp(928)	lbp(928)
# of objects	2000	169	1474	2386
# of classes	10	6	7	20

purity (PUR). For each measure, higher value indicates better performance [17]. In comparison experiments, following [21], the average rank of the performance obtained by each method is also reported across all datasets.

3.3 Parameter Analysis

In this subsection, parameter analysis is conducted to show the effect of the three parameters α, γ and k. The first parameter to be analyzed is k, which determines the value of β^v. By fixing $\alpha = 1$ and $\gamma = 0.001$, we tune the value of k in range $[5, 70]$ with step 5. The performance in terms of all three measures on the four datasets are reported in Fig. 3. From the figure, we find that the method performs not so well when k is too small due to the failure of preserving the neighborhood structures. As the value of k increases, the performance will gradually increase. After reaching the highest point (often around $k = 30$), the value of curve will gradually decrease. Although the method may perform not so well with relatively larger k, it produces acceptable results. The main reason is that by introducing the idea of representative, which transforms the original data into a more suitable state for proximity learning, the negative impact caused by the noisy neighbors will be alleviated.

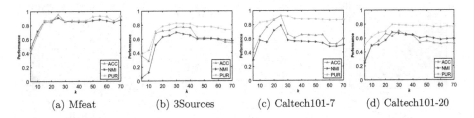

(a) Mfeat   (b) 3Sources   (c) Caltech101-7   (d) Caltech101-20

Fig. 3. Parameter analysis on number of neighbors k.

Next we analyze the effect of α and γ by setting $k = 30$. According to the properties of datasets, different ranges of γ are used for different datasets while the same range of α is used for all datasets. The experimental results are shown in Figs. 4, 5, 6 and 7 respectively. From the figures, we find our method has

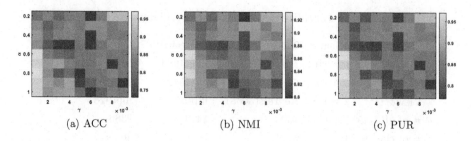

(a) ACC   (b) NMI   (c) PUR

Fig. 4. Parameter analysis on α and γ on Mfeat.

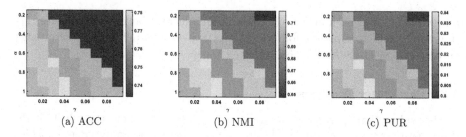

(a) ACC   (b) NMI   (c) PUR

Fig. 5. Parameter analysis on α and γ on 3Sources.

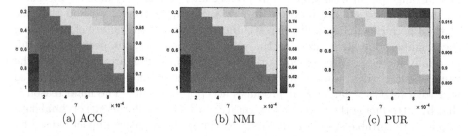

(a) ACC   (b) NMI   (c) PUR

Fig. 6. Parameter analysis on α and γ on Caltech101-7.

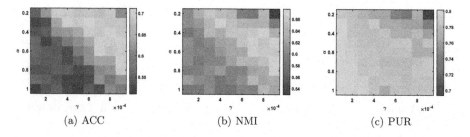

(a) ACC   (b) NMI   (c) PUR

Fig. 7. Parameter analysis on α and γ on Caltech101-20.

similar performance with similar γ/α. What is more, the value of α should not be set too large since it may lead to information loss in terms of topological structure. In practice, user can select the value of α in $[0.5, 1]$ and the value of γ from $\{0.01, 0.001, 0.0001\}$ by which satisfactory performance can be obtained.

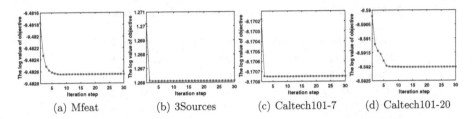

| (a) Mfeat | (b) 3Sources | (c) Caltech101-7 | (d) Caltech101-20 |

Fig. 8. Convergence analysis of optimization.

3.4   Convergence Analysis

In this subsection, convergence analysis is conducted to explore the convergence property of the proposed iterative algorithm by setting $\alpha = 1$, $\gamma = 0.001$ and $k = 30$. Figure 8 plots the log value of objective as a function of iteration step. From the subfigures, we find that the log values of objective decrease rapidly during the iterations on all four datasets. Usually, the algorithm will converge within 30 times of iteration.

3.5   Comparison Experiment

In this subsection, we compare the proposed MVPL method with several state-of-the-art algorithms. Two types of clustering methods are used for comparison, namely the traditional single-view clustering methods and the multi-view clustering methods. For the single-view methods, three representative algorithms are selected, namely k-means (KM) [27], normalized cut (NCut) [19] and robust continuous clustering (RCC) [21]. These single-view methods operate on each individual views from which the best results are reported. For multi-view clustering methods, five state-of-the-art algorithms are used, namely multi-view k-means (MVKM) [28], multi-view spectral clustering (MVSC) [12], co-training multi-view clustering (CoTrn) [13], co-regularized multi-view clustering (CoReg) [15] and multi-view learning with adaptive neighbors (MLAN) [29]. Following [13], for the methods that generate multiple view-specific clustering results (i.e. CoTrn, CoReg and MVPL), prior knowledge is used to select the most informative view for comparison purpose. For all the spectral-clustering-like compared methods, we use the method mentioned in Sect. 2.2 to construct the proximity matrices, which is shown to be a good method for proximity construction [30]. And the sparsity of the proximity matrices is determined by the number of neighbors k. We will tune k in the range of $[10, 50]$ to select the best proximity according to the three measures for all the methods. For all spectral-clustering-like methods and k-means-like methods, we set the number of clusters c as the ground-truth number. Besides, for all the methods involving k-means, we run each algorithm 50 times in the same parameter setting and select the results with the smallest objective as the result for this setting. For all the methods, the best parameters are tuned as suggested by the authors.

Table 2. Clustering results in terms of ACC on all datasets.

Method	Mfeat	Caltech101-7	Caltech101-20	3Sources	Rank
KM	0.729	0.463	0.466	0.527	8.5
NCut	0.753	0.646	0.486	0.746	5.0
RCC	0.779	0.761	0.597	0.420	4.6
MVKM	0.738	0.754	0.516	0.775	4.0
MVSC	0.834	0.556	0.445	0.645	6.8
CoTrn	0.833	0.588	0.473	0.734	5.5
CoReg	0.844	0.586	0.497	0.675	4.8
MLAN	0.750	0.707	0.475	0.757	5.0
MVPL	**0.970**	**0.926**	**0.719**	**0.781**	**1.0**

Table 3. Clustering results in terms of NMI on all datasets.

Method	Mfeat	Caltech101-7	Caltech101-20	3Sources	Rank
KM	0.685	0.459	0.582	0.506	7.8
NCut	0.742	0.521	0.564	0.679	6.0
RCC	0.790	0.621	0.588	0.344	5.5
MVKM	0.650	0.616	0.619	0.587	5.3
MVSC	0.819	0.473	0.551	0.619	6.3
CoTrn	0.846	0.555	0.597	0.696	2.8
CoReg	0.830	0.489	0.596	0.690	4.3
MLAN	0.815	0.544	0.464	0.613	6.3
MVPL	**0.932**	**0.789**	**0.677**	**0.720**	**1.0**

Table 4. Clustering results in terms of PUR on all datasets.

Method	Mfeat	Caltech101-7	Caltech101-20	3Sources	Rank
KM	0.729	0.875	0.786	0.757	7.3
NCut	0.774	0.891	0.783	0.834	5.3
RCC	0.836	0.876	**0.866**	0.716	4.8
MVKM	0.738	0.899	0.810	0.781	4.8
MVSC	0.834	0.868	0.764	0.811	6.0
CoTrn	0.857	0.896	0.803	**0.846**	2.3
CoReg	0.844	0.860	0.788	0.811	5
MLAN	0.778	0.857	0.665	0.793	7.8
MVPL	**0.970**	**0.929**	0.803	0.840	**2.0**

The comparison results obtained by all the methods on the four real-world datasets in terms of ACC, NMI and PUR are reported in Tables 2, 3 and 4 respectively. In the tables, the best performance among all the methods is highlighted in bold. From the tables, we find that the proposed MVPL method outperforms all the other methods on ACC and NMI. In particular, our method has achieved on average 14% percent improvement in terms of ACC and 12% improvement in terms of NMI on all the datasets. For PUR, although our method cannot reach the highest PUR on all the datasets, it still ranks the first on average. Overall, the comparison results have demonstrated the effectiveness of the proposed method.

4   Conclusion

In this paper, we propose a novel proximity learning method for multi-view clustering, called multi-view proximity learning. Through the method, proximities between data objects with multiple views of features can be obtained, which are suitable for multi-view clustering. Accordingly, our method adopts two criteria to fulfill the task, namely intra-view criterion and inter-view criterion. For the intra-view part, we not only make use of the relations between data objects but also take cluster structures into account within individual views. For the interview part, we model the correlations between views based on spectral embedding, which utilizes the view consistency and complement properties such that the learning performance is improved. Extensive experiments conducted on both synthetic and real-world datasets demonstrate the effectiveness of our method.

Acknowledgments. This work was supported by NSFC (61502543), Guangdong Natural Science Funds for Distinguished Young Scholar (2016A030306014), and Tip-top Scientific and Technical Innovative Youth Talents of Guangdong special support program (2016TQ03X542).

References

1. Gao, Y., Gu, S., Li, J., Liao, Z.: The multi-view information bottleneck clustering. In: Kotagiri, R., Krishna, P.R., Mohania, M., Nantajeewarawat, E. (eds.) DASFAA 2007. LNCS, vol. 4443, pp. 912–917. Springer, Heidelberg (2007). https://doi.org/10.1007/978-3-540-71703-4_78
2. Müller, E., Assent, I., Sánchez, I.P., Mülle, Y., Böhm, K.: Outlier ranking via subspace analysis in multiple views of the data. In: 12th IEEE International Conference on Data Mining, pp. 529–538. IEEE (2012)
3. Chen, C., He, J., Bliss, N., Tong, H.: Towards optimal connectivity on multi-layered networks. IEEE Trans. Knowl. Data Eng. **29**(10), 2332–2346 (2017)
4. Blum, A., Mitchell, T.: Combining labeled and unlabeled data with co-training. In: Proceedings of the 11th Annual Conference on Computational Learning Theory, pp. 92–100 (1998)
5. Amini, M., Usunier, N., Goutte, C.: Learning from multiple partially observed views-an application to multilingual text categorization. Adv. Neural Inf. Process. Syst. **22**, 28–36 (2009)
6. Xu, Y.M., Wang, C.D., Lai, J.H.: Weighted multi-view clustering with feature selection. Pattern Recogn. **53**, 25–35 (2016)
7. Zhang, G.Y., Wang, C.D., Huang, D., Zheng, W.S.: Multi-view collaborative locally adaptive clustering with Minkowski metric. Expert Syst. Appl. **86**, 307–320 (2017)

8. Huang, L., Chao, H.Y., Wang, C.D.: Multi-view intact space clustering. In: Proceedings of the 4th Asian Conference on Pattern Recognition, pp. 500–505 (2017)
9. Xu, R., Wunsch, D.C.: Survey of clustering algorithms. IEEE Trans. Neural Netw. **16**(3), 645–678 (2005)
10. Xu, C., Tao, D., Xu, C.: A survey on multi-view learning. CoRR abs/1304.5634 (2013)
11. Xia, T., Tao, D., Mei, T., Zhang, Y.: Multiview spectral embedding. IEEE Trans. Syst. Man Cybern. Part B **40**(6), 1438–1446 (2010)
12. Tzortzis, G., Likas, A.: Kernel-based weighted multi-view clustering. In: 12th IEEE International Conference on Data Mining, pp. 675–684 (2012)
13. Kumar, A., Daumé, H.: A co-training approach for multi-view spectral clustering. In: Proceedings of the 28th International Conference on Machine Learning, pp. 393–400 (2011)
14. Son, J.W., Jeon, J., Lee, A., Kim, S.J.: Spectral clustering with brainstorming process for multi-view data. In: Proceedings of the 31st AAAI Conference on Artificial Intelligence, pp. 2548–2554 (2017)
15. Kumar, A., Rai, P., Daumé, H.: Co-regularized multi-view spectral clustering. In: Advances in Neural Information Processing Systems, pp. 1413–1421 (2011)
16. Lu, C., Yan, S., Lin, Z.: Convex sparse spectral clustering: single-view to multi-view. IEEE Trans. Image Process. **25**(6), 2833–2843 (2016)
17. Wang, C.D., Lai, J.H., Yu, P.: Multi-view clustering based on belief propagation. IEEE Trans. Knowl. Data Eng. **28**(4), 1007–1021 (2016)
18. Xia, R., Pan, Y., Du, L., Yin, J.: Robust multi-view spectral clustering via low-rank and sparse decomposition. In: Proceedings of the 28th AAAI Conference on Artificial Intelligence, pp. 2149–2155 (2014)
19. Luxburg, U.V.: A tutorial on spectral clustering. Stat. Comput. **17**(4), 395–416 (2007)
20. Zelnik-Manor, L., Perona, P.: Self-tuning spectral clustering. Adv. Neural Inf. Process. Syst. **17**, 1601–1608 (2005)
21. Shah, S.A., Koltun, V.: Robust continuous clustering. Proc. Nat. Acad. Sci. U.S.A. **114**(37), 9814 (2017)
22. Nie, F., Wang, X., Huang, H.: Clustering and projected clustering with adaptive neighbors. In: Proceedings of the 23rd ACM SIGKDD International Conference on Knowledge Discovery and Data Mining, pp. 977–986 (2014)
23. Wang, W., Carreira-Perpinán, M.A.: Projection onto the probability simplex: an efficient algorithm with a simple proof, and an application. CoRR abs/1309.1541 (2013)
24. Bache, K., Lichman, M.: UCI machine learning repository (2013). http://archive.ics.uci.edu/ml/index.php
25. Li, F.F., Fergus, R., Perona, P.: Learning generative visual models from few training examples: an incremental Bayesian approach tested on 101 object categories. In: IEEE Conference on Computer Vision and Pattern Recognition Workshops, CVPR Workshops 2004, p. 178 (2004)
26. Li, Y., Nie, F., Huang, H., Huang, J.: Large-scale multi-view spectral clustering via bipartite graph. In: Proceedings of the 29th AAAI Conference on Artificial Intelligence, pp. 2750–2756 (2015)
27. MacQueen, J.: Some methods for classification and analysis of multivariate observations. In: Proceedings of the 5th Berkeley Symposium on Mathematical Statistics and Probability, vol. 1, pp. 281–297 (1967)

28. Cai, X., Nie, F., Huang, H.: Multi-view k-means clustering on big data. In: Proceedings of the 23rd International Joint Conference on Artificial Intelligence, pp. 2598–2604 (2013)
29. Nie, F., Cai, G., Li, X.: Multi-view clustering and semi-supervised classification with adaptive neighbours. In: Proceedings of the 31st AAAI Conference on Artificial Intelligence, pp. 2408–2414 (2017)
30. Nie, F., Wang, X., Jordan, M.I., Huang, H.: The constrained Laplacian rank algorithm for graph-based clustering. In: Proceedings of the 30th AAAI Conference on Artificial Intelligence, pp. 1969–1976 (2016)

Extracting Label Importance Information for Multi-label Classification

Dengbao Wang[1], Li Li[1]([✉]), Jingyuan Wang[1], Fei Hu[1], and Xiuzhen Zhang[2]

[1] College of Computer and Information Science, Southwest University,
Chongqing, China
{dbwang,wjykim,jackyetz}@email.swu.edu.cn, lily@swu.edu.cn
[2] School of Computer Science and Information Technology, RMIT University,
Melbourne, Australia
xiuzhen.zhang@rmit.edu.au

Abstract. Existing multi-label learning approaches assume all labels in a dataset are of the same importance. However, the importance of each label is generally different in real world. In this paper, we introduce *multi-label importance* (MLI) which measures label importance from two perspectives: label predictability and label effects. Specifically, label predictability and label effects can be extracted from training data before building models for multi-label learning. After that, the *multi-label importance* information can be used in existing approaches to improve the performance of multi-label learning. To prove this, we propose a classifier chain algorithm based on *multi-label importance* ranking and a improved kNN-based algorithm which takes both feature distance and label distance into consideration. We apply our algorithms on benchmark datasets demonstrating efficient multi-label learning by exploiting *multi-label importance*. It is also worth mentioning that our experiments show the strong positive correlation between label predictability and label effects.

1 Introduction

Traditional single-label classification aims at building a classifier that can tag each instance with a single label. Both binary classification and multi-class classification belong to this learning framework. However, multi-label classification is a more general learning framework. In multi-label learning, each instance in the dataset is associated with a set of labels, and the task of multi-label problem is to output a label set whose size is unknown for each test instances.

Multi-label problems are ubiquitous in the real world, for example, in image categorization, each image can be associated with multiple labels, such as *sea, desert* and *mountain* [1]; in text categorization, each text may belong to a set of topics, such as *economics, poetry* and *health* [2,21]; in bioinformatics, a gene may be related to multiple functions, such as *metabolism* and *protein synthesis* [3].

© Springer International Publishing AG, part of Springer Nature 2018
J. Pei et al. (Eds.): DASFAA 2018, LNCS 10828, pp. 424–439, 2018.
https://doi.org/10.1007/978-3-319-91458-9_26

Existing multi-label classification methods usually assume that all labels in a dataset have the same importance. However, because of the different attributes carried by labels, the importance of each label is generally different in real world. Intuitively, the label importance can be described from two perspectives: label predictability and label effects.

- The predictability of labels in multi-label learning is generally different. Given a multi-label dataset, there may be some labels can get the high accuracy prediction, while some labels cannot. This issue can affect the performances of some approaches (e.g. Classifier Chains) since the propagation of error information.
- Label effects represent the influence power of each label in multi-label data. The presence or absence of different labels have different effects on the overall label structure. We can extract Label effects from training data before building classifiers for multi-label learning.

In this paper, we introduce *multi-label importance* combining label predictability and label effects. Specifically, label predictability can be gained utilizing existing approaches as the base classifiers, and label effects are extracted from label structure of train data. To prove the usefulness of MLI information, we propose a classifier chain algorithm based on multi-label importance ranking and an improved ML-kNN algorithm considering both feature distance and label distance. Our experiments show the significantly improved performance by exploiting *multi-label importance* and the positive correlation between label predictability and label effects.

The paper is organized as follows. In Sect. 2, we review previous work on multi-label learning. In Sect. 3, we introduce *multi-label importance* (MLI). Section 4 present two approaches by exploiting MLI. The experiment results on real-world datasets are given in Sect. 5. Finally, we conclude and propose future work in Sect. 6.

2   Multi-label Learning

We denote $\mathcal{X} = \mathcal{R}^d$ as the d-dimensional feature space an $\mathcal{Y} = \{0,1\}^L$ as the label space with L possible labels, then the goal of multi-label classifier is to learn a function $f: \mathcal{X} \mapsto \mathcal{Y}$. Given a multi-label dataset, we can divide it into feature space X and label space Y. An instance in multi-label problem is associated with a subset of labels $Y_i \subseteq \mathcal{Y}$ (finite set of labels), and a multi-label dataset is composed of m examples $(x_1, Y_1), (x_2, Y_2), \dots, (x_n, Y_n)$ [4,5].

Given a multi-label learning task, it can be transformed into other well-established learning tasks. This way is formally defined as *Problem Transforma-tion*. In this way, we can decompose a multi-label problem into multiple single-label problems, and each single-label problems can be tackled by a binary classifier. Another way to tackle multi-label classification problems is *Algorithm Adaptation* [8]. This category of approaches tackle multi-label learning problem by adapting existing learning approaches to deal with multi-label problem directly.

Problem Transformation is widely used in multi-label learning problems owing to its greater flexibility. The multi-label classification function can be represented in another form $\boldsymbol{f} = \{f_1, f_2, \ldots, f_L\}$ in this way. Binary Relevance (BR) [1] is the most well known first-order approach in multi-label learning. It decomposes a multi-label problem into multiple binary problems, and for each binary classification problem, several existing algorithms such as k Nearest Neighbor (kNN), Support Vector Machine and Logistic Regression can be employed. The second-order approach Calibrated Label Ranking [7] transforms the multi-label problem into the label ranking problems.

The high-order approach Classifier Chains (CC) [4] transform the multi-label classifier into a chain of binary classifier, and exploit label correlations by extend the feature space using the outputs of previous binary classifiers in the chain (see in Fig. 1). It is obvious that the order of the chain itself has an effect on prediction accuracy. Ensembles of Classifier Chains (ECC) solve the issue by using an ensemble framework with a different random chain ordering for each iteration, but this strategy takes a lot of time in prediction phase. After introducing *multi-label importance*, the order of classifier chains can be determined by the *multi-label importance* ranking.

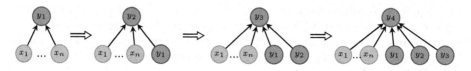

Fig. 1. Classifier Chains (CC) transform the multi-label classifier into a chain of binary classifier.

Algorithm Adaptation method adapt popular learning approaches such as AdaBoost, kNN or Neural Networks to deal with the multi-label problems directly. Adaboost.MR [2] is a improved boosting algorithm proposed to solve multi-label text and speech categorization task. MP-MLL [10] is proposed to tackle the multi-label problems on functional genomics and text categorization by employing a novel error function to capture the characteristics of multi-label learning, i.e. the labels belonging to one instance should be ranked higher than those not belonging to this instance.

ML-kNN [9] is the first lazy learning approach for multi-label learning, which is derived from the traditional k-nearest neighbor (kNN) algorithm. It is modified based on the classic kNN algorithm by utilizing the maximum a posteriori rule. The application of maximum a posteriori principle is the kernel part of ML-kNN:

$$\begin{aligned}
\boldsymbol{y}_t(l) &= \operatorname*{arg\,max}_{b \in \{0,1\}} P(H_b^l | E_{C_t(l)}^l) \\
&= \operatorname*{arg\,max}_{b \in \{0,1\}} P(E_{C_t(l)}^l | H_b^l) P(H_b^l)
\end{aligned} \tag{1}$$

where $\boldsymbol{y}_t$ is the label vector for a test sample t. $\boldsymbol{C}_t(l)$, $E_{C_t(l)}^l$ and H_b^l are the same as described in [9]. However, it ignores the correlation between labels, i.e., it does not consider label distance as the base of measurement when identifying neighbors for a test sample. In Sect. 4, we will address this problem by exploiting label distance.

3   Multi-label Importance

Existing multi-label learning algorithms assume that the importance of all labels is equal. However, the importance of each label in label space is generally different in real world. As the intrinsic correlations interact between different labels, a label can effect the prediction of other labels. Intuitively speaking, different labels have different degrees of effect on the label structure, and the more important the label is, the more effect it has on entire label structure. Another factor that determines the label importance is the predictability of each label. Given a multi-label dataset, there may be some labels can get the high prediction accuracy, while some labels cannot. This issue can effect the performances of some approaches like Classifier Chains since the disseminating of error information. Based on the above assumption, we introduce *multi-label importance* (MLI) in this section.

Note that our proposed *multi-label importance* is different from *relative labeling importance* (RLI) proposed in [11]. The former is used to reveal the influence power of a label in whole label space. The latter represents the degree to which a label l describes an instance, similar to *label distribution* in *Label Distribution Learning* [12].

Different predictive accuracy of multiple labels imply that the predictability of each label is different from each other. The predictability of each label can be measured by utilizing existing approaches as the base classifiers (we need further divide the training data into two parts). In our work, we respectively use two popular approaches SVM and ML-kNN to measure the predictability of labels. The rest of this section describes how to extract label effects from label space.

Label effects represent the influence power of each label in multi-label data. The presence or absence of different labels have a different influence power on the overall label structure. Let $\boldsymbol{E}$ be the effect relation matrix, where E_{ij} stands the effect of label i on label j.

$$E_{ij} = \frac{1}{N(j)} \left| \frac{1}{|\boldsymbol{u}_0^i|} \sum_{l \in \boldsymbol{u}_0^i} Y_l^j - \frac{1}{|\boldsymbol{u}_1^i|} \sum_{l \in \boldsymbol{u}_1^i} Y_l^j \right| \tag{2}$$

where

$$N(j) = \sum_{i \in \boldsymbol{C}_j} \left| \frac{1}{|\boldsymbol{u}_0^i|} \sum_{l \in \boldsymbol{u}_0^i} Y_l^j - \frac{1}{|\boldsymbol{u}_1^i|} \sum_{l \in \boldsymbol{u}_1^i} Y_l^j \right| \tag{3}$$

is the normalizing constant which ensures that the effects of all other labels on label j sum up to 1. And $\boldsymbol{C}_j$ is the complementary set of label j. We divide training samples into two parts by the positive and negative values of label i when calculating E_{ij}, formulated as follows: $\boldsymbol{u}_0^i = \{s | Y_s^i = 0, 0 < s < n\}$, and

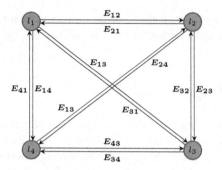

Fig. 2. E_{ij} stands the effect of label i on label j. Given label size L, there are $L(L-1)$ edges in the graph. After introducing the normalizing constant $N(j)$, the sum of the effects of all labels on label j is 1 (for label l_1, $E_{21} + E_{31} + E_{41} = 1$).

$u_1^i = \{s|Y_s^i = 1, 0 < s < n\}$. The hypothesis is that each label in training sets has both positive and negative samples, so u_0^i and u_1^i will not be empty sets.

After calculating E_{ij} for each label l_i and l_j, the interaction effect relationships of all label pairs can be shown in effect relation matrix $\boldsymbol{E}$. Note that the diagonal elements of the effect relation matrix is 0. We use a directed graph to demonstrate the relationship between labels in Fig. 2.

We use v_l to denote the effect degree of label l on label structure, and the degree vector $\boldsymbol{v} = [v_1, v_2, \ldots, v_L]^T$. We can calculate the effect degree vector as following:

$$\boldsymbol{v}^{(t)} = \boldsymbol{E}\boldsymbol{v}^{(t-1)}$$
$$\boldsymbol{v}^{(1)} = \boldsymbol{E}\boldsymbol{u}^T \tag{4}$$

Combining these two equations:

$$\boldsymbol{v}^{(t)} = \boldsymbol{E}^t\boldsymbol{u}^T \tag{5}$$

where $\boldsymbol{u} = [1, 1, \ldots, 1]$ is the all-1 vector to initiate the effect degree vector. As shown in (4) and (5), we calculate the effect degree using an iterative process. Because of the transfers of influence powers between different labels, the effect degree vector $\boldsymbol{v}$ will dynamically update at each iteration.

As we described above, we have introduced a normalizing constant to ensure the effects of all other labels on label j sum up to 1. Thus effect relation matrix $\boldsymbol{E}$ has two properties:

(1) $\boldsymbol{E}_{ij} \geqslant 0$ for $i, j \in \mathcal{Y}$ (All entries greater than or equal to zero).
(2) $\sum_{i=1}^{L} \boldsymbol{E}_{ij} = 1$ for $j \in \mathcal{Y}$ (All columns add to 1).

Then according to the properties of Markov matrices, we know that the effect vector $\boldsymbol{v}$ will converge to a stable state as long as the number of iterations is large enough. Actually, for our specific requirements, the convergence usually occurs within a few dozens of iterations.

Algorithm 1. v =get-$Degree(Y, t)$

---

1   Obtaining size of label space: $[n, L] = size(Y)$

1   Initializing $\boldsymbol{E} = zeros(L, L)$

2   **for** $i \in \mathcal{Y}$ **do**:

3   Divide training samples into two parts:

(i) $\boldsymbol{u}_0^i = \{s | Y_s^i = 0, 0 < s < n\}$

(ii) $\boldsymbol{u}_0^i = \{s | Y_s^i = 0, 0 < s < n\}$

4   **for** $j \in \mathcal{C}_i$ **do**:

5            $E_{ij} = \left| \frac{1}{|u_0^i|} \sum_{l \in u_0^i} Y_l^j - \frac{1}{|u_1^i|} \sum_{l \in u_1^i} Y_l^j \right|$

6   **end**

7   **end**

8   Normalizing the columns of $\boldsymbol{E}$

9    $\boldsymbol{v} = \boldsymbol{E}^t \boldsymbol{u}$

11   **return** $\boldsymbol{v}$

We show how to extract the effect degrees from label structure in Algorithm 1. The meanings of the input arguments Y and the output argument $\boldsymbol{v}$ are the same as described previously. The input arguments t is the number of iterations.

Given a multi-label dataset, we can extract the effect power degrees from label space Y of training data. For label l, the degree v_l can be determined by measuring its effect power on other labels, i.e., its influence power can be measured by the effect it makes to the rest of the label set. Obviously, the higher the value of v_l, the more influence power conveyed by label l.

After extracting the label effects information, we can determine the *multi-label importance* for multi-label learning by combining the label predictability and label effects. Although there may exist several possible ways to coordinate the contribution ratios of both label predictability and label effects, we simply use the pointwise product of these two terms as the MLI degree vector $\mathcal{I}$ to show its usefulness in our works.

$$\mathcal{I} = \boldsymbol{v} \cdot \boldsymbol{p} \tag{6}$$

where $\boldsymbol{p} = [p_1, p_2, ..., p_L]^T$ denotes the predictive accuracy vector which measured by the base classifiers, and each p_i is a real number which is greater than 0 and less than 1.

4   Exploiting Multi-label Importance Information

To prove the usefulness of *multi-label importance* extraction, we apply *multi-label importance* to optimize the label ordering in multi-label Classifier Chains and propose a improved ML-kNN algorithm considering both feature distance and label distance.

4.1   Optimizing the Label Ordering in Classifier Chains

The Classifier Chains (CC) is one of the most popular methods for multi-label learning since its flexibility and effective exploiting of label correlations. Nevertheless, the label ordering is still a challenging issue of CC. In the basic CC model, the label ordering is decided at random. It is clear that the inadequate order may cause a significant decrease in predictive accuracy since the propagation of error information. The authors of CC solve the issue by using an ensemble framework with a different random chain ordering for each iteration. [14] proposed a genetic algorithm for optimizing the label ordering in a chain of classifiers. However, all these approaches caused great computational complexity while optimizing the label ordering in Classifier Chains.

Based on the analysis of *multi-label importance*, we solve the issue using a heuristics by exploiting the *multi-label importance* ranking. Before building the classifiers for multi-label learning, we firstly extract the MLI information from training data. Specifically, the predictability of each label can be measured by the base classifier built for each label (we need further divide the training data into two parts), and the label effects can be obtained by Algorithm 1. Then we use the product of these two terms as the MLI degree. Then we can determine the label ordering of Classifier Chains by ranking the MLI degrees.

Our strategy offers important advantages for multi-label Classifier Chains.

- First, it considers both the label predictability and label effects. That means it can reduce the error propagation along the classifier chain while effectively exploiting the label correlations.
- Second, it has relatively low computational complexity and can be parallelized when determining the label ordering.

4.2   Considering Label Distance in ML-kNN

ML-kNN has become one of the most influential lazy learning approach for multi-label data. Nevertheless, it only considers feature distance between two samples when identifying neighbors, which means it ignores the implicit correlation between different labels. To achieve effective multi-label learning, the label distance can also be considered when identifying neighbors for each label [19]. In this section, we firstly discuss the distance metric between different label sets. Then we introduce a novel algorithm MLLD-kNN (ML-kNN by considering label distance).

Label Distance. Label distance indicates the dissimilarity degree between two label vectors (binary vector), and it can be measured by Hamming distance. The Hamming distance [13] is a metric expressing the distance between two objects by the number of mismatches among their pairs of variables. In particular, Hamming distance between two label vectors of equal length is the number of positions at which the corresponding label are different.

$$\mathcal{D}_h(i,j) = \sum_{k=1}^{L} [\![Y_i^k \neq Y_j^k]\!] \qquad (7)$$

However, the Hamming distance treats each label equally, which is not appropriate for our problem. Therefore, a weighted Hamming distance is presented for effectively measuring the dissimilarity between label vectors [16]. The *multi-label importance* degrees can be used as a weight vector to calculate the weighted Hamming distance.

$$\mathcal{D}_h^*(i,j) = \sum_{k=1}^{L} \mathcal{I}_k [\![Y_i^k \neq Y_j^k]\!] \qquad (8)$$

where $\mathcal{I}_k$ stands the MLI degree of label k. As shown in Fig. 3, the difference between Hamming distance and our proposed weighted Hamming distance can be illustrated by two 3-bit binary cubes. The Hamming distance $\mathcal{D}_h(i,j)$ between points can be indicated by the first cube (see in Fig. 3 (a)), where each side length is 1. The weighted Hamming distance $\mathcal{D}_h^*(i,j)$ can be indicated by the second cube (see in Fig. 3(b)), the side lengths of which are 1, 1.5 and 0.5.

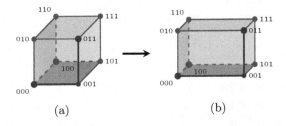

(a)   (b)

Fig. 3. $\mathcal{D}_h(i,j)$ and $\mathcal{D}_h^*(i,j)$ can be illustrated by these two 3-D cubes. The minimum rectilinear distance between any two vertices indicates distance between two binary strings. For example, $000 \mapsto 011$ has distance 2 in (a) and 2.5 in (b).

MLLD-kNN Algorithm. Before prediction phase, we need to determine MLI for given multi-label dataset. In prediction phase, given a test instance, we firstly output a label set using ML-kNN algorithm, then we can identify its k nearest neighbors by measuring both feature distance and label distance. Finally, we obtain the result label set after multiple iterations. The output label set $\boldsymbol{y}_t$ will be dynamically updated in each iteration.

Algorithm 2 shows the test procedure of our proposed algorithm. The meanings of input arguments S, k, t and the output argument $\boldsymbol{y}_t$ are the same as described in Sect. 2. While the input argument T is the maximum numbers of iterations. As discussed previously, we need to extract the MLI vector $\boldsymbol{\mathcal{I}}$ (step-14 in Algorithm 2) before implementing the iteration steps (steps from 15 to 17).

The kernel function MLLD-kNN is modified on the basis of ML-kNN algorithm. The measurements of distance between two samples are different in these two algorithms. In particular, the distance used in MLLD-kNN can be defined as:

$$\mathcal{D}(i,j) = \mathcal{D}_f(i,j) + \lambda \mathcal{D}_h^*(i,j) \tag{9}$$

where $\mathcal{D}_h^*(i,j)$ is the weighted Hamming distance between the label sets of two samples as defined above. λ is used to determine the contribution ratios of both feature distance and label distance. $\mathcal{D}_f(i,j)$ is the feature distance which can be measured by different metrics. As a commonly used metric, Euclidean distance can be employed.

$$\mathcal{D}_f^2(i,j) = \mathcal{D}_{euc}^2(i,j) = \sum_{k=1}^{L}(X_i^k - X_j^k)^2 \tag{10}$$

Algorithm 2. MLLD-*k*NN Algorithm

---

2   **MLLD-*k*NN**$(S, t, k, \boldsymbol{\mathcal{I}}, \boldsymbol{y}_t)$

3   **if** $\boldsymbol{y}_t$ is *null*

4   Identify $\mathcal{N}(t)$ only considering feature distance

5   **else**

6   Identify $\mathcal{N}(t)$ based on $\mathcal{D}(i,j)$
                where $\mathcal{D}(i,j) = \mathcal{D}_f(i,j) + \lambda \mathcal{D}_h^*(i,j)$

7   **for** l in range($\mathcal{Y}$) **do:**

8   get $\boldsymbol{y}_t(l)$ using maximum a posteriori principle

9   **end**

10   **return** $\boldsymbol{y}_t$

12   **main**(S, k, t, T)

13   get Y and L from dataset S

14   Extracting the MLI vector $\boldsymbol{\mathcal{I}}$

15   **for** i in $range(T)$ **do:**

16           $\boldsymbol{y}_t$=**MLLD-*k*NN**$(S, t, k, \boldsymbol{\mathcal{I}}, \boldsymbol{y}_t)$

17   **end**

18   output $\boldsymbol{y}_t$

---

5   Experimental Results

5.1   Evaluation Metrics

In multi-label learning, the evaluation is more complicated than that in single-label learning. Various evaluation metrics have been proposed to measure the performance of multi-label classifier [17]. There are five commonly used metrics: *hamming lass,ranking loss, coverage, one error* and *average precision.*

$$\boldsymbol{hloss}(H) = \frac{1}{n}\sum_{i=1}^{n}\frac{1}{L}|H(x_i)\Delta Y_i| \tag{11}$$

where H is the multi-label classifier which outputs a binary set for each instance and Δ stands for the symmetric difference between two sets.

$$\boldsymbol{one\text{-}error}(f) = \frac{1}{n}\sum_{i=1}^{n}[\![\arg\max_{l\in L} f(x_i, y)] \notin Y_i]\!] \tag{12}$$

where f is ranking function which outputs a real-value (between 0 and 1) set, so we can get the ranking of predicted labels of each instance. For any predicate π, $[\![\pi]\!]$ equals 1 if π holds and 0 otherwise. This evaluation criteria indicates the probability of that the classifier failed to get even one of the labels correct.

$$rloss(f) = \frac{1}{n} \sum_{i=1}^{n} \frac{|R_i|}{|Y_i||\bar{Y}_i|} \tag{13}$$

where $\bar{Y}_i$ denotes the complementary set of Y_i in $\mathcal{Y}$. For instance (x_i, Y_i), $|R_i|$ is the number of label pairs that are reversely ordered by ranking function f, and R_i defined as: $R_i = \{(l_1, l_2) | f(x_i, l_1) \leq f(x_i, l_2), \ (l_1, l_2) \in Y_i \times \bar{Y}_i\},$.

$$coverage(f) = \frac{1}{n} \sum_{i=1}^{n} |C(x_i)| - 1 \tag{14}$$

where $C(x_i) = \{l | f(x_i, l) \geq f(x_i, l_i^{min}), l \in \mathcal{Y}\}$ and $l_i^{min} = \arg\min_{y \in S_i} f(x_i, y)$.

$$ave\text{-}prec(f) = \frac{1}{n} \sum_{i=1}^{n} \frac{1}{|Y_i|} P(x_i) \tag{15}$$

where

$$P(x_i) = \sum_{y \in Y_i} \frac{|\{l | f(x_i, l) \geq f(x_i, y), l \in Y_i\}|}{|\{l | f(x_i, l) \geq f(x_i, y), l \in L\}|}. \tag{16}$$

These above five metrics evaluate the performance of a multi-label classifier from different horizon. Hamming loss is based on the multi-label classifier H, and others are based on the real-valued ranking function f, which concern the ranking quality of different labels.

5.2   Experiments

Datasets. We evaluate the performance of our approaches on four multi-label classification datasets from different domains[1]:

- *Emotions* come from the music domain, which consists of 593 songs with six clusters of music emotions [18].
- *Scene* is a benchmark for image classification containing 2407 natural scene images and six possible labels: *Beach*, *Sunset*, *FallFoliage*, *Field*, *Mountain* and *Urban* [1].
- *Yeast* is a dataset for predicting the gene functional [3]. It has 2,417 instances that each instance in the dataset represents a yeast gene and there are 14 possible labels indicating gene functional groups.
- *Enron* is a benchmark for text classification. It is a subset of the Enron email corpus [15], including 1702 emails with 53 possible labels.

---

[1] Data sets were downloaded from http://mulan.sourceforge.net/datasets.html and http://meka.sourceforge.net/#datasets.

- The dataset *Slashdot* was collected from the Slashdot web page and consists of article blurbs labelled with subject categories.
- *Ohsumed* was constructed from a collection of peer-reviewed medical articles and labelled with the appropriate disease categories.

Table 1. Multi-label datasets used in experiments.

Name	Domain	Instances	Labels	Cardinality	Density
emotions	music	593	6	1.869	0.311
scene	image	2407	6	1.074	0.179
yeast	biology	2417	14	4.237	0.303
enron	text	1702	53	3.378	0.064
slashdot	text	3782	22	1.181	0.054
ohsumed	text	13929	23	1.663	0.072

There are two measures for evaluating the characteristics of a dataset: cardinality and density [6]. The cardinality of a dataset S is the mean of the number of labels of the instances that belong to S, defined by (16), and the density of S is the mean of the number of labels of the instances that belong to S divided by L, defined by (17).

$$cardinality(S) = \frac{1}{n} \sum_{i=1}^{n} |Y_i| \tag{17}$$

$$density(S) = \frac{1}{n} \sum_{i=1}^{n} \frac{|Y_i|}{L} \tag{18}$$

Results. To prove the usefulness of *multi-label imp-ortance* extraction, we apply *multi-label importance* in CC and ML-kNN. We first compare CC-MLI with the basic CC model and BR method (all these three method based on SVM). For each dataset, we select 5 random label ordering to build CC models respectively. Results show that our strategy performs significantly better than most CC models. In Table 2, CC* is the best one in these 5 (random ordering) CC models. Note that although our strategy can significantly improve the performance of Classifier Chains, it exhibits a predictive accuracy inferior to BR method on *yeast*.

Then we compare MLLD-kNN with the ML-kNN. The 6th and 7th columns of Table 2 shows the comparison results on four datasets. As can be seen from the comparison of the experimental results, our proposed approach achieves effective classification on *emotions*, *scene* and *enron*. Similar to the results of CC-MLI, the improved ML-kNN algorithm exhibits a worse performance on *yeast*. We think it is due to the error propagation after considering label correlations [20].

Table 2. We evaluate the performance of our approaches on four multi-label classification datasets, and five evaluation metrics (•(∘) indicates our approaches is better (worse) than the corresponding approach).

Datasets	Metrics	CC-MLI	CC*	BR	MLLD-kNN	ML-kNN
emotions	*hloss*	0.217	0.213∘	0.213∘	0.196	0.196•
	one-error	0.306	0.301∘	0.282∘	0.247	0.272•
	rloss	0.187	0.191•	0.165∘	0.150	0.160•
	coverage	1.930	1.965•	1.850∘	1.811	1.815•
	ave-prec	0.788	0.787•	0.803∘	0.819	0.802•
scene	*hloss*	0.111	0.115•	0.112•	0.093	0.088∘
	one-error	0.285	0.301•	0.287•	0.235	0.248•
	rloss	0.106	0.109•	0.107•	0.085	0.089•
	coverage	0.636	0.646•	0.629∘	0.526	0.551•
	ave-prec	0.825	0.816•	0.820•	0.856	0.850•
yeast	*hloss*	0.203	0.216•	0.198∘	0.198	0.196∘
	one-error	0.253	0.257•	0.242∘	0.233	0.235•
	rloss	0.196	0.219•	0.174∘	0.191	0.168∘
	coverage	7.060	7.660•	6.476∘	6.870	6.286∘
	ave-prec	0.733	0.718•	0.752∘	0.749	0.762∘
enron	*hloss*	0.051	0.051•	0.056•	0.051	0.052•
	one-error	0.317	0.309∘	0.359•	0.284	0.313•
	rloss	0.087	0.090•	0.115•	0.090	0.093•
	coverage	12.674	12.775•	15.582•	13.250	13.091∘
	ave-prec	0.655	0.652•	0.578•	0.636	0.626•
slashdot	*hloss*	0.049	0.052•	0.054•	0.052	0.051∘
	one-error	0.432	0.502•	0.511•	0.646	0.669•
	rloss	0.181	0.189•	0.184•	0.181	0.180∘
	coverage	4.344	4.304∘	4.645•	4.164	4.284•
	ave-prec	0.623	0.601•	0.579•	0.522	0.495•
ohsumed	*hloss*	0.066	0.072•	0.068•	0.069	0.074•
	one-error	0.373	0.386•	0.388•	0.631	0.656•
	rloss	0.143	0.154•	0.122∘	0.230	0.231•
	coverage	6.384	7.435•	4.255∘	7.112	6.984∘
	ave-prec	0.638	0.636•	0.655∘	0.517	0.487•

Since *multi-label importance* provides important prior information for the following training and testing process, our proposed algorithms applied on benchmark datasets demonstrate efficient multi-label learning by exploiting multi-label importance. Based on the analysis of comparison results, we conclude that the

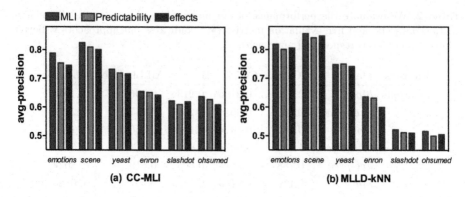

Fig. 4. The results demonstrate the performances of MLI-based approaches better others (consider just label predictability or label effect).

multi-label importance information extracted from training data is helpful to improve the performance of some multi-label classification algorithms.

In the above experiments we considered both label predictability and label effects. It is also important to observe what would happen if we consider just label predictability or just label effect. As shown in Fig. 4, we conduct the comparison experiments and the results demonstrate that the MLI-based approaches (considering both label predictability and label effects) have superior performance than others (considering just label predictability or just label effect).

Another key observation of our experiments shows that there is a strong positive correlation between the label predictability and label effects. As we described above, different predictive accuracy of multiple labels imply that the predictability of each label is different from each other. We determine the label predictability by ranking the predictive accuracy of labels on test data utilizing SVM (in fact, the label predictability ranking measured by different models are approximately the same). More concretely, for *emotions*, we can get the accuracy of each label using SVM: $\{0.787, 0.707, 0.663, 0.891, 0.806, 0.861\}$. Then we obtain the label predictability ranking: $\{4, 6, 5, 1, 2, 3\}$. The effect degrees extracted from label structure are $\{0.868, 0.705, 1.01, 1.209, 1.109, 1.096\}$ and the effect power ranking can be obtained:$\{4, 5, 6, 1, 3, 2\}$. Thus the Spearman correlation coefficient of these two ranking vectors can be calculated (as shown in Table 3). All values are greater than 0 and close to 1 in Table 3, which demonstrate the positive correlation between those two ranking vectors (if there is no correlation between two vectors, the correlation coefficient is about 0).

We can also use the Euclidean distance to measure the dissimilarity of the vectors. We compare the Euclidean distance between two ranking vectors against that between random arrays (see in Fig. 5). For each data set, we select 3 groups of arrays and each group has 5 random arrays (the length of arrays is equal to the label size L). For each group, we calculate the Euclidean distance of each array pair and average the results. As shown in Fig. 5, the distances between "acc & effects" are significantly shorter than corresponding comparative groups.

Table 3. We calculate the Spearman correlation coefficient of predictive accuracy and effect power of labels.

Datasets	emotions	scene	yeast	enron	slashdot	ohsumed
ρ	0.886	0.771	0.481	0.207	0.335	0.567

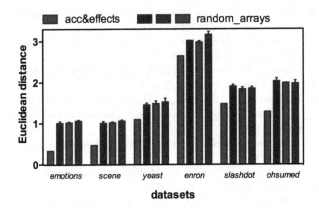

Fig. 5. The comparison of Spearman correlation coefficients.

Experimental results show the strong positive correlation between label predictability and label effects. In other words, for a multi-label data, more predicable labels have more effect on entire label structure.

6   Conclusion

We introduced *multi-label importance* (MLI) in this paper and addressed the issue of label importance extraction. To prove the usefulness of *multi-label importance* we applied it to two existing multi-label algorithms. The experimental results explicitly demonstrate the efficient multi-label learning by exploiting *multi-label importance*. Another contribution of this work is that our experiments show the strong positive correlation between label predictability and label effects. We think this finding is helpful for exploring the effective strategies for multi-label learning.

In future work, we plan to conduct more experiments on other large multi-label datasets to fully prove the usefulness of *multi-label importance* extraction. Applying *multi-label importance* extraction to more multi-label classification approaches will be another interesting work to be investigated.

Acknowledgement. It was supported by NSF Chongqing China (cstc2017zdcy-zdyf0366).

References

1. Boutell, M.R., Luo, J., Shen, X., Brown, C.M.: Learning multi-label scene classification. Pattern Recognit. **37**(9), 1757–1771 (2004)
2. Singer, Y., Schapire, R.E.: BoosTexter: a boosting-based system for text categorization. Mach. Learn. **39**, 135–168 (2000)
3. Elisseeff, A., Weston, J.: A kernel method for multi-labelled classification. In: Proceedings of NIPS, vol. 14, pp. 681–687 (2001)
4. Read, J., Pfahringer, B., Holmes, G., Frank, E.: Classifier chains for multi-label classification. Mach. Learn. **85**(3), 333 (2011)
5. Zhang, M.L., Zhou, Z.H.: A review on multi-label learning algorithms. IEEE Trans. Knowl. Data Eng. **26**(8), 1819–1837 (2014)
6. Tsoumakas, G., Katakis, I., Vlahavas, I.: Mining multi-label data. In: Maimon, O., Rokach, L. (eds.) Data Mining and Knowledge Discovery Handbook, pp. 667–685. Springer, Boston (2009)
7. Brinker, K.: Multilabel classification via calibrated label ranking. Mach. Learn. **73**(2), 133–153 (2008)
8. Tsoumakas, G., Katakis, I., Taniar, D.: Multi-label classification: an overview. Int. J. Data Warehous. Min. **3**(3), 1–13 (2007)
9. Zhang, M.L., Zhou, Z.H.: Ml-KNN: a lazy learning approach to multi-label learning. Pattern Recogn. **40**(7), 2038–2048 (2007)
10. Zhang, M.L., Zhou, Z.H.: Multilabel neural networks with applications to functional genomics and text categorization. IEEE Trans. Knowl. Data Eng. **18**(10), 1338–1351 (2006)
11. Li, Y.K., Zhang, M.L., Geng, X.: Leveraging implicit relative labeling-importance information for effective multi-label learning. In: IEEE International Conference on Data Mining, vol. 6, pp. 251–260. IEEE (2016)
12. Geng, X., Ji, R.: Label distribution learning. In: IEEE International Conference on Data Mining Workshops, pp. 377–383. IEEE Computer Society (2013)
13. Hamming, R.W.: Error detecting and error correcting codes. Bell Syst. Tech. J. **29**(2), 147–160 (1982)
14. Plastino, A., Freitas, A.A.: A genetic algorithm for optimizing the label ordering in multi-label classifier chains. In: IEEE International Conference on TOOLS with Artificial Intelligence, pp. 469–476. IEEE Computer Society (2013)
15. Klimt, B., Yang, Y.: Introducing the Enron Corpus. In: Conference on Email and Anti-Spam. DBLP (2004)
16. Zhang, L., Zhang, Y., Tang, J., Lu, K., Tian, Q.: Binary code ranking with weighted hamming distance. In: 2013 IEEE Conference on Computer Vision and Pattern Recognition (CVPR), vol. 9, pp. 1586–1593. IEEE (2013)
17. Wu, X. Z., Zhou, Z. H.: A Unified View of Multi-Label Performance Measures. arXiv preprint arXiv: 1609.00288 (2016)
18. Trohidis, K., Tsoumakas, G., Kalliris, G., Vlahavas, I.: Multilabel classification of music into emotions. Blood **90**(9), 3438–3443 (2008)
19. Yu, Y., Pedrycz, W., Miao, D.: Multi-label classification by exploiting label correlations. Expert Syst. App. **41**(6), 2989–3004 (2014)

20. Senge, R., del Coz, J.J., Hüllermeier, E.: On the problem of error propagation in classifier chains for multi-label classification. In: Spiliopoulou, M., Schmidt-Thieme, L., Janning, R. (eds.) Data Analysis, Machine Learning and Knowledge Discovery. SCDAKO, pp. 163–170. Springer, Cham (2014). https://doi.org/10.1007/978-3-319-01595-8_18

21. Hu, F., Xu, X., Wang, J., Yang, Z., Li, L.: Memory-enhanced latent semantic model: short text understanding for sentiment analysis. In: Candan, S., Chen, L., Pedersen, T.B., Chang, L., Hua, W. (eds.) DASFAA 2017. LNCS, vol. 10177, pp. 393–407. Springer, Cham (2017). https://doi.org/10.1007/978-3-319-55753-3_25

Exploiting Instance Relationship for Effective Extreme Multi-label Learning

Feifei Li[1], Hongyan Liu[2], Jun He[1(✉)], and Xiaoyong Du[1]

[1] Key Laboratory of DEKE (MOE), School of Information,
Renmin University of China, Beijing, China
{pangxiaobi,hejun,duyong}@ruc.edu.cn
[2] School of Economics and Management, Tsinghua University, Beijing, China
hyliu@tsinghua.edu.cn

Abstract. Extreme multi-label classification is an important data mining technique, which can be used to label each unseen instance with a subset of labels from a large label set. It has wide applications and many methods have been proposed in recent years. Existing methods either seek to compress label space or train a classifier based on instances' features, among which tree-based classifiers enjoy the advantages of better efficiency and accuracy. In many real world applications, instances are not independent and relationship between instances is very useful information. However, how to utilize relationship between instances in extreme multi-label classification is less studied. Exploiting such relationship may help improve prediction accuracy, especially in the circumstance that feature space is very sparse. In this paper, we study how to utilize the similarity between instances to build more accurate tree-based extreme multi-label classifiers. To this end, we introduce the utilization of relationship between instances to state-of-the-art models in two ways: feature engineering and collaborative labeling. Extensive experiments conducted on three real world datasets demonstrate that our proposed method achieves higher accuracy than the state-of-the-art models.

1 Introduction

Multi-label classification is to label each instance with the most relevant subset of labels from a label set. But when label set is very large, traditional multi-label classification methods don't work well. Therefore, extreme multi-label classification has been studied to deal with this situation in recent years. Lots of applications have a large number of labels today. For instance, in e-commerce platforms, each item such as product or service has labels such as categories or tags to describe it. In social network platforms, users have several tags to express their preferences. In app stores, there are tags describing functions of apps. In these scenarios, the number of labels is large. How to automatically and accurately assign most relevant labels to those without labels is very challenging and essential for many other applications such as information retrieval

© Springer International Publishing AG, part of Springer Nature 2018
J. Pei et al. (Eds.): DASFAA 2018, LNCS 10828, pp. 440–456, 2018.
https://doi.org/10.1007/978-3-319-91458-9_27

and recommendation. For instance, Agrawal et al. [1] treated search queries as labels to train a classifier that could automatically tag a new web page with the most relevant queries sorted by their relevance. Jain et al. [9] showed how to treat product recommendation as an extreme multi-label classification problem, where items bought by a user can be seen as user labels. These methods provide a fresh way to re-think recommendation and ranking problem. Also, if items and users are accurately labeled with tags, traditional ranking algorithms and recommendation algorithms can also be improved by taking labels into consideration, as labels can better describe items' characteristics or users' preferences.

To deal with the extreme multi-label classification problem, existing methods either seek to compress label space through embedding approach or train tree-based classifiers based on instances' features. Among existing tree-based methods, state-of-the-art methods are FastXML [16], PfastreXML [9] and PLTs [11], which are usually more efficient and accurate than other methods. The quality of the classifiers built by these methods relies heavily on the features that are used to describe each instance. In many cases, the number of features is big while the feature space is very sparse. For example, most features come from description of instances, where the total number of terms occurred in all descriptions is big and quality of the descriptions is not good. In this case, accuracy of the classifiers built is usually low. To solve these problems, we study how to leverage instance relationship to improve accuracy, which is less studied in extreme multi-label classification model. Usually, instances are not independent but associated with each other. Therefore, relationship between instances can be exploited to improve classification. Specifically, in this paper, we propose to make use of the similarity among instances to modify existing tree-based extreme multi-label classification models, including FastXML, PfastreXML and PLTs.

The intuition behind our proposed method is that similar instances are more likely to have common labels. This kind of relationship is universally available in many applications and we can infer this relationship in different ways. For example, in app store, users usually browse and compare some apps before downloading one of them. In this case, the browsed apps in one short session are usually similar in functions thus having common labels. In Twitter, users usually follow internet celebrities according to their interests. If two celebrities are followed by the same users, they are very likely to be in the same field or have similar interests. Therefore, we can infer that they are similar and share similar tags.

The proposed method, incorporating **R**elationship into **T**ree-based **C**lassifiers (**RTC** for short), can utilize instance relationship in different ways. We propose two ways: feature engineering and collaborative labeling, which can be used separately or combined in different scenarios. By feature engineering, we mean that we transform the relationship between instances into features. By collaborative labeling, we mean that the relationship is used during tree building process and prediction process, where similar instances collectively help label the target instance, acting like collaborative labeling.

To summarize, we make the following contributions in this paper:

- To the best of our knowledge, we are the first to take instance relationship into consideration for extreme multi-label learning. We propose different ways to infer and represent the relationship.
- We propose methods to modify existing tree-based extreme multi-label classification models including FastXML, PfastreXML and PLTs to leverage the relationship.
- We present extensive experiments conducted on several real world datasets to evaluate the proposed method. Experimental results demonstrate that our proposed method performs better than state-of-the-art original models.

2   Related Work

2.1   Extreme Multi-label Learning

Existing extreme multi-label learning methods are usually embedding based or tree based ones.

Traditional methods like one-vs-all [12], which train classifier for each label, cannot be used for extreme multi-label classification due to high time complexity. Embedding based methods compress the dimension of labels based on the correlation between labels, usually through a projection matrix. Many methods have been proposed to tackle extreme multi-label learning problem in embedding based ways [3,4,10,13,17]. Methods are different in compression and decompression techniques, including SVP [10], SVD [17], etc. Predictions can be made based on k nearest neighbors in embedding space. Embedding based methods enjoy strong theoretical foundation, ease of implementation and simplicity. But usually they cannot perform better than one-vs-all methods in terms of prediction accuracy [16].

Tree based methods like FastXML [16] and PfastreXML [9] usually learn a hierarchy to recursively partition the training dataset into subsets in child nodes. The intuition is that only a small number of labels are present or active in each local region. So we can estimate the unseen instance's label distribution according to training instances within the region. PfastreXML introduces propensity to FastXML to deal with missing labels and tail labels. Others like PLTs [11] are similar to conditional probability estimation trees. Each internal node of the tree decides whether the path should continue or not, and each leaf node corresponds to a label. The product of conditional probability of the path from root to leaf is the probability whether a label should be assigned to an instance.

Tree based methods usually enjoy better efficiency and accuracy [16]. Therefore, in this paper, we choose the competing tree based classifiers to improve further.

2.2   Relational Learning

Relational learning [7] has been used in classification for a long time. Relation between instances is usually represented by a graph or network, where nodes usually share some common characteristics with its linked neighbors, as nodes with similar labels are more likely to be connected [15].

Many relational classification models have been proposed, which usually learn models by using the correlation between labels of linked nodes in the network. The relational neighbor classifier [14] is a probabilistic model, where an instance's class labels depend on its neighbors' class labels. This model mainly solves the single-label classification problems. SCRN [20] extends relational neighbor classifier to solve multi-label problems. The model represents each edge of the network as a vector of 0 and 1, where the edges' linked nodes are 1 s and the others are 0s. Then the nodes' features are constructed based on edge cluster IDs. Predictions are made by the nearest neighbors in embedding space. SocDim [18,19] learns latent social dimensions on social network and uses SVM or logistic regression to train a classifier.

3   Preliminary

In this section we overview three state-of-the-art tree-based extreme multi-label classification models, FastXML, PfastreXML and PLTs, which are the base models of our proposed method. We also use them as comparison baseline in Sect. 5.

3.1   FastXML

Based on the intuition that only a small number of labels are active in each region of instances' feature space, FastXML learns a hierarchy over the feature space. Labels that are active in each region are the union of labels of all instances in the region. So the objective is to partition regions that can recognize similar instances correctly based on instances' features.

FastXML considers both feature and label distributions to build the tree model, which recursively partitions a parent's feature space between its children, directly optimizing a normalized Discounted Cumulative Gain (nDCG) [21] loss function, which is sensitive to both ranking and relevance. An effective method is also proposed to solve the optimization problem. Through experiments, the proposed model is shown to be more accurate than other competing methods.

When the tree is built, prediction is made based on the label distribution of all the instances within the leaf node where the unseen instance goes to, starting from the root of the tree.

3.2   PfastreXML

PfastreXML is a modified FastXML model. Apart from following the same way of node-splitting and prediction, it takes missing labels into consideration, using a propensity to weigh each label. After making label prediction as used in FastXML, it ranks the predicted labels by a classifier designed specially for tail labels. This model performs well in dealing with missing label and tail label.

3.3   Partial Label Trees

PLTs is based on conditional probability trees [2] and probabilistic classifier chains [5]. Different from FastXML, the structure of PLTs is predefined or randomly assigned. Each node of the tree stands for a conditional probability, and each leaf node corresponds to one label. The product of the conditional probability of each node along a path from root to a leaf node is calculated as the probability that an instance has the corresponding label. During training time, instances traverse each node from root to every leaf to train a linear estimator for the node, which is used to measure the conditional probability in feature space at the node.

At prediction time, the final probability of whether a label should be assigned to an instance is determined by the product of conditional probability of nodes in the path from root to some leaf nodes. If the product is greater than a threshold, the label corresponding to that leaf will be assigned to the instance. If the probability at some nodes within the path is less than a threshold, the path will not be checked further.

4   The RTC Approach

In this section, we present our proposed method, RTC, which takes the relationship between instances into consideration for extreme multi-label learning task, aiming to achieve higher accuracy and overcome the problem of feature sparsity. Incorporating the idea of RTC to models FastXML, PfastreXML and PLTs, we introduce the first method to get RTC-FastXML, RTC-PfastXML and RTC-PLTs respectively. After that, we introduce the second method, feature engineering, for extreme multi-label classification models.

In many applications, we can infer similarity relationship between instances. For example, we can infer similarity relationship between instances (items) by analyzing user behavior sequence. Taking user's behavior in app store as an example, when we want to download an app, we usually first search and browse many relevant apps and then download some of them. Those apps searched, browsed or downloaded together by many users are usually very similar in function. Therefore, we can infer similarity based on user behavior. Based on this observation, we incorporate instance relationship into the competing extreme multi-label classification models. By doing this, for those instances we don't have high quality features to describe, similar instances can collectively help to label them.

We propose two ways to utilize instances' relationship information, collaborative labeling and feature engineering, which can be used separately or combined in different scenarios.

Through experiments we find that both ways can improve classification accuracy compared to the original model, and combining the two ways in RTC achieves better performance.

We adopt the same symbols used in FastXML to define the mining problem. Training dataset consists of a set of instances, $\mathcal{B} = \{(\boldsymbol{x}_i, \boldsymbol{y}_i)|i = 1, 2, \ldots, N\}$, where $\boldsymbol{x}_i$ is feature vector for training instance i ($\boldsymbol{x}_i \in \mathcal{R}^D$), and $\boldsymbol{y}_i$ is a binary label vector with L dimensions. If label l is assigned to instance i, $y_{il} = 1$, otherwise $y_{il} = 0$. In other words, there are N training instances and the number of distinct labels is L. Specially, in each node of a tree, Sd_i is the set of instances similar to instance i, s_{ij} is the strength of similarity relationship between instances i and j, which could be measured in different ways in different scenarios, $S_i = \sum_{j \in Sd_i} s_{ij}$.

4.1   RTC-FastXML

A. Training Model. Model FastXML builds a classification tree based on feature space of instances. The quality of instance's feature influences the performance of the tree. In many cases, the feature space is huge and sparsity is a major problem. To alleviate this problem, we introduce the collaboration of similar instances into the tree building process, and the modified model is called RTC-FastXML. It builds a binary tree recursively. Initially, all training instances belong to a root node. Then, training instances in each node are split into two children nodes, with one child node called positive node and the other negative node. The key point is how to split the current node. For each instance i present in current node, let δ_i in $\{-1, +1\}$ denote whether it is assigned to the positive child node ($\delta_i = +1$) or the negative child node ($\delta_i = -1$). To get the value of δ_i, a linear separator $\boldsymbol{w}$ is learnt to partition the current node through optimizing an objective function, which is shown below.

$$
\min \|\boldsymbol{w}\|_1 + \sum_i C_\delta(\delta_i) log(1 + e^{-\delta_i \boldsymbol{w}^T \boldsymbol{x}_i}) - C_r \sum_i \frac{1}{2}(1 + \delta_i)\mathcal{L}_{nDCG@L}(\boldsymbol{r}^+, \boldsymbol{y}_i)
$$

$$
- C_r \sum_i \frac{1}{2}(1 - \delta_i)\mathcal{L}_{nDCG@L}(\boldsymbol{r}^-, \boldsymbol{y}_i) - C_s \sum_i \sum_{j \in Sd_i} \delta_i \delta_j log(\frac{s_{ij}}{S_i} + 1)
$$

$$
w.r.t \quad \boldsymbol{w} \in \mathcal{R}^D, \boldsymbol{\delta} \in \{-1, +1\}^n, \boldsymbol{r}^+, \boldsymbol{r}^- \in \Pi(1, L)
$$

$$(1)$$

where i is the index of the training instances present at the node being partitioned, j is an instance of the similar instance set Sd_i and n is the number of instances in the node. $\Pi(1, L)$ is the set of all permutations of $\{1, 2, \ldots, L\}$ with each permutation representing a rank of L labels, $\boldsymbol{r}^+$ and $\boldsymbol{r}^-$ are the predicted label rankings for the positive and negative nodes respectively. Suppose $\boldsymbol{r}^+ = (r_1, r_2, \ldots, r_L)$, then label r_1 is the most relevant predicted label in the positive node. $\boldsymbol{y}_i$ is the ground truth label vector with L dimensions for instance i. C_r and C_s are hyper-parameters, determining the relative importance of the three terms. $C_\delta(\delta_i)$ is a function of δ_i to allow different misclassification penalties for positive and negative nodes. $C_\delta(+1) = n_{pos}/(n_{pos} + n_{neg})$, $C_\delta(-1) = n_{neg}/(n_{pos} + n_{neg})$. n_{pos} is the number of instances assigned to the positive child at the current node, and n_{neg} is that of instances assigned to the negative node. s_{ij} measures the relationship strength between instances i and j.

Algorithm 1. RTC-FastXML: SPLIT-NODE($\{x_i, y_i\}_{i=1}^{N}, node$)

1: **Input:** $\{x_i, y_i\}_{i=1}^{N}, node$
2: **Output:** $w[t], node^+, node^-$
3: $Id \leftarrow node.Id$
4: $Sd \leftarrow node.Sd$
5: $\delta_i[0] \sim \{-1, 1\}, \forall i \in Id$
6: $w[0] \leftarrow 0, t \leftarrow 0, t_w \leftarrow 0, \mathcal{W}_0 \leftarrow 0$
7: **repeat**
8:     $r^{\pm}[t+1] \leftarrow rank_L(\sum_{i \in Id} \frac{1}{2}(1 \pm \delta_i[t]) I_L(y_i) y_i)$
9:     **for** $i \in Id$ **do**
10:         $v_i^{\pm} \leftarrow C_\delta(\pm 1) log(1 + e^{\mp w[t]^T x_i}) - C_r I_L(y_i) \sum_{l=1}^{L} \frac{y_{ir_l^{\pm}[t+1]}}{log(1+l)} \mp C_s \sum_{j \in Sd[i]} \delta_j[t] log(\frac{s_{ij}}{S_i} +$
   1)   ▷ consider relationship
11:         **if** $v^+ = v^-$ **then**
12:             $\delta_i[t+1] = \delta_i[t]$
13:         **else**
14:             $\delta_i[t+1] = sign(v^- - v^+)$
15:     **if** $\delta[t+1] = \delta[t]$ **then**
16:         $w[t+1] \leftarrow \arg\min_{w} \|w\|_1 + C_\delta(\delta_i[t]) \sum_{i \in Id} log(1 + e^{-\delta_i[t]w^T x_i})$
17:         $\mathcal{W}_{t_w+1} \leftarrow t+1$
18:         $t_w \leftarrow t_w + 1$
19:     $t \leftarrow t+1$
20: **until** $\delta[\mathcal{W}_{t_w}] = \delta[\mathcal{W}_{t_w-1}]$
21: $node^+ \leftarrow$ new node, $node^- \leftarrow$ new node
22: $node^+.Id \leftarrow \{i \in Id : w[t]^T x_i > 0\}$
23: $node^-.Id \leftarrow \{i \in Id : w[t]^T x_i \leqslant 0\}$

The normalized Discounted Cumulative Gain for the top k predictions $\mathcal{L}_{nDCG@k}$ in (1) is defined in (2).

$$\mathcal{L}_{nDCG@k}(r, y_i) = I_k(y_i) \sum_{l=1}^{k} \frac{y_{ir_l}}{log(1+l)} \qquad (2)$$

$I_k(y_i)$ is the inverse of the $DCG@k$ of the ideal ranking for y_i defined in (3).

$$I_k(y_i) = \frac{1}{\sum_{l=1}^{\min(k, \|y_i\|_0)} \frac{1}{log(1+l)}} \qquad (3)$$

The first term in (1) is a regularization. The second one is based on feature values to determine the separator and δ. The third and fourth items consider the label distributions of the partition. We add the fifth term to make use of relationship between instances to facilitate the learning of the separator. The intuition behind this term is that similar instances are more likely to go to the same child node. The higher the similarity between instances i and j, the more likely they will be distributed to the same child node. The tree building process is shown in Algorithm 1, where *node* represents the splitting node.

B. Prediction. When the separator w is learned in each node of the tree, for an unseen instance i, we use formula (4) to determine which node it goes to starting from the root node. If the result of (4) is greater than zero, then it goes

Algorithm 2. RTC-FastXML: PREDICT($\{\mathcal{T}_1, ..., \mathcal{T}_T\}, \boldsymbol{x}_i$)

1: **Input:**$\{\mathcal{T}_1, ..., \mathcal{T}_T\}, \boldsymbol{x}_i$
2: **Output:** $r(\boldsymbol{x}_i)$
3: **for** $l = 1, ..., T$ **do**
4:      $node \leftarrow \mathcal{T}_l.root$
5:      **while** $node$ is not a leaf **do**
6:          $w \leftarrow node.w$
7:          $Id \leftarrow node.Id$
8:          $Sd \leftarrow node.Sd$
9:          **if** $\boldsymbol{w}^\mathrm{T}\boldsymbol{x}_i + C_p \sum_{j \in Sd[i]} \delta_j log(\frac{s_{ij}}{S_i} + 1) > 0$ **then**
10:             $node \leftarrow node.left\_child$
11:         **else**
12:             $node \leftarrow node.right\_child$ ▷ consider relationship
13:     $\boldsymbol{p}_l^{leaf}(\boldsymbol{x}_i) \leftarrow node.\boldsymbol{P}$
14: $r(\boldsymbol{x}_i) = rank_k(\frac{1}{T} \sum_{l=1}^{T} \boldsymbol{P}_l^{leaf}(\boldsymbol{x}_i))$

to the positive child node. Otherwise, it goes to the negative one.

$$\boldsymbol{w}^\mathrm{T}\boldsymbol{x}_i + C_p \sum_{j \in Sd_i} \delta_j log(\frac{s_{ij}}{S_i} + 1) \tag{4}$$

j refers to similar instance of instance i in the training set belonging to the present node. C_p is another hyper-parameter, indicating the importance of relationship to the classification task. The added second term in (4) implies that which node instance i goes to depends on not only its feature values but also its similar instances. If more similar instances go to the positive node, node instance i is more likely go to the same node. Vice versa. Prediction is made based on the label distribution of the training instances within the leaf node where instance i goes to. We assign top k labels to it, as shown in line 12 of Algorithm 2. T is the number of trees we have trained and $\boldsymbol{P}$ is label distribution of the leaf.

C. Optimization. In our proposed model, we need to optimize $\boldsymbol{w}$, $\boldsymbol{r}$ and $\boldsymbol{\delta}$. We optimize one of them while keeping the other two fixed. $\boldsymbol{w}$ and $\boldsymbol{r}$ are optimized in the same way as in FastXML.

We optimize with respect to $\boldsymbol{\delta}$ while keeping $\boldsymbol{r}$ and $\boldsymbol{w}$ fixed.

Each δ_i can be optimized by checking whether it is optimized by $\delta_i^\star = +1$ or -1 while keeping others fixed. This yields

$$\delta_i^\star = sign(v_i^- - v_i^+) \tag{5}$$

$$v_i^\pm = C_\delta(\pm 1)log(1 + e^{\mp \boldsymbol{w}^\mathrm{T}\boldsymbol{x}_i}) - C_r I_L(\boldsymbol{y}_i) \sum_{l=i}^{L} \frac{y_{ir_l^\pm}}{log(1 + l)} \mp C_s \sum_{j \in Sd_i} \delta_j log(\frac{s_{ij}}{S_i} + 1) \tag{6}$$

$I_L(\boldsymbol{y}_i)$ is defined in (3), and the other variables are the same as those in (1).

4.2   RTC-PfastreXML

In real world applications, an instance may not have the complete set of correct labels due to many reasons. For example, for an app, it is not easy to find all of appropriate tags, as it is very time consuming to go over each tag to find suitable ones. PfastreXML assumes that a relevant label l is observed for an instance i by a marginal probability p_{il}. It is estimated by a sigmoidal function and the label's observation times.

The objective function of RTC-PfastreXML is the same as that of RTC-FastXML as shown in formula (1), with a replacement of $\mathcal{L}_{nDCG_{@k}}(\boldsymbol{r}, \boldsymbol{y}_i)$ with $\mathcal{L}_{PSnDCG_{@k}}(\boldsymbol{r}, \boldsymbol{y}_i)$:

$$\mathcal{L}_{PSnDCG_{@k}}(\boldsymbol{r}, \boldsymbol{y}_i) = I_k(\boldsymbol{y}_i) \sum_{l=1}^{k} \frac{y_{ir_l}}{p_{ir_l} log(1+l)} \tag{7}$$

where p_{ir_l} is the marginal probability and other variables are the same as Eq. (2).

Similar to RTC-FastXML, the fifth term is added to the original objective function with the same intuition. The prediction procedure and optimization method are the same as those in RTC-FastXML. After prediction, we also rank the labels again by a classifier designed specially for tail labels, which is the same as PfastreXML.

4.3   RTC-PLTs

A. Training model. To build a multi-label classification tree $\mathcal{T}$, we need to learn a classifier in each node. Each leaf in the tree corresponds to a label. We denote a set of leaves of a (sub) tree rooted in node t by $L(t)$, $pa(t)$ stands for the parent node of t and $ch(t)$ stands for the set of child nodes of t. The training process is shown in Algorithm 3.

The objective of PLTs is to minimize the following loss function for each training instance i:

$$L(f|\boldsymbol{x}_i) = \sum_{j=1}^{L} \sum_{t \in path(j)} |\eta_T(\boldsymbol{x}_i, t) - \hat{\eta}_T(\boldsymbol{x}_i, t)| \tag{8}$$

In this equation, $\boldsymbol{x}_i$ is the feature of instance i. $\eta_T(\boldsymbol{x}_i, t)$ is the real conditional probability at node t of instance i, and $\hat{\eta}_T(\boldsymbol{x}_i, t)$ is the predicted value, measured by a real valued function $f_t(\boldsymbol{x}_i)$. $\hat{\eta}_T(\boldsymbol{x}_i, t) = \sigma(f_t(\boldsymbol{x}_i))$. σ is sigmoid function, mapping $f_t(\boldsymbol{x}_i)$ from $\mathbb{R}$ to $[0, 1]$. In PLTs, $f_t(\boldsymbol{x}_i) = \boldsymbol{w}\boldsymbol{x}_i$, where $\boldsymbol{w}$ is weight in node t. Variable j refers to each of the L labels and t stands for each node along the path from root to the leaf corresponding to label j. If instance i has several labels and these labels correspond to some leaves in the tree, for the nodes along the paths to these leaves, $\eta_T(\boldsymbol{x}_i, t)$ is 1. For other nodes, $\eta_T(\boldsymbol{x}_i, t)$ is 0, as shown in line 5 of Algorithm 3.

In RTC-PLTs, we have the same loss function as Eq. (8) and to utilize instance relationship information, we add one more term to $f_t(\boldsymbol{x}_i)$:

$$f_t(\boldsymbol{x}_i) = \boldsymbol{w}\boldsymbol{x}_i + C_s \sum_{j \in Sd_i} log(\frac{s_{ij}}{S_i} + 1)(\hat{\eta}_T(\boldsymbol{x}_j, t) - \frac{1}{2}) \tag{9}$$

Algorithm 3. RTC-PLTs: TRAIN($\{x_i, y_i\}_{i=1}^N, \mathcal{T}$)

1: **Input:** $\{x_i, y_i\}_{i=1}^N, \mathcal{T}$
2: **Output:** w
3: **for** each node $t \in \mathcal{T}$ **do**
4:      Initialize w_t randomly
5:      **for** $i = 1 \to N$ **do**
6:          **if** t is root or $\eta_T(x_i, pa(t)) = 1$ **then**
7:              **if** $\sum_{j \in L(t)} y_{ij} \geq 1$ **then** $\eta_T(x_i, t) = 1$ **else** $\eta_T(x_i, t) = 0$
8:              $w_t \leftarrow \min \sum_{j \in L(t)} |\eta_T(x_i, t) - \hat{\eta}_T(x_i, t)|$
9:              $\hat{\eta}_T(x_i, t) = \sigma(w_t x_i + C_s \sum_{j \in Sd_i} \log(\frac{s_{ij}}{S_i} + 1)(\hat{\eta}_T(x_j, t) - \frac{1}{2}))$

Algorithm 4. RTC-PLTs: PREDICT($\{x_i\}, \mathcal{T}$)

1: **Input:** $\{x_i\}, \mathcal{T}$
2: **Output:** y_i
3: $\hat{\eta}_T(x_i, root(\mathcal{T})) = \sigma(w_{root} x_i + C_p \sum_{j \in Sd_i} \log(\frac{s_{ij}}{S_i} + 1)(\hat{\eta}_T(x_j, root(\mathcal{T})) - \frac{1}{2}))$
4: $Q = \varnothing, y_i = 0, Q.add(root(\mathcal{T}), \hat{\eta}_T(x_i, root(\mathcal{T})))$
5: **while** $Q \neq \varnothing$ **do**
6:      $(t, p_t) = pop(Q)$
7:      **if** $p_t \geq$ threshold **then**
8:          **if** t is a leaf node **then**
9:              $\hat{y}_t = 1$
10:          **else**
11:              **for** $c \in ch(t)$ **do**
12:                  $\hat{\eta}_T(x_i, c) = \sigma(w_c x_i + C_p \sum_{j \in Sd_i} \log(\frac{s_{ij}}{S_i} + 1)(\hat{\eta}_T(x_j, c) - \frac{1}{2}))$
13:                  $Q.add(c, p_t \cdot \hat{\eta}_T(x_i, c))$

where $\hat{\eta}_T(x_j, t)$ is the conditional probability of instance j calculated in the last training epoch at node t. s_{ij}, S_i, Sd_i and C_s have the same meanings as those in Eq. (1).

Adding the second term, we aim to leverage similar instances' help to judge if instance i should go further along the path. If the similar instances of i are more likely to continue the path at the current node, i.e. $\hat{\eta}_T(x_j, t) > \frac{1}{2}$, then instance i is more likely to continue, too. Vice versa. The more similar two instances are, the more likely they continue the same paths and have the same labels.

B. Prediction. After we get w at each node, at prediction time, the conditional probability at each node is calculated based on Eq. (9), where C_s is replaced by another hyper-parameter C_p. $\hat{\eta}_T(x_j, t)$ here is the conditional probability of training instance j calculated finally in the training process at node t. The intuition is that

when we predict labels of an instance i, we use both the estimator w and the similar instances of i in the training set to calculate the conditional probability at each node. If the similar instances in the training set are more likely to continue at a node, then instance i is also more likely to continue at that node.

Finally, the product of $\hat{\eta}_T(x_i, t)$ of all nodes along a specific path is used to decide whether the label represented by the leaf nodes should be assigned to instance i. The predict process is shown in Algorithm 4. We initialize a queue Q to perform a breadth first traversal.

C. Optimization. The second term we added in Eq. (9) is a constant value for w, so we can directly optimize w in the same way as done by PLTs with a two-phase gradient descent step [6].

4.4   Feature Engineering

A simple way to utilize instance relationship is to transform relationship information into features. We regard each instance as a feature. As for instance i, those instances similar to it have a nonzero weight as their values, and others have value of zero. The weight can be calculated based on the strength of the relationship. Taking the app downloading scenario for example, we can use tf-idf measure as the weight. Suppose a user u's behavior sequence is denoted by $q_u = <session_1^u, session_2^u, \dots >$, where each session $session_i^u$ is a sequence of apps (instances) browsed and downloaded by the user in a short period session. Let N be the number of instances (app) in the training dataset, M be the number of instances in the testing dataset. Then, for each instance i in training dataset or testing dataset, we use a $N + M$ dimensional vector z_i to depict its relationship with other instances, where the jth element is defined in (10), which is similar to tf-idf measure.

$$z_{ij} = \frac{s_{ij}}{S_i} \times log\frac{N + M}{sf_j} \tag{10}$$

where s_{ij} is the number of times instances i and j co-occur in a same session of all of users' behavior sequences, sf_j is the number of distinct instances co-occurred with instance j.

We directly concatenate z_i to the raw feature x_i as the feature of instance i in RTC related methods.

5   Experiments

In this section, we present experiments conducted on three different kinds of real datasets to evaluate the performance of our proposed method.

5.1   Datasets

A. Mobile App Dataset. This dataset contains app description and user behavior sequence, including browsing and downloading behavior. We use the co-occurrence of apps in the behavior sequence of a session to infer the similarity between apps, and use the tf-idf of description terms as the original features. The tags of apps are labels, such as *Game, Camera, Chat*, etc. This dataset is obtained from a famous mobile app store.

B. Amazon Category Dataset. The Amazon Category dataset contains the description of each item sold on Amazon and a set of similar items of each one. We use the similar items as relationship between items directly. It is published in *Stanford Large Network Dataset Collection*[1]. We use the tf-idf of the item

---

[1] http://snap.stanford.edu/data/#amazon.

description words as the original feature, published by Probhu et al. [9]. The labels here are the hierarchical categories of the items, such as *Books, Bible, Family Bibles*, etc.

Table 1. Dataset statistics

Dataset	Train N	Features D	Labels L	Test M	Avg. labels instance	Avg. instances per label	Avg. similar instances per instance
Mobile App	60406	768631	20118	20136	2.65	10.60	3.45
Amazon	57168	203882	14385	16856	11.68	60.11	1.63
Twitter	2442	17829	159	1221	12.5	288.8	49.77

C. Twitter Dataset. The Twitter dataset contains the description of each twitter celebrities and each celebrity has several tags to describe his or her interest. Normal users may follow the celebrities due to some aspects, so we exploit relationship of celebrities in the following list of normal users. That is to say, two celebrities are similar to some extent if they co-occur in the same following list, and we use the co-occurrence frequency as s_{ij} to measure the similarity between i and j. The dataset is published by He et al. [8]

The detail of the three datasets is shown in Table 1.

5.2 Hyper-parameters

RTC related classification models have two common hyper-parameters C_s, C_p. To choose appropriate values of these parameters, we calculate precision of the predicted top 1 tag on the Mobile App dataset under different parameters, which is illustrated in Fig. 1. In Fig. 1.(a), we set $C_p = 1.0$, and C_s varies from 0.01 to 0.1. In Fig. 1.(b), we set C_s as the best of Fig. 1.(a), and C_p varies from 0.1 to 1.5. As we can see from the figures, when C_s and C_p change in this range, precisions of three models don't change obviously. In the following experiments, we set $C_s = 0.03$ and $C_p = 0.8$ for RTC-FastXML, $C_s = 0.1$ and $C_p = 0.6$ for RTC-PfastreXML, $C_s = 0.07$ and $C_p = 0.9$ for RTC-PLTs. Besides, we choose Huffman tree with $32°$ in PLTs.

The other hyper-parameters are the same as used in the original models published by the authors. The number of trees in FastXML and PfastreXML is both 50, and epoch in PLTs is 30.

For the mobile app dataset, we split the behavior sequence into sessions by 2 min' gap. That is to say, we regard the behaviors of one user within 2 min as a session.

5.3 Experimental Results

Implementations of FastXML, PfastreXML and PLTs are provided by their authors. We evaluate the performance of various algorithms based on measures of

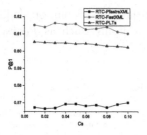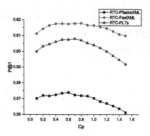

Fig. 1. Parameter sensitivity study on Mobile App dataset

precision at k and nDCG at k (k=1, 3, 5). The baseline algorithms are FastXML, PfastreXML, PLTs, SLEEC [3], and PDSparse [22].

Table 2 compares the performance using precision. *time* in the table means testing time on the test dataset in seconds. *difference* in the table means the accuracy difference between RTC related algorithms and the original models. Figures in bold are the highest precision among six algorithms for each dataset. This result corresponds to the combination of the two ways of relationship utilization: collaborative labeling and feature engineering. We also conduct experiments on the datasets to compare the two ways. We find that they both can improve classification accuracy, and their combination is the best. In the following, we only show the results of combination due to space limitation. Table 3 is the result in terms of nDCG. Our proposed models also perform better than corresponding original models.

As can be seen from Tables 2 and 3, the proposed RTC related algorithms lead to significantly better prediction accuracies compared to the state-of-the-art methods on all the three datasets. Overall, RTC-FastXML achieves the highest accuracy, and FastXML works better than PLTs and PfastreXML on the three datasets.

As for the efficiency, PLTs needs the least time. PDsparse is performed in a multi-thread method while the other models are all performed in single thread. PDSparse and SLEEC need more time to train the model and consume much more memory than the others. Amazon dataset needs more time than the other two datasets, because instances have more labels on average and the scale of the dataset is bigger. RTC related algorithms need a little more seconds to run than the original models, mainly because the number of features for each instance becomes bigger by adding new features based on feature engineering method. But the extra time to predict all test instances is less than 10 s.

In addition, from the experiments we also find that the quality of the relationship information is important. On Twitter dataset, we use the information of the frequency of two celebrities co-occurred in the same following list of a normal user to measure the similarity. We may follow people due to different reasons, and there may be some noise. Hence, the information is not so strong as that used in the first two datasets. Therefore, the improvement of precision is relatively smaller than the other two datasets.

5.4   Case Study

In this section, we showcase the experimental results with several cases, which can help understand the proposed model better. We take FastXML and RTC-FastXML on Amazon dataset as examples. Table 4 shows the top 10 categories (labels) predicted by them on Amazon dataset. The first column of this table shows the name of each item (underlined) and terms extracted from its description. The fifth column shows the similar items (underlined) and their categories. Categories shown in bold in the third, fourth and fifth column are those matching with the ground truth. The first two cases in the table suffer the problem of feature sparsity or limitation of description information.

From the table, it is easy to see that RTC-FastXML can predict more accurate labels than FastXML, especially in the first two cases. For example, the first case is about item *The Sound of the Mountain*, which has three terms in its description: mountain, sound and text. FastXML assigns label *music* to it, possibly due to the fact that the term *sound* appears in its description text. As a matter of fact, it is a Japanese novel. With the help of similar items, RTC-FastXML can successfully predict all the right categories for this item. Therefore, when description terms of the item are not enough to represent its characteristics, similar items are especially important to improve the accuracy.

Table 2. Precision@k of different algorithms (%) and the time for testing in seconds

Algorithm	Mobile app				Amazon				Twitter			
	P@1	P@3	P@5	time	P@1	P@3	P@5	time	P@1	P@3	P@5	time
FastXML	57.07	40.28	29.56	19.66	93.38	89.21	79.53	67.24	76.00	65.33	59.62	2.31
RTC-FastXML	**61.78**	**43.70**	31.70	29.87	**94.36**	**91.30**	**82.79**	70.98	76.25	67.13	59.80	3.09
PfastreXML	51.00	39.11	29.36	42.37	77.71	78.12	72.96	80.12	73.96	63.69	56.18	1.10
RTC-PfastreXML	57.39	43.12	**31.74**	48.96	82.50	82.09	77.05	84.5	75.51	65.74	58.97	3.30
PLTs	54.27	38.07	28.04	2.97	66.89	24.00	14.49	4.10	75.10	66.09	59.51	0.08
RTC-PLTs	60.88	43.12	31.22	3.53	70.20	46.22	34.91	4.82	**76.82**	**67.26**	**60.36**	0.10
PDSparse	51.22	35.72	25.95	3.33	88.10	82.81	74.41	4.42	65.38	57.39	51.86	0.08
SLEEC	53.33	37.03	26.71	37.67	93.73	88.31	76.84	70.12	75.38	66.02	58.72	2.30

Table 3. nDCG@k of different algorithms (%)

Algorithm	Mobile app			Amazon			Twitter		
	nDCG@1	nDCG@3	nDCG@5	nDCG@1	nDCG@3	nDCG@5	nDCG@1	nDCG@3	nDCG@5
FastXML	57.07	56.11	58.30	93.38	90.12	83.44	76.00	71.65	60.16
RTC-FastXML	**61.78**	60.70	62.68	**94.36**	**91.98**	**86.17**	76.25	71.18	61.50
PfastreXML	51.00	53.25	56.23	77.71	78.01	74.68	73.96	67.78	64.26
RTC-PfastreXML	57.39	57.55	60.82	82.50	82.17	78.88	75.51	69.77	66.83
PLTs	54.27	63.10	65.42	66.89	33.56	24.41	75.10	86.82	83.35
RTC-PLTs	60.88	**70.86**	**72.91**	70.20	55.21	45.32	**76.82**	**88.00**	**84.47**
PDSparse	51.22	50.90	53.50	88.10	84.13	78.19	65.38	61.91	60.02
SLEEC	53.33	51.84	53.31	93.73	89.52	81.50	75.38	74.47	72.84

Table 4. Top 10 categories predicted by FastXML and RTC-FastXML on Amazon Category dataset

Item and description	Ground truth	Categories predicted by different models		Similar items and labels
		FastXML	RTC-FastXML	
The Sound of the Mountain: mountain sound text	Books; Subjects; General; Literature & Fiction; World Literature; Literary; Japanese	*General*; Music; Styles; Specialty Stores; Indie Music; Country; Bluegrass; Contemporary; Folk;Traditional	*Books*; *Subjects*; *General*; *Literature & Fiction*; *Literary*; *World Literature*; *Japanese*; Contemporary; Short Stories; Authors, A-Z	1.First Snow on Fuji: *Books*; *Subjects*; *Literature & Fiction*; *General*; *Literary*;Short Stories; *World Literature*; *Japanese*; 2.Snow Country: *Books*; *Subjects*; *Literature & Fiction*; *General*; *Literary*; *World Literature*; *Japanese*; 3.Beauty and Sadness: *Books*; *Subjects*; *Literature & Fiction*; *General*; *Literary*; *World Literature*; *Japanese*
Adrenalize: cassette	General; Music; Styles; Classic Rock; Album-Oriented Rock (AOR); Hard Rock & Metal; Hard Rock; Pop Metal	*Styles*; *Music*; Categories; *General*; Amazon.com Stores; Travel; Formats; Music Outlet; Bargains; Today's Deals in Music	*Styles*; *Music*; *General*; *Classic Rock*; *Hard Rock & Metal*; *Album-Oriented Rock (AOR)*; *Hard Rock*; *Pop Metal*; Rock; British Metal	1.Retro Active: *Music*; *Styles*; *Hard Rock & Metal*; *General*; *Pop Metal*; *Classic Rock*; *Album-Oriented Rock (AOR)*; *Hard Rock*; 2.Euphoria: *Music*; *Styles*; *Hard Rock & Metal*; *General*; *Pop Metal*; *Classic Rock*; *Album-Oriented Rock (AOR)*; *Hard Rock*; 3.Slang: *Music*;*Styles*; *Hard Rock & Metal*; British Metal; *General*; *Pop Metal*; *Classic Rock*; *Album-Oriented Rock (AOR)*
Planning Your Addition: adding addition architect architectural assembled book builder color design detailed explore house ideas illustrations involved photos planning plans point show source starting use	Books; Subjects; Home & Garden; General; Home Design; Remodeling & Renovation; Design & Construction; House Plans	*Books*; *Subjects*; *General*; Amazon.com Stores; Business & Investing Books; Home & Office; Business & Investing; Marketing & Sales; *Home & Garden*; Marketing	*Books*; *Subjects*; *General*; *Home & Garden*; *Home Design*; *Remodeling & Renovation*; *Design & Construction*; Interior Design; How-to & Home Improvements; Professional & Technical	1. Additions: Your Guide to Planning and Remodeling (Better Homes and Gardens) : *Books*; *Subjects*; *Home & Garden*; *Home Design*; *Remodeling & Renovation*; *General*

6   Conclusions

In this paper we study how to leverage similarity relationship among instances to improve tree-based extreme multi-label learning. We propose to incorporate instance similarity relationship information into state-of-the-art models: FastXML, PfastreXML and PLTs. Two methods, collaborative labeling and feature engineering, are developed to utilize the relationship, which can be used individually or combined. Experiments were conducted on three real world datasets.

Results show that combining collaborative labeling and feature engineering with existing tree-based extreme multi-label classification models can improve classification performance significantly. Case study further helps us analyze the experimental results, from which we can see that exploiting instance relationship does help find more accurate labels, especially when the original feature space is sparse.

Acknowledgment. This work was supported in part by National Natural Science Foundation of China under grant No. U1711262, 71771131, 71272029, 71490724 and 61472426.

References

1. Agrawal, R., Gupta, A., Prabhu, Y., Varma, M.: Multi-label learning with millions of labels: recommending advertiser bid phrases for web pages. In: Proceedings of the 22nd International Conference on World Wide Web, pp. 13–24. ACM (2013)
2. Beygelzimer, A., Langford, J., Lifshits, Y., Sorkin, G., Strehl, A.: Conditional probability tree estimation analysis and algorithms. Eprint Arxiv, pp. 51–58 (2009)
3. Bhatia, K., Jain, H., Kar, P., Varma, M., Jain, P.: Sparse local embeddings for extreme multi-label classification. In: Cortes, C., Lawrence, N.D., Lee, D.D., Sugiyama, M., Garnett, R. (eds.) Advances in Neural Information Processing Systems vol. 28, pp. 730–738. Curran Associates, Inc. (2015)
4. Bi, W., Kwok, J.: Efficient multi-label classification with many labels. In: International Conference on Machine Learning, pp. 405–413 (2013)
5. Dembczynski, K., Cheng, W., Hllermeier, E.: Bayes optimal multilabel classification via probabilistic classifier chains. In: International Conference on Machine Learning, pp. 279–286 (2010)
6. Duchi, J., Singer, Y.: Efficient online and batch learning using forward backward splitting. J. Mach. Learn. Res. **10**(18), 2899–2934 (2009)
7. Getoor, L.: Introduction to Statistical Relational Learning. MIT press, Cambridge (2007)
8. He, W., Liu, H., He, J., Tang, S., Du, X.: Extracting interest tags for non-famous users in social network. In: Proceedings of the 24th ACM International on Conference on Information and Knowledge Management, pp. 861–870. ACM (2015)
9. Jain, H., Prabhu, Y., Varma, M.: Extreme multi-label loss functions for recommendation, tagging, ranking & other missing label applications. In: Proceedings of the 22nd ACM SIGKDD International Conference on Knowledge Discovery and Data Mining, pp. 935–944. ACM (2016)
10. Jain, P., Meka, R., Dhillon, I.S.: Guaranteed rank minimization via singular value projection. In: Advances in Neural Information Processing Systems, pp. 937–945 (2010)
11. Jasinska, K., Ski, K.D., Busa-Fekete, R., Pfannschmidt, K., Klerx, T., Llermeier, E.H.: Extreme f-measure maximization using sparse probability estimates. In: ICML (2016)
12. Lewis, D.D., Yang, Y., Rose, T.G., Li, F.: RCV1: a new benchmark collection for text categorization research. J. Mach. Learn. Res. **5**(Apr), 361–397 (2004)
13. Lin, Z., Ding, G., Hu, M., Wang, J.: Multi-label classification via feature-aware implicit label space encoding. In: ICML, pp. 325–333 (2014)

14. Macskassy, S.A., Provost, F.: A simple relational classifier. Technical report, DTIC Document (2003)
15. McPherson, M., Smith-Lovin, L., Cook, J.M.: Birds of a feather: homophily in social networks. Ann. Rev. Sociol. **27**(1), 415–444 (2001)
16. Prabhu, Y., Varma, M.: Fastxml: a fast, accurate and stable tree-classifier for extreme multi-label learning. In: Proceedings of the 20th ACM SIGKDD International Conference on Knowledge Discovery and Data Mining, KDD 2014, pp. 263–272. ACM, New York (2014)
17. Tai, F., Lin, H.T.: Multilabel classification with principal label space transformation. Neural Comput. **24**(9), 2508–2542 (2012)
18. Tang, L., Liu, H.: Relational learning via latent social dimensions. In: Proceedings of the 15th ACM SIGKDD International Conference on Knowledge Discovery and Data Mining, pp. 817–826. ACM (2009)
19. Tang, L., Liu, H.: Leveraging social media networks for classification. Data Min. Knowl. Discov. **23**(3), 447–478 (2011)
20. Wang, X., Sukthankar, G.: Multi-label relational neighbor classification using social context features. In: Proceedings of the 19th ACM SIGKDD International Conference on Knowledge Discovery and Data Mining, pp. 464–472. ACM (2013)
21. Wang, Y., Wang, L., Li, Y., He, D., Liu, T.Y., Chen, W.: A theoretical analysis of NDCG type ranking measures. arXiv preprint arXiv:1304.6480 (2013)
22. Yen, I.E.H., Huang, X., Ravikumar, P., Zhong, K., Dhillon, I.: PD-sparse: a primal and dual sparse approach to extreme multiclass and multilabel classification. In: International Conference on Machine Learning, pp. 3069–3077 (2016)

Exploiting Ranking Consistency Principle in Representation Learning for Location Promotion

Siyuan Zhang[1]([✉]), Yu Rong[2], Yu Zheng[3], Hong Cheng[1], and Junzhou Huang[2]

[1] The Chinese University of Hong Kong, Hong Kong, China
{syzhang,hcheng}@se.cuhk.edu.hk
[2] Tencent AI Lab, Shenzhen, China
{royrong,joehhuang}@tencent.com
[3] Microsoft Research, Beijing, China
yuzheng@microsoft.com

Abstract. Location-based services, which use information of people's geographical position as service context, are becoming part of our daily life. Given the large volume of heterogeneous data generated by location-based services, one important problem is to estimate the visiting probability of users who haven't visited a target Point of Interest (POI) yet, and return the target user list based on their visiting probabilities. This problem is called the *location promotion problem*. The location promotion problem has not been well studied due to the following difficulties: (1) the cold start POI problem: a target POI for promotion can be a new POI with no check-in records; and (2) heterogeneous information integration. Existing methods mainly focus on developing a general mobility model for all users' check-ins, but ignore the ranking utility from the perspective of POIs and the interaction between geographical and preference influence of POIs.

In order to overcome the limitations of existing studies, we propose a unified representation learning framework called *hybrid ranking and embedding*. The core idea of our method is to exploit the ranking consistency principle into the representation learning of POIs. Our method not only enables the interaction between the geographical and preference influence for both users and POIs under a ranking scheme, but also integrates heterogeneous semantic information of POIs to learn a unified preference representation. Extensive experiments show that our method can return a ranked user list with better ranking utility than the state-of-the-art methods for both existing POIs and new POIs. Moreover, the performance of our method with respect to different POI categories is consistent with the hierarchy of needs in human life.

Keywords: Location promotion · Ranking consistency
Graph embedding

© Springer International Publishing AG, part of Springer Nature 2018
J. Pei et al. (Eds.): DASFAA 2018, LNCS 10828, pp. 457–473, 2018.
https://doi.org/10.1007/978-3-319-91458-9_28

1   Introduction

Location-based services, such as Foursquare, Yelp and Facebook Place, are becoming increasingly popular these days. The data generated in a typical location-based service consist of two parts: the check-in records of users and the profiles of Points of Interest (POIs). Many methods have been developed for applications such as POI recommendation [2,6,9–11,13,14,19,21,22] and friend recommendation in location-based social networks [15,18]. Besides providing user-centric services as mentioned above, mining data from location-based services can also help local companies to promote their business more effectively. For example, a new restaurant owner at Pittsburgh would like to know the target users who are more likely to have dinner at his restaurant according to their check-in histories so that he can distribute coupons to them.

Different from existing studies on POI recommendation [2,6,9–11,13,14,19, 21,22], which recommends POIs to a target user as a user-centric task, we need to recommend users to a target POI in the coupon distribution scenario. In this paper, we define the task of ranking users according to their visiting probabilities to a target POI as a **location promotion problem**. There are two major challenges for this problem: (1) the cold start POI problem, i.e., a target POI for promotion can be a new POI with no check-in records; and (2) heterogeneous information integration, as there are both geographical and semantic information associated with POIs. Existing POI recommendation methods [10,11,13,22] cannot be directly applied to solve the location promotion problem with satisfactory performance.

Existing solutions [21,28] for location promotion is unsatisfactory due to two reasons. *First*, they build their models by maximizing the likelihood of observing all check-in records. They ignore "unobserved" POI-user pairs, i.e., those users with no check-in at certain POIs, in their models. But such "null" relationship can be combined with the check-in records to help infer the ranking of potential users to specific POIs. *Second*, they consider modeling interaction between users and POIs in only one space. [28] only considers the geographical proximity between users and POIs, i.e., geographical space. [21] only considers representing users and POIs in one POI latent space. However, using only one space is not enough for integrating heterogenous information in location-based service because different information may have different interaction patterns.

In order to overcome the limitations of existing studies, we propose a unified representation learning framework called **hybrid ranking and embedding (HRE)** to solve the location promotion problem. The core idea of our method is to **exploit the ranking consistency principle in the representation learning framework of POIs**. The *ranking consistency principle* states that users with more check-ins at the POI should be ranked higher than users with fewer check-ins or no check-in at the POI. With the ranking consistency principle, we can use the unobserved POI-user pairs to alleviate the data sparsity issue in check-in records. To measure the interaction between users and POIs, we use geographical embeddings and preference embeddings to represent both users and POIs. The geographical embeddings measure the geographical influence to users and POIs in each region. The preference embeddings measure the similarity of check-in patterns between users and between POIs. The final ranking score of users w.r.t. the target POI is estimated by their embeddings in both geographical space and preference space.

To learn the embeddings of new POIs in preference space as well as improve the ranking performance of existing POIs, we build different types of weighted bipartite POI-semantic graphs to capture and integrate heterogeneous semantic information of POIs. Our intuition is that POIs with similar semantic information should be close in preference space. Five kinds of information, including sequential visiting POIs, temporal check-in number in each time slot, region proximity, tags, and neighborhood visiting users, are considered for learning the preference embeddings for all POIs through multiple graph embeddings. A joint learning algorithm is proposed to train a unified preference representation for each POI.

Compared with existing solutions, our method has four advantages. *First*, we can gain better ranking performance for location promotion since we optimize from the POIs' perspective and integrate heterogeneous semantic information for POIs. *Second*, our hybrid model can learn preference embeddings for new POIs by sharing POI preference embeddings in different POI-semantic graphs. New POIs can utilize the collective check-in data of existing POIs to achieve better ranking performance. *Third*, the embeddings of users are divided into two spaces, the geographical and preference spaces, and trained by a joint learning algorithm, which enables the interaction between geographical influence and preference influence as well as preserving the heterogenous interaction pattern in different spaces. *Fourth*, our method can be easily extended to incorporate other kinds of semantic information such as the photos and comments posted by users at POIs.

We have made the following contributions in this paper.

- To the best of our knowledge, we are the first to incorporate the ranking consistency information into the representation learning framework of POIs, which has been ignored by current literatures.
- Our method can integrate both the geographical and semantic information of POIs. Existing POIs and new POIs can have a unified embedding in both geographical space and preference space.
- Extensive experiment results show that our method can boost the ranking performance measured by AUC and Tau for both existing POIs and new POIs, compared to several baselines and state-of-the-art methods. On average, we increase the AUC by 3.0% and Tau by 7.7% for all POIs in four cities of United States, compared to the second best solution.
- We conduct a case study to demonstrate that the performance of our method with respect to different POI categories is consistent with the hierarchy of needs in human life, which has not been reported by previous studies.

The remainder of this paper is organized as follows: Sect. 2 gives a formal definition of our problem and performs data analysis for geographical space embedding. Section 3 describes the hybrid ranking and embedding method and the parameter learning algorithm. Section 4 reports the experimental results. Section 5 reviews related work. Finally Sect. 6 concludes this paper.

2   Preliminaries

2.1   Problem Description

Assume we have a user set $U = \{u_1, u_2, \ldots, u_N\}$ and a POI set $L = \{l_1, l_2, \ldots, l_M\}$. A check-in record can be defined as a triple $c = (u, l, t)$ which indicates that user u performs a check-in action on POI l at time t. We denote $C_u = \{c_1, c_2, \ldots\}$ as the user u's check-in records, C_{ul} as the set of check-in records which are performed by user u on POI l, and U_l as the set of users that have check-in at l. Furthermore, we denote a POI as $l = (\omega_l, \tau_l)$, where ω_l is the coordinate of l and τ_l is the tag set of l. The *location promotion problem* can be defined as:

Definition 1 *(Location Promotion Problem). Given the user set* $U = \{u_1, u_2, \ldots, u_N\}$, *all users' check-in records* $C = \{C_1, \ldots, C_N\}$ *and a target POI* l, *return a rank list of candidate users who have not visited POI* l *yet, i.e.,* $u \in U \setminus U_l$, *in descending order of their probability of visiting* l.

2.2   Data Analysis

Different from the traditional recommendation problem, user's check-in records not only contain the user's preference on POIs but also indicate the user's geographical preference. For example, the check-in records of a user usually cluster around his/her workplace and home [4].

In order to find a reasonable geographical embedding method for users and POIs, we perform data analysis on the check-in records from two cities Los Angeles and San Diego. We first use k-means algorithm to cluster POIs in a city into K regions according to their coordinates. Then we consider all sequential check-in records in pairs made by the same user, in the form of (l_{j-1}, l_j). Denote the region center where l_{j-1} belongs to as ω_k. We calculate the distance between ω_k and the next POI l_j. Then we plot the distribution of such distances from a region center to the next check-in POI in Fig. 1. We observe that the visiting probability decreases as the distance between the region center and next check-in POI increases.

To further choose a suitable distribution to describe the relationship, we try to fit the data with three alternatives: Pareto distribution, Exponential distribution and Log-normal distribution, and show the loglikelihood of different distributions in Table 1. We find that the Pareto distribution has the largest loglikelihood in modeling the distance distribution between the region center and the next check-in POI. Thus, we decide to use Pareto distribution in POIs' geographical embeddings.

Table 1. Loglikelihood of different distance distribution in Los Angeles and San Diego

City	Pareto	Exponential	Log-normal
Los Angeles	**−2.476**	−3.186	−2.879
San Diego	**−2.153**	−3.199	−2.839

3   Hybrid Ranking and Embedding Method

In this section, we first introduce the two building blocks of our unified model: ranking consistency model and graph based POI embedding. Then we describe our joint learning model.

3.1   Learning Geographical and Preference Embedding with Ranking Consistency

Based on the analysis on Sect. 2.2, we utilize two latent spaces $\mathcal{G}$ and $\mathcal{V}$ to model the geographical factor and the preference factor respectively.

Geographical Factor Embedding. We propose a geographical embedding method to encode the spatial proximity between candidate users and target POIs. For POI l_j, we denote its geographical feature by an embedding vector $\mathbf{g}_j \in \mathcal{G}$ as:

$$\mathbf{g}_j = [f(d(\omega_j, \omega_1)), \ldots, f(d(\omega_j, \omega_K))]^T. \tag{1}$$

In Eq. (1), $d(\cdot, \cdot)$ represents the distance between two coordinates and the k-th entry of $\mathbf{g}_j$ denotes the probability that a user moves from the k-th region to l_j. In general, $f(\cdot)$ can be any function that can output a probability. In our proposed framework, we adopt the Pareto distribution based on the analysis in Sect. 2.2. Following the setting in [28] where a similar observation is reported between sequentially visited POIs, the specific form of $\mathbf{g}_j$ is:

$$\mathbf{g}_j = [(1 + d(\omega_j, \omega_1))^{-\alpha}, \ldots, (1 + d(\omega_j, \omega_K))^{-\alpha}]^T. \tag{2}$$

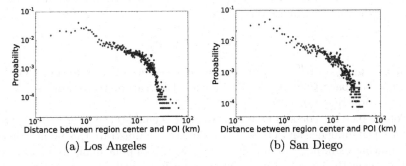

(a) Los Angeles   (b) San Diego

Fig. 1. The distance distribution between the sequential check-in region center and POI in two cities

α is a shape parameter in Pareto distribution and is estimated by maximum likelihood estimation.

The geographical embedding for user u_i can be defined as:

$$\mathbf{g}_i = [\gamma_{i,1}, \gamma_{i,2}, \ldots, \gamma_{i,K}]^T, \tag{3}$$

where $\gamma_{i,k}$ denotes the expected number of check-ins in the k-th region by u_i.

Using the geographical embedding of POIs and users, we define the ranking score y_{ij} that u_i will visit l_j as:

$$y_{ij} = \mathbf{g}_i \cdot \mathbf{g}_j. \tag{4}$$

Equation (4) can be considered as a two-layer model, where the inner product encompasses the weighted sum of the expected check-in number from user u_i's activity regions denoted by $\mathbf{g}_i$. The weight factor $\mathbf{g}_j$ of a target POI l_j only depends on its distance to each region center, which implies the spatial influence from each region to l_j.

Preference Factor Embedding. Though the geographical factor has a large impact on determining the ranking score of candidate users, there are still portions of check-in records that cannot be explained by the geographical factor. For example, when people choose a certain store they like to check-in in a shopping mall, the distance factor has less impact compared to users' preference. If we represent all users' check-in records with a POI-user matrix where each entry is the number of check-ins from a specific user at a POI, we can infer a user's preference by comparing his/her check-in records with those of other users who have similar check-in patterns.

To model the preference, we introduce an embedding vector $\mathbf{v}_i \in \mathcal{V}$ for u_i, and $\mathbf{v}_j \in \mathcal{V}$ for l_j to represent the preference factors of user u_i and POI l_j respectively. If u_i has check-ins at l_j, $\mathbf{v}_i \cdot \mathbf{v}_j$ should be larger than that of another user $u_{i'}$ who has no check-in at l_j. Under this assumption, two users who visit many common POIs should have close preference embeddings. On the other hand, POIs that share many common users should have close preference embeddings in $\mathcal{V}$. To combine the information from geographical coordinates and POI-user matrix, we define the final ranking score y_{ij} as:

$$y_{ij} = \mathbf{v}_i \cdot \mathbf{v}_j + \mathbf{g}_i \cdot \mathbf{g}_j. \tag{5}$$

In Eq. (5), $\mathbf{g}_i, \mathbf{g}_j \in \mathbb{R}_+^K$ indicate the feature vectors of user u_i and POI l_j in geographical space $\mathcal{G}$; $\mathbf{v}_i, \mathbf{v}_j \in \mathbb{R}^D$ indicate the feature vectors of user u_i and POI l_j in preference space $\mathcal{V}$.

Combining the scores in two separate spaces enables the interaction between geographical influence and preference influence. To illustrate this point, consider the following example. For a target shop in a shopping mall, users who have check-ins near the shopping mall are more likely to walk by and visit the target shop, i.e., from the perspective of the geographical embedding. On the other hand, users who have the similar preference to the target shop are also more likely to visit the target shop, i.e., from the perspective of preference embedding.

In our solution, the geographical embedding $\mathbf{g}_i$ for users and the preference embedding $\mathbf{v}_i$ for users and $\mathbf{v}_j$ for POIs are all trained using the check-in records because the influence of geographical embeddings and preference embeddings on check-in behaviors varies from user to user. Later we will describe how to learn the vectors $\mathbf{g}_i, \mathbf{v}_i, \mathbf{v}_j$ for all users and POIs in our inference algorithm.

Ranking Consistency Model. Given the definition of ranking score, we propose our ranking consistency model to learn the latent representation of users and POIs in both geographical and preference spaces simultaneously. The core idea behind our model lies in the *ranking consistency principle*, which indicates that the ranking score of different users should be consistent with their check-in records. Concretely, given a target POI l_j, the following constraints should be satisfied:

- A user who has performed check-in at l_j should be ranked higher than those who have not performed check-in at l_j.
- A user with more check-in records at l_j should be ranked higher than those with less check-in records at l_j.

Based on the *ranking consistency principle*, given POI l_j, our model can be defined as:

$$f(l_j|\Theta) = \prod_{u_i \in U_{l_j}} \prod_{u_{i'} \in U_{<|C_{ij}|}} P((y_{ij} - y_{i'j}) > 0|\Theta). \tag{6}$$

In Eq. (6), $U_{<|C_{ij}|} = \{u_{i'} | |C_{i'j}| < |C_{ij}|\}$ is the set of users whose number of check-in records at POI l_j is less than that of user u_i. $\Theta = \{\mathbf{v}_i, \mathbf{v}_j, \mathbf{g}_i | u_i \in U, l_j \in L\}$ is the parameter set. $P((y_{ij} - y_{i'j}) > 0|\Theta)$ is the probability which indicates that the ranking score of user u_i is higher than that of user $u_{i'}$. Following the Bayesian personalized ranking scheme [16], we apply the logistic function $\sigma(x) = 1/(1 + \exp(-x))$ to output the probability $P((y_{ij} - y_{i'j}) > 0|\Theta)$:

$$P((y_{ij} - y_{i'j}) > 0|\Theta) = \frac{1}{1 + e^{-(y_{ij} - y_{i'j})}}. \tag{7}$$

Therefore, given a POI set L, the log-likelihood function $\mathcal{F}_{RC}(L, \Theta)$ is written as:

$$\mathcal{F}_{RC}(L, \Theta) = \sum_{l_j \in L} \log f(l_j|\Theta) - \lambda ||\Theta||^2. \tag{8}$$

In Eq. (8), $\lambda ||\Theta||^2$ is the Gaussian prior for regularization. Compared with the methods that directly approximate the check-in frequency like rating based recommendation problems, Bayesian personalized ranking criterion learns the ranking models based on pairwise comparison of users such that the area under the ROC curve (AUC) can be maximized [16]. On the other hand, it alleviates the data sparsity problem in modeling check-in records by fully utilizing information of the number of check-in records of users at a POI. This happens to meet our goal for location promotion.

3.2   Learning POI Semantic with Graph Based Embedding

Although the ranking based embedding learning method can alleviate the data sparsity problem in check-in data, it cannot handle new POIs without any check-in record. Other than the geographical and user preference information, POIs in location-based services also contain rich semantic information, such as tags, spatial check-in relations, temporal check-in patterns and neighborhoods.

In order to handle new POIs and enhance the representation power of our POI embeddings, based on the intuition that *POIs with similar semantic information would share common potential target users*, we utilize the POI semantic information to infer the preference embeddings of all POIs. This would be particularly useful for new POIs with no check-in records.

POI-semantic Graph Construction. We first extract the semantic information for POIs by constructing POI-semantic graphs. A POI-semantic graph is a bipartite graph carrying weights on the edges. According to different semantics, we design five types of POI-semantic graphs as follows.

- POI-POI graph: It is denoted as $G_{LL} = (L, L, E_{LL}, W_{LL})$ and designed for capturing POI check-in sequential relationship. The two node sets are both the POI set L. If there exists a user who visits two POIs $l_j, l_k \in L$ sequentially and the time gap between the two visits is less than a threshold ΔT, we add an edge $e_{jk} \in E_{LL}$ from l_j to l_k. The weight $w_{jk} \in W_{LL}$ of e_{jk} is defined as the number of such sequential visits between l_j and l_k in all check-in records.
- POI-Time graph: It is denoted as $G_{LT} = (L, T, E_{LT}, W_{LT})$ and designed for capturing POI temporal visit patterns. One node set is the POI set L, and the other node set is T representing different time slots. We first divide all the check-in timestamps into 24 time slots and denote each time slot as a node in T. For a POI $l_j \in L$ and a time slot $t_k \in T$, an edge $e_{jk} \in E_{LT}$ denotes that there are check-ins in l_j at t_k. The weight $w_{jk} \in W_{LT}$ for e_{jk} is the number of check-ins in l_j at t_k.
- POI-Tag graph: It is denoted as $G_{LW} = (L, W, E_{LW}, W_{LW})$ and designed for capturing functions of POIs. One node set is the POI set L, and the other node set is W representing different tags (such as "*Chinese Restaurant*" and "*Coffee*") of POIs. For a POI $l_j \in L$ and a tag $w_k \in W$, an edge $e_{jk} \in E_{LW}$ exists if l_j has tag w_k, and the weight $w_{jk} \in W_{WL}$ is defined as the *tf.idf* value.
- POI-Region graph: It is denoted as $G_{LR} = (L, R, E_{LR}, W_{LR})$ and designed for capturing region influence to POIs. One node set is the POI set L, and the other node set is R representing the K regions defined in Sect. 3.1. For a POI $l_j \in L$ and a region $r_k \in R$, an edge $e_{jk} \in E_{LR}$ with a unit weight denotes that l_j belongs to r_k.

- POI-Neighborhood User graph: It is denoted as $G_{LU} = (L, U, E_{LU}, W_{LU})$ and designed for capturing the neighborhood visit patterns. One node set is the POI set L, and the other node set is the user set U. A user $u_i \in U$ may not have check-in records in $l_j \in L$, but has check-ins in the k-nearest neighbor POIs of l_j. Then we consider u_i as l_j's neighborhood user, i.e., whose check-ins are near l_j, and an edge $e_{ij} \in E_{LU}$ is added. The edge weight $w_{ij} \in W_{LU}$ is the number of check-ins at l_j's neighbor POIs by u_i.

Overall, there are five types of semantic information as described above, thus we use a semantic set $S = \{L, T, R, W, U\}$ to denote them collectively. It is noted that the POI-POI, POI-Time and POI-Tag graphs are in the same form as defined in [21].

To solve the location promotion problem within a city, we further define POI-Region graph and POI-Neighborhood User graph to model different levels of geographical proximity between POIs. The construction of the two graphs helps us to learn a more geo-aware preference embedding for POIs and improve the ranking performance.

Learning POI and Semantic Embeddings. Based on the semantic information captured by the above POI-semantic graphs, we can learn the preference embeddings for new POIs. For the ease of presentation, we use a generic notation $G_{LS} = (L, S, E_{LS}, W_{LS})$ to denote a POI-semantic graph, which can be any specific type of the above five POI-semantic graphs. For the node set S in the POI-semantic graph, we call the nodes $s_1, s_2, \ldots, s_m \in S$ *semantic nodes*. Our target is to map these semantic nodes as well as the POIs to the preference latent space $\mathcal{V}$. We define the empirical conditional probability that s_k can be represented by l_j as:

$$\widehat{p}(s_k|l_j) = \frac{w_{jk}}{\sum_{s_{k'} \in S_{l_j}} w_{jk'}}, \tag{9}$$

where $S_{l_j} \subseteq S$ is the semantic node set related to POI l_j.

In the preference latent space $\mathcal{V}$, the embedding vector of s_k is $\mathbf{v}_k$. We use the softmax function to model the conditional probability that s_k can be represented by l_j in $\mathcal{V}$:

$$p(s_k|l_j) = \frac{\exp(\mathbf{v}_k \cdot \mathbf{v}_j)}{\sum_{s_{k'} \in S_{l_j}} \exp(\mathbf{v}_{k'} \cdot \mathbf{v}_j)}. \tag{10}$$

Since we have $\sum_{s_k \in S} \widehat{p}(s_k|l_j) = 1$ and $\sum_{s_k \in S} p(s_k|l_j) = 1$, given the POI l_j, the conditional probability over semantic information S can be treated as a distribution which is denoted as $\mathcal{P}_{l_j}$ (empirical distribution as $\widehat{\mathcal{P}}_{l_j}$). The objective of the embedding is to make the conditional distribution $\mathcal{P}_{l_j}$ close to the empirical distribution $\widehat{\mathcal{P}}_{l_j}$ for all POIs. We use the KL-divergence $KL(.,.)$ to measure the distance between two distributions. Thus the objective function for the semantic information S is written as:

$$\mathcal{F}(S, L | \Theta_S) = - \sum_{l_j \in L} \sigma_j KL(\widehat{\mathcal{P}}_{l_j}, \mathcal{P}_{l_j})$$

$$= \sum_{e_{jk} \in E_{LS}} \omega_{jk} \log p(s_k | l_j). \tag{11}$$

In Eq. (11), $\Theta_S = \{\mathbf{v}_k | s_k \in S\}$ is the embedding vector set of semantic nodes in S. $\sigma_j = \sum_{s_k \in S} \omega_{jk}$ is the importance of node l_j.

As there are five types of POI-semantic graphs, we construct a combinational graph-based POI embedding scheme. Given the semantic information set $S = \{L, T, R, W, U\}$, the objective function can be written as:

$$\mathcal{F}_{EMB}(\mathcal{S}, L | \Theta_S) = \sum_{S \in \mathcal{S}} \mathcal{F}(S, L | \Theta_S). \tag{12}$$

In Eq. (12), each POI in L has a unique preference embedding among different POI-semantic graphs. Therefore, the learned POI preference embeddings integrate heterogeneous information related to POIs and improve the ranking performance for all POIs as confirmed in experiments.

3.3 The Unified Model

We propose to jointly learn the embeddings of all elements simultaneously by sharing the preference embedding of POIs. The final objective function of our unified model is written as:

$$\mathcal{F}(L, \mathcal{S}, \Theta, \Theta_S) = \mathcal{F}_{RC}(L, \Theta) + \beta \mathcal{F}_{EMB}(\mathcal{S}, L | \Theta_S), \tag{13}$$

where $\Theta = \{\mathbf{v}_i, \mathbf{v}_j, \mathbf{g}_i | u_i \in U, l_j \in L\}$ and $\Theta_S = \{\mathbf{v}_k | s_k \in S, S \in \mathcal{S}\}$. After learning the embeddings of POIs, users and all kinds of semantic information, we can calculate the visiting score for each unobserved POI-user pair and rank the unvisited users w.r.t. a POI according to Eq. (5). Our model can be trained efficiently by a stochastic gradient descent method with mixed unified and alias sampling. The training time for the largest data set of LA is less than 10 min with 4 threads.

4 Experimental Results

In this section, we conduct extensive experiments on the real data sets. Firstly we evaluate the overall performance of our algorithms and the other baselines. Specifically, we compare the performance of new POIs to demonstrate the effectiveness of our algorithms in cold start situation. Then we analyze the parameter sensitivity. At last, we conduct a case study on the real world data sets to show the explanation power of our algorithms.

4.1   Settings

Datasets. We conduct experiments on a foursquare data set collected from 4 cities in USA[1]. We list the statistics of the data set in Table 2. For each city, we only consider users with check-ins at more than 4 POIs as active users. For each user, we first aggregate the check-ins at each POI and sort the POIs in ascending order according to the first check-in timestamp. Then we select the earliest 80% POIs as training data and the remaining 20% as test data. As a result, the numbers of test POIs in four cities are 3686 in LA, 4035 in SF, 2149 in SD, and 1392 in NY.

Table 2. Statistics of our data set

City	#POIs	#Users	#Check-ins
Los Angeles (LA)	7986	1128	49616
San Francisco (SF)	7907	1204	58547
San Diego (SD)	4770	393	25618
New York City (NY)	3631	365	8523

Evaluation Metrics. We use two metrics to evaluate the performance of our model: AUC and Kendall's Tau Coefficient.

AUC. For a target POI l_j and candidate users $U \backslash U_{l_j}$, we consider candidate users that have check-ins at l_j in the test set as positive users and other candidate users as negative users. Then we can plot the ROC curve according to the predicted ranking scores of a model and calculate the Area Under ROC Curve (AUC) as $AUC(l_j)$. We compare the average AUC of all POIs in the test set produced by different models.

Kendall's Tau Coefficient. Kendall's Tau is used to measure the overall ranking accuracy when we consider the number of check-ins in the test data. For a target POI l_j and two candidate users $u_i, u_{i'} \in U \backslash U_{l_j}$, we can get user ranking scores $y_{ij}, y_{i'j}$ and check-in numbers $|C_{ij}|, |C_{i'j}|$. Then, the user pair $(u_i, u_{i'})$ is said to be *concordant*, if both $y_{ij} > y_{i'j}$ and $|C_{ij}| > |C_{i'j}|$, or if both $y_{ij} < y_{i'j}$ and $|C_{ij}| < |C_{i'j}|$. On the other hand, if both $y_{ij} > y_{i'j}$ and $|C_{ij}| < |C_{i'j}|$, or if both $y_{ij} < y_{i'j}$ and $|C_{ij}| > |C_{i'j}|$, $(u_i, u_{i'})$ is said to be *discordant*. We define #*cons* and #*disc* as the number of concordant and discordant user pairs in candidate users $U \backslash U_{l_j}$, and $Tau = \frac{\#cons - \#disc}{\#cons + \#disc}$. We compare the average Tau of all POIs in the test set produced by different models.

---

[1] The data set is publicly available at https://sites.google.com/site/dbhongzhi/.

Baseline Methods. Our method is denoted as **HRE**, which stands for *hybrid ranking and embedding*. We compare HRE with the following baselines.

- **User Popularity (POP)** ranks each candidate user according to the number of check-ins by that user at all POIs in the training data.
- **Distance-based Mobility Model (DMM)** [28] considers the probability that a candidate user moves from his/her visited POIs to a target POI. It estimates the probability of moving from a visited POI to the target POI by a Pareto distribution from the distances between sequentially visited POIs.
- **Graph Embedding (GE)** [21] is the state-of-the-art model for POI recommendation that learns the embedding of POIs by considering four types of information: POI sequential visit pattern, temporal visit pattern, regional proximity and content similarity. The user embedding in GE is the sum of POI embedding in a user's visited records. We rank each candidate user by the dot product of user embedding and the target POI embedding.
- **Multi-Context Embedding (MC)** [26] is also the state-of-the-art model for POI recommendation that incorporates user-level, trajectory-level, location-level, and temporal contexts for learning embeddings of users and POIs. However, this model cannot learn the embeddings for new POIs.

Parameter Settings. The major parameters for our method HRE include: (1) the dimension D for preference embeddings of users and POIs. D is also the dimension of semantic embeddings for time slots, regions, tags and neighborhood users; (2) the number of regions K. By default, we set $D = 50$, $K = 50$. We set the regularization parameter $\lambda = 0.001$ in Eq. (8), the number of nearest neighbor POIs $k = 30$ in constructing POI-neighborhood user graph, and the weight $\beta = 1.0$ in Eq. (13). We set the temporal threshold ΔT for constructing POI-POI graph to be 25 days as [21]. We conduct parameter sensitivity test in Sect. 4.4 to study the influence of parameters on the ranking performance.

4.2   Overall Ranking Performance

We report the performance of different methods for all test POIs in the four cities. The results for AUC and Tau are in Fig. 2. From Fig. 2, we observe that:

- Our method HRE has the best performance in terms of both AUC and Tau among all comparison methods in four cities, because HRE is designed to optimize the ranking consistency and incorporate the heterogeneous information. Compared with the second best method DMM, HRE increases the AUC by *3.0%* and Tau by *7.7%* on average.
- To our surprise, the ranking performance of two existing POI recommendation methods GE and MC are lower than those of DMM and HRE. The reasons are three-folds. *First,* GE and MC are not designed for optimizing ranking consistency. *Second,* GE and MC ignore the geographical proximity between regions, which are captured by geographical embedding and POI-Neighborhood User graph in HRE. *Third,* GE and MC ignore the interaction between geographical and preference influence for modeling user mobility.

- Regarding the results of four cities, we find that LA, SF and SD have much larger AUC than NY. This is because the hometowns of all users in our data set are located in California. These users' check-in behaviors become less regular and predictable when they travel to a new city such as New York City, which was also reported in [19]. The performance gain of HRE in LA and SF is larger than the gain in SD and NY, this is because there are more check-ins in LA and SF which contain more collaborative information between POIs for POI preference embedding enrichments.

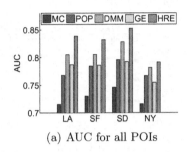
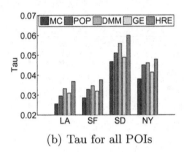

(a) AUC for all POIs   (b) Tau for all POIs

Fig. 2. Performance on all POIs

4.3 Performance on New POIs

We consider POIs with no check-in users in the training set as new POIs in our evaluation. Since MC cannot be applied to new POIs, we only report the performance of the four methods for new POIs in Fig. 3. Again we can observe that HRE has the largest AUC and Tau among all methods, which confirms that incorporating the ranking information in the user-POI interaction can enrich the representation of POIs without check-in information.

4.4 Parameter Sensitivity Test

We evaluate the performance of HRE w.r.t. four parameters: the dimension D of preference embedding, the dimension K of geographical embedding, the number of k-nearest POIs, k, in selecting neighborhood users, and the coefficient β for tuning weight of graph-based POI embedding. The AUC values for all POIs and new POIs in all four cities are reported in Fig. 4. We observe that the ranking performance of our method is not very sensitive to the parameter change. This shows the robustness of our method HRE.

4.5 A Case Study

We report the AUC for each category of POIs in the test set. The category information is obtained from tags of POIs in the data set, where each POI may have

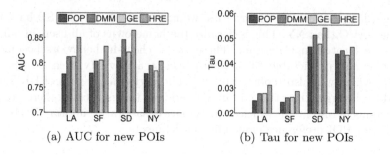

(a) AUC for new POIs

(b) Tau for new POIs

Fig. 3. Performance on new POIs

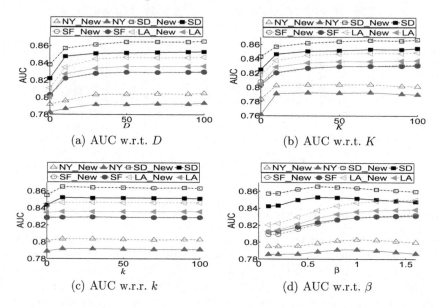

(a) AUC w.r.t. D

(b) AUC w.r.t. K

(c) AUC w.r.r. k

(d) AUC w.r.t. β

Fig. 4. AUC w.r.t. different parameters in HRE

more than one tag because it may have multiple functions. We calculate the average AUC of one tag by aggregating the POIs in four cities. Then we sort the tags based on their AUC produced by our method in descending order. We plot a heat map recording the AUC produced by five methods in Fig. 5. We observed that the AUC rankings w.r.t. category produced by HRE are more consistent with the hierarchy of needs for human life. The top three categories with the largest AUC, which are "Food & Drink Shop", "Shop & Service", "Residence", stand for more fundamental needs. On the other hand, categories such as "Arts & Entertainment", "Office" and "Airport" stand for higher-level human needs related to spirit and self-actualization [5]. In contrast, ranking categories based on other methods' AUC fail to reveal the needs order of human due to the lack of user-POI interaction information in learning POIs' preference embeddings. For example, the AUC of "Clothing" in GE and DMM are close to "Art & entertainment". In fact, "Clothing" is a more fundamental need than "Arts & Entertainment".

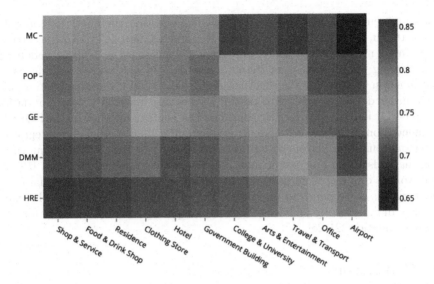

Fig. 5. AUC w.r.t. category

5   Related Works

Influential Users Selection in Location Based Social Network. Unlike traditional influence maximization problems in online social networks [1,7,17], users' locations need to be considered in finding influential users in location-based social networks [3,24,27]. [8] identifies a set of users who are most influential to other users in a specific region. [20] considers the user-POI distance in calculating the influence spread. Both works assume that the user's location is fixed and ignore the user mobility. [28] first studies the location promotion problem where the activation probability of each edge is determined by users' probability to visit a target location. They propose two mobility models for setting the users' visiting probability where the distance-based mobility model (DMM) is reported to have the largest AUC under different activation thresholds. As mentioned in Sect. 1, DMM only considers check-in records of one user in deriving his/her visiting probability, while our method considers heterogeneous information such as POI tags and neighborhood user visiting patterns in check-in records of all users. Besides, our method focuses on maximizing the ranking consistency on setting user visiting score so that it has the highest ranking performance as shown in the experiments.

User Mobility Modeling and POI Recommendation. Studies [4,12,25] related to user mobility modeling focus on maximizing the likelihood of observing all check-in records of individual users. [4] finds that most of the human movement is based on periodic behaviors. They propose a Periodic Mobility Model (PMM) that consists of two Gaussian distributions to denote user's states at work and home respectively. In POI recommendation [2,6,9–11,13,14,19,21,22], the search engine is asked to return a set of POIs that the query user may be interested in. Most of the works on POI recommendation [9,11,13,22] are based on a fused model that considers geographical influence as well as collaborative filtering information from check-in records. [10] first uses a ranking based loss function to optimize the precision and recall of top-k recommendation, but their method cannot deal with new POIs because they only use information of user-POI matrix. [21] further considers the context information in POI recommendation and proposes a multi-graphs embedding method for integrating heterogeneous information. It outperforms other cold start POI recommendation methods based on content information and geographical locations [19,23]. Our work is different from user mobility modeling and POI recommendation as our target is to recommend users for a target POI. We also confirm the effectiveness of our ranking based methods by comparing with the methods for POI recommendation [21,26].

6   Conclusion

In this paper, we study the location promotion problem in location-based services. In order to return a target user list w.r.t. a target POI, we propose a unified representation learning framework called hybrid ranking and embedding. Our framework maximizes the ranking consistency and integrates heterogeneous information of POIs, which alleviates the data sparsity problem in users' check-in data and solves the cold-start POI problem. Experiments on four cities of the United States show that our method has better ranking performance than the state-of-the-art methods.

Acknowledgement. The work is supported by a Microsoft Research Asia Collaborative Research Grant.

References

1. Arora, A., Galhotra, S., Ranu, S.: Debunking the myths of influence maximization: an in-depth benchmarking study. In: SIGMOD, pp. 651–666 (2017)
2. Bao, J., Zheng, Y., Wilkie, D., Mokbel, M.: Recommendations in location-based social networks: a survey. GeoInformatica **19**(3), 525–565 (2015)
3. Bouros, P., Sacharidis, D., Bikakis, N.: Regionally influential users in location-aware social networks. In: SIGSPATIAL, pp. 501–504 (2014)
4. Cho, E., Myers, S.A., Leskovec, J.: Friendship and mobility: user movement in location-based social networks. In: SIGKDD, pp. 1082–1090 (2011)
5. Fu, Y., Xiong, H., Ge, Y., Yao, Z., Zheng, Y., Zhou, Z.H.: Exploiting geographic dependencies for real estate appraisal: a mutual perspective of ranking and clustering. In: SIGKDD, pp. 1047–1056 (2014)
6. Kefalas, P., Symeonidis, P., Manolopoulos, Y.: Recommendations based on a heterogeneous spatio-temporal social network. WWW **21**(2), 345–371 (2018)
7. Kempe, D., Kleinberg, J., Tardos, É.: Maximizing the spread of influence through a social network. In: SIGKDD, pp. 137–146 (2003)

8. Li, G., Chen, S., Feng, J., Tan, K.L., Li, W.S.: Efficient location-aware influence maximization. In: SIGMOD, pp. 87–98 (2014)
9. Li, H., Ge, Y., Hong, R., Zhu, H.: Point-of-interest recommendations: learning potential check-ins from friends. In: SIGKDD, pp. 975–984 (2016)
10. Li, X., Cong, G., Li, X.L., Pham, T.A.N., Krishnaswamy, S.: Rank-GeoFM: a ranking based geographical factorization method for point of interest recommendation. In: SIGIR, pp. 433–442 (2015)
11. Lian, D., Zhao, C., Xie, X., Sun, G., Chen, E., Rui, Y.: GeoMF: joint geographical modeling and matrix factorization for point-of-interest recommendation. In: SIGKDD, pp. 831–840 (2014)
12. Lichman, M., Smyth, P.: Modeling human location data with mixtures of kernel densities. In: SIGKDD, pp. 35–44 (2014)
13. Liu, B., Fu, Y., Yao, Z., Xiong, H.: Learning geographical preferences for point-of-interest recommendation. In: SIGKDD, pp. 1043–1051 (2013)
14. Liu, Y., Pham, T.A.N., Cong, G., Yuan, Q.: An experimental evaluation of point-of-interest recommendation in location-based social networks. VLDB 10(10), 1010–1021 (2017)
15. Pham, H., Shahabi, C., Liu, Y.: EBM: an entropy-based model to infer social strength from spatiotemporal data. In: SIGMOD, pp. 265–276 (2013)
16. Rendle, S., Freudenthaler, C., Gantner, Z., Schmidt-Thieme, L.: BPR: bayesian personalized ranking from implicit feedback. In: UAI, pp. 452–461 (2009)
17. Tang, Y., Shi, Y., Xiao, X.: Influence maximization in near-linear time: a martingale approach. In: SIGMOD, pp. 1539–1554 (2015)
18. Wang, H., Li, Z., Lee, W.C.: PGT: measuring mobility relationship using personal, global and temporal factors. In: ICDM, pp. 570–579 (2014)
19. Wang, W., Yin, H., Chen, L., Sun, Y., Sadiq, S., Zhou, X.: Geo-SAGE: a geographical sparse additive generative model for spatial item recommendation. In: SIGKDD, pp. 1255–1264 (2015)
20. Wang, X., Zhang, Y., Zhang, W., Lin, X.: Distance-aware influence maximization in geo-social network. In: ICDE, pp. 1–12 (2016)
21. Xie, M., Yin, H., Wang, H., Xu, F., Chen, W., Wang, S.: Learning graph-based poi embedding for location-based recommendation. In: CIKM, pp. 15–24 (2016)
22. Ye, M., Yin, P., Lee, W.C., Lee, D.L.: Exploiting geographical influence for collaborative point-of-interest recommendation. In: SIGIR, pp. 325–334 (2011)
23. Yin, H., Zhou, X., Shao, Y., Wang, H., Sadiq, S.: Joint modeling of user check-in behaviors for point-of-interest recommendation. In: CIKM, pp. 1631–1640 (2015)
24. Zhang, C., Shou, L., Chen, K., Chen, G., Bei, Y.: Evaluating geo-social influence in location-based social networks. In: CIKM, pp. 1442–1451 (2012)
25. Zhang, C., Zhang, K., Yuan, Q., Zhang, L., Hanratty, T., Han, J.: GMove: group-level mobility modeling using geo-tagged social media. In: SIGKDD, pp. 1305–1314 (2016)
26. Zhou, N., Zhao, W.X., Zhang, X., Wen, J.R., Wang, S.: A general multi-context embedding model for mining human trajectory data. TKDE 28(8), 1945–1958 (2016)
27. Zhou, T., Cao, J., Liu, B., Xu, S., Zhu, Z., Luo, J.: Location-based influence maximization in social networks. In: CIKM, pp. 1211–1220 (2015)
28. Zhu, W.Y., Peng, W.C., Chen, L.J., Zheng, K., Zhou, X.: Modeling user mobility for location promotion in location-based social networks. In: SIGKDD, pp. 1573–1582 (2015)

Patent Quality Valuation with Deep Learning Models

Hongjie Lin, Hao Wang, Dongfang Du, Han Wu,
Biao Chang, and Enhong Chen[✉]

Anhui Province Key Laboratory of Big Data Analysis and Application,
University of Science and Technology of China, Hefei, China
{hjlin,wanghao3,dfdu,wuhanhan,chbiao}@mail.ustc.edu.cn,
cheneh@ustc.edu.cn

Abstract. Patenting is of significant importance to protect intellectual properties for individuals, organizations and companies. One of practical demands is to automatically evaluate the quality of new patents, i.e., patent valuation, which can be used for patent indemnification and patent portfolio. However, to solve this problem, most traditional methods just conducted simple statistical analyses based on patent citation networks, while ignoring much crucial information, such as patent text materials and many other useful attributes. To that end, in this paper, we propose a Deep Learning based Patent Quality Valuation (DLPQV) model which can integrate the above information to evaluate the quality of patents. It consists of two parts: Attribute Network Embedding (ANE) and Attention-based Convolutional Neural Network (ACNN). ANE learns the patent embedding from citation networks and attributes, and ACNN extracts the semantic representation from patent text materials. Then their outputs are concatenated to predict the quality of new patents. The experimental results on a real-world patent dataset show our method outperforms baselines significantly with respect to patent valuation.

Keywords: Patent quality valuation · Attribute network embedding
Convolutional Neural Network · Patent citation network

1 Introduction

With regard to industry research and development, patent application is one of the most significant sources of key technologies for protecting intellectual properties. And patenting is also an important asset for companies in the knowledge economy. Over the past few decades, with the rapid development of multifaceted technology in different application domains, a large amount of patents are applied and authorized. They serve as one of the crucial intellectual property components for individuals, organizations and companies. Many companies, especially burgeoning firms, apply several thousands of patents each year[1]. The granted patent information is open to public and can be available for professional organizations in various countries or regions around the world. For instance, World Intellectual Property Organization (WIPO)[2] reported over 2 million total patent applications authorized worldwide within a year [15]. The researches which contrapose patent information are more and more important in order to make fair and credible valuation results available to investors.

In fact, questions involving patent mining have intrigued scholars for decades, and there have been many influential academic researches in this area, including patent retrieval [7], patent classification [4], patent visualization and cross-language patent mining [8] and patent valuation [1,10]. In this work, we devoted to exploring this deeper and hope to make further support for the last topic, patent valuation, a common process of evaluating the quality of patent documents.

Indeed, assessing the value of a patent is crucial both at the licensing stage and during the resolution of a patent infringement lawsuit [20], and it is undeniable that business community have paid much concern about this question because of its considerable significance, so they might hire many professional patent analysts engaging in this. Obviously, patent valuation is a non-trivial task which requires tremendous amount of human efforts. What's more, it is necessary for patent analysts to have a certain degree of expertise in different research domains, including information retrieval, data mining, domain-specific technologies, and business intelligence [32]. As a result, it is of great significance to evaluate the potential value of a given patent automatically, which is the goal of this work.

However, there are many challenges to solve this question. First of all, different from general text analysis, patent document contains dozens of special features, including structured items and unstructured items [32]. The structured items are uniform in semantics and format (such as patent number, inventor, assignee, application date, grant date and classification code) and the unstructured ones consist of text content in different length(including claims, abstracts, and descriptions of the invention.). Second, there contains much useful information in patent citation network, but how to model it and make it effectively contribute to patent valuation is kind of difficult, which is one of the technicality goal of our framework modeling.

As mentioned above, there are indeed previous works focusing on patent valuation, while most of them just focus on one aspect of patent value, such as statistical analysis [28] and text mining [13]. As far as concerned, none of the existing works [13,20] takes into account both the patent text materials and the citation networks in terms of finding more valuable patents. To solve all these problems with addressing the challenges above, we propose Deep Learn-

---

[1] http://www.ificlaims.com/, in 2015, IBM received 8,088 granted U.S. patents, followed by Samsung (5,518), Canon (3,665), Qualcomm (2,897), Google (2,835), Intel (2,784), LG (2,428), Microsoft (2,398).

[2] http://www.wipo.int.

ing based Patent Quality Valuation (DLPQV) model to evaluate patent quality, which extracts the patent attribute network embedding by Attribute Network Embedding (ANE) and analyzes patent text materials by Attention-based Convolutional Neural Network (ACNN).

Specially, given the text materials, citation relations and meta features of patents, we first design an unified CNN-based and ANE-based architecture to exploit the semantic representations and network embedding for all patents. Then we qualify the quality valuation contribution of each sentence to the title by utilizing an attention strategy. Next, train DLPQV and generate the quality valuation prediction for each patent. Finally, extensive experiments on a large-scale real-world dataset validate both the effectiveness and explanatory power of our proposed framework. The main contributions of this paper could be summarized as:

(1) We are the first one to apply deep learning method to patent document analysis, which is an ingenious piece of work combining the strength of deep learning and patent characteristics.
(2) We present novel attribute network embedding for learning the low-dimensional vectors of patent citation networks, which is one of the most important components of patent valuation.
(3) We propose a unified framework to combine attribute network embedding and deep learning based CNN methods, which allows jointly modeling patent information for patent quality valuation.
(4) The extensive experiments in a real patent dataset show the proposed method outperforms baselines significantly.

2   Related Work

Generally, the related work can be classified into the following two parts, i.e., patent citation network studies in patent quality valuation and text mining techniques for patent analysis.

2.1   Patent Citation Network in Patent Quality Valuation

Many scholars have suggested that patent citation counts are strongly relevant to patent value or patent quality [1,10,12,22]. Sterzi [29], who proposed that the number of times a patent has been cited by other patents is significantly associated with the value of the patent, trying to solve data truncation problems by using year dummies; these dummies represented the period from the priority year up to 3 years, the period from the priority year up to 6 years, and the period from 7 years to the search year. Fischer and Henkel [6] used the natural logarithms of the number of forward citations +1 to reduce the skewness of the distribution of patent citation counts. The number of citations made by other firms or researchers in a similar field for up to 5 years after the publication date showed a considerable association with economic patent value [19,29], and late citations those made after 5 years since a patent was granted showed a strong relationship with the market value of a firm [11,29]. In addition, Karki [17] considered the number of citations to reflect a patents technological influence on subsequent inventions. The number of backward citations signifies references that are quoted by the relevant patent, and a variety of technological information is expected to contribute to high patent quality [2]. Based on all the previous works, we can tell that the number of patent citations can reflect patent value in terms of novelty.

However, the common limitation of these works is that these methods are usually based on statistical analysis using historical citation information in order to explore some specific relationships between patent citation count and patent value, and there still need extensive and unified approaches to synthetically measuring patent quality, which is what we devote to. Different from them, our study adopts both the citation networks with the patent meta features and abundant patent documents to predict the potential patent value, trying to reveal more deeper insights in this problem using attribute network embedding method.

2.2 Text Mining Techniques for Patent Analysis

One of the crucial steps in our framework is the understanding and representations of patent text materials, which aims at automatically processing patent document inputs and producing textual outputs. Most of the previous researches in this area are based on bag-of-words or LDA. Hasan et al. [13] built a patent ranking software, named COA (Claim Originality Analysis) that rates a patent based on its value by measuring the recency and the impact of the important phrases. Shaparenko et al. [27] discovered important documents in a document collection, which are clustered by their word bags. They find that a document is important if it has fewer similar documents published before it, and has more similar documents published after it. Specifically, Tang et al. [31] designed and implement a general topic-driven framework for analyzing and mining the heterogeneous patent network. Besides, to assess the technology prospecting of a company, Jin et al. [16] proposed an Assignee-Location-Topic (ALT) Model to extract emerging technology terms from patent documents of different companies, which are also based on LDA method.

However, these existing methods fail to display the relationships among words or sentences in patent documents, which is exactly the strengths of deep learning methods in NLP (Natural Language Processing) field.

Combining the above two points, in this work, we adopt both the citation networks with the patent meta features and abundant patent documents and propose a novel framework (DLPQV) of patent quality valuation, consisting of Attribute Network Embedding (ANE) and Attention-based Convolutional Neural Network (ACNN), which mixes patent text materials, meta features and citation network together to carry our point of comprehensive valuation of given patents.

3   Deep Learning Based Patent Quality Valuation (DLPQV) Framework

In this section, we first detailedly introduce the Patent Quality Analysis task, and then we introduce the technical details of DLPQV. The DLPQV model consists of Attribute Network Embedding (ANE) and Attention-based Convolutional Neural Network (ACNN).

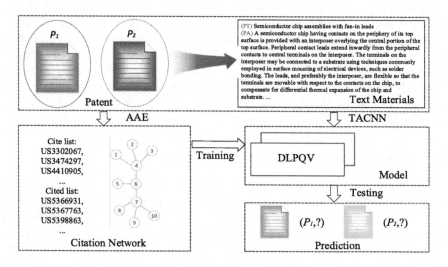

Fig. 1. The flowchart overview of our work

3.1   Problem and Study Overview

Traditional patent citation analysis can work on different applications for patents. For instance, if a patent have a high citation count, the cited patent probably have high chance to be a foundation of the citing patents. That is to say, highly-cited patents are possibly more important compared with those less ones. Therefore, we regard forward citation within two decades after authorization as patent quality with normalization.

Definition 1. *Formally, given a set of patents with corresponding text materials including title (PT), abstract (PA), citation networks and patent meta features. And each patent has a quality valuation record obtained from cited amount with normalization (see Table 1). Our goal is to leverage the information of patent P_i available to train a prediction model $\mathcal{M}$ (i.e., DLPQV), which can be effectively used to valuate the quality of patents in the new granted patents.*

As is shown in Fig. 1, our solution is a two-stage framework, which includes a training stage and a testing stage: (1) In the training stage, given patent features including text materials, citation network and patent meta features

Table 1. Examples of patent instances with text, citation and attributive information.

Quality (Q)	Patent (P)	Text materials		Citation network		Meta features			
		Title (PT)	Abstract (PA)	Cite	Cited	WIPO	Claims	Nber	...
0.8761	US5148265	Semiconductor chip...	A semiconductor chip having contacts on...	US3302067...	US5367763...	A	39	Meth	
0.1205	US5366931	Fabrication method...	This invention relates to a technology...	US4177480...	US5895966...	B2	5	Meth	
?	US5477611	Method of forming...	A method for creating an interface...	US4079511...	US5776796...	B1	39	Elec	

(see Table 1), we propose DLPQV to represent the text materials of each patent P_i and embedding the attribute network so as to evaluate patent quality Q_i. (2) In the testing stage, after the training of DLPQV is completed, for each new granted patent, DLPQV could estimate its quality with the available patent features.

Our DLPQV detailed framework is showed in Fig. 2, and we will introduce the model specifically in the following description, which covers Attribute Network Embedding (ANE) and Attention-based Convolutional Neural Network (ACNN).

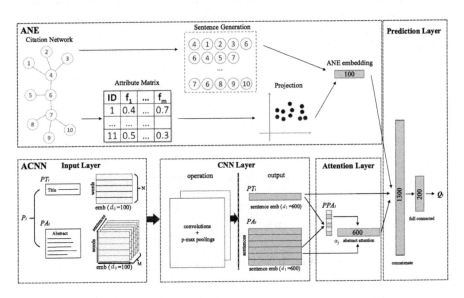

Fig. 2. Deep Learning based Patent Quality Valuation (DLPQV) framework

3.2   Attribute Network Embedding for Citation Network

Definition 2 (*ANE for Citation Network*). *Treating the granted patents as nodes and citation relations among them as edges respectively, we construct a citation network and use the proposed attribute network embedding method to learn the patent representation. Our citation network representation problem is formalized as follows. Given a citation network $G = (V, E, F)$, where V is the sets of nodes, E is the sets of edges and $F = \{f_1, f_2, ..., f_{|V|}\}$ represents the sets of features of size m for each node. We aim to learn a low-dimensional vector representation $u_v \in R^d$ for each node $v \in V$ in G, where d is much smaller than $|V|$.*

Attribute Network Embedding Framework. For citation network, we propose a Attribute Network Embedding model (*ANE*) that incorporates the node attributes, whose framework is shown in Fig. 3. Firstly, different from the *sentences generation* (like *word2vec*) method used in previous work, we propose the *sentences generation* method based on nodes' neighbors. We can preserve the citation network structure based on these sentences, so that nodes with the similar neighborhoods will have the similar citation network embedding. Then, in order to incorporate the attributes of nodes into citation network embedding, we take nodes' attributes as the initial input and utilize the mapping function to project it into the node embedding space. Finally, through the optimization of the model, we obtain the citation network embedding which can simultaneously preserve the citation network structure and reflect the similarity of node attributes. In the following section, we will introduce our model in detail.

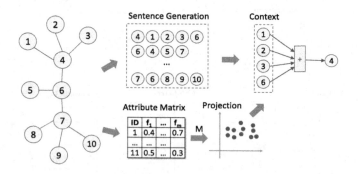

Fig. 3. ANE model framework

Sentences Generation. In previous network representation learning research works, there are two main ways to learn the network structure information. Like Deepwalk [25], node2vec [9] etc., they sample uniformly a random node $v \in G$ as the root of random walk and generate a truncated random walk sequences as the training *Sentences* to learn the node embedding. They are based on the assumption that the node is similar to the surrounding nodes in window size k, which we think is too strong for some network structures, like the network in Fig. 3. The other is to learn the network embedding by preserving the First-order Proximity or the Second-order Proximity, like [3,30], etc. However, these methods only consider the similarity between the node and their neighborhoods and don't consider the similarity between nodes' neighborhoods. In order to alleviate these problems, inspired by [26], we proposed the *sentences generation* method based on nodes' neighborhoods as follows:

We use each node as the root once, and take the random permutations of the root node's neighborhoods into the sentence. Each generated sentences has the form: $[v_{root}, v_1, ..., v_n]$, where $\forall 1 \leq i \leq n, v_i$ is the neighborhood of v_{root}. Take the node 2 in Fig. 3 as an example, [3, 4, 1] is a permutation of the node 2's neighborhoods and [2, 3, 4, 1] is an instance of sentence generation of node 2. Also, it is important to note that the nodes in root node's neighborhoods should be no explicit order. So we set the number of permutations of root node's neighborhood to be N^P. The larger the value N^P is selected, the more evenly distributed root node's neighborhoods are in generated sentences.

ANE Model Formulation. Here, we describe how the ANE model incorporates the node attributes into citation network embedding. For each node in generated sentences, the ANE model predicts the center node v_i given a representation of the surrounding context nodes $v \in \{v_{i-k}, ..., v_{i+k}\} \setminus \{v_i\}$, where k is the window size of context nodes. The objective function of ANE model is to maximize the average log probability of the center node v_i given the context nodes $context(v_i)$ for all the sentences $s \in S$, which is defined as following:

$$L = -\frac{1}{|s|} \sum_{i=1}^{|s|} \log p(v_i|context(v_i)) = -\frac{1}{|s|} \sum_{i=1}^{|s|} \log \frac{\exp u'^T_i u_{context(i)}}{\sum_{j=1}^{|V|} \exp u'^T_j u_{context(i)}} \quad (1)$$

where u'_i is 'output' vector representation of node v_i, $u_{context(i)}$ is vector representation of context words of node v_i and $|V|$ is the number of citation network nodes as well as the number of patents.

In order to make full use of the nodes' own attributes, as shown in Fig. 3, we take the nodes' attributes as the initial input of the model. Then we transform it to node embedding space with the use of transformation matrix M, where we have:

$$u_i = M^T f_i \quad (2)$$

where u_i is the 'input' vector representation of the node v_i, f_i is attribute value of node v_i. And $M \in R^{m \times d}$, where m is the node attributes dimension, and d is the dimension of u_i.

Furthermore, we defined $u_{context(i)}$ as weighted average of the 'input' vector representation of context nodes:

$$u_{context(i)} = \frac{1}{2k} \sum_{j \in [i-k, i+k] \setminus \{i\}} u_j \quad (3)$$

Finally, by minimizing Eq. (1), we obtain 'input' representation u_i and 'output' vector representation u_i' for node $v_i \in V$, and both of them can be regraded as low-dimensional representation of node. Therefore, we utilize the concatenation of them as the citation network embedding, and each patent is represented by a citation network embedding.

Model Optimization. Next, we introduce the details of how to use the Stochastic Gradient Descent method(SGD) to train the ANE model. Then we present the algorithm framework and time complexity of the model.

Approximation by Negative Sampling: Optimizing the Eq. (1) is computationally expensive, because the denominator of $p(v_i|context(v_i))$ requires summation over all the nodes in citation network, which the number of node is usually very large. To address this problem, we adopt the approach of negative sampling proposed in [23], which select negative samples according to the noisy distribution $P_n(v)$ for each node contexts. As a result, the $\log p(v_i|context(v_i))$ in Eq. (1) is replaced by the following objective function:

$$L_i = \log \sigma(u_i'^T u_{context(i)}) + \sum_{t=1}^{neg} E_{v_t \sim P_n(v)}[\log \sigma(-u_t'^T u_{context(i)})] \qquad (4)$$

where $\sigma(x) = 1/(1 + exp(-x))$, neg is the number of negative samples. And we set the node noisy distribution $P_n(v) \propto d_v^{3/4}$ as proposed in [23], where d_v is the out-degree of node v.

We employ the widely used Adaptive Moment Estimation (Adam) algorithm [18] to optimize the Eq. (4). In each step, the Adam algorithm samples a mini-batch of training instance(center-context) and then update the model parameter by walking along the descending gradient direction,

$$u_i'^{t+1} = u_i'^t - \eta \cdot \frac{\partial L_i}{\partial u_i'} \qquad (5)$$

where u' is the 'output' vector representation of the node v_i, and t is the iteration times. η is the learning rate, which is automatically adjusted in Adam algorithm.

3.3   Attention-Based Convolutional Neural Network

Through ANE for citation network, a patent P_i as a node in citation network is represented as a representation vector u_i', which is expressed as PU_i in the following description. In this subsection, we will introduce the specific components of ACNN of DLPQV, which deals with text materials to obtain the representation of patents. As shown in Fig. 2, ACNN can be divided into four components, i.e., *Input Layer, CNN Layer, Attention Layer and Prediction Layer*. The following will cover concrete content about the four layer, especially CNN Layer and Attention Layer.

Definition 3 (*ACNN of DLPQV*). *Given a dataset of patents with text materials including patent titles (PT), patent abstracts (PA) and patent attribute citation network embedding (PU), and each patent P_i has a quality valuation Q_i (e.g., 0.8761) obtained from the normalized cited amount (see Table 1), we aim at leveraging the information of patents to train a prediction model based on ACNN, which can estimate the qualities of patents.*

Input Layer. The input to ACNN is the title text and all abstract text of a patent Pi, i.e., title(PT_i) and abstract(PA_i). Specifically, the abstract text PA_i is expressed to a sequence of sentences $PA_i = \{s_1, s_2, ..., s_M\}$ where M is the sequence length. And the title PT_i is an individual sentence. Considered to sentence constituents, each sentence consists of a sequence of words $s = \{w_1, w_2, ..., w_N\}$ where $w_i \in \mathbb{R}^{d_0}$ is obtained from d_0-dimensional pre-trained word embedding and N is the length of sentence. Finally, the title of a patent is translated into a matrix $PT_i \in \mathbb{R}^{N \times d_0}$, and the abstract of a patent is represented by a tensor $PA_i \in \mathbb{R}^{M \times N \times d_0}$.

CNN Layer. We aim at learning each sentence representation from word embedding in CNN Layer. Reasonably, we choose CNN-based model to learn sentence embedding with following reasons: (1) Because of convolution-pooling operations, CNN works better on considering dominated information of each sentence from local to global views. Usually, sentence is well represented by local key words. (2) CNN leverages shared convolution filter for training model, so it can reduce the complexity compared with other deep learning architecture, such as DNN or RNN [21]. (3) CNN is suitable for learning the interactions between words and deeply mining the semantic representations for sentences.

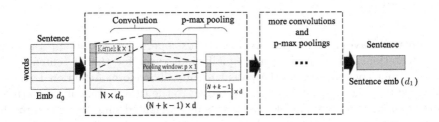

Fig. 4. CNN Layer, which contains several layers of convolution and p-max pooling.

As shown in Fig. 2, we design CNN Layer as a traditional model [5] that selects several layers of convolution and p-max pooling. Then each sentence is represented as a fixed length vector. Next, we will introduce the detail of the convolution-pooling operation in CNN Layer.

Specifically, we analyze the first convolution-pooling operation, and the other more operations are similar to that. In Input Layer, we transform a sentence into a sentence matrix input $s \in \mathbb{R}^{N \times d_0}$ as the input of CNN Layer (showed in Fig. 4), then the wide convolution operates on a sliding window of every k words with a kernel $k \times 1$. Through the first convolution operation, the input sentence $s = w_1, w_2, ..., w_N$ is transformed to a new hidden sequence, i.e., $e^c = \{\vec{e}_1^c, \vec{e}_2^c, ..., \vec{e}_{N+k-1}^c\}$, where:

$$\vec{e}_i^c = ReLU(\mathbf{G} \cdot [w_{i-k+1} \oplus ... \oplus w_i] + \mathbf{b}), \tag{6}$$

here, $\mathbf{G} \in \mathbb{R}^{d \times kd_0}, \mathbf{b} \in \mathbb{R}^d$ are the convolution parameters, and d is the output dimension of the convolution operation. $ReLU(x)$ is a nonlinear activation function which is equal to $ReLU(x) = max(0, x)$. "$\oplus$" is in order to concatenate k word vectors into a long vector.

After the convolution operation, we obtain a local semantic representation by convoluting sequential k words. Next, we leverage p-max pooling operation to transform the convolution sequence e^c into a new global hidden sequence, i.e., $e^{cp} = \{\vec{e}_1^{cp}, \vec{e}_2^{cp}, ..., \vec{e}_{\lfloor (N+k-1)/p \rfloor}^{cp}\}$, where:

$$\vec{e}_i^{cp} = \left[max \begin{bmatrix} e_{i-p+1,1}^c \\ ... \\ e_{i,1}^c \end{bmatrix}, ..., max \begin{bmatrix} e_{i-p+1,d}^c \\ ... \\ e_{i,d}^c \end{bmatrix} \right]. \tag{7}$$

Similar to the first convolution-pooling operation, more layers of convolution-pooling processes are merged into the ACNN model to gradually express the global semantic information of sequential words in a sentence. Finally, a sentence consisted of N word embedding is transformed to a vectorial representation $s \in \mathbb{R}^{d_1}$, where d_1 is the output dimension of CNN Layer.

Through CNN Layer, the title of a patent is transformed into a vector $PT_i \in \mathbb{R}^{d_1}$. Meanwhile, the abstract of a patent which contains M sentences is represented by a matrix $PA_i \in \mathbb{R}^{M \times d_1}$. The output form of CNN Layer is showed in Fig. 2.

Attention Layer. After the previous layers' operation, we obtain sentence representation. However, it is not equally important for the M sentences of the abstract contributing to the patent quality. Therefore, Attention Layer is designed to assign different weights according to the title. Detailedly, the attention representations are modeled as vectors by a weighted sum aggregated result of the sentence representations from abstract perspectives. For example, the abstract attention score PAA_i of a specific patent P_i is represented as follows:

$$PAA_i = \sum_{j=1}^{M} \alpha_j s_j^{PA_i}, \ \alpha_j = cos(s_j^{PA_i}, s^{PT_i}), \tag{8}$$

here, $s_j^{PA_i}$ is the j-th sentence in PA_i, s^{PT_i} is the sentence representation of patent title PT_i; *Cosine similarity* α_j is denoted as the attention score for measuring the weight of the sentence s_j in abstract PA_i for patent P_i, which means the importance of the contribution to the patent quality.

Prediction Layer. The last layer of ACNN is Prediction Layer, which aims at predicting the quality $\widetilde{Q}_i$ of patent Pi considered the abstract-attention representation PAA_i, the title representation s^{PT_i} and the attribute network embedding PU_i. To be specific, we first merge those three representation vectors into a long vector by concatenation operation, then use a classical full-connected network [14] to learn the overall valuation representation o_i, then predict the quality $\widetilde{Q}_i$ by LeakyReLU function, which we will discuss detailedly in Sect. 4:

$$o_i = ReLU\left(W_{ReLU} \cdot [PAA_i \oplus s^{PT_i} \oplus PU_i] + b_{ReLU}\right), \tag{9}$$

$$\widetilde{Q}_i = LeakyReLU(W_{LeakyReLU} \cdot o_i + b_{LeakyReLU}), \tag{10}$$

where $W_{ReLU}, b_{ReLU}, W_{LeakyReLU}, b_{LeakyReLU}$ are parameters to tune the network.

And we formulate the function by minimizing the least square loss with a l_2-regularization term:

$$\mathcal{J}(\Phi) = \sum_{P_i} (Q_i - \widetilde{Q}_i)^2 + \lambda_\Phi ||\Phi_\mathcal{M}||^2, \tag{11}$$

where $\mathcal{M}$ represents the DLPQV that transforms text materials, citation relation and attribute information of patent P_i into predicted patent quality $\widetilde{Q}_i$ (Eq. (10)). $\Phi_\mathcal{M}$ denotes all parameters in DLPQV and λ_Φ is the regularization hyperparameter.

4   Experiments

In this section, we first introduce our DLPQV framework settings, then compare the performance of DLPQV against the baseline approaches on patent quality valuation task. At last, we provide a *case study* to visualize the explanatory power of DLPQV.

4.1   Dataset Description

The experimental dataset is supplied by United States Patent and Trademark Office (USPTO)[3], which grants US patents to inventors and assignees all over the world since 1976. Patents are classified according to the technical features of patented invention. These classification are mapped to broader, more easily understood technology fields.

For data pre-processing, we extract 51224 patents from USPTO dataset as our experimental dataset including the titles, abstracts, citation relation and meta features. Text materials are cleaned by deleting stop words, and meta features contain WIPO document kind codes, number of claims, categories by National Bureau of Economic Research, authorization year, assign information

---

[3] http://www.patentsview.org.

and so on, which are also transformed into one-hot form (8035 dimensions). Lastly, the cited amount of a patent within two decades after granted is normalized as the patent quality.

4.2 Experimental Setup

Word Embedding. The word embedding in Input Layer of ACNN are trained on a large-scale *gigaword* corpus using public *word2vec* tool [23] with the dimension 100.

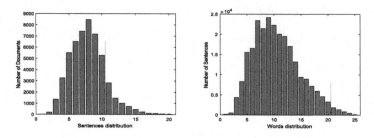

Fig. 5. Statistics of sentence and word distribution.

DLPQV Setting. In ANE of DLPQV, we set patent citation network embedding dimension as 100, and negative sampling number is set as 4 when the maximal length of sentence generation path is 40. In ACNN of DLPQV, we set maximum length N(M) of words (sentences) in sentences (abstracts) as 10 (20) (zero padded when necessary) according to our statistics in Fig. 5, i.e., around 90% sentences (abstracts) contains less than 10 (20) words (sentences). There are four layers of convolution consisted of three wide convolutions and one narrow convolutions and max-pooling. And they are employed for CNN Layer in ACNN to accommodate the sentence length N, where the numbers of the feature maps for four convolutions are (200, 400, 600, 600) respectively. Meanwhile, the kernel size k is set as 3 for all four convolution layers and the pooling window p is set as (2, 2, 2, 1) for each max pooling respectively. We notice that LeakyReLU performances better in the patent quality valuation task, due to the property that it can not only preserve the advantage of ReLU like fast convergence speed, but also retain the informance in the negative axis. $LeakyReLU(x)$ denotes x when $x > 0$, and αx when $x \leqslant 0$. Further, we choose the value of α as 0.1 by conducting several experiments.

Training Setting. On the basis of the operation [24], we randomly initialize all vector and matrix parameters in ACNN with uniform distribution in the range between $-\sqrt{6/(nin + nout)}$ and $\sqrt{6/(nin + nout)}$, where nin and $nout$ are the numbers of the input and output matrix feature sizes. To measure the performance of DLPQV, we use the widely used *Root Mean Squared Error* (RMSE) for the comparison of patent quality valuation precision. Overall speaking, the smaller the RMSE is, the better performance the result has.

$$RMSE = \sqrt{\frac{\sum_{i=1}^{n}(Q_i - \widetilde{Q_i})^2}{n}} \qquad (12)$$

4.3 Baseline Approaches

To our best knowledge, this is the first work based on deep learning for predicting patent quality valuation based on cited amount, which integrated text materials, citation network and patent meta features, so we verify the effectiveness of each component of DLPQV. The details of comparison are as follows:

- *ANE*: ANE is a framework without ACNN part, and only use citation network embedding PU_i as the patent embedding to predict the patent quality Q_i.
- *ACNN*: ACNN only consider text materials without citation relation and patent meta features.
- *CNN*: CNN is a framework with attention-ignored strategy compared with ACNN. Here, the attention-ignored strategy means the attention parameters α in Eq. (8) are the same for all sentences.
- *ANE-CNN*: ANE-CNN is a framework with attention-ignored strategy compared with DLPQV.

Both DLPQV and baselines are all implemented by Tensorflow and all experiments are run on a Tesla K20m GPU.

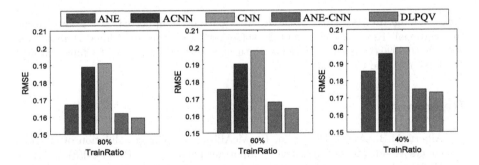

Fig. 6. Overall performance on the patent quality valuation task.

4.4 Experimental Results

Overall Results. To observe the several models' performance for different data sparsity, we randomly select 80%, 60%, 40% of the extracted patent dataset as the training sets, and the rests as testing sets, respectively. In Fig. 6, we summarize the patent quality valuation results of all models. Obviously, we can see that DLPQV model performs best. Concretely, DLPQV performs better than

ANE, which indicated that the semantic representation from ACNN can provide patents' content features to improve the patent quality valuation accuracy rate. Both attribute information and patent effects are well integrated to enhance the network embedding so that Fig. 6 shows DLPQV beats ACNN, which indicates that ANE is also significant to DLPQV. Meanwhile, ACNN beats CNN as well as DLPQV beats ANE-CNN, which qualifies the contributions of texts with attention strategy. In summary, DLPQV has a best performance in different scale of training sets, and each part of DLPQV provides an important role for enhancing patent quality's forecast accuracy.

5   Conclusions

In this paper, we proposed a novel *Deep Learning based Patent Quality Valuation* (DLPQV) framework to predict patent quality. It is the first one to apply deep learning method to patent quality valuation problem with attribute network embedding and CNN method combined. We firstly design ANE to learn the patent embedding from attribute citation networks. Then, in order to represent text materials, we use a CNN-based architecture for exploiting sentence representations with attention strategy. And we qualified the contributions of abstract sentences to the patent valuation by an attention strategy. Finally, we mix citation network embedding and text representation to generate the patent quality prediction value. Experiments on real-world dataset supplied by USPTO proved that our framework could effectively predict the patent quality. In the future, we will focus on the patent quality variation tendency over time based deep learning method.

Acknowledgements. This research was partially supported by grants from the National Key Research and Development Program of China (Grant No. 2016YFB1000904), and the National Natural Science Foundation of China (Grants No. U1605251 and 61727809).

References

1. Albert, M.B., Avery, D., Narin, F., McAllister, P.: Direct validation of citation counts as indicators of industrially important patents. Res. Policy **20**(3), 251–259 (1991)
2. Burke, P.F., Reitzig, M.: Measuring patent assessment quality-analyzing the degree and kind of (in) consistency in patent offices' decision making. Res. Policy **36**(9), 1404–1430 (2007)
3. Chang, S., Han, W., Tang, J., Qi, G.-J., Aggarwal, C.C., Huang, T.S.: Heterogeneous network embedding via deep architectures. In: Proceedings of the 21th ACM SIGKDD International Conference on Knowledge Discovery and Data Mining, pp. 119–128. ACM (2015)
4. Chen, Y.-L., Chang, Y.-C.: A three-phase method for patent classification. Inf. Process. Manag. **48**(6), 1017–1030 (2012)

5. Collobert, R., Weston, J., Bottou, L., Karlen, M., Kavukcuoglu, K., Kuksa, P.: Natural language processing (almost) from scratch. J. Mach. Learn. Res. **12**(Aug), 2493–2537 (2011)

6. Fischer, T., Henkel, J.: Patent trolls on markets for technology-an empirical analysis of NPEs' patent acquisitions. Res. Policy **41**(9), 1519–1533 (2012)

7. Fujii, A., Iwayama, M., Kando, N.: Overview of the patent retrieval task at the NTCIR-6 workshop. In: NTCIR (2007)

8. Fujii, A., Utiyama, M., Yamamoto, M., Utsuro, T.: Evaluating effects of machine translation accuracy on cross-lingual patent retrieval. In: Proceedings of the 32nd International ACM SIGIR Conference on Research and Development in Information Retrieval, pp. 674–675. ACM (2009)

9. Grover, A., Leskovec, J.: node2vec: scalable feature learning for networks. In: Proceedings of the 22nd ACM SIGKDD International Conference on Knowledge Discovery and Data Mining, pp. 855–864. ACM (2016)

10. Guellec, D., de la Potterie, B.V.P.: Applications, grants and the value of patent. Econ. Lett. **69**(1), 109–114 (2000)

11. Hall, B.H., Jaffe, A., Trajtenberg, M.: Market value and patent citations. RAND J. Econ. **36**, 16–38 (2005)

12. Harhoff, D., Narin, F., Scherer, F.M., Vopel, K.: Citation frequency and the value of patented inventions. Rev. Econ. Stat. **81**(3), 511–515 (1999)

13. Hasan, M.A., Spangler, W.S., Griffin, T., Alba, A.: Coa: finding novel patents through text analysis. In: Proceedings of the 15th ACM SIGKDD International Conference on Knowledge Discovery and Data Mining, pp. 1175–1184. ACM (2009)

14. Hecht-Nielsen, R., et al.: Theory of the backpropagation neural network. Neural Netw. **1**(Supplement–1), 445–448 (1988)

15. Hsia, W.L.: The value of patents toward business. Intellect. Prop. Manag. **16**, 20–21 (1998)

16. Jin, B., Ge, Y., Zhu, H., Guo, L., Xiong, H., Zhang, C.: Technology prospecting for high tech companies through patent mining. In: 2014 IEEE International Conference on Data Mining (ICDM), pp. 220–229. IEEE (2014)

17. Karki, M.M.S.: Patent citation analysis: a policy analysis tool. World Patent Inf. **19**(4), 269–272 (1997)

18. Kingma, D., Ba, J:. Adam: a method for stochastic optimization. arXiv preprint arXiv:1412.6980 (2014)

19. Lanjouw, J.O., Schankerman, M.: The quality of ideas: measuring innovation with multiple indicators. Technical report, National bureau of economic research (1999)

20. Liu, X., Yan, J., Xiao, S., Wang, X., Zha, H., Chu, S.M.: On predictive patent valuation: forecasting patent citations and their types. In: AAAI, pp. 1438–1444 (2017)

21. Ma, L., Lu, Z., Li, H.: Learning to answer questions from image using convolutional neural network. In: AAAI, vol. 3, p. 16 (2016)

22. Martinez-Ruiz, A., Aluja-Banet, T.: Toward the definition of a structural equation model of patent value: PLS path modelling with formative constructs. REVSTAT-Stat. J. **7**(3), 265–290 (2009)

23. Mikolov, T., Sutskever, I., Chen, K., Corrado, G.S., Dean, J.: Distributed representations of words and phrases and their compositionality. In: Advances in Neural Information Processing Systems, pp. 3111–3119 (2013)

24. Montavon, G., Orr, G.B., Müller, K.-R. (eds.): Neural Networks: Tricks of the Trade. LNCS, vol. 7700. Springer, Heidelberg (2012). https://doi.org/10.1007/978-3-642-35289-8

25. Perozzi, B., Al-Rfou, R., Skiena, S.: Deepwalk: online learning of social represen-
    tations. In: Proceedings of the 20th ACM SIGKDD International Conference on
    Knowledge Discovery and Data Mining, pp. 701–710. ACM (2014)
26. Pimentel, T., Veloso, A., Ziviani, N.: Unsupervised and scalable algorithm for
    learning node representations (2017)
27. Shaparenko, B., Caruana, R., Gehrke, J., Joachims, T.: Identifying temporal pat-
    terns and key players in document collections. In: Proceedings of the IEEE ICDM
    Workshop on Temporal Data Mining: Algorithms, Theory and Applications (TDM
    2005), pp. 165–174 (2005)
28. Stephens, J.C., Jiusto, S.: Assessing innovation in emerging energy technologies:
    socio-technical dynamics of carbon capture and storage (CCS) and enhanced
    geothermal systems (EGS) in the USA. Energy Policy **38**(4), 2020–2031 (2010)
29. Sterzi, V.: Patent quality and ownership: an analysis of UK faculty patenting. Res.
    Policy **42**(2), 564–576 (2013)
30. Tang, J., Qu, M., Wang, M., Zhang, M., Yan, J., Mei, Q.: Line: large-scale infor-
    mation network embedding. In: Proceedings of the 24th International Conference
    on World Wide Web, pp. 1067–1077. International World Wide Web Conferences
    Steering Committee (2015)
31. Tang, J., Wang, B., Yang, Y., Hu, P., Zhao, Y., Yan, X., Gao, B., Huang, M.,
    Xu, P., Li, W., et al.: Patentminer: topic-driven patent analysis and mining. In:
    Proceedings of the 18th ACM SIGKDD International Conference on Knowledge
    Discovery and Data Mining, pp. 1366–1374. ACM (2012)
32. Zhang, L., Li, L., Li, T.: Patent mining: a survey. ACM SIGKDD Explor. Newsl.
    16(2), 1–19 (2015)

Learning Distribution-Matched Landmarks for Unsupervised Domain Adaptation

Mengmeng Jing, Jingjing Li$^{(\boxtimes)}$, Jidong Zhao, and Ke Lu

School of Computer Science and Engineering, University of Electronic Science
and Technology of China, Chengdu, China
poxiaoge77@foxmail.com, lijin117@yeah.net, {jdzhao,kel}@uestc.edu.cn

Abstract. Domain adaptation is widely used in database applications, especially in data mining. The basic assumption of domain adaptation (DA) is that some latent factors are shared by the source domain and the target domain. Revealing these shared factors, as a result, is the core operation of many DA approaches. This paper proposes a novel approach, named Learning Distribution-Matched Landmarks (LDML), for unsupervised DA. LDML reveals the latent factors by learning a domain-invariant subspace where the two domains are well aligned at both feature level and sample level. At the feature level, the divergences of both the marginal distribution and the conditional distribution are mitigated. At the sample level, each sample is evaluated so that we can take full advantage of the pivotal samples and filter out the outliers. Extensive experiments on two standard benchmarks verify that our approach can outperform state-of-the-art methods with significant advantages.

Keywords: Domain adaptation · Transfer learning
Landmark selection

1 Introduction

In real-world multimedia databases and data mining applications, cross-domain contents are often involved, such as videos, audios, texts and so on [3,5,29]. Naturally, there exists the need for multi-domain knowledge transferring among these applications. For example, if a biologist wants to develop an application that can accurately index or retrieve an endangered breed of bird, numerous labeled images of the birds are required if he or she employs the conventional machine learning algorithms. Unfortunately, it is impossible to collect plenty of such images and consume high labor-cost to label them manually. In recent years, DA algorithms get a rapid development for their advantage on solving the cross-domain knowledge transfer problems. Domain adaptation has been widely used in computer vision and database applications, such as image classification [18, 19], object recognition [11], text analysis [29] and recommendation systems [16].

© Springer International Publishing AG, part of Springer Nature 2018
J. Pei et al. (Eds.): DASFAA 2018, LNCS 10828, pp. 491–508, 2018.
https://doi.org/10.1007/978-3-319-91458-9_30

The principle of existing DA algorithms is to learn the latent shared factors between the source and target domains. These factors are at either feature level or sample level. Most existing DA algorithms, therefore, can be roughly classified into two groups: feature-based methods and sample-based methods.

Methods in the first group aim to reduce the gaps between domains at the feature level. The idea of these methods is to learn a new feature representation, typically a latent subspace, to minimize distribution divergence between domains. For example, Gopalan et al. [11] find intermediate representations by learning subspaces along the geodesic path that connects the source subspace and the target one on the Grassmann manifold. Fernando et al. [7] align subspaces of the two domains by directly using one linear mapping matrix so that the source and target subspaces can move closer. LRDE [18] constructs a novel graph structure under the graph embedding framework so that it can preserve geometric information in both the ambient instance space and the embedding feature space. JGSA [28] learns two coupled projections that project the source domain and the target domain data onto low-dimensional subspaces where the geometric structures are preserved and the distribution divergence is reduced. For a better understanding, we show the main idea of these methods in Fig. 1(a) and (b).

Methods in the second group employ landmark selection at the sample level. Landmarks are defined as a subset of the training samples which can bridge the two domains. Figure 1(c) illustrates the importance of landmark selection. After projecting data onto a low-dimensional subspace, discrepancies among samples remain. Some cross-domain pivotal samples are close to each other and can reduce the distribution divergence between domains, while other samples are far from the ones in the other domain and can increase the gaps between domains. The process of landmark selection takes full advantage of pivotal samples that align the source and target domains well, and filters out outliers. For instance, Gong et al. [9] discover landmarks at multiple granularities and construct auxiliary tasks correspondingly to compose domain-invariant features. LSSA [1] computes a quality measure for each sample using the Gaussian kernel. If the measure is above a threshold, the corresponding sample is selected as a landmark. CDLS [15] selects landmarks that match cross-domain data distributions well by using the empirical Maximum Mean Discrepancy (MMD) [12].

However, most of the existing approaches are based on either features or samples. They consider only one of the two factors independently. We observe that features and samples are factors at two different levels, and they can adapt two domains at different granularities: coarse-grained adaptation at the sample level and fine-grained adaptation at the feature level. It is obvious that both of the factors are interrelated and can reinforce each other. Therefore, in this paper, we propose a novel approach, named Learning Distribution-Matched Landmarks (LDML), for unsupervised DA. LDML learns a domain-invariant subspace where the two domains are well aligned at both feature level and sample level. Technically, at the feature level, we use MMD as the measure of the marginal and conditional distribution divergences. Then, minimizing MMD is employed to

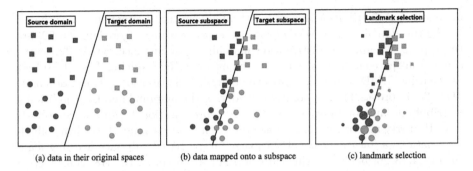

(a) data in their original spaces   (b) data mapped onto a subspace   (c) landmark selection

Fig. 1. The main idea of LDML. Points with the same color stand for samples that belong to the same domain and points with the same shape stand for samples that belong to the same class. The size of a point represents the weight of a sample. (a) the distribution divergence of data in their original spaces is large; (b) the distribution divergence between domains is reduced after mapping data onto a latent shared subspace; (c) select landmarks that can bridge the source and target subspaces.

align the two domains. Furthermore, we construct a graph under the graph embedding framework to preserve the locality structures in both domains. At the sample level, we evaluate each sample and put different weights on them. As a result, the pivotal samples are selected automatically in the iteration process and can be used to train a robust model. The contributions of this paper can be summarized as follows:

(1) We propose a multi-granularity DA method which considers factors at not only fine-grained feature level but also coarse-grained sample level. Therefore, our method can not only reduce the distribution divergence between domains but also be robust to outliers.
(2) Different from the conventional domain adaptation methods which learn only one projection matrix, we learn two projection matrices, one for each domain, so that the proposed method is more generalized and can be easily extended to handle the heterogeneous domain adaptation problems where the two domains may have different dimensionalities and features.
(3) Extensive experiments on two standard benchmarks with three features verify that our algorithm outperforms state-of-the-art algorithms with a significant advantage.

2 Related Work

As stated above, existing DA methods can be roughly classified as feature-based methods and sample-based methods. We now give a brief review of these methods from both categories.

Feature-based methods can be further categorized as distribution matching, e.g., subspace learning [17,20,22], and property preserving, e.g., geometric structures preserving [8,28]. Distribution matching algorithms reduce marginal or conditional distribution divergence through learning a latent subspace shared by the two domains. The MMD-based subspace learning algorithms [20,22] are the typical examples of distribution matching. For instance, TCA [22] learns some transfer components across domains in a Reproducing Kernel Hilbert Space using MMD, which mitigates the marginal distribution between domains. JDA [20] improves TCA by jointly considering the marginal and the conditional distributions and building a new feature representation to guarantee robustness. Property preserving algorithms preserve important properties when embedding data to a low-dimensional subspace. Graph-based subspace learning algorithms [4,18,26] belong to this category. For example, LRDE [18] constructs a novel graph structure under the graph embedding framework. Specifically, LRDE builds a within-class graph to encourage data in the same class to move closer, and a between-class graph to make data between classes far away. JGSA [28] is a feature matching algorithm which learns two coupled projections that project the two domains onto low-dimensional subspaces where the geometric structures are preserved and the distribution divergence is reduced.

Landmark [30] selection is a DA method at the sample level. For instance, TJM [21] employs instance reweighting by imposing the $\ell_{2,1}$-norm regularizer on the source projection matrix. CDLS [15] selects landmarks that have higher matching degree across domains by minimizing MMD. LSSA [1] uses the Gaussian kernel to compute quality measures of all samples and selects landmarks by a quality threshold.

The most related work is JGSA [28]. However, our method is significantly different from JGSA in at least two aspects:

(1) JGSA is merely a feature-based method which does not consider discrepancies among samples. When the two domains are not closely related, JGSA still enforces samples close to each other though they are far away. This transfer may degrade the performance of DA, even lead to the negative transfer [23]. LDML reweights samples to ensure that the pivotal samples are taken full advantage of and outliers are filtered out. The advantage of landmark selection is illustrated in Fig. 1.

(2) To preserve the discriminative information, JGSA employs the LDA criterion on the source domain data, which assumes the data are sampled from Gaussian distribution and makes the algorithm sensitive to outliers [27]. LDML applies the graph embedding strategy to preserve the local and global geometric information, which is more robust.

Experimental results in Sect. 4.4 show that our method can achieve better performance compared with JGSA.

3   Learning Distribution-Matched Landmarks

3.1   Problem Definition

Definition 1. A domain D is defined by a feature space χ and its probability distribution $P(X)$, where $X \in \chi$. For a specific domain, a classification task T

consists of class information y and a classifier $f(x)$, that is $T = \{y, f(x)\}$. We use subscripts s and t to indicate the source domain and the target domain, respectively. This paper focuses on the following problem:

Problem 1. Given a labeled source domain $\{X_s, y_s\}$ and an unlabeled target domain $\{X_t, y_t\}$, where X_s and X_t are the source and target domain samples, y_s and y_t are labels for the corresponding domains respectively, y_t is unknown, $P(X_s) \neq P(X_t)$ and $P(y_s|X_s) \neq P(y_t|X_t)$, project the source and target domains onto a subspace by projection matrices A and B so that the common latent features shared by involved domains are uncovered, the data manifold structure is preserved, and the domain difference is minimized.

3.2 Problem Formulation

Distribution Matching. LDML reveals the latent factors by learning a domain-invariant subspace where the two domains are well aligned at both feature level and sample level. Previous work [14,20] only learn one projection matrix, which has a major limitation when the two domains have different dimensionalities. To gain strong generalization ability, we learn two different projection matrices, one for each domain. We deploy these two matrices to project the source and target domain data onto a latent shared subspace. Specifically, let A be the projection matrix for the source domain and B for the target domain. $X_s \in \mathbb{R}^{m \times n_s}$ and $X_t \in \mathbb{R}^{m \times n_t}$ are the source and target domain data respectively, where n_s and n_t are the total numbers of the corresponding domain samples, m is the original dimensionality. $A \in \mathbb{R}^{m \times d}$ can project the source domain data onto a d-dimensional subspace, where $d \ll m$. Then, the low-dimensional data can be represented by $A^T X_s$. Similarly the low-dimensional representation of the target domain data X_t is $B^T X_t$. Considering the substantial distribution divergence, we align both the marginal and the conditional distribution between the two domains by adopting MMD. It is worth noting that the learned subspace should be shared by two domains. Therefore, we employ the Frobenius norm to minimize the distance between the two domains. Then, our objective function can be written as:

$$\min_{A,B} E_M(X_s, X_t, A, B) + E_C(X_s, X_t, A, B) + \lambda \|A - B\|_F^2, \tag{1}$$

where $\lambda > 0$ is the regularization parameter, E_M and E_C match the marginal and the conditional cross-domain data distributions, respectively. For simplicity, we set E_M and E_C with the same coefficient. E_M can be calculated as:

$$E_M = \left\| \frac{1}{n_s} \sum_{i=1}^{n_s} A^T x_s^i - \frac{1}{n_t} \sum_{j=1}^{n_t} B^T x_t^j \right\|_F^2, \tag{2}$$

E_C can be calculated as:

$$E_C = \sum_{c=1}^{C} \left(\left\| \frac{1}{n_s^c} \sum_{i=1}^{n_s^c} A^T x_s^{i,c} - \frac{1}{n_t^c} \sum_{j=1}^{n_t^c} B^T x_t^{j,c} \right\|_F^2 + \frac{1}{n_s^c n_t^c} \sum_{i=1}^{n_s^c} \sum_{j=1}^{n_t^c} \left\| A^T x_s^{i,c} - B^T x_t^{j,c} \right\|_F^2 \right),$$

$$(3)$$

where C is the number of classes, n_s^c and n_t^c denote the total numbers of the source and target domain samples in class c, respectively. Since labels in the target domain are not available, we use the pseudo labels to classify the target domain samples.

Landmark Selection. At the sample level, we learn two weight vectors α and β for the source and target domains, respectively. Each entry of the vector is a weight of the corresponding sample. Then the objective function can be rewritten as:

$$\min_{A,B} E_M(\alpha, \beta, X_s, X_t, A, B) + E_C(\alpha, \beta, X_s, X_t, A, B) + \lambda \|A - B\|_F^2,$$

$$\text{s.t. } \{\alpha_i^c, \beta_i^c\} \in [0,1], \frac{\alpha^{c^T} \mathbf{1}_{n_s^c}}{n_s^c} = \frac{\beta^{c^T} \mathbf{1}_{n_t^c}}{n_t^c} = \delta,$$

$$(4)$$

where $\alpha = [\alpha^1; \cdots; \alpha^c; \cdots; \alpha^C] \in \mathbb{R}^{n_s}$ and $\beta = [\beta^1; \cdots; \beta^c; \cdots; \beta^C] \in \mathbb{R}^{n_t}$ are the weights of data in the source and target domains respectively, $\alpha^c = [\alpha_1^c; \cdots; \alpha_{n_s^c}^c]$, $\beta^c = [\beta_1^c; \cdots; \beta_{n_t^c}^c]$, $\mathbf{1}_{n_s^c} \in \mathbb{R}^{n_s^c}$ and $\mathbf{1}_{n_t^c} \in \mathbb{R}^{n_t^c}$ are column vectors with all ones. $\delta \in [0,1]$ controls the average weight of the whole source or target domain samples.

The E_M and E_C in (4) can be updated as:

$$E_M = \left\| \frac{1}{\delta n_s} \sum_{i=1}^{n_s} \alpha_i A^T x_s^i - \frac{1}{\delta n_t} \sum_{j=1}^{n_t} \beta_j B^T x_t^j \right\|_F^2,$$

$$E_C =$$
$$\sum_{c=1}^{C} \left(\left\| \frac{1}{\delta n_s^c} \sum_{i=1}^{n_s^c} \alpha_i A^T x_s^{i,c} - \frac{1}{\delta n_t^c} \sum_{j=1}^{n_t^c} \beta_j B^T x_t^{j,c} \right\|_F^2 + \frac{1}{\delta^2 n_s^c n_t^c} \sum_{i=1}^{n_s^c} \sum_{j=1}^{n_t^c} \left\| \alpha_i A^T x_s^{i,c} - \beta_j B^T x_t^{j,c} \right\|^2 \right),$$

Equation (4) can be further transformed to its matrix form as follows:

$$\min_{A,B} \mathrm{Tr} \left([A^T B^T] \begin{bmatrix} M_{ss} + \lambda I & M_{st} - \lambda I \\ M_{ts} - \lambda I & M_{tt} + \lambda I \end{bmatrix} \begin{bmatrix} A \\ B \end{bmatrix} \right),$$

$$(5)$$

where

$$M_{ss} = X_s H_{ss} X_s^T, \qquad M_{tt} = X_t H_{tt} X_t^T,$$
$$M_{st} = X_s H_{st} X_t^T, \qquad M_{ts} = M_{st}^T.$$

Each entry $(H_{ss})_{ij}$ in $H_{ss} \in \mathbb{R}^{n_s \times n_s}$ denotes the coefficient associated with $x_s^{i^T} x_s^j$. Similar remarks can be applied to $H_{tt} \in \mathbb{R}^{n_t \times n_t}$ and $H_{st} \in \mathbb{R}^{n_s \times n_t}$. Detailed derivations are similar to that in CDLS [15]. For the conciseness, we omit the regular mathematical derivations in this paper.

Locality Structure Preservation. In general, a sample tends to have the same label with its k-nearest neighbors. This locality property is crucial in many computer vision tasks [4,18]. Therefore, we propose to preserve the geometric structures in the source domain so that the discriminative information can be maximized and then be transferred to the target domain. On the other hand, the geometric structures in the target domain should be preserved as well, so that the manifold structures in the original space can be retained. So we introduce two Laplacian graph terms, one for each domain:

$$\min \frac{\mathrm{Tr}(A^T X_s L_w^s X_s^T A)}{\mathrm{Tr}(A^T X_s L_b^s X_s^T A)} = \min \frac{\mathrm{Tr}(A^T S_w^s A)}{\mathrm{Tr}(A^T S_b^s A)}, \tag{6}$$

$$\min \frac{\mathrm{Tr}(B^T X_t L_w^t X_t^T B)}{\mathrm{Tr}(B^T X_t L_b^t X_t^T B)} = \min \frac{\mathrm{Tr}(B^T S_w^t B)}{\mathrm{Tr}(B^T S_b^t B)}, \tag{7}$$

where

$$S_b^s = X_s L_b^s X_s^T, \quad S_w^s = X_s L_w^s X_s^T,$$
$$S_b^t = X_t L_b^t X_t^T, \quad S_w^t = X_t L_w^t X_t^T,$$

where $L = D - W$, D is a diagonal matrix and its diagonal entry is $D_{ii} = \sum_{j \neq i} W_{ij}$. L_w^s and L_b^s are the Laplacian matrices of the intrinsic graph and the penalty graph for the source domain. Similarly, L_w^t and L_b^t are the Laplacian matrices for the target domain. W_w and W_b are the weight matrices for the intrinsic graph and the penalty graph, respectively. In this paper, we follow the following two criteria to construct W_w and W_b.

(a) Construct the intrinsic weight matrix W_w: For each sample x, connect the nearest neighbor v with x where v has the same class information with x.
(b) Construct the penalty weight matrix W_b: For each domain, connect the k-nearest vertex pairs where samples in each pair belong to different classes.

By applying (a), samples from the same class can be more compact and the local manifold structure can be preserved. By deploying (b), samples from the same domain but different classes can be more separable and the global discriminative information can be retained. We apply the heat kernel method to get W_w and W_b, e.g., if two samples x_i and x_j are connected, then the weight of them is $\exp(-((\|x_i - x_j\|^2)/(2\sigma^2)))$, otherwise it is 0.

Overall Objective Function. Considering all the above discussions, we get the overall objective function:

$$\min_{A,B} \frac{\mathrm{Tr}\left([A^T \ B^T] \begin{bmatrix} M_{ss} + \gamma S_w^s + \lambda I & M_{st} - \lambda I \\ M_{ts} - \lambda I & M_{tt} + \gamma S_w^t + (\lambda + \mu)I \end{bmatrix} \begin{bmatrix} A \\ B \end{bmatrix}\right)}{\mathrm{Tr}\left([A^T \ B^T] \begin{bmatrix} \gamma S_b^s & 0 \\ 0 & \gamma S_b^t + \mu S_h^t \end{bmatrix} \begin{bmatrix} A \\ B \end{bmatrix}\right)}, \tag{8}$$

where

$$S_h^t = X_t(I_t - \frac{1}{n_t} \mathbf{1}_{n_t} \mathbf{1}_{n_t}^T) X_t^T \tag{9}$$

is the covariance matrix of the target domain, γ, μ and λ are trade-off parameters for the graph-embedding term, the target variance term and $||A - B||_{\mathrm{F}}^2$, respectively.

3.3   Problem Optimization

Optimizing A and B. To optimize (8), we rewrite$[A; B]$ as P. Thus, the objective function can be rewritten as:

$$\min_{P} \frac{\mathrm{Tr}\left(P^{\mathrm{T}} \begin{bmatrix} M_{\mathrm{ss}} + \gamma S_{\mathrm{w}}^{\mathrm{s}} + \lambda I & M_{\mathrm{st}} - \lambda I \\ M_{\mathrm{ts}} - \lambda I & M_{\mathrm{tt}} + \gamma S_{\mathrm{w}}^{\mathrm{t}} + (\lambda + \mu)I \end{bmatrix} P\right)}{\mathrm{Tr}\left(P^{\mathrm{T}} \begin{bmatrix} \gamma S_{\mathrm{b}}^{\mathrm{s}} & 0 \\ 0 & \gamma S_{\mathrm{b}}^{\mathrm{t}} + \mu S_{\mathrm{h}}^{\mathrm{t}} \end{bmatrix} P\right)}. \tag{10}$$

We can reformulate (10) as:

$$\max_{P} \mathrm{Tr}\left(P^{\mathrm{T}} \begin{bmatrix} \gamma S_{\mathrm{b}}^{\mathrm{s}} & 0 \\ 0 & \gamma S_{\mathrm{b}}^{\mathrm{t}} + \mu S_{\mathrm{h}}^{\mathrm{t}} \end{bmatrix} P\right), \tag{11}$$

$$\text{s.t.} \quad \mathrm{Tr}\left(P^{\mathrm{T}} \begin{bmatrix} M_{\mathrm{ss}} + \gamma S_{\mathrm{w}}^{\mathrm{s}} + \lambda I & M_{\mathrm{st}} - \lambda I \\ M_{\mathrm{ts}} - \lambda I & M_{\mathrm{tt}} + \gamma S_{\mathrm{w}}^{\mathrm{t}} + (\lambda + \mu)I \end{bmatrix} P\right) = 1 .$$

The Lagrange function of (11) is

$$\begin{aligned} L = {} & \mathrm{Tr}\left(P^{\mathrm{T}} \begin{bmatrix} \gamma S_{\mathrm{b}}^{\mathrm{s}} & 0 \\ 0 & \gamma S_{\mathrm{b}}^{\mathrm{t}} + \mu S_{\mathrm{h}}^{\mathrm{t}} \end{bmatrix} P\right) \\ & + \mathrm{Tr}\left(\left(P^{\mathrm{T}} \begin{bmatrix} M_{\mathrm{ss}} + \gamma S_{\mathrm{w}}^{\mathrm{s}} + \lambda I & M_{\mathrm{st}} - \lambda I \\ M_{\mathrm{ts}} - \lambda I & M_{\mathrm{tt}} + \gamma S_{\mathrm{w}}^{\mathrm{t}} + (\lambda + \mu)I \end{bmatrix} P - I\right) \Phi\right), \end{aligned} \tag{12}$$

where $\Phi = \mathrm{diag}(\phi_1, \cdots, \phi_d)$ and $(\phi_1, \cdots, \phi_d)$ are the d largest eigenvalues of the following eigendecomposition problem:

$$\begin{bmatrix} \gamma S_{\mathrm{b}}^{\mathrm{s}} & 0 \\ 0 & \gamma S_{\mathrm{b}}^{\mathrm{t}} + \mu S_{\mathrm{h}}^{\mathrm{t}} \end{bmatrix} P = \begin{bmatrix} M_{\mathrm{ss}} + \gamma S_{\mathrm{w}}^{\mathrm{s}} + \lambda I & M_{\mathrm{st}} - \lambda I \\ M_{\mathrm{ts}} - \lambda I & M_{\mathrm{tt}} + \gamma S_{\mathrm{w}}^{\mathrm{t}} + (\lambda + \mu)I \end{bmatrix} P \Phi. \tag{13}$$

As a result, P consists of the corresponding d eigenvectors of the above problem. Once the transformation matrix P is obtained, the subspace A and B can be obtained easily.

Optimizing α and β. By regarding A and B as constants, the optimization of our objective function can be written as:

$$\min_{\alpha, \beta} \frac{1}{2} \alpha^{\mathrm{T}} K_{\mathrm{ss}} \alpha - \alpha^{\mathrm{T}} K_{\mathrm{st}} \beta, \tag{14}$$

$$\text{s.t.} \ \{\alpha_i^c, \beta_i^c\} \in [0, 1], \frac{\alpha^{c\mathrm{T}} \mathbf{1}}{n_{\mathrm{s}}^c} = \frac{\beta^{c\mathrm{T}} \mathbf{1}}{n_{\mathrm{t}}^c} = \delta ,$$

where $(K_{ss})_{i,j}$ in $K_{ss} \in \mathbb{R}^{n_s \times n_s}$ is the coefficient associated with $(A^T x_s^i)^T A^T x_s^i$, and $(K_{st})_{i,j}$ in $K_{st} \in \mathbb{R}^{n_s \times n_t}$ is the coefficient associated with $(A^T x_s^i)^T B^T x_t^j$. Limited by space, the detailed derivations are omitted since they are similar to that in CDLS [15].

With the above formulation, we can apply Quadratic Programming (QP) solvers to optimize the equivalent problem:

$$\min_{z_i \in [0,1],\, Z^T V = G} \frac{1}{2} Z^T B Z, \tag{15}$$

where

$$Z = \begin{pmatrix} \alpha \\ \beta \end{pmatrix}, B = \begin{pmatrix} K_{ss} & -K_{st} \\ -K_{st}^T & 0 \end{pmatrix}, G \in \mathbb{R}^{1 \times 2C} \text{ with}$$

$$(G)_c = \begin{cases} \delta n_s^c & \text{if } c \leq C \\ \delta n_t^{c-C} & \text{if } c > C \end{cases}, V = \begin{bmatrix} V_s & \mathbf{0}_{n_s \times C} \\ \mathbf{0}_{n_t \times C} & V_t \end{bmatrix} \in \mathbb{R}^{(n_s + n_t) \times 2C} \text{ with}$$

$$(V_s)_{ij} = \begin{cases} 1 & \text{if } x_s^i \in \text{class } j \\ 0 & \text{otherwise} \end{cases}, (V_t)_{ij} = \begin{cases} 1 & \text{if } x_t^i \text{ predicted as class } j \\ 0 & \text{otherwise} \end{cases}.$$

Finally, we use the weight sensitive libsvm [2] to train a classifier so that the learned weights can be fully exploited. We show the procedure of LDML in Algorithm 1.

3.4 Computational Complexity

Now we analyze the computational complexity of Algorithm 1 by the big O notation. As stated above, $X_s \in \mathbb{R}^{m \times n_t}$, $X_t \in \mathbb{R}^{m \times n_s}$, $X = [X_s\ X_t] \in \mathbb{R}^{m \times n}$, $A \in \mathbb{R}^{m \times d}$, $B \in \mathbb{R}^{m \times d}$, where m is the original dimensionality, d is the dimensionality of the subspace, $n = n_s + n_t$ is the number of all samples. We note the number of classes as C and the number of iterations as T. The time cost of Algorithm1 consists of the following three parts:

(1) Solving the eigendecomposition problem in step 2 is $O(Tdm^2)$.
(2) Solving the equality constrained QP problems in step 5 is $O(Tn^3)$.
(3) Computing the MMD matrices and the graph embedding matrices in step 6 are both $O(TCn^2)$.

Then, the overall computational complexity of Algorithm 1 is $O(Tdm^2 + Tn^3 + TCn^2)$. In Sect. 4.5, we show that the number of iterations T is usually set as 5, which is enough to guarantee convergence. Besides, the typical values of d are not greater than 200, so $T \ll \min(m, n)$, $d \ll \min(m, n)$. Therefore, the computational complexity of Algorithm 1 depends mainly on the number of samples n and the dimensionality m.

Algorithm1: Learning Distribution-Matched Landmarks

Input: source and target domain data: X_s, X_t; labels for source domain data: y_s;
    Parameters: $\delta = 0.5$, d, λ, μ, γ
Output: Predicted labels y_t for target domain data
begin
    1: Initialize pseudo labels of target domain $\hat{y}_t$ using PCA; Construct S_h^t, M_{ss}, M_{tt}, M_{st},
        M_{ts}, S_b^s, S_w^s, S_b^t, S_w^t according to (9)(5)(6)(7);
    while not converge **do**
    2: Solve the generalized eigendecomposition problem in (13) and select d corresponding
        eigenvectors of d largest eigenvalues as the transformation P, and obtain transformation
        A and B;
    3: Map the original data to respective subspace to get the embeddings: $Z_s = A^T X_s$,
        $Z_t = B^T X_t$;
    4: Train a SVM classifier on α, Z_s, y_s to update pseudo labels in target domain $\hat{y}_t$;
    5: Update landmark weights α, β by (15);
    6: Update $M_{ss}, M_{tt}, M_{st}, M_{ts}, S_b^t, S_w^t$ by (5)(6)(7);
    end while
end

---

4 Experiments

4.1 Datasets

Office + Caltech: Office [24] consists of three different datasets: **A**mazon (images downloaded from amazon.com), **W**ebcam (low-resolution images taken by a web camera), **DSLR** (high-resolution images taken by a digital SLR camera). **Caltech** [13] is a dataset which has 256 classes and 30, 607 images. Ten common classes of all four datasets are selected [10] and each dataset is regarded as a domain. As a result, we have four domains: C (Caltech), A (Amazon), W (Webcam), D (DSLR) and 12 DA evaluations by selecting two different domains as the source and target domains respectively. We consider two types of features: SURF and DeCAF$_6$ [6]. SURF features are encoded with the 800-bin histogram with codebooks trained from a subset of Amazon images. DeCAF$_6$ are activation features of the 6th fully connected layer of a convolutional network constructed by [6].

PIE: PIE [25] is a face recognition benchmark which has 68 individuals with 41, 368 face images. These individuals have different poses, expressions and illuminations. Five subsets of PIE are selected to conduct the face recognition experiments: C05 (left pose), C07 (upward pose), C09 (downward pose), C27 (frontal pose) and C29 (right pose). All face images are cropped and resized to 32 × 32 pixels and then converted to grayscale. Each subset is regarded as a domain, and each time one subset plays the role of the source domain and the other one plays the role of the target domain. Then, we can generate 20 DA evaluations.

4.2 Compared Baselines

We compare our LDML with the following six state-of-the-art baselines: transfering with SVM, subspace alignment (SA) [7], geodesic flow kernel (GFK) [10], joint distribution analysis (JDA) [20], transfer joint matching (TJM) [21], joint geometrical and statistical alignment (JGSA) [28].

4.3 Experimental Settings

Following the previous work [10,21], we normalize all of the data to have zero mean and unit standard deviation in each dimension.

For all the baselines, we report the best results we can achieve. For LDML, we fix the average weight $\delta = 0.5$. Both of the numbers of neighbors in the intrinsic graph and the penalty graph are 5. The number of iterations T is usually set to 5 except for testing the convergence of our method. For different experimental tasks, we set different hyperparameters to gain good performance. Specifically, when evaluating our method on Office + Caltech with SURF features, we set $\lambda = 0.5$, $\mu = 0.1$, $\gamma = 0.001$, $d = 40$. On Office + Caltech with DeCAF$_6$ features, we set $\lambda = 0.05$, $\mu = 0.5$, $\gamma = 0.01$, $d = 40$. On PIE dataset, we set $\lambda = 1$, $\mu = 0.1$, $\gamma = 0.02$, $d = 120$.

To exploit the learned weights for samples, our LDML uses the weight sensitive SVM to train a classifier. For fairness, we also use SVM to train classifiers on some baselines, e.g., SA, JDA, TJM and JGSA.

In this paper, we follow the previous work [10,22] and use the following equation as the classification accuracy:

$$\frac{|x : x \in X_t \wedge \hat{y}_t = y_t|}{|x : x \in X_t|},$$

where x is a sample in target domain, y_t is the real label for x, and $\hat{y}_t$ is the predicted label for x.

4.4 Experimental Results

The results of LDML and other baseline methods on all the datasets are reported in Tables 1, 2 and 3.

On Office + Caltech with SURF features, LDML achieves the best classification accuracy on 8 out of 12 evaluations. Besides, the average accuracy of LDML is 55.81%, gaining an improvement of 3.43% compared with the best baseline JGSA. JGSA is a method that can reduce not only the distribution divergence but also the geometrical divergence simultaneously when projecting data onto low-dimensional subspaces. However, since the domain difference in Office + Caltech is substantially large, there may exist samples in one domain that are irrelevant to samples in the other domain even in subspaces learned by JGSA. LDML handles limitation of JGSA by selecting pivotal samples in both of the source and target domains, which contributes to the better performance compared with JGSA.

On Office + Caltech with DeCAF$_6$ features, all of the methods achieve better performance than that on Office + Caltech with SURF features, especially our LDML. LDML achieves the best classification accuracy on 10 out of 12 evaluations. It is also observed that LDML, JGSA and JDA achieve the best three average classification accuracies. The common point of these methods is that they all reduce both the marginal and conditional distribution divergences

Table 1. Accuracy (%) on the Office + Caltech dataset with SURF features.

X_s	X_t	SVM	GFK	SA	JDA	TJM	JGSA	LDML
C	A	54.18	41.02	51.04	54.28	51.77	58.87	**59.60**
	W	44.41	40.68	40.34	47.80	44.41	54.58	**58.98**
	D	43.95	38.85	45.86	45.86	49.68	45.22	**54.14**
A	C	**45.41**	40.25	44.79	43.46	43.99	40.69	45.24
	W	36.61	38.98	38.64	46.10	45.08	**60.34**	55.25
	D	37.58	36.31	39.49	40.76	45.22	**58.60**	52.87
W	C	33.04	30.72	35.71	32.23	35.89	31.79	**37.76**
	A	34.55	29.75	36.95	38.41	38.10	40.40	**43.11**
	D	84.08	80.89	71.97	84.08	82.17	84.71	**91.72**
D	C	30.01	30.28	34.11	33.30	**35.80**	34.11	35.44
	W	32.57	32.05	35.39	35.60	37.16	38.62	**44.47**
	A	73.90	75.59	76.61	82.03	84.75	80.68	**91.19**
Avg.		45.86	42.95	45.91	48.65	49.50	52.38	**55.81**

Table 2. Accuracy (%) on the Office + Caltech dataset with DeCAF$_6$ features.

X_s	X_t	SVM	GFK	SA	JDA	TJM	JGSA	LDML
C	A	89.46	89.04	90.61	90.60	90.92	92.07	**92.90**
	W	77.28	87.80	82.37	80.68	86.10	83.05	**89.49**
	D	80.25	85.99	86.62	84.71	87.90	88.54	**89.81**
A	C	81.21	78.45	84.06	85.04	82.46	85.57	**87.71**
	W	74.58	81.02	82.71	**87.46**	86.10	86.78	87.12
	D	82.80	80.25	86.62	**89.17**	87.90	87.90	87.26
W	C	66.34	72.31	76.58	82.37	79.52	85.40	**87.00**
	A	75.68	81.84	84.45	90.61	89.04	91.34	**91.44**
	D	99.36	**100**	99.36	99.36	98.73	99.36	**100**
D	C	64.74	75.96	73.73	72.31	73.91	74.98	**88.07**
	W	76.93	77.35	84.45	89.46	88.52	88.94	**93.11**
	A	97.29	97.29	94.24	97.29	94.58	**99.66**	99.66
Avg.		80.49	83.94	85.48	87.42	87.14	88.63	**91.13**

simultaneously by using MMD. TJM only considers the marginal distribution divergence, which leads to its relatively poor performance.

On PIE dataset, LDML outperforms state-of-the-art methods on all of the 20 evaluations. The average classification accuracy of LDML is 82.36%, which overwhelmingly has a 19.62% improvement compared with the best baseline JGSA. On PIE dataset, the Euclidean distance between samples from one domain but different classes may be smaller than that of samples within one class but from

Table 3. Accuracy (%) on the PIE dataset.

X_s	X_t	SVM	GFK	SA	JDA	TJM	JGSA	LDML
C05	C07	33.52	45.49	35.54	54.08	34.87	52.73	**80.11**
	C09	43.69	50.31	46.38	60.54	44.36	51.84	**72.49**
	C27	61.28	65.82	63.62	85.97	60.74	73.72	**95.31**
	C29	36.46	41.97	39.58	49.33	34.93	52.39	**64.64**
C07	C05	42.05	46.91	44.24	60.41	38.75	64.26	**80.34**
	C09	41.85	56.74	44.06	55.45	49.82	58.88	**75.49**
	C27	65.64	70.86	66.45	82.43	63.41	70.71	**94.86**
	C29	34.13	41.85	35.48	47.37	35.23	49.02	**69.73**
C09	C05	49.58	48.65	51.80	58.64	43.04	64.89	**78.75**
	C07	42.91	56.23	46.41	49.60	38.74	59.91	**79.19**
	C27	67.98	74.95	68.76	67.80	65.33	72.63	**96.00**
	C29	42.40	50.61	45.22	46.69	41.12	57.72	**74.69**
C27	C05	66.96	71.43	69.57	79.26	63.39	74.73	**95.98**
	C07	62.06	81.34	64.03	77.59	60.59	76.24	**96.13**
	C09	70.71	86.34	72.06	77.45	74.39	67.89	**94.24**
	C29	54.23	59.87	55.94	58.88	50.61	63.05	**84.74**
C29	C05	46.16	39.98	48.50	51.02	37.18	63.99	**77.46**
	C07	34.81	38.80	35.30	42.79	29.28	54.02	**69.06**
	C09	47.98	48.96	49.39	41.85	42.77	59.87	**77.63**
	C27	59.12	54.73	59.51	62.51	46.77	66.39	**90.36**
Avg.		50.17	56.59	52.09	60.48	47.77	62.74	**82.36**

different domains. The illustration can be seen in Fig. 2. To handle this, LDML constructs an intrinsic graph to preserve the local manifold structure, and a penalty graph to make samples with different labels more separable. This novel graph structure enhances the discriminability of the model and leads to the best performance of LDML on PIE dataset. We also notice that TJM performs poorly on PIE dataset, even not as good as SVM. TJM imposes $\ell_{2,1}$ norm on the source projection matrix to get the "row sparsity". On PIE dataset, $\ell_{2,1}$ norm reinforces the modal information among the faces with the same pose but from different people, and weaken the classification information among the faces from one person but with different poses. This results in low performance of TJM.

Above all, our LDML obviously performs the best on almost all evaluations, especially on PIE dataset. The only two exceptions are on A → W and A → D. As stated in [10], Amazon is greatly different from Webcam and DSLR geometrically and statistically. Besides, we also notice that Amazon is the unique dataset where images usually have no backgrounds. In this case, the MMD distances of almost all cross-domain sample pairs are so far that the selected landmarks have no obvious discrepancies with other samples. Even so, LDML still

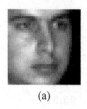

(a)   (b)   (c)

Fig. 2. Three faces are chosen from the CMU PIE dataset. Profile and frontal faces can be regarded as two domains. The faces of different people represent different classes. Figure 2(a) and (b) are two faces from the same domain but with different labels. Figure 2(b) and (c) are two faces with the same label but from different domains. We simply compute the Euclidean distance of two images. One is between the profile face of two classes in (a) and (b), the other is between two domains of one class in (b) and (c). We get that the distance of (a) and (b) is much smaller than that of (b) and (c), which is a common phenomenon in PIE dataset.

achieves the second-best performance on A → W(SURF), A → D (SURF) and A → W (DeCAF$_6$), which reveals the robustness of LDML.

4.5   Parameter Sensitivity and Convergence

To analyze the parameter sensitivity, we evaluate four hyperparameters: λ, μ, γ and d. When one of them is being evaluated, we set the others the same as parameters in Sect. 4.3. We conduct sensitivity analysis on C05 → C07 and C → A. The results are shown in Fig. 3.

In Fig. 3(a), when the dimensionality of subspace d varies from 20 to 200, the classification accuracies on Office + Caltech with both SURF and DeCAF$_6$ features fluctuate in small ranges. We observe that the classification accuracy on PIE is sensitive to d and the optimal range of d on PIE is between 100 and 120. In Fig. 3(b), it is obvious that the classification accuracy will reduce on all datasets if $\gamma > 0.01$. In Fig. 3(c), we can observe that the optimal λ for Office + Caltech is around 0.1, while the optimal λ for PIE is around 1. In Fig. 3(d), μ affects the performance on Office + Caltech dataset little when $\mu > 0.01$, while the optimal μ on PIE is around 0.1.

To show the convergence of our method, we report the values of the objective function on C → W (SURF), C → W (DeCAF$_6$) and C05 → C07 with different iterations. The results are illustrated in Fig. 4. It is obvious that the value of the objective function is monotonically decreasing when $T \leq 5$ and keeps stable when $T > 5$. This result reveals that LDML can converge in 5 iterations.

4.6   Effectiveness Analysis

To evaluate the effectiveness of landmark selection, we conduct experiments on Office + Caltech (SURF) with two settings: one is LDML with the same weight, the other is LDML with landmark selection. The results are reported

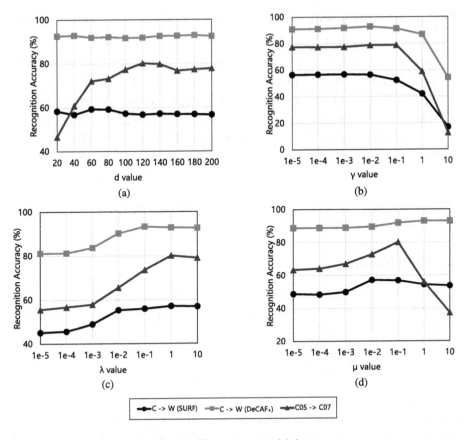

Fig. 3. Parameter sensitivity.

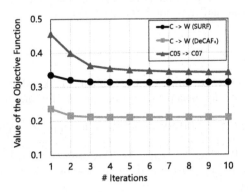

Fig. 4. Convergence analysis.

in Table 4. We observe that LDML with landmark selection achieves the better performance on 10 out of 12 evaluations. Besides, the average accuracy of LDML with landmark selection gains an improvement of 2.13% compared with that of

Table 4. Effectiveness analysis: accuracies (%) of two settings on Office + Caltech (SURF).

Method	C → A	C → W	C → D	A → C	A → W	A → D	-
No landmarks	58.35	**61.02**	**54.14**	43.01	50.51	47.77	-
With landmarks	**59.60**	58.98	**54.14**	**45.24**	**55.25**	**52.87**	-
Method	W → C	W → A	W → D	D → C	D → A	D → W	**Avg.**
No landmarks	30.81	41.02	**92.99**	31.26	42.07	**91.19**	53.68
With landmarks	**37.76**	**43.11**	91.72	**35.44**	**44.47**	**91.19**	**55.81**

LDML with the same weight. These results reveal the effectiveness of landmark selection in our method.

5   Conclusion

In this paper, we propose a novel method for unsupervised domain adaptation, named Learning Distribution-Matched Landmarks (LDML). LDML aligns the source and target domains by reducing the distribution divergence and selecting landmarks that bridge two domains. Comprehensive experiments on two real-world datasets verify the effectiveness of the proposed method. In the future, we plan to extend our LDML to be capable of handling the heterogeneous DA.

Acknowledgment. This work was supported in part by the National Postdoctoral Program for Innovative Talents under Grant BX201700045, China Postdoctoral Science Foundation under Grant 2017M623006, the Applied Basic Research Program of Sichuan Province under Grant 2015JY0124, and the Fundamental Research Funds for the Central Universities under Grant ZYGX2016J089.

References

1. Aljundi, R., Emonet, R., Muselet, D., Sebban, M.: Landmarks-based kernelized subspace alignment for unsupervised domain adaptation. In: CVPR, pp. 56–63 (2015)
2. Chang, C.C., Lin, C.J.: LIBSVM: a library for support vector machines. ACM TIST **2**(3), 27 (2011)
3. Chattopadhyay, R., Sun, Q., Fan, W., Davidson, I., Panchanathan, S., Ye, J.: Multisource domain adaptation and its application to early detection of fatigue. KDD **6**(4), 18 (2012)
4. Chen, H.T., Chang, H.W., Liu, T.L.: Local discriminant embedding and its variants. In: CVPR, vol. 2, pp. 846–853. IEEE (2005)
5. Ding, Z., Shao, M., Fu, Y.: Deep low-rank coding for transfer learning. In: IJCAI, pp. 3453–3459 (2015)
6. Donahue, J., Jia, Y., Vinyals, O., Hoffman, J., Zhang, N., Tzeng, E., Darrell, T.: DeCAF: a deep convolutional activation feature for generic visual recognition. In: ICML, pp. 647–655 (2014)

7. Fernando, B., Habrard, A., Sebban, M., Tuytelaars, T.: Unsupervised visual domain adaptation using subspace alignment. In: ICCV, pp. 2960–2967 (2013)
8. Ghifary, M., Balduzzi, D., Kleijn, W.B., Zhang, M.: Scatter component analysis: a unified framework for domain adaptation and domain generalization. IEEE TPAMI **39**(7), 1414–1430 (2016)
9. Gong, B., Grauman, K., Sha, F.: Connecting the dots with landmarks: discriminatively learning domain-invariant features for unsupervised domain adaptation. In: ICML, pp. 222–230 (2013)
10. Gong, B., Shi, Y., Sha, F., Grauman, K.: Geodesic flow kernel for unsupervised domain adaptation. In: CVPR, pp. 2066–2073. IEEE (2012)
11. Gopalan, R., Li, R., Chellappa, R.: Domain adaptation for object recognition: an unsupervised approach. In: ICCV, pp. 999–1006. IEEE (2011)
12. Gretton, A., Borgwardt, K.M., Rasch, M., Schölkopf, B., Smola, A.J.: A kernel method for the two-sample-problem. In: NIPS, pp. 513–520 (2007)
13. Griffin, G., Holub, A., Perona, P.: Caltech-256 Object Category Dataset (2007)
14. Gu, Q., Li, Z., Han, J.: Joint feature selection and subspace learning. In: IJCAI, vol. 22, p. 1294 (2011)
15. Hubert Tsai, Y.H., Yeh, Y.R., Frank Wang, Y.C.: Learning cross-domain landmarks for heterogeneous domain adaptation. In: CVPR, pp. 5081–5090 (2016)
16. Li, J., Lu, K., Huang, Z., Shen, H.T.: Two birds one stone: on both cold-start and long-tail recommendation. In: ACM Multimedia, pp. 898–906. ACM (2017)
17. Li, J., Lu, K., Zhu, L., Li, Z.: Locality-constrained transfer coding for heterogeneous domain adaptation. In: Huang, Z., Xiao, X., Cao, X. (eds.) ADC 2017. LNCS, vol. 10538, pp. 193–204. Springer, Cham (2017). https://doi.org/10.1007/978-3-319-68155-9_15
18. Li, J., Wu, Y., Zhao, J., Lu, K.: Low-rank discriminant embedding for multiview learning. IEEE TCYB **47**(11), 3516–3529 (2017)
19. Li, J., Zhao, J., Lu, K.: Joint feature selection and structure preservation for domain adaptation. In: IJCAI, pp. 1697–1703 (2016)
20. Long, M., Wang, J., Ding, G., Sun, J., Yu, P.S.: Transfer feature learning with joint distribution adaptation. In: ICCV, pp. 2200–2207 (2013)
21. Long, M., Wang, J., Ding, G., Sun, J., Yu, P.S.: Transfer joint matching for unsupervised domain adaptation. In: CVPR, pp. 1410–1417 (2014)
22. Pan, S.J., Tsang, I.W., Kwok, J.T., Yang, Q.: Domain adaptation via transfer component analysis. IEEE TNN **22**(2), 199–210 (2011)
23. Pan, S.J., Yang, Q.: A survey on transfer learning. IEEE TKDE **22**(10), 1345–1359 (2010)
24. Saenko, K., Kulis, B., Fritz, M., Darrell, T.: Adapting visual category models to new domains. In: Daniilidis, K., Maragos, P., Paragios, N. (eds.) ECCV 2010. LNCS, vol. 6314, pp. 213–226. Springer, Heidelberg (2010). https://doi.org/10.1007/978-3-642-15561-1_16
25. Sim, T., Baker, S., Bsat, M.: The CMU pose, illumination, and expression (PIE) database. In: FG, pp. 53–58. IEEE (2002)
26. Sugiyama, M.: Local fisher discriminant analysis for supervised dimensionality reduction. In: ICML, pp. 905–912. ACM (2006)
27. Yan, S., Xu, D., Zhang, B., Zhang, H.J., Yang, Q., Lin, S.: Graph embedding and extensions: a general framework for dimensionality reduction. IEEE TPAMI **29**(1), 40–51 (2007)

28. Zhang, J., Li, W., Ogunbona, P.: Joint geometrical and statistical alignment for visual domain adaptation. In: CVPR (2017)
29. Zhou, M., Chang, K.C.: Unifying learning to rank and domain adaptation: enabling cross-task document scoring. In: SIGKDD, pp. 781–790. ACM (2014)
30. Zhu, L., Shen, J., Jin, H., Zheng, R., Xie, L.: Content-based visual landmark search via multimodal hypergraph learning. IEEE TCYB **45**(12), 2756–2769 (2015)

Factorization Meets Memory Network: Learning to Predict Activity Popularity

Wen Wang, Wei Zhang[✉], and Jun Wang

Shanghai Key Laboratory of Trustworthy Computing,
East China Normal University, Shanghai, China
51164500120@stu.ecnu.edu.cn, zhangwei.thu2011@gmail.com,
jwang@sei.ecnu.edu.cn

Abstract. We address the problem, i.e., early prediction of activity popularity in event-based social networks, aiming at estimating the final popularity of new activities to be published online, which promotes applications such as online advertising recommendation. A key to success for this problem is how to learn effective representations for the three common and important factors, namely, activity organizer (who), location (where), and textual introduction (what), and further model their interactions jointly. Most of existing relevant studies for popularity prediction usually suffer from performing laborious feature engineering and their models separate feature representation and model learning into two different stages, which is sub-optimal from the perspective of optimization. In this paper, we introduce an end-to-end neural network model which combines the merits of Memory netwOrk and factOrization moDels (MOOD), and optimizes them in a unified learning framework. The model first builds a memory network module by proposing organizer and location attentions to measure their related word importance for activity introduction representation. Afterwards, a factorization module is employed to model the interaction of the obtained introduction representation with organizer and location identity representations to generate popularity prediction. Experiments on real datasets demonstrate MOOD indeed outperforms several strong alternatives, and further validate the rational design of MOOD by ablation test.

Keywords: Popularity prediction · Event-based social network
Memory network · Factorization model

1 Introduction

In recent years, a growing body of studies have explored the problem of popularity prediction for user-generated content [1], which finds a wide range of real applications, including online advertising [2], recommender system [3], and trend detection [4], to name a few. In this paper, we present a new variant of general popularity prediction problem, titled *early prediction of activity popularity*. Activity is the fundamental component in event-based social networks [5–8],

© Springer International Publishing AG, part of Springer Nature 2018
J. Pei et al. (Eds.): DASFAA 2018, LNCS 10828, pp. 509–525, 2018.
https://doi.org/10.1007/978-3-319-91458-9_31

an increasingly popular social media linking online and offline worlds. This problem focuses on predicting the ultimate number of participants given activities to be published online with respect to three important types of factors, namely, organizer (**Who** organize the activities?), location (**Where** are the activities held?), and textual introduction (**What** are the activities about?). It is significant for both activity organizers to understand whether their activities will be attractive in advance and ordinary users to avoid information overload and filter unappealing activities (see Fig. 1).

Many efforts have been devoted to different popularity prediction problems in the literature [9–12]. Among them, textual based static popularity modeling approaches [13–15], which have no need of targets' existing popularity dynamics over time, are relevant to our study. However, most of these approaches suffer from heavy engineering cost to pursue effective representations of different factors for the studied targets, especially for unstructured textual data. Such complicated feature design limits its generalization ability. Moreover, feature representation and model learning are separated into two stages, which is suboptimal from the perspective of optimization as the pre-specified feature representation might not be very suitable for the prediction object. These limitations pose a major challenge for this study: how can we learn multiple effective feature representations and model these representations jointly to generate accurate popularity prediction in an end-to-end fashion?

Proposed Model. To address the challenge, we develop an end-to-end neural network approach which fuses *Memory netwOrk with factOrization moDels* (MOOD), inspired by recent advances of attention and memory mechanisms for natural language processing [16,17]. The central idea is to endow MOOD with the ability of learning effective textual representation through powerful memory network and jointly modeling representations of multiple factors by tensor factorization. More specifically, MOOD first builds a memory network module to learn the representation of activity introduction. Organizer and location are leveraged as contextual information when performing attention to capture the importance of each word in the activity introduction. Through this way, the same word associated with different organizers and locations might have different contributions to build the activity introduction representation, enabling the representation being personalized. Afterwards, a tensor factorization module with pairwise interaction is employed to model the activity introduction representation, organizer representation, and location representation jointly and generate an integrated representation for the final prediction. An end-to-end learning framework ensures the representation learning more focuses on the target of prediction, which is promising to achieve better performance.

Contributions. To sum up, the main contributions of this paper lie in three aspects:

- We formulate the problem of early prediction of activity popularity in event-based social networks, a variant of existing popularity prediction problems.
- We present a neural network approach called MOOD, which is able to learn effective representation for text through memory network and jointly model

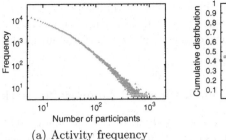
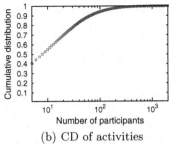

(a) Activity frequency   (b) CD of activities

Fig. 1. Analysis of activity participants. The data used in this figure comes from Douban Event (https://beijing.douban.com/events/future-all). Figure (a) shows the activity frequency of different number of participants and Figure (b) describes the corresponding cumulative distribution (CD). We observe many activities have only a few participants and about more than 90% activities have less than 100 participants, revealing that many activities are not very appealing and it is necessary to provide them with less attention than hot activities.

multiple representations by tensor factorization. Its key novelty is to combine the merits of memory network and factorization model by a unified deep learning framework.

- We conduct comprehensive experiments on real datasets to demonstrate the benefits of our model over several strong alternatives and verify the rationality of the model design by ablation test. To make our model repeatable, we make the code of MOOD and the dataset available[1].

2 Related Work

2.1 Popularity Prediction

According to whether considering existing sequential patterns about popularity dynamics of targets, the methods for popularity prediction can be categorized into dynamic popularity modeling [10,12,18,19] and static popularity modeling [9,13–15,20,21]. Although the former methods behave well as reported in their experiments, they have to collect enough records of popularity dynamics before performing prediction and thus lack of timeliness. Furthermore, it might not be easy to obtain popularity dynamics due to restricted access of third-parties [22], which limits the scope of application. Therefore, we consider the research direction of the latter methods.

Some of the static popularity modeling based methods [9,20,21,23] are carefully designed for domain-specific tasks and could not be easily generalized to our problem setting. The most related studies to us are [13,14], both of which consider textual content and the publishers' influence on popularity. The first study obtains the representations of tweets by topic modeling [24] and then incorporates them into a non-negative matrix factorization framework. Consequently,

---

[1] https://github.com/Autumn945/MOOD.

its learning involves a two-stage process. The latter proposes diverse features relevant to user and text. However, the efforts of feature engineering might be tedious and not so necessary. In the experiments, we compare our model MOOD with them to validate its effectiveness.

2.2 Deep Learning for Personalization and Memory Network

Deep learning methodologies have flourished since [25] and made great success in many domains including computer vision and natural language processing. In this paper, we pay attention to deep learning for personalization and memory network, related to the model we proposed. On the one hand, deep learning for personalization is promising for recommender system. It is employed to model attributes of items [26] or replace simple inner product between factors [27]. However, most deep learning methods are not designed for text popularity prediction problem. On the other hand, memory network [16], with recurrent attention to basic memory units, has shown new progress in natural language processing. It exploits interactions between query and text to perform representation learning and improve the performance of textual question answering [17], sentiment classification [28], etc. In the pursuit of learning effective representation from textual modality, we enhance basic memory network with both organizer and location attentions to capture importance of each word.

3   Problem Definition

Activity is the most essential component in event-based social networks (EBSNs) [5]. Each activity is associated with an organizer who can be a user or an institution, a textual introduction to describe what it is about, and a location denoting where it will be held. Organizers usually publish activities online and other users in EBSNs can register to participate offline activities. The above three types of factors are most critical for the popularity of each activity. Besides, we also know the starting time of each activity. As no one can attend an activities after it start, we can determine the corresponding ultimate number of participants. However, we do not consider the time information when building models, due to the reason that time seems to be not a significant factor to influence activity popularity, which is discussed in later experiments. We leave how to model the time information effectively as future work.

Specifically, we assume $\mathcal{A}$, $\mathcal{U}$, $\mathcal{Q}$, and $\mathcal{V}$ to be activity, organizer, location, and vocabulary sets, respectively. The vocabulary set consists of a large quantity of words, $\mathcal{V} = \{w_v\}_{v=1}^{v=|\mathcal{V}|}$, where $|\mathcal{V}|$ is the size of $\mathcal{V}$. For an activity $a \in \mathcal{A}$, we denote its organizer as $u_a \in \mathcal{U}$, location as $q_a \in \mathcal{Q}$, and its ultimate number of participants as $\bar{r}_a \in \mathbb{Z}_0^+$. To suppress large variance of participants for different activities, we predict a rescaled version of r_a just like [19,23], which can be regarded as the popularity score of activity a and is defined as follows,

$$r_a = \log(\bar{r}_a + 1). \tag{1}$$

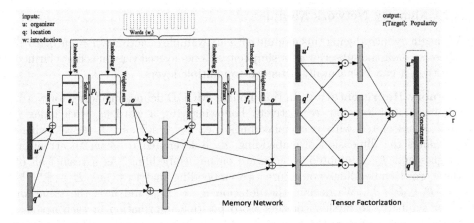

Fig. 2. The graphical representation of MOOD. In this figure, the gray rectangles represent intermediate representations. Memory network is plotted as two-layered versions. $\oplus$ means element-wise addition of corresponding embeddings while $\odot$ denotes element-wise multiplication of connected embeddings.

Moreover, the activity a has a textual introduction, denoted as $d_a = \{w_1^a, \ldots, w_i^a, \ldots, w_n^a\}$, where n is the length of d_a. For the ease of later clarification, we further denote the training, validation, and test parts of the activity set as $\mathcal{A}^{tn}$, $\mathcal{A}^{vd}$ and $\mathcal{A}^{tt}$, respectively. In a nutshell, we have $\{u_a, q_a, d_a, r_a\}$ for each activity $a \in \mathcal{A}$. With these preliminaries, we can formally define the studied problem as below,

*Problem 1 (**Early Prediction of Activity Popularity**).* For a new activity a to be published in event-based social networks, given its organizer u_a, location q_a, and textual description d_a, the goal is to predict the popularity r_a of this activity.

4   Computational Model

This section first presents the overview of the proposed model. Afterwards, it goes deeper into the details of the model to clarify it.

4.1   Model Overview

Our model MOOD is an end-to-end learning framework which takes textual introduction, organizer, and location of the target activity as input and output its predicted popularity score. Graphical illustration of MOOD is shown in Fig. 2. Essentially, the cores of the model are the memory network and tensor factorization modules. In each module, the model designs specific organizer and location embeddings with different roles: attention embedding for the memory network module and interaction embedding for the tensor factorization module. We also consider bias embedding when generating the final prediction.

4.2   Memory Network Module

We begin by introducing this module with an example of activity, $a = \{u, q, d, r\}$, where we omit the subscript a for simplicity. A one-layered version is first clarified and then it can be naturally extended to multiple layers.

Memory Representation: In this module, MOOD defines memories for word, organizer, and location, respectively. Following [16], a word w_v with index v in $\mathcal{V}$ is associated with two embeddings, i.e., $e_v \in \mathbb{R}^k$ and $f_v \in \mathbb{R}^k$, where k denotes the dimension of embedding. e_v is leveraged to generate attention weights and f_v is adopted to generate output embedding. As a result, all of these word embeddings constitute two embedding matrices, i.e., $E \in \mathbb{R}^{k \times |\mathcal{V}|}$ and $F \in \mathbb{R}^{k \times |\mathcal{V}|}$. We declare the notation $\hat{e}_v = E_{:,v}$ and it is the same for other symbols. To further consider word position information in each activity introduction, we follow the idea of [29] by incorporating two absolute position encoding matrix $E^p \in \mathbb{R}^{k \times L}$ and $F^p \in \mathbb{R}^{k \times L}$ into basic word embeddings, where L is the length of the document.

Analogously, MOOD defines attention embedding matrices for both organizer and location, $U^A \in \mathbb{R}^{k \times |\mathcal{U}|}$ and $Q^A \in \mathbb{R}^{k \times |\mathcal{Q}|}$. Without losing generality, we assume the dimension of word embedding equals to those of organizer and location attention embeddings. Likewise, we have $u_u^A = U_{:,u}^A$ for organizer u and $q_q^A = Q_{:,q}^A$ for location q.

Attention for Memory: In the original vocabulary space, a textual introduction can be represented as a sequence of one-hot vectors. For the j-th word w_j in the introduction d, the one-hot vector is expressed as $\hat{w}_j \in \{0, 1\}^{|\mathcal{V}|}$. Assume $v(j)$ represents the index of w_j in the vocabulary, we can obtain the embedding $e_{v(j)} = E\hat{w}_j + E_{:,j}^p$. In a similar fashion, $f_{v(j)}$ can be acquired as well.

Based on these embeddings, MOOD computes the attention weight of the organizer u and location q to the word $v(j)$ through the follow equation,

$$\omega_{v(j)}^{u,q} = (u_u^A + q_q^A)^{\mathrm{T}} e_{v(j)}. \tag{2}$$

Intuitively speaking, larger $\omega_{v(j)}^{u,q}$ denotes word $v(j)$ is more relevant to its corresponding organizer and location, and thus it could be more important for representing the introduction. The addition of organizer and location embeddings ensures the joint influence on word embedding, which shares the similar idea adopted in the matrix factorization approach for modeling multiple factors [30].

Output Representation of Text: The central goal of the memory network module is to obtain better representation for activity introduction. We first calculate $p_{v(j)}^{u,q} = \mathrm{softmax}(\omega_{v(j)}^{u,q})$, in which the probability denoting the importance of $v(j)$ to represent d. We regard the representation of d learned from the one-layered memory network as o. It could be computed by cumulative sum of each word output embedding $f_{v(j)}$ as follows,

$$o = \sum_j p_{v(j)}^{u,q} f_{v(j)}. \tag{3}$$

Multi-layered Extension: Analogous to the common strategy adopted in deep memory network [16], MOOD updates organizer and location embeddings between each layer. Assume the embeddings of the organizer u and location q in the k-th layer are expressed as $\boldsymbol{u}_u^{A,k}$ and $\boldsymbol{q}_q^{A,k}$, respectively. For the first layer, we have $\boldsymbol{u}_u^{A,1} = \boldsymbol{u}_u^A$ and $\boldsymbol{q}_q^{A,1} = \boldsymbol{q}_q^A$. On the basis of the output from Eq. 3, iterative updates can be formulated as below,

$$\begin{aligned} \boldsymbol{u}_u^{A,k+1} &= \boldsymbol{u}_u^{A,k} + \boldsymbol{o}^k \\ \boldsymbol{q}_q^{A,k+1} &= \boldsymbol{q}_q^{A,k} + \boldsymbol{o}^k. \end{aligned} \tag{4}$$

Using the updated organizer and location embeddings, this module iteratively calculates attention weights until determining final introduction representation. We have tried other more complex updating manners such as fusing these embeddings through matrix transformation. However, they do not improve performance notably while increasing the complexity of the model.

Suppose the total number of layers is K, then the output embedding of this module, denoted as $\boldsymbol{o}_d$, is formally defined as following,

$$\boldsymbol{o}_d = \boldsymbol{o}^K + \boldsymbol{u}_u^{A,K} + \boldsymbol{q}_q^{A,K} \tag{5}$$

where $\boldsymbol{o}_d$ is then fed into the tensor factorization module introduced below. From the Eqs. 3 and 5, we can see that the learned textual representation $\boldsymbol{o}_d$ is deeply personalized. Even if two introductions have the same text, their representations could be different for different organizers and locations.

4.3 Tensor Factorization Module

In the tensor factorization module, despite the input of the introduction embedding from the memory network module, MOOD defines interaction embeddings for both organizer and location. It models the three types of embeddings together to capture their joint influence on activity popularity. The interaction embeddings are expressed as $\boldsymbol{u}_u^I \in \mathbb{R}^k$ for organizer u and $\boldsymbol{q}_q^I \in \mathbb{R}^k$ for location q.

[31] suggests a tensor factorization model with pairwise factor interaction to calculate multiple factors and obtain scalar values, denoting the preference of users to items under specific context. Inspired by this idea, we define the following formula to get an integrated vector representation $\boldsymbol{\psi}$,

$$\boldsymbol{\psi}_d^{u,q} = \boldsymbol{u}_u^I \odot \boldsymbol{q}_q^I + \boldsymbol{u}_u^I \odot \boldsymbol{o}_d + \boldsymbol{q}_q^I \odot \boldsymbol{o}_d \tag{6}$$

Alternative factorization models include PARAFAC and Tucker decomposition [32]. However, we choose the one in Eq. 6 for its simplicity and good performance in the experiments.

It is common that each organizer or location has popularity bias, regardless of whatever activity introduction is actually about. Based on this intuition, we introduce bias embeddings $\boldsymbol{u}_u^B \in \mathbb{R}^k$ for user u and $\boldsymbol{q}_q^B \in \mathbb{R}^k$ for location q.

We further concatenate the bias embeddings and the integrated embedding, and associate them with a fully connected layer to calculate the popularity $\hat{r}$,

$$\hat{r} = \boldsymbol{\theta}^{\mathrm{T}} \sigma(\boldsymbol{W}_1^{\mathrm{T}} [\boldsymbol{\psi}_d^{u,q}; \boldsymbol{u}_u^B; \boldsymbol{q}_q^B] + \boldsymbol{b}_1) + b \tag{7}$$

where σ denotes the ReLU (Rectified Linear Unit) with the form ReLU(x) = max$(0, x)$, $\boldsymbol{W}_1$ and $\boldsymbol{b}_1$ are the parameters of the first full connected hidden layer, and $\boldsymbol{\theta}$ and b are the parameters of the output layer. We adopt only one hidden fully connected layer due to its already good experimental results.

4.4   Training

Now based on the above formulations, we define the objective function for later optimization. For an activity a with the known popularity score r_a in training data, suppose $\hat{r}_a$ is the corresponding prediction generated by our model for the activity. We then choose square error, usually adopted in regression tasks, as the target to be optimized,

$$\mathcal{L} = \sum_{a \in \mathcal{A}^{tn}} (r_a - \hat{r}_a)^2. \tag{8}$$

We train the model by taking the first-order gradients of all model parameters through back-propagation, and adopt Adagrad [33] to learn the parameters.

5   Experimental Setup

5.1   Datasets

We adopt the Douban event dataset [34,35] as the experimental data. Douban is a very popular website, containing a large user base and various types of rich data. Thus some previous studies have conducted experiments using the datasets created from Douban. The Douban dataset we used has totally more than 350k activities which cover a long time range and are held in many cities. Activities are locally constrained by cities and different cities have different number of candidate participants. For this reason, we first segment all the activities by their cities. Then we choose the largest two cities, Beijing and Shanghai in China, which contain more than 40% activities to build the two datasets we used in the experiments.

We perform Chinese word segmentation and sparse word filtering for activity introduction, and keep activities with the length of introduction more than five. Moreover, following the common filtering step in personalization modeling [30], we keep organizers and locations with more than four activities.

For later comparison, we divide the datasets into training, validation and test sets in chronological order for each user and location. Specifically, the training set is composed of the first half of activities for both organizers and locations. Then, we randomly select one-third of the remaining activities as the validation set and the remaining activities are regarded as the test set. The basic statistics of the processed datasets are summarized in Table 1. As mentioned in the first section, we make the source code of MOOD and anonymous data publicly available.

Table 1. Experimental data statistics.

Data	Activity	Organizer	Location	Word	Training	Validation	Testing
Beijing	33,923	882	1,767	79,212	21,851	4,024	8,048
Shanghai	26,133	714	1,376	65,618	16,525	3,202	6,406

5.2   Baselines

To validate the advantages of MOOD, we compare it with several alternative baselines, some of which have strong performances.

- **GloAve, OrgAve, LocAve.** The three simple methods are just based on popularity average in training data. The former one takes all activities into computation while the later two consider them for each organizer and location, respectively.
- **HF-NMF** [13] **and HF-NTF.** The hybrid factor non-negative matrix factorization (HF-NMF) model is proposed to estimate the number of retweets given textual content of original tweets and their authors. It utilizes the topics learned from latent Dirichlet allocation (LDA) [24] as textual features and incorporates them into a non-negative matrix factorization framework. The original HF-NMF model only considers two types of factors, i.e. text and user. To better adapt it to our problem setting, we extend it with pairwise interaction tensor factorization, ensuring the fairness of performance comparison.
- **PoissonMF** [36] **and PoissonTF.** This model utilizes the benefit of Poisson distribution to generate count data by regarding the result of matrix factorization as the expected mean of this distribution. Following the methodology exploited in HF-NTF, we extend PoissonMF to PoissonTF to handle multiple factors.
- **FeaReg** [14]. This method needs hand-crafted features to describe how the three types of factors influence final activities' popularity. As with the study [14], we have designed features such as one-hot representation and TF-IDF to characterize organizer, location, and textual description. We have tried several standard statistical regression models (random forest, ridge regression, etc.) and choose ridge linear regression due to its better performance.

In later experiments, we also conduct ablation test to verify the contribution of each factor considered. We denote FeaReg (D+U) as the one only modeling introduction (D) and organizer (U), and FeaReg (D+Q) as the one only modeling introduction (D) and location (Q). Other notations such as MOOD (D+U) are determined in a similar fashion.

5.3   Variants of MOOD

To verify the design rationality of the proposed model, we present two variants of MOOD, which can be utilized to demonstrate the benefits of the proposed two modules.

- **LSTM-TF.** This model chooses Long Short-Term Memory (LSTM) Network [37] instead of memory network. Since LSTM performs well in many text modeling tasks in recent years, we compare it with MOOD to verify the benefit of leveraging the memory network module.
- **DMN.** It just feeds the output of the memory network module into the final prediction. In other words, this model does not consider tensor factorization and the interaction embeddings of organizer and location. It can be utilized to demonstrate the effectiveness of modeling interaction embedding with tensor factorization module.

5.4   Implementation Details and Evaluation Metrics

We set the dimension of all the embeddings used in our model and baselines to be 128. We set the hyper-parameters of Adagrad to be the default ones shown in [33] and the batch size is 128. The number of layers in the memory network module is set to 2 which performs better. L2 regularization is adopted to reduce overfitting.

We adopt mean square error (MSE) and mean absolute error (MAE), which are employed by many previous studies for popularity prediction [13, 20, 23, 38]. Moreover, MSE is consistent with the optimization target we adopted for learning MOOD and other competitors, and MAE often acts as a complement to MSE. All the models mentioned above are run five times and the average of their results are reported.

6   Experimental Results

In this section, we present the detailed experimental results and some intuitive analysis to first answer the following core research questions:

Q1: Does the proposed model MOOD indeed outperform all the other competitors in terms of the evaluation metrics? Does the memory network module reveal its advantages over some alternatives? Can the tensor factorization module really benefit the studied problem?

Q2: What is the relative importance of each type of the three factors we consider for the activity popularity prediction problem? Does joint modeling all the three factors achieve better performance?

On this basis, we further provide some necessary experimental discussions about (1) the number of layers in the memory network module, (2) activity time information, and (3) case study of the visual attention results.

6.1   Model Performance Comparison (Q1)

The overall results are shown in Table 2 with MSE and MAE metrics. By first comparing GloAve, OrgAve, and LocAve, we can observe that both OrgAve

Table 2. Comparisons of different models on activity popularity prediction.

Models	Beijing		Shanghai	
	MSE	MAE	MSE	MAE
Traditional approaches				
GloAve	2.0070	1.1897	1.6727	1.0850
OrgAve	0.8151	0.6605	0.9851	0.7378
LocAve	0.8573	0.6642	0.9487	0.7173
HF-NMF	0.9619	0.7288	1.0703	0.7764
HF-NTF	1.0023	0.7221	1.0564	0.7629
PoissonMF	1.0056	0.7298	1.1147	0.7875
PoissonTF	0.7779	0.6437	0.8753	0.6963
FeaReg	0.6690	0.6028	0.7739	0.6574
Deep learning models (MOOD and its variants)				
LSTM-TF	0.7388	0.6322	0.8555	0.6889
DMN	0.6896	0.6109	0.7999	0.6658
MOOD (Ours)	**0.6536**	**0.5850**	**0.7505**	**0.6360**

and LocAve improve GloAve by a large margin as GloAve considers no factor of activities. It is surprising that HF-NFM and HF-NTM behave obviously worse than OrgAve, which reflects that directly utilizing them for our studied problem is not suitable. Although PoissonMF shows no good results as well, its extension to tensor factorization presents obviously better results, indicating the benefits of considering the three factors to some extent. However, the results are still far from satisfactory, compared with the methods discussed below. One of the reasons may be that textual feature representation and model learning are separated into two stages, which is not very optimal.

Based on hand-crafted features, FeaReg performs best among traditional approaches, and even better than the variants of our model, i.e., LSTM-TF and DMN. Finally, MOOD not only performs better than DMN and LSTM-TF on the two datasets, but also better than FeaReg. Through the above comparisons, we can find that the integration of memory network and tensor factorization can complement each other and achieve the best results among all the adopted models, which can answer the question Q1.

6.2   Factor Contribution (Q2)

We investigate how the three types of factors contribute to the popularity prediction in integrated models. To achieve this, ablation test is adopted by removing one type of factor each time from textual introduction, organizer, and location. We choose FeaReg and our model MOOD, which achieve the best performance among traditional approaches and deep learning models, respectively.

Table 3 shows MOOD outperforms FeaReg on almost every combination of introduction, organizer, and location for the MAE metric, which further demonstrates the advantages of MOOD. We observe that both FeaReg (U+Q) and MOOD (U+Q) obtain better results than other models which also consider two types of factors. This phenomenon is rational since structured organizer and location information are easier to be modeled than unstructured text.

Table 3. Ablation test for factor contribution.

Models	Beijing		Shanghai	
	MSE	MAE	MSE	MAE
FeaReg (U+Q)	0.7006	0.6170	0.8031	0.6728
FeaReg (D+Q)	0.7588	0.6453	0.8397	0.6873
FeaReg (D+U)	0.7339	0.6258	0.8519	0.6826
FeaReg	0.6690	0.6028	0.7739	0.6574
MOOD (U+Q)	0.6653	0.5884	0.7761	0.6501
MOOD (D+Q)	0.7832	0.6367	0.8343	0.6749
MOOD (D+U)	0.7234	0.6168	0.8329	0.6756
MOOD	**0.6536**	**0.5850**	**0.7505**	**0.6360**

Finally, we notice that modeling three types of factors jointly can achieve better performances consistently in the two datasets, no matter which of the two methods is selected. This phenomenon may reveal that the three factors may be complementary to each other for the activity popularity prediction problem. In summary, we can answer question Q2 through the above discussions.

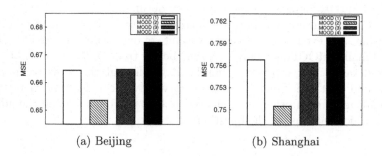

(a) Beijing   (b) Shanghai

Fig. 3. Results of MOOD with different number of layers.

6.3 Impact of Number of Layers

We investigate how the number of layers in the memory network module impacts prediction performance. We analyze the memory network module with different number of layers, and the corresponding results are introduced in Fig. 3. Obviously, the results of the module with two layers achieve the best performance across the two datasets. We also show the visualizations of attention values given sampled examples later, which indicate that the second layer is more focused than the first layer. In summary, setting the number of layers to be two is a rational choice.

6.4 Impact of Activity Time

We consider time information of activities and present a simple average-based method called TimeAve to test it. We first discretize the continuous time space into fixed-length time periods, similar to some previous studies [39]. We regard one week as a cycle and one hour as a period, and get 7×24 periods. Afterwards, we calculate the popularity average for each period in the training dataset and generate prediction according to which period the target activity belongs to. Figure 4 shows TimeAve is only slightly better than GloAve, but much worse than OrgAve and LocAve, which reveals that time information is not easily to be modeled to improve performance. The reason might be that registering online for participating activities mainly reflects users' preference, but not their final decision to participate. Thus the time factor is not very important to be considered by users, which is also empirically verified in [34].

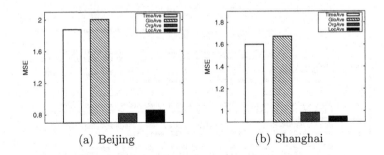

(a) Beijing   (b) Shanghai

Fig. 4. Results of average-based methods.

6.5 Case Study for Attention Visualization

We study the difference of word attention weights in different layers of MOOD and how the attention weights change when we associate the organizers and locations with the introduction text not really belonging to them.

We use deeper colors to denote larger attention weights for different words in Fig. 5. Each word is followed by an English translation and a number indicating the normalized value of attention weight. As different introduction have different length, making the average attention weights not the same. To enable the visualization of attention weight comparison for different introduction, we adopt a simple strategy by multiplying each word's attention weight with the

Layer-1	摄影-photography(1.173)/ 与-and(0.787)/ 魔术-magic(1.193)/ 的(0.396)/ 碰撞-impact(1.212)/ 让-made(0.891)/ 你-you(0.824)/ 体验-experience(1.001)/ 与众不同-different(1.242)/ 的(0.387)/ 视觉-visual(1.210)/ 盛宴-feast (1.222)
Layer-2	摄影-photography(1.271)/ 与-and(0.442)/ 魔术-magic(1.423)/ 的(0.082)/ 碰撞-impact(1.494)/ 让-made(0.650)/ 你-you(0.544)/ 体验-experience(0.882)/ 与众不同-different(1.602)/ 的(0.076)/ 视觉-visual(1.478)/ 盛宴-feast(1.520)
Layer-1	各类-various types(1.071)/ 酒会-reception(1.072)/ 上-on(1.004)/ 你-you(0.963)/ 懂-know(1.073)/ 酒-wine(1.088)/ 并-and(1.011)/ 能-can(1.040)/ 优雅-gracefully(1.075)/ 地(1.056)/ 品酒-wine tasting(1.089)/ 那么-then(1.061)/ 你-you(0.972)/ 就是-are(1.042)/ 酒会-reception(1.078)/ 上-on(0.997)/ 的(0.780)/ 亮点-highlights(1.080)
Layer-2	各类-various types(1.339)/ 酒会-reception(1.326)/ 上-on(0.793)/ 你-you(0.659)/ 懂-know(1.326)/ 酒-wine(1.452)/ 并-and(0.861)/ 能-can(1.066)/ 优雅-gracefully(1.369)/ 地(1.190)/ 品酒-wine tasting(1.407)/ 那么-then(1.245)/ 你-you(0.706)/ 就是-are(1.101)/ 酒会-reception(1.386)/ 上-on(0.758)/ 的(0.612)/ 亮点-highlights(1.412)

Fig. 5. Attention weights in different attention layers.

length of the corresponding introduction. Through this way, the average attention weight of all words equals to 1 and an attention weight less than 1 means less attention to the corresponding word and vice versa. We observe in Fig. 5 that more meaningful words have larger attention weights on both layers. Besides, the attention weights in layer-2 are more centralized than those in layer-1. In a nutshell, we qualitatively indicate the multi-layered recurrent attention mechanism is beneficial for our model.

Layer-1	歌剧-opera(1.215)/团-group(1.051)/聚集-assemble(1.177)/着-with(0.906)/众多-numerous(1.224)/优秀-excellent(1.181)/的-of(0.272)/歌剧-opera(1.202)/表演艺术家-performing artist(1.240)/他们-they(0.929)/大都-mostly(1.142)/曾-already(1.026)/获得-obtain(1.072)/过(1.023)/国内外-at home and abroad(1.176)/声乐-vocal music(1.272)/大赛-competition(1.169)/的-of(0.280)/重要-important(1.139)/奖项-awards(1.197)
Layer-2	歌剧-opera(1.349)/团-group(1.031)/聚集-assemble(1.264)/着-with(0.765)/众多-numerous(1.344)/优秀-excellent(1.259)/的-of(0.170)/歌剧-opera(1.294)/表演艺术家-performing artist(1.375)/他们-they(0.754)/大都-mostly(1.211)/曾-already(0.977)/获得-obtain(1.048)/过(0.919)/国内外-at home and abroad(1.258)/声乐-vocal music(1.454)/大赛-competition(1.258)/的-of(0.183)/重要-important(1.189)/奖项-awards(1.313)
Layer-1	歌剧-opera(0.973)/团-group(0.960)/聚集-assemble(0.966)/着-with(0.936)/众多-numerous(0.968)/优秀-excellent(0.975)/的-of(1.659)/歌剧-opera(0.955)/表演艺术家-performing artist(0.972)/他们-they(0.923)/大都-mostly(0.964)/曾-already(0.953)/获得-obtain(0.962)/过(0.934)/国内外-at home and abroad(0.969)/声乐-vocal music(0.968)/大赛-competition(0.958)/的-of(1.681)/重要-important(0.967)/奖项-awards(0.975)
Layer-2	歌剧-opera(1.229)/团-group(1.026)/聚集-assemble(1.170)/着-with(0.821)/众多-numerous(1.211)/优秀-excellent(1.170)/的-of(1.033)/歌剧-opera(1.158)/表演艺术家-performing artist(1.228)/他们-they(0.766)/大都-mostly(1.145)/曾-already(0.982)/获得-obtain(1.028)/过(0.891)/国内外-at home and abroad(1.164)/声乐-vocal music(1.269)/大赛-competition(1.157)/的-of(1.114)/重要-important(1.125)/奖项-awards(1.213)

Fig. 6. Different attentions to the same introduction.

We further randomly sample an activity to get its textual introduction, and calculate the corresponding attention weights with different organizers and locations. The visualization of attention weights are shown in Fig. 6. The first part of the figure adopts the organizer and location which the introduction belongs to, while the second part uses an arbitrary pair of organizer and location. Each Chinese word is followed by an English translation and a value corresponding to its attention weight. As the figure shows, the attentions of the first part seem to be better and more meaningful than the second, which reveals that the attention mechanism adopted by MOOD is personalized.

7 Conclusion

We formulate the problem of early predicting the ultimate popularity for a new activity given its organizer, location, and introduction. To avoid tedious feature engineering and fuse the two separate stages of feature representation and model learning for popularity prediction, we present MOOD, a deep learning approach which combines memory network with tensor factorization in a unified end-to-end learning framework. It is endowed with the ability of acquiring effective representations for text and jointly modeling the three types of considered factors effectively. We conduct experiments on real datasets and validate the advantages and rationalities of the proposed model.

Acknowledgements. This work was supported in part by NSFC (61702190), Shanghai Sailing Program (17YF1404500), SHMEC (16CG24), NSFC-Zhejiang (U1609220), and NSFC (61672231, 61672236).

References

1. Szabó, G., Huberman, B.A.: Predicting the popularity of online content. J. Commun. ACM **53**(8), 80–88 (2010)
2. Figueiredo, F., Benevenuto, F., Almeida, J.M.: The tube over time: characterizing popularity growth of youtube videos. In: WSDM, pp. 745–754 (2011)
3. Chang, B., Zhu, H., Ge, Y., Chen, E., Xiong, H., Tan, C.: Predicting the popularity of online serials with autoregressive models. In: CIKM, pp. 1339–1348 (2014)
4. Agarwal, N., Liu, H., Tang, L., Yu, P.S.: Identifying the influential bloggers in a community. In: WSDM, pp. 207–218 (2008)
5. Liu, X., He, Q., Tian, Y., Lee, W., McPherson, J., Han, J.: Event-based social networks: linking the online and offline social worlds. In: SIGKDD, pp. 1032–1040 (2012)
6. Zhang, W., Wang, J., Feng, W.: Combining latent factor model with location features for event-based group recommendation. In: SIGKDD, pp. 910–918 (2013)
7. Du, R., Yu, Z., Mei, T., Wang, Z., Wang, Z., Guo, B.: Predicting activity attendance in event-based social networks: content, context and social influence. In: UbiComp, pp. 425–434 (2014)
8. She, J., Tong, Y., Chen, L.: Utility-aware social event-participant planning. In: SIGMOD, pp. 1629–1643 (2015)
9. Khosla, A., Sarma, A.D., Hamid, R.: What makes an image popular? In: WWW, pp. 867–876 (2014)
10. Zhao, Q., Erdogdu, M.A., He, H.Y., Rajaraman, A., Leskovec, J.: SEISMIC: a self-exciting point process model for predicting tweet popularity. In: SIGKDD, pp. 1513–1522 (2015)
11. Xiao, S., Yan, J., Li, C., Jin, B., Wang, X., Yang, X., Chu, S.M., Zha, H.: On modeling and predicting individual paper citation count over time. In: IJCAI, pp. 2676–2682 (2016)

12. Rizoiu, M., Xie, L., Sanner, S., Cebrián, M., Yu, H., Hentenryck, P.V.: Expecting to be HIP: hawkes intensity processes for social media popularity. In: WWW, pp. 735–744 (2017)
13. Cui, P., Wang, F., Liu, S., Ou, M., Yang, S., Sun, L.: Who should share what?: item-level social influence prediction for users and posts ranking. In: SIGIR, pp. 185–194 (2011)
14. Martin, T., Hofman, J.M., Sharma, A., Anderson, A., Watts, D.J.: Exploring limits to prediction in complex social systems. In: WWW, pp. 683–694 (2016)
15. Dimitrov, D., Singer, P., Lemmerich, F., Strohmaier, M.: What makes a link successful on Wikipedia? In: WWW, pp. 917–926 (2017)
16. Sukhbaatar, S., Szlam, A., Weston, J., Fergus, R.: End-to-end memory networks. In: NIPS, pp. 2440–2448 (2015)
17. Kumar, A., Irsoy, O., Ondruska, P., Iyyer, M., Bradbury, J., Gulrajani, I., Zhong, V., Paulus, R., Socher, R.: Ask me anything: dynamic memory networks for natural language processing. In: ICML, pp. 1378–1387 (2016)
18. Shen, H., Wang, D., Song, C., Barabási, A.: Modeling and predicting popularity dynamics via reinforced poisson processes. In: AAAI, pp. 291–297 (2014)
19. Wu, B., Mei, T., Cheng, W., Zhang, Y.: Unfolding temporal dynamics: predicting social media popularity using multi-scale temporal decomposition. In: AAAI, pp. 272–278 (2016)
20. Chen, J., Song, X., Nie, L., Wang, X., Zhang, H., Chua, T.: Micro tells macro: predicting the popularity of micro-videos via a transductive model. In: MM, pp. 898–907 (2016)
21. Zhang, W., Wang, W., Wang, J., Zha, H.: User-guided hierarchical attention network for multi-modal social image popularity prediction. In: WWW, pp. 1277–1286 (2018). https://dl.acm.org/citation.cfm?id=3186026
22. He, X., Gao, M., Kan, M., Liu, Y., Sugiyama, K.: Predicting the popularity of web 2.0 items based on user comments. In: SIGIR, pp. 233–242 (2014)
23. Li, C., Ma, J., Guo, X., Mei, Q.: DeepCas: an end-to-end predictor of information cascades. In: WWW, pp. 577–586 (2017)
24. Blei, D.M., Ng, A.Y., Jordan, M.I.: Latent dirichlet allocation. JMLR **3**, 993–1022 (2003)
25. Hinton, G.E., Salakhutdinov, R.R.: Reducing the dimensionality of data with neural networks. Science **313**(5786), 504–507 (2006)
26. Wang, H., Wang, N., Yeung, D.Y.: Collaborative deep learning for recommender systems. In: SIGKDD, pp. 1235–1244. ACM (2015)
27. He, X., Liao, L., Zhang, H., Nie, L., Hu, X., Chua, T.: Neural collaborative filtering. In: WWW, pp. 173–182 (2017)
28. Tang, D., Qin, B., Liu, T.: Aspect level sentiment classification with deep memory network. In: EMNLP, pp. 214–224 (2016)
29. Vaswani, A., Shazeer, N., Parmar, N., Uszkoreit, J., Jones, L., Gomez, A.N., Kaiser, L., Polosukhin, I.: Attention is all you need. In: NIPS, pp. 6000–6010 (2017)
30. Aizenberg, N., Koren, Y., Somekh, O.: Build your own music recommender by modeling internet radio streams. In: WWW, pp. 1–10 (2012)
31. Rendle, S., Schmidt-Thieme, L.: Pairwise interaction tensor factorization for personalized tag recommendation. In: WSDM, pp. 81–90. ACM (2010)
32. Cichocki, A., Zdunek, R., Phan, A.H., Amari, S.: Nonnegative Matrix and Tensor Factorizations - Applications to Exploratory Multi-way Data Analysis and Blind Source Separation. Wiley, Hoboken (2009)
33. Duchi, J.C., Hazan, E., Singer, Y.: Adaptive subgradient methods for online learning and stochastic optimization. JMLR **12**, 2121–2159 (2011)

34. Zhang, W., Wang, J.: A collective Bayesian poisson factorization model for cold-start local event recommendation. In: SIGKDD, pp. 1455–1464 (2015)
35. Yin, H., Hu, Z., Zhou, X., Wang, H., Zheng, K., Nguyen, Q.V.H., Sadiq, S.: Discovering interpretable geo-social communities for user behavior prediction. In: ICDE, pp. 942–953. IEEE (2016)
36. Ma, H., Liu, C., King, I., Lyu, M.R.: Probabilistic factor models for web site recommendation. In: SIGIR, pp. 265–274. ACM (2011)
37. Hochreiter, S., Schmidhuber, J.: Long short-term memory. Neural Comput. 9(8), 1735–1780 (1997)
38. Ahmed, M., Spagna, S., Huici, F., Niccolini, S.: A peek into the future: predicting the evolution of popularity in user generated content. In: WSDM, pp. 607–616 (2013)
39. Yuan, Q., Zhang, W., Zhang, C., Geng, X., Cong, G., Han, J.: PRED: periodic region detection for mobility modeling of social media users. In: WSDM, pp. 263–272 (2017)

Representation Learning for Large-Scale Dynamic Networks

Yanwei Yu[1,2(✉)], Huaxiu Yao[1], Hongjian Wang[1],
Xianfeng Tang[1], and Zhenhui Li[1]

[1] College of Information and Science Technology, Pennsylvania State University,
State College, USA
{yuy174,huaxiuyao,hxw186,xianfeng,jessieli}@ist.psu.edu
[2] School of Computer and Control Engineering, Yantai University, Yantai, China
yuyanwei@ytu.edu.cn

Abstract. Representation leaning on networks aims to embed networks into a low-dimensional vector space, which is useful in many tasks such as node classification, network clustering, link prediction and recommendation. In reality, most real-life networks constantly evolve over time with various kinds of changes to the network structure, e.g., creation and deletion of edges. However, existing network embedding methods learn the representation vectors for nodes in a static manner, which are not suitable for dynamic network embedding. In this paper, we propose a dynamic network embedding approach for large-scale networks. The method incrementally updates the embeddings by considering the changes of the network structures and is able to dynamically learn the embedding for networks with millions of nodes within a few seconds. Extensive experimental results on three real large-scale networks demonstrate the efficiency and effectiveness of our proposed methods.

1 Introduction

Networks are ubiquitous in our daily life, such as social networks, communication networks, biological networks, academic networks and the World Wild Web. People have studied many important data mining problems on networks, including network visualization [21], node classification [3], community detection [12], link prediction [20] and recommendation [37]. A typical way to tackle these problems is based on hand-crafted features of networks, which requires a lot of manual efforts on feature engineering and usually is constricted to a specific problem. Network embedding techniques provide an alternative way to learn features automatically. The basic idea of network embedding is to learn the low-dimensional representation of nodes by preserving the network structure. Following the initial ideas in network embedding [2,10,31], recent techniques such as DeepWalk [25] and node2vec [13] learn node representation using random walks sampled in the network. A limitation of such random walk based method is the high computational cost. To scale up to large-scale network with millions of nodes, LINE [28]

© Springer International Publishing AG, part of Springer Nature 2018
J. Pei et al. (Eds.): DASFAA 2018, LNCS 10828, pp. 526–541, 2018.
https://doi.org/10.1007/978-3-319-91458-9_32

utilizes edge sampling in the network to learn representations that preserve the first-order and the second-order proximities.

However, existing studies mostly focus on static networks. In the real world, the networks could vary over time with creation and deletion of edges [1]. For instance, in social networks, users may add a user as a new friend or unfriend a user who used to be a friend. In co-author network, people build new co-author relationships over time. In co-location graph where edges as two people being within certain distance, people gather and depart dynamically.

Though dynamic networks widely exist, the studies of representation learning on dynamic networks are limited. A naive method is to re-run the embedding methods on the whole network when the network is updated. However, learning network representation is costly, especially for the large-scale network. Re-computing on the whole network with every batch of updates may not be feasible in the real-world setting. Naturally, we ask the question, "can we learn network embedding dynamically in a more efficient way?"

To address these challenges, in this paper, we propose an efficient embedding method, DLNE, for dynamic network embedding. Our intuition is that, we update previous embeddings by considering the changes of the networks, i.e., newly added (or deleted) edges. The method will be much more efficient compared with re-computing on the whole network because the changes in a large network could be relatively small. More specifically, our proposed method is built based on the LINE method [28], which has been shown to be significantly faster than other embedding methods. We use LINE to obtain the initial representations on the current network. With the new batch of updates on the network structure, we update the representations of corresponding affected nodes by optimizing the loss function defined based on first-order and second-order proximities.

Our method is validated on three large-scale real world networks, including social networks and citation networks. We conduct extensive experiments to verify the effectiveness and efficiency of our method by comparing with state-of-the-art methods via a multi-label classification task. The results suggest that DLNE is able to incrementally update the representations for nodes on a dynamic network with millions of edges in time scale of seconds.

To summarize, the major contributions of this paper are as follows:

- We formally study the problem of dynamic large-scale network embedding. To the best of our knowledge, we are the first method to efficiently learn embeddings dynamically on a large-scale network.
- We propose a novel method DLNE to solve the problem of dynamic large-scale network embedding. We design the loss function to learn the updated representations by considering the changes of the network structure.
- We conduct extensive evaluations through a multi-label classification task on three real-world networks. Experimental results demonstrate the effectiveness and efficiency of our proposed method.

The rest of paper is organized as follows. Related work is discussed in Sect. 2. Then, we formally state our problem definition in Sect. 3 and the proposed online embedding method is described in Sect. 4. Experimental results are reported in Sect. 5. Finally, we conclude the paper in Sect. 6.

2   Related Work

Dimension reduction or low-dimension graph representation learning have been studied extensively in the literature. Many methods are proposed in various fields, such as multidimensional scaling [10], IsoMap [31], LLE [27], and Laplacian Eigenmaps [2]. Chen et al. [8] propose the network embedding for directed network. They use Markov random walks to measure the locality link structure of directed networks. Following Chen's work and motivated by the success of word2vec technique [22,23], Perozzi et al. [25] propose DeepWalk for social network embedding. They use a truncated random walk to construct the context of a vertex, then they employ word2vec to learn latent representations for all vertices in social network. Grover and Leskovec [13] further propose node2vec, which improve DeepWalk by enabling a controlled random walk. Tang et al. [28] propose LINE to learn embedding for both undirected and directed large-scale information networks with unweighted or weighted edges, which is particularly designed to preserve both the first-order and second-order proximities. Cao et al. [4] extend LINE to support high-order graph representation learning by capturing different k-step local relational information. Chen and Wang [9] propose an heterogeneous information network embedding that considers local and global semantic among multi-typed entities. In addition, other deep learning based approaches [5,32] are proposed to enhance the network representation.

Another line of work aims to learn the graph embedding while considering additional information other than graph connectivity. For example, Yang et al. [36] propose a matrix factorization based method to learn network representations that incorporates text feature into network structure. Chen et al. [7] incorporate group information to learn network embedding. Most recently, Huang et al. [14] propose LANE framework for learning representation vectors for attributed networks. They aim to learn better feature representation incorporating label information into network embedding while preserving their correlations. Xu et al. [35] propose Embedding of Embedding (EOE) framework for coupled heterogeneous networks. They incorporate a harmonious embedding matrix to further embed the embeddings that only encode intra-network edges. Wang et al. [34] propose a Modularized Nonnegative Matrix Factorization (MNMF) model to incorporate the community structure into network embedding. Network embedding techniques have attracted more and more attentions in network science community. Li et al. [17] study on how to leverage node order information and annotation data to improve network embedding in a semsupervised manner. *However, all the approaches mentioned so far only handle static networks. They are not applicable to learn representations on dynamic evolving information networks.*

In practice, many real world networks (e.g., social networks and co-occurrence networks) are dynamic networks whose edges and vertices change over time. In dynamic network analysis, Ning et al. [24] propose an incremental spectral clustering for evolving networks by updating the eigenvalue system continuously. Chen and Tong [6] propose an online approach to track the eigen-functions of adjacency matrix for a dynamic network. Li et al. [19] propose a unsupervised feature selection for dynamic networks, which leverages the temporal evolution property of dynamic networks to update the feature results incrementally. *However, such work focus on online analysis of network structure change in dynamic networks, while our work aims to online representation learning for dynamic evolving networks.*

Wang and Li [33] propose a graph embedding method to learn the temporal dynamics of urban region graph. *Although they study a dynamic temporal graph, the embedding learning is not conducted in an incremental manner. Instead, they construct the whole evolving network, and learn different embedding for regions at different timestamp simultaneously. In our method, the embedding vectors of networks are updated continuously and efficiently, rather than re-learned from scratch.*

Most recently, Li et al. [18] propose dynamic network embedding for attributed network. They first use an off-line Laplacian Eigenmaps-based model to learn graph embeddings. Then they update the embedding by updating the top eigenvectors and eigenvalues according to the updated graph matrix. Jian et al. [15] propose an online network embedding algorithm for node classification on streaming network. They use same Laplacian Eigenmaps-based model to update embedding representations for newly arrived nodes. *However, the graph factorization-based method only considers the one-hop relationships in adjacency matrix, and it is difficult to scale because of the use of laplacian eigenmaps. Meanwhile, our proposed method considers both local and global network structure and is able to handle large-scale dynamic networks with millions of vertices and edges in an online fashion.*

3 Problem Definition

In this section, we first introduce some concepts used in this paper. Then we define the problem of dynamic network embedding.

Definition 1 (Network). *A network is denoted as* $\mathcal{G} = (V, E)$, *where* $V = \{v_1, v_2, \ldots, v_n\}$ *is the set of vertices, and* $E = \{e_{ij}\}$, *where* $i, j \in \{1, 2, \cdots, n\}$, *is the set of edges. Each edge* e_{ij} *connects two vertices* v_i *and* v_j, *and the weight of this edge is* w_{ij}.

In practice, networks can be categorized as unweighted (e.g., social networks) or weighted (e.g., word co-occurrence network) networks. And networks can also be directed (e.g., citation networks) or undirected (e.g., co-author networks) networks. In unweighted network, $w_{ij} = 1$ if e_{ij} exists, while w_{ij} takes continuous values in weighted network. In undirected network $e_{ij} = e_{ji}$ with the same weight, while $e_{ij} \neq e_{ji}$ and $w_{ij} \neq w_{ji}$ in a directed network.

The structure of networks often evolves over time by adding or deleting edges and vertices. Without loss of generaliy, we partition the time dimension into discrete timestamps $t = \{1, 2, \cdots, T\}$ with fixed interval τ. We use $\mathcal{G}_t$ to denote the dynamic evolving network at time t. Correspondingly, the set of edges and vertices are denoted as E_t and V_t.

For simplicity, we track network updates by the addition and deletion of edges, because the addition (deletion) of vertex could be implemented by the addition (deletion) of edges. More specifically, one vertex is deleted when all edges connecting to this vertex are deleted. Also, if a vertex is added to into current network, at least one edge should be created to connect the new vertex with an existing vertex. To this end, within each time interval τ, we use E_a and E_d to denote the set of edges that are added and deleted, respectively.

Network embedding aims to represent each vertex of the networks as a low dimensional vector, and vertices should have similar embedding vectors if they are connected. To achieve such an embedding result, the network structures must be preserved. Next, we formally define the first-order proximity and the second-order proximity to preserve the local and global network structure, respectively.

Definition 2 (First-Order Proximity). *The first-order proximity in a network is defined as the local pairwise proximity between two directly connected vertices. For each pair of vertices u and v, if there exists $e_{uv} \in E$, the weight w_{uv} indicates the first-order proximity between them. Otherwise, the first-order proximity between u and v is 0.*

Definition 3 (Second-Order Proximity). *The second-order proximity between a pair of vertices is defined to account for their neighborhood structure. Let $\mathcal{N}_u = \{v_{u1}, v_{u2}, v_{u3}, \cdots\}$ denote the set of direct neighbors of vertex u. The second-order proximity between u and v is determined by the similarity of two sets $\mathcal{N}_u$ and $\mathcal{N}_v$.*

In this paper, we aim to learn embedding vectors to preserve the first-order proximity and the second-order proximity among vertices, meanwhile the learned embeddings could be updated to account for the temporal dynamics of networks. The formal definition of dynamic network embedding problem is given as follows:

Problem 1 (Dynamic Network Embedding). *Given a dynamic network $\mathcal{G}_t$, the current embeddings Φ_t of all the vertices V_t in $\mathcal{G}_t$, the problem of dynamic network embedding aims to efficiently calculate the embeddings Φ_{t+1} for all vertices in $\mathcal{G}_{t+1}$ from Φ_t. $\Phi : V \rightarrow \mathbb{R}^d$ can also be regarded as the mapping function from vertices to d-dimension vector representations.*

4   DLNE: Dynamic Large-Scale Network Embedding

In this section, we provide details for our proposed Dynamic Large-scale Network Embedding (DLNE). First, we present the technical details on how to update the embeddings according to the addition and deletion edge sets E_a and E_d.

More specifically, we account for first-order and second-order proximity information while update the embedding learning. Finally, we give the algorithm and optimization steps.

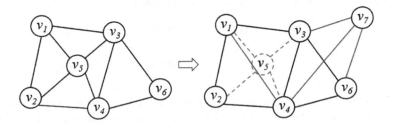

Fig. 1. An example of evolving network: solid red edges denote the added edges E_a, ane dashed grey edges denote the deleted edges E_d. (Color figure online)

4.1 Overall Framework

In Fig. 1 we show an example of evolving network, where the network evolves from left to the right. In the given time window, vertex v_5 is deleted, which is equivalent to delete edges $\{e_{15}, e_{25}, e_{35}, e_{45}\}$. Similarly, the addition of v_7 is equivalent to addition of $\{e_{37}, e_{47}, e_{67}\}$. We use Φ_l to denote the embedding function for the graph on the left. The goal of our problem is to calculate Φ_r with the addition edge set E_a and deletion edge set E_d.

Intuitively, the edge embedding is used to account for the local network structure information. Given each edge within the updated edge sets E_a and E_d, the network structure only changes locally. Driven by this intuition, we propose to update the vertex embedding locally. Namely, for each edge $e_{ij} \in E_a \cup E_d$, we only update the embedding vecter $\Phi(v)$, where v is in the local structure of v_i and v_j. The problem becomes how to define local structure, and update the embeddings accordingly. In this paper, we define the local structure with the first-order and the second-order proximity. In the following sections, we give the technical details on how to update the embedding to account for the first-order and the second-order proximity, respectively.

DLNE with First-Order Proximity. First, we consider the first-order proximity when there is an edge added into current network. For each edge $e_{ij} \in E_a$, the probability of the connection between vertex v_i and v_j is formulated as follows:

$$p_1(v_i, v_j) = \sigma(\Phi(v_j)^T \cdot \Phi(v_i)), \tag{1}$$

where $\sigma(x) = 1/(1 + \exp(-x))$ is the sigmoid function and $\Phi(v_i) \in \mathbb{R}^d$ is the embedded vector of vertex v_i in the current timestamp. Namely, the probability of v_j being a neighbor of vertex v_i is correlated to the similarity of their potential representation vectors.

However, the influence of added and deleted edge on the corresponded vertices is different. For each newly added edge $e_{ij} \in E_a$, in order to maintain the influence of both connected and non-connected vertices, we utilize negative edge sampling from the noise distribution $P_n(V)$ to model the influence of non-connected vertices and then the loss function is:

$$
\begin{aligned}
\mathcal{L}_{fa}(v_i, v_j) = & - \log \sigma(\Phi(v_j)^T \cdot \Phi(v_i)) - \\
& \sum_{x=1}^{k} \mathbb{E}_{v_x \sim P_n(V)} \Big[\log \sigma(-\Phi(v_x)^T \cdot \Phi(v_i)) \Big],
\end{aligned}
\tag{2}
$$

where k is the number of sampled non-connected edges. Based on the loss function, we maximize the probability of v_j being a neighbor of vertex v_i and minimize the probability of each negative vertex v_x being a neighbor of vertex v_i. We set $P_n(V) \propto d_v^{3/4}$ as suggested in [23], where d_v is the degree of vertex v.

For each deleted edge $e_{kh} \in E_d$, we use the negative edge sampling shown as Eq. (3) to minimize the probability of vertex v_k being a neighbor of v_h, which reduces the similarity between v_k and v_h in the latent representation.

$$
\mathcal{L}_{fd}(v_k, v_h) = -\log(1 - \sigma(\Phi(v_k)^T \cdot \Phi(v_h))).
\tag{3}
$$

Then, in order to update the embedding vectors of all vertices which correspond to the created and deleted edges, we maximize the joint probability over all evolved vertices. The loss function for preserving the first-order proximity is formulated as follows:

$$
\mathcal{L}_1 = \sum_{e_{ij} \in E_a} w_{ij} \mathcal{L}_{fa}(v_i, v_j) + \sum_{e_{kh} \in E_d} w_{kh} \, \mathcal{L}_{fd}(v_k, v_h),
\tag{4}
$$

where w_{ij} is the weight of edge e_{ij}, representing the importance of edge e_{ij} in constructing the embeddings of v_i and v_j.

DLNE with Second-Order Proximity. Second, the second-order proximity is determined by the similarity of neighbors between two vertices. The intuition is that two vertices are more similar if they share more common neighbors. The second-order proximity has been demonstrated to be a good metric to measure the similarity of a pair of vertices, even if they are not connected [20]. In this work, we employ the conditional probability of "context" v_j linked with vertex v_i [28]:

$$
p_2(v_j | v_i) = \frac{exp(\Psi(v_j)^T \cdot \Phi(v_i))}{\sum_{k=1}^{|V|} exp(\Psi(v_k)^T \cdot \Phi(v_i))},
\tag{5}
$$

where $|V|$ is the number of neighbors of v_i, and $\Psi(v_j) \in \mathbb{R}^d$ is an auxiliary vector of v_j that needs to be learned when v_j is treated as "context". We can see that $p_2(\cdot | v_i)$ defines a conditional distribution of vertex v_i among its contexts.

Similar to the first-order proximity, the influence of created and deleted edges are different. For each added edge $e_{ij} \in E_a$, we also use the negative edge sampling method to model the influence of the vertices that are not in the context set of v_i, and then the loss function is defined as:

$$\mathcal{L}_{sa}(v_i, v_j) = -\log\sigma(\Psi(v_j)^T \cdot \Phi(v_i)) -$$
$$\sum_{x=1}^{k} \mathbb{E}_{v_x \sim P_n(V)} \Big[\log\sigma(-\Psi(v_x)^T \cdot \Phi(v_i))\Big]. \tag{6}$$

The loss function $\mathcal{L}_{sa}(v_i, v_j)$ is to maximize the log-probability of observing the context of each vertex v_i that connect with the created edges.

For each deleted edge $e_{kh} \in E_d$, one negative edge sampling process is used to model the influence of edge e_{kh} and then the loss function $\mathcal{L}_{sd}$ is defined as:

$$\mathcal{L}_{sd}(v_k, v_h) = -\log(1 - \sigma(\Psi(v_h)^T \cdot \Phi(v_k))). \tag{7}$$

By combining the influence of each added and deleted edge, the loss function for preserving the second-order proximity is defined as:

$$\mathcal{L}_2 = \sum_{e_{ij} \in E_a} w_{ij}\mathcal{L}_{sa}(v_i, v_j) + \sum_{e_{kh} \in E_d} w_{kh}\mathcal{L}_{sd}(v_k, v_h). \tag{8}$$

The representation Φ and Ψ can be learned by training the empirical distribution $p_2(v_j|v_i) = \frac{w_{ij}}{\sum_{k \in \mathcal{N}_i} w_{ik}}$ that can be observed in the network, where the denominator is the out-degree of vertex v_i.

Finally, we jointly consider the influence of the first-order proximity and the second-proximity, and the joint loss function is defined as follows:

$$\mathcal{L} = \mathcal{L}_1 + \mathcal{L}_2. \tag{9}$$

4.2   Algorithm and Optimization

The Algorithm 1 is the pseudo-code for our DLNE that preserves both first-order and second-order proximities. We first perform edge sampling in the set of added edges E_a to generate the training vertices that are associated with the sampled edges. In particular, we use negative sampling to implement our loss function (Eqs. (4) and (8)) for each sampled new edge. Then, we process the deleted edges by edge sampling in E_d. The sampled deleted edges is addressed in form of a negative edge in consistent with Eq. (3) for the first-order proximity and Eq. (7) for the second-order proximity. Note that AddedEdgeSample(E_a) and DeletedEdgeSample(E_d) only perform one edge sampling from the set of added edges E_a and the set of deleted edges E_d, respectively. $NEG_k(v_i)$ represents the set of k sampled negative vertices w.r.t v_i. To optimize the loss function, we employ the asynchronous stochastic gradient algorithm (ASGD) [26]. The learning rate η for ASGD is initially set to 0.025 and then decreased linearly with the number of vertices that have been trained.

Algorithm 1. DLNE: Dynamic Large-scale Network Embedding

Input: Dynamic network $\mathcal{G}_t$; embeddings Φ_t and auxiliary vectors Ψ_t of $\mathcal{G}_t$; network
  updates E_a and E_d; embedding dimension d; the number of edge sampling s_n; the
  number of negative sampling k; learning rate η.
Output: Updated embedding vectors Φ_{t+1} of $\mathcal{G}_{t+1}$.
 1: $num_s \leftarrow 0$;
 2: **while** $(num_s < s_n)$ **do**
 3:     $e_{ij} \leftarrow$ AddedEdgeSample(E_a);
 4:     **for all** $v \in v_j \cup NEG_k(v_i)$ **do**
 5:         $\Phi(v) \leftarrow \Phi(v) - \eta\frac{\partial \mathcal{L}_1}{\partial \Phi(v)}$;
 6:         $\Psi(v) \leftarrow \Psi(v) - \eta\frac{\partial \mathcal{L}_2}{\partial \Psi(v)}$;
 7:     **end for**
 8:     $\Phi(v_i) = \Phi(v_i) - \eta(\frac{\partial \mathcal{L}_1}{\partial \Phi(v_i)} + \frac{\partial \mathcal{L}_2}{\partial \Phi(v_i)})$;
 9:     $e_{kh} \leftarrow$ DeletedEdgeSample(E_d);
10:     $\Phi(v_h) \leftarrow \Phi(v_h) - \eta\frac{\partial \mathcal{L}_1}{\partial \Phi(v_h)}$;
11:     $\Phi(v_k) \leftarrow \Phi(v_k) - \eta(\frac{\partial \mathcal{L}_1}{\partial \Phi(v_k)} + \frac{\partial \mathcal{L}_2}{\partial \Phi(v_k)})$;
12:     $\Psi(v_h) \leftarrow \Psi(v_h) - \eta\frac{\partial \mathcal{L}_2}{\partial \Psi(v_h)}$;
13:     $num_s + +$;
14: **end while**

The gradients in Algorithm 1 for edge $e_{ij} \in E_a$ are calculated as follows:

$$\frac{\partial \mathcal{L}_1}{\partial \Phi(v)} = \begin{cases} -w_{ij}(\sigma(-\Phi(v)^T\Phi(v_i)))\Phi(v_i) & v = v_j \\ w_{ij}(\sigma(\Phi(v)^T\Phi(v_i)))\Phi(v_i) & v \in NEG_k(v_i) \end{cases}$$

$$\frac{\partial \mathcal{L}_1}{\partial \Phi(v_i)} = -w_{ij}(\sigma(-\Phi(v_j)^T\Phi(v_i)))\Phi(v_j) + w_{ij}\sum_{x=1}^{k}(\sigma(\Phi(v_x)^T\Phi(v_i)))\Phi(v_x)$$

$$\frac{\partial \mathcal{L}_2}{\partial \Psi(v)} = \begin{cases} -w_{ij}(\sigma(-\Psi(v)^T\Phi(v_i)))\Phi(v_i) & v = v_j \\ w_{ij}(\sigma(\Psi(v)^T\Phi(v_i)))\Phi(v_i) & v \in NEG_k(v_i) \end{cases}$$

$$\frac{\partial \mathcal{L}_2}{\partial \Phi(v_i)} = -w_{ij}(\sigma(-\Psi(v_j)^T\Phi(v_i)))\Psi(v_j) + w_{ij}\sum_{x=1}^{k}(\sigma(\Psi(v_x)^T\Phi(v_i)))\Psi(v_x)$$

The gradient for edge $e_{kh} \in E_d$ are calculated as follows:

$$\frac{\partial \mathcal{L}_1}{\partial \Phi(v_h)} = w_{kh}(\sigma(\Phi(v_h)^T\Phi(v_k)))\Phi(v_k)$$

$$\frac{\partial \mathcal{L}_1}{\partial \Phi(v_k)} = w_{kh}(\sigma(\Phi(v_h)^T\Phi(v_k)))\Phi(v_h)$$

$$\frac{\partial \mathcal{L}_2}{\partial \Psi(v_h)} = w_{kh}(\sigma(\Psi(v_h)^T\Phi(v_k)))\Phi(v_k)$$

$$\frac{\partial \mathcal{L}_2}{\partial \Psi(v_k)} = w_{kh}(\sigma(\Psi(v_h)^T\Phi(v_k)))\Phi(v_h)$$

We use the same strategy as proposed in [28] to sample edges with probabilities in proportional to the original edge weights in the created or deleted slides.

Intuitively, the edges with large weight would be sampled more times. In this way, the embedding method can support the weighted graph. Specifically, we use the alias table method [16] to draw an edge sample, which only takes $O(1)$ time. Therefore, we are able to efficiently update the embedding vectors for each vertex in the current window.

It is worthy to mention that if the previous embedding Φ_t and Ψ_t are not available, our method can still apply. In this case, we treat previous graph G_t as an empty graph, and randomly initialize Φ and Ψ. By adding all edges in E_a and runs Algorithm 1, we are able to learn the dynamic network embeddings.

5   Experiments

5.1   Data Description

We use three real world networks to evaluate our method:

- **YouTube**[1] [30] is a video-sharing website on which users can upload, view, and share videos. Both the user social network and group membership information are included in the dataset. The group is defined by common video genres (e.g. anime and wresting) that the user followed. We use such group information as user labels.
- **DBLP** is an author-paper network[2] [29]. We use the DBLP dataset to construct two citation networks, which are DBLP (Paper) and DBLP (Author). DBLP (Paper) is a directed network, which represents the citation relationships among papers. As a directed weighted network, DBLP (Author) represents the citation relationships among authors, where edge weight is the number of cited papers. The labels of DBLP (Author) and DBLP (Paper) are the research areas of the published papers and authors. We choose 10 research areas in the field of computer science, including AI, computer networks, information security, high-performance computing, software engineering, computer graphics and multimedia, theoretical computer science, human computer interaction and ubiquitous computing, interdisciplinary studies, and database, data mining and information retrieval.

The detailed statistics of these networks are summarized in Table 1. Each network contains at least half million vertices and millions of edges.

We random assign timestamp to edges in YouTuBe dataset due to lack of time information. We rank the edges in the DBLP datasets by the publish time. Without loss of generality, we add the same number of edges E_a and delete the same number of E_d within each time interval.

---

[1] Available at http://socialcomputing.asu.edu/pages/datasets.
[2] Available at https://aminer.org/citation.

Table 1. Statistics of the information networks

Name	YouTube	DBLP (Paper)	DBLP (Author)
$|V|$	1,138,499	781,109	524,061
$|E|$	2,990,443	4,191,677	20,580,238
Average degree	5.25	10.73	78.54
# labels	47	10	10
# train	31,703	61,257	117,934

5.2   Baselines and Evaluation Metrics

We compare our method with the following baselines:

- **DeepWalk.** This approach learns low-dimensional feature representations for each vertex in the social networks by simulating truncated random walks [25].
- **LINE.** LINE [28] is an approach for large-scale information network embedding. LINE preserves the first-order (LINE(1st)) and the second-order (LINE(2nd)) proximities and supports both weighted and directed networks.
- **node2vec.** node2vec [13] extends DeepWalk by proposing a flexible neighbor selection method for vertices instead of simple random walk.

Note that we did not compare with DANE [18], because DANE employs both network adjacency matrix and node attribute matrix. While it is possible to apply DANE only on network matrix, called DANE-N, it requires to calculate the eigenvalues and thus cannot be applied in the large-scale networks with millions of vertices used in the paper.

To facilitate the comparison between DLNE and baselines, we perform a supervised task – multi-label classification on the embedding results. Specifically, we randomly sample a portion of the labeled vertices as training data, and the rest for testing. We employ a one-vs-rest logistic regression classifier implemented by LibLinear[3] [11]. We repeat this process 10 times, and report the average performance results in terms of *Micro-F1* and *Macro-F1*, which are defined as follows:

$$Micro - F1 = \frac{2 \times Precision \times Recall}{Precision + Recall}, \tag{10}$$

$$Macro - F1 = \frac{\sum_{i=1}^{D} F1(i)}{D}, \tag{11}$$

where D is the number of categories and $F1(i)$ is the *Micro-F1* in the ith category. The *Precision* and *Recall* are calculated on all categories. We evaluate the efficiency of our method on a machine with Core^{TM} i7-6700 (3.4 GHz) CPU and 16 GB memory. In this experiment, the dimensional d of embedding is set as 128. For each baseline (Deepwalk, LINE and node2vec), we follow the parameter settings used in their original papers.

---

[3] Available at http://www.csie.ntu.edu.tw/~cjlin/liblinear/.

Table 2. Performance of algorithms w.r.t. $\frac{|E_a|}{|E|}$ on incremental networks that only consider newly added edges

Algorithm	$\frac{|E_a|}{|E|}$	YouTube			DBLP(Paper)			DBLP(Author)		
		Mic-F1	Mac-F1	Time(s)	Mic-F1	Mac-F1	Time(s)	Mic-F1	Mac-F1	Time(s)
DeepWalk	1	44.50	35.46	24653	54.41	45.19	16125	61.55	56.72	11997
node2vec	1	44.66	35.94	27756	57.59	46.18	19524	61.91	57.23	15354
LINE	1	45.11	36.21	682	61.16	48.62	340	63.33	59.18	688
DLNE	0.01	41.07	33.07	3.5	52.42	39.23	3.9	56.42	52.89	5.5
	0.05	42.53	34.65	15.5	57.31	45.83	18.5	59.94	55.95	29
	0.1	43.85	35.76	27	60.11	47.73	35	62.08	57.83	61
	0.5	44.58	36.17	125	61.03	48.14	144	62.97	58.76	355
	1	45.35	36.53	255	61.12	48.57	335	63.22	59.13	680

Note: Mic-F1 and Mac-F1 mean Micro-F1 and Macro-F1 scores, and Time(s) denotes runtime (in seconds) of each update. $|E_a|$ is the number of newly added edges at each timestamp.

5.3   Performance Comparison

Comparison on Incremental Networks. We first evaluate the effectiveness and efficiency of DLNE on multi-label classification compared with baselines in an incremental environment. Namely, we assume edges are only added into and never deleted from the network.

We simulate different evolving networks by changing the value of $|E_a|/|E|$ from 0.01 to 1, where $|E_a|$ is the number of newly added edges with each time interval and $|E|$ is the number of edges in the dataset. Semantically, $|E_a|/|E|$ defines the evolving speed of a dynamic network. For example, when $|E_a|/|E| = 0.01$, DLNE adds 1% edges of whole network at each timestamp. For the compared baselines, the representation is learned based on the whole network (i.e., $|E_a|/|E| = 1$). Additionally, since the first-order proximity is not applicable on directed graph [28], the evaluation of DLNE on DBLP(Author) and DBLP(Paper) only consider the second-order proximity. On YouTuBe dataset DLNE is able to consider both the first-order and the second-order proximities, and thus LINE(1st+2nd) is compared on YouTuBe dataset only.

We evaluate the final embedding of the dynamic graph with a multi-class classification task. In this task, we choose 20% of the labeled vertices as training, and report the micro-F1 and macro-F1 on the testing data in Table 2. We report the running time for different embedding methods as well.

It is clear that DLNE achieves similar performance compared with other methods, while DLNE is 10–25 times faster than the best baseline. These observations are consistent on all three datasets. Learning the representation on the entire network should achieve the best performance, since more information could be captured, although at the cost of longer running time. We also observe that as $|E_a|/|E|$ increases, the DLNE achieves better prediction performance with longer running time. The reason is that incrementally updating a small portion of newly added edges saves time but loses some information in the process.

Additionally, on each network, we can see that DLNE and LINE run faster than DeepWalk and node2vec. This is because the latter two baselines need to sample and train a huge number of random walks, which is computationally expensive. The superior performance of DLNE and LINE also indicates that the first-order and second-order proximities effectively captures the local network structure.

Furthermore, on YouTube dataset, although both DLNE and Line consider the first-order and second-order proximities, DLNE achieves better Micro-F1 and Macro-F1 than LINE. The potential reason is that LINE learns the embedding of each vertex by preserving the first-order and second-order proximities separately and then concatenate them. Meanwhile, DLNE accounts for the first-order and second-order proximities by optimizing the joint loss function, which automatically balances the effect of these two proximities.

Table 3. Performance of algorithms on dynamic networks that consider both newly added and deleted edges

$\frac{|E_0|}{|E|}$	Method	YouTube			DBLP(Paper)			DBLP(Author)		
		Mic-F1	Mac-F1	Time(s)	Mic-F1	Mac-F1	Time(s)	Mic-F1	Mac-F1	Time(s)
0.9	Deepwalk	41.75	33.51	21687	53.32	44.41	17162	58.41	55.68	14497
	node2vec	41.85	33.85	23980	57.26	45.89	18771	58.26	55.54	17816
	LINE	42.43	35.09	640	60.16	47.72	340	57.95	54.27	688
	DLNE	**42.91**	**35.37**	**25**	**60.79**	**48.01**	**33**	**59.35**	**56.38**	**58**
0.7	Deepwalk	38.71	28.04	15143	45.68	30.49	13645	54.52	50.37	12425
	node2vec	40.15	29.34	16250	45.86	31.56	14350	54.94	50.73	14265
	LINE	41.97	32.76	568	52.66	34.56	274	54.82	50.28	532
	DLNE	**42.70**	**33.85**	**22**	**53.53**	**36.74**	**34**	**56.45**	**52.26**	**50**
0.5	Deepwalk	37.36	27.39	13254	43.62	26.19	12816	52.56	48.36	10797
	node2vec	39.04	28.12	14650	44.94	26.91	13010	53.43	48.45	11971
	LINE	40.12	28.47	410	51.74	32.79	193	51.97	46.62	398
	DLNE	**42.49**	**32.08**	**20**	**52.85**	**34.96**	**32**	**55.06**	**50.20**	**53**

Note: Mic-F1 and Mac-F1 mean Micro-F1 and Macro-F1 scores, and Time(s) denotes running time (in seconds).

Comparison on Dynamic Networks. Next, we evaluate DLNE on a dynamic network, where edges are added and deleted simultaneously. In this experiment, we vary the edge size $|E_0|$ of the initial graph G_0 from $0.5|E|$ to $0.9|E|$, and the numbers of added edges and deleted edges at each timestamp are fixed as $|E_a| = |E_d| = 0.1|E|$. Similar to last experiment, we compare the DLNE output after the last timestamp T with the baseline embeddings of the last snapshot graph G_T.

We report the comparison results in Table 3. The DLNE consistently runs about 6–25 times faster than LINE, because the incremental updating embeddings are time efficient. We also see that DLNE achieves better classification performance than all baselines on both micro-F1 and macro-F1. The reason is

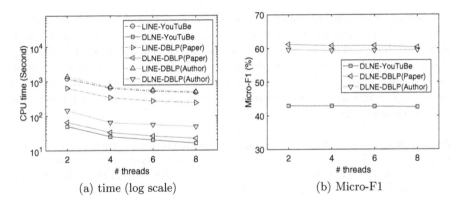

Fig. 2. Performance w.r.t #threads.

that DLNE updates the embeddings of vertices over evolving networks, where the historical network information are also captured. Other baselines only capture the network structure of G_T and overlook the temporal dependency.

Furthermore, we compare the results under different window sizes. Table 3 shows that our proposed method DLNE outperforms other methods consistently with different window size, which demonstrates that our method is robust. In particular, we can see that the performance gap of DLNE over LINE increases (e.g. Macro-F1 gaps are 0.8%, 3.3%, 12.7% on YouTube when $|E_0|/|E| = 0.9, 0.7, 0.5$), when the size of initial graph G_0 decreases. Smaller initial graph size means the networks evolves more rounds. When the network evolves more rounds, there are more historical information, which means stronger temporal dependency in the dynamic evolving network. Our DLNE captures more information in evolving networks since it updates the representation incrementally, while other methods only learn the representation on the snapshot. This also demonstrates the superiority of our DLNE since most real-world networks continuously evolve.

5.4   Parallel Computing

Finally, we evaluate the scalability of our method by running DLNE with different number of threads. The embedding learning parameters are exactly the same as previous experiment. Figure 2 shows the performance comparison w.r.t the number of threads on three datasets. We can see that the speed up of DLNE is stable with the increase of thread number. At the same time, increasing the number of threads does not affect the classification performance, as shown in Fig. 2(b). In summary, DLNE exhibits strong parallelism potential in handling large scale network dynamics.

6   Conclusion

This paper proposes an incremental model DLNE to learn the representation of large-scale dynamic networks. DLNE efficiently update the representation of network in a dynamic environment. The model preserves the first-order and the second-order proximities by optimizing the joint loss function. Extensive evaluations on three real-world networks demonstrate the effectiveness and efficiency of our method. In the future, we plan to explore how to learn the representation on dynamic networks with the changing of the weights of edges. Besides, it is interesting to investigate the efficient deep representation learning for dynamic large-scale networks.

Acknowledgments. This work is partially supported by the National Natural Science Foundation of China under grant Nos. 61773331 and 61403328, the National Science Foundation under grant Nos. 1544455, 1652525, and 1618448, and the China Scholarship Council under grant No. 201608370018.

References

1. Aggarwal, C., Subbian, K.: Evolutionary network analysis: a survey. ACM Comput. Surv. (CSUR) **47**(1), 10 (2014)
2. Belkin, M., Niyogi, P.: Laplacian eigenmaps and spectral techniques for embedding and clustering. In: Proceedings of NIPS, pp. 585–591 (2002)
3. Bhagat, S., Cormode, G., Muthukrishnan, S.: Node classification in social networks. In: Aggarwal, C. (ed.) Social Network Data Analytics, pp. 115–148. Springer, Boston (2011). https://doi.org/10.1007/978-1-4419-8462-3_5
4. Cao, S., Lu, W., Xu, Q.: GraRep: learning graph representations with global structural information. In: Proceedings of CIKM, pp. 891–900. ACM (2015)
5. Chang, S., Han, W., Tang, J., Qi, G.J., Aggarwal, C.C., Huang, T.S.: Heterogeneous network embedding via deep architectures. In: Proceedings of SIGKDD, pp. 119–128. ACM (2015)
6. Chen, C., Tong, H.: Fast eigen-functions tracking on dynamic graphs. In: Proceedings of SDM, pp. 559–567. SIAM (2015)
7. Chen, J., Zhang, Q., Huang, X.: Incorporate group information to enhance network embedding. In: Proceedings of CIKM, pp. 1901–1904. ACM (2016)
8. Chen, M., Yang, Q., Tang, X.: Directed graph embedding. In: IJCAI, pp. 2707–2712 (2007)
9. Chen, Y., Wang, C.: HINE: heterogeneous information network embedding. In: Candan, S., Chen, L., Pedersen, T.B., Chang, L., Hua, W. (eds.) DASFAA 2017. LNCS, vol. 10177, pp. 180–195. Springer, Cham (2017). https://doi.org/10.1007/978-3-319-55753-3_12
10. Cox, T.F., Cox, M.A.: Multidimensional Scaling. CRC Press, Boca Raton (2000)
11. Fan, R.E., Chang, K.W., Hsieh, C.J., Wang, X.R., Lin, C.J.: Liblinear: a library for large linear classification. J. Mach. Learn. Res. **9**(Aug), 1871–1874 (2008)
12. Fortunato, S.: Community detection in graphs. Phys. Rep. **486**(3), 75–174 (2010)
13. Grover, A., Leskovec, J.: node2vec: scalable feature learning for networks. In: Proceedings of SIGKDD, pp. 855–864. ACM (2016)
14. Huang, X., Li, J., Hu, X.: Label informed attributed network embedding. In: Proceedings of WSDM, pp. 731–739. ACM (2017)
15. Jian, L., Li, J., Liu, H.: Toward online node classification on streaming networks. Data Mining Knowl. Discov. **32**, 1–27 (2017)
16. Li, A.Q., Ahmed, A., Ravi, S., Smola, A.J.: Reducing the sampling complexity of topic models. In: Proceedings of SIGKDD, pp. 891–900. ACM (2014)

17. Li, C., Li, Z., Wang, S., Yang, Y., Zhang, X., Zhou, J.: Semi-supervised network embedding. In: Candan, S., Chen, L., Pedersen, T.B., Chang, L., Hua, W. (eds.) DASFAA 2017. LNCS, vol. 10177, pp. 131–147. Springer, Cham (2017). https://doi.org/10.1007/978-3-319-55753-3_9

18. Li, J., Dani, H., Hu, X., Tang, J., Chang, Y., Liu, H.: Attributed network embedding for learning in a dynamic environment. arXiv preprint arXiv:1706.01860 (2017)

19. Li, J., Hu, X., Jian, L., Liu, H.: Toward time-evolving feature selection on dynamic networks. In: Proceedings of ICDM, pp. 1003–1008. IEEE (2016)

20. Liben-Nowell, D., Kleinberg, J.: The link-prediction problem for social networks. J. Assoc. Inf. Sci. Technol. **58**(7), 1019–1031 (2007)

21. Maaten, L.V.D., Hinton, G.: Visualizing data using t-SNE. J. Mach. Learn. Res. **9**(Nov), 2579–2605 (2008)

22. Mikolov, T., Chen, K., Corrado, G., Dean, J.: Efficient estimation of word representations in vector space. arXiv preprint arXiv:1301.3781 (2013)

23. Mikolov, T., Sutskever, I., Chen, K., Corrado, G.S., Dean, J.: Distributed representations of words and phrases and their compositionality. In: Proceedings of NIPS, pp. 3111–3119 (2013)

24. Ning, H., Xu, W., Chi, Y., Gong, Y., Huang, T.: Incremental spectral clustering with application to monitoring of evolving blog communities. In: Proceedings of SDM, pp. 261–272. SIAM (2007)

25. Perozzi, B., Al-Rfou, R., Skiena, S.: DeepWalk: online learning of social representations. In: Proceedings of SIGKDD, pp. 701–710. ACM (2014)

26. Recht, B., Ré, C., Wright, S.J., Niu, F.: HOGWILD: a lock-free approach to parallelizing stochastic gradient descent. In: Proceedings of NIPS, pp. 693–701 (2011)

27. Roweis, S.T., Saul, L.K.: Nonlinear dimensionality reduction by locally linear embedding. Science **290**(5500), 2323–2326 (2000)

28. Tang, J., Qu, M., Wang, M., Zhang, M., Yan, J., Mei, Q.: Line: large-scale information network embedding. In: Proceedings of WWW, pp. 1067–1077. ACM (2015)

29. Tang, J., Zhang, J., Yao, L., Li, J., Zhang, L., Su, Z.: ArnetMiner: extraction and mining of academic social networks. In: Proceedings of SIGKDD, pp. 990–998. ACM (2008)

30. Tang, L., Liu, H.: Scalable learning of collective behavior based on sparse social dimensions. In: Proceedings of CIKM, pp. 1107–1116. ACM (2009)

31. Tenenbaum, J.B., De Silva, V., Langford, J.C.: A global geometric framework for nonlinear dimensionality reduction. Science **290**(5500), 2319–2323 (2000)

32. Wang, D., Cui, P., Zhu, W.: Structural deep network embedding. In: Proceedings of SIGKDD, pp. 1225–1234. ACM (2016)

33. Wang, H., Li, Z.: Region representation learning via mobility flow. In: Proceedings of CIKM. ACM (2017)

34. Wang, X., Cui, P., Wang, J., Pei, J., Zhu, W., Yang, S.: Community preserving network embedding. In: Proceedings of AAAI, pp. 203–209 (2017)

35. Xu, L., Wei, X., Cao, J., Yu, P.S.: Embedding of embedding (EOE): joint embedding for coupled heterogeneous networks. In: Proceedings of WSDM, pp. 741–749. ACM (2017)

36. Yang, C., Liu, Z., Zhao, D., Sun, M., Chang, E.Y.: Network representation learning with rich text information. In: IJCAI, pp. 2111–2117 (2015)

37. Yu, X., Ren, X., Sun, Y., Gu, Q., Sturt, B., Khandelwal, U., Norick, B., Han, J.: Personalized entity recommendation: a heterogeneous information network approach. In: Proceedings of WSDM, pp. 283–292. ACM (2014)

Multi-view Discriminative Learning via Joint Non-negative Matrix Factorization

Zhong Zhang, Zhili Qin, Peiyan Li, Qinli Yang, and Junming Shao[(✉)]

School of Computer Science and Engineering, Big Data Reserach Center,
University of Electronic Science and Technology of China, Chengdu 611731, China
{zhongzhang,zhiliqin,peiyanli}@std.uestc.edu.cn,
{qinli.yang,junmshao}@uestc.edu.cn
http://dm.uestc.edu.cn

Abstract. Multi-view learning attempts to generate a classifier with a better performance by exploiting relationship among multiple views. Existing approaches often focus on learning the consistency and/or complementarity among different views. However, not all consistent or complementary information is useful for learning, instead, only class-specific discriminative information is essential. In this paper, we propose a new robust multi-view learning algorithm, called DICS, by exploring the **D**iscriminative and non-discriminative **I**nformation existing in **C**ommon and view-**S**pecific parts among different views via joint non-negative matrix factorization. The basic idea is to learn a latent common subspace and view-specific subspaces, and more importantly, discriminative and non-discriminative information from all subspaces are further extracted to support a better classification. Empirical extensive experiments on seven real-world data sets have demonstrated the effectiveness of DICS, and show its superiority over many state-of-the-art algorithms.

Keywords: Multi-view learning · Matrix factorization · Classification

1 Introduction

Many real-world entities are often represented with different views such as web pages [1,33], multi-lingual news [2,8,16] and neuroimaging [22–24]. Consistency and complementarity, as the bridges to link all views together, are the two main assumptions in current multi-view learning [30]. The consistency assumption suggests that there is consistent information shared by all views [3,18,31]. Apparently, it is insufficient to exploit multi-view data using only consistent information since each view also contains complementary knowledge that other views do not have [1,9,19]. Therefore, investigating the complementarity of views is another important paradigm to learn multi-view data.

However, a question comes to our mind: whether the derived consistent and (or) complementary information really always support a better classification performance? Our answer is: *no*, since empirical pre-experiments indicate that

J. Pei et al. (Eds.): DASFAA 2018, LNCS 10828, pp. 542–557, 2018.
https://doi.org/10.1007/978-3-319-91458-9_33

prediction performance on multi-view data can be even worse than using single-view data in some real-world data sets. The main reason is that the consistent or complementary information does not learn discriminative information directly. The classifier constructed by multi-view data may give an even worse classification performance if the learned consistent and (or) complementary information contains no clear discriminative information.

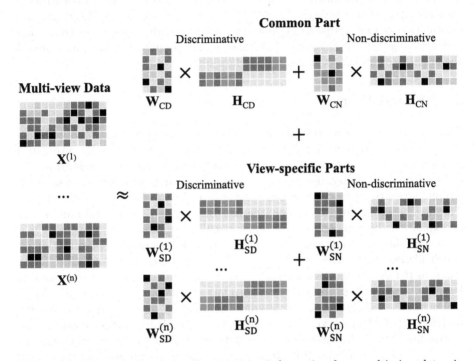

Fig. 1. Illustration of extracting discriminative information from multi-view data via joint non-negative matrix factorization. Each view of the data matrix is a superposition of four different parts: common discriminative part, common non-discriminative part, specific discriminative part and specific non-discriminative part.

In this paper, towards robust multi-view learning, we examine both discriminative and non-discriminative information existing in the consistent and complementary parts, and use only discriminative information for learning. Following this idea, we propose a new multi-view learning algorithm, called **DICS**, by exploring the **D**iscriminative and non-discriminative **I**nformation existing in **C**ommon and view-**S**pecific parts among different views via joint non-negative matrix factorization (NMF). Specifically, as usual, multi-view data is factorized into common part shared across views and view-specific parts existing within each view. Beyond, for both common part and each view-specific part, they are further factorized into two parts (discriminative part and non-discriminative part). To better obtain the discriminative parts, a supervised constraint is added to guide the joint NMF factorization. For illustration, Fig. 1 gives a simple example to illustrate the decomposition. Here, each view of data is factorized into four parts: the common discriminative, common non-discriminative, specific discriminative and specific non-discriminative part, respectively. To find the optimal decomposition, we follow the block coordinate descent (BCD) framework [14] to solve the objective function of DICS. Finally, only the derived discriminative parts from common part and view-specific parts are used to construct a classifier. Experimental results show that DICS allows extracting discriminative information as well as discarding non-discriminative information effectively, and supports a gained classification performance, which outperforms many state-of-the-art algorithms on seven real-world data sets.

2   Related Work

The most simplest way to deal with multi-view data is to concatenate all feature vectors of different views into one single long feature vector. However, such method ignores the relationships among multiple views and may suffer from the curse of dimensionality. To present, many advanced multi-view learning algorithms have been proposed, which can be broadly categorized into two types: The first category aims to exploit the consistency, and the second one focuses on exploiting the complementarity among multiple views.

Studies in exploiting consistency generally seek a common representation on which all views have minimum disagreement. For instance, Canonical Correlation Analysis (CCA) related algorithms [3,6,11,12,26] project two or more views into latent subspaces by maximizing the correlations among projected views. Spectral methods [5,16,20,29,33] use weighted summation to merge graph Laplacian matrices from different views into one optimal graph for further clustering or embedding. Matrix factorization based methods [8,18,27] jointly factorize multi-view data into one common centroid representation by minimizing the overall reconstruction loss of different views. In addition, multiple kernel learning (MKL) [7] can also be considered as exploiting the consistency across different views, where each view is mapped into a new space (e.g. kernel Hilbert space) using kernel trick, and then combines all kernel matrices into one unified kernel by minimizing a pre-defined objective function.

Another paradigm of multi-view learning is to explicitly preserve complementary information of different views. Co-training style algorithms [1,15,28,32] treat each view as complementarity. Generally speaking, it iteratively trains two classifiers on two different views, and each classifier generates its complementary information to help the other classifier to train in the next iteration. Beyond, the Co-EM algorithm [21] can be considered as a probabilistic version of co-training. Subspace related methods are also adopted to learn the complementarity. For instance, [9,10,13,19,25] learn one shared latent factor and view-specific latent factors to simultaneously capture the consistency and complementarity.

In summary, most existing multi-view learning algorithms mainly focus on learning consistency and complementarity from multi-view data. However, discriminative information existing in consistency and complementarity is not fully investigated, which is actually the direct factor to dominate the learning performance.

3   The Proposed Method

3.1   Preliminaries

Given a non-negative matrix $\mathbf{X} \in \mathbb{R}_+^{m \times n}$, where each column represents a data point. NMF aims to approximately factorize the data matrix into two non-negative matrix $\mathbf{W} \in \mathbb{R}_+^{m \times k}$ and $\mathbf{H} \in \mathbb{R}_+^{n \times k}$, so that,

$$\min_{\mathbf{W},\mathbf{H}} ||\mathbf{X} - \mathbf{W}\mathbf{H}^T||_F^2$$
$$\text{s.t. } \mathbf{W}, \mathbf{H} \geq 0 \tag{1}$$

where $|| \cdot ||_F$ denotes the Frobenius norm. Note that the original data matrix is a linear combination of all column vectors in $\mathbf{W}$ with weights of corresponding column vectors in $\mathbf{H}$. Therefore, $\mathbf{W}$ and $\mathbf{H}$ are often called the basis matrix and the coefficient matrix respectively.

For multi-view data, NMF-based approaches often take either $\mathbf{W}$ or $\mathbf{H}$ as a common factor. One of the representative formulation is as follows.

$$\min_{\mathbf{W},\mathbf{H}} \sum_{v=1}^{n_v} ||\mathbf{X}^{(v)} - \mathbf{W}^{(v)}\mathbf{H}^T||_F^2 + \Phi(\mathbf{W}, \mathbf{H})$$
$$\text{s.t. } \mathbf{W}, \mathbf{H} \geq 0 \tag{2}$$

where n_v denotes the number of views, and $\mathbf{W}^{(v)}$ denotes the basis matrices corresponding to different views. $\mathbf{H}$ denotes the common coefficient matrix shared across views, and $\Phi(\cdot)$ are some regularization terms on $\mathbf{W}$ and $\mathbf{H}$. It assumes that different views of one identical object are generated from distinct subspaces, and all views share with one centroid latent representation. This paradigm considers the consistency shared by all views, however, it ignores the complementary knowledge existing in each view.

3.2   Discriminant Learning on Multi-view Data

As multiple views have their commonality and distinctiveness, we first decompose the multi-view data into two parts: common part and view-specific parts, like many existing approaches [9,10,13,19]. Formally, let $\mathbf{W}_C$ represents the common subspace shared by all views and $\mathbf{W}_S^{(v)}$ represents the distinct subspace corresponding to each specific view. Therefore, each view of data matrix can be written as $\mathbf{X}^{(v)} = \mathbf{W}_C\mathbf{H}_C^T + \mathbf{W}_S^{(v)}\mathbf{H}_S^{(v)T}$. To derive the common and view-specific information, we thus can formulate our objective function as follows.

$$\min_{\mathbf{W},\mathbf{H}} \sum_{v=1}^{n_v} \left|\left| \mathbf{X}^{(v)} - \begin{bmatrix} \mathbf{W}_C & \mathbf{W}_S^{(v)} \end{bmatrix} \begin{bmatrix} \mathbf{H}_C^T \\ \mathbf{H}_S^{(v)T} \end{bmatrix} \right|\right|_F^2 + \Phi(\mathbf{W}, \mathbf{H})$$
$$\text{s.t. } \mathbf{W}, \mathbf{H} \geq 0 \tag{3}$$

To learn the discriminative information existing in multi-view data, we further leverage the available label information to guide joint matrix factorization in a supervised way. Specifically, we first divide the common part and each view-specific part into the discriminative part and the non-discriminative part, respectively. Namely,

$$\widetilde{\mathbf{W}} = \begin{bmatrix} \mathbf{W}_{\mathrm{CD}} & \mathbf{W}_{\mathrm{CN}} & \mathbf{W}_{\mathrm{SD}}^{(v)} & \mathbf{W}_{\mathrm{SN}}^{(v)} \end{bmatrix} \tag{4}$$

$$\widetilde{\mathbf{H}} = \begin{bmatrix} \mathbf{H}_{\mathrm{CD}} & \mathbf{H}_{\mathrm{CN}} & \mathbf{H}_{\mathrm{SD}}^{(v)} & \mathbf{H}_{\mathrm{SN}}^{(v)} \end{bmatrix} \tag{5}$$

where $\mathbf{W}_{\mathrm{CD}}$ and $\mathbf{W}_{\mathrm{CN}}$ indicate the common discriminative as well as the non-discriminative part of matrix $\widetilde{\mathbf{W}}$, respectively. Similarly, $\mathbf{W}_{\mathrm{SD}}^{(v)}$ and $\mathbf{W}_{\mathrm{SN}}^{(v)}$ indicate the view-specific parts. It is the same for $\widetilde{\mathbf{H}}$.

Afterwards, we impose the supervised constraint on the latent coefficient matrix $\mathbf{H}$. Here, it is worth noting that we only add the constraint on the discriminative part of $\mathbf{H}$ to derive discriminability. In addition, we should notice that the discriminative information not only exists in the common part, but also in each view-specific part. Therefore, the objective function is further reformulated as follows.

$$\min_{\mathbf{W},\mathbf{H},\mathbf{B}} \sum_{v=1}^{n_v} \left|\left| \mathbf{X}^{(v)} - \widetilde{\mathbf{W}}\widetilde{\mathbf{H}}^T \right|\right|_F^2 + \Phi(\mathbf{W},\mathbf{H})$$

$$+ \gamma \left|\left| \mathbf{Y} - \begin{bmatrix} \mathbf{B}_{\mathrm{CD}} & \mathbf{B}_{\mathrm{SD}}^{(v)} \end{bmatrix} \begin{bmatrix} \mathbf{H}_{\mathrm{CD}}^T \\ \mathbf{H}_{\mathrm{SD}}^{(v)T} \end{bmatrix} \right|\right|_F^2 \tag{6}$$

$$\text{s.t. } \mathbf{W}, \mathbf{H} \geq 0, \; ||(\mathbf{W})._{,i}||_2 = 1$$

where $\mathbf{Y} \in \mathbb{R}^{c \times n}$ is the label matrix, c is the number of classes, and n is the number of data instances. $y_{i,j} = 1$ if the instance j belong to class i and 0 otherwise. $\mathbf{B} = [\mathbf{B}_{\mathrm{CD}} \; \mathbf{B}_{\mathrm{SD}}^{(v)}] \in \mathbb{R}^{c \times (k1+k3)}$ is a linear projection matrix which maps the latent representation into label space. Subscript "C" and "S" represent "common" and "specific" respectively. "D" and "N" represent "discriminative" and "non-discriminative" respectively. For example, $\mathbf{W}_{\mathrm{CD}}$ denotes the common discriminative subspace. We normalize each column vector of $\mathbf{W}$ to ensure a unique solution. The supervised regularization term is imposed on $\mathbf{H}_{\mathrm{D}} = [\mathbf{H}_{\mathrm{CD}} \; \mathbf{H}_{\mathrm{SD}}^{(v)}]$ to make the derived patterns discriminative.

3.3   Regularization Terms

To further enhance the discriminative power of latent subspaces, we impose a $\ell_{1,1}$ norm constraint on $\mathbf{W}_{\mathrm{D}}$ as $||\mathbf{W}_{\mathrm{D}}^T\mathbf{W}_{\mathrm{D}}||_{1,1}$, where $\mathbf{W}_{\mathrm{D}} = [\mathbf{W}_{\mathrm{CD}} \; \mathbf{W}_{\mathrm{SD}}^{(v)}]$. This term can be factorized into two parts: $||\mathbf{W}_{\mathrm{D}}^T\mathbf{W}_{\mathrm{D}}||_{1,1} = \sum_i \mathbf{w}_{\mathrm{D}i}^T\mathbf{w}_{\mathrm{D}i} + \sum_{i \neq j} \mathbf{w}_{\mathrm{D}i}^T\mathbf{w}_{\mathrm{D}j}$. The first term is used to prevent overfitting. The second term encourages basis vectors to be as orthogonal as possible, which reduces the redundancy of discriminative bases. At last, we impose a $\ell_{1,1}$ norm constraint on $\mathbf{H}_{\mathrm{D}}$, which encourages the discriminative coefficients to be sparse. The reason is that data points of different classes should not possess identical latent concepts (i.e. basis vectors). It is reasonable that a latent concept only appears in a certain class but not in the others. With such intuition, a discriminative latent representation vector $\mathbf{h}_{\mathrm{D}i}$ should be sparse in the elements which are corresponding to the latent concepts that $\mathbf{h}_{\mathrm{D}i}$ doesn't posses. Finally, putting all terms together, the objective function of DICS is formulated as follows.

$$
\min_{\mathbf{W},\mathbf{H},\mathbf{B}} \sum_{v=1}^{n_v} ||\mathbf{X}^{(v)} - \widetilde{\mathbf{W}}\widetilde{\mathbf{H}}^T||_F^2 + \alpha||\mathbf{W}_{\mathrm{D}}^T\mathbf{W}_{\mathrm{D}}||_{1,1}
$$

$$
+ \beta||\mathbf{H}_{\mathrm{D}}||_{1,1} + \gamma\left\|\mathbf{Y} - \begin{bmatrix}\mathbf{B}_{\mathrm{CD}} & \mathbf{B}_{\mathrm{SD}}^{(v)}\end{bmatrix}\begin{bmatrix}\mathbf{H}_{\mathrm{CD}}^T \\ \mathbf{H}_{\mathrm{SD}}^{(v)T}\end{bmatrix}\right\|_F^2 \tag{7}
$$

$$
\text{s.t. } \mathbf{W},\mathbf{H} \geq 0, \ ||(\mathbf{W})_{.,i}||_2 = 1
$$

where α, β, γ are non-negative parameters to balance the regularization terms.

3.4   Optimization

The objective function Eq. (7) is not convex over both variables $\mathbf{W}$ and $\mathbf{H}$. Therefore, it is impractical to find the global optimum. We follow the general BCD framework to divide the objective function Eq. (7) into several convex subproblems corresponding to each column of $\mathbf{W}$ and $\mathbf{H}$, then solve each subproblem successively by fixing the others. In this way, the global convergence and local minimum solutions can be obtained [4].

Firstly, we represent $\mathbf{W}\mathbf{H}^T$ as the sum of rank-1 outer products. We can equivalently reformulate the objective function Eq. (7) as follows.

$$
f(\mathbf{W},\mathbf{H},\mathbf{B}) =
$$

$$
\sum_{v=1}^{n_v}\left\|\mathbf{X}^{(v)} - \sum_{i=1}^{k1}\mathbf{w}_{\mathrm{CD}i}\mathbf{h}_{\mathrm{CD}i}^T - \sum_{i=1}^{k2}\mathbf{w}_{\mathrm{CN}i}\mathbf{h}_{\mathrm{CN}i}^T - \right.
$$

$$
\left. \sum_{i=1}^{k3}\mathbf{w}_{\mathrm{SD}i}^{(v)}\mathbf{h}_{\mathrm{SD}i}^{(v)T} - \sum_{i=1}^{k4}\mathbf{w}_{\mathrm{SN}i}^{(v)}\mathbf{h}_{\mathrm{SN}i}^{(v)T}\right\|_F^2 +
$$

$$
\alpha(\sum_{i=1}^{k1}\sum_{j=1}^{k1}\mathbf{w}_{\mathrm{CD}i}^T\mathbf{w}_{\mathrm{CD}j} + \sum_{i=1}^{k3}\sum_{j=1}^{k3}\mathbf{w}_{\mathrm{SD}i}^{(v)T}\mathbf{w}_{\mathrm{SD}j}^{(v)} + \tag{8}
$$

$$
2\sum_{i=1}^{k1}\sum_{j=1}^{k3}\mathbf{w}_{\mathrm{CD}i}^T\mathbf{w}_{\mathrm{SD}j}^{(v)}) + \beta\mathbf{1}_{1\times n}(\sum_{i=1}^{k1}\mathbf{h}_{\mathrm{CD}i} + \sum_{i=1}^{k3}\mathbf{h}_{\mathrm{SD}i}^{(v)})
$$

$$
\gamma\left\|\mathbf{Y} - \sum_{i=1}^{k1}\mathbf{b}_{\mathrm{CD}i}\mathbf{h}_{\mathrm{CD}i}^T - \sum_{i=1}^{k3}\mathbf{b}_{\mathrm{SD}i}^{(v)}\mathbf{h}_{\mathrm{SD}i}^{(v)T}\right\|_F^2
$$

where $\mathbf{w}_{\mathrm{CD}i}$, $\mathbf{w}_{\mathrm{CN}i}$, $\mathbf{w}_{\mathrm{SD}i}^{(v)}$, $\mathbf{w}_{\mathrm{SN}i}^{(v)}$, $\mathbf{h}_{\mathrm{CD}i}$, $\mathbf{h}_{\mathrm{CN}i}$, $\mathbf{h}_{\mathrm{SD}i}^{(v)}$, $\mathbf{h}_{\mathrm{SN}i}^{(v)}$ are the i-th column vectors of $\mathbf{W}_{\mathrm{CD}}$, $\mathbf{W}_{\mathrm{CN}}$, $\mathbf{W}_{\mathrm{SD}}^{(v)}$, $\mathbf{W}_{\mathrm{SN}}^{(v)}$, $\mathbf{H}_{\mathrm{CD}}$, $\mathbf{H}_{\mathrm{CN}}$, $\mathbf{H}_{\mathrm{SD}}^{(v)}$, $\mathbf{H}_{\mathrm{SN}}^{(v)}$ respectively. $\mathbf{1}_{1\times n}$ is a row vector of length n with all elements 1.

By fixing all column vectors except the one we want to update, we can obtain the convex subproblem respect to it, then solve it based on the BCD framework. Note that we use $[\cdot]_+$ to denote $\max(0, \cdot)$, which projects the negative value to the boundary of feasible region of zero. Finally, we give the update rules as follows.

$$\mathbf{w}_{\mathrm{CD}i} = \mathbf{w}_{\mathrm{CD}i} + \left[\frac{\sum_{v=1}^{n_v}(\mathbf{R}^{(v)}\mathbf{h}_{\mathrm{CD}i} - \alpha(\mathbf{W}_{\mathrm{CD}}\mathbf{1}_{k1\times1} + \mathbf{W}_{\mathrm{SD}}^{(v)}\mathbf{1}_{k3\times1}))}{n_v(\mathbf{h}_{\mathrm{CD}i}^T\mathbf{h}_{\mathrm{CD}i} + \alpha)} \right]_+ \tag{9}$$

$$\mathbf{w}_{\mathrm{CN}i} = \mathbf{w}_{\mathrm{CN}i} + \left[\frac{\sum_{v=1}^{n_v}\mathbf{R}^{(v)}\mathbf{h}_{\mathrm{CN}i}}{n_v(\mathbf{h}_{\mathrm{CN}i}^T\mathbf{h}_{\mathrm{CN}i})} \right]_+ \tag{10}$$

$$\mathbf{w}_{\mathrm{SD}i}^{(v)} = \mathbf{w}_{\mathrm{SD}i}^{(v)} + \left[\frac{\mathbf{R}^{(v)}\mathbf{h}_{\mathrm{SD}i}^{(v)} - \alpha(\mathbf{W}_{\mathrm{CD}}\mathbf{1}_{k1\times1} + \mathbf{W}_{\mathrm{SD}}^{(v)}\mathbf{1}_{k3\times1})}{\mathbf{h}_{\mathrm{SD}i}^{(v)T}\mathbf{h}_{\mathrm{SD}i}^{(v)} + \alpha} \right]_+ \tag{11}$$

$$\mathbf{w}_{\mathrm{SN}i}^{(v)} = \mathbf{w}_{\mathrm{SN}i}^{(v)} + \left[\frac{\mathbf{R}^{(v)}\mathbf{h}_{\mathrm{SN}i}^{(v)}}{\mathbf{h}_{\mathrm{SN}i}^{(v)T}\mathbf{h}_{\mathrm{SN}i}^{(v)}} \right]_+ \tag{12}$$

$$\mathbf{h}_{\mathrm{CD}i} = \mathbf{h}_{\mathrm{CD}i} + \left[\frac{\sum_{v=1}^{n_v}(\mathbf{R}^{(v)T}\mathbf{w}_{\mathrm{CD}i} - \frac{\beta}{2}\mathbf{1}_{n\times1} + \gamma\mathbf{Q}^{(v)T}\mathbf{b}_{\mathrm{CD}i})}{n_v(\mathbf{w}_{\mathrm{CD}i}^T\mathbf{w}_{\mathrm{CD}i} + \gamma\mathbf{b}_{\mathrm{CD}i}^T\mathbf{b}_{\mathrm{CD}i})} \right]_+ \tag{13}$$

$$\mathbf{h}_{\mathrm{CN}i} = \mathbf{h}_{\mathrm{CN}i} + \left[\frac{\sum_{v=1}^{n_v}\mathbf{R}^{(v)T}\mathbf{w}_{\mathrm{CN}i}}{n_v(\mathbf{w}_{\mathrm{CN}i}^T\mathbf{w}_{\mathrm{CN}i})} \right]_+ \tag{14}$$

$$\mathbf{h}_{\mathrm{SD}i}^{(v)} = \mathbf{h}_{\mathrm{SD}i}^{(v)} + \left[\frac{\mathbf{R}^{(v)T}\mathbf{w}_{\mathrm{SD}i}^{(v)} - \frac{\beta}{2}\mathbf{1}_{n\times1} + \gamma\mathbf{Q}^{(v)T}\mathbf{b}_{\mathrm{SD}}^{(v)}}{\mathbf{w}_{\mathrm{SD}i}^{(v)T}\mathbf{w}_{\mathrm{SD}i}^{(v)} + \gamma\mathbf{b}_{\mathrm{SD}i}^{(v)T}\mathbf{b}_{\mathrm{SD}i}^{(v)}} \right]_+ \tag{15}$$

$$\mathbf{h}_{\mathrm{SN}i}^{(v)} = \mathbf{h}_{\mathrm{SN}i}^{(v)} + \left[\frac{\mathbf{R}^{(v)T}\mathbf{w}_{\mathrm{SN}i}^{(v)}}{\mathbf{w}_{\mathrm{SN}i}^{(v)T}\mathbf{w}_{\mathrm{SN}i}^{(v)}} \right]_+ \tag{16}$$

where $\mathbf{R}^{(v)}$ and $\mathbf{Q}^{(v)}$ are

$$\mathbf{R}^{(v)} = \mathbf{X}^{(v)} - \mathbf{W}_{\mathrm{CD}}\mathbf{H}_{\mathrm{CD}}^T - \mathbf{W}_{\mathrm{CN}}\mathbf{H}_{\mathrm{CN}}^T - \mathbf{W}_{\mathrm{SD}}^{(v)}\mathbf{H}_{\mathrm{SD}}^{(v)T} - \mathbf{W}_{\mathrm{SN}}^{(v)}\mathbf{H}_{\mathrm{SN}}^{(v)T} \tag{17}$$

$$\mathbf{Q}^{(v)} = \mathbf{Y} - \mathbf{B}_{\mathrm{CD}}\mathbf{H}_{\mathrm{CD}}^T - \mathbf{B}_{\mathrm{SD}}^{(v)}\mathbf{H}_{\mathrm{SD}}^{(v)T} \tag{18}$$

Note that we extract the common factors $\mathbf{R}^{(v)}$ and $\mathbf{Q}^{(v)}$ from the equations just for saving the writing space. However, it is not efficient for implementation, since the computation orders i.e. $(\mathbf{WH}^T)\mathbf{h}_i$ and $\mathbf{W}(\mathbf{H}^T\mathbf{h}_i)$ largely affect the computational complexity. The former takes $mn(k+1)$ multiply operations, the later takes $(m+n)k$ multiply operations. Obviously the later form is much more efficient in implementation.

In addition, when the other variables are fixed, the projection matrices $\mathbf{B}_{\mathrm{CD}}$ and $\mathbf{B}_{\mathrm{SD}}^{(v)}$ can be solved in a closed form as follows.

$$\mathbf{B}_{\mathrm{CD}} = \frac{\sum_{v=1}^{n_v}(\mathbf{Y} - \mathbf{B}_{\mathrm{SD}}^{(v)}\mathbf{H}_{\mathrm{SD}}^{(v)T})}{n_v}\mathbf{H}_{\mathrm{CD}}(\mathbf{H}_{\mathrm{CD}}^T\mathbf{H}_{\mathrm{CD}} + \lambda\mathbf{I})^{-1} \qquad (19)$$

$$\mathbf{B}_{\mathrm{SD}}^{(v)} = (\mathbf{Y} - \mathbf{B}_{\mathrm{CD}}\mathbf{H}_{\mathrm{CD}}^T)\mathbf{H}_{\mathrm{SD}}^{(v)}(\mathbf{H}_{\mathrm{SD}}^{(v)T}\mathbf{H}_{\mathrm{SD}}^{(v)} + \lambda\mathbf{I})^{-1} \qquad (20)$$

where $\mathbf{I}$ is the identity matrix, λ is a small positive number.

Initialization. Since the NMF objective function is non-convex and has many local minima, a proper initialization is beneficial to improve learning performance. We develop a heuristic approach to initialize the basis matrix. DICS encourages the discriminative bases to achieve a degree of orthogonality, thus we try to initialize them as orthogonal as possible. To initialize $\mathbf{W}_{\mathrm{C}}$, we first calculate the mean of multi-view data, i.e. $\bar{\mathbf{X}} = \frac{1}{n_v}\sum_v^{n_v}\mathbf{X}^{(v)}$. Afterwards, we clustering $\bar{\mathbf{X}}$ into $k1 + k2$ clusters and obtain the corresponding centroids. Then we compute the pairwise linear correlation coefficients between each pair of centroids, and sort them in an ascending order. At last, we select $k1$ centroids corresponding to the top $k1$ correlation coefficients to initialize $\mathbf{W}_{\mathrm{CD}}$, and use the rest $k2$ centroids to initialize $\mathbf{W}_{\mathrm{CN}}$. It is same to initialize each $\mathbf{W}_{\mathrm{S}}^{(v)}$ by replacing $\bar{\mathbf{X}}$ with $\mathbf{X}^{(v)}$.

Time Complexity. The computational complexity of DICS is the same as solving standard NMF problem via hierarchical alternating least squares (HALS) algorithm under the BCD framework [14]. It is $O(\sum_v m_v n k)$ in the multi-view case, where m_v is the dimension of the v-view feature. Finally, the pseudocode of DICS is given in Algorithm 1.

4   Experiment

In this section, we first experimentally evaluate the proposed algorithm DICS in classification task on seven real world multi-view data sets. Then we empirically investigate that whether the extracted discriminative information from the common and the view-specific parts are really helpful for improving the learning performance. At last, the sensitivity of parameters and the convergence of DICS are analyzed.

Algorithm 1. DICS Algorithm

Input:
   Multi-view data matrices $\mathbf{X}^{(1)}, \mathbf{X}^{(2)}, ..., \mathbf{X}^{(n_v)}$, label matrix $\mathbf{Y}$, parameters α, β, γ, number of latent factors $k1$, $k2$, $k3$, $k4$.

Output:
   Basis matrices $\mathbf{W} = \{\mathbf{W}_{\mathrm{CD}}, \mathbf{W}_{\mathrm{CN}}, \mathbf{W}_{\mathrm{SD}}^{(v)}, \mathbf{W}_{\mathrm{SN}}^{(v)}\}$,
   Coefficient matrices $\mathbf{H} = \{\mathbf{H}_{\mathrm{CD}}, \mathbf{H}_{\mathrm{CN}}, \mathbf{H}_{\mathrm{SD}}^{(v)}, \mathbf{H}_{\mathrm{SN}}^{(v)}\}$,
   Projection matrices $\mathbf{B} = \{\mathbf{B}_{\mathrm{CD}}, \mathbf{B}_{\mathrm{SD}}^{(v)}\}$.

1: Initialize $\mathbf{W}$, $\mathbf{H}$, and $\mathbf{B}$.
2: **repeat**
3:     Update each column of $\mathbf{W}_{\mathrm{CD}}$ using Eq. (9)
4:     Update each column of $\mathbf{W}_{\mathrm{CN}}$ using Eq. (10)
5:     **for** $v = 1$ to n_v **do**
6:         Update each column of $\mathbf{W}_{\mathrm{SD}}^{(v)}$ using Eq. (11)
7:     **end for**
8:     **for** $v = 1$ to n_v **do**
9:         Update each column of $\mathbf{W}_{\mathrm{SN}}^{(v)}$ using Eq. (12)
10:    **end for**
11:    Update each column of $\mathbf{H}_{\mathrm{CD}}$ using Eq. (13)
12:    Update each column of $\mathbf{H}_{\mathrm{CN}}$ using Eq. (14)
13:    **for** $v = 1$ to n_v **do**
14:        Update each column of $\mathbf{H}_{\mathrm{SD}}^{(v)}$ using Eq. (15)
15:    **end for**
16:    **for** $v = 1$ to n_v **do**
17:        Update each column of $\mathbf{H}_{\mathrm{SN}}^{(v)}$ using Eq. (16)
18:    **end for**
19:    Update $\mathbf{B}_{\mathrm{CD}}$ using Eq. (19)
20:    **for** $v = 1$ to n_v **do**
21:        Update $\mathbf{B}_{\mathrm{SD}}^{(v)}$ using Eq. (20)
22:    **end for**
23: **until** convergence or max no. iterations reached;

4.1   Data Sets

Four popular real-world multi-view data sets are used in the experiment, including WebKB, Reuters, YaleFace and BBC, where the WebKB data set can be further divided into four sub data sets, namely Cornell, Texas, Washington, Wisconsin. Therefore, finally seven data sets are used to evaluate the performance of the proposed algorithm in this study. The statistics of data sets are summarized in Table 1.

4.2   Selection of Comparison Algorithms

We compare DICS algorithm with several single-view and multi-view algorithms to demonstrate its effectiveness. For fair comparison, the source codes of all comparing algorithms are directly downloaded from the author's website or requested from the author by email. The parameters of all algorithms are selected within

Table 1. Statistics of the data sets

Data sets	Data size	# of views	# of classes	# of dimensions
Reuters[a]	1200	5	6	2000 for all
Cornell[a]	195	2	5	1703/585
Texas[a]	187	2	5	1703/561
Washington[a]	230	2	5	1703/690
Winsconsin[a]	265	2	5	1703/795
YaleFace[b]	256	2	8	2016 for all
BBC[c]	685	4	5	4659/4633/4665/4684

[a] http://lig-membres.imag.fr/grimal/data.html
[b] http://vision.ucsd.edu/~iskwak/ExtYaleDatabase/ExtYaleB.html
[c] http://mlg.ucd.ie/datasets/segment.html

the range that the author suggested, which are listed in the following. Also, the source code of our proposed DICS algorithm can be acquired from Dropbox[1].

- *KNN.* We use the KNN algorithm (Set $k = 1$) as the baseline algorithm since all NMF-based algorithms can be regarded as a preprocessing before KNN. We apply KNN on all single views and report the best performance on the view. Also we apply the KNN algorithm on the concatenated feature vector (i.e. KNNcat).
- *NMF.* We apply the standard NMF algorithm on each of the single view data and the concatenated feature vector (i.e. NMFcat), as another baseline algorithm.
- *SSNMF.* This is a supervised NMF variant proposed in [17], which incorporates a linear classifier to encode the supervised information. We select the regularization parameter λ within the range of [0.5:0.5:3].
- *GNMF*[2]. This is a manifold regularized version of NMF [2], which preserves the local similarity by imposing a graph Laplacian regularization. We use the normalized dot product (cosine similarity) to construct the affinity graph, and select the regularization parameter λ within the set of $\{10^0, 10^1, 10^2, 10^3, 10^4\}$.
- *multiNMF*[3]. This is a well-known multi-view NMF algorithm proposed in [18]. We select the regularization parameter λ within the set of $\{10^{-3}, 10^{-2}, 10^{-1}, 10^0\}$.
- *MVCC*[4]. MVCC incorporates the local manifold regularization for multi-view learning [27]. We set parameter α to 100, and select β and γ within the set of $\{50, 100, 200, 500, 1000\}$.
- *MCL.* This is a semi-supervised multi-view NMF variant with graph regularized constraint [8]. We select parameter α within the range of [100:50:250],

[1] https://www.dropbox.com/s/guohn1zhq073x9f/DICS.zip?dl=0.
[2] http://www.cad.zju.edu.cn/home/dengcai/Data/GNMF.html.
[3] http://jialu.cs.illinois.edu.
[4] https://github.com/vast-wang/Clustering.git.

β within the set of {0.01, 0.02, 0.03}, and set gamma to 0.005 as author suggested.
- *DICS.* This is the proposed algorithm. We select parameters: α, β and γ within the set of $\{10^{-2}, 10^{-1}, 10^{0}, 10^{1}, 10^{2}\}$.

4.3   Classification on Real-World Data Sets

For DICS and all comparing algorithms, we first perform a five-folds cross validation to select the parameters, then we run ten times 10-folds cross validation with the selected parameters to obtain the final average classification accuracy and standard deviation. For all comparing NMF-based methods, we don't fix the number of latent factors k a global constant number, considering different algorithms may prefer different ks. Thus, we select k within the range of [5:5:100] for each algorithm. As for DICS, we need to set the number of four latent factors $k1$, $k2$, $k3$, $k4$ respectively. To avoid searching too large parameter space, we first select $k_i(i = 1, 2, 3, 4)$ within the range of [5:5:20], then we select the regularization parameters by fixing all k_i.

For classification, we first obtain latent representations from different NMF-based approaches, then we use KNN($k = 1$) for classification. Specifically, for unsupervised algorithms including NMF, GNMF, multiNMF, MCL and MVCC, we first apply algorithms on the data sets to obtain the latent representations $\mathbf{H}$, then we use $\mathbf{H}$ for further training and testing. For supervised method like DICS, we first obtain the discriminative basis $\mathbf{W}_D$ on training data, then we use the Moore-Penrose Pseudoinverse of $\mathbf{W}_D$ as projection matrix to obtain new data representation, namely $\widetilde{\mathbf{X}}^{(v)} = (\mathbf{W}_D^T \mathbf{W}_D)^{-1} \mathbf{W}_D^T \mathbf{X}^{(v)}$. Then we concatenate $\widetilde{\mathbf{X}}^{(v)}$ as the input for KNN.

Table 2 summarizes the classification results of different multi-view learning algorithms, where the numbers in the parentheses of the table denote the standard deviation. The best result on each data set is highlighted in boldface. As we can see from the results, the proposed DICS outperforms the other comparison algorithms on all seven data sets. DICS is slightly better than other algorithms on Reuters, YaleFace and BBC. But it achieves remarkably promising performance on four WebKB sub data sets, where it outperforms the second best algorithm up to 9.01% on Texas especially. The amazing result may result from twofold: (a) DICS not only explores the common and the view-specific information, but more importantly, the discriminative information existing in these parts is further extracted, which thus supports a gained prediction performance. (b) By filtering out the non-discriminative information from common part and view-specific parts, and adding the supervised constraints on encoding coefficients, the extracted discriminative information is much more effective for classification.

4.4   Empirical Study of DICS Algorithm

DICS assumes that multi-view data can be decomposed into the common part and the view-specific parts, and only the discriminative information in them

Table 2. Multi-view classification performance on real-world data sets

Method	ACC (%)						
	Reuters	Cornell	Texas	Washington	Wisconsin	YaleFace	BBC
KNN	43.3 (0.5)	61.6 (1.4)	66.4 (1.2)	74.4 (1.3)	60.2 (1.4)	89.1 (1.1)	36.7 (0.6)
KNNcat	37.3 (0.5)	62.5 (1.5)	65.5 (1.1)	70.4 (0.7)	61.7 (1.4)	74.9 (1.1)	22.7 (1.0)
NMF	59.2 (1.1)	64.0 (2.9)	72.8 (3.5)	74.5 (1.0)	74.9 (2.1)	93.3 (0.7)	69.8 (1.9)
NMFcat	59.6 (0.9)	65.9 (1.9)	73.7 (2.1)	75.3 (2.5)	77.1 (1.7)	92.5 (1.4)	87.9 (1.4)
SSNMF	64.0 (0.7)	66.5 (2.5)	69.3 (2.6)	73.4 (1.1)	73.9 (2.6)	93.8 (0.8)	81.6 (0.7)
GNMF	50.0 (1.0)	49.0 (2.0)	59.8 (1.8)	58.2 (2.0)	67.0 (1.8)	14.9 (1.3)	46.6 (1.7)
multiNMF	61.1 (0.8)	54.7 (1.4)	67.4 (2.8)	59.0 (2.8)	61.5 (2.8)	90.7 (1.2)	89.3 (1.4)
MCL	64.4 (0.8)	69.9 (2.2)	70.0 (2.5)	74.7 (2.1)	79.6 (2.4)	90.0 (0.2)	90.0 (0.8)
MVCC	55.0 (1.5)	64.9 (1.7)	71.0 (3.4)	70.8 (2.8)	76.4 (2.7)	29.5 (3.6)	70.2 (9.6)
DICS	**66.9 (1.6)**	**75.5 (2.3)**	**82.7 (2.0)**	**78.7 (1.3)**	**84.3 (1.1)**	**94.1 (1.0)**	**91.9 (0.7)**

is essential. To verify this assumption, we first construct the following subspaces: $\mathbf{W}_D = [\mathbf{W}_{CD} \; \mathbf{W}_{SD}^{(v)}]$, $\mathbf{W}_N = [\mathbf{W}_{CN} \; \mathbf{W}_{SN}^{(v)}]$, $\mathbf{W}_C = [\mathbf{W}_{CD} \; \mathbf{W}_{CN}]$ and $\mathbf{W}_S = [\mathbf{W}_{SD}^{(v)} \; \mathbf{W}_{SN}^{(v)}]$, denoting as the "Discriminative", "Non-discriminative", "Common" and "Specific" subspace. Afterwards, we project the original data onto these subspaces to obtain the corresponding components of data. We perform classification on each component, and the results are given in Fig. 2. The classification performance of the "Common" part is much worse than the "Specific" part, which suggests that only using the consistent information of

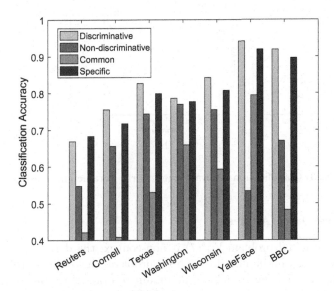

Fig. 2. Classification accuracy of DICS on different extracted components of multi-view data.

multi-view data is not enough to capture the whole discriminative information. Also, performance on the "Discriminative" part is better than all other parts in all data sets except Reuters. It suggests that extracting the discriminative information from the common as well as the view-specific parts, and discarding the non-discriminative parts do help improve the learning performance.

4.5   Parameter Study

There are three regularization parameters in DICS, i.e. α, β and γ. α controls the orthogonality degree of discriminative bases $\mathbf{W_D}$, β controls the degree of sparsity of discriminative latent representation $\mathbf{H_D}$, and γ balances the importance of supervised regularization term. To investigate how these parameters affect the final classification accuracy, we vary one parameter at a time within the set of $\{10^{-4}, 10^{-3}, 10^{-2}, 10^{-1}, 10^0, 10^1, 10^2, 10^3\}$, and fix the others to 10^{-3}. Figure 3 shows the variation trend of classification accuracy over different parameters on four typical data sets. The classification accuracy is relatively stable when α and β are less than 1, then drops sharply after α and β are increasing. As for parameter γ, the classification accuracy on BBC largely increases after γ is greater than 10^{-2}, and has become steady after γ is greater than 1. It is similar to other data sets except YaleFace, classification accuracy on YaleFace starts to decrease after γ is greater than 1. Based on the observation, we suggest selecting parameters α and β within a small range of $[0\ 1]$, and simply set the parameter $\gamma = 1$ for practical use.

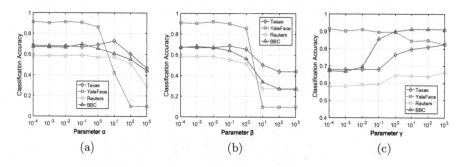

Fig. 3. Classification accuracy curve w.r.t. parameters α, β and γ.

4.6   Convergence Analysis

Though the original problem Eq. (7) is non-convex, the derived updating rules can achieve optimal minimum for each subproblem, the original problem Eq. (7) will eventually converge to a local minimum solution. In order to empirically investigate the convergence property of DICS, we plot the convergence curve and the corresponding classification accuracy curve on four typical data sets (see Fig. 4). From all four plots, we can observe that the objective values drop sharply and meanwhile the classification accuracies increase rapidly within about the first 10 iterations. After that, convergence curves and the accuracy curves begin to grow/decrease mildly, then it converges eventually. Usually, DICS will converge in no more than 50 iterations, while the corresponding classification accuracy becomes stable.

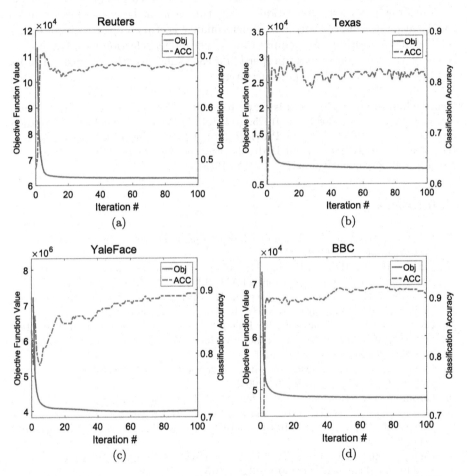

Fig. 4. Convergence and the corresponding classification accuracy curve of DICS on four typical data sets.

5   Conclusion

In this paper, we propose a novel multi-view learning algorithm, called DICS, by exploiting the discriminative information existing in multi-view data. To this end, a joint non-negative matrix factorization is employed to factorize multi-view data into a common part and view-specific parts. Beyond, the discriminative and non-discriminative information in these parts are further extracted in a supervised way. In contrast to existing multi-view learning approaches focusing on consistent and (or) complementary information, our new approach, offers an intuitive and effective way to improve classification performance based on the direct discriminative information. The high discriminative power of derived distinct patterns, further demonstrates the effectiveness of DICS on seven multi-view real-world data sets. Although DICS has several desirable properties, it has its own drawbacks. One limitation is that tuning k_i in DICS is quite troublesome, since inferring the subspace dimensionality is still an open problem for all NMF-based algorithms. We simply tune k_i via model selection with traditional strategy. However, once we set proper k_i for each subspace, the promising results can be obtained as we have demonstrated.

Acknowledgments. This work is supported by the National Natural Science Foundation of China (61403062, 41601025, 61433014,), Science-Technology Foundation for Young Scientist of SiChuan Province (2016JQ0007), State Key Laboratory of Hydrology-Water Resources and Hydraulic Engineering (2017490211), National key research and development program (2016YFB0502300).

References

1. Blum, A., Mitchell, T.: Combining labeled and unlabeled data with co-training. In: COLT. pp. 92–100 (1998)
2. Cai, D., He, X., Han, J., Huang, T.S.: Graph regularized nonnegative matrix factorization for data representation. TPAMI **33**(8), 1548–1560 (2011)
3. Chaudhuri, K., Kakade, S.M., Livescu, K., Sridharan, K.: Multi-view clustering via canonical correlation analysis. In: ICML, pp. 129–136 (2009)
4. Chu, M., Diele, F., Plemmons, R., Ragni, S.: Optimality, computation, and interpretation of nonnegative matrix factorizations. SIMAX (2004). http://users.wfu.edu/plemmons/papers/chu_ple.pdf
5. De Sa, V.R.: Spectral clustering with two views. In: ICML Workshop on Learning with Multiple Views, pp. 20–27 (2005)
6. Farquhar, J.D., Hardoon, D.R., Meng, H., Shawe-Taylor, J., Szedmak, S.: Two view learning: SVM-2K, theory and practice. In: NIPS, pp. 355–362 (2005)
7. Gönen, M., Alpaydın, E.: Multiple kernel learning algorithms. JMLR **12**(July), 2211–2268 (2011)
8. Guan, Z., Zhang, L., Peng, J., Fan, J.: Multi-view concept learning for data representation. TKDE **27**(11), 3016–3028 (2015)
9. Gupta, S.K., Phung, D., Adams, B., Tran, T., Venkatesh, S.: Nonnegative shared subspace learning and its application to social media retrieval. In: KDD, pp. 1169–1178 (2010)
10. Gupta, S.K., Phung, D., Adams, B., Venkatesh, S.: Regularized nonnegative shared subspace learning. DMKD **26**(1), 57–97 (2013)
11. Hardoon, D.R., Szedmak, S., Shawe-Taylor, J.: Canonical correlation analysis: an overview with application to learning methods. Neural Comput. **16**(12), 2639–2664 (2004)
12. Kan, M., Shan, S., Zhang, H., Lao, S., Chen, X.: Multi-view discriminant analysis. TPAMI **38**(1), 188–194 (2016)

13. Kim, H., Choo, J., Kim, J., Reddy, C.K., Park, H.: Simultaneous discovery of common and discriminative topics via joint nonnegative matrix factorization. In: KDD, pp. 567–576 (2015)
14. Kim, J., He, Y., Park, H.: Algorithms for nonnegative matrix and tensor factorizations: a unified view based on block coordinate descent framework. JGO **58**(2), 285–319 (2014)
15. Kumar, A., Daumé, H.: A co-training approach for multi-view spectral clustering. In: ICML, pp. 393–400 (2011)
16. Kumar, A., Rai, P., Daume, H.: Co-regularized multi-view spectral clustering. In: NIPS, pp. 1413–1421 (2011)
17. Lee, H., Yoo, J., Choi, S.: Semi-supervised nonnegative matrix factorization. IEEE Sig. Process. Lett. **17**(1), 4–7 (2010)
18. Liu, J., Wang, C., Gao, J., Han, J.: Multi-view clustering via joint nonnegative matrix factorization. In: SDM, pp. 252–260 (2013)
19. Liu, J., Jiang, Y., Li, Z., Zhou, Z.H., Lu, H.: Partially shared latent factor learning with multiview data. TNNLS **26**(6), 1233–1246 (2015)
20. Nie, F., Li, J., Li, X.: Parameter-free auto-weighted multiple graph learning: a framework for multiview clustering and semi-supervised classification. In: IJCAI (2016)
21. Nigam, K., McCallum, A.K., Thrun, S., Mitchell, T.: Text classification from labeled and unlabeled documents using EM. Mach. Learn. **39**(2), 103–134 (2000)
22. Shao, J., Meng, C., Tahmasian, M., Brandl, F., Yang, Q., Luo, G., Luo, C., Yao, D., Gao, L., Riedl, V., et al.: Common and distinct changes of default mode and salience network in schizophrenia and major depression. Brain Imaging Behav. 1–12 (2018). https://doi.org/10.1007/s11682-018-9838-8
23. Shao, J., Myers, N., Yang, Q., Feng, J., Plant, C., Böhm, C., Förstl, H., Kurz, A., Zimmer, C., Meng, C., et al.: Prediction of Alzheimer's disease using individual structural connectivity networks. Neurobiol. Aging **33**(12), 2756–2765 (2012)
24. Shao, J., Yang, Q., Wohlschlaeger, A., Sorg, C.: Discovering aberrant patterns of human connectome in Alzheimer's disease via subgraph mining. In: ICDMW, pp. 86–93 (2012)
25. Shao, J., Yu, Z., Li, P., Han, W., Sorg, C., Yang, Q.: Exploring common and distinct structural connectivity patterns between schizophrenia and major depression via cluster-driven nonnegative matrix factorization. In: ICDM (2017)
26. Sharma, A., Kumar, A., Daume, H., Jacobs, D.W.: Generalized multi-view analysis: a discriminative latent space. In: CVPR, pp. 2160–2167 (2012)
27. Wang, H., Yang, Y., Li, T.: Multi-view clustering via concept factorization with local manifold regularization. In: ICDM, pp. 1245–1250 (2016)
28. Wang, W., Zhou, Z.H.: A new analysis of co-training. In: ICML, pp. 1135–1142 (2010)
29. Xia, T., Tao, D., Mei, T., Zhang, Y.: Multiview spectral embedding. IEEE Trans. Syst. Man Cybern. Part B (Cybern.) **40**(6), 1438–1446 (2010)
30. Xu, C., Tao, D., Xu, C.: A survey on multi-view learning. arXiv preprint arXiv:1304.5634 (2013)
31. Ye, H.J., Zhan, D.C., Miao, Y., Jiang, Y., Zhou, Z.H.: Rank consistency based multi-view learning: a privacy-preserving approach. In: CIKM, pp. 991–1000 (2015)
32. Zhang, M.L., Zhou, Z.H.: CoTrade: confident co-training with data editing. IEEE Trans. Syst. Man Cybern. Part B (Cybern.) **41**(6), 1612–1626 (2011)
33. Zhou, D., Burges, C.J.: Spectral clustering and transductive learning with multiple views. In: ICML, pp. 1159–1166 (2007)

Efficient Discovery of Embedded Patterns from Large Attributed Trees

Xiaoying Wu[1] and Dimitri Theodoratos[2(✉)]

[1] Computer School, Wuhan University, Wuhan, China
xiaoying.wu@whu.edu.cn
[2] New Jersey Institute of Technology, Newark, USA
dth@njit.edu

Abstract. Discovering informative patterns deeply hidden in large tree datasets is an important research area that has many practical applications. Many modern applications and systems represent, export and exchange data in the form of trees whose nodes are associated with attributes. In this paper, we address the problem of mining frequent embedded attributed patterns from large attributed data trees. Attributed pattern mining requires combining tree mining and itemset mining. This results in exploring a larger pattern search space compared to addressing each problem separately. We first design an interleaved pattern mining approach which extends the equivalence-class based tree pattern enumeration technique with attribute sets enumeration. Further, we propose a novel layered approach to discover all frequent attributed patterns in stages. This approach seamlessly integrates an itemset mining technique with a recent unordered embedded tree pattern mining algorithm to greatly reduce the pattern search space. Our extensive experimental results on real and synthetic large-tree datasets show that the layered approach displays, in most cases, orders of magnitude performance improvements over both the interleaved mining method and the attribute-as-node embedded tree pattern mining method and has good scaleup properties.

1 Introduction

Trees are used for representing data in a plethora of applications ranging from computational biology, genome and chemistry compound analysis to ontologies, scientific workflows and business process management. Because of their flexibility in representing data they have been promoted into the standard format for exporting, exchanging and integrating data on the web (e.g., JSON) and as the core data model for databases including NoSQL databases (e.g., MongoDB). The need to analyze data has triggered the last years extensive research on mining patterns from tree data. Over the years the extracted tree patterns have evolved from *induced* patterns [2,5] to *embedded* patterns [16–18]. Embedded patterns generalize induced patterns: while induced patterns involve parent-child edges and are mapped to the data tree using isomorphisms, embedded patterns involve

© Springer International Publishing AG, part of Springer Nature 2018
J. Pei et al. (Eds.): DASFAA 2018, LNCS 10828, pp. 558–576, 2018.
https://doi.org/10.1007/978-3-319-91458-9_34

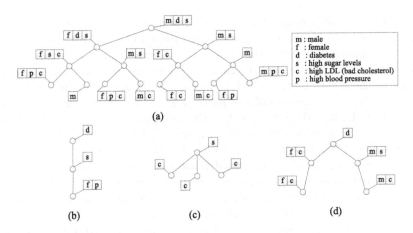

Fig. 1. (a) An attributed tree T (b), (c), (d) embedded attributed frequent patterns on T (minsup = 2).

ancestor-descendant edges and are mapped to the data tree using embeddings. As such, embedded patterns are able to extract relationships "hidden" (or embedded) deeply within data trees which might be missed by induced patterns [17,18]. However, mining embedded patterns is computationally more challenging than mining induced patterns.

Besides structural information such as nodes and edges, tree structured data often contains also attributes. A node represents an entity (or an object) and is associated with a number of attributes (or features) which represent properties of this entity. We refer to these data trees as *attributed trees*. Attributed trees and graphs are useful in many application domains. For instance, in social networks, individuals have many characteristics. In webpage browsing networks, webpages are characterized by a number of keywords. In publication networks, various features like author, title and keywords are recorded for every publication. A simplified example of an attributed tree is shown in Fig. 1(a) where a data tree records diseases and medical conditions (diabetes, high blood pressure, high LDL etc.) and other properties for a person and his ancestors. Every node represents a person and the medical conditions and diseases of the person are shown as attributes of the node. If attributed trees are to be mined, the embedded patterns to be extracted need to be defined appropriately so that attributes in the patterns are associated with nodes which have these attributes. Figures 1(b), (c) and (d) show three examples of frequent embedded patterns extracted from the attributed tree of Fig. 1(a) assuming that the frequency threshold is two. Figure 1(b) says that it is frequent to have a patient with diabetes who has an ancestor with high sugar levels who, in turn, has a female ancestor with high LDL. Similarly the pattern of Fig. 1(c) says that a patient with high sugar levels who has three ancestors with high LDL is also a frequent case. Note that edges in the patterns are ancestor-descendant edges, that is, they can be mapped to paths in the data tree.

As the attributes associated with a node in a tree pattern constitute an itemset, extracting patterns from attributed trees involves two types of mining: structural mining (for extracting the tree patterns) and itemset mining (for extracting the attribute sets of the nodes) [1]. Most previous approaches address the one or the other problem. Real life application though, like those mentioned above, require the extraction of tree patterns from attributed trees whose nodes are associated with attribute sets. Clearly, this problem is more difficult since it combines the identification of frequent embedded tree structures with the identification of frequent itemsets associated with the nodes.

Algorithms which mine embedded patterns from trees cannot be directly applied to attributed trees by representing attributes as (another type of) nodes. Indeed, as the attributes of a node in a pattern should be associated with the node that has these attributes and not with an ancestor of this node, these algorithms do not have a way to enumerate attributed tree patterns. A number of papers address the problem of mining attributed patterns from attributed graphs [11,12]. However, graph mining algorithms cannot efficiently mine patterns from trees: as they are designed for the more general setting of graphs, they cannot take advantage of the specificities of trees. In addition, these algorithms mine not only embedded tree patterns, but also non-tree graph patterns (tree patterns extended with transitive edges). As they generate and output too many useless frequent patterns, their performance on trees is degraded.

In this paper, we present novel algorithms for mining unordered embedded patterns from attributed trees. The main contributions of the paper are the following:

- We formally define the problem of mining unordered embedded tree patterns from large attributed tree data. Our formalization of the problem covers the extraction of patterns from a collections of small attributed trees but also from a single large attributed tree (Sect. 2).
- We design an interleaved approach for mining attributed patterns, which extends the well-known equivalence-class based tree pattern enumeration technique with attribute set enumeration to systematically generate candidate attributed tree patterns (Sect. 4).
- We further propose a novel layered approach, which, instead of explicitly enumerating candidate attributed tree patterns, discovers the frequent attributed patterns in stages. This approach seamlessly integrates the itemset mining technique with a recent unordered embedded tree pattern mining algorithm to greatly reduce the pattern search space (Sect. 5).
- We run extensive experiments to evaluate the performance and scalability of our approaches on real and synthetic datasets. The experimental results show that the layered approach mines embedded attributed patterns up to several orders of magnitude faster than both the interleaved mining method and the attribute-as-node embedded tree pattern mining method at low frequency thresholds. Further, the layered approach scales smoothly in terms of execution time and space consumption empowering the extraction of patterns from large attributed datasets (Sect. 6).

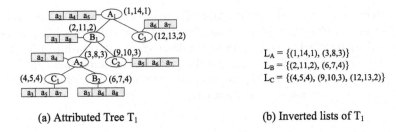

(a) Attributed Tree T₁   (b) Inverted lists of T₁

$L_A = \{(1,14,1), (3,8,3)\}$
$L_B = \{(2,11,2), (6,7,4)\}$
$L_C = \{(4,5,4), (9,10,3), (12,13,2)\}$

Fig. 2. An attributed tree and its inverted lists.

2   Definitions and Problem Statement

Trees. Let $\mathcal{A}$ denote a set of attributes $\{a_1, a_2, \ldots, a_n\}$, where each attribute takes its value from a finite domain. A *rooted attributed tree*, $T = (V, E)$, is a directed acyclic connected graph consisting of a set of nodes V and a set of edges $E \subseteq V \times V$ satisfying the following properties: (1) there is a distinguished node called the *root* that has no incoming edges; (2) there is a unique path from the root to any other node; (3) there is a labeling function *lb* mapping nodes to labels; and (4) there is a function *av* assigning a subset of $\mathcal{A}$ to each node of V. A tree is called *ordered* if it has a predefined left-to-right ordering among the children of each node. Otherwise, it is *unordered*. The *size* of a tree is defined as the number of its nodes.

In this paper, unless otherwise specified, a tree is a rooted, labeled, unordered, attributed tree.

Data Tree Encoding Scheme. We assume that the input data tree T is preprocessed and the position of every node is encoded following the regional encoding scheme [3]. According to this scheme, every node in T is associated with its positional representation which is a *(begin, end, level)* triple. The fields *begin* and *end* in the triple of a node correspond to the order of the first and the last visit of this node in a depth first traversal of T. The region encoding allows efficiently checking structural relationships between two nodes. For instance, node n_1 is an ancestor of node n_2 iff $n_1.begin < n_2.begin$, and $n_2.end < n_1.end$; node n_1 is the parent of node n_2 iff $n_1.begin < n_2.begin$, $n_2.end < n_1.end$, and $n_1.level = n_2.level-1$. For every label A in T, an inverted list L_A of the positional representations of the nodes with label A is produced, ordered by their *begin* field. Figure 2(a) shows an attributed data tree T_1 and the positional representation of its nodes. Subscripts are used in the labels of nodes in T_1 to distinguish between nodes with the same label (e.g., nodes A_1 and A_2 whose label is A). Figure 2(b) shows the inverted lists of T_1's labels. In the following, and depending on the context, we use the same symbol T to refer interchangeably to the tree T and to its set of inverted lists.

Tree Patterns. A tree pattern is a tree. There are two types of mined tree patterns in the literature: patterns whose edges represent child relationships (*child edges*) and patterns whose edges represent descendant relationships (*descendant edges*). The latter are more general, as a child relationship subsumes a descendant relationship between two nodes. In this paper, we focus on embedded patterns (defined next) which have descendant edges.

Tree Morphisms. A tree morphism determines if a tree pattern is included in a tree. Given a pattern P and a tree T, an *embedding* from P to T is an injective function m mapping nodes of P to nodes of T, such that: (1) for any node $x \in P$, $lb(x) = lb(m(x))$; (2) (x, y) is a child edge in P *iff* $m(x)$ is a child of $m(y)$ in T, while x is an ancestor of y in P *iff* $m(x)$ is an ancestor of $m(y)$ in T; and (3) for any node $x \in P$, $av(x) \subseteq av(m(x))$.

Patterns with descendant edges mined using embeddings are called *embedded* patterns [17,18]. Patterns with child edges mined using isomorphisms are qualified as *induced* [2,5].

Support. The *support* of a pattern P on a data tree T is defined as the minimum cardinality of the image sets of the pattern nodes under all possible embeddings of P to T. This support definition is popularly adopted when mining a single large graph/tree [7]. It enjoys also the antimonotonicity property that is central in pattern mining algorithms for a substantial pruning of the pattern search space.

Problem Statement. Given a large rooted labeled attributed tree T and a minimum support threshold *minsup*, our goal is to mine all the frequent unordered embedded attributed patterns.

3   Mining Embedded Tree Patterns from Non-attributed Trees

In this section, we briefly describe an algorithm called *embTM* presented in [16]. Algorithm *embTM* mines embedded patterns from rooted labeled trees without attributes. We will discuss in the following sections how to extend *embTM* to mine embedded attributed patterns from attributed trees. We have chosen *embTM* as extension basis since it displays orders of magnitude performance improvements over a state-of-the-art embedded tree mining algorithm *sleuth* [17].

Algorithm *embTM* works by iterating between the candidate generation phase and the support counting phase. In the first phase, it uses a systematic way to generate candidate patterns that are potentially frequent. In the second phase, it incrementally computes the support of candidate patterns. The main techniques used in the two phases are briefly reviewed below.

3.1   Candidate Pattern Generation

Concepts and Notation. Let P be a tree pattern. Each node of P is identified by its *depth-first position* in the tree, determined through a depth-first traversal of P, by sequentially assigning numbers to the first visit of the node. The *rightmost leaf* of P, denoted $rml(P)$, is the node with the highest depth-first position. The *immediate prefix* of P is the sub-pattern of P obtained by deleting $rml(P)$. A sub-pattern of P obtained by a sequence of deletions of $rml(P)$ is called *prefix* of P.

Equivalence Classes. In order to systematically generate candidate patterns, *embTM* adopts a method called the *equivalence class-based pattern generation* introduced in [17,18]. It is briefly outlined below.

Let the term *k-pattern* refer to a pattern of size k. Given a (k-1)-pattern P ($k \geq 1$), its *equivalence class* is the set of all k-patterns that have P as immediate prefix. The equivalence class of P is denoted as $[P]$. The notation P_X^i denotes a k-pattern in $[P]$ formed by adding a child node labeled by x to the node with position i in P as the rightmost leaf node.

Pattern Join Operations. The equivalence class-based pattern generation works by joining patterns from the same equivalence class. A join operation can produce one or two result patterns. One option is to make the right operand's rightmost leaf as the rightmost cousin/sibling of the left operand's rightmost leaf. The other option is to make the rightmost leaf of the right operand as the rightmost child of the left operand's rightmost leaf. Formally, let P_X^i and P_Y^j denote any two elements in $[P]$. The join operation $P_X^i \otimes P_Y^j$ is defined as follows:

- if $j = i$, return the pattern Q_Y^{k-1} where $Q = P_X^i$.
- if $j \leq i$, return the pattern Q_Y^j where $Q = P_X^i$.

The join operation is not defined if $i < j$. The two result patterns Q_Y^{k-1} and Q_Y^j are called *child* and *cousin* expansions of P_X^i by P_Y^j, respectively. Notice that for each result pattern its immediate prefix is its left-parent, and its rightmost leaf is the rightmost leaf of its right-parent.

Equivalence Class-Based Pattern Generation. The main idea of this expansion is to *join* each pattern $P_X^i \in [P]$ with any other pattern in $[P]$, including itself (self expansion), to produce the patterns of the equivalence class $[P_X^i]$. The equivalence class expansion method correctly generates all possible labeled, ordered, candidate tree patterns and each candidate is generated at most once [18].

Canonical Form. An unordered pattern may have multiple alternative iso-morphic representations. To avoid the redundant generation of the isomorphic representations of the same pattern, a further step is needed to check if the newly generated candidate is in canonical form. The concept of canonical form of a tree is used by pattern mining algorithms as an ordered representative of the corresponding unordered pattern [16,17]. A detailed study of various canonical representations of trees can be found in [6].

3.2   Support Computation

Core Idea. In order to compute the support of P on T, Algorithm $embTM$ first computes the homomorphic matches of P under all possible homomorphisms of P to T, and then, filters out non-embedded matches to retain the embedded ones. Embeddings are special cases of homomorphisms. Unlike embeddings, homomor-phisms do not require the mapping from nodes of P to nodes of T to be injective, and they allow two distinct sibling nodes in P mapping to two nodes on the same path in T. Because of the relaxations, finding a homomorphism from P to T can be done in PTIME [10], whereas the problem of finding an unordered embedding of P to T is NP-Complete [8].

In more detail, in the first phase, $embTM$ incrementally computes the homo-morphisms of a pattern using the materialized homomorphisms of its parent patterns and other previously computed patterns. Instead of materializing the homomorphic matches of previously considered patterns, $embTM$ materializes the homomorphic instances for each node of the pattern. In the second phase, $embTM$ uses a polynomial procedure to filter out non-embedded pattern matches from the homomorphic ones. The 2-phase procedure was further extended to prune non-embedded node instances of the nodes of the pattern without explic-itly generating the homomorphic matches of the pattern.

3.3   Algorithm $embTM$

Algorithm $embTM$ is outlined in Fig. 3. It starts by finding the set F_1 of frequent 1-patterns. Set F_1 and the inverted lists of the dataset are used to compute the set F_2. Then, using the equivalence class-based pattern expansion method, the algorithm recursively expands all the patterns in F_2. The pattern search space is traversed in a depth-first manner.

In more detail, the main loop starts by calling the procedure $MineEmbPat$-$terns$ on every frequent 2-pattern (lines 3–4). Before expanding a pattern P_X^i, $MineEmbPatterns$ makes sure that P_X^i is in canonical form (line 2). It tries to expand P_X^i with every pattern $P_Y^j \in [P]$ and computes the support of each possible expansion outcome using function $IsFrequent$ (lines 4–6). The inputs of $IsFrequent$ are the pattern under consideration, and the occurrence list sets (defined in the next section) of its parent patterns on the input data tree. The detailed description of function $IsFrequent$ can be found in [16].

Any new pattern that turns out to be frequent is added to the new class $[P_X^i]$ (line 7). After all patterns P_Y^j have been joined with P_X^i, the class $[P_X^i]$ contains

---

Input: inverted lists $\mathcal{L}$ of tree T and *minsup*.
Output: all the frequent embedded tree patterns in T.

1. $F_1 := \{\text{frequent 1-patterns}\}$;
2. $F_2 := \{\text{equivalent classes } [P] \text{ of frequent 2-patterns}\}$;
3. **for** (every $[P] \in F_2$) **do**
4. $MineEmbPatterns([P])$;

Procedure $MineEmbPatterns$(Equivalence class $[P]$)
1. **for** (each $P_X^i \in [P]$) **do**
2.   **if** (P_X^i is in canonical form) **then**
3. $[P_X^i] := \emptyset$;
4.   **for** (each $P_Y^j \in [P]$) **do**
5.   **for** (each expansion outcome Q of $P_X^i \otimes P_Y^j$) **do**
6.   **if** (IsFrequent(Q, $OL(P_X^i)$, $OL(P_Y^j)$)) **then**
7.   add Q to $[P_X^i]$;
8. $MineEmbPatterns([P_X^i])$

Fig. 3. Algorithm *embTM* for mining embedded tree patterns.

all the frequent tree patterns sharing the immediate prefix P_X^i. The procedure is then called on $[P_X^i]$ to discover larger frequent tree patterns having P_X^i as prefix (line 8). The recursive process is repeated until no more frequent patterns can be generated.

4 Mining Embedded Attributed Patterns: An Interleaved Approach

Based on *embTM*, we design an embedded attributed pattern mining algorithm called *embATM-inter*. Figure 4 shows the outline of Algorithm *embATM-inter*. The main difference between *embATM-inter* and *embTM* is on their candidate pattern enumeration. To generate candidate attributed patterns, *embATM-inter* extends the equivalence class-based pattern generation with attribute sets enumeration. The idea is to interleave the pattern expansion with the enumeration of attribute sets for pattern nodes during the pattern mining process.

Without loss of generalization, in the following discussion, we assume that each node in the data tree T is associated with a non-empty subset of the attribute set $\mathcal{A}$.

Concepts and Notation. We call pattern nodes that are associated with an non-empty attribute set as *attributed* pattern nodes. A pattern is called *attributed* pattern if every node is attributed. Let the term *k-pattern* refer to a pattern with k nodes; F_k and AF_k, $k \geq 2$, denote respectively the set of the equivalence classes of frequent k-patterns and attributed k-patterns.

Given a pattern P and a data tree T, an *occurrence* of P on T is a tuple indexed by the nodes of P whose values are the images of the corresponding nodes in P under an embedding of P to T. Remember that an embedding needs to satisfy the three conditions given in the definition (Ref. Sect. 2). The set of occurrences of P under all possible embeddings of P to T is a relation OC whose schema is the set of nodes of P.

If X is a node in P labeled by label A, the *occurrence list of X on T* is a sublist L_X of the inverted list L_A containing only those nodes that occur in the column for X in OC. The set of all the occurrence lists of the nodes of P on T is denoted by OL, that is, $OL = \{L_X \mid X \in nodes(P)\}$.

Let X be a node in P which is not yet attributed, let also L_X be its occurrence list. After attaching an non-empty attribute set $S\,(\subseteq \mathcal{A})$ to X, the *occurrence list of S w.r.t X on T*, denoted by $L_{X(S)}$, is $\{x \mid x \in L_X, S \subseteq av(x)\}$. We call S a *frequent attribute set* of X, if $|L_{X(S)}|$ is not less than *minsup*.

Attributed Pattern Nodes Enumeration. We first discuss how to generate AF_k, where $k \geq 2$. Let P_X^i be a candidate k-pattern ($k \geq 2$) in class $[P]$. Every pattern in class $[P]$ has the common property that, every node except for the rightmost leaf has been attributed. *embATM-inter* iteratively computes attribute sets from all the attributes to which the rightmost leaf node X of P_X^i is mapped in T (Lines 2–3 in Procedure *MineAttrEmbPatterns*). Occurrence list of attribute sets are computed based on occurrence list L_X via a call to function *GetAttrOccList*. We will talk about the efficient computation of attribute occurrence lists in the next section. Each computed attribute set that is frequent is attached to node X of P_X^i. The procedure for checking whether an attributed set is frequent is not shown for simplicity. Any non-frequent attribute sets of X can be discarded, since by the antimonotonicity property of the pattern support definition, an attributed pattern with an infrequent attribute set must also be infrequent. For each canonical attributed pattern of P_X^i (we will discuss next how to check canonicity for attributed patterns), *embATM-inter* computes its frequency on T (Line 4). The same frequency computation function *IsFrequent* from Algorithm *embTM* can be applied here, using occurrence list set OL of P_X^i and occurrence list $L_{X(S)}$ as the inputs, where S is the current frequent attribute set under consideration. Each frequent attributed pattern $P_{X(S)}^i$ is added to AF_k (Line 5).

Attributed Pattern Expansion. *embATM-inter* proceeds to expand the newly obtained frequent attributed pattern $P_{X(S)}^i$ with candidates P_Y^j from $[P]$ to generate *(k+1)-patterns* (Lines 6–7). Here the equivalence class-based pattern generation method of *embTM* is applied. Each frequent *(k+1)-pattern* is added to class $[P_{X(S)}^i]$ (Lines 8–9). Notice that *embATM-inter* keeps the invariant that only the rightmost leaf node of every *(k+1)-pattern* is not yet attributed. Once all P_Y^j have been processed, *embATM-inter* recursively explores the new class $[P_{X(S)}^i]$ in a depth-first manner (Line 10).

Obtaining AF_1 and F_2. Frequent attributed *1-patterns* in AF_1 can be obtained by first augmenting frequent *1-patterns* with frequent attribute sets and then

Input: Inverted lists $\mathcal{L}$ of tree T, attribute set $\mathcal{A}$ and *minsup*.
Output: All the frequent embedded attributed tree patterns in T.

1. $F_1 := \{\text{frequent 1-patterns}\}$;
2. $AF_1 := GetAF_1(F_1)$;
3. $F_2 := \{\text{equivalent classes } [P] \text{ of frequent 2-patterns}\}$;
4. **for** (every $[P] \in F_2$) **do**
5. $MineAttrEmbPatterns([P], 2)$;

Procedure $GetAF_1$(Frequent 1-patterns F_1)
1. **for** (each $X \in F_1$) **do**
2.   **for** (each subset S of $\mathcal{A}$) **do**
3. $L_{X(S)} := GetAttrOccList(S, X)$;
4.   **if** ($|L_{X(S)}| \geq minsup$) **then**
5.   add $X(S)$ to AF_1;

Function $GetAttrOccList$(Attribute set S, pattern node X)
1. **for** (each $x \in L_X$ in its preorder appearance in T) **do**
2.   **if** ($S \subseteq av(x)$) **then**
3.   add x to $L_{X(S)}$;
4. **return** $L_{X(S)}$;

Procedure $MineAttrEmbPatterns$(Equivalence class $[P]$, size k)
1. **for** (each $P_X^i \in [P]$) **do**
2.   **for** (each subset S of $\mathcal{A}$) **do**
3. $L_{X(S)} := GetAttrOccList(S, X)$;
4.   **if** ($P_{X(S)}^i$ is canonical and IsFrequent($P_{X(S)}^i$, $OL(P_X^i)$, $L_{X(S)}$)) **then**
5.   add $P_{X(S)}^i$ to AF_k and set $[P_{X(S)}^i]$ to be empty;
6.   **for** (each $P_Y^j \in [P]$) **do**
7.   **for** (each expansion outcome Q of $P_{X(S)}^i \otimes P_Y^j$) **do**
8.   **if** (IsFrequent(Q, $OL(P_{X(S)}^i)$, $OL(P_Y^j)$)) **then**
9.   add Q to $[P_{X(S)}^i]$;
10. $MineAttrEmbPatterns([P_{X(S)}^i], k+1)$

Fig. 4. Algorithm *embATM-inter* for mining embedded attributed tree patterns.

computing their occurrence lists (Procedure $GetAF_1$). Frequent *2-patterns* in F_2 can then be computed by joining AF_1 and F_1. We omit the details due to space limitation.

Checking Canonicity of Attributed Patterns. By the antimonotonicity property of the pattern support definition, each frequent attributed k-pattern ($k \geq 2$) must have a subset of AF_1 as its attributed nodes. Based on this observation, we assign each *1-pattern* in AF_1 a unique id and use those ids to identify pattern nodes in attributed k-patterns. By giving an ordering to those ids, we can apply the canonical form checking method for unordered tree patterns to check canonicity of attributed patterns. This way, no explicit definition of a canonical form for attributed patterns is needed. An attributed pattern canonical form that considers labels of both entity nodes and attributes is given in [13].

5   Mining Embedded Attributed Patterns: A Layered Approach

One difficulty with the interleaved approach *embATM-inter* for mining attributed patterns is that it has to maintain a large amount of states to keep track of attribute sets enumerated during the recursive mining process. This will consume significant amount of memory when the size of attribute domains is large, confirmed by the experimental results in Sect. 6. We provide in this section an improved method for finding frequent attributed patterns which successfully addresses this issue. The new method does not alternate between the pattern node expansion and the attribute set enumeration, but instead, it takes a layered approach to generate attributed patterns in stages.

The layered approach is called *embATM-layer*. It first computes a set of frequent *singleton* patterns, which are frequent 1-patterns each of which has a single attribute. Based on the singleton patterns, it adapts an itemset mining algorithm to find the set of frequent attributed 1-patterns. In the final step, it feeds attributed 1-patterns and their occurrence lists into the embedded pattern mining algorithm *embTM* to produce all the possible attributed k-patterns ($k \geq$ 2). The processing flow of *embATM-layer* is summarized in Fig. 5. We describe the three steps in more detail below.

Layer 1: Singleton Pattern Generation. Singleton patterns and their occurrence lists can be obtained in the data preprocessing phase. Recall that before a pattern mining process starts, the input data tree T is preprocessed and every node is associated with a (*begin, end, level*) triple encoding the node's depth-first position in T (Ref. Sect. 2). Also, for every label A in T, an inverted list L_A of triples of the node with label A is produced, ordered by their *begin* field. All the node triples and inverted lists of node labels can be obtained by performing one depth-first traversal of T.

Each singleton pattern is composed of a node label in the input data tree T and an attribute from $\mathcal{A}$. During the depth-first traversal of T, we record each distinct pair of node labels and attributes appearing in T. Each distinct pair is a singleton pattern. Occurrence lists of singleton patterns are also produced

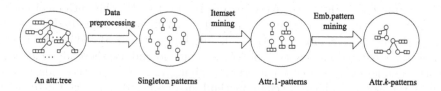

Fig. 5. The processing flow of algorithm *embATM-layer*.

during the depth-first traversal. For each singleton pattern instance encountered during the depth-first traversal, we add the corresponding node triple to the occurrence list of the singleton pattern.

Layer 2: Attributed 1-Pattern Generation. By making each singleton pattern as an item, and each node in T as a transaction identified by its triplet, the problem of mining frequent attributed 1-patterns reduces to the problem of mining frequent itemsets from the database of transactions. Any frequent itemset mining algorithm developed in the literature can be applied here. Once the set of frequent attributed 1-patterns is obtained, we compute the occurrence list of each 1-pattern by intersecting occurrence lists of singleton patterns contained in that 1-pattern.

Layer 3: Attributed k-Pattern Generation. Each k-pattern ($k \geq 2$) is a tree formed by 1-patterns. By making attributed 1-patterns, instead of node labels, as the information unit of pattern mining, we can use the embedded pattern mining algorithm *embTM* to find all the frequent attributed k-patterns. Here, the inputs to *embTM* are attributed 1-patterns and their occurrence lists. No changes need to be made on *embTM* for this purpose.

6   Experimental Evaluation

In this section, we study the performance of the two proposed approaches for mining embedded attributed tree patterns in terms of execution time, memory consumption and scalability. We implemented and compared the following algorithms that mine patterns from large tree data: (1) two versions of the recent embedded tree patterns algorithm which mines all the frequent embedded tree patterns from tree data [16], one which ignores attributes in the data, denoted as *embTM*, and another one which treats attributes as tree nodes, denoted as *embTM(AttrAsNode)*; and (2) the two embedded attributed pattern mining algorithms presented in this paper, one using the interleaved strategy, denoted as *embATM-inter*, and another one using the layered strategy, denoted as *embATM-layer*. All four algorithms employ the techniques described in [16] for computing pattern support and materializing pattern occurrences.

Our implementation was coded in Java. All the experiments reported here were performed on a workstation having an Intel Xeon CPU 3565 @3.20 GHz processor with 8 GB memory running JVM 1.7.0 in Windows 7 Professional. The Java virtual machine memory size was set to 4 GB.

Datasets. We ran experiments on three real and synthetic datasets with different structural properties.

$D10^1$ is a synthetic dataset generated by the tree generation program provided by Zaki [18]. To generate attribute sets for nodes in the tree, we used the

---

[1] http://www.cs.rpi.edu/~zaki/software/.

IBM synthetic dataset generator (See footnote 1) for itemsets and sequences. The results presented here were based on setting which associates each tree node with an attribute set of size ranging from 1 to 5, and each attribute takes its value from a domain of 20 different values in total. We also ran experiments on the dataset with different attribute settings. The results are similar and are omitted due to space limitation.

XMark[2] is an XML benchmark dataset modeling an auction website. The dataset is deep and has many regular structural patterns. It includes a few recursive elements (elements on one root-to-leaf path with the same label).

DBLP[3] is a real XML dataset proving bibliographic information on major CS journals and proceedings. We used a DBLP XML fragment including 50,000 publications provided as a duplicate detection benchmark for XML [14].

When we parse an XML document for mining attributed patterns, we generate the positional representation, that is, the triplets (Ref. Sect. 2) only for elements having attributes and/or subelements, which are called *entity nodes* here. For instance, *article* elements in DBLP are parsed as entity nodes. Empty elements or elements having only PCDATA as their contents are parsed as attributes, such as *year* elements in DBLP. Entity nodes having no attributes are assigned a special (dummy) attribute. During the data preprocessing step, we record also the positional representation of entity nodes for each of their associated attributes. The dataset D10 is processed in the same way.

Value nodes in XMark and DBLP are processed as follows: for XMark, we ignored its *text* elements, whose contents are taken from Shakespeare's plays. For DBLP, in order to mine informative patterns from its contents, we used the Apache OpenNLP Tools[4] to extract noun phrases from the contents of *title* elements, and made each extracted phrase as a value associated with the title attribute of the corresponding publication. The main characteristics of the above three datasets are summarized in Table 1. Columns 2 and 3 show the total number of elements/entity nodes in datasets before and after the data preprocessing step, respectively. The total number of attributes shown in the table does not count dummy attributes.

Table 1. Dataset statistics.

Dataset	#elements	#entity nodes	#attributes	#labels	Max/Avg depth	#paths
D10	272407	27371	51089	100	11/2.4	100K trees
XMark	78314	33208	71375	29964	13/6.4	138840
DBLP	260401	49240	395442	227467	5/3.5	1033285

---

[2] http://xml-benchmark.org.

[3] http://dblp.uni-trier.de/xml/.

[4] https://opennlp.apache.org/.

Time Performance. Figures 6(a), 7(a), and 8(a) show the time spent by *embTM, embTM(AttrAsNode), embATM-inter*, and *embATM-layer* on the D10, XMark, and DBLP, respectively, under different support thresholds. In all the cases, we excluded the data preprocessing time. Notice that, in all the figures, a logarithmic scale is used on the Y-axis. We stopped testing algorithms which were unable to finish within 6 h for the support levels under examination are below certain values on each dataset. Table 2 compares the number of candidates and final frequent patterns generated by the four algorithms under different support thresholds on the three datasets. We have the following observations.

Algorithm *embTM(AttrAsNode)* runs slower than others by at least one order of magnitude in most applicable cases. Also, its runtime increase rate is much sharper than others as the support level decreases. This can be explained as follows. The pattern computation method of [16] used by all the four algorithms is based on the structural join operation [3,15] among entity nodes. Labels of attribute values generally have more variety than labels of entity nodes. When treating attributes as entity nodes, *embTM(AttrAsNode)* clearly needs to perform more structural joins compared to algorithms that compute structural joins among entity nodes only. Further, the larger number of labels can substantially increase the pattern search space. As shown in Table 2, *embTM(AttrAsNode)* not only evaluates substantially more candidates but also produces more frequent patterns. Many of those generated patterns include false information, as they associate the attributes of pattern nodes with their ancestors.

Understandably, Algorithm *embTM* has a better performance than *embTM(AttrAsNode)* in all the testing cases, and in particular on D10, since it only considers entity nodes for mining patterns. This reduces both the pattern search space and the number of structural joins. But on the two XML datasets XMark and DBLP, *embTM* still runs much slower than the two attributed pattern mining algorithms. This can be explained by the following two remarks: (1) *embTM* (and *embTM(AttrAsNode)*) treats every XML element as a tree node. As shown in Table 1, the number of elements is usually larger than the number of entity nodes obtained after the preprocessing step. (2) Attribute labels usually have smaller number of occurrences in the data than tree node labels. When attributes in the data have high selectivity, both the pattern search space and the structural joins can be greatly reduced when mining attributed patterns.

Algorithm *embATM-inter* performs poorly when the size of attribute domains is large. It is because the number of candidate attribute sets enumerated for tree pattern nodes and hence the number candidate patterns considered by *embATM-inter* grows explosively during the mining process. This negatively affects the time performance of *embATM-inter*. For instance, on DBLP, the domain size of attributes such as *title* and *author* is expectedly very large, *embATM-inter* has to evaluates 544 times more candidates than *embATM-layer* when $minsup = 650$ (Table 2).

In contrast, *embATM-layer* is able to mine attributed patterns in reasonable time even at very low support thresholds on each dataset. For instance, it is able to mine 21950 frequent attributed patterns in 433 s when $minsup = 3$, whereas all other approaches fail before $minsup$ reaches 499, either due to prohibitively long

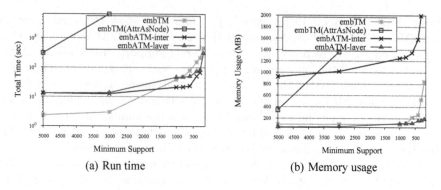

Fig. 6. Performance comparison on D10.

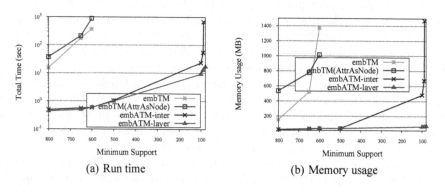

Fig. 7. Performance comparison on XMark.

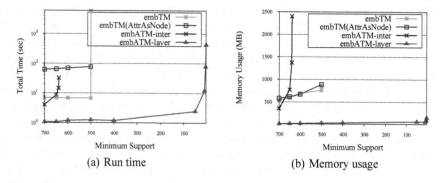

Fig. 8. Performance comparison on DBLP.

execution time or out-of-memory error. This clearly demonstrates the advantages of layered strategy used by *embATM-layer* for mining attributed patterns. The produced frequent patterns on DBLP at low support thresholds provide useful information on tasks such as detecting duplicated publications, searching recent hot research topics, or finding frequent co-authorship patterns.

Table 2. Statistics for frequent patterns mined from the three datasets.

dataset	algorithms	tot.cand. patterns	tot.freq. patterns	dataset	algorithms	tot.cand. patterns	tot.freq. patterns
D10 (minsup=3000)	embTM	498	76	XMark (minsup=600)	embTM	146287	13480
	embTM(AttrAsNode)	9673	1109		embTM(AttrAsNode)	319248	27008
	embATM-inter	754446	107		embATM-inter	264	24
	embATM-layer	8569	107		embATM-layer	189	24
XMark (minsup=90)	embTM	n/a	n/a	DBLP (minsup=650)	embTM	293	55
	embTM(AttrAsNode)	n/a	n/a		embTM(AttrAsNode)	235797	1016
	embATM-inter	16530918	2403		embATM-inter	196606	36
	embATM-layer	30333	2403		embATM-layer	292	36

Memory Usage. Figures 6(b), 7(b), and 8(b) show the memory consumption of the four algorithms on the D10, XMark, and DBLP, respectively, under different support thresholds. Overall, *embATM-layer* has the best memory performance, consuming substantially less memory than the other three algorithms in almost all the test cases.

The large memory consumption of *embTM* and *embTM(AttrAsNode)* is mainly due to the larger number of entity nodes processed as well as the larger amount of candidate patterns evaluated and final frequent patterns produced during the mining process.

Algorithm *embATM-inter* consumes substantially more memory than *embATM-layer*, especially when the attribute domain size is large. It aborted due to out-of-memory error when the support level reaches certain values on each dataset. This is mainly because *embATM-inter* has to maintain a large amount of states to keep track of attribute sets enumerated during the recursive mining process.

Scalability. We studied the scalability of the four algorithms, as we increase the size of input data on XMark. We generated 6 XMark trees by setting $factor = 0.05, 0.06, \ldots, 0.1$. Figure 9 shows the scalability results of the four algorithms when $minsup = 1200$.

As we increase the input data size, the growth of the running time of both *embTM* and *embTM(AttrAsNode)* is much sharper than that of *embATM-inter* and *embATM-layer*. The memory usage of *embTM* and *embTM(AttrAsNode)* is also larger than the *embATM-inter* and *embATM-layer*. The latter two algorithms have similar time performance and memory usage.

We also show in the same figure the scalability results of *embATM-inter* and *embATM-layer* when $minsup = 136$. In this case, both *embATM-inter* and *embATM-layer* were unable to finish within 12 h even on the smallest sized XMark tree. The running time and memory consumption of *embATM-inter* grow more sharply than *embATM-layer*. The latter runs up to 63 times faster than *embATM-inter*, while consuming up to 12 times less memory.

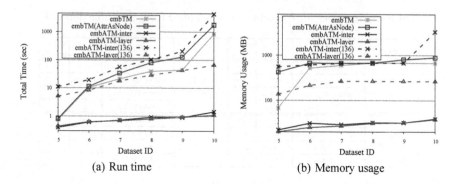

(a) Run time

(b) Memory usage

Fig. 9. Scalability comparison for mining embedded patterns on XMark with increasing size. The solid curves correspond to $minsup = 1200$, the dashed curves correspond to $minsup = 136$.

7   Related Work

The problem of mining tree patterns has been studied since the last decade. Existing work has focused almost exclusively on mining induced and embedded patterns from a set of small trees. The problem of mining unordered embedded tree mining is computationally more challenging than mining induced patterns or ordered embedded patterns. Among the many tree mining algorithms studied in the literature [4–6,16–18], only few mine unordered embedded patterns [16,17].

There is a growing interest in mining itemsets organized in structures. Among the works on mining frequent patterns from attributed graphs [9,11–13], *FAT-miner* [9] and *IMIT* [13] mine attributed tree patterns from a collection of small attributed trees. While *FAT-miner* [9] aims at mining ordered embedded attributed patterns, *IMIT* [13] focuses on mining ordered and unordered induced attributed patterns. Both works proposed a candidate attributed pattern generation method that is based on two operations: itemset expansion and tree expansion. The enumeration method extends the rightmost path expansion [2,6,17] with itemset enumeration. The interleaved approach proposed in this paper also requires explicitly enumerating attributed candidate patterns, but it extends the most recent equivalence class-based pattern generation [16] with itemset enumeration. The existing works on mining graph patterns from attributed graphs [11,12] combine itemset expansion with graph expansion and explicitly enumerate candidate attributed patterns. In contrast to these approaches, our proposed layered approach discovers attributed patterns in stages and avoids enumerating attributed candidate patterns explicitly.

8   Conclusion

We have addressed the important problem of mining unordered embedded tree patterns from large attributed tree data. To the best of our knowledge, none of the previous approaches can mine embedded attributed tree patterns from large data trees.

To cope with this pattern mining problem, we have designed an interleaved approach, which extends the widely-used equivalence-class based tree pattern enumeration technique with attribute sets enumeration. We have further designed a novel layered approach, which, instead of explicitly enumerating candidate attributed tree patterns, discovers all frequent attributed patterns in stages. The layered approach combines different techniques from frequent itemset mining and embedded tree pattern mining to greatly reduce the pattern search space. Our experimental results on real and synthetic datasets show that the layered approach displays orders of magnitude performance improvement over both the interleaved mining method and the attribute-as-node embedded tree pattern mining method for mining embedded attributed tree patterns.

The number of frequent embedded attributed patterns for a given support threshold can be too large for users to understand and analyze. Our future work will focus on incorporating user-specified constraints to the proposed approach in order to enable constraint-based attributed pattern mining. We are also investigating summarization techniques for mining compact sets of frequent attributed tree patterns.

References

1. Aggarwal, C.C., Han, J. (eds.): Frequent Pattern Mining. Springer, Cham (2014). https://doi.org/10.1007/978-3-319-07821-2
2. Asai, T., Abe, K., Kawasoe, S., Arimura, H., Sakamoto, H., Arikawa, S.: Efficient substructure discovery from large semi-structured data. In: SDM (2002)
3. Bruno, N., Koudas, N., Srivastava, D.: Holistic twig joins: optimal XML pattern matching. In: SIGMOD (2002)
4. Chehreghani, M.H., Bruynooghe, M.: Mining rooted ordered trees under subtree homeomorphism. Data Min. Knowl. Discov. **30**(5), 1249–1272 (2016)
5. Chi, Y., Muntz, R.R., Nijssen, S., Kok, J.N.: Frequent subtree mining - an overview. Fundam. Inform. **66**(1–2), 161–198 (2005)
6. Chi, Y., Yang, Y., Muntz, R.R.: Canonical forms for labelled trees and their applications in frequent subtree mining. Knowl. Inf. Syst. **8**(2), 203–234 (2005)
7. Elseidy, M., Abdelhamid, E., Skiadopoulos, S., Kalnis, P.: GRAMI: frequent subgraph and pattern mining in a single large graph. PVLDB **7**(7), 517–528 (2014)
8. Kilpeläinen, P., Mannila, H.: Ordered and unordered tree inclusion. SIAM J. Comput. **24**(2), 340–356 (1995)
9. Knijf, J.D.: FAT-miner: mining frequent attribute trees. In: SAC, pp. 417–422 (2007)
10. Miklau, G., Suciu, D.: Containment and equivalence for a fragment of XPath. J. ACM **51**(1), 2–45 (2004)
11. Miyoshi, Y., Ozaki, T., Ohkawa, T.: Frequent pattern discovery from a single graph with quantitative itemsets. In: ICDM Workshops, pp. 527–532 (2009)
12. Pasquier, C., Flouvat, F., Sanhes, J., Selmaoui-Folcher, N.: Attributed graph mining in the presence of automorphism. Knowl. Inf. Syst. **50**(2), 569–584 (2017)

13. Pasquier, C., Sanhes, J., Flouvat, F., Selmaoui-Folcher, N.: Frequent pattern mining in attributed trees: algorithms and applications. Knowl. Inf. Syst. **46**(3), 491–514 (2016)

14. Weis, M., Naumann, F., Brosy, F.: A duplicate detection benchmark for xml (and relational) data (2006)

15. Wu, X., Souldatos, S., Theodoratos, D., Dalamagas, T., Sellis, T.K.: Efficient evaluation of generalized path pattern queries on XML data. In: WWW (2008)

16. Wu, X., Theodoratos, D.: Leveraging homomorphisms and bitmaps to enable the mining of embedded patterns from large data trees. In: Renz, M., Shahabi, C., Zhou, X., Cheema, M.A. (eds.) DASFAA 2015. LNCS, vol. 9049, pp. 3–20. Springer, Cham (2015). https://doi.org/10.1007/978-3-319-18120-2_1

17. Zaki, M.J.: Efficiently mining frequent embedded unordered trees. Fundam. Inform. **66**(1–2), 33–52 (2005)

18. Zaki, M.J.: Efficiently mining frequent trees in a forest: algorithms and applications. IEEE Trans. Knowl. Data Eng. **17**(8), 1021–1035 (2005)

Classification Learning from Private Data in Heterogeneous Settings

Yiwen Nie$^{(\boxtimes)}$, Shaowei Wang, Wei Yang, Liusheng Huang, and Zhenhua Zhao

University of Science and Technology of China, Hefei, China
{nyw2016,wangsw,hzq}@mail.ustc.edu.cn, {qubit,lshuang}@ustc.edu.cn

Abstract. Classification is useful for mining labels of data. Though well-trained classifiers benefit many applications, their training procedures on user-contributed data may leak users' privacy.

This work studies methods for private model training in heterogeneous settings, specially for the Naïve Bayes Classifier (NBC). Unlike previous works focusing on centralized and consistent datasets, we consider the private training in two more practical settings, namely the local setting and the mixture setting. In the local setting, individuals directly contribute training tuples to the untrusted trainer. In the mixture setting, the training dataset is composed of individual tuples and statistics of datasets from institutes. We propose a randomized response based NBC strategy for the local setting. To cope with the privacy of heterogeneous data (single tuples and the statistics) in the mixture setting, we design a unified privatized scheme. It integrates respective sanitization strategies on the two data types while preserving privacy. Besides contributing error bounds of estimated probabilities constituting NBC, we prove their optimality in the minimax framework and quantify the classification error of the privately learned NBC. Our analyses are validated with extensive experiments on real-world datasets.

Keywords: Differential privacy · Classification

1 Introduction

With the vast penetration of Internet and mobile devices, massive and diverse data are generated. Classification, as a useful learning tool, has been universally applied on these data to build relationship among data attributes (e.g., intrusion detection). Among many types of classification methods, the Naïve Bayes Classifier is a simple but effective one. By assuming that data features are conditionally independent on a class, NBC estimates the relevance between feature values and classes on the training set and uses Bayes theorem to classify unlabeled data. Yet its training procedure usually relies on individual data, which may compromise people's privacy (e.g., locations and medical history) since the trainer is usually curious and not always trustful.

© Springer International Publishing AG, part of Springer Nature 2018
J. Pei et al. (Eds.): DASFAA 2018, LNCS 10828, pp. 577–585, 2018.
https://doi.org/10.1007/978-3-319-91458-9_35

Recently, local differential privacy (LDP) [10] has emerged into a de facto notion to handle data privacy in the local setting. LDP privatizes data independently on the individual side before sharing, and ensures that snoopers cannot distinguish users' true data by observing perturbed ones. The practical setting and protective strength of LDP motivate many related private statistical methods (e.g., [1,3,6]). However, most of existing LDP works concern about basic estimations (e.g., mean) and seldom design strategies for widely-used learning tools (e.g., classifiers) which are actually more close to individual privacy.

Moreover, the local setting is not always feasible in actual scenarios, since except for individuals, entities with raw datasets (e.g., census department) also offer data for analyses. Hence, the mixture setting which combines the distributed setting and the local setting is considerable in actual scenarios. Current distributed privacy schemes [9,12] generally exploit secure multiparty computation (SMC) [8]. It has high computational/interactive costs which cannot be afforded by resource-constrained individuals. Due to the data type difference, existing LDP approaches cannot be extended on datasets' statistics either.

In this paper, we design accurate and efficient schemes for the private training of NBC in both the local setting and the mixture setting. In the local setting, a randomized response strategy is applied to collect individual tuples and train NBC privately. In the mixture setting, a general strategy is proposed, filling in the gap caused by high computational/interactive cost and data type difference. By rigorous proofs, all these strategies are demonstrated optimal on utility in the minimax framework.

In summary, our contributions are listed as follows:

- *Effective private training strategies.* In the local setting, we give a *Utility-First Strategy* (UFS) to privately train NBC on individual categorical tuples. In the mixture setting, we propose a simple but effective method, namely *Expanded Geometric Strategy* (EGS). To the best of our knowledge, it is the first time that different data types are simultaneously considered in one private scheme of the model learning.
- *Provable optimality guarantees and quantified classification error.* By the quantified ℓ_2-norm error, we prove the optimality of UFS and EGS in the minimax framework. Following Bayes theorem, classification errors of UFS and EGS are formally presented. This error provides a reasonable estimator to measure the utility loss of a classifier resulting from a private mechanism.
- *Experimental validation.* The performances of our strategies are evaluated over a simulative dataset and two real-world datasets. The results experimentally confirm theoretical conclusions about two strategies.

2   System Model

2.1   Problem Definition

In the local setting, there are N individuals; in the mixture setting, there are N_1 individuals and N_2 institutes. The institutes are trustworthy that they own

Table 1. Notations

Symbol	Description
$\mathcal{X}_j, M_j, t$	The j-th feature domain $\{a_1^j, a_2^j, \ldots, a_{M_j}^j\}$ ($|\mathcal{X}_j| = M_j, 1 \leqslant j \leqslant t$)
$\mathcal{Y}$	The class domain $\{c_1, c_2, \ldots, c_K\}$ and $\mathcal{Y} = K$
$\mathcal{U}_{c_k}$	The set of users with label c_k
$\mathbf{x}_i[j]$	The j-th feature of user i
$\mathbf{binx}_i^j$	The binary basis vector of length M_j representing $\mathbf{x}_i[j]$
N_{c_k}	The amount of data with label c_k in the training set
$\mathbf{p}_{jk}[s_j]$	The true conditional probability of $\mathbf{x}[j] = a_{s_j}^j$ on the class c_k

datasets of exact individuals' tuples. The NBC trainer is assumed to be curious that it would not attack data providers for more information but would snoop on the received data as much as possible. The data formation is described as $d_i = (\mathbf{x}_i, Y_i)$, where $\mathbf{x}_i$ is the feature vector and Y_i is the label. We privatize the feature vector and leave the label public in this paper.

Individuals and institutes send sanitized tuples and privatized statistics of datasets to the trainer. By the trained results (i.e., $P(\mathbf{x}[j]|Y = c_k)$), the trainer classify the unlabeled data with $Y = \arg\max_{c_k} P(Y = c_k|\mathbf{x})$, where $P(Y = c_k|\mathbf{x}) = \frac{P(Y = c_k) \prod_j P(\mathbf{x}[j]|Y = c_k)}{\sum_k P(Y = c_k) \prod_j P(\mathbf{x}[j]|Y = c_k)}$. The denotations used in this paper are presented in Table 1.

2.2 Privacy Model

We adopt two forms of differential privacy (DP) [4] as definitions in this paper; the classical notion of DP for the statistics of datasets and LDP for a data tuple.

Definition 1 (DP). *A is a randomized mechanism over any neighboring datasets D' and D. Let z be a possible output of A. A is said to satisfy ϵ-differential privacy, if $\frac{Pr[\mathcal{A}(D) = z]}{Pr[\mathcal{A}(D') = z]} \leqslant e^\epsilon$.*

Neighboring datasets used in the definition refers to two datasets D and D' that D' can be obtained from D by adding, removing or modifying a tuple.

Definition 2 (LDP [10]) *A randomized mechanism $\mathcal{A}$ is said to satisfy the ϵ-LDP, if for any pair of tuples $v, v' \in V$, and for any output z, $\frac{Pr[\mathcal{A}(v) = z]}{Pr[\mathcal{A}(v') = z]} \leqslant e^\epsilon$.*

Theorem 1 (Post-processing theorem [5]). *For any method ψ which works on the output of a ϵ-differentially private mechanism $\mathcal{M}$ without accessing the raw data, the integrated procedure $\psi \circ \mathcal{M}$ remains ϵ-differentially private.*

Algorithm 1. Utility-First User (UFU)

Require: User data $\{\mathbf{binx}_i^j\}_{j=1}^t$, ϵ.
Ensure: Sanitized vector set $< \{\mathbf{z}_i^j\}_{j=1}^t, Y_i >$.
1: **for** $j = 1$ to t and $s = 1$ to M_j **do**
2:     Compute every bit of the privatized vector $\mathbf{z}_i^j$ independently with Eq. (1);
3: **end for**
4: Send $< \{\mathbf{z}_i^j\}_{j=1}^t, Y_i >$ to the trainer.

Algorithm 2. Utility-First Trainer (UFT)

Require: $\bigcup_{i=1}^N < \{\mathbf{z}_i^j\}_{j=1}^t, Y_i >$, ϵ.
Ensure: Classifier $\mathcal{C} = \bigcup_{\mathbf{x}[j]\in\mathcal{X}_j}^{Y\in\mathcal{Y}} \widehat{P}(\mathbf{x}[j]|Y)$.
1: Group tuples by the label Y;
2: **for** $k = 1$ to K and $j = 1$ to t **do**
3:     $\widehat{\mathbf{p}}_{jk}^{UF} = |\mathcal{U}_{c_k}|^{-1} \sum_{u\in\mathcal{U}_{c_k}} [\mathbf{z}_u^j - \mathbb{1}(e^{\frac{\epsilon}{2}} + 1)^{-1}](e^{\frac{\epsilon}{2}} + 1)(e^{\frac{\epsilon}{2}} - 1)^{-1}$
        $\triangleright$ $\mathbb{1}$ represents an all-one vector of length M_j;
4:     **for** $s = 1$ to M_j **do**
5:         Add $\widehat{\mathbf{p}}_{jk}^{UF}[s]$ to $\mathcal{C}$;
6:     **end for**
7: **end for**

2.3   Utility Metrics

To describe the utility loss of a private strategy, for final classification results, we adopt variance $Var[\cdot]$; for the estimated conditional probability, we use *Mean Squared Error* $MSE(\widehat{\mathbf{p}}) = \mathbb{E}[||\widehat{\mathbf{p}}-\mathbf{p}||_2^2]$ and ℓ_1-norm error $Dis_1(\widehat{\mathbf{p}}) = \mathbb{E}[||\widehat{\mathbf{p}}-\mathbf{p}||_1]$, where $\widehat{\mathbf{p}}$ and $\mathbf{p}$ denote the estimated probability and the true one, respectively.

The error bounds of private strategies are discussed in high privacy regime $\epsilon \in [0, 1]$ and their optimality are measured within the minimax framework.

Theorem 2 ([3]). *The optimal minimax error of private multinomial estimation, w.r.t. $\epsilon \in [0,1]$, are bounded by $||\cdot||_2^2 \leqslant c\min\{1, \frac{d}{N\epsilon^2}\}$ and $||\cdot||_1 \leqslant c\min\{1, \frac{d}{\sqrt{N\epsilon^2}}\}$, where N, d and c denote the sample size, the size of data domain and a constant.*

3   Private Naïve Bayes Classifier in the Local Setting

In this section, we present a private strategy for NBC in the local setting, and provide the bound of its utility loss. Due to the space limitation, some proofs of our conclusions are omitted in this paper.

3.1   Utility-First Strategy (UFS)

We use RR based mechanism introduced in [3] as the basic private block for UFS. For any user data $\mathbf{binx}_i^j$, its corresponding sanitized output $\mathbf{z}_i^j$ is also a binary vector, each bit of which is independently set by

$$\mathbf{z}_i^j[s] = \begin{cases} \mathbf{binx}_i^j[s] & \text{with probability } e^{\frac{\epsilon}{2}}(1+e^{\frac{\epsilon}{2}})^{-1} \\ 1 - \mathbf{binx}_i^j[s] & \text{with probability } (1+e^{\frac{\epsilon}{2}})^{-1} \end{cases}. \tag{1}$$

Algorithm 1 shows the workflow of UFS on the user side, and Algorithm 2 shows the training process. Since LDP is satisfied by RR and the post-processing property, the ϵ-LDP can be guaranteed for UFS.

3.2 Theoretical Analysis

Theorem 3. *The MSE of $\widehat{\mathbf{p}}_{jk}^{UF}$ is $N_{c_k}^{-1}[e^{\frac{\epsilon}{2}} M_j(e^{\frac{\epsilon}{2}} - 1)^{-2} + (1 - \sum_{s=1}^{M_j} \mathbf{p}_{jk}^2[s])]$.*

When $\epsilon \in [0,1]$, it holds that $e^{\frac{\epsilon}{2}}(e^{\frac{\epsilon}{2}} - 1)^{-2} \leqslant (4+\epsilon)\epsilon^{-2} \leqslant 5\epsilon^{-2}$. Therefore, the ℓ_2-norm error of $\widehat{\mathbf{p}}_{jk}^{UF}$ is bounded as $MSE(\widehat{\mathbf{p}}_{jk}^{UF}) \leqslant \min\{2, 5\frac{M_j}{N_{c_k}\epsilon^2}\}$. With the Cauchy-Schwarz inequality that $||\widehat{\mathbf{p}}_{jk}^{UF} - \mathbf{p}_{jk}||_1 \leqslant \sqrt{M_j}||\widehat{\mathbf{p}}_{jk}^{UF} - \mathbf{p}_{jk}||_2$, we have $Dis_1(\widehat{\mathbf{p}}_{jk}^{UF}) \leqslant \min\{1, \frac{\sqrt{5}M_j}{\epsilon\sqrt{N_{c_k}}}\}$. Thus, UFS is optimal in the minimax framework.

To measure the accuracy of the private classifier, the error bound of classification probabilities is given below.

Theorem 4. *The error bound of the classification probability of UFT is $Var[\widehat{P}^{UF}(Y = c_k|\mathbf{x})] \leqslant \{N_{c_k}^{-1}[e^{\frac{\epsilon}{2}}(e^{\frac{\epsilon}{2}} - 1)^{-2} + 1] + 1\}^t - 1$.*

4 Private Naïve Bayes Classifier in the Mixture Setting

In this section, we introduce an Expanded Geometric Strategy (EGS) for the mixture setting.

4.1 Mechanism Description

EGS is a localized expansion of the traditional Geometric mechanism, which was first proposed in [7] for databases in the centralized setting. It abstracts noise from two-sided geometric distribution $Pr[Z = z] = \frac{1-\alpha}{1+\alpha}\alpha^{|z|}$ whose probability exponentially decrease on integers, and satisfies $\alpha^{\Delta q}$-DP, where Δq is the sensitivity of a query. We set $\alpha = e^{-\frac{\epsilon}{2}}$.

For trusted institutes, EGS groups the original dataset by data labels, counts the histogram $\mathbf{Hist}^{jk}$ of j-th feature in the group c_k, and privatizes $\mathbf{Hist}^{jk}$ by $\mathbf{Histz}^{jk} = \mathbf{Hist}^{jk} + < \mathbf{Geo}(e^{-\frac{\epsilon}{2}}) >^{M_j}$. Then it sends the size of each group $\{g^k\}_{k=1}^K$ and $\mathbf{Histz}^{jk}$ to the trainer. For individual data providers, the workflow of privatization is similar as in Algorithm 2 with two-sided geometric noise.

The trainer aggregates data from institutes and individuals, and estimates the probability $\widehat{\mathbf{p}}_{jk}^{EG} = N_{c_k}^{-1}(\sum_{i=1}^{N_2} \mathbf{Histz}_i^{jk} + \sum_{u \in \mathcal{U}_{c_k}} \mathbf{z}_u^j)$, where $N_{c_k} = \sum_{i=1}^{N_2} g_i^k + |\mathcal{U}_{c_k}|$ and g_i^k is the group size of the k-th label provided by the i-th institute.

Theorem 5. *EGS satisfies ϵ-DP and ϵ-LDP.*

4.2   Theoretical Analysis

Geometric mechanism is utility-optimal on datasets [7]. To illustrate the optimality of EGS in the mixture setting, we prove its effectiveness on a tuple.

Theorem 6. *The MSE of the estimator $\widehat{p}_{jk}^{EG}$ is $2e^{\frac{\epsilon}{2}}VM_j[N_{c_k}(e^{\frac{\epsilon}{2}}-1)]^{-2}$, where $V = N_2 + |\mathcal{U}_{c_k}|$.*

Following this result, in the pure local setting where N_{c_k} individuals provide training data with label c_k, the ℓ_2-norm error of the conditional probability estimated by EGS is $2e^{\frac{\epsilon}{2}}M_jN_{c_k}^{-1}(e^{\frac{\epsilon}{2}}-1)^{-2} \leqslant \min\{2, \frac{10M_j}{N_{c_k}\epsilon^2}\}$. Therefore, EGS is utility-optimal in the minimax framework on individual data tuples.

The classification error of EGS has the same formation as UFS; that is:

Theorem 7. *The error bound of the classification probability of EGS in the mixture setting is $Var[\widehat{P}^{EG}(Y = c_k|\boldsymbol{x})] \leqslant (2Ve^{\frac{\epsilon}{2}}[N_{c_k}(e^{\frac{\epsilon}{2}}-1)]^{-2}+1)^t - 1$, where $V = N_2 + |\mathcal{U}_{c_k}|$.*

5   Experiment

In this section, we use MSE and Dis_1 of the estimated conditional probability and the correct rate (CR) of the final classification result to evaluate the performance of proposed strategies.

Settings. We use two real-world datasets and one simulative dataset for modeling NBC. Two real-world datasets are bank dataset [13] and car dataset [11]. Bank dataset contains 17 features and two classes (i.e., a term deposit subscriber or not), and has 45,307 instances, 41,188 of which are used for training. Car dataset contains 5 features and 4 types of labels describing car quality, and has 1728 instances, 1228 of which are used for training. The simulative dataset has 5 features and 8 labels. Feature values are generated with independently conditional probability on labels. This dataset has 4,000,000 instances and 3,000,000 of them are used as training set.

Results. The experimental performances of private NBCs in the local setting are illustrated in Fig. 1. The error trends of UFS and EGS on the conditional probability estimation are similar w.r.t. the privacy budget ϵ. When the privacy level is relatively high ($\epsilon \leqslant 1$), UFS slightly outperforms EGS. These experimental phenomena validate theoretical analyses that the proposed private strategies share the same error bound but have different constant factors. The CRs of UFS and EGS are averages of 1000 times experiments and presented in Fig. 2. The black dashed lines denote average CRs of the original NBC without privatization. On the bank dataset and simulative dataset, the experimental results are mostly above the original CR. This is because that the perturbation from private strategies relaxes the overfitting of the non-private model and avoids the

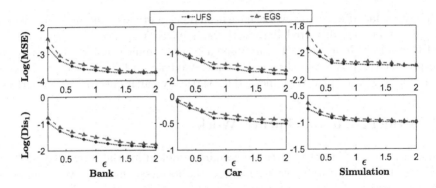

Fig. 1. Estimation errors on different datasets

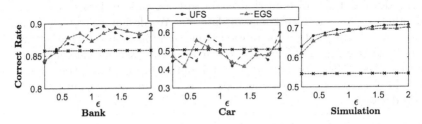

Fig. 2. Correct rate

error caused by $\mathbf{p}^{jk} = 0$. However, the CRs of three strategies fluctuate around non-private one on the car dataset. Since the simulative dataset and car dataset are both multi-label datasets, this fluctuation may be caused by the insufficient training data.

6   Related Work

Differential Privacy. The definition of Differential Privacy (DP) and relevant mechanisms are first presented in [4]. They are used in various fields, such as location service [15] and machine learning [14].

Then, the traditional definitions of DP are extended into distributed scenarios by many works (e.g., [9,12]). Most of them adopting SMC [8] and cryptosystem to assist distributed DP, which has high computational/interactive costs and cannot be afforded by resource-constrained individuals.

Local privacy is proposed by Kasiviswanathan et al. [10], to cope with data privacy with no trustful data curators. Duchi et al. [3] characterized the trade-off between the utility and LDP for various types of estimation problems. Based on [3], Bassily et al. proposed an efficient protocol for succinct histograms in [1]. In practical applications, Erlingsson [6] proposed a scheme RAPPOR to privately collect crowdsourcing data. Depending on [1], Chen et al. [2] provided a personalized LDP model to estimate the distribution of users over a certain area.

Private Classifier. Many works have been done in private training of classifiers. Zhang et al. [18] distorted data with randomized response strategy to train NBC privately. Yang et al. [17] proposed a cryptographic classification approach to protect the privacy in the local setting. Vaidya and Clifton [16] proposed a decision tree based privatized method for vertically partitioned data over parties.

7   Conclusion and Future Work

In this paper, we study the private training problem for the Naïve Bayes Classifier (NBC) in two different settings. In the local setting, we give a randomized response (RR) based strategy *Utility-First Strategy* (UFS). To cover more real-world data aggregation scenarios, we formally present *the mixture setting*. It includes two types of data that commonly appears in data collections, i.e., individual tuples and histograms of datasets. We design a strategy *Expanded Geometric Strategy* (EGS) to train NBC privately in this setting. Measured by common error estimators (e.g., ℓ_1, ℓ_2-norm error) and specific classification error for the private NBC, we demonstrate the optimality of these strategies in the minimax framework, and our conclusions are experimentally confirmed on real-world datasets.

Acknowledgments. This work was supported by the National Natural Science Foundation of China (No. 61572456), the Anhui Province Guidance Funds for Quantum Communication and Quantum Computers and the Natural Science Foundation of Jiangsu Province of China (No. BK20151241).

References

1. Bassily, R., Smith, A.: Local, private, efficient protocols for succinct histograms. In: STOC (2015)
2. Chen, R., Li, H., Qin, A., Kasiviswanathan, S.P., Jin, H.: Private spatial data aggregation in the local setting. In: ICDE (2016)
3. Duchi, J., Wainwright, M., Jordan, M.: Minimax optimal procedures for locally private estimation. arXiv preprint (2016)
4. Dwork, C.: Differential privacy. In: ICALP (2006)
5. Dwork, C., McSherry, F., Nissim, K., Smith, A.: Calibrating noise to sensitivity in private data analysis. In: Halevi, S., Rabin, T. (eds.) TCC 2006. LNCS, vol. 3876, pp. 265–284. Springer, Heidelberg (2006). https://doi.org/10.1007/11681878_14
6. Erlingsson, Ú., Pihur, V., Korolova, A.: RAPPOR: randomized aggregatable privacy-preserving ordinal response. In: CCS (2014)
7. Ghosh, A., Roughgarden, T., Sundararajan, M.: Universally utility-maximizing privacy mechanisms. SICOMP **41**, 1673–1693 (2012)
8. Goldreich, O.: Secure multi-party computation (1998)
9. Hong, Y., Vaidya, J., Lu, H., Karras, P., Goel, S.: Collaborative search log sanitization: toward differential privacy and boosted utility. TDSC **12**, 504–518 (2015)
10. Kasiviswanathan, S.P., Lee, H.K., Nissim, K., Raskhodnikova, S., Smith, A.: What can we learn privately? SICOMP **40**, 793–826 (2011)

11. Lichman, M.: UCI machine learning repository (2013). http://archive.ics.uci.edu/ml
12. Mohammed, N., Alhadidi, D., Fung, B.C., Debbabi, M.: Secure two-party differentially private data release for vertically partitioned data. TDSC **11**, 59–71 (2014)
13. Moro, S., Cortez, P., Rita, P.: A data-driven approach to predict the success of bank telemarketing. Decis. Support Syst. **62**, 22–31 (2014)
14. Shokri, R., Shmatikov, V.: Privacy-preserving deep learning. In: CCS (2015)
15. To, H., Ghinita, G., Shahabi, C.: A framework for protecting worker location privacy in spatial crowdsourcing. VLDB **7**, 919–930 (2014)
16. Vaidya, J., Clifton, C.: Privacy-preserving decision trees over vertically partitioned data. In: Jajodia, S., Wijesekera, D. (eds.) DBSec 2005. LNCS, vol. 3654, pp. 139–152. Springer, Heidelberg (2005). https://doi.org/10.1007/11535706_11
17. Yang, Z., Zhong, S., Wright, R.N.: Privacy-preserving classification of customer data without loss of accuracy. In: SDM (2005)
18. Zhang, P., Tong, Y., Tang, S., Yang, D.: Privacy preserving Naive Bayes classification. In: Li, X., Wang, S., Dong, Z.Y. (eds.) ADMA 2005. LNCS (LNAI), vol. 3584, pp. 744–752. Springer, Heidelberg (2005). https://doi.org/10.1007/11527503_88

Multimedia Data Processing

Fusing Satellite Data and Urban Data for Business Location Selection: A Neural Approach

Yanan Xu, Yanyan Shen[(✉)], Yanmin Zhu[(✉)], and Jiadi Yu

Shanghai Jiao Tong University, Shanghai, China
{xuyanan2015,shenyy,yzhu,jiadiyu}@sjtu.edu.cn

Abstract. Business location selection is of great importance in practice, but is a long and costly process. Traditional approaches to selecting optimal business locations consider complex factors such as foot traffic, neighborhood structure, space rent and available workforce, which are typically hard to acquire or measure. In this paper, we propose to exploit the highly available satellite data (e.g., satellite images and nighttime light data) as well as urban data for business location selection. We first perform an empirical analysis to evaluate the direct relationship between satellite features and business locations. We then propose a novel regression-and-ranking combined neural network model, to collectively predict the popularity and ranking for each location. Our model fuses the heterogeneous yet discriminative features from satellite and urban data and captures feature interactions effectively. We conduct experiments to compare our approach with various baselines. The results verify the effectiveness of the extracted satellite features and the superior performance of our model in terms of four metrics.

Keywords: Satellite data · Nighttime light · Satellite images
Business location selection

1 Introduction

Selecting a good location for starting a new business is essential for entrepreneurs. A good location would lead to business prosperity, while an unsuitable one may result in serious business risk and even the failure of the business. More importantly, many mistakes for starting a new business can be corrected later on, but a poor choice of the location is very difficult, if not impossible, to be repaired.

Business location selection is typically a long and costly process, during which one of the most important considerations is the *potential popularity* of locations. To evaluate the popularity of a location, great efforts have to be denoted to collecting and assessing factors such as foot traffic, neighborhood structure, space rent, available workforce and quality partners in the vicinity. While those factors

ⓒ Springer International Publishing AG, part of Springer Nature 2018
J. Pei et al. (Eds.): DASFAA 2018, LNCS 10828, pp. 589–605, 2018.
https://doi.org/10.1007/978-3-319-91458-9_36

are able to disclose the popularity of a location, they are very difficult to acquire or measure in practice. Recently, a number of research studies explore the data from location-based services for business location selection, including check-in data [11], wifi connections and search engine queries [23]. However, most of these data are not publicly available due to the privacy concerns, e.g., check-in records contain sensitive and personal mobility trajectories.

We find that satellite data and urban data are widely available, and they contain effective indicators for determining the popularity of a location. The satellite data mainly include three pieces of information: *nighttime light intensity, visible and infrared radiometer (VIRR) data*, and *satellite images*. Intuitively, the nighttime light intensity reflects the population and business concentrations of an area; the VIRR data collected by visible and infrared radiometer sensors contain Land Surface Temperature (LST), Normalized Difference Vegetation Index (NDVI) and emissivity of ground that indicate the distribution of vegetation and building constructions, for example, business centers usually have high temperature and few plants; and the satellite images capture important transportation infrastructures, such as roads and rivers. All the above information is valuable in predicting the potential popularity of a business location (see empirical analysis in Sect. 3). As for the urban data including POIs, road networks and taxi trajectories, they have been widely used in many urban computing methods [8,11], because they capture important spatial characteristics that are correlated with location popularity. For example, the number of POIs show the prosperity of locations and its diversity implies the completeness of the serving facilities; road networks reflect traffic convenience; and taxi trajectories retain the mobility patterns of people. Intuitively, if properly analyzed, the urban data can be a rich source to improve the performance of estimating location popularity for business purpose.

Inspired by our intuitions, in this paper, we propose to exploit both satellite data and urban data for business location selection. Our goal is to predict the popularity of any location in a city, and we use the number of check-ins as the ground-truth popularity score of a known business location. The key challenges come from three aspects. First, the satellite and urban data are very heterogeneous, and we need to mine useful features from both low-level image pixels and various spatial-temporal data. Second, the identified features from different data sources have to be fused in a complex yet effective way in order to achieve high prediction accuracy. Last but not least, we observe that a popularity ranking order among a set of locations is more valuable than individual popularity scores, as people always choose the rank-1 location with the highest potential popularity. However, it is worth noting that a small error in popularity prediction may cause a huge ranking error, and it is extremely difficult to predict accurate popularity scores that preserve the correct ranking order, as pointed out in [18].

To address the above challenges, we first extract useful features from heterogeneous satellite and urban data using different machine learning methods. A deep empirical analysis is then performed to evaluate the relationship between extracted features and popularity scores of locations. To predict location popularity, we propose a novel neural network model, named R^2Net, which effectively identifies discriminative latent features and captures feature interactions via fully connected layers and convolutional operations. With R^2Net, we are able to fuse heterogeneous features from both satellite and urban data in a unified way. To preserve ranking order among different locations, our neural network model acts in two roles by using an objective function that combines popularity regression and ranking order prediction simultaneously. The regression part focuses on predicting location popularity with high accuracy, while the ranking part regularizes the regression part to predict popularity scores with the preserved ranking order. We conduct extensive experiments using a real-world dataset. The results verify the superior popularity prediction performance of our proposed approach, compared with several baseline methods.

To summarize, the main contributions of this paper are the following.

- To the best of our knowledge, we are the first to exploit both satellite data and urban data for business location selection problem. In this paper, we consider three kinds of satellite data, i.e., nighttime light intensity, VIRR and satellite images.
- We introduce various feature extraction methods to identify important features from heterogeneous satellite and urban data according to different data characteristics. We measure the correlation between the features and the location popularity, and perform empirical analysis to evaluate the effectiveness of the proposed features in depth.
- We propose a novel neural network model, named R^2Net, for predicting location popularity. Our model employs fully connected layers and convolutional operations to fuse satellite and urban features in a unified manner, and captures latent feature interactions automatically. A regression-and-ranking combined objective function is adopted to predict location popularities with high accuracy as well as preserve ranking order among different locations.
- Extensive experiments are conducted with a real-world dataset. The results show that (1) our proposed approach outperforms other baseline methods in terms of different metrics and (2) the features from satellite data are effective for identifying the locations with high popularities.

2   Problem and Framework

In this section, we first define the satellite data and urban data explored in this paper. We then present the problem statement and the proposed framework.

Definition 1 (Nighttime Light Intensity). *Nighttime light intensity is a map which indicates the intensity of light generated by human at the nighttime. We uniformly sample light intensity points from the light intensity map (one sampling point for every 50 m) and each point p contains light intensity in a location, i.e., (lon, lat, intensity).*

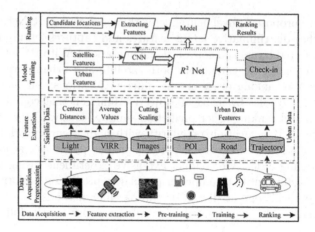

Fig. 1. Framework of our approach

Definition 2 (VIRR Data). *VIRR data are collected by the visible and infrared radiometer (VIRR) sensors on the satellites. Each record in VIRR data consists of seven fields including a timestamp, geographic coordinates, emissivity, reflectance, land surface temperature (LST), vegetation coverage (NDVI), i.e.,* $(time, lon, lat, emi, ref, lst, ndvi)$.

Definition 3 (Satellite Imagery). *Satellite imagery contains earth images collected by the imaging satellites. Each record in satellite imagery includes an image, the geographic coordinates of the place in the image center, and the geographic range of the image, i.e.,* (img, lon, lat, r).

Definition 4 (POI). *A Point of Interest (POI) is a venue (e.g., bus stop) in the city associated with name, category, and geographic coordinates.*

Definition 5 (Road Network). *A road network consists of a set of road segments. Each segment is represented by an identifier, its length, type, and a list of points forming the shape of the road segment.*

Definition 6 (Taxi Trajectory). *A taxi trajectory τ is a sequence of geographical points with the corresponding timestamps, i.e.,* $\{p_1, p_2, ..., p_n\}$.

Problem Statement. We consider a set L of locations. Following the definition in [11], we define each *location* as a region centered at a geographical point with a radius r (e.g., 200 m). Given both satellite data and urban data as defined above, our aim is to predict a popularity score $y(l)$ for each location $l \in L$ and output a ranking list for all the locations in L based on the estimated popularities. In this paper, we use the total number of check-in records crawled from Dianping, the largest online review platform in China, as the gold standard popularity score for a location. The number of check-in records is positively correlated with the number of customers for business, and it is also applied as the business popularity in other works [11, 25].

Framework. Figure 1 provides the framework of our approach to the business location selection problem, which consists of four major components as follows.

Data Preprocessing. Given the satellite data and urban data, we remove anomalous values, e.g., satellite data influenced by clouds and trajectory points far away from roads or adjacent points. We conduct a map-matching algorithm to reduce the noises in trajectories.

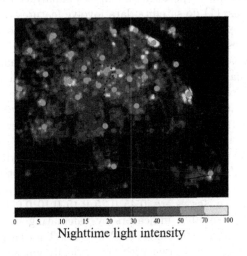

Nighttime light intensity

Fig. 2. The nighttime light intensity map and light centers

Feature Extraction. We extract useful features for each location from satellite and urban data. Given a location l, we consider the data in its *affecting region* (i.e., within certain distance). This is because the popularity of a location is typically determined by its surrounding environment. The satellite features contain the average light intensity, the distances to centers whose light intensities are higher than their surrounding area, and the average values in June and December for emissivity, reflectance, NDVI, and LST. The urban features include density and entropy for POIs, length and number of intersections for roads, visiting times and average speed for taxi trajectories.

Model Training. We first pre-train a convolutional neural network to learn latent features from satellite images. The image features are then concatenated with other features to learn the feature interactions in the proposed R^{2}Net model. Our model employs convolutional layers to handle the numeric satellite and urban features, and uses a fully connected merging layer to combine features from the satellite data and urban data. Deep hidden layers learn a latent representation that models feature interactions well. R^{2}Net leverages a regression-and-ranking combined loss function that predicts location popularity with preserved ranking order among locations.

Ranking. The ranking component predicts the popularity score for each location in L and outputs locations in descending order of their estimated popularities.

3   Empirical Data Analysis

A novel attempt of this paper is to utilize satellite data for optimal store location selection. In this section, we perform an empirical analysis over satellite data, to answer two questions: (1) do satellite data involve features that effectively indicate the business popularity of a location? (2) how are these features correlated with location business popularity?

To answer the above two questions, we focus on Shanghai, the largest metropolis in China. We first divide the whole Shanghai area into grids of $1\,km \times 1\,km$ and each grid is treated as one location. The ground-truth popularity for each location is measured by the total number of check-ins within the location crawled from Dianping, a premier online review website in China. For the ease of illustration, we categorize locations into five classes according to their popularity. As we are interested in studying the distribution characteristics of locations with high popularity, the locations are not divided uniformly. Instead, the number of the most popular locations is small and the percentages of locations for the five classes are $[0.45, 0.25, 0.15, 0.10, 0.05]$.

Next, we identify satellite indicators by calculating the average values of light intensity, emissivity, reflectance, NDVI and LST for each grid. In addition, we observe that higher light intensities are typically gathered in several small areas, as shown in Fig. 2. We thus group light intensity points and identify light centers in the city. For each location, we compute its distance to the closest light center as one satellite indicator for its popularity. Figure 3 provides the correlation analysis results between satellite indicators and location popularity. We summarize two key observations as follows.

Observation 1: *Popular locations tend to have higher nighttime light intensities and smaller distances to the light centers.* Figure 3(a) shows the distributions of two light intensity indicators for the locations in five classes. We also provide the scatter plot for each location with respect to two indicators. In the figure, each row/column denotes one feature and a plot denotes the distribution of locations over two features. We can see that the locations with higher popularity (the red hexagons) have higher light intensities and smaller distances to their nearest light centers than other locations. The reason may be that high light intensities indicate the existence of a residence area or flourishing business center.

Observation 2: *Popular locations tend to have lower reflectance and vegetation coverage, higher land surface temperature and medium emissivity.* Figure 3(b) and (c) show the correlation matrices between the VIRR related features and location popularity. From the figure, we can see that the locations with higher

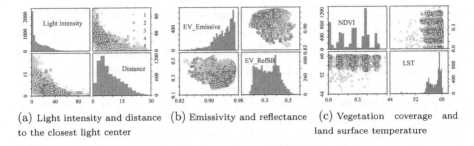

(a) Light intensity and distance to the closest light center   (b) Emissivity and reflectance   (c) Vegetation coverage and land surface temperature

Fig. 3. The correlation between satellite features and location popularity

popularity seem to have medium emissivity, low reflectance and vegetation coverage, high land surface temperature. One possible explanation is that popular locations are typically more crowded with human and buildings that easily cause the biased distributions of the corresponding satellite features. For instance, popular city business centers often have low vegetation coverage and human activities in the centers lead to high land surface temperature.

4 Location Popularity Appraisal Using Neural Networks

4.1 Urban Context and Satellite Feature Extraction

Before introducing our proposed model, we first describe the features extracted from satellite and urban data in this study.

Satellite Features. We employ convolutional neural network (CNN) to extract latent features from satellite images[1] (see details in Sect. 4.2). Here, we mainly describe the features extracted from nighttime light intensity and VIRR data for any location l: (1) average light intensity f^n: we uniformly sample light intensity points from the light pollution map[2] and compute the average light intensity by aggregating sampled intensities within the location; (2) distances to light centers f^{dis}: we cluster the light intensity points using DBSCAN [6] to find light centers, and then compute the distances from the location center to light centers; (3) distance to the closest light center f^{mdis}. The VIRR data[3] are collected every 10 days and we consider the following features: average emissivity f^e, reflectance f^r, vegetation coverage f^v, land surface temperature of the location over the recent two months.

Urban Features. Urban data have been widely used for urban computing [11, 24, 26]. In this study, we extract the following urban features in any location l. For POI data, we compute: (1) POI category frequency f^{pf}: the number

---

[1] http://map.tianditu.com/map/index.html.
[2] https://www.lightpollutionmap.info/.
[3] http://satellite.nsmc.org.cn/portalsite/default.aspx.

of POIs in each category within the location; (2) POI density f^{pd}: the total number of POIs in all categories; (3) POI entropy f^{pe}: this is computed by $f^{pe}(l) = -\sum_i \frac{f_i^{pf}(l)}{f^{pd}(l)} \times \log \frac{f_i^{pf}(l)}{f^{pd}(l)}$, where i is a POI category. For road network data, we consider features: (1) road length vector f^{rl}: the length of roads in each type within the location, which is also used in [3]; (2) total road length f^{rs}; (3) the number of road intersections f^{rc}. For taxi trajectory data, we divide one day into 24 time slots and extract the following features: (1) trajectory density f^{tn}: a vector recording the number of GPS points within the location in each time slot; (2) the number of visits f^{tv}: a vector recording the number of trajectories that enter the location per time slot; (3) average moving speed f^{ts}: a vector for average moving speed of trajectories per time slot. The values of trajectory features are averaged over different days.

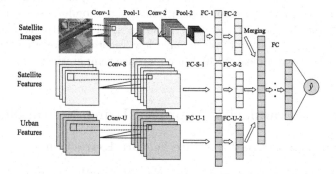

Fig. 4. Structure of R^{2}Net (Color figure online)

4.2 R^{2}Net: Proposed Model

We propose a regression-and-ranking combined neural network model, named R^{2}Net, for predicting location popularity. Figure 4 shows the structure of our model. The input to our model consists of satellite images, extracted satellite features and urban features for a particular location l. Due to the heterogeneity of different features, we leverage three sub-nets (in blue, green and red colors) to capture the corresponding feature interactions separately. We further concatenate the latent representations learned from three sub-nets and use fully connected layers to obtain a unified representation for predicting the final popularity score $y(l)$. The details of the components in R^{2}Net are described as follows.

(1) Extracting Satellite Image Features. We use I to denote one satellite image and feed it into a convolutional neural network. The first four layers are convolutional layers and pooling layers for extracting features from the pixels. The outputs of the two pooling layers are as follows.

$$z_1^c = P_{max}(\sigma(W_1^c * I + b_1^c)), \qquad z_2^c = P_{max}(\sigma(W_2^c * z_1^c + b_2^c)), \tag{1}$$

where $*$ denotes the convolutional operator. The W_i^c and b_i^c are kernel and bias of the ith convolutional layer, respectively. σ is the $ReLU$ activation function. $P_{max}(\cdot)$ denotes the max-pooling function. We then use two fully connected layers as Eq. (2) for learning the interactions of features from convolutional layers.

$$z^I = h_2^c(h_1^c(z_2^c)), \tag{2}$$

where the $h(\cdot)$ is one fully connected layer, i.e., $h(u) = \sigma(Wu + b)$.

(2) Learning Interactions Among Satellite/Urban Features. Inspired by [15], for each location, we also extract features for its neighbor locations in order to capture the spatial influences, e.g., the popularity scores of nearby locations typically change continuously. Hence, we organize location l and its neighbor locations (within a given window size λ) into a map and employ a convolutional layer to capture spatial influences. As satellite features and urban features typically encode different information, we use separate CNN layers to learn feature interactions for each of them. Consider satellite features as an example.

$$\phi = [x^l, x^N], \tag{3}$$

where x^l is the feature vector of location l and $x^N = \{x^{l'}|l' \in N(l)\}$. The input feature vector ϕ is organized as a $\lambda \times \lambda \times d$ tensor T. Similar to an image, the first and second dimension indicate the position of one location in the window. The target location l is at the center of the window. The last dimension indicates different channels, i.e., different kinds of features in this paper. Then, with the convolution layer, our model learns the interactions between different features and influences of neighbors.

$$z_0 = \sigma(F * T + b_0), \tag{4}$$

where F is the kernel and b_0 is the bias vector.

Two hidden layers follow the convolutional layer to learn the feature interactions in a further step, i.e., $z = h_2(h_1(z_0))$. We use z^s and z^u to denote the output vectors of satellite data and urban data respectively.

(3) Location Popularity Prediction. The three vectors, i.e., z^I, z^s, and z^u, are merged and as the input of a neural network with several fully connected layers to learn combinatorial features with data from different sources. For simplicity, the three vectors are concatenated together for merging. The output $\hat{y}$ is as Eq. (5). It should be noticed that the fully connected hidden layers can be replaced by a deeper residual network as [19].

$$\hat{y} = h^N(...h^2(h^1([z^I, z^s, z^u]))) \tag{5}$$

(4) Regression-and-Ranking Combined Objective Function. Considering that the ranking score of traditional learning to rank model usually just indicates the order of locations without estimating the true location popularity which indicates the number of potential customers and is concerned by entrepreneurs. On the contrary, the regression methods may yield arbitrary poor ranking performance [18]. In this paper, we combine popularity regression and ranking order prediction simultaneously to reserve the correct ranking order and predict accurate popularity. The loss function of our approach consists of two parts, i.e., ranking loss L^p and regression loss L^r, as shown in Eq. (6).

$$L(D, \theta) = \alpha L^p(D, \theta) + (1 - \alpha)L^r(D, \theta), \tag{6}$$

where D is the training data set and θ denotes parameters of our R^{2}Net.

Loss Function for Ranking. Similar to the RankNet [1], we adopt the pairwise loss function for ranking. Given two input samples x_i and x_j, and location i is more popular than j, we let $\hat{o}_{ij} = \hat{y}_i - \hat{y}_j$. The cross-entropy cost function is defined as the following equation.

$$L^p = \sum_{i,j} C(\hat{y}_i - \hat{y}_j) = \sum_{i,j} -P_{ij} \log \hat{P}_{ij} - (1 - P_{ij}) \log(1 - \hat{P}_{ij}), \tag{7}$$

where $\hat{P}_{ij} = \frac{1}{1+e^{-\hat{o}_{ij}}}$ which indicates the probability that location i should be placed ahead of location j. P_{ij} is the desired target values. $P_{ij} = \frac{1}{1+e^{-o_{ij}}}$. And $o_{ij} = y_i - y_j$.

Loss Function for Popularity Regression. We add the following squared-error loss function for minimizing the regression errors.

$$L^r = \sum_i (y_i - \hat{y}_i)^2. \tag{8}$$

4.3   Model Training and Optimization

For the CNN, we can add an output layer, i.e., $y^t = h^t(z^I)$ and pre-train it with the location popularity as it has a large number of parameters to learn. Then the component is trained together with the R^{2}Net.

Algorithm 1 shows the training steps of our R^{2}Net model. At first, the parameters of the model are initialized. Then, the model is trained with ranking samples and regression samples until the model converges. We adopt the mini-batch Adagrad [4] to optimize the loss function as it achieves faster convergence than the SGD. To prevent overfitting, we adopt dropout [20] on each hidden layer as the regularization. The dropout ratio is denoted by ρ.

ALGORITHM 1. R^2Net model training

Input: training data D, tradeoff parameter α, iterations t

1 Initialize the parameters of the neural network ;
2 $i = 0$;
3 **repeat**
4 | pick r from a uniform distribution between 0 and 1 ;
5 | **if** $r < \alpha$ **then**
6 | | $((x_a, y_a), (x_b, y_b)) \leftarrow$ RandomPair(D) ;
7 | | Take a gradient step with $((x_a, y_a), (x_b, y_b))$ using Eq. (7) ;
8 | **end**
9 | **else**
10 | | $(x, y) \leftarrow$ RegressionExample(D) ;
11 | | Take a gradient step with (x, y) with loss function Eq. (8) ;
12 | **end**
13 | $i \leftarrow i + 1$;
14 **until** $i > t$ *or convergence;*

---

5 Experiments

5.1 Datasets

We use the datasets of Shanghai to evaluate the performance of the proposed approach. The POI dataset is collected with the API of Baidu map[4]. The road network data are downloaded from the OpenStreetMap website[5]. In addition, we crawled the check-in data from Dianping[6] as the ground truth. The statistics of the experimental data are shown in Table 1. Similar to the Sect. 3, we divide the area of Shanghai into disjoined grids of the same size. The total number of check-ins in a grid is treated as its business popularity. After preprocessing the raw data and extracting features for grids, we randomly choose about 80% grids as the training data and the rest as the test data.

5.2 Performance Metrics

We adopt the following three kinds of metrics to assess the quality of ranking results.

Normalized Discounted Cumulative Gain. We choose the $NDCG@k$ defined in [11] to measure the extent to which the top-k locations with the highest popularity are actually highly ranked in the predicted list. The Discounted Cumulative Gain measure is defined as $DCG@k = \sum_{i=1}^{k} \frac{2^{rel(l_i)} - 1}{\log_2(i+1)}$, where $rel(l_i)$ is the score of the relevance of an instance at the position i in the predicted ranking list. The measure is normalized by the DCG value of the ideal ranking,

---

[4] http://lbsyun.baidu.com/index.php?title=jspopular.
[5] http://www.openstreetmap.org/.
[6] http://www.dianping.com.

Table 1. Details of the datasets

Data	Properties	Statistics
Light intensity	Resolution	50 m
VIRR data	Time period	June and december 2016
	Resolution	0.01 degree
Satellite image	Number of images	24,505
POI	Number of POIs	486,822
Road	Number of roads	229,398
	Road length	95,000 km
Trajectory	Number of GPS points	2,750,033
	Time period	July 2014
	Number of cars	4,373

i.e., the instances are sorted by the real relevance. We use the relative position in the actual ranking list as the relevance, i.e., $rel(l_i) = \frac{|L| - \overline{rank}(l_i) + 1}{|L|}$.

Kendall's Tau Coefficient. Kendall's Tau Coefficient (Tau for short) measures the ranking quality over the whole list [12]. For a location pair $<i, j>$, it is said to be concordant, if both $y_i > y_j$ and $\hat{y}_i > \hat{y}_j$ or if both $y_i < y_j$ and $\hat{y}_i < \hat{y}_j$. They are said to be discordant if $y_i > y_j$ and $\hat{y}_i < \hat{y}_j$ or if both $y_i < y_j$ and $\hat{y}_i > \hat{y}_j$. The Tau is defined as $Tau = \frac{\#conc - \#disc}{\#conc + \#disc}$.

Precision and Recall. We select the top N locations with the highest popularity. Given a top-k location list L_k sorted in a descending order of the predicted ranking scores, the precision and recall are defined as $Precision@k = \frac{L_k \cap L_N}{k}$ and $Recall@k = \frac{L_k \cap L_N}{N}$ where L_N is the list of the top N locations with the greatest number of check-ins. In the experiment, we set N to 50.

5.3   Baseline Methods

We compare our proposed approach R²Net with the following methods.

- *Lasso.* Lasso is a linear regression method using L1 norm regularization for selecting features.
- *SVR* [9]. SVR is an extension of SVM for solving regression problems.
- *MART* [7]. It is a boosted tree method and linearly combines the outputs of a set of regression trees.
- *RankNet* [1]. RankNet is a neural network based ranking method and adopts a pairwise objective function.
- *ListNet* [2]. ListNet is a list-wise learning to rank model. It achieves the best performance among all baseline methods.

For the two regression methods, i.e., Lasso and SVR, we use the implementation of skicit-learn library[7]. For SVR method, we employ the polynomial kernel

---
[7] http://scikit-learn.org.

which performs the best and the degree is 3. For the learning to rank methods, we use RankLib[8]. We set the number of trees to 1000, the number of leaves of each tree to 10, the learning rate to 0.1 for MART method. We set the number of hidden layers to 3, the number of nodes for each layer to 50 for RankNet. As for ListNet, the number of epochs is set to 1500 and the learning rate is 0.0001.

For our model, we set the learning rate $= 0.001$, the window size $\lambda = 3$. The α is set to 0.9. The dropout ratio ρ is 0.1. The optimal hyperparameters are chosen with the 10-fold cross-validation on the training dataset.

5.4 Performance Comparison of Different Approaches

We compare our approach with five methods mentioned above and report the comparison results in Fig. 5. Our approach outperforms the baseline methods for all metrics. Specifically, our method achieves 0.22 Tau and increases it by about 15% compared with the ListNet. The NDCG metrics are above 0.85 for different ks. We also observe that our approach has better performances than the RankNet method which has a same pairwise loss function as ours. We think it is because our approach adds one regression loss function to estimate the location popularity and fine tunes the parameters of the CNN component. In addition, we compare the regression results of our approach with that of the two regression baseline methods, i.e., Lasso and SVR. Our approach improves the performance by about 18% and 5% in terms of RMSE compared with Lasso and SVR respectively.

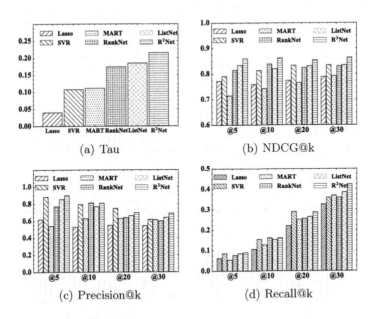

Fig. 5. Performance comparison of our approach and baselines

---

[8] http://sourceforge.net/p/lemur/wiki/RankLib/.

Table 2. Impact of $\frac{1-\alpha}{\alpha}$ on the ranking performance.

Metric	0.001	0.01	0.1	1.0	10
NDCG@5	0.8388	0.8829	0.8322	0.8171	0.8250
NDCG@10	0.8547	0.8655	0.8337	0.8333	0.8231
Tau	0.2333	0.2206	0.2011	0.2124	0.1633

5.5   Impact of Combination of Ranking and Regression

To prove the effectiveness of combing the ranking and regression, we vary the α to change the weights of the two parts in the loss function. It should be noticed that if α is bigger, ranking loss has a greater weight. Table 2 shows the performances of our approach with different ratio $\frac{1-\alpha}{\alpha}$. We find that when the ratio is 0.01, R^{2}Net achieves the best performance in terms of NDCG. As the ratio grows, i.e., regression loss has greater weights, the performances drop for metric Tau. We think it is because that the regression objective function ignores the ranking information among locations.

5.6   Feature Evaluation

We explore which kind of data is more effective for the business location selection. The results for the features from six data sources are shown in Fig. 6 in terms of four metrics. It seems that POI, satellite images, and nighttime light intensity

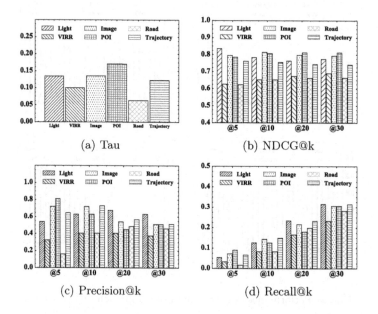

(a) Tau

(b) NDCG@k

(c) Precision@k

(d) Recall@k

Fig. 6. Performances of features from different sources

features achieve better performance for almost all cases. Specifically, the POI features perform the best w.r.t. the Tau and NDCG. We think it is because that some POIs are highly related to the businesses and bring many customers, e.g., bus stops. The light intensity and satellite images are useful for identifying the places with a large number of citizens and lead to good performances.

6   Related Work

Business location placement problem has been studied in recent years. Various optimization approaches [13,21] have been proposed to solve the problem. Xiao et al. focused on selecting a location to build a new facility in the road network by minimizing a specific cost function [21]. Li et al. tried to find the optimal locations for ambulance stations by minimizing the average travel time [13]. All these optimization techniques are clearly different from our approach which mines various features for location popularity prediction.

LBS (Location Based Service) attracts increasing attention as mobile phones have been used widely in recent years. Check-in data and POI data are utilized to analysis business area and find the optimal business locations in [11,17]. Fu et al. propose an estate ranking predictor to rank residential real estates leveraging features extracted from online user reviews and offline trajectories [8]. Lin et al. seek to explore the user check-ins, type of business, and business locations data from Facebook to evaluate locations for the business of a specific type [14]. Eravci et al. propose to find mobility patterns of customers and their habits from check-in data and identify a suitable place to open new business venues [5]. Besides the check-in data, some other location related data are explored. Xu et al. take advantage of the query log data from Baidu Maps and predict the demand for specific business to find the optimal places for opening new business venues [23]. As mentioned earlier, some of these data may not be easy to obtain or incur privacy concerns.

Satellite data have been explored in many traditional sectors. NASA[9], NSMC (National Satellite Meteorological Center). Some government organizations publish a great amount of satellite data which provide opportunities for urban computing. Many studies have explored satellite data for new applications [10,16,22]. Xie et al. try to predict poverty of Africa with satellite imagery and solve the data sparsity problem with transfer learning method and nighttime light intensities [10,22]. Pinkovskiy et al. use the nighttime light data to estimate the business development of a region [16].

7   Conclusion

This paper has focused on the problem of selecting promising locations for business and is inspired by the appealing advantages of satellite data including ease

---

[9] https://www.nasa.gov/.

of collection, wide coverage, valuable information on assessing the business popularity of an area. The empirical analysis confirms the predictive power of the extracted features. We proposed a novel model R^2Net for selecting promising business locations by leveraging the indicative features. With real-world satellite data and urban data, our experimental results demonstrate that our proposed approach outperforms other baseline methods and the features extracted from satellite data are effective for the business location selection. We also envisage that the satellite data features may benefit a variety of other applications, such as real estate price analysis and urban area development analysis.

Acknowledgment. This research is supported in part by 973 Program (No. 2014CB340303), NSFC (No. 61772341, 61472254, 61170238, 61602297 and 61472241), and Singapore NRF (CREATE E2S2). This work is also supported by the Program for Changjiang Young Scholars in University of China, the Program for China Top Young Talents, and the Program for Shanghai Top Young Talents.

References

1. Burges, C., Shaked, T., Renshaw, E., Lazier, A., Deeds, M., Hamilton, N., Hullender, G.: Learning to rank using gradient descent. In: ICML. ACM (2005)
2. Cao, Z., Qin, T., Liu, T.Y., Tsai, M.F., Li, H.: Learning to rank: from pairwise approach to listwise approach. In: ICML. ACM (2007)
3. Cheng, W., Shen, Y., Zhu, Y., Huang, L.: A neural attention model for urban air quality inference: learning the weights of monitoring stations. In: AAAI (2018)
4. Duchi, J., Hazan, E., Singer, Y.: Adaptive subgradient methods for online learning and stochastic optimization. J. Mach. Learn. Res. **12**(Jul), 2121–2159 (2011)
5. Eravci, B., Bulut, N., Etemoglu, C., Ferhatosmanoğlu, H.: Location recommendations for new businesses using check-in data. In: ICDMW. IEEE (2016)
6. Ester, M., Kriegel, H.P., Sander, J., Xu, X., et al.: A density-based algorithm for discovering clusters in large spatial databases with noise. In: KDD (1996)
7. Friedman, J.H.: Greedy function approximation: a gradient boosting machine. Ann. Stat. **29**, 1189–1232 (2001)
8. Fu, Y., Ge, Y., Zheng, Y., Yao, Z., Liu, Y., Xiong, H., Yuan, J.: Sparse real estate ranking with online user reviews and offline moving behaviors. In: ICDM. IEEE (2014)
9. Hearst, M.A., Dumais, S.T., Osuna, E., Platt, J., Scholkopf, B.: Support vector machines. IEEE Intell. Syst. Appl. **13**(4), 18–28 (1998)
10. Jean, N., Burke, M., Xie, M., Davis, W.M., Lobell, D.B., Ermon, S.: Combining satellite imagery and machine learning to predict poverty. Science **353**(6301), 790–794 (2016)
11. Karamshuk, D., Noulas, A., Scellato, S., Nicosia, V., Mascolo, C.: Geo-spotting: mining online location-based services for optimal retail store placement. In: KDD. ACM (2013)
12. Kendall, M.G.: A new measure of rank correlation. Biometrika **30**, 81–93 (1938)
13. Li, Y., Zheng, Y., Ji, S., Wang, W., Gong, Z., et al.: Location selection for ambulance stations: a data-driven approach. In: SIGSPATIAL. ACM (2015)
14. Lin, J., Oentaryo, R., Lim, E.P., Vu, C., Vu, A., Kwee, A.: Where is the goldmine?: finding promising business locations through Facebook data analytics. In: Proceedings of the 27th ACM Conference on Hypertext and Social Media. ACM (2016)

15. Liu, Z., Shen, Y., Zhu, Y.: Inferring dockless shared bike distribution in new cities. In: WSDM. ACM (2018)
16. Pinkovskiy, M., Sala-i Martin, X.: Lights, camera... income! illuminating the national accounts-household surveys debate. Q. J. Econ. **131**(2), 579–631 (2016)
17. Qu, Y., Zhang, J.: Trade area analysis using user generated mobile location data. In: WWW. ACM (2013)
18. Sculley, D.: Combined regression and ranking. In: KDD. ACM (2010)
19. Shan, Y., Hoens, T.R., Jiao, J., Wang, H., Yu, D., Mao, J.: Deep crossing: web-scale modeling without manually crafted combinatorial features. In: KDD. ACM (2016)
20. Srivastava, N., Hinton, G.E., Krizhevsky, A., Sutskever, I., Salakhutdinov, R.: Dropout: a simple way to prevent neural networks from overfitting. J. Mach. Learn. Res. **15**(1), 1929–1958 (2014)
21. Xiao, X., Yao, B., Li, F.: Optimal location queries in road network databases. In: ICDE. IEEE (2011)
22. Xie, M., Jean, N., Burke, M., Lobell, D., Ermon, S.: Transfer learning from deep features for remote sensing and poverty mapping. arXiv preprint (2015)
23. Xu, M., Wang, T., Wu, Z., Zhou, J., Li, J., Wu, H.: Store location selection via mining search query logs of Baidu maps. arXiv preprint (2016)
24. Xu, Y., Zhu, Y.: When remote sensing data meet ubiquitous urban data: fine-grained air quality inference. In: BigData. IEEE (2016)
25. Yu, Z., Tian, M., Wang, Z., Guo, B., Mei, T.: Shop-type recommendation leveraging the data from social media and location-based services. TKDD **11**(1), 1 (2016)
26. Zheng, Y., Liu, F., Hsieh, H.P.: U-air: when urban air quality inference meets big data. In: KDD. ACM (2013)

Index and Retrieve Multimedia Data: Cross-Modal Hashing by Learning Subspace Relation

Luchen Liu[1], Yang Yang[1(✉)], Mengqiu Hu[1], Xing Xu[1], Fumin Shen[1], Ning Xie[1], and Zi Huang[2]

[1] Center for Future Media and School of Computer Science and Engineering, University of Electronic Science and Technology of China, Chengdu, China
lucchenliu@gmail.com, dlyyang@gmail.com, mqhu2018@gmail.com, xing.xu@uestc.edu.cn, fumin.shen@gmail.com, seanxiening@gmail.com
[2] School of Information Technology and Electrical Engineering, The University of Queensland, Brisbane, Australia
huang@itee.uq.edu.au

Abstract. Hashing methods have been extensively applied to efficient multimedia data indexing and retrieval on account of explosion of multimedia data. Cross-modal hashing usually learns binary codes by mapping multi-modal data into a common Hamming space. Most supervised methods utilize relation information like class labels as pairwise similarities of cross-modal data pair to narrow intra-modal and inter-modal gap. In this paper, we propose a novel supervised cross-modal hashing method dubbed Subspace Relation Learning for Cross-modal Hashing (SRLCH), which exploits relation information in semantic labels to make similar data from different modalities closer in the low-dimension Hamming subspace. SRLCH preserves the discrete constraints and nonlinear structures, while admitting a closed-form binary codes solution, which effectively enhances the training efficiency. An iterative alternative optimization algorithm is developed to simultaneously learn both hash functions and unified binary codes, indexing multimedia data in an efficient way. Evaluations in two cross-modal retrieval tasks on three widely-used datasets show that the proposed SRLCH outperforms most cross-modal hashing methods.

Keywords: Multimedia index · Retrieval · Cross-modal hashing
Discrete optimization

1 Introduction

With the development of multimedia technologies, the quantity of multimedia data on the Internet such as images and text has increased rapidly [19,34]. Information retrieval techniques are not restricted to a single modality, thus, cross-modal retrieval, which means using an example from one modality as the

© Springer International Publishing AG, part of Springer Nature 2018
J. Pei et al. (Eds.): DASFAA 2018, LNCS 10828, pp. 606–621, 2018.
https://doi.org/10.1007/978-3-319-91458-9_37

query to retrieve relevant items from other modalities draws more attention in multimedia area [21,26,29]. For example, users can directly search an image by a textual sentence that describes the semantic content of the image. The challenge is how to retrieve multimedia data quickly from large-scale databases due to explosion of data. To achieve this problem in cross-modal retrieval, it is necessary to index vast multimedia data in an efficient way. In consideration of this, hashing, which aims to learn binary codes to reduce the storage and enhance the retrieval speed via bit operations, shows effectiveness and flexibility in indexing and retrieving items in a database [16,27,28].

Earlier hashing techniques for multimedia index and retrieval usually focus on uni-modal data [10,11,17,24], these methods tried to achieve the problem of intra-modal *semantic gap* [32,33], which means the differences of data contents, via a common Hamming space. However, in addition to semantic gap, the key challenge in cross-modal hashing is solving inter-modal *heterogeneous gap*, which demands to measure the similarity of between different modalities in an effective way. To bridge the two gaps, cross-modal hashing targets to preserve both intro-modal and inter-modal correlations in common Hamming space. Existing cross-modal hashing can be divided into two main categories of them based on whether the labels are exploited: unsupervised methods and supervised ones. Unsupervised cross-modal hashing usually learns binary codes via a predefined metric to measure similarities between modalities without class labels. These methods [1,3,15,23,25,39] can be applied to the data that lack label information to support hashing training. Existing unsupervised representative methods includes Predictable Dual-view Hashing (PDH) [23], Inter-media Hashing (IMH) [25], Collective Matrix Factorization Hashing (CFMH) [3], Fusion Similarity Hashing (FSH) [15] and so on.

Unlike unsupervised hashing, supervised methods [12–14,31,36–38] make full use of the class labels to enhance the hash codes learning. The relation information in labels is well exploited so that codes generated are more discriminative. Representative supervised methods includes Semantic Correlation Maximization (SCM) [36], Semantics-Preserving Hashing (SePH) [13], Coupled Dictionary Hashing (DCDH) [35], Discrete Cross-modal Hashing (DCH) [31] and so on. These supervised methods show different approaches in interpreting semantic information of labels. To be general, supervised schemes usually achieve better performance than unsupervised ones due to the relation information in class labels.

Moreover, Deep Neural Networks (DNN) based hashing methods have been studied recently inspired by the success of deep learning [30]. Such methods as Deep Visual-semantic Hashing (DVH) [9] and Deep Cross-modal Hashing (DCMH) [8] merge feature learning and hash function learning in a unified end-to-end framework, however, training DNN is really difficult and time-consuming.

There are still some problems in both supervised and unsupervised approaches. Most of the existing methods generate hash functions which use projection matrices to map the original features into the Hamming space [14], as a result, some significant feature structures are discarded and the learned hash codes lose some key information. In addition, since the discrete hash codes lead to difficulties in training, most hashing methods relax the original discrete constraints then solve the objective function in a continuous way to accelerate the algorithm [12], thus causing a lot of quantization error and the accuracy of learned hash codes decreases. Some supervised method like DCH develops a discrete optimization algorithm to solve original objective function without any relaxation but generated hash codes in a inefficient bit-by-bit way. Furthermore, how to effectively exploit inter-modal and intra-modal relation information to learn discriminative hash code in a faster way is still under study.

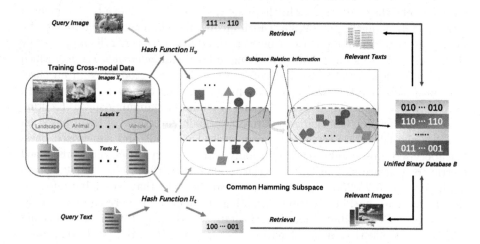

Fig. 1. Framework of the proposed SRLCH.

To overcome those drawbacks, we propose a novel supervised hashing approach, termed *Subspace Relation Learning for Cross-modal Hashing* (SRLCH), which learns binary codes via exploiting the relation information in the Hamming subspace. Our major contributions can be summarized as follows:

– Our method directly exploits class labels in a Hamming subspace, which can utilize transformed class labels to subspace relation information to jointly learn more discriminative hash codes and hash functions by optimizing the distances between hash codes and the relation information. A symmetrical framework is designed to generate unified binary codes retrieval database and learn hash function for query item synchronously. Kernel mapping and subspace projection is used in our hash functions, in this way, nonlinear structures in original features are preserved which contribute a lot to improve the performance significantly.
– We also develop an efficient discrete optimization hashing algorithm to solve the objective function without relaxing discrete constraints, even in our algorithm, we finally obtain a closed-form solution. And the binary codes can be generated in a single step, which contributes a lot to large-scale training.

– Our proposed method is evaluated on three popular datasets in two typical cross-modal retrieval tasks, experimental results show it outperforms several state-of-the-art methods.

2   Related Work

A lot of studies in hashing have been done to narrow the semantic gap in the same modality and heterogeneous gap in different modalities. Both uni-modal and cross-modal hashing methods focus on learning binary representation for data objectives, specifically, uni-modal hashing aims at data with homogeneous features and cross-modal hashing needs to fuse heterogeneous data to achieve similarity preservation across modalities. Related work to our method in this paper can be roughly divided into two main areas: how to embed data into a subspace and how to interpret label semantic information to bridge heterogeneous data.

The former one is uni-modal hashing usually does. Representative hashing methods for single modality [4,10,24] project homogeneous data into a Hamming subspace via linear projection or nonlinear modeling. Based on such subspace learning, some hashing methods enhance the accuracy and robustness via new similarity-measurement or regularization metrics [7,18], some methods adjust the loss function to obtain faster training speed [5]. As for cross-modal hashing, the goal of subspace learning is to learn a common Hamming subspace from two modalities where specific features can be matched well, in this way, some classical methods like Canonical Correlation Analysis (CCA) [6] are applied.

Supervised hashing methods utilize diverse class labels in subspace, which refers to the second area we introduced. The goal of supervised methods is to map samples belonging to same class as close as possible while different-class instances far away from each other [29], thus semantic labels are integrated into hashing learning procedure to excavate inter-modal and intra-modal correlations. Generally, in most supervised cross-modal hashing methods [12,14,36], a semantic similarity matrix is pre-constructed before training, which shows pairwise similarities of all training data. To be detailed, SCM [36] reconstructs the correlation matrix through sequential hash codes learning with relation of orthogonality constraints, what's more, in Supervised Matrix Factorization Hashing (SMFH) [14], collective matrix factorization with the label consistency is utilized to generate unified hash codes, Linear Subspace Ranking Hashing (LSRH) [12] uses rank order of features. Other hashing methods like DCDH [35] preserves sematic label information via graph model and dictionary learning. In some small-scale dataset it works well, however, for large-scale dataset, constructing correlation matrices or graphs causes much memory occupation. Unlike these methods, DCH [31] learns unified binary codes without measuring the similarities between different modalities. Inspired by the uni-modal hashing method Discrete Supervised Hashing (SDH) [24], DCH directly utilizes labels by predicting the class information of hash codes during learning, specifically, DCH learns a classifier of binary codes via class labels and the binary codes are more discriminative.

In this paper, our SRLCH integrates semantic labels into hashing framework, preserving both intra-modal and inter-modal similarities when embedding data into common Hamming subspace. Advantages are well demonstrated in experimental results.

3   Proposed Method

3.1   Notations

Suppose we have n training instances in the dataset, each instance is an image-text pair. For the i-th instance, an image is denoted by $x_v^i \in \mathbb{R}^a$ and a textual sentence is denoted by $x_t^i \in \mathbb{R}^b$, where a and b are the numbers of dimensions in each specific feature space. Class label vector $y^i = [y_1^i, y_2^i, \ldots, y_c^i]^\top \in \mathbb{R}^c$ is also available for each instance, where c denotes the category number, and $y_k^i = 1$ if the instance pertains to the k-th class and otherwise $y_k^i = 0$. Moreover, we denote $X_v = [x_v^1, x_v^2, \ldots, x_v^n]^\top \in \mathbb{R}^{n \times a}$ as the image visual feature matrix, $X_t = [x_t^1, x_t^2, \ldots, x_t^n]^\top \in \mathbb{R}^{n \times b}$ the text semantic feature matrix and $Y = [y^1, y^2, \ldots, y^n]^\top \in \mathbb{R}^{n \times c}$ the label matrix respectively. We denote L as the length of hash codes, also the dimension number of Hamming subspace.

3.2   Hash Functions

Hashing models can be divided into two categories: linear model and nonlinear one. As images and texts are typical unstructured data, latent structure in original feature vectors cannot be well preserved by the former one. Therefore, we learn nonlinear hash functions via the combination of kernel mapping and linear projection. Kernel functions can be adopted to better express the nonlinear intra-modal correlations among original features [11,17]. We define the kernel function $\phi(\cdot)$ via the RBF kernel mapping [10,11,17]. Specifically, for each instance, the kernelized feature of each image vector can be expressed as $\phi_v(x_v)$ and the text vector is $\phi_t(x_t)$, where for a specific feature vector x,
$\phi(x) = \left[exp(\frac{\|x-a_1\|}{2\sigma^2}), \ldots, exp(\frac{\|x-a_m\|}{2\sigma^2})\right]^\top$, $\{a_j\}_{j=1}^m$ denotes the randomly chosen m anchor samples and σ is the width.

To generate hash codes for each instance, the mapped feature vectors should be transformed from their specific space to a common Hamming subspace. Thus, to achieve the transformation for image and text modalities, we define two hash functions as

$$H_v(x_v) = sgn(\phi_v(x_v)P_v),$$
$$H_t(x_t) = sgn(\phi_t(x_t)P_t), \tag{1}$$

where $P_v \in \mathbb{R}^{m \times L}$ and $P_t \in \mathbb{R}^{m \times L}$ are matrices that project specific mapped features into the low-dimensional Hamming subspace, and the sign function $sgn(\cdot)$ outputs $+1$ for positive numbers and -1 otherwise.

3.3   Subspace Relation Learning for Cross-Modal Hashing

The hash functions defined have similar form but aim to different modalities respectively, though features are embedded to common Hamming space, binary codes generated by each function are not discriminative. To reduce heterogeneous differences and match data from same category, class labels should be utilized in subspace. As semantic label information is a bridge for both homogeneous and heterogeneous data that can use categories to distinguish and map embedded features. Therefore, we try to use a new projection matrix $W \in \mathbb{R}^{c \times L}$, to map class label vectors into the low-dimension subspace as *subspace relation information*, which can be considered as a "datum line". Note that the goal is to learn better subspace relation information and binary codes to preserve both inter-modal and intra-modal similarities, making more similar data mapped closer in Hamming space. Finally, projection matrix W and hash functions can be learned by jointly minimizing the distance between binary codes and the "datum line". This regression task is conducted in our hashing scheme to learn more discriminative binary codes. Our frame is illustrated in Fig. 1, and based on our frame, we proposed a novel hashing model as:

$$\min_{H_v, H_t} \mu_v \|H_v(X_v) - YW\|_F^2 + \mu_t \|H_t(X_t) - YW\|_F^2, \tag{2}$$

where μ_v and μ_t are the model parameters used to balance two-modality features, and $\|\cdot\|_F^2$ denotes the Frobenius-norm, and λ represents the regularization parameter.

Since the hash function in Eq. (2) is nonlinear, original problem cannot be solved trivially by an off-the-shelf solver. To solve the model, we introduce the binary codes matrix $B = [b_1, b_2, \ldots, b_n]^\top \in \mathbb{R}^{n \times L}$ for all training data, where L is the length of the binary codes and each binary code $b_i \in \{-1, 1\}^L$. For the feature matrix X from a specific modality and the corresponding hash function $H(X) = sgn(\phi(X)P)$, original quantization loss $\|H(X) - YW\|_F^2$ can be transformed as:

$$\|B - YW\|_F^2 + \|B - \phi(X)P\|_F^2. \tag{3}$$

Note that Eq. 3 keeps the discrete constrains, in this way, solving the hash functions H_v and H_t can be regarded as solving the projection matrices P_v and P_t. What's more, let B_v and B_t be binary codes generated by the two hash function respectively, to fit our modality-consistent condition, it also has a latent constraint, which requires unified binary codes $B = B_v = B_t$ instead of minimized distance $\|B_v - B_t\|_F$ as prior methods do, because it is very difficult to effectively require minimized distance between two sparse binary codes and it will introduce an unnecessary trade-off term. Combining the discrete constrains and unified B, we rewrite Eq. 2 as:

$$\min_{P_v, P_t, B, W} \|B - YW\|_F^2 + \nu_v \|B - \phi_v(X_v)P_v\|_F^2 + \nu_t \|B - \phi_t(X_t)P_t\|_F^2$$
$$+ \Omega(W, P_v, P_t), \ s.t. \ B \in \{-1, 1\}^{n \times L}, \tag{4}$$

where $\Omega(W, P_v, P_t) = \lambda \|W\|_F^2 + \alpha \|P_v\|_F^2 + \beta \|P_t\|_F^2$ is a newly-added penalty term with L_2 regularization to enhance the stability of the model, λ, α and β are

---

Algorithm 1. Subspace Relation Cross-modal Hashing

Input:

  Training data X_v, X_t for image and text feature matrices respectively;
  Label matrix Y;
  Model parameters ν_v, ν_t;
  Penalty parameters λ, α, β.

Output:

  Unified binary codes matrix B;
  Specific projection matrices in hash function P_v, P_t.

1: Normalize training data X_v, X_t by L_2 norm with zero mean on each row.
2: Obtain $\phi(X_v)$ and $\phi(X_t)$ by RBF kernel mapping.
3: Initialize P_v, P_t, W randomly, and B is initialized as a random $\{-1, 1\}^{n \times L}$ matrix.
4: **repeat**
5:    Compute P_v, P_t according to Eq. (6).
6:    Calculate W by Eq. (9).
7:    Update B using Eq. (12).
8: **until** Objective function converges.
9: **return** P_v, P_t and B

---

penalty parameters. Both ν_v and ν_t are the new model parameters of the two modalities respectively, where $\nu_v = \frac{\mu_v}{\mu_v + \mu_t}$ and $\nu_v = \frac{\mu_t}{\mu_v + \mu_t}$. In the new model formulated in Eq. (4), we can directly learn the linear projection matrices and the unified retrieval binary database B, which simplifies the computation. We call the proposed method which exploits subspace relation information in class labels to learn unified binary codes as *Subspace Relation Learning for Cross-modal Hashing* (SRLCH).

3.4   Optimization Algorithm

The formula shown in Eq. (4) is still non-convex and difficult to get the local optimal solution. We propose an iterative alternative optimization algorithm to solve the problem with respect to one variable while keeping others fixed. In this way, we can iteratively get the local optimal solution of each variable one by one with the following three steps until convergence. Note that throughout the optimization, discrete constraints are preserved well without relaxation. As conventional alternative optimization algorithm does, the whole optimization process is decomposed into three sub-steps.

Step-1. Firstly, we fix B and W to update the projection matrices P_v and P_t respectively. Since the two matrices are independent, the problem can be rewritten as:

$$\min_{P_v} \nu_v \|B - \phi_v(X_v)P_v\|_F^2 + \alpha \|P_v\|_F^2,$$

$$\min_{P_t} \nu_t \|B - \phi_t(X_t)P_t\|_F^2 + \beta \|P_t\|_F^2, \tag{5}$$

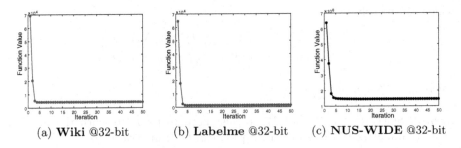

(a) **Wiki** @32-bit   (b) **Labelme** @32-bit   (c) **NUS-WIDE** @32-bit

Fig. 2. Convergence curves.

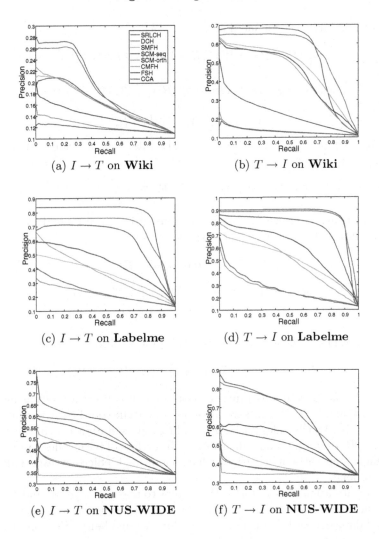

(a) $I \rightarrow T$ on **Wiki**   (b) $T \rightarrow I$ on **Wiki**

(c) $I \rightarrow T$ on **Labelme**   (d) $T \rightarrow I$ on **Labelme**

(e) $I \rightarrow T$ on **NUS-WIDE**   (f) $T \rightarrow I$ on **NUS-WIDE**

Fig. 3. Precision-recall curves @32-bit.

which can be solved as:

$$P_v = (\phi_v(X_v)^\top \phi_v(X_v) + \eta_v I)^{-1} \phi_v(X_v)^\top B,$$
$$P_t = (\phi_t(X_t)^\top \phi_t(X_t) + \eta_t I)^{-1} \phi_t(X_t)^\top B, \tag{6}$$

where $\eta_v = \frac{\alpha}{\nu_v}$ and $\eta_t = \frac{\beta}{\nu_t}$.

Step-2. Since we get P_v and P_t, we just need to fix B to learn the relation projection matrix W. To solve W, Eq. (4) can be transformed to

$$\min_W \|B - YW\|_F^2 + \lambda \|W\|_F^2, \tag{7}$$

which can be further formulated as:

$$\min_W Tr\left((B - YW)^\top (B - YW) + \lambda Tr(W^\top W)\right) \tag{8}$$

where $Tr(\cdot)$ denotes the trace norm and $B \in \{-1, 1\}^{n \times L}$. Hence, we get the solution as

$$W = (\lambda I + Y^\top Y)^{-1} Y^\top B. \tag{9}$$

Step-3. When P_v, P_s and W are fixed, in this step, we can compute B by rewriting Eq. (4) as:

$$\min_B \|B - YW\|_F^2 + \nu_v \|B - \phi_v(X_v)P_v\|_F^2 + \nu_t \|B - \phi_t(X_t)P_t\|_F^2, \tag{10}$$

where two penalty terms of P_v and P_t are discarded since P_v and P_t have been fixed. Equation (10) can be further expressed as:

$$\min_B Tr\left((YW - B)^\top (YW - B)\right)$$
$$+\nu_v Tr\left((\phi_v(X_v)P_v - B)^\top (\phi_v(X_v)P_v - B)\right) \tag{11}$$
$$+\nu_t Tr\left((\phi_t(X_t)P_t - B)^\top (\phi_t(X_t)P_t - B)\right)$$

As $B \in \{-1, 1\}^{n \times L}$, hence, we get the closed-form solution of B as:

$$B = sgn\left(\nu_v \phi_v(X_v)P_v + \nu_t \phi_t(X_t)P_t) + YW\right). \tag{12}$$

Since B has discrete constraints, conventional discrete optimization algorithm in SDH [24] and DCH [31] uses iterative discrete cyclic coordinate descent (DCC) to solve optimal hash codes bit by bit, leading to lots of iterations in the sub-step of solving B. In our method, when solving B, all bits of hash codes can be solved in a single step without iteration, which shows high efficiency especially when codes are long.

In each iteration, we can follow three steps to update P_v, P_t, W and B respectively until the model converges. Finally, we can learn the unified binary codes as retrieval database and the hash functions for each modality. The complete optimization algorithm of our SRLCH is presented in Algorithm 1. Figure 2 shows the rapid convergence of our algorithm on different datasets.

Table 1. Statistics of datasets

Dataset	Features		Categories	Size	Retrieval/train	Query/test
	Image	Text				
Wiki	128	10	8	2866	2173	693
Labelme	512	245	10	2688	2016	672
NUS-WIDE	500	1000	10	186577	184577	2000

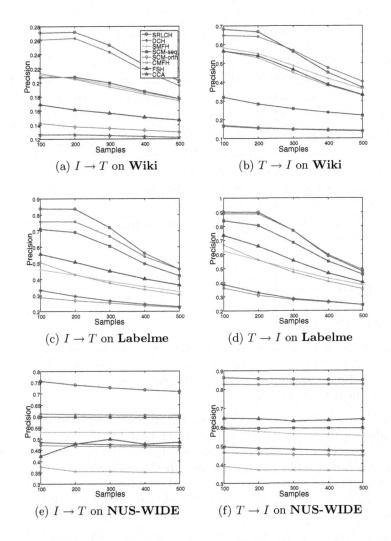

(a) $I \rightarrow T$ on **Wiki**   (b) $T \rightarrow I$ on **Wiki**

(c) $I \rightarrow T$ on **Labelme**   (d) $T \rightarrow I$ on **Labelme**

(e) $I \rightarrow T$ on **NUS-WIDE**   (f) $T \rightarrow I$ on **NUS-WIDE**

Fig. 4. Comparative precision evaluation with top 500 retrieved samples @32-bit.

3.5   Indexing and Cross-Modal Retrieval

After learning P_v, P_t and B according to Algorithm 1, we can get specific hash functions H_v and H_t from Eq. (1). In training process, unified binary codes matrix B as retrieval database is also obtained. Note that B contains both image and text modalities information, in which all the pairwise image-text data are indexed. For a given query example, either an image feature vector x'_v or a textual feature vector x'_t, we can use the learned specific hash functions to generate a binary-code query vector b'. We can conduct several bit operations to return relevant query results, which enhances the retrieval efficiency.

Table 2. mAP scores on Wiki dataset

Method	$Img \rightarrow Txt$				$Txt \rightarrow Img$			
	16-bit	32-bit	64-bit	128-bit	16-bit	32-bit	64-bit	128-bit
CCA [4]	0.1669	0.1519	0.1495	0.1472	0.1587	0.1392	0.1272	0.1211
CMFH [3]	0.2098	0.2295	0.2333	0.2431	0.4784	0.5312	0.5287	0.5443
FSH [15]	0.2333	0.2596	0.2557	0.2613	0.2230	0.2572	0.2594	0.2622
SMFH [14]	0.1879	0.2389	0.2519	0.2638	0.4130	0.5454	0.6012	0.6122
SCM-orth [36]	0.1538	0.1402	0.1303	0.1289	0.1540	0.1373	0.2485	0.1224
SCM-seq [36]	0.2341	0.2410	0.2462	0.2566	0.2257	0.2459	0.1258	0.2528
DCH [31]	0.3294	0.3383	**0.3756**	0.3726	0.6855	0.7086	0.7107	0.7050
SRLCH	**0.3526**	**0.3648**	0.3537	**0.3829**	**0.7356**	**0.7251**	**0.7343**	**0.7354**

4   Experiments

4.1   Datasets

We use three widely-used multi-modal datasets and there are two small-scale dataset: Wiki [22] and Labelme [20,39], and one large-scale dataset: NUS-WIDE [2]. Table 1 shows the statistics of each dataset in detail and we briefly describes the datasets as follows:

Wiki. There are 2,866 data in the Wiki [22] dataset, which contain image and text information annotated with 10 different class labels. Each image in this dataset is represented as a 128-dimensional feature vector extracted by SIFT, each text is generated by Latent Dirichlet Allocation (LDA) as a 10-dimensional feature vector. We randomly choose around 20% of the total data as the test data to query and the remaining data are selected as training data for retrieval database.

Labelme. The Labelme [20,39] dataset consists of 2,688 images along with textual data. Each image is a 512-dimensional GIST feature vector, and each textual vector is extracted as a 245-dimensional feature vector. The training and test sizes are the same as Wiki dataset.

NUS-WIDE. Unlike Wiki and Labelme, NUS-WIDE [2] is a large-scale dataset containing 186,577 images and corresponding textual tags, which are divided into 10 categories. 1% data of NUS-WIDE dataset are taken as query set to test, the rest are used to training hash functions and building binary codes retrieval database. Each image is represented by 500-dimensional manual features, and each text by 1000-dimensional BoW features.

As is shown in Table 1, all experiments are performed under the same settings.

Table 3. mAP scores on Labelme dataset

Method	$Img \rightarrow Txt$				$Txt \rightarrow Img$			
	16-bit	32-bit	64-bit	128-bit	16-bit	32-bit	64-bit	128-bit
CCA [4]	0.3199	0.2735	0.2326	0.2029	0.3534	0.2965	0.2560	0.2278
CMFH [3]	0.3873	0.4197	0.4539	0.3039	0.5003	0.5599	0.5995	0.3026
FSH [15]	0.4517	0.5405	0.6234	0.6528	0.5310	0.6532	0.7262	0.7601
SMFH [14]	0.3202	0.4542	0.5411	0.6368	0.4047	0.5767	0.6947	0.7815
SCM-orth [36]	0.3587	0.2902	0.2503	0.2321	0.3831	0.2697	0.2080	0.1681
SCM-seq [36]	0.6684	0.7026	0.6564	0.7354	0.7865	0.8104	0.7638	0.8381
DCH [31]	0.7663	0.8226	0.8150	0.8293	0.9054	0.9053	0.9143	0.9156
SRLCH	**0.8607**	**0.8705**	**0.8855**	**0.8896**	**0.9178**	**0.9202**	**0.9203**	**0.9247**

4.2 Compared Methods and Evaluation Metrics

The proposed SRLCH has been compared with six cross-modal hashing methods. There are three unsupervised methods including CCA [4], CMFH [3] and FSH [15], and three supervised methods SMFH [14] SCM [36] and DCH [31]. Note that SCM has two approaches: SCM-orth and SCM-seq, which are based on orthogonal projection learning and sequential learning respectively. To get the best performance of each compared approach, all the parameters are set default in the codes available or as the original papers suggest.

The retrieval is conducted on two retrieval tasks: $Img \rightarrow Txt$ and $Txt \rightarrow Img$. To test the performance of different methods, we use precision at top K samples (pre@K), mean Average Precision (mAP) and precision-recall as metrics.

4.3 Implementation

For kernel mapping method, our experiments show that the parameter σ is not sensitive to the results when its value is not very small, and we set it equaling to 0.6 in both kernel functions. However, the number of anchor samples m is set to different values, in small-scale datasets like Wiki and Labelme, m is set to 500, in large-scale dataset like NUS-WIDE, m is set to 5000.

In our method, there are two main model parameters ν_v and ν_t, and three regularizer parameters λ, α and β. To fit features for different modalities, we apply linear search for ν_v and ν_t over $\{10^{-6}, 10^{-5}, \ldots, 10^{-1}\}$, and empirically set the ν_v and ν_t to 10^{-1} and 10^{-5} respectively. As for λ, α and β, to simplify computation, we replace α and β by η_v and η_t in Eq. (6), then fix them to 0.1, 0.1 and 0.01 for all experiments based on experience.

Table 4. mAP scores on NUS-WIDE dataset

Method	$Img \rightarrow Txt$				$Txt \rightarrow Img$			
	16-bit	32-bit	64-bit	128-bit	16-bit	32-bit	64-bit	128-bit
CCA [4]	0.3861	0.3729	0.3612	0.3531	0.3809	0.3689	0.3589	0.3519
CMFH [3]	0.3478	0.3446	0.3446	3435	0.3546	0.3499	0.3507	0.3507
FSH [15]	0.5115	0.5172	0.5204	0.5191	0.4816	0.4851	0.4885	0.4920
SMFH [14]	0.4268	0.4361	0.4433	0.4470	0.4008	0.4097	0.4151	0.4177
SCM-orth [36]	0.3986	0.3790	0.3645	0.3570	0.3861	0.3694	0.3586	0.3533
SCM-seq [36]	0.5172	0.5379	0.5547	0.5554	0.4849	0.5038	0.5148	0.5170
DCH [31]	**0.5968**	0.6002	0.6023	0.5912	**0.7232**	0.7392	0.7305	0.7036
SRLCH	0.5964	**0.6416**	**0.6263**	**0.6559**	0.7038	**0.7556**	**0.7561**	**0.7899**

4.4   Experimental Results

Comparison with Baselines

We set various binary code length levels from 16-bit to 128-bit. The mAPs of SRLCH and all the baselines on small-scale datasets Wiki and Labelme are recorded in Tables 2 and 3. The mAP values of our method is higher than most supervised and unsupervised approaches on two database for two different tasks. In detail, on Wiki dataset, our SRLCH improves averagely by 10% than the best approach and on Labelme dataset, SRLCH obtains the best performance of all compared methods. As Table 4 illustrates, we can observe that our proposed method shows good mAP scores on large-scale dataset NUS-WIDE with the code length more than 32-bit while DCH only outperforms with 16-bit code length. For larger code length, our SRLCH has higher mAP scores, for example, with 128-bit length, SRLCH improves by 12% than the best baseline DCH.

We also sketch the precision-recall curves in Fig. 3. Apparently, our proposed method always maintains the best for both tasks on Wiki and Labelme with the precision-recall metric. Figure 4 illustrates the variation of precision values at from 100 to 500 samples for different tasks with 32-bit code length. It can be observed that our method achieve a very preferable performance on two datasets for both $I \rightarrow T$ and $T \rightarrow I$ tasks. To be specific, some methods are only

Table 5. Training time (second) of SRLCH and baselines on **Labelme**.

Code length	Method							
	SRLCH	DCH [31]	SCM-seq [36]	SCM-orth [36]	SMFH [14]	FSH [15]	CMFH [3]	CCA [4]
16-bit	**0.1305**	0.4102	12.0327	0.6945	0.6881	4.3859	1.7778	1.0567
32-bit	**0.1290**	0.4447	23.1629	0.7008	0.7261	4.6017	1.4875	1.0322
64-bit	**0.1344**	0.6595	26.6193	0.7008	0.8947	4.7865	1.7859	1.0326
128-bit	0.2110	5.6338	91.7365	0.6962	1.2453	5.3416	**0.1345**	1.0366

competitive in certain tasks and datasets, for example, DCH shows a preferable pre@K only when K is more than 300 for the $T \rightarrow I$ task on Wiki, however, the pre@K is not so good as the $I \rightarrow T$ task on the same dataset. What's more, on other datasets including Labelme and NUS-WIDE, for both tasks, our SRLCH can beat all baselines outright throughout 100–500 retrieval samples.

In addition, there are also some interesting findings from the experimental results. Firstly, with the increase of the hash code length, the mAP value of our SRLCH also increases. Secondly, both the pre@K curves and precision-recall curves of our SRLCH have slower trends than most methods, which shows the stability of our model.

Training Speed

We test the training time (second) of each method on Labelme dataset with different code lengths, and record the time in Table 5. All of the algorithms run on the same 64-bit system with Intel Core i7-4790 @ 3.6 GHz and 8 GB of memory. Notably, the unsupervised method CMFH has a very competitive speed at 128-bit, since its specific matrix factorization algorithm causes the abnormally fast convergence at that code length level. However, at other code length levels, our SRLCH has higher training speed than all baselines, even about 3 to 5 times faster than the best baselines. At long code length level (128-bit), SRLCH can achieve similar training speed with those on other levels.

5   Conclusions

In this paper, we proposed a novel supervised cross-modal hashing method, referred to as *Subspace Relation Learning for Cross-modal Hashing* (SRLCH), for efficient multimedia indexing and retrieval. In detail, to interpret semantic labels and integrate them into our scheme, we projected class labels of two modalities into a subspace, meanwhile, we exploited the subspace relation information to jointly learn the unified binary codes and hash functions for two different modalities in a common Hamming space. The subspace relation information can be regarded as a "datum line" to map similar features together. Thanks to the optimization algorithm we developed, the objective binary codes were generated in a single step, which accelerated the model training to some extent. As a result, we learned more discriminative binary codes in less time using SRLCH, enhancing the efficiency of cross-modal retrieval. The experimental results demonstrated that SRLCH outperforms many other methods in accuracy and speed for two cross-modal retrieval tasks.

Acknowledgments. This work was supported in part by the National Natural Science Foundation of China under Project 61572108, Project 61632007 and Project 61502081.

References

1. Bronstein, M.M., Bronstein, A.M., Michel, F., Paragios, N.: Data fusion through cross-modality metric learning using similarity-sensitive hashing. In: Proceedings of CVPR, pp. 3594–3601 (2010)
2. Chua, T., Tang, J., Hong, R., Li, H., Luo, Z., Zheng, Y.: NUS-WIDE: a real-world web image database from National University of Singapore. In: Proceedings of CVIR (2009)
3. Ding, G., Guo, Y., Zhou, J.: Collective matrix factorization hashing for multimodal data. In: Proceedings of CVPR, pp. 2083–2090 (2014)
4. Gong, Y., Lazebnik, S.: Iterative quantization: a procrustean approach to learning binary codes. In: Proceedings of CVPR, pp. 817–824 (2011)
5. Gui, J., Liu, T., Sun, Z., Tao, D., Tan, T.: Fast supervised discrete hashing. IEEE TPAMI **40**(2), 490–496 (2018)
6. Hardoon, D.R., Szedmák, S., Shawe-Taylor, J.: Canonical correlation analysis: an overview with application to learning methods. Neural Comput. **16**(12), 2639–2664 (2004)
7. Hu, M., Yang, Y., Shen, F., Xie, N., Shen, H.T.: Hashing with angular reconstructive embeddings. IEEE TIP **27**(2), 545–555 (2018)
8. Jiang, Q., Li, W.: Deep cross-modal hashing. In: Proceedings of CVPR, pp. 3270–3278 (2017)
9. Kang, Y., Kim, S., Choi, S.: Deep learning to hash with multiple representations. In: Proceedings of ICDM, pp. 930–935 (2012)
10. Kulis, B., Darrell, T.: Learning to hash with binary reconstructive embeddings. In: Proceedings of NIPS, pp. 1042–1050 (2009)
11. Kulis, B., Grauman, K.: Kernelized locality-sensitive hashing for scalable image search. In: Proceedings of ICCV, pp. 2130–2137 (2009)
12. Li, K., Qi, G., Ye, J., Hua, K.A.: Linear subspace ranking hashing for cross-modal retrieval. IEEE TPAMI **39**(9), 1825–1838 (2017)
13. Lin, Z., Ding, G., Hu, M., Wang, J.: Semantics-preserving hashing for cross-view retrieval. In: Proceedings of CVPR, pp. 3864–3872 (2015)
14. Liu, H., Ji, R., Wu, Y., Hua, G.: Supervised matrix factorization for cross-modality hashing. In: Proceedings of IJCAI, pp. 1767–1773 (2016)
15. Liu, H., Ji, R., Wu, Y., Huang, F., Zhang, B.: Cross-modality binary code learning via fusion similarity hashing. In: Proceedings of CVPR, pp. 6345–6353 (2017)
16. Liu, J., Wang, R., Gao, X., Yang, X., Chen, G.: Anglecut: a ring-based hashing scheme for distributed metadata management. In: Proceedings of DASFAA, pp. 71–86 (2017)
17. Liu, W., Wang, J., Ji, R., Jiang, Y., Chang, S.: Supervised hashing with kernels. In: Proceedings of CVPR, pp. 2074–2081 (2012)
18. Luo, Y., Yang, Y., Shen, F., Huang, Z., Zhou, P., Shen, H.T.: Robust discrete code modeling for supervised hashing. PR **75**, 128–135 (2018)

19. McNamara, Q., de la Vega, A., Yarkoni, T.: Developing a comprehensive framework for multimodal feature extraction. In: Proceedings of ACM SIGKDD, pp. 1567–1574 (2017)
20. Oliva, A., Torralba, A.: Modeling the shape of the scene: a holistic representation of the spatial envelope. IJCV **42**(3), 145–175 (2001)
21. Peng, Y., Huang, X., Zhao, Y.: An overview of cross-media retrieval: concepts, methodologies, benchmarks and challenges. CoRR abs/1704.02223 (2017)
22. Rasiwasia, N., Pereira, J.C., Coviello, E., Doyle, G., Lanckriet, G.R.G., Levy, R., Vasconcelos, N.: A new approach to cross-modal multimedia retrieval. In: Proceedings of ACM MM, pp. 251–260 (2010)
23. Rastegari, M., Choi, J., Fakhraei, S., III, H.D., Davis, L.S.: Predictable dual-view hashing. In: Proceedings of ICML, pp. 1328–1336 (2013)
24. Shen, F., Shen, C., Liu, W., Shen, H.T.: Supervised discrete hashing. In: Proceedings of CVPR, pp. 37–45 (2015)
25. Song, J., Yang, Y., Yang, Y., Huang, Z., Shen, H.T.: Inter-media hashing for large-scale retrieval from heterogeneous data sources. In: Proceedings of ACM SIGMOD, pp. 785–796 (2013)
26. Wang, B., Yang, Y., Xu, X., Hanjalic, A., Shen, H.T.: Adversarial cross-modal retrieval. In: Proceedings of ACM MM, pp. 154–162 (2017)
27. Wang, J., Shen, H.T., Song, J., Ji, J.: Hashing for similarity search: a survey. CoRR abs/1408.2927 (2014)
28. Wang, J., Zhang, T., Song, J., Sebe, N., Shen, H.T.: A survey on learning to hash. CoRR abs/1606.00185 (2016)
29. Wang, K., Yin, Q., Wang, W., Wu, S., Wang, L.: A comprehensive survey on cross-modal retrieval. CoRR abs/1607.06215 (2016)
30. Wang, W., Yang, X., Ooi, B.C., Zhang, D., Zhuang, Y.: Effective deep learning-based multi-modal retrieval. VLDB J. **25**(1), 79–101 (2016)
31. Xu, X., Shen, F., Yang, Y., Shen, H.T., Li, X.: Learning discriminative binary codes for large-scale cross-modal retrieval. IEEE TIP **26**(5), 2494–2507 (2017)
32. Xu, Y., Yang, Y., Shen, F., Xu, X., Zhou, Y., Shen, H.T.: Attribute hashing for zero-shot image retrieval. In: Proceedings of ICME, pp. 133–138 (2017)
33. Yang, Y., Luo, Y., Chen, W., Shen, F., Shao, J., Shen, H.T.: Zero-shot hashing via transferring supervised knowledge. In: Proceedings of ACM MM, pp. 1286–1295 (2016)
34. Yang, Z., Li, Q., Liu, W., Ma, Y.: Learning manifold representation from multimodal data for event detection in flickr-like social media. In: Gao, H., Kim, J., Sakurai, Y. (eds.) DASFAA 2016. LNCS, vol. 9645, pp. 160–167. Springer, Cham (2016). https://doi.org/10.1007/978-3-319-32055-7_14
35. Yu, Z., Wu, F., Yang, Y., Tian, Q., Luo, J., Zhuang, Y.: Discriminative coupled dictionary hashing for fast cross-media retrieval. In: Proceedings of ACM SIGIR, pp. 395–404 (2014)
36. Zhang, D., Li, W.: Large-scale supervised multimodal hashing with semantic correlation maximization. In: Proceedings of AAAI, pp. 2177–2183 (2014)
37. Zhen, Y., Yeung, D.: Co-regularized hashing for multimodal data. In: Proceedings of NIPS, pp. 1385–1393 (2012)
38. Zhen, Y., Yeung, D.: A probabilistic model for multimodal hash function learning. In: Proceedings of ACM SIGKDD, pp. 940–948 (2012)
39. Zhou, J., Ding, G., Guo, Y.: Latent semantic sparse hashing for cross-modal similarity search. In: Proceedings of ACM SIGIR, pp. 415–424 (2014)

Deep Sparse Informative Transfer SoftMax for Cross-Domain Image Classification

Hanfang Yang[1], Xiangdong Zhou[1(✉)], Lan Lin[2], Bo Yao[1], Zijing Tan[1], Haocheng Tang[1], and Yingjie Tian[3]

[1] School of Computer Science, Fudan University, Shanghai, China
{hfyang14,xdzhou,16110240008,zjtan,15210240090}@fudan.edu.cn
[2] School of Electronics and Information Engineering,
Tongji University, Shanghai, China
linlan@tongij.edu.cn
[3] State Grid Shanghai Municipal Electric Power Company, Shanghai, China
13901712348@163.com

Abstract. In many real applications, it is often encountered that the models trained on source domain cannot fit the related target images very well, due to the variants and changes of the imaging background, lighting of environment, viewpoints and so forth. Therefore cross-domain image classification becomes a very interesting research problem. Lots of research efforts have been conducted on this problem, where many of them focus on exploring the cross-domain image features. Recently transfer learning based methods become the main stream. In this paper, we present a novel transfer SoftMax model called Sparse Informative Transfer SoftMax (SITS) to deal with the problem of cross-domain image classification. SITS is a flexible classification framework. Specifically, the principle eigenvectors of the target domain feature space are introduced into our objective function, hence the informative features of the target domain are exploited in the process of the model training. The sparse regularization for feature selection and the SoftMax classification are also employed in our framework. On this basis, we developed Deep SITS network to efficiently learn informative transfer model and enhance the transferable ability of deep neural network. Extensive experiments are conducted on several commonly used benchmarks. The experimental results show that comparing with the state-of-the-art methods, our method achieves the best performance.

Keywords: Transfer learning · Neural network · Deep learning
Image classification · Sparse regularization

1 Introduction

With the fast and widespread proliferation of image and video sensors, the technology of image classification becomes very popular and necessary in many

© Springer International Publishing AG, part of Springer Nature 2018
J. Pei et al. (Eds.): DASFAA 2018, LNCS 10828, pp. 622–637, 2018.
https://doi.org/10.1007/978-3-319-91458-9_38

practical applications such as target detection, web image search and video surveillance etc. However, it is often the case that in some applications the pre-trained models do not work very well in the real running of classifications, where the testing domain is different from the training domain. Figure 1 shows some examples of cross-domain image datasets. One solution for this problem is to collect data from the testing domain and train a new model. But it is labor expensive and tedious to build a new dataset with human annotations. Therefore, the cross-domain image classification which aims to use the source training data to learn a related target domain classifier becomes an important research topic.

In this work, we propose a novel cross-domain classification framework called Sparse Informative Transfer SoftMax (SITS). To explore the cross-domain features, we exploit principle eigenvectors of the target domain feature space in our objective function. We further employ the sparse regularization method to select and preserve these informative features. The SoftMax with the aforementioned regularization approach is applied as the SITS layer in the deep neural networks and brings significant improvement on the cross-domain classification problem. Our main contribution can be summarized as:

1. We propose a novel cross-domain classification model SITS to explore the target domain using the principal eigenvectors and sparse regularization to fine tune the decision hyperplane.
2. We propose the Deep SITS network which is a combination of convolutional layers, SITS layers and fully connected layers. SITS layer exploits transferable information from the target domain to assist final connected layers training.
3. Our experiments show that SITS is more effective and powerful combined with the pre-processing cross-domain methods and bring further performance improvement.

Fig. 1. Image examples and domains' profile of cross-domain benchmark Office-Caltech datasets. Office-Caltech dataset contains the 10 overlapping categories between the Office dataset (includes three domains: Amazon, Webcam and Dslr) and Caltech256 dataset.

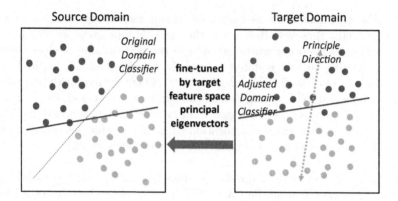

Fig. 2. SITS classifier: the decision boundary is the red line (Red dotted line represents original boundary trained by labeled data, while red solid line is adjusted boundary fitting for target data by using unlabeled target informative principle eigenvectors). Grey dotted line is the first principle direction of target domain feature space. (Color figure online)

The experiments are conducted on several commonly used benchmark datasets. The previous pre-processing methods, such as JDA [15] and TCA [17], and the state-of-the-art transfer learning based approaches [14,16,22,27] are adopted for performance comparisons. The experimental results show that our methods achieve the best performance on all the datasets in our experiments. Especially, integrating with the pre-processing and Convolutional Neural Network (CNN), our method achieves significant improvement compared with other combination approaches.

The paper is organized as follows. The review of related work is given in Sect. 2. In Sect. 3, our SITS model is given and the sparse regularization and informative factors are also introduced. In Sect. 4, we introduce the benchmark data and our experimental settings and compare our method with the baselines and the state-of-the-art methods in terms of accuracy for evaluation. In Sect. 5 we conclude our work.

2   Related Work

Most of the cross-domain classification methods can be categorized into two classes: the feature pre-processing based methods [3,4] and the transfer learning based methods [9,15,17]. The feature pre-processing based methods focus on exploring the image features with cross-domain characteristics. Since the most domain bias/difference is often related with the bias of sample selections or the shifting of means/covariances, several linear and nonlinear transformation approaches are exploited to reduce the distribution difference. Transfer learning models make use of domain knowledge to tackle the related target classification problem. The cross-domain image classification can be regarded as a kind of transfer learning problem. Hence the transfer learning techniques are extensively applied for dealing with the problem of cross-domain classification. Multiple Kernel Learning (MKL) [6] and distance metric learning [6,23,24] techniques are also applied to cross-domain image classification. Zhu et al. [27] exploit the first principle component of the target feature space in their classification model training.

It is noticed that with the fast development of deep learning technique, the performance of transfer classification has obtained a significant boost with deep structure, such as deep low rank coding [5] and deep features training [7]. Besides, Yosinski et al. [26] comprehensively explore the deep CNN transferability through manifest invariant factors across domains with sufficient target labels. Tzeng et al. [20,21] improve the adaptation ability of deep networks with domain invariance and discriminative feature learning. Due to the limited labeled data in reality, domain-invariant features with deep CNNs [14,22] and domain-adversarial deep architectures [8] have been proposed. Deep domain confusion (DDC) [22] is proposed for learning a domain-invariant representation by adding network adaptation layer and dataset shift loss. Deep adaptation network (DAN) [14] focuses on increasing transferability in deep feed-forward networks with kernel Hilbert space reproducing and discrepancy reducing. In our work, the deep features of the source domain are adapted with the target domain by PCA and L1 regularization based feature selection. In our experiment comparing with several previous deep CNN transfer learning models [13,14,22,25] on the standard benchmarks, our method achieves the best performance.

3   Sparse Informative Transfer SoftMax Model

In this section, we introduce the proposed Sparse Informative Transfer SoftMax (SITS) model in details.

3.1   Problem Formulation

In the cross-domain classification, we are given $\mathbf{M}$ classes from two specific domains $\mathbf{D_{src}}$ and $\mathbf{D_{tar}}$, where $\mathbf{D_{src}}$ denotes the labeled source domain while $\mathbf{D_{tar}}$ denotes the unlabeled target domain. Let $\{\mathbf{x_{src_i}}, \mathbf{y_{src_i}}\}_{i=1}^{n_{src}}$ denote the training instances and labels in the source domain $\mathbf{D_{src}}$, and $\{\mathbf{x_{tar_i}}\}_{i=1}^{n_{tar}}$ denote the samples in the target domain without labels. Differ from the traditional classification problem, the IID (Identical Independent Distribution) assumption between the training (source) and testing (target) data sets is not guaranteed in the scenario of cross-domain classification. Generally, in the cross-domain classification problem, $\mathbf{D_{src}}$ and $\mathbf{D_{tar}}$ have different input spaces/distributions. Due to the distribution discrepancy between the source domain and the target domain, a classifier $\mathbf{f_{src}}$ trained on the source domain $\mathbf{D_{src}}$ usually does not work very well on the target domain $\mathbf{D_{tar}}$.

In cross-domain classification, our aim is to predict the labels of the samples of the target domain $\mathbf{D_{tar}}$ by training the model $\mathbf{f_{src}}$ on the source domain training data and exploring the target data distribution. We optimize a multi-nominal logistic regression with Elastic Net regularization and informative factors of the target domain. SoftMax function is used to define the n-class logistic classifier model in our experiment which is given by:

$$P(y_i = Y|x_i; \omega) = f_{src}(y_i = Y|x_i, \omega_1, \omega_2, \ldots, \omega_M) = \frac{e^{\omega_Y^T x_i}}{\sum_{k=1}^{M} e^{\omega_k^T x_i}}. \tag{1}$$

Here $\omega_1, \omega_2, \ldots, \omega_M$ are the parameters of the classifier and the term $\sum_{k=1}^{M} e^{\omega_k^T x_i}$ normalizes the distribution. The model is trained by maximum-likelihood and we rewrite the regularized form of the negative log-likelihood as:

$$L_{src}(\omega) = -\sum_{i=1}^{n_{src}} \sum_{j=1}^{M} 1\{y_i = j\} log \frac{e^{\omega_{y_i}^T x_i}}{\sum_{k=1}^{M} e^{\omega_k^T x_i}}, \tag{2}$$

where M is the number of training classes.

However, f_{src} often leads to the overfitting in transfer learning problem due to the discrepancy between the source and target domain distributions. With the deep network image features such as CNN features being widely applied, it is reasonable to assume that the feature space is often high-dimensional, which reduces the overfitting problem. Therefore to deal with the problem, we propose to fit the generalized linear model f_{src} by minimizing the negative log-likelihood with sparse penalty $L_{spr}(\omega)$. $L_{spr}(\omega)$ is a regulation factor for SoftMax regression, which scales well to large problems. We combines lasso and ridge as $L_{spr}(\omega)$ regression, also known as Elastic Net in our model training.

To explore the target domain input space, we distill informative properties by using a numbers of principle eigenvectors for regularization $L_{inf}(\omega)$ to bridge the gap between the source and target feature space. It helps to fine tune the direction of the decision hyperplane to more suitable for target domain as shown in Fig. 2

Hence the loss function is reconstructed as follows:

$$L_{tar}(\omega) = L_{src}(\omega) + \mu L_{inf}(\omega) + \lambda L_{spr}(\omega), \tag{3}$$

where the hyper-parameters λ and μ determine the weight of sparse penalty and informative factors. $L_{spr}(\omega)$ is the sparse constraint for SoftMax regression and $L_{inf}(\omega)$ is a combination of multiple informative transfer factors helping to understand target datasets, which will be described in details in the following sections.

3.2   Multiple Informative Transfer Factor

In this subsection we introduce the idea of maximizing the information representation of the target domain data into the sparse SoftMax model training. With our objective function, the informative features of target domain can be preserved and utilized to adjust the classification hyperplane in the model training which consequently leads to performance improvement in the testing.

PCA is powerful for extracting unlabeled data structure and obtaining its intrinsic variability. In PCA the data samples are projected onto the new basis of principal component (PC) and each PC is a linear combination of the original input space. PCA is often used to reduce the number of dimensions of high-dimensional data in traditional data processing. For the covariance matrix or the correlation matrix that PCA applied, the eigenvectors correspond to the principal components and represent the general directions of the variability of the input space.

In our SITS model, with the Elastic Net regularization multiple principal eigenvectors of the target domain are selected to fine-tune the decision hyper-plane. This makes the classification model more discriminative in the larger variance directions of the target domain and finally brings performance improvement. The informative factor is given as follows.

$$L_{inf}(\omega) = \sum_{k=1}^{M} \frac{\|\omega_k v_1\|^2 + \frac{1}{2!}\|\omega_k v_2\|^2 + \ldots + \frac{1}{t!}\|\omega_k v_t\|^2}{1 + \frac{1}{2!} + \ldots + \frac{1}{t!}}, \tag{4}$$

where ω is the SoftMax classifier parameter. In the above equation, $v_1, v_2, \ldots, v_t$ are the first t principle eigenvalues of the target domain data. In this equation, we use the reciprocal of factorial as decay penalty to weight each principal direction. As illustrated in Fig. 2, where t equals to 1 and $\omega_k^T v_i$ makes classifier parameter ω adjusting to vertical to the first principal direction of the target domain. In our experiment, we use multiple informative factors to exploit more principal directions of the target data.

3.3   Sparse SoftMax Model

In a lot of multinomial learning methods, SoftMax transformation is a key component used as activation function, encompassing multinomial logistic regression in neural networks and reinforcement learning as multi-instance classifier. Soft-Max with simple log-likelihood loss function and appealing efficiency is widely used in deep neural network.

The sparse SoftMax model adds a sparsity constraint to the objective function. The sparsity constraint is given by the reconstruction error of the Elastic Net regularization $\mathbf{L_{spr}}(\omega)$:

$$L_{spr}(\omega) = \sum_{k=1}^{M} \lambda_1 \|\omega_k\|^2 + \lambda_2 \|\omega_k\|_1, \tag{5}$$

where the first term tends to decrease the magnitude of the weights to prevent overfitting. The first parameter λ_1 controls the relative importance of the term and λ_2 controls how sparse the model is.

In our sparse softmax model, lasso penalty shrinks the linear regression coefficients towards zero and generate sparse models, providing a computationally feasible model selection way. However ridge penalty helps extend the number of non-zero coefficients, while lasso does not perform well when the number of predictors is much larger than the observations or the predictors are highly correlated, Elastic Net regularization is proposed integrating both advantages [11].

3.4   Parameter Estimation

Here we give the solution for our objective function, which is shown as follows:

$$min_{(\lambda,\mu)}[L_{src}(\omega) + L_{inf}(\omega) + L_{spr}(\omega)]. \tag{6}$$

The combining of Elastic Net regularization $\mathbf{L_{spr}}(\omega)$ implies an automatic selection of the principle eigenvectors in $\mathbf{L_{inf}}(\omega)$.

The solution for SoftMax with sparse regularization has mature theory and algorithm. Gradient descent and coordinate descent are the most commonly used methods to solve the problem. In this paper, our objective function is minimized by using batch gradient descent.

The gradient of the loss function is:

$$G(\omega) = \nabla L_{tar}(\omega) = \nabla L_{src}(\omega) + \mu\nabla L_{inf}(\omega) + \lambda\nabla L_{spr}(\omega), \tag{7}$$

which denoted by the $\mathbf{m} \times \mathbf{n}$ indicator response matrix ($\mathbf{m}$ is the number of dimensions of the input space and $\mathbf{n}$ is the classes number). Then we can write the partial derivative of the SoftMax as:

$$\nabla_{\omega_k} L_{src}(\omega) = -\sum_{i=1}^{n_{src}}[x_i(1\{y_i = k\} - P(y_i = k|x_i; \omega))], \tag{8}$$

where $\nabla_{\omega_k} L_{src}$ helps for grouping the k-th category responses for each value $\mathbf{x_i}$. By solving the minimization problem with sparse penalty, the absolute value function does not have a derivative at zero. Thus we introduce a positive tiny factor ϵ in solving the derivatives.

$$\nabla_{\omega_k} L_{spr}(\omega) = 2 * \lambda_1 * \omega_k + \lambda_2 sign(\omega_k)(|\omega_k| - \epsilon)_+, \tag{9}$$

where for each element of ω_k

$$sign(\omega_k)(|\omega_k| - \epsilon)_+ = \begin{cases} \omega_{k,i} - \epsilon & \text{if } \omega_{k,i} > \epsilon \\ 0 & \text{if } |\omega_{k,i}| \le \epsilon \\ \omega_{k,i} + \epsilon & \text{if } \omega_{k,i} < -\epsilon \end{cases}. \tag{10}$$

And the gradient of the $\nabla\mathbf{L_{inf}}(\omega)$ is expand as:

$$\nabla_{\omega_k} L_{inf}(\omega) = 2\omega_k * \frac{v_1^2 + \frac{1}{2!}v_2^2 + \ldots + \frac{1}{t!}v_t^2}{1 + \frac{1}{2!} + \ldots + \frac{1}{t!}}. \tag{11}$$

Typically, determining the number of principle eigenvectors is a trade-off: poor transfer efficiency versus noise information. We will give more details in the Section of Experiment.

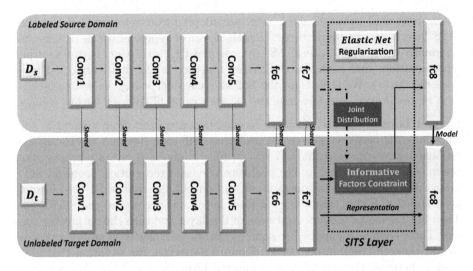

Fig. 3. Overall Caffe-based Deep SITS Architecture. As deep features transit from general to specific along the network, convolutional layer conv1–conv5 and fully connected layers fc6–fc7 share the hyper-parameters of pre-trained deep neural network. SITS layer generates target domain constraint and sparse regularization to the final layer. The classifier layer fc8 is tailored to task-specific structures with SITS model that consider informative factors constraint from target domain and elastic net regularization with prior layer representation.

3.5 Deep SITS Network

In this section, we give an overview of Deep SITS network architecture depicted in Fig. 3. Deep SITS network takes source domain images, source domain labels and target domain images as input. The network first processes the source images and target images with several convolutional (conv) and fully connected layers individually. Then, the SITS layer learns informative factors from the final convolutional layer or one of fully connected layers of target network. The informative factors work for the source network classifier layer (i.e. the final fully connected layer) training to fine-tunes the connected weights with target informative properties. In addition, the classifier layer weights are constrained by sparse Elastic Net at the same time. The output weights of classifier layer, combined with the SITS layer informative factors constraint and Elastic Net regulation, is more tailored to target domain classification task. Deep SITS network aims to construct a deep neural network which is able to train a transferable classifier that bridges the cross-domain discrepancy. More details of our deep network are given as follows:

Network Initialization: Deep SITS networks include five convolutional layers as deep image feature extractors, two fully connected layers, one SITS layer and one classifier layer as class label predictor. We experiment with two pre-trained AlexNet weights based on ImageNet datasets [13] for source and target domain respectively. The layers of $conv1 - fc7$ are adopted from released CaffeNet [12]. SITS layer takes as input target domain features representations to generate informative constraint. After SITS layer calculating, two distinct network output representations and configs are gathered together in the final connected layer (classifier layer). We then train the SoftMax classifier using the source labeled data representation and fine-tune the network using SITS layer output configs.

Layer Choose: The deep features representation transits from general to specific along the network. The recent researches [19] indicates that the last convolutional layer representation and several fully connection layers representation are more suited for tackling the problems of visual recognition problem. SITS layer takes the representation as input and output the configs for the classifier layer training. Either target layer is feasible to be connected to the SITS layer. Then SITS achieves the representation of this layer and then the output the configs. These configs are applied to source network classifier layer training process to add network transferable ability. Figure 3 illustrates the situation that SITS layer is between the second fully connected layer (fc7) and the classifier layer (fc8). In our experiment, we investigate which layer representation from conv5 to fc8 is most effective for Deep SITS framework. More details are given in the following experiments.

4   Experiments

This paper compare SITS model with the state-of-the-art transfer learning methods and evaluate it on two standard benchmark datasets: Office-31 [18] and 10 common classes Office-Caltech [9].

4.1   DataSet

Caltech-256 [10] and Office [9,18] are increasingly popular benchmark for the evaluation of image cross-domain classification, where Office datasets includes three real world object domains: Amazon, Webcam and DSLR. One challenge of the dataset is that both Office datasets and Caltech datasets are imbalance. We adopt 10 classes Caltech-Office datasets and encode SUFT features to 800-bin histogram with bag-of-word method [2].

Therefore, with these two benchmarks, we have 4 different domains images with common classes labels. The details of the four domains images are described as follows: The Caltech domain: 256 object classes and 30, 607 images, downloaded from Google. The Amazon domain: 31 classes, each of which includes different object instances seen from one canonical viewpoint, closely monitored environment, studio lighting conditions, large intra-class variations, downloaded from online merchants. The DSLR domain: 31 categories and 4652 images, acquired with a digital SLR camera, high-resolution images, realistic environment, natural light. The Webcam domain: a similar environment as the DSLR ones, low-resolution images, contain significant noise.

4.2   Experimental Settings

We adopt two baselines in our experiment: Principal Component Analysis (PCA) as data pre-processing method with 1-NN classification and the Support Vector Machine (SVM) based method. PCA projects input data points into a new orthogonal space and is often used for dimensional reduction. It is combined with 1-Nearest Neighbor Classifier (NN) for cross-domain image classification [9,17].

Multiple classifiers like SVM and ILR are trained with the labeled data from the source domain and used to predict the data samples of the target domain. We also compare our method extensively to a number of transfer learning methods listed as follows:

1. TCA [17] is a conventional transfer learning method based on MMD-regularized PCA.
2. GFK [9] is a widely-adopted method which interpolates across intermediate subspaces to bridge the source and target.
3. LapCNN [25] is a semi-supervised variant of CNN based on Laplacian graph regularization.
4. CNN [13] is the leading method in the ImageNet 2012 competition, and it turns out to be a strong model for learning transferable features [26].
5. JDA [15] is a powerful pre-processing method for cross-domain problem extending MMD by jointly adapting both the marginal distribution and conditional distribution.
6. JD-CNN is the strong baseline in our experiment. It combines pre-processing the CNN feature with JDA [15] models, and then trained by nearest-neighbor. Specifically, experiment adopt supervised CNN based fc7 layer as feature representation, which trained in the ImageNet using the released CaffeNet [12] weights.
7. ILR [27] introduces the target data distribution constraints into logistic regression classification model to keep informative transfer information.
8. TJM [16] is also a practical pre-procession transfer learning model especially in large domain difference problem. TJM is jointly matching the features and reweighting the instances across domain in a principled dimensionality reduction procedure.
9. DDC [22] is proposed for learning a domain-invariant representation by adds an adaptation layer between the fc7 and fc8 layers that is regularized by single-kernel MMD to the deep CNN.
10. DAN [14] generalizes a new deep CNN to domain adaptation that embedding layers to reproducing kernel Hilbert space and reducing discrepancy using an optimal multi-kernel selection method.

In the experiments, we evaluate two variants of sparse informative transfer softmax (SITS): SITS and JD-SITS, by using the SURF embedding feature and deep representations. JD-SITS is the extended model of SITS, which is a combination methods with the joint adaptation pre-processing analysis [15]. For both Office-Caltech and Office-31 datasets, we follow the standard evaluation for cross-domain classification.

We examine the influence of deep representations for cross-domain classification by employing the AlexNet [13] from the open source Caffe [12] package, which is pre-trained on the ImageNet dataset. The contrast methods, including CNN, LapCNN, JD-CNN, DDC, and DAN methods, adopt the same network configuration settings with our proposed methods.

In order to study the effect of various layer features, we evaluate variants layer feature combined with our SITS transfer learning model classification accuracy. To determine the optimal values of the hyper-parameters we perform a grid search over the parameter space of the number of units and the optimal weight decay was found to be: $\lambda_1 = 1 \times 10^{-4}$, $\lambda_2 = 5 \times 10^{-4}$ and $\mu = 5 \times 10^{-4}$. The first 10 principle eigenvectors are adopted in our model training in both SURF-based and deep-based experiments.

4.3   Experimental Results

We utilize two kinds of visual features: SURF features encoded by the bag-of-words model and the deep convolutional neural network features on both of Caltech and Office datasets. 10 common classes of these two datasets are adopted as usual for the classification task. In addition, the Office dataset with 31 classes is also employed for deep feature based evaluation.

Our experiment is divided into two parts according to which kind of feature being employed: SURF feature and deep feature. For the experiment with SURF feature, Table 1 shows the experimental results of our models comparing with the baselines, the state-of-the-art pre-processing methods and the transfer models respectively. In contrast to ILR [27], which exploits a single pre-calculated sparse principle component of the target domain for transfer learning, our SITS model employs multiple salient direction factors to fine-tune the decision hyperplane.

Table 1. Accuracy (%) on 10 classes Office-Caltech cross-domain images datasets with SURF feature.

Datasets		Baseline		Prior transfer learning methods					Our methods	
Source	Target	PCA	SVM	GFK [9]	TCA [17]	ILR [27]	JDA [15]	TJM [16]	SITS	JD-SITS
Caltech	Amazon	36.95	55.64	41.02	38.20	55.11	44.78	46.76	57.93	**58.14**
Caltech	Webcam	32.54	43.73	40.68	38.64	48.81	41.69	38.98	45.08	**49.15**
Caltech	DSLR	38.22	45.22	38.85	41.40	**46.50**	45.22	44.59	44.59	**46.50**
Amazon	Caltech	34.73	45.77	40.25	37.76	42.39	39.36	39.45	**45.24**	42.48
Amazon	Webcam	35.59	39.66	38.98	37.63	37.29	37.97	**42.03**	41.36	**42.03**
Amazon	DSLR	27.39	42.04	36.31	33.12	42.04	39.49	**45.22**	42.68	41.40
Webcam	Caltech	26.36	26.62	30.72	29.30	37.79	31.17	30.19	32.21	**34.64**
Webcam	Amazon	31.00	29.39	29.75	30.06	35.80	32.78	29.96	37.89	**38.62**
Webcam	DSLR	77.07	63.39	80.89	87.26	84.08	89.17	**89.17**	85.35	**89.17**
DSLR	Caltech	29.65	34.76	30.28	31.70	37.27	31.52	31.43	34.11	**35.35**
DSLR	Amazon	32.05	31.43	32.05	32.15	33.57	33.09	32.78	36.01	**42.28**
DSLR	Webcam	75.93	82.80	75.59	86.10	81.36	**89.49**	85.42	84.07	86.78
Average		39.79	45.04	42.94	43.61	48.50	46.31	47.10	48.96	**50.55**

Table 2. Accuracy (%) on 10 classes Office-Caltech datasets with deep convolutional neural network

Datasets		Baseline deep methods			Recent methods		Our methods	
Source	Target	LapCNN [25]	CNN [13]	JD-CNN [13,15]	DDC [22]	DAN [14]	SITS	JD-SITS
Amazon	Caltech	83.6	83.8	84.2	84.3	86.0	85.2	**87.0**
Webcam	Caltech	77.8	76.1	85.1	76.9	81.5	79.8	**88.0**
DSLR	Caltech	80.6	80.8	86.0	80.5	82.0	82.9	**87.0**
Average		80.7	80.2	85.1	80.6	83.1	82.7	**87.3**
Caltech	Amazon	92.1	91.1	89.8	91.3	92.0	91.8	**92.7**
Caltech	Webcam	81.6	83.1	86.8	85.5	92.0	87.1	**93.6**
Caltech	DSLR	87.8	89.0	84.7	89.1	**90.5**	89.8	**90.5**
Average		87.2	87.7	87.1	88.6	91.5	89.6	**92.3**
Overall average		83.9	84.0	86.1	84.6	87.3	86.1	**89.8**

Table 3. Accuracy (%) on Office-31 datasets with deep convolutional neural network

Datasets		Baseline deep methods			Recent methods		Our methods	
Source	Target	LapCNN [25]	CNN [13]	JD-CNN [13,15]	DDC [22]	DAN [14]	SITS	JD-SITS
Amazon	Webcam	60.4	61.6	63.9	61.8	68.5	62.5	**68.9**
DSLR	Webcam	94.7	95.4	**96.9**	95.0	96.0	95.6	96.0
Webcam	DSLR	99.1	99.0	98.0	98.5	99.0	**99.4**	96.4
Amazon	DSLR	63.1	63.8	61.5	64.4	67.0	64.7	**68.3**
DSLR	Amazon	51.6	51.1	58.8	52.1	54.0	51.9	**60.0**
Webcam	Amazon	48.2	49.8	53.5	52.2	53.1	50.3	**57.9**
Average		69.5	70.1	72.1	70.6	72.9	70.7	**74.6**

The experimental results show that our model JD-SITS achieves the best performance, 50.55% on average accuracy and followed by our SITS model 48.96% and ILR model 48.5%.

For the experiment with deep feature, we compare our methods against the baseline of CNN and the state-of-the-art CNN transfer models, such as DAN [14] and DDC [22]. Table 2 shows the experimental results on the 10 common classes of Office and Caltech datasets. It shows that our JD-SITS model achieves the best performance on average and outperform 5.8% comparing with CNN. The experimental results also shown that our framework is more effective for model integration. It is also observed that with the integration of JD pre-processing method, our JD-SITS improves the accuracy by 3.7%, from 86.1% to 89.8%. In contrast, the strong baseline JD-CNN, which is an integration of JD pre-processing with CNN method improves the accuracy from 84% to 86.1%, the margin is 2.1%. Hence, the experimental results show that our SITS model is more effective of integrating with pre-processing transfer learning than the traditional ones.

Table 3 shows the experimental results on the Office datasets with 31 classes. In this experiment, our models are compared with serval deep learning models, such as CNN, LapCNN, DDC and DAN. It is shown that our deep feature based

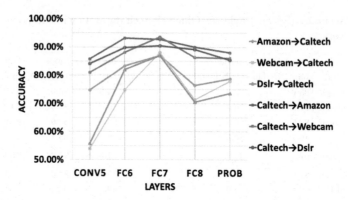

Fig. 4. Joint method (JD-SITS) evaluation based on various CNN layers representation over 10 classes Office-Caltech datasets

SITS model with pre-process (JD-SITS) achieves the best accuracy 74.6% on average. Specifically, our JD-SITS model achieves 3.9% (from 70.7% to 74.6%) accuracy improvement comparing with SITS, while the JD-CNN rises the accuracy 2.1% (from 70.1% to 72.1%) comparing CNN.

From Tables 2 and 3, it can be observed that our method JD-SITS outperforms 2.5% and 1.7% on average comparing with the state-of-the-art method DAN [14], respectively. All the experimental results demonstrate that our SITS model and JD-SITS model are more effective to boost the transferability compared with the baselines and the state-of-the-art methods.

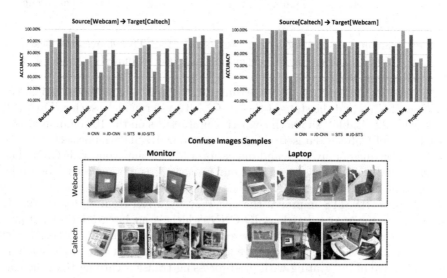

Fig. 5. Classification accuracy (%) comparison on each classes of 10 classes Ofce-Caltech datasets and confuse intra-class images samples.

4.4   Discussion

In this section we give more details of our experiment to illustrate the effectiveness of our proposed method on cross-domain image classification. Figure 5 shows the classification accuracies on ten classes in both of Office and Caltech datasets. The 10 classes labels are: *projector, mug, mouse, monitor, laptop, keyboard, headphones, calculator, bike, backpack*. Where CNN and our SITS models are compared by using deep features. From the results, it can be observed that comparing with CNN, SITS can enhance the transfer ability significantly by exploring the informative and significant directions of the target domain. From Table 2, it can be observed that our SITS model outperforms 2.5% (from 80.2% to 82.7%) than the CNN method with Caltech dataset as the target domain. When the Office dataset (amazon, webcam and dslr) is utilized as the target domain, our SITS model outperforms 1.9% (from 87.7% to 89.6%) than the CNN method. We notice that the accuracy improvements are apparently different when we exchange the target and the source domains ($Caltech \rightarrow Office$ VS $Office \rightarrow Caltech$). With the observation of the images in these datasets, we find that the background of Caltech images is more complex than the Office dataset. It partially indicates that with the PCA and sparse regularization approach our SITS model can exploit the complex target images more effectively and finally brings further improvements. Although in most of the cases our SITS model achieves the best performance, there is some exceptions in Fig. 5, for instance on the class of "*monitor*", our STIS model shows some worse performance than the CNN method with the Caltech dataset bing the target domain. By investigating the images of Caltech and Office, it shows that in the target domain (Caltech) the images of "*monitor*" class are very complex and prone to be confused with other class such as "Laptop", please refer to the "confuse Image Samples" in Fig. 5 for more details. Therefore like the idiom said "*A coin has two sides*", in this situation the PCA method in our SITS may not always select the correct informative features and finally leads to the even worse performance comparing with the CNN model. While in general the experimental results show that our STIS and JD-STIS models work quite good and achieve the best performance comparing with CNN methods.

Our experimental results also demonstrate that deep feature brings significant improvement on cross-domain image classification accuracy comparing with SURF feature. However, how many layers of the deep network should be used in the cross-domain image classification is not a well studied problem. Previous work [1] concludes that the first fully connected layer (fc6) works best for most visual recognition problem. While for the datasets of Office and Caltech in our experiment, we observe that the deep feature of the second fully connection layer (fc7) brings the best cross-domain image classification accuracy on average. Please refer to Fig. 4 for more details. In our experiment the first fully connection layer (fc6) brings less performance than fc7 by 4.7% on average. Therefore, we adopt fc7 deep features with all the deep learning methods in our experiment.

5   Conclusions

In this paper, we have proposed a novel transfer learning classifier model SITS. It integrates target domain distribution information to classifier training process to fine-tune the decision hyperplane. The deep SITS framework is also introduced to improve the classification accuracy. As a novel classification model, it is flexible and can be easily integrated with the pre-processing approach to bring further improvements. Our experimental results demonstrate that our model achieves significant improvement over the state-of-the-art methods.

Acknowledgment. This work was supported by the National High Technology Research and Development Program (863 Program) of China (2015AA050203), NSFC grant no. 61370157, NSFC grant no. 61373106, NSFC grant no. 61572135 and State Grid Shanghai Company Project No. 52094016001Z.

References

1. Azizpour, H., Razavian, A.S., Sullivan, J., Maki, A., Carlsson, S.: From generic to specific deep representations for visual recognition. CoRR abs/1406.5774 (2014)
2. Borgwardt, K.M., Gretton, A., Rasch, M.J., Kriegel, H.-P., Schölkopf, B., Smola, A.J.: Integrating structured biological data by kernel maximum mean discrepancy. Bioinformatics **22**(14), e49–e57 (2006)
3. Bruzzone, L., Marconcini, M.: Domain adaptation problems: a DASVM classification technique and a circular validation strategy. IEEE Trans. Pattern Anal. Mach. Intell. **32**(5), 770–787 (2010)
4. Chen, M., Weinberger, K.Q., Blitzer, J: Co-training for domain adaptation. In: Advances in Neural Information Processing Systems, pp. 2456–2464 (2011)
5. Ding, Z., Shao, M., Fu, Y.: Deep low-rank coding for transfer learning. In: IJCAI, pp. 3453–3459 (2015)
6. Lixin, D., Tsang, I.W., Xu, D.: Domain transfer multiple kernel learning. IEEE Trans. Pattern Anal. Mach. Intell. **34**(3), 465–479 (2012)
7. Ganin, Y., Lempitsky, V.: Unsupervised domain adaptation by backpropagation. arXiv preprint arXiv:1409.7495 (2014)
8. Ganin, Y., Ustinova, E., Ajakan, H., Germain, P., Larochelle, H., Laviolette, F., Marchand, M., Lempitsky, V.: Domain-adversarial training of neural networks. J. Mach. Learn. Res. **17**(59), 1–35 (2016)
9. Gong, B., Shi, Y., Sha, F., Grauman, K.: Geodesic flow kernel for unsupervised domain adaptation. In: 2012 IEEE Conference on Computer Vision and Pattern Recognition (CVPR), pp. 2066–2073. IEEE (2012)
10. Griffin, G., Holub, A., Perona, P.: Caltech-256 object category dataset (2007)
11. Hastie, T., Tibshirani, R., Wainwright, M.: Statistical Learning with Sparsity. CRC Press, Boca Raton (2015)
12. Jia, Y., Shelhamer, E., Donahue, J., Karayev, S., Long, J., Girshick, R., Guadarrama, S., Darrell, T.: Caffe: convolutional architecture for fast feature embedding. In: Proceedings of the 22nd ACM International Conference on Multimedia, pp. 675–678. ACM (2014)
13. Krizhevsky, A., Sutskever, I., Hinton, G.E.: Imagenet classification with deep convolutional neural networks. In: Advances in Neural Information Processing Systems, pp. 1097–1105 (2012)

14. Long, M., Cao, Y., Wang, J., Jordan, M.I.: Learning transferable features with deep adaptation networks. In: ICML, pp. 97–105 (2015)
15. Long, M., Wang, J., Ding, G., Sun, J., Yu, P.S.: Transfer feature learning with joint distribution adaptation. In: Proceedings of the IEEE International Conference on Computer Vision, pp. 2200–2207 (2013)
16. Long, M., Wang, J., Ding, G., Sun, J., Yu, P.S.: Transfer joint matching for unsupervised domain adaptation. In: Proceedings of the IEEE Conference on Computer Vision and Pattern Recognition, pp. 1410–1417 (2014)
17. Pan, S.J., Tsang, I.W., Kwok, J.T., Yang, Q.: Domain adaptation via transfer component analysis. IEEE Trans. Neural Netw. **22**(2), 199–210 (2011)
18. Saenko, K., Kulis, B., Fritz, M., Darrell, T.: Adapting visual category models to new domains. In: Daniilidis, K., Maragos, P., Paragios, N. (eds.) ECCV 2010. LNCS, vol. 6314, pp. 213–226. Springer, Heidelberg (2010). https://doi.org/10.1007/978-3-642-15561-1_16
19. Razavian, A.S., Azizpour, H., Sullivan, J., Carlsson, S.: CNN features off-the-shelf: an astounding baseline for recognition. In: Proceedings of the IEEE Conference on Computer Vision and Pattern Recognition Workshops, pp. 806–813 (2014)
20. Tzeng, E., Hoffman, J., Darrell, T., Saenko, K.: Simultaneous deep transfer across domains and tasks. In: Proceedings of the IEEE International Conference on Computer Vision, pp. 4068–4076 (2015)
21. Tzeng, E., Hoffman, J., Saenko, K., Darrell, T.: Adversarial discriminative domain adaptation. In: ICLR Workshop (2017)
22. Tzeng, E., Hoffman, J., Zhang, N., Saenko, K., Darrell, T.: Deep domain confusion: maximizing for domain invariance. arXiv preprint arXiv:1412.3474 (2014)
23. Wang, H., Wang, W., Zhang, C., Xu, F.: Cross-domain metric learning based on information theory. In: AAAI, pp. 2099–2105 (2014)
24. Wang, W., Wang, H., Zhang, C., Xu, F.: Transfer feature representation via multiple kernel learning. In: AAAI, pp. 3073–3079 (2015)
25. Weston, J., Ratle, F., Mobahi, H., Collobert, R.: Deep learning via semi-supervised embedding. In: Montavon, G., Orr, G.B., Müller, K.-R. (eds.) Neural Networks: Tricks of the Trade. LNCS, vol. 7700, pp. 639–655. Springer, Heidelberg (2012). https://doi.org/10.1007/978-3-642-35289-8_34
26. Yosinski, J., Clune, J., Bengio, Y., Lipson, H.: How transferable are features in deep neural networks? In: Advances in Neural Information Processing Systems, pp. 3320–3328 (2014)
27. Zhu, G., Yang, H., Lin, L., Zhou, G., Zhou, X.: An informative logistic regression for cross-domain image classification. In: Nalpantidis, L., Krüger, V., Eklundh, J.-O., Gasteratos, A. (eds.) ICVS 2015. LNCS, vol. 9163, pp. 147–156. Springer, Cham (2015). https://doi.org/10.1007/978-3-319-20904-3_14

Sitcom-Stars Oriented Video Advertising via Clothing Retrieval

Haijun Zhang, Yuzhu Ji[(✉)], Wang Huang, and Linlin Liu

Department of Computer Science, Shenzhen Graduate School,
Harbin Institute of Technology, Shenzhen, China
andrewchiyz@stu.hit.edu.cn

Abstract. This paper introduces a novel learning-based framework for video content-based advertising, DeepLink, which aims at linking sitcom-stars and online stores with clothing retrieval by using state-of-the-art deep convolutional neural networks (CNNs). Concretely, several deep CNN models are adopted for composing multiple sub-modules in DeepLink, including human-body detection, human-pose selection, face verification, clothing detection and retrieval from advertisements (ads) pool that is constructed by clothing images collected from real-world online stores. For clothing detection and retrieval from ad images, we firstly transfer the state-of-the-art deep CNN models to our data domain, and then train corresponding models based on our constructed large-scale clothing datasets. Extensive experimental results demonstrate the feasibility and efficacy of our proposed clothing-based video-advertising system.

Keywords: Video advertising · Deep learning · Object detection
Face verification · Image retrieval · Clothing detection

1 Introduction

According to the increasing online video traffic and its growing revenue, video advertising has a huge potential of business opportunities in the online video market [1]. However, the widely used advertising approach to most video sites is still relying on directly inserting an advertisement (ad) in the beginning or the middle of a video. The ad is usually unrelated with the video content. On the other hand, undifferentiated advertising for all the users will increase the advertising cost. Thus, a tradeoff between reducing the impact on users viewing experience and keeping the advertising revenue should be considered jointly.

Typical video advertising systems work on finding appropriate locations for ads in sports videos, personalized ads delivery in interactive digital television (IDTV) [2], text-based advertising, video segment-level advertising [3], and object-level video advertising [4]. Although the ad selection in IDTV considers a users preference, such as viewer information, or current or past users activities [2], these advertising approaches deliver ads without using the content-relevance

© Springer International Publishing AG, part of Springer Nature 2018
J. Pei et al. (Eds.): DASFAA 2018, LNCS 10828, pp. 638–646, 2018.
https://doi.org/10.1007/978-3-319-91458-9_39

of the ads and target video. Widely-used text-based advertising methods, such as AdSense and AdWords[1]. Segment-level video advertising, e.g. vADeo and VideoSense [3], works by recognizing scene changes in a video and locate a related ad in a suitable place of the screen. Moreover, an in-stream video advertising strategy was designed by considering the emotional impact of the videos as well as ads [5]. In recent work, an object-level video advertising (OLVA) framework was developed [4]. Despite the encouraging results obtained by OLVA, only low-level features (i.e. HOG) were used in OLVA and its effectiveness depends heavily on the performance of the object detection method.

In this paper, we introduce a novel learning-based framework for content-based video advertising, which aims to link soapstars and online shops with clothing retrieval by using state-of-the-art deep learning models. We call the proposed framework as DeepLink in the following context. Our framework was tested on a famous American sitcom, *The Big Bang Theory*. Experimental results demonstrate the feasibility and efficacy of our framework. The contribution of this paper is three-fold: (1) A new learning-based video-advertising framework, DeepLink, is presented by using several state-of-the-art deep learning models. (2) The performance of deep CNNs, which is experiencing increased popularity in both academia and industry, has not yet been explored in clothing-related video-advertising tasks. We provide an empirical study on the performance of state-of-the-art deep CNN. (3) Three large-scale datasets for clothing-detection, pose selection, and clothing retrieval, were constructed for clothing-related research.

2   Models and Implementation Details

Overview of DeepLink. The main idea of our proposed clothing-related advertising is based on accurate object detection and image retrieval for ads recommendation by using state-of-the-art CNNs. The whole DeepLink system consists of several sub-modules, including human-body detection, pose selection, leading character verification, clothing detection, and ads image retrieval (see Fig. 1).

Learning to Detect the Human Body. Since we consider the content-based video-advertising system from the perspective of detecting the clothing of leading characters, detecting human bodies from video frames is the initial step. By considering the scalability of our framework, we trained a model for multiple-object detection. The module will be activated if a human body is detected.

(1) **Dataset:** PASCAL VOC 2012[2] contains a total of 11,530 images, categorized into 20 classes with 27,450 ROI annotated objects. Approximately 4,087 images contain human bodies and annotated location data. We trained our networks on this dataset for human-body detection.

(2) **Evaluation result:** We evaluated five deep CNN models, i.e., Faster R-CNN with ZF model [6], Faster R-CNN with VGGNet-16 [7], Faster R-CNN with VGG_CNN_M_1024, SSD300 with VGGNet-16, and SSD300 with ResNet101

---

[1] http://www.google.com/adwords/.
[2] http://host.robots.ox.ac.uk/pascal/VOC/voc2012/index.html.

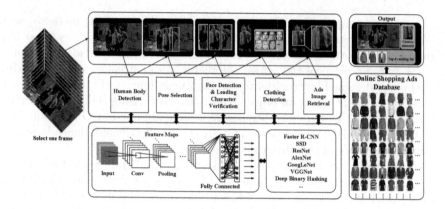

Fig. 1. Overview of our proposed DeepLink framework.

[8]. All of the deep CNN models were implemented by using Caffe library[3]. Experimental results are shown in Table 1. Based on this evaluation, we chose SSD300 with ResNet101 for human-body detection in our implementation. By using such model, we cropped out 208,519 human body regions from the raw video-frame dataset.

Table 1. Comparison results of human-body detection.

Network	Faster R-CNN (ZF)	Faster R-CNN (VGG_CNN_M_1024)	Faster R-CNN (VGG-16)	SSD300 (VGG-16)	SSD300 (ResNet)
Acc.	0.753	0.841	0.671	0.836	0.853

Learning to Select Pose. In our framework, we are actually concern with whether or not the body is in a good pose to facilitate clothing detection, instead of concerning what pose the detected body is in. For evaluating pose selection, we selected two widely-used deep CNN structures, including AlexNet [9], and GoogLeNet [10]. All of the above two network structures were modified with a binary classifier in order to transfer the deep CNN model to our data domain.

(1) Dataset: We collected 23,167 human-body images, in which 11,097 images were annotated as positive pose samples with frontal full body, and 12,070 images as negative ones with half bodies, and bodies with deformed clothing. In the dataset, 13,190 images were collected from the videos, and 9,977 images were cropped from the Street-shop dataset [11]. All of the cropped body images were manually cleaned and labeled. We randomly held out 50% of each class of images as training/validation set, and the rest 50% of images were used for testing.

---

[3] http://caffe.berkeleyvision.org/.

(2) **Evaluation Result:** According to our experiment, GoogLeNet achieved better performance, with 92.2% accuracy on pose selection in comparison to AlexNet (88.3%). Thus, GoogLeNet was selected to perform pose selection. According to the human-body detection result in the previous step, 208,519 human-body images will be fed into the pose selection module in this part. As a result, 84,415 human-body images were determined by our model as positive samples for the next part of sitcom-star face detection and verification.

Learning to Recognize Sitcom-Stars. To simplify this task, we performed face recognition by verification on a relatively small set of faces by collecting the faces of seven leading stars of *The Big Bang Theory*. Concretely, we adopted an open source face recognition engine, SeetaFace [12], to conduct face detection, face alignment and feature extraction. Then 2,048-dimension features were extracted for similarity calculation. Cosine similarity between leading roles and detected faces were calculated. By empirically setting the verification threshold to 0.66, a binarized verification vector was obtained. To verify a detected face, vectors with all-zero elements and multiple ones located in multiple intervals of leading characters were regarded as non-leading character samples, which will be strictly filtered out.

In our implementation, among the 84,415 positive human-pose samples, around 60,312 faces were detected, and 22,786 face images were verified as leading characters. As a result, among the 22,786 verified leading role faces, 412 faces of other characters were verified as leading roles. Thus the precision for leading character verification with respect to the verified detected faces is 98.19%, which suggests the reliability of our verification procedure. In practice, the corresponding 22,786 pose regions were preserved for clothing detection in the next step.

Table 2. Comparative detection results on our established clothing dataset.

Network	gallus-w	sweater-w	shirt-w	hoodie-m	tshirt-w	tshirt-m	suit-m	skirts-w	dress-w	shirt-m	mAP
Faster R-CNN(ZF)	0.906	0.881	0.873	0.907	0.863	0.984	0.908	0.907	0.973	0.907	0.911
Faster R-CNN(VGG_CNN_M_1024)	0.906	0.885	0.884	0.905	0.870	0.977	0.908	0.907	0.907	0.907	0.905
Faster R-CNN(VGGNet-16)	0.906	0.888	**0.893**	0.907	0.883	0.909	**0.981**	0.907	**0.985**	0.908	0.917
SSD300(VGGNet-16)	**0.934**	**0.915**	0.884	0.915	**0.890**	0.984	0.938	**0.937**	0.973	**0.937**	**0.931**
SSD300(ResNet101)	0.924	0.894	0.893	**0.923**	0.883	0.962	0.917	0.915	0.909	0.916	0.914

Learning to Detect Clothing. After we obtain a candidate region with leading stars' body, clothing-detection module will be executed to locate the potential clothes' sub-regions. Since the ultimate goal of this proposed advertising pipeline is clothing retrieval, we are focused on whether or not the clothing can be detected in a frame without the necessity to consider the category of the clothing.

(1) **Dataset:** We established a dataset by collecting clothing images from Amazon.com and Taobao.com. The constructed clothing dataset contains a total of 14,812 images, categorized into 10 classes. For clarity, we organized our dataset structure in the format of the PASCAL VOC dataset.

(2) **Evaluation Result:** We evaluated five deep CNN models, i.e., Faster R-CNN with ZFnet, Faster R-CNN with VGG_CNN_M_1024, Faster R-CNN with VGGNet-16, SSD with VGGNet-16, and SSD with ResNet101. For dataset partition, we randomly selected 50% of each class of images as training/validation set, and the remaining 50% images for testing. Experimental results are shown in Table 2. As a result, SSD300 with VGGNet-16 achieved 93.1% mAP (mean Average Precision), which is higher than the other networks. Therefore, this model was selected for clothing detection in our implementation. By using such clothing-detection model, 18,471 clothing candidate sub-images were cropped from the previous 22,786 images in a good pose with leading characters.

3   DeepLink in the Wild

Clothing Retrieval. The training and image retrieval process by transferring the original network [13] to our data domain can be summarized as follows. The whole framework consists of three modules. For clarity and without loss of generality, we took AlexNet as an example. The first module aims at learning rich image representations. In the second module, hash-like representations of the images were learned by adding an extra latent layer between the layer FC_7 and FC_8. Finally, a hierarchical search strategy was designed for coarse-to-fine image retrieval by utilizing the learned hash-like binary codes and feature representations from layer FC_7.

Dataset. A total of 263,865 clothing images were collected from Amazon and Taobao, and those images were categorized into 22 common classes. Training samples were generated automatically by using the trained clothes detection model. For each classes, 80% of images in each class were chosen as training samples, and the rest of the images were treated as validation samples.

Query Redundancy Removal by Clustering. Since similar query images may obtain similar retrieval results, a clustering process was used to filter out the redundancy queries in this step. Concretely, the density peaks clustering algorithm (DPCA) [14] was utilized as the pre-processing stage of image retrieval, and we set the hyper-parameter, the percentage of average number of neighbors, to 2%, and normalized the threshold of ρ and δ to 0.5 and 0.2, respectively. Image features are extracted from the FC_7 layer in our trained deep binary hash model with AlexNet. As a result, for 18,471 samples, the DPCA automatically generated 6,875 clusters under such hyper-parameter settings. Among them, 4,233 clusters with only one sample were regarded as outliers. Centroid samples in the remaining 2,642 clusters were selected as finally queries.

Experimental Setup. We employed three models under the binary hash code framework with different networks: AlexNet, VGGNet-16, and GoogLeNet. The length of the binary hash code of the latent layer was set to 20, and we modified the last FC layer with 22-way softmax output for 22 categories prediction. For clothing retrieval, we extracted the output of the softmax layer, and binary hash code from the latent layer for coarse-level searching, and used feature representations from the FC layer before for fine-level searching. We also extracted three traditional features for image retrieval, i.e., local binary pattern (LBP) [15], local texture pattern (LTP) [16], and local tetra patterns (LTrps) [17].

Table 3. Quantitative results of different methods on the baseline set Baidu (%).

Features	MRR	S@10	S@20	S@50	NDCG@20	NDCG@50	NDCG@100
AlexNet	57.71	**81.76**	88.64	95.53	30.78	**39.95**	**46.67**
GoogLeNet	**58.98**	78.84	84.94	91.52	**31.62**	39.38	45.37
VGGNet-16	47.92	78.88	**88.91**	**96.97**	23.32	32.01	39.15
LBP	15.47	36.03	50.53	73.24	4.97	7.87	11.07
LTP	16.38	37.85	53.22	74.22	5.26	8.36	11.53
LTrps	15.69	33.88	49.05	71.88	4.43	6.72	9.35

Evaluation Design. Many search engines have provided APIs for image search. Such a service provides a possible solution to an objective evaluation on our framework. Therefore, we have constructed such a ground-truth retrieval list by using Baidu and Google search engines. After clothing detection from the previous steps in DeepLink, 2,642 clothing images remained for constructing a query set. As a result, 2,642 queries and 2,301 queries were obtained retrieval results returned by Baidu and Google, respectively. For each query, we collected the top 20 images searched from Baidu and Google. Specifically, for the entire query set, Baidu and Google returned 52,840 and 46,020 images respectively. Sub-regions with the highest probability of clothing detection were cropped. We added these cropped images, used as positive samples, into our candidate dataset.

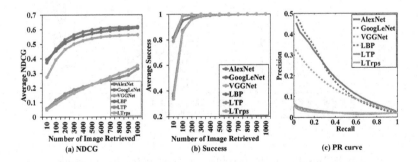

Fig. 2. Results on the baseline set constructed from Baidu.

For performance evaluation, we adopted five commonly used metrics in recommendation, retrieval and ranking systems, including: Mean Reciprocal Rank

(MRR), Success at rank k (S@k), and Normalized Discounted Cumulative Gain at rank k (NDCG@k), as well as precision, and recall. Implementation details of these metrics can be found in [18].

Table 4. Quantitative results of different methods on the baseline set Google (%).

Features	MRR	S@10	S@20	S@50	NDCG@20	NDCG@50	NDCG@100
AlexNet	**54.56**	**76.31**	**83.62**	**90.74**	**42.40**	**53.37**	**59.56**
GoogLeNet	49.89	70.80	78.53	86.40	37.48	48.76	55.71
VGGNet-16	28.19	51.89	65.80	83.05	25.46	37.39	46.63
LBP	23.93	44.59	56.58	73.66	10.51	14.97	19.52
LTP	21.82	41.76	54.50	73.01	9.71	14.09	18.62
LTrps	20.50	40.29	52.50	68.62	8.94	12.90	16.74

Experimental Result. (1) Positive Samples from Baidu: Quantitative results of different models and features on the baseline set constructed by Baidu are summarized in Table 3. It shows that deep CNN models deliver better results than other traditional features. In particular, GoogLeNet can achieve over 1.2% and 11.06% MRR improvement in comparison to AlexNet and VGGNet-16, respectively. On the contrary, AlexNet produces superior performance on S@10, but slightly lower on S@20 and S@50 when comparing with VGGNet. Figure 2(a)–(b) visually illustrates the results against the number of retrieved images. Figure 2(c) presents the precision-recall curve. According to our observation, GoogLeNet performs the best on precision. However, AlexNet achieved better results with respect to precision and recall when increasing the number of retrieved images.

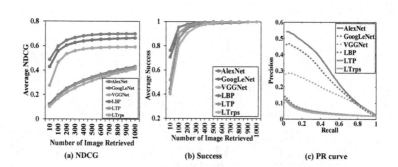

(a) NDCG   (b) Success   (c) PR curve

Fig. 3. Results on the baseline set constructed from Google.

(2) Positive Samples from Google: Comparative results of different models on the baseline set constructed by Google are given in Table 4. It shows that AlexNet outperforms other methods and traditional features consistently. Figure 3(a)–(b)

visually illustrates the results against the number of retrieved images. It is clear that AlexNet achieves better NDCG consistently compared with other methods. From the precision-recall curve shown in Fig. 3(c), AlexNet outperforms other models and features to a large extent.

4   Conclusion

This paper introduced a learning-based framework for video content-based advertising, DeepLink, which aims at linking sitcom-stars with online stores via clothing retrieval by using state-of-the-art deep CNN models. To the best of our knowledge, this research constitutes the first attempt to implement such a system for mining fashion data from videos. Extensive experimental results demonstrated the feasibility and efficiency of our proposed clothing-based system by linking the clothing of sitcom-stars in videos with similar items from a large-scale dataset in the real world.

Acknowledgments. This work was supported in part by the Natural Science Foundation of China under Grant 61572156 and in part by the Shenzhen Science and Technology Program under Grant JCYJ20170413105929681.

References

1. Li, Y., et al.: Real time advertisement insertion in baseball video based on advertisement effect. In: Proceedings of ACMMM, pp. 343–346. ACM (2005)
2. Redondo, R.P.D., et al.: Bringing content awareness to web-based IDTV advertising. IEEE Trans. Syst. Man Cybern. Part C (Appl. Rev.) **42**(3), 324–333 (2012)
3. Mei, T., et al.: VideoSense: a contextual in-video advertising system. IEEE Trans. Circuits Syst. Video Technol. **19**(12), 1866–1879 (2009)
4. Zhang, H., et al.: Object-level video advertising: an optimization framework. IEEE Trans. Ind. Inform. **13**(2), 520–531 (2017)
5. Yadati, K., Katti, H., Kankanhalli, M.: CAVVA: computational affective video-in-video advertising. IEEE Trans. Multimedia **16**(1), 15–23 (2014)
6. Zeiler, M.D., Fergus, R.: Visualizing and understanding convolutional networks. In: Fleet, D., Pajdla, T., Schiele, B., Tuytelaars, T. (eds.) ECCV 2014. LNCS, vol. 8689, pp. 818–833. Springer, Cham (2014). https://doi.org/10.1007/978-3-319-10590-1_53
7. Simonyan, K., Zisserman, A.: Very deep convolutional networks for large-scale image recognition (2014). arXiv preprint arXiv:1409.1556
8. He, K., et al.: Deep residual learning for image recognition. In: Proceedings of CVPR, pp. 770–778 (2016)
9. Krizhevsky, A., et al.: Imagenet classification with deep convolutional neural networks. In: Advances in Neural Information Processing Systems, pp. 1097–1105 (2012)
10. Szegedy, C., et al.: Going deeper with convolutions. In: Proceedings of CVPR, pp. 1–9 (2015)
11. Kiapour, M.H., et al.: Where to buy it: matching street clothing photos in online shops. In: Proceedings of ICCV, pp. 3343–3351 (2015)

12. Liu, X., et al.: Front. Comput. Sci. VIPLFaceNet: an open source deep face recognition SDK **11**, 208–218 (2017)
13. Lin, K., et al.: Deep learning of binary hash codes for fast image retrieval. In: Proceedings of CVPR Workshops, pp. 27–35 (2015)
14. Rodriguez, A., Laio, A.: Clustering by fast search and find of density peaks. Science **344**(6191), 1492–1496 (2014)
15. Ojala, T., et al.: Multiresolution gray-scale and rotation invariant texture classification with local binary patterns. IEEE Trans. Pattern Anal. Mach. Intell. **24**(7), 971–987 (2002)
16. Tan, X.Y., Triggs, B.: Enhanced local texture feature sets for face recognition under difficult lighting conditions. IEEE Trans. Image Process. **19**(6), 1635–1650 (2010)
17. Murala, S., et al.: Local tetra patterns: a new feature descriptor for content-based image retrieval. IEEE Trans. Image Process. **21**(5), 2874–2886 (2012)
18. Zhang, H., et al.: Organizing books and authors by multilayer SOM. IEEE Trans. Neural Netw. Learn. Syst. **27**(12), 2537–2550 (2016)

Distributed Computing

Efficient Snapshot Isolation
in Paxos-Replicated Database Systems

Jinwei Guo[1], Peng Cai[1,2(✉)], Bing Xiao[1], Weining Qian[1], and Aoying Zhou[1]

[1] School of Data Science and Engineering, East China Normal University,
Shanghai 200062, People's Republic of China
{guojinwei,bingxiao}@stu.ecnu.edu.cn,
{pcai,wnqian,ayzhou}@dase.ecnu.edu.cn
[2] Guangxi Key Laboratory of Trusted Software,
Guilin University of Electronic Technology, Guilin 541004, People's Republic of China

Abstract. Modern database systems are increasingly deployed in a cluster of commodity machines with Paxos-based replication technique to offer better performance, higher availability and fault-tolerance. The widely adopted implementation is that one database replica is elected to be a leader and to be responsible for transaction requests. After the transaction execution is completed, the leader generates transaction log and commit this transaction until the log has been replicated to a majority of replicas. The state of the leader is always ahead of that of the follower replicas since the leader commits the transactions firstly and then notifies other replicas of the latest committed log entries in the later communication. As the follower replica can't immediately provide the latest snapshot, both read-write and read-only transactions would be executed at the leader to guarantee the strong snapshot isolation semantic. In this work, we design and implement an efficient snapshot isolation scheme. This scheme uses adaptive timestamp allocation to avoid frequently requesting the leader to assign transaction timestamps. Furthermore, we design an early log replay mechanism for follower replicas. It allows the follower replica to execute a read operation without waiting to replay log to generate the required snapshot. Comparing with the conventional implementation, we experimentally show that the optimized snapshot isolation for Paxos-replicated database systems has better performance in terms of scalability and throughput.

1 Introduction

Replication is a key technique to achieve better scalability, availability and fault-tolerance in distributed systems, and its challenge is how to keep the consistency between replicas. Many traditional DBMS products adopt primary-backup technique including *eager* or *lazy* schemes to replicate writes from a primary replica node to multiple backup nodes [12]. Eager replication has bad performance as the transaction can't be committed until its updates have been synchronously installed at all replicas. On the other hand, lazy replication sacrifices consistency

J. Pei et al. (Eds.): DASFAA 2018, LNCS 10828, pp. 649–665, 2018.
https://doi.org/10.1007/978-3-319-91458-9_40

especially when the primary crashed and failed to replicate the logs of committed transaction. Replication protocols based on group communication presented by Kemme and Alonso [15] utilize total order broadcast primitives to ensure that updates are applied in the same order at all replicas. This technique can maintain the scalability of replication without violating consistency. However, the cost of its synchronization is still high due to the fact that a write is synchronized to all replicas in the group [26]. Furthermore, systems using group communication need to rely on an external highly available component due to the assumption that only a primary group is able to continue when network is partitioned [14].

Paxos [17] has been widely used to build a highly available and consistent distributed system containing unreliable servers and asynchronous network. Therefore, using Paxos to replicate log is a popular choice in database systems, such as IBM's Spinnaker [24], Google's MegaStore [3] and Spanner [9]. To reduce network overhead, Spanner replicates transactional log entries using multi-Paxos [18] in which a replica is elected to be the leader and the first phase in the classic Paxos is not necessary. In common cases, the leader replica commits a log entry after synchronizing it to a majority of Paxos members referred to as followers. Then, the leader notifies the followers of the latest commit information in the later leader-follower communication.

Snapshot isolation (SI) [4], a well-known multi-version concurrency control method, has been widely available in many DBMS engines like Oracle, PostgreSQL and Microsoft SQL Server due to the non-blocking read processing. In the past decade, there are many research works [5–7,10,13,19] focusing on the combination of SI and various replication schemes for distributed database systems. Strong snapshot isolation (strong-SI), which is regarded as one-copy SI, is friendly to application programmers, and is also used to naturally resolve the problem of transaction inversions in lazy replication [19]. It needs strongly consistent read to guarantee the *recency property* from the strong consistency. Unfortunately, it's still non-trivial to achieve strongly consistent read when reading the data from any replica. Paxos-based protocols require a write to be visible only when its corresponding log entry has been persisted in a majority of Paxos members. There is non-negligible data version difference between the leader and follower nodes since the leader always commits a transaction first and then informs the followers to commit. Until the log is replayed in the followers, the latest updated data can not be observed from these replicas. This procedure leads to the result that the data on the follower nodes are less fresh than that on the leader. If the application requires strongly consistent read, the database system must process read operations only in the leader which owns the latest state of database. In this case, the leader node has the potential risk of suffering from overload and this makes a significant impact on the throughput and response time. Therefore, many system designs relax the strong consistency requirement to achieve high performance via weakly consistent read.

In this paper, we propose early log replay (ELR) algorithm, by which we can achieve efficient snapshot isolation (ESI) in Paxos-replicated database systems. The main goal of our approach is to avoid read blocking and read failure in the conventional implementation of strong-SI. To decrease the overhead of leader replica, we also present an adaptive timestamp allocation (ATA) mechanism. ATA effectively reduces the number of timestamp requests to the leader. The following is the list of our main contributions.

- We give a basic implementation of SI in a Paxos-replicated database system, analyze its transaction read execution and figure out the root causes of read blocking, read failure and leader overloading.
- We propose early log replay (ELR) mechanism to implement an efficient snapshot isolation (ESI) in Paxos-replicated database systems. ELR can avoid read blocking effectively. To guarantee the correctness of data, we give the recovery mechanism for ELR.
- We present adaptive timestamp allocation (ATA) to reduce the leader's overhead imposed by frequently processing timestamp requests. ATA allows the leader's timestamp to be embedded into response messages (e.g., write/commit response messages). By means of ATA, the leader only handles timestamp requests in a few rare cases.
- We implement the efficient snapshot isolation in an open source database system OceanBase. Experimental results demonstrate the effectiveness of our method in terms of scalability and throughput.

The reminder of the paper is organized as follows: The background of SI is introduced in Sect. 2. We give and analyze SI in Paxos-replicated database systems in Sect. 3. Section 4 presents an efficient version of SI for Paxos replication systems. In Sect. 5, we introduce the adaptive timestamp allocation, and experimental results are presented in Sect. 6. Finally, related work and conclusion are presented in Sects. 7 and 8 respectively.

2   Background

Snapshot isolation (SI), which is one kind of multi-version concurrency control (MVCC), was proposed by Berenson et al. [4]. Under snapshot isolation, the transaction manager assigns a transaction T a start timestamp ($T.sts$) when it receives a start transaction request. The transaction T reads data from the latest snapshot of database containing the data committed before its start timestamp. When a transaction T is ready to commit, it gets a commit timestamp ($T.cts$) which is larger than any existing start timestamps or commit timestamps.

It's not straight-forward to extend snapshot isolation, originally defined over centralized database, to replicated database systems. The main reason is the "latest" snapshot is not well defined in distributed environment. Generalized snapshot isolation (GSI) allows the transaction to execute over an old local snapshot of database [11]. As the read may not get the last committed write in GSI, it violates the recency guarantee of strong consistency [2].

To achieve the same snapshot isolation semantics of centralized DBMS, a simple approach is to globally order all transactions to maintain the partial order of operations from different clients. In other words, when a database replica receives a start transaction request from a client, it ensures the latest snapshot is allocated to this transaction. It's referred to as the concept of strong snapshot isolation (strong-SI) and the formal definition is defined below [10]:

Definition 1 (Strong Snapshot Isolation). *A history $\mathcal{H}$ of transactions satisfies strong snapshot isolation, it has the following property: for any pair of transactions T_i and T_j, if the database replica receives the* commit *of T_i before the* start *of T_j, then $T_i.cts \leq T_j.sts$.*

Owing to the recency guarantee, strong-SI in distributed database systems is also regarded as one-copy SI [19], where the effect of transactions performed on the database replicas is the same as that in a single centralized database. Strong-SI is friendly to application developers. Therefore, in this paper, our target is to implement strong-SI in a Paxos-replicated database system.

3   SI in Paxos-Replicated Database Systems

3.1   System Architecture

For ease of description, we assume that the replicated database is a main-memory key-value store. The simplified architecture contains two components:

- **Request Processing Nodes (RP-Node)**: RP-Node is the bridge between the clients and the database. Its task is to parse the SQL, generate the logical/physical plan, and forward the generated plan to the transaction processing nodes. It should be noted that RP-Node is stateless.
- **Transaction Processing Nodes (TP-Node)**: A set of TP-nodes constitute a Paxos group, and each one maintains a full copy of the database. TP-node is responsible for concurrency control and log replication. To reduce the consensus cost, the group adopting multi-Paxos consists of only one leader as primary replica and multiple followers as backups.

3.2   Transaction Execution

In the architecture described above, all start-transaction, read/write and commit-transaction requests need to be processed by the leader TP-node, which is responsible for transaction scheduling and log replication. When the leader receives the commit-transaction request of an update transaction, it generates a log entry including the transaction's writes and then uses Paxos-based replication protocol to synchronize the log to other TP-nodes. As shown in Fig. 1, the classical Paxos-based log replication is divided into two phases:

Phase 1: The leader TP-Node sends the log entry to other TP-nodes. After the leader confirms the log has been successfully replicated to the majority of TP-Nodes, it can commit this transaction and then respond to the client;

Phase 2: The leader firstly updates its local *commit point*, and then asynchronously notifies other TP-Node of the latest committed logs in the later

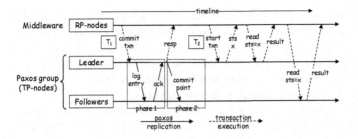

Fig. 1. Transaction execution in a Paxos-replicated database system.

communication. It should be noted that the commit point is represented by log sequence number (LSN), and log entries are persisted and replayed in the order of LSN. When a follower TP-Node receives the latest commit point from leader, it can refresh its local commit point and replay these logs with LSN prior to and including this commit point.

As described in Phase 2, the data committed at leader is not immediately visible at follower until it receives the commit point and completes the log replaying. To obtain the scalability property of replication, after a transaction gets its *start timestamp* from the leader, it's expected that subsequent read requests can be forwarded to leader or followers, as illustrated in Fig. 1. To handle a read request under strong-SI, a follower needs to hold a multi-version storage engine, and a read operation of T_j can access data only when the replica includes all T_i's writes, where $T_i.cts \leq T_j.sts$. In this work, the *commit timestamp* of a transaction is embedded into its log entry. The state of followers reflecting data recency is represented by three variables. (1) **flush_cts** (**f_cts**): The commit timestamp in the last log entry flushed by this follower TP-Node; (2) **commit_cts** (**c_cts**): The commit timestamp in the log entry whose LSN is equal to the latest received commit point; (3) **publish_cts** (**p_cts**): The commit timestamp in the log entry with maximal LSN in all replayed log entries.

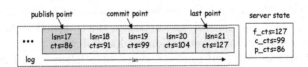

Fig. 2. An example of a follower's log and state. The boxes with light gray are used to denote the log entries that are not replayed. The *publish point* and *last point* are used to denote LSN of the last replayed entry and LSN of the last persisted entry, respectively.

To further demonstrate how the state of a follower influences transaction execution, an example is presented in Fig. 2. The commit point and last point in the follower are 19 and 21, respectively. Therefore, the corresponding *c_cts* and

f_cts are 99 and 127, respectively. Since the log entries before LSN 18 have been replayed, p_cts is equal to 86. Based on the follower's state, there are four cases on how a read request of transaction T is handled according to strong-SI:

Case 1: $f\_cts < T.sts$. The follower realizes that the RP-node requests a non-existent snapshot. *In other words, its log lags behind leader's.* Therefore, it immediately returns a failure response message.

Case 2: $c\_cts < T.sts \leq f\_cts$. This indicates that the follower is waiting for the latest commit point from the leader. To avoid waiting a long time, a failure response for the request is returned directly.

Case 3: $p\_cts < T.sts \leq c\_cts$. The follower learns that the expected version of data will be available soon. Accordingly, *the request is blocked until $p\_cts \geq T.sts$.* Finally, the follower returns a response message with expected data.

Case 4: $T.sts \leq p\_cts$. The follower directly gets the data of expected version from local database snapshot and returns a successful response including the data to the RP-node.

If the followers handle read requests according to the four cases, it's referred to as the *basic* implementation of strong-SI in Paxos-replicated DBMS. Although the simplest method to achieve the Strong-SI is to let the leader TP-Node handle all read/write operations, it loses much scalability. Even under the basic implementation of Strong-SI, the read requests handled by followers may still be failed or blocked. We make an analysis on this problem in the next subsection.

3.3   Problem Analysis

Recall that Paxos-based replication has two properties: (1) The leader can commit a log entry if it receives a majority of acknowledgments. In other words, a follower TP-node may not have the latest log. In Fig. 3, the TP-node 3 has not the entry with LSN 22. (2) The follower can replay a log entry only when receiving the commit information about the entry. In Fig. 3, the commit point is sent asynchronously to the follower TP-node 2. The visibility of a write in the TP-node 2 is always later than that in the leader. To guarantee the recency property, a transaction is assigned a newest start timestamp from the leader TP-node. As shown in Fig. 3, when the leader receives a start-transaction request, it assigns the $p\_cts(128)$ as the *start timestamp* to the transaction. We summarize the issues in the execution of transaction read as follows:

Issue 1: A transaction read can be rejected by a follower. In Fig. 3, the RP-node sends a read request of a transaction T with $sts(128)$ to TP-node 3. Since $f\_cts(127) < T.sts(128)$, the read is rejected directly by TP-node 3.

Issue 2: A transaction read can be blocked in a follower. In Fig. 3, we find that a read request of a transaction T with $sts(128)$ is blocked by TP-node 2. Although TP-node 2 has the latest log, the log is being replayed and the data of expected version is invisible.

Issue 3: Under Strong-SI, the start-transaction request of each transaction is always forwarded to the leader replica. Therefore, the leader may be faced with a large number of requests, which can negatively impact the system's performance.

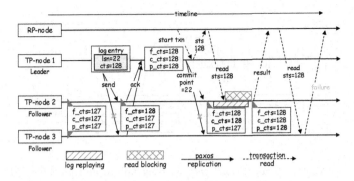

Fig. 3. An example of transaction read with blocking or failure. This is 3-way Paxos replication system, where TP-node 1 is the leader. Assuming that all variables (i.e., f_cts, c_cts and p_cts) in each TP-node's server state are 127 at the beginning.

Both issue 1 and 2 decrease the read performance of Paxos-replicated database systems. To address these issues, we introduce a design of efficient snapshot isolation utilizing early log replay mechanism. For issue 3, we present adaptive timestamp allocation for read-only transactions in Sect. 5.

4 Efficient Snapshot Isolation

4.1 Overview

We have introduced a basic implementation of SI in Paxos-replicated database system in Sect. 3. In this section, we present an *efficient* version. To implement an efficient snapshot isolation, our design target has two sides: (1) A RP-node is required to forward a read request to the TP-node having the latest log; (2) A follower can replay its local log without waiting leader's commit point.

An example of transaction read in efficient snapshot isolation is illustrated in Fig. 4. For the first target, the leader responds with a message including the available followers information (TP-node 2). This is because the leader ensures that TP-node 2 has the latest log. RP-node can forward subsequent read requests according to the response. For the second target, when the follower TP-node 2 receives a log entry with *commit timestamp* ($cts = 128$) from the leader, it replays the entry immediately without waiting the corresponding commit point. Then, the follower can update its p_cts to 128 after replaying successfully. Owing to the efforts for both targets, a transaction read with $sts(128)$ can be processed without blocking in TP-node 2.

The first target can be easily achieved. To guarantee the correctness of implementation for the second target, we introduce early log replay (ELR) mechanism in the following subsections.

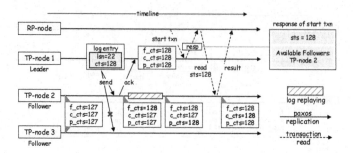

Fig. 4. An example of transaction read avoiding failure and blocking. All variables (i.e., f_cts, c_cts and p_cts) in each TP-node's server state are 127 at start.

4.2 Early Log Replay

Owing to the multi-version storage of snapshot isolation for main-memory database, each object can utilize a committed list (c_list) to store multiple versions in the order of committed timestamps [27]. In a follower replica, each version—which contains a value and a timestamp (cts of the transaction creating the version)— can be constructed from the corresponding log entry. We design a *centralized uncommitted list* (uc_list) to store the pointers to all of uncommitted versions. uc_list is used for data recovery when failure happens (see Sect. 4.4).

To safely replay log without commit point in followers, ELR decomposes the conventional log replay into two phases: *early log replay* and *log commit*. The pseudocode of functions used by ELR is showed in Algorithm 1. Assuming that a log entry e contains only one updated object. Accordingly, we use $e.key$ and $e.value$ to denote the object's key and value, respectively. Next, we detailedly describe the transition of follower state using ELR.

When a follower receives a log entry message from the leader, it invokes the function `LogReceiver`. The function flushes the entry into the non-volatile storage and then refreshes the local f_cts from the log entry (lines 2–3). Next, the follower appends the entry to a queue (line 4), which stores the log entries that are ready to replay. We call it *log_queue*. Finally, the follower responds to the leader (line 5).

Early Log Replay Phase: There is a thread running `EarlyLogReplayer`, which is responsible for early log replay phase. If the thread gets a log entry from the *log_queue*, it parses the entry and gets the corresponding object from the local (line 11). Then, it generates a new version and appends it to the object's c_list (lines 12–13). Next, the follower adds the pointer of the new version to the uc_list (line 14). Finally, it updates the local p_cts using the cts of the entry (line 15).

Log Commit Phase: There is a background thread which is in charge of the log commit phase. It periodically invokes the function `LogCommitter` using c_cts.

Algorithm 1. Early log replay algorithm

```
 1 Function LogReceiver(entry)
 2 │    flush entry to disk;
 3 │    f_cts = entry.cts;
 4 │    log_queue.enqueue(entry);
 5 │    response to the leader;
 6 end
   /* early log replay phase                                          */
 7 Function EarlyLogReplayer()
 8 │    while true do
 9 │    │    if ¬log_queue.isEmpty() then
10 │    │    │    entry = log_queue.dequeue();
11 │    │    │    obj = getObjectByKey(entry.key);
12 │    │    │    version = new Version(entry.value, entry.cts);
13 │    │    │    obj.c_list.add(version);
14 │    │    │    uc_list.add(version);
15 │    │    │    p_cts = max(p_cts, entry.cts);
16 │    │    end
17 │    end
18 end
   /* log commit phase                                                */
19 Function LogCommitter(c_cts)
20 │    while ¬uc_list.isEmpty() ∧ c_cts > uc_list.get(0).cts do
21 │    │    uc_list.remove(0);
22 │    end
23 end
```

It gets the head in the *uc_list*. If the *cts* of the version is not larger than *c_cts*, the follower removes the pointer from the *uc_list* (lines 20–21).

4.3   Transaction Read Execution

When a follower TP-node receives a read request, it needs to return the data satisfying the expected version. This processing flow is similar to that described in Sect. 3.2, excepting that the follower's *c_cts* is not used to determine whether this read request can be processed. Since a read is forwarded to a follower having the latest log, it can be served without blocking due to ELR mechanism.

We note that a read request contains the *start timestamp*, which represents a required snapshot. In other words, the log entry whose timestamp is not larger than the *start timestamp* is committed. *Therefore, the follower can update local c_cts to the request's sts and invoke the function* LogCommitter *to handle the versions in the uncommitted list* uc_list.

4.4 Recovery

If the leader TP-Node is corrupted, Paxos group will leverage election mechanism to achieve automatic fault tolerance. A typical method is that a TP-node wins the election if its logs are not older than a majority of TP-nodes. When a new leader is elected, any component in the system needs to take efforts to guarantee strong-SI services:

- **New Leader**: The new leader must do some work for takeover. It is required to ensure that all writes in the log are committed, i.e., it synchronized local log to at least a majority of TP-nodes. After log synchronization, it can empty the uc_list and apply all log entries. Finally, the leader returns to normal and can receives requests from RP-nodes.
- **Follower**: When a TP-node detects that a new leader is not itself, it becomes a follower and needs to take some measures to ensure the validation of local data. Due to invalid versions in the uc_list (i.e., a version does not exist in the new leader), the follower needs to ask the leader to check the local log to find the invalid log entries. Then, the follower traverses the uc_list, removes the pointer of committed version directly and deletes the invalid versions in the objects' c_list.
- **RP-node**: When a RP-node is informed of the crash of leader, it notes that all writes are blocked until the new leader is elected and returns to normal. On the other hand, the RP-node can issue these unfinished read-only transaction to other TP-node's.

If a follower TP-Node recovers from a failure, it only checks the local log and applies the correct entries to local due to the main memory storage engine.

5 Adaptive Timestamp Allocation

Recall from Sect. 3 that a RP-node needs to ask the leader to get a latest *start timestamp* for each start-transaction request from the clients. As a result, the leader may be faced with tremendous pressure. In this section, we introduce adaptive timestamp allocation (ATA) mechanism to reduce the number of timestamp requests.

Note that if a transaction is a read-only one, it doesn't need to be registered in the leader replica. Therefore, in order to reduce the overhead caused by the *start timestamp* requests in the leader, we can adopt batch processing technique in the RP-node for the read-only transactions. More specifically, the RP-node uses a buffer to keep a batch of start-transaction requests from clients and send only one timestamp request to the leader. The RP-node asks the leader for a new start timestamp every d ms. In real application, it's difficult to determine which is the optimal value of d. If d is too small, the leader will receive an enormous amount of start requests; if d is too large, the delay of clients' requests will be increased.

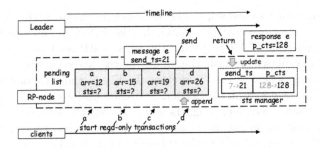

Fig. 5. An example of assigning start timestamps for clients' read-only transactions without violating recency guarantee. A read-only transaction is allocated an arrival time (arr) and appended into *pending list*. Then it is waiting for a suitable start timestamp (sts), which is triggered by the update of *sts manager*.

The idea of ATA is to allow the RP-node to embed the timestamp request into a write or commit request issued by other transaction. Now we analyze the cost of timestamp request in terms of how many requests sent to the leader. We assume that the arrival time of read/write request is uniformly distributed, and a RP-node needs $1000/d$ timestamp requests per second. In one second, a RP-node receives n requests. If the percentage of writes is w in the workload, there are $n \cdot w$ writes per second that need to be forwarded to the leader. Equation 1 shows the number of timestamp requests sent to the leader:

$$f(d, n, w) = \max(0, \frac{1000}{d} - n \cdot w) \tag{1}$$

Assume that $w = 5\%$, which is a typical value in the read-intensive workload. According to Eq. 1, we can see that if the extra timestamp requests are not required, the optimal d can be set to less than 4 ms when $n \geq 5,000$. If n is small, it indicates that the overhead of the leader is not heavy. In this kind of case, the RP-node can send additional timestamp requests to the leader to further decrease the delay of clients' requests. It is clear that the average delay of a client's request is incremented by $d/2$ ms. We can see that if $n = 10,000$, the request delay is only incremented by 1 ms.

Although the leader TP-Node embeds its state of related timestamps into a sequence of message for responding many write/commit requests, it should be noted that not all returned timestamps can be served as the start-timestamp of a transaction. *To guarantee the recency property of strong snapshot isolation, only the returned timestamps can be taken as valid start timestamp of a transaction if this message is responded to the write/commit request sent by the RP-Node after the start-transaction request arrived.* Therefore, the RP-node is required to record the sent time *send_ts* for each request. In order to efficiently allocate the start timestamps, RP-node utilizes two components *pending list* and *sts manager*, which are illustrated in Fig. 5. The FIFO *pending list* is used to store the pending read-only transactions in the order of their arrival timestamps. We can see that there are four transactions a, b, c and d in Fig. 5, which are waiting for the suitable start timestamps. The *sts manager* manages the latest *send_ts* of a message whose response is received by the RP-node and the *p_cts* in the corresponding response. More specifically, when a RP-node receives a response from the leader, it will refresh the values in the *sts manager* if the *send_ts* of corresponding send message is greater. In Fig. 5, the RP-node receives the response of message *e* from the leader, and then updates *send_ts* and *p_ts* to 21 and 128, respectively. This triggers an event that the RP-node allocates the *p_ts*(128) to transactions in the *pending list*, whose arrival timestamps are not larger than the *send_ts*(21), i.e., the transactions *a*, *b* and *c*.

There is a background thread which is responsible for checking the *sts manager* periodically. If the values in the manager are not updated in *d* ms, the thread will send a start timestamp request to the leader.

6 Experiments

We implemented efficient snapshot isolation in OceanBase 0.4.2 [1], which is a scalable open source RDBMS developed by Alibaba. We conducted an experimental study to evaluate the performance of the proposed efficient snapshot isolation. Experimental setup and the benchmark used in this evaluation are given below.

Cluster Platform: We deployed a 3-way replication database system including RP-nodes and TP-nodes on a cluster of 18 machines, and each machine is equipped with a 2-socket Intel Xeon E5606 @2.13 GHz (a total of 8 physical cores), 96 GB RAM and 100 GB SSD while running CentOS version 6.5.

Competitors: We use SINGLE to denote the system containing only one TP-node without any log replication. Its experimental results will be helpful for understanding the behavior with other strategies. We use BASIC and ESI to denote the implementation of basic version and efficient version of SI, respectively. The framework of generalized snapshot isolation, which allows a transaction read over an arbitrary old data version, is denoted as WEAK.

Benchmark: First, we adopt YCSB [8] to evaluate our implementation. We use the workloads YCSB A and B (abbr. workload-A and workload-B), which have a read/write ratio of 50/50 and 95/5 respectively, and each transaction contains only one operation. Second, to investigate the performance of complicated transaction workload, we use five read/write operations to generate a transaction with multiple operations. The size of each update is about 100 bytes.

6.1 Scalability

Figure 6(a) illustrates the system throughput over the increasing number of clients under the workload-B with read-intensive operations. Because all requests were forwarded to the single node, the performance of SINGLE is the worst. Owing to the serviceability of followers, WEAK has the highest throughput, compared with SINGLE by about 2.5× when the number of clients was more

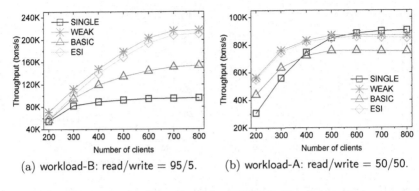

(a) workload-B: read/write = 95/5.   (b) workload-A: read/write = 50/50.

Fig. 6. Throughput for YCSB workloads.

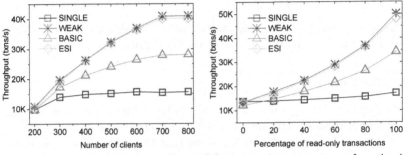

(a) Varing the number of clients (read- (b) Varing the percentage of read-only
only transactions account for 80%).   transactions (clients = 600).

Fig. 7. Throughput for transaction workloads, where each read-only transaction contains five read operations and each update transaction contains three reads and two writes.

than 600. Recall from Sect. 3.3 that there are some issues in the basic version of GSI. Therefore, the results of BASIC was about two third of that of WEAK. Since an uncommitted write can be published in ELR, a read could be processed quickly without blocking in any alive replica. This advantage makes the performance of ESI be very similar to WEAK. Accordingly, ESI maintains the goodness of replication under read intensive workload.

Figure 6(b) shows the results under workload-A with write-intensive operations. The trend of the results is similar to that in Fig. 6(a). Due to the overhead for synchronizing large amount of updates log by the leader node, the throughput of the 3-way replication system was limited and less than SINGLE when the number of clients is more than 640. It's worthwhile noting that although ESI provides strong-SI, it has a similar performance to WEAK. Furthermore, we also observed that ESI outperforms BASIC, since the read requests are always not blocked in the followers adopting early log replay.

In the transaction workload, a read-only transaction contains multiple read operations. As ESI reduces the timestamp requests dramatically and allows the

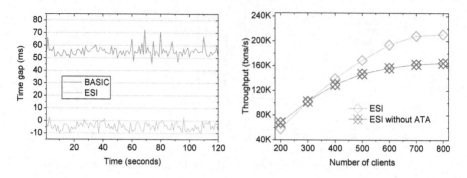

Fig. 8. Effectiveness of ELR: workload-A.   **Fig. 9.** Effectiveness of ATA: workload-B.

RP-nodes to quickly forward read operations to a qualified follower node. Thus, the results of ESI are nearly the same as that of WEAK, which is showed in Fig. 7.

6.2   Effectiveness

To investigate the effectiveness of early log replay (ELR) mechanism, we compare ESI against BASIC. Recall from Sect. 3.2 that p_cts is used to denote the latest server state a client can access. Assuming that the leader ℓ and the follower f update their p_cts to the same timestamp at physical time t_ℓ and t_f, respectively. We use the result of $(t_f - t_\ell)$ to denote the time gap between the same visible state of leader l and one follower f.

Figure 8 shows the visibility difference of all p_cts's between a follower and the leader over 120 s for the workload-B. The number of clients is fixed to 240, where the system can output stable results. We can see that the visibility difference of BASIC exceeds 50 ms, which suggests that a read expecting a latest version may not be satisfied in followers. And the visibility difference of ESI was close to the zero, which indicates that a log entry has been replayed successfully in followers when it finishes the commit phase in the leader. ELR ensures that the expected data can be returned immediately for the transaction read at the follower replica.

To investigate the effectiveness of adaptive timestamp allocation (ATA), we evaluate the performance of ESI without ATA, where RP-nodes send a start timestamp request for each read. Figure 9 illustrates the experimental results under workload-B. It is clear that the delay of a start transaction request in RP-nodes may impact the system performance. We can see that ESI without ATA exceeds ESI when the workload is low (i.e., the leader TP-node has the enough capacity to process each request). However, as the number of clients increases, the leader is faced with increasing overhead. When workload is high, due to the reduction of requests to the leader, ESI significantly outperforms ESI without ATA. Therefore, in the case of heavy workload, ATA has a positive effect on system performance.

7   Related Work

Replication: Replication is an effective mechanism to provide scalability, high availability and fault tolerance in distributed systems. State machine replication (SMR) [25], a fundamental approach to fault-tolerant services, can ensure that the replica is consistent with each other only if the operations are executed in the same order on all replicas. Eager or lazy replication [12] has been a standard criteria for database systems. But these schemes can not satisfy the requirement of both performance and consistency. Kemme and Alonso [15] presented a replication protocol based on group communication, which can maintain the scalability of replication without violating consistency. However, the assumption that high throughput network and an external highly available service are required limits the use of group communication. Wiesmann et al. [23] compared and summarized the replication techniques from database and distributed system communities.

Snapshot Isolation for Replication: Snapshot isolation [4], which allows write-skew anomaly but offers greater performance, is widely available in DBMS engines like Oracle and PostgreSQL. In the past decade, there are many research works [5–7,10,13,19] focusing on the combination of replication and SI for distributed database systems. Lin et al. [19] presented a middleware-based replication scheme which provides strongly global SI. Daudjee and Salem [10] showed how snapshot isolation can be maintained in lazy replicated systems while taking full advantage of each replica's concurrency control. Jung et al. [13] proposed replicated serializable snapshot isolation, which can guarantee 1-copy serializable global execution. ConfluxDB [7] determine a total snapshot isolation order for update transactions over multiple master sites without requiring global coordination. Binnig et al. [5] further defined distributed snapshot isolation and accordingly proposed a criteria to efficiently implement snapshot isolation. To our best knowledge, there is no snapshot isolation implementation over Paxos replicated main-memory database systems.

Paxos Replication: To provide highly available services, modern database systems often adopt Paxos protocol—which is first described by Lamport in [17,18]—to replicate data from primary to backup replicas. Raft [22] is a famous variant of Paxos, which is widely used in open-source databases. Google has developed MegaStore [3] and Spanner [9], which utilize Paxos for log replication. Spanner implements distributed transaction processing and provides external consistency based on TrueTime API. Nonetheless, the non-blocking transaction read can only be forwarded to the leader replica. Spinnaker [24] builds a scalable, consistent and highly available datastore using Paxos-based replication. However, a read request processed by a backup node only offer weak consistency. Moraru et al. [20] introduced the method of Paxos quorum leases to allow strongly consistent local read be performed at replicas which are lease holders. The lease mechanism have small negative impact on the availability.

Spanner implemented concurrency control mechanism and Paxos based replication at different layers, and both of them have consistency requirements. To reduce the coordination cost, [16,21,28] consolidate concurrency and Paxos census to decrease the network round-trips.

8 Conclusion

In this work we presented an efficient snapshot isolation (ESI), an optimized implementation of strong-SI in Paxos-replicated database systems. By analyzing the transaction execution of basic version, we proposed two effective mechanisms for ESI, i.e., early log replay (ELR) and adaptive timestamp allocation (ATA). ELR avoids to block or fail the execution of transaction read in followers. ATA relieves the leader's overhead by reducing the number of timestamp requests sent to the leader. Experimental results demonstrate the effectiveness of ELR and ATA.

Acknowledgments. This work is partially supported by National High-tech R&D Program (863 Program) under grant number 2015AA015307, NSFC under grant numbers 61432006 and 61332006, and Guangxi Key Laboratory of Trusted Software (kx201602).

References

1. OceanBase website. https://github.com/alibaba/oceanbase/
2. Bailis, P., Davidson, A., Fekete, A., et al.: Highly available transactions: virtues and limitations. PVLDB **7**(3), 181–192 (2013)
3. Baker, J., Bond, C., Corbett, J.C., et al.: Megastore: providing scalable, highly available storage for interactive services. In: CIDR, pp. 223–234 (2011)
4. Berenson, H., Bernstein, P., Gray, J., et al.: A critique of ANSI SQL isolation levels. SIGMOD Rec. **24**(2), 1–10 (1995)
5. Binnig, C., Hildenbrand, S., et al.: Distributed snapshot isolation: global transactions pay globally, local transactions pay locally. VLDB J. **23**(6), 987–1011 (2014)
6. Bornea, M.A., Hodson, O., Elnikety, S., Fekete, A.: One-copy serializability with snapshot isolation under the hood. In: ICDE, pp. 625–636 (2011)
7. Chairunnanda, P., Daudjee, K., Özsu, T.M.: ConfluxDB: multi-master replication for partitioned snapshot isolation databases. In: VLDB, pp. 947–958 (2014)
8. Cooper, B.F., Silberstein, A., Tam, E., Ramakrishnan, R., Sears, R.: Benchmarking cloud serving systems with YCSB. In: SoCC, pp. 143–154 (2010)
9. Corbett, J.C., Dean, J., Epstein, M., et al.: Spanner: Google's globally distributed database. TOCS **31**(3), 8 (2013)
10. Daudjee, K., Salem, K.: Lazy database replication with snapshot isolation. In: VLDB, pp. 715–726 (2006)
11. Elnikety, S., Zwaenepoel, W., Pedone, F.: Database replication using generalized snapshot isolation. In: SRDS, pp. 73–84. IEEE Computer Society (2005)
12. Gray, J., Helland, P., O'Neil, P., Shasha, D.: The dangers of replication and a solution. SIGMOD Rec. **25**(2), 173–182 (1996)
13. Jung, H., Han, H., Fekete, A., Röhm, U.: Serializable snapshot isolation for replicated databases in high-update scenarios. In: VLDB, pp. 783–794 (2011)
14. Kemme, B., Alonso, G.: A suite of database replication protocols based on group communication primitives. In: ICDCS, pp. 156–163 (1998)

15. Kemme, B., Alonso, G.: Database replication: a tale of research across communities. PVLDB **3**(1), 5–12 (2010)
16. Kraska, T., Pang, G., Franklin, M.J., et al.: MDCC: multi-data center consistency. In: EuroSys, pp. 113–126 (2013)
17. Lamport, L.: The part-time parliament. TOCS **16**(2), 133–169 (1998)
18. Lamport, L.: Paxos made simple. ACM SIGACT News **32**(4), 18–25 (2001)
19. Lin, Y., Kemme, B., Patiño Martínez, M., Jiménez-Peris, R.: Middleware based data replication providing snapshot isolation. In: SIGMOD, pp. 419–430 (2005)
20. Moraru, I., Andersen, D.G., Kaminsky, M.: Paxos quorum leases: fast reads without sacrificing writes. In: SOCC, pp. 22:1–22:13 (2014)
21. Mu, S., Nelson, L., Lloyd, W., Li, J.: Consolidating concurrency control and consensus for commits under conflicts. In: OSDI, pp. 517–532 (2016)
22. Ongaro, D., Ousterhout, J.K.: In search of an understandable consensus algorithm. In: ATC (2014)
23. Pedone, F., Wiesmann, M., Schiper, A., Kemme, B., Alonso, G.: Understanding replication in databases and distributed systems. In: ICDCS, pp. 464–474 (2000)
24. Rao, J., Shekita, E.J., Tata, S.: Using Paxos to build a scalable, consistent, and highly available datastore. In: VLDB, pp. 243–254 (2011)
25. Schneider, F.B.: Implementing fault-tolerant services using the state machine approach: a tutorial. CSUR **22**(4), 299–319 (1990)
26. Wiesmann, M., Schiper, A.: Comparison of database replication techniques based on total order broadcast. TKDE **17**(4), 551–566 (2005)
27. Wu, Y., Arulraj, J., Lin, J., et al.: An empirical evaluation of in-memory multi-version concurrency control. Proc. VLDB Endow. **10**(7), 781–792 (2017)
28. Zhang, I., Sharma, N.K., Szekeres, A., et al.: Building consistent transactions with inconsistent replication. In: SOSP, pp. 263–278. ACM (2015)

Proof of Reputation: A Reputation-Based Consensus Protocol for Peer-to-Peer Network

Fangyu Gai, Baosheng Wang, Wenping Deng$^{(\boxtimes)}$, and Wei Peng

School of Computer, National University of Defense Technology, Changsha, China
wpdeng@nudt.edu.cn

Abstract. The advent of blockchain sheds light on addressing trust issues of peer-to-peer networks by providing a distributed tamper-resistant ledger. Beyond cryptocurrencies, it is believed that blockchain can also be used to protect other properties such as reputation. Most of the existing studies of enhancing reputation systems using blockchains are built on top of the bitcoin-like blockchains, so they are inherently constrained by the low-efficiency and high-consumption of the underlying blockchain. To fill this gap, we present a reputation-based consensus protocol called Proof of Reputation (PoR), which guarantees the reliability and integrity of transaction outcomes in an efficient way. In PoR, we let reputation serves as the incentive for both good behavior and block publication instead of digital coins, therefore no miners are needed. We also implement a prototype and our scalability experiments show that our protocol can scale to over a thousand participants in a peer-to-peer network with throughput of hundreds of transactions per second.

Keywords: Blockchain · Consensus protocol · Reputation system
Peer-to-peer network

1 Introduction

Since 2009 Bitcoin was proposed by Nakamoto [1], more than 250 similar altcoins have been proposed. Blockchain, the core technology behind Bitcoin, is considered to benefit not only economic but also politics, healthcare, supply chain and scientific domains [2]. A blockchain is a distributed ledger that cryptographically links a chain of blocks, which records a set of time-ordered transactions. With this emerging technology, a trustless environment can be created for distributed applications. Most blockchain applications running depends on coins as the incentive for miners to produce blocks, such as Bitcoin and Namecoin [3]. In these applications, the security of the blockchain will be affected if miners are not paid [4]. In fact, miners and coins are not necessary, as long as the provision of appropriate incentives to maintain the security of underlying blockchain.

Reputation can be defined as the rating of a member's trustworthiness by others [5]. In peer-to-peer (p2p) networks, reputation systems are applied to drive

© Springer International Publishing AG, part of Springer Nature 2018
J. Pei et al. (Eds.): DASFAA 2018, LNCS 10828, pp. 666–681, 2018.
https://doi.org/10.1007/978-3-319-91458-9_41

the ability for each participant to trust one another and facilitate a successful interaction [6]. The existing studies can be divided into two kinds: the centralized and the distributed, both of which have obvious drawbacks. While centralized reputation systems (i.e. eBay online auction site [7]) can grasp the overall reputation of each participant in the context of their own use case, the single-point-failure problem is inevitable. On the other hand, in distributed methods, participants only catch partial reputation evidence. To assess the trustworthiness, they need both direct experience and indirect opinion of its peers, which is inefficient. Worsely, the effective communication, and sharing of unmodified reputation evidence remain unsolved [8].

Actually, reputation and blockchain can be a good combination: reputation serves as the incentive and the blockchain keeps the reputation records safe in turn. To achieve this purpose, we propose a reputation-based consensus protocol called *Proof of Reputation* (PoR), which provides a distributed ledger of reputation. We hope to build a decentralized protocol that records a history of transaction outcomes without central third parties involved. The protocol is used in the permissioned blockchain, where an access control layer is built into the blockchain nodes. Participants can authenticate with each other using asymmetric cryptography [9]. The reason why we choose permissioned blockchains is that reputation is inherently tied to identity and needs time to accumulate. In PoR, each participant maintains a distributed ledger of reputation evidence, which achieves consistent with the consensus algorithm. Compared with the *Proof of Work* (PoW), the consensus algorithm of Bitcoin, our protocol uses the reputation as the incentive, which is cost-efficient since there is no miners or hash power being consumed during block competition. The cryptographical nature of blockchain can protect the integrity and reliability of the reputation evidence. Theoretically, any p2p applications and decentralized reputation systems (e.g. second-hand dealing, social networking, service sharing etc.) can utilize our protocol as a reputation layer to objectively and securely record transactional history, based on which each participant's reputation can be evaluated without being manipulated by third parties.

The rest of this paper is organized as follows: In Sect. 2, we review the earlier work related to our work. We define the problem and threat model in Sect. 3. The rationale for its design and details are presented in Sect. 4, and the implementation of the protocol and experiment results are discussed in Sect. 5. Finally, Sect. 6 concludes the paper with a discussion of the future work.

2   Related Works

The reputation systems in p2p networks have been studied in the literature for decades. One of the first reputation systems of p2p networks is proposed by Gupta et al. [5]. It is also the most complete and effective solution, although the drawbacks are evident. Other studies also made great contributions to reputation systems [10–12], but none of them are truly decentralized. In recent few years, integrating reputation with blockchain sheds light on this area and is under active research.

Carboni [13] describes how a decentralized feedback management system can be built on top of the Bitcoin blockchain. The author explains that online reputation has the same requirements of electronic money: (1) can be expressed as a numerical variable; (2) its value is agreed by every participant and cannot be manipulated by third parties. The proposed system records the interaction feedback during payment, leaving the computation of the actual reputation to third-party applications. Dennis et al. [14,15] propose a novel reputation system based on the blockchain, which aims to solve the majority of issues remaining in current reputation systems. They built an entirely new blockchain to store reputation data and utilized merge mining from Bitcoin network to prevent 51% attack. However, their blockchain needs miners to produce and verify the blocks, and a possible drawback is requiring the users to be online for the miners to verify the transaction. Buechler et al. [16] propose a reputation algorithm named net flow convergence and a decentralized system that records reputation evidence. The algorithm is to detect fraudulent behavior by looking at the network flow of the Bitcoin. The system, which is implemented in three layers, analyzes the structure of the underlying transaction network and builds a history of transaction outcomes by utilizing smartcontract. Schaub et al. [17] present a trustless, decentralized, and anonymity preserving reputation system based on blockchain for e-commerce applications. The system utilizes Proof of Stake blockchain to keep the reputation system consensus, and meanwhile, it allows customers to submit ratings as well as textual reviews.

All of these studies use blockchain as a decentralized database storing reputation evidence. Meanwhile, these systems rely on blockchain mining to motivate miners to publish blocks, during which a great amount of computing power is consumed. Furthermore, they are subject to the underlying blockchain, and therefore they are facing the same challenges as the Bitcoin blockchain, such as limits on data storage, slow writes, limited bandwidth, and endless ledger [17].

A different study called *Trustchain* is proposed by Otte et al. [18], which is to build a permission-less tamper-proof and Sybil-resistant blockchain for storing transaction records of participants. Compared to traditional blockchains, each block in Trustchain contains only a single transaction record and Trustchain blocks together form a directed acyclic graph (DAG). They also propose the NetFlow accounting mechanism to prevent Sybil attack. Similar to our work, there are no miners in Trustchain, where two agents in a transaction produce their block and store it locally. Nevertheless, participants in Trustchain have less motivation to produce blocks and they do not share a global picture of reputation in the network.

3 Problem and Threat Model

3.1 Problem Definition

The aim of this study is to solve the problem of reputation agreement in p2p networks by providing a distributed ledger of reputation. Unlike Bitcoin's consensus protocol, known as PoW, there is no miner nor bitcoin in PoR. Therefore, nodes do not have to compete for the right to write the block. In our protocol, the node who writes the block is the one who has the highest trust value in the pre-committed block. Reputation serves as the incentive, because in order to increase its overall trustworthiness, the participant urges to write the block into the blockchain when it has the highest trust value in this block.

Noticeably, during the consensus process, there are no complex mathematical problems to be solved, which means our protocol is cost-efficient. Furthermore, we do not have to worry about the double-spending problem, because reputation is an overall status of a node after a number of transactions, which can not be spent or transferred. We make three assumptions about the underlying network and nodes involved in the network:

1. **Enrolment Control:** Our consensus protocol requires an access control layer built into the blockchain nodes, which is known as "permissioned blockchain". Candidates need to go through an enrolment process before joining the network. Each candidate has a pair of cryptographical key (like Bitcoin) used for authentication and digital signature. The public key will be submitted to the registry as the participant's identity, the hash of which will be considered as its ID, and then the registry will broadcast this key so that other participants can authenticate.
2. **Secure Channel:** For simplicity, we assume a secure broadcast channel avoiding man-in-the-middle attack (MITM) [19]. It means that no third parties can intercept or modify messages. Each pair of participants can authenticate each other reliably.
3. **Quick Bootstrap:** In Bitcoin, a new node needs to take more than 3 days to download the whole blockchain from its adjacent peers, verify it for bootstrap. Contrarily, a new PoR node can boot up instantly by requesting a trust ranking list from the registry. Details will be discussed in Sect. 4.6.

On the other hand, we do not make any assumption that nodes are reliable during protocol runs. That means the impact of Byzantine nodes is not within the scope of our discussion.

We formalize the problem of designing a reputation consensus protocol for permissioned blockchains as follows. Assume N participants have registered themselves to join the network, and an individual participant is presented by $p_i, i \in N$. Each participant stores others' public keys locally so that it can authenticate identities and verify transactions. Each transaction generated by p_i toward p_j is represented by an real number $x_i^j \in \mathbb{R}$ signed by p_i's private key, denoted as $Sig(x_i^j)$. The transaction from the same pair can only be included once in one block, which means each block contains $N(N-1)$ transactions at most, and a threshold $\lambda \leq N(N-1)$ can be adjusted to limit the number of transactions in one block.

3.2   Threat Model

Although we assume that nodes communicate through a reliable authenticated point-to-point channel, the network can still be damaged by selfish behaviors, malicious attacks, and even unintentional misconfiguration. Especially, the reputation-based protocol itself can be an attractive target for attackers. Here we discuss several Potential attacks [20].

1. **Bad-mouthing attack** is to provide dishonest recommendations to deframe good nodes, which is the most straightforward attack [21]. In our protocol, malicious nodes can continuously give bad comments to a specific node or all other nodes to improve his trustworthiness ranking in a block in return.
2. **Replay attack** attempts to reuse transactions and replay them in order to increase the impact of the same transaction. By carrying out this attack, a malicious participant can claim he has been involved in a transaction that is profitable for him multiple times. It can also be used to undermine hostile participants.
3. **On-off attack** is referred to irregular behaviors of attackers, which means that malicious nodes can perform well or badly alternatively in order to remain undetected while causing damage.
4. **Sybil attack and newcomer attack**, which was first described by Douceur [22], is harmful to almost all p2p networks. It can be described that attackers "legally" create more than a single ID. If one ID gets low reputation by performing bad behaviors, it switches to a new ID and starts over.

Further, since every participant can publish arbitrary blocks, malicious nodes can rewrite all the transactions inside or publish a fabricated block. Nonetheless, provided proper security mechanisms, high attack cost, and low reward, there is little incentive for attackers to attack a reputation system [23].

4   The Proof of Reputation Protocol

In this section, we present the Proof of Reputation (PoR) protocol, which serves to provide global reputation agreement in a p2p environment.

4.1   Design Overview

The core idea of PoR is to ensure that every one of participants holds an agreed ledger, recording the rate of each transaction. To achieve this purpose, three questions must be answered:

1. How to keep the ledgers consensus? The answer to this question is similar to that of Bitcoin. Bitcoin applies the PoW protocol, where the first processor that has solved the mathematical problem gains the right to publish the block, and others can openly verify the block. Likewise, in the PoR protocol, the one who has the highest trust value in a group of transactions can package them into a block and publish it. The verification of the block is also open to other participants.

2. How to motivate participants to publish blocks? Compared with Bitcoin, there is no block reward or transaction fee as incentives in the PoR. We use trustworthiness instead. However, trustworthiness, which is entirely different with coins, can not be spent or transferred. It is a kind of reputation which indicates that participants with high trust value can provide better services. In each round, only the one that has the highest trust value can publish the block. Publishing the block can undoubtedly increase the overall trust rank of the publisher.

3. How to prevent transactions or blocks from being modified? The Bitcoin network is entirely open to every machine that has a client installed on it. Modifying the transactions consumes huge hash power, which is known as *51% attack* [24]. On the contrast, the PoR protocol is applied in the permissioned blockchain, where all participants are registered with their public keys with private keys stored locally, and every participant has a copy of all public keys. Therefore, if some malicious nodes attempt to fabricate identities, it will be detected instantly. Also, the Merkle Tree [25] is used in forming the transactions in the block, which can protect the integrity of transactions.

In each round, the PoR protocol runs the following 3 steps:

1. *Broadcasting Transactions.* At the end of each interaction, a piece of feedback will be generated by the service requestor, recording the rate of the service. Then the requestor will broadcast this message together with its signature. Other nodes that have received these messages will verify and store them in the memory.
2. *Building Block.* When the number of transactions reaches the threshold, the node stops receiving transactions and starts to calculate a ranking list according to each service providers' scores recorded in this set of transactions. If the top of the list is the node itself, it constructs a block and publishes it after signing with its private key.
3. *Verifying Block.* Nodes receiving a block will recalculate the ranking list to check if the sender has the highest trust value. They also need to verify the signature of each transaction using the signer's public key. After verification, the nodes will append the block into the blockchain and prepare for the next round.

4.2   Broadcasting Transactions

We define the participant that provides a certain service during an interaction as the *provider*, while another participant that rates the service as the *rater*. During the interaction, the *provider* sends the requested service, such as a bunch of data, signed by its private key, along with the hash of the data and a timestamp. After verifying the integrity of the data by checking the recalculating the hash, the *rater* produces a *transaction* consisting of the *reputation score*, a timestamp, the hash of the received data, and its digital signature. Next, the *rater* broadcasts the *transaction* to the network. A diagram of the format of a *transaction* is

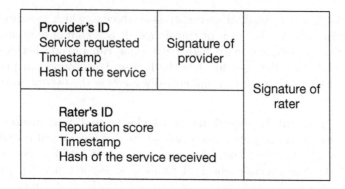

Provider's ID Service requested Timestamp Hash of the service	Signature of provider	
Rater's ID Reputation score Timestamp Hash of the service received		Signature of rater

Fig. 1. A diagram of the format of a *transaction*.

illustrated in Fig. 1. The *reputation score*, represented by x_i^j, can be expressed either in a binary way ($x_i^j \in \{0,1\}$, i.e., p_i rates 1 if it is satisfied with the service and 0 otherwise), or a real number in a continuous range ($x_i^j \in [0,1]$) to represent the degree of satisfaction.

An example of this step is shown in Fig. 2, where p_1 is the *rater* and p_5 is the provider. At the end of the interaction, p_1 is satisfied with the service and rates it with 1. Then the *transaction*, which is represented as $Sig(p_1 \rightarrow p_5 = 1)$, of this interaction is generated, and sent to the other participants by p_1.

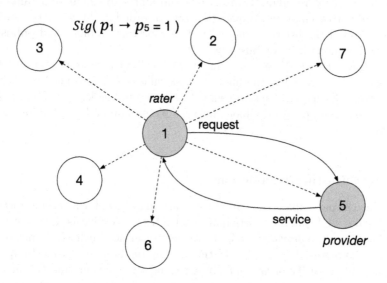

Fig. 2. Broadcasting transactions step of p_1 rating the service of p_5.

Algorithm 1. Transaction Filter Algorithm

    Input: A set of *tansactions*, represented as $T = \{t_1, t_2, t_3, ..., t_m\}$ where m is
        the number of the set.
    Output: Filtered *transactions*
1 $FilteredNumber = 0$;
2 $FilteredTransactions = \{\}$;
3 **for** t_i **in** T **do**
4   **if** *VerifySig(t_i)* **then**
5   **if** *the rater and the provider of t_i*
6   *have appeared in FilteredTransactions* **then**
7   update(*FilteredTransactions*);
8   **else**
9   **if** *FilteredNumber* $< \lambda$ **then**
10   FilteredTransactions.append(t_i);
11   FilteredNumber+=1;
12   **else**
13   break;
14   **else**
15   drop(t_i);

4.3 Transactions Filter

Malicious nodes can spread forged *transactions* and *transactions* with dishonest rating. To prevent this, we need filter *transactions* before packaging them up.

Our algorithm to filter *transactions* is depicted in Algorithm 1. More specifically, when a *transactions* comes, its signature will be tested whether it comes from an authenticated participant. Next step is to check if the same pair of *rater* and *provider* have appeared in the current block. If so, the *transaction* will be replaced by the later one. For example, assume a *transaction* $Sig(x_5^1)$ has already been verified, then if another $Sig(x_5^1)$ comes, it will replace the former one. The purpose of keeping only one *transaction* of the same pair in one block is to reduce the impact of *bad-mouthing attack*. Finally, when the number of filtered *transaction* reaches the threshold λ, this set of *transactions* would be packaged up to conduct an alternative block, while other *transactions* will remain in the memory pool for the next round.

4.4 Block Publication

With a set of verified *transactions*, there would be two steps to construct an alternative block.

1. Trustworthiness Evaluation. From the set of verified *transactions*, a participant can extract each *provider*'s action log, which can be represented as $l_j = s_1, s_2, s_3, ..., s_n$ for the *provider* p_j, where s_i is the ith *score* of l_j and n

is the number of the *scores*. We can use a certain trustworthiness evaluation model to squash the log into a trust value. For simplicity, we use the sigmoid function to evaluate the trustworthiness in our protocol (Other trust evaluation algorithms can also be used, i.e., EigenTrust [12]). Assume that $s_i \in \{1, 0, -1\}$ for *good, general, bad* actions respectively. Then the trustworthiness of p_j can be calculated as follows:

$$trust^j = \frac{1}{1 + e^{-\phi \sum_{i=1}^n s_i}} \tag{1}$$

Accordingly, ϕ is simply the adjustable parameter. With this measure, each *provider* will be assigned a trust value, and a *ranking list* is generated. Then the participant needs to check if itself is the *provider* and it has the highest trust value. If so, it moves on to the second step to publish the block, otherwise, it has to give up this round and prepare the next round. The reason why only the *provider* with the highest trust value can publish the block in each round is that this mechanism can motivate participants that have the highest trust value in each round to publish the block since it will increase their trust ranks.

2. Block Construction. The way we consider to construct a block is comparable to Bitcoin, but with different data structure. Details are shown in Fig. 3. Merkle tree is also introduced when constructing the block to facilitate transaction searching. The threshold is to control the number of *transactions* contained in a single block. In addition, the participants who publish this block need to digitally sign the block and add its signature into the blockheader for authentication.

After these two steps above, the block will be committed and spread to the network.

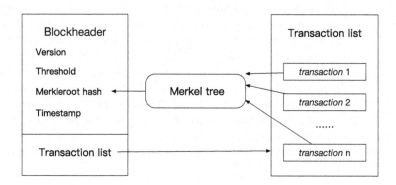

Fig. 3. Data structure of the block.

4.5   Block Verification

Every block received by participants need to be verified before appended to the blockchain. The verification process is presented in Algorithm 2. First, the signature of the block should be from an authenticated participant, so be the signatures of each *transactions* in the block. Next, a trustworthiness ranking list will be recalculated based on the checked transactions. Finally, it will be checked that whether the provider on top of the list is same as the sender of the block.

Algorithm 2. Block Verification Algorithm

Input: *currentBlock*
Output: block validity
1 *rankingList* = {};
2 **if** !*verify*(*currentBlock.signature*) **then**
3 | return false;
4 **if** !*verify*(*currentBlock.transactionList.signatures*) **then**
5 | return false;
6 *rankingList* = *calculateRankingList*(currentBlock.transactionList);
7 **if** *(currentBlock.signature is not matched to the public key of rankingList.top)*
   then
8 | return false;
9 return true;

After verification, the block it will be cryptographically linked to the blockchain and wrote to the disk. If any *transactions* in the memory pool also exist in the newly written block, they would be removed.

4.6   Cost Analysis and Quick Bootstrap

The space cost of a single block is closely related to the number of *transactions* controlled by λ, which can also be utilized to control the average time a block is generated according to the number of participants in the network. A block with no *transactions* would be about 80 bytes, which is close to that of Bitcoin. A single *transaction* costs about 100 bytes. If we suppose $\lambda = 100$, a single block costs no more than 10KB. Since blockchain is an append-only data structure, the space cost grows linearly with time. However, because the number of participants is limited in permissioned blockchains, the space cost is far less than Bitcoin blockchain.

To boost setup, new participants do not have to download the whole blockchain. Instead, they can simply get a ranking list from the registry. This ranking list is similar to that described in Sect. 4.4 except that it is calculated from the whole blockchain. The ranking list is signed by the registry and expressed in JSON format. An example of it is shown below.

```
1  {
2    "number":128
3    "rankingList": {
4      "rankingitem": [
5        {"id": "62923F...CF0629", "value": "0.84"},
6        {"id": "730F75...82848F", "value": "0.74"},
7        ...
8        {"id": "27B2C1...B93255", "value": "0.92"}
9      ]
10   }
11   "rankingListHash":"7520149...60265B"
12   "rankingListHashSignature":{
13     "r":"F0AB42...1D6875"
14     "s":"4C556E...31C42B"
15   }
16   "time":1496374331
17 }}
```

4.7 Security Analysis

After the details of the protocol design, we provide security analysis for how PoR prevents potential threats and works securely. Since we have already assumed that our protocol is deployed in permissioned blockchains where participants can interact in a secure channel and can authenticate with each other, but the trust-based protocol itself is easily targeted by potential attackers.

Malicious participants may perform *bad-mouthing attack* by rating the *provider* dishonestly. In PoR, the *transaction* of the same pair of a *rater* and a *provider* can only appear once in one block. Therefore, even though the *rater* is malicious, its dishonest recommendation has limited impact on the *provider*. Moreover, it can effectively resist *Replay attack*. In *on-off attack*, attackers have irregular behaviors. In our protocol, the distributed ledger of reputation can provide a trust ranking list, which is calculated from all the *transactions* recorded in the blockchain. This ranking list indicates the overall trustworthiness of each participant, so the attackers performing *on-off attack* will be reported by the *rater* resulting in a lower rank. If a participant has a low rank, they may want to switch to another ID and start over, which is known as *Sybil attack* or *Newcomer attack*. The access control layer built on the blockchain can increase the cost of creating a new ID to prevent this kind of attack.

5 Experiments and Evaluation

In this section, we carry out several experiments and performance measurements based on a prototype implementation of the PoR protocol. The goal of our evaluation is to quantify the overhead and scalability of our protocol when deploying it in a p2p network.

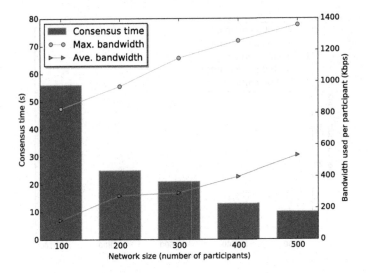

Fig. 4. Consensus time and bandwidth of PoR with different network sizes.

5.1 Experiment Setting

We developed a prototype implementation of PoR in the Python 2.7 programming language. Then we virtualized participants by running the protocol on Docker [26], which is an open source software technology providing containers. We also used Rancher [27], which is a container management platform, to manage our containers so that we can easily create p2p networks of different sizes and measure the performance of our protocol. We vary the number of participants in the network from 100 to 500, using 3 servers with load balancing strategy. Each server has an E5-2640 CPU with 16 cores and 64 GB of memory.

5.2 Performance Evaluation

The ability to log *transactions* in a light-weight, scalable and efficient manner is key to a transactional p2p network. In our experiments, we measure the bandwidth consumptions per node of PoR, consensus time, production time, as well as the throughput in different settings.

Scalability of PoR. We start with a network of 100 participants, then we adjust the network size 4 times, from 100 participants to 500 participants. Meanwhile, we set the number of transactions in each block fixed to 1000. We quantify the average time to reach consensus, maximum and average bandwidth consumed at each participant. The experimental results are shown in Fig. 4.

The maximum bandwidth refers to the bandwidth used when the participant receiving or delivering a published block, while the average bandwidth is the bandwidth consumed when processing transactions. The results show that both the maximum and average bandwidth increase linearly (from 820 Kbps to

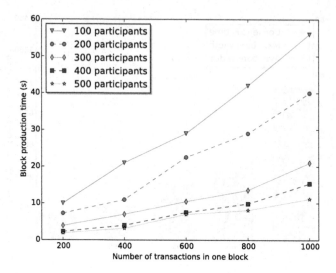

Fig. 5. Average time to produce a block with different block sizes.

1380 Kbps and 160 Kbps to 470 Kbps, respectively) as we enlarge the network size (100 to 500 accordingly). This is because more participants joining the network, more messages being transferred in the network. In contrast, the consensus time is shorter (e.g., 56 s in 100 participants to less than 10 s in 500 participants) since more transactions are generated, each participant requires less time to publish the block. These results show the average time of PoR to reach consensus is less than a minute, which is much shorter than that of Bitcoin blockchain (usually 1 day). In addition, the average bandwidth is less than 1 Mbps, which is negligible in most p2p networks.

Production Time. We next vary the number of transactions in a single block from 200 to 1000 to measure the average time it takes to produce a block with different network sizes. In our experiment, we measured the production time in different network sizes ranging from 100 to 500 participants.

Figure 5 depicts that longer time is consumed if the block size is larger since a participant needs to receive more transactions to package them up. Meanwhile, we observe that smaller network has longer block production time and it grows faster as the block size goes up. This is because smaller network generates fewer transactions per unit time. When the block size is set to 1000, the production time of the network with 100 participants is less than a minute, and in the network with 500 participants, the production time is less than 10 seconds. This indicates that the PoR protocol is far more efficient than bitcoin-like ones, where it takes about 10 min to produce a block. In addition, the results show that we can adjust the time a single block consumed by changing the block size according to the requirement of upper applications.

Throughput. We have the same definition of throughput in [28], where it is defined as the number of transactions committed per unit of time. In this experiment, we also vary the block size from 200 to 1000 transactions and the network size from 100 to 500 participants. The results are plotted in Fig. 6.

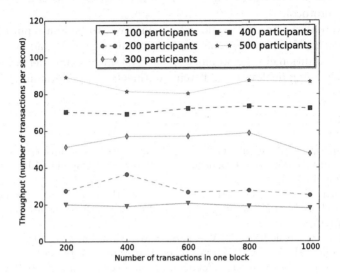

Fig. 6. Throughput with different block sizes.

From Fig. 6 we can see that the throughput has an increment when more participants joining the network, the reason of which is obvious. Meanwhile, the throughput fluctuates slightly around a certain value (i.e., 20, 30, 50, 70, 85 transactions per second respectively in different network sizes) as the block size increases. The most likely explanation for this is that although a participant can commit more transactions in larger-size blocks, it takes more time to produce one. These results indicate that the throughput of PoR hinges on the number of participants involved in the network. Even in a thousand-level network, it is possible to process hundreds of transactions in one second, which is far more efficient than the throughput of bitcoin-like protocols.

In summary, the experiments confirm the expected scalability of PoR's transaction throughput and bandwidth usage. The performance of our protocol largely depends on the network size and the block size, which gives it flexibility to adjust various p2p environment. In addition, the proposed protocol can also be deployed on IoT devices since low bandwidth and low computation resources are consumed.

6   Conclusion and Future Work

We present PoR, a reputation consensus protocol for p2p networks, which is derived from the idea of blockchain technology. The contribution of our study is to provide a distributed ledger of reputation so that each participant can share a global view of reputation. We have implemented a prototype of our protocol and experimental results demonstrate that PoR can be a suitable component to for transactional applications to make reputation based decisions. More generally, our study has shown the potential of reputation to serve as the incentive in a consensus protocol.

As future directions, we would extend our work to build a reputation based system which contains access control, identity management, and other security strategies. Additionally, we would test our system in a broader area, such as participants being deployed in different continents, for performance evaluation.

References

1. Nakamoto, S.: Bitcoin: a peer-to-peer electronic cash system (2008)
2. Swan, M.: Blockchain: Blueprint for a New Economy. O'Reilly Media Inc., Sebastopol (2015)
3. Kalodner, H.A., Carlsten, M., Ellenbogen, P., Bonneau, J., Narayanan, A.: An empirical study of Namecoin and lessons for decentralized namespace design. In: WEIS (2015)
4. Carlsten, M., Kalodner, H., Weinberg, S.M., Narayanan, A.: On the instability of bitcoin without the block reward. In: Proceedings of the 2016 ACM SIGSAC Conference on Computer and Communications Security, pp. 154–167. ACM (2016)
5. Gupta, M., Judge, P., Ammar, M.: A reputation system for peer-to-peer networks. In: Proceedings of the 13th International Workshop on Network and Operating Systems Support for Digital Audio and Video, pp. 144–152. ACM (2003)
6. Selcuk, A.A., Uzun, E., Pariente, M.R.: A reputation-based trust management system for P2P networks. In: IEEE International Symposium on Cluster Computing and the Grid, CCGrid 2004, pp. 251–258. IEEE (2004)
7. Resnick, P., Zeckhauser, R.: Trust among strangers in internet transactions: empirical analysis of ebay's reputation system. In: The Economics of the Internet and E-Commerce, pp. 127–157. Emerald Group Publishing Limited (2002)
8. Dewan, P., Dasgupta, P.: Securing reputation data in peer-to-peer networks. In: Proceedings of International Conference on Parallel and Distributed Computing and Systems, PDCS (2004)
9. Shrier, D., Wu, W., Pentland, A.: Blockchain & infrastructure (identity, data security). Technical report (2016). http://cdn.resources.getsmarter.ac/wp-content/uploads/2016/06/MIT_Blockain_Whitepaper_PartThree.pdf. Accessed 27 Nov 2016
10. Wang, Y., Vassileva, J.: Trust and reputation model in peer-to-peer networks. In: Proceedings of the Third International Conference on Peer-to-Peer Computing (P2P 2003), pp. 150–157. IEEE (2003)
11. Schiffner, S., Clauß, S., Steinbrecher, S.: Privacy, liveliness and fairness for reputation. In: Černá, I., Gyimóthy, T., Hromkovič, J., Jefferey, K., Královič, R., Vukolić, M., Wolf, S. (eds.) SOFSEM 2011. LNCS, vol. 6543, pp. 506–519. Springer, Heidelberg (2011). https://doi.org/10.1007/978-3-642-18381-2_42
12. Kamvar, S.D., Schlosser, M.T., Garcia-Molina, H.: The Eigentrust algorithm for reputation management in P2P networks. In: Proceedings of the 12th International Conference on World Wide Web, pp. 640–651. ACM (2003)

13. Carboni, D.: Feedback based reputation on top of the bitcoin blockchain. arXiv preprint arXiv:1502.01504 (2015)

14. Dennis, R., Owen, G.: Rep on the block: a next generation reputation system based on the blockchain. In: 2015 10th International Conference for Internet Technology and Secured Transactions (ICITST), pp. 131–138. IEEE (2015)

15. Dennis, R., Owenson, G.: Rep on the roll: a peer to peer reputation system based on a rolling blockchain. Int. J. Digit. Soc. (IJDS) **7**(1), 1123–1134 (2016)

16. Buechler, M., Eerabathini, M., Hockenbrocht, C., Wan, D.: Decentralized reputation system for transaction networks. Technical report, University of Pennsylvania (2015)

17. Schaub, A., Bazin, R., Hasan, O., Brunie, L.: A trustless privacy-preserving reputation system. In: Hoepman, J.H., Katzenbeisser, S. (eds.) SEC 2016. IFIP AICT, vol. 471, pp. 398–411. Springer, Cham (2016). https://doi.org/10.1007/978-3-319-33630-5_27

18. Otte, P., de Vos, M., Pouwelse, J.: TrustChain: a Sybil-resistant scalable blockchain. Future Gener. Comput. Syst. (2017)

19. Callegati, F., Cerroni, W., Ramilli, M.: Man-in-the-middle attack to the HTTPS protocol. IEEE Secur. Priv. **7**(1), 78–81 (2009)

20. Sun, Y., Han, Z., Ray Liu, K.J.: Defense of trust management vulnerabilities in distributed networks. IEEE Commun. Mag. **46**(2), 112–119 (2008)

21. Dellarocas, C.: Mechanisms for coping with unfair ratings and discriminatory behavior in online reputation reporting systems. In: Proceedings of the Twenty-First International Conference on Information Systems, ICIS 2000, Brisbane, Australia, 10–13 December 2000, pp. 520–525 (2000)

22. Douceur, J.R.: The Sybil attack. In: Druschel, P., Kaashoek, F., Rowstron, A. (eds.) IPTPS 2002. LNCS, vol. 2429, pp. 251–260. Springer, Heidelberg (2002). https://doi.org/10.1007/3-540-45748-8_24

23. Jøsang, A., Golbeck, J.: Challenges for robust trust and reputation systems. In: Proceedings of the 5th International Workshop on Security and Trust Management (SMT 2009), Saint Malo, France, p. 52 (2009)

24. Eyal, I., Sirer, E.G.: Majority is not enough: bitcoin mining is vulnerable. In: Christin, N., Safavi-Naini, R. (eds.) FC 2014. LNCS, vol. 8437, pp. 436–454. Springer, Heidelberg (2014). https://doi.org/10.1007/978-3-662-45472-5_28

25. Szydlo, M.: Merkle tree traversal in log space and time. In: Cachin, C., Camenisch, J.L. (eds.) EUROCRYPT 2004. LNCS, vol. 3027, pp. 541–554. Springer, Heidelberg (2004). https://doi.org/10.1007/978-3-540-24676-3_32

26. Docker. https://www.docker.com

27. Rancher. http://rancher.com

28. Miller, A., Xia, Y., Croman, K., Shi, E., Song, D.: The honey badger of BFT protocols. In: Proceedings of the 2016 ACM SIGSAC Conference on Computer and Communications Security, pp. 31–42. ACM (2016)

Incremental Materialized View Maintenance on Distributed Log-Structured Merge-Tree

Huichao Duan, Huiqi Hu[(✉)], Weining Qian, Haixin Ma, Xiaoling Wang, and Aoying Zhou

East China Normal University, Shanghai, China
stevenduan@stu.ecnu.edu.cn, {hqhu,wnqian,ayzhou}@dase.ecnu.edu.cn,
xlwang@sei.ecnu.edu.cn

Abstract. Modern database systems are in need of supporting highly scalable transactions of data updates and efficient queries over data simultaneously for the real-time applications. One solution to reach the demand is to implement query optimization techniques on the online transaction processing (OLTP) systems. The materialized view is considered as a panacea to improve query latency. However, it also involves a significant cost of maintenance which trades away transaction performance. In this paper, we develop materialized views on a distributed log-structured merge-tree (LSM-tree), which is a well-known structure adopted to improve data write performance. We examine the design space and conclude several design features for the implementation of view on LSM-tree. An asynchronous approach with two optimizations are proposed to decouple the view maintenance with transaction process. Under the asynchronous update, we also provide consistency query for views. Experiments on TPC-H benchmark show our method achieves better performance than straightforward methods on different workloads.

1 Introduction

Databases have always been facing challenges, where an important one is that it requires supporting highly scalable transactions of data updates and efficient queries over massive data simultaneously for the real-time applications. To meet the demand, a practical solution is to leverage query optimization techniques on scalable OLTP systems. In term of query optimization, materialized view can significantly facilitate query by reducing execution time. The technique succeeds in many data warehouses and decision support systems which mainly update data by importing other data source via Extract-Transform-Load (ETL) tools. It can meet the analytical demand by avoiding the impact of online date updates, but the accessed data may be out of date. It's much more challenging to develop materialized views on OLTP systems than on those data warehouses. In OLTP systems, data updates frequently happen on data tables. To get an up-to-second query result, the materialized view must be updated along with the base table.

© Springer International Publishing AG, part of Springer Nature 2018
J. Pei et al. (Eds.): DASFAA 2018, LNCS 10828, pp. 682–700, 2018.
https://doi.org/10.1007/978-3-319-91458-9_42

However, views are usually expensive to maintain since they are mainly constructed for schemas associated with a bunch of data records. The improvement of query performance and the cost of maintenance have put the system into a dilemma for whether to use materialized views.

The answer to the question may be depended on the workload loaded on the system. But it is apparently beneficial to optimize the view maintenance for those systems that support high-speed update. In this paper, we aim to support views in a distributed LSM-tree architecture. Many systems choose LSM-tree since it offers high write throughput. Following the notations used in [9,15], LSM-tree organizes records in multiple components: a memtable and several sstables. The memtable is a memory-based structure for data writes and the sstable is a disk-based structure, offering large storage capacity and servicing read requests only. The LSM-tree structure has been widely implemented by distributed systems such as BigTable [9] and Cassandra [13], where the memtable and sstables are kept in the main memory and distributed file system (e.g. GFS [10]) respectively.

We conclude the major problems for view maintenance on distributed LSM-tree are: (1) the structure is distributed. Thus a record of base tables and its related records in view table may be located on different machines. If an update happens on the record of the base table, it takes many costs to identify these related updates in view. (2) In a LSM-tree system, the memtable is responsible for high-speed data write. The resources of the memtable's server are usually precious and the burden on it is heavy, hence we don't want to add many additional overheads of view updates on the server. (3) The system serves both the transactional update and view query at the same time. If they access same data, we need to guarantee the consistency of query. (4) The updates of base table produce plenty of updates in view table. However, if queries and updates fall on different data, we just need to update the related records in view instead of updating all of them directly. We can take advantage of this to make better use of system resources, which requires fine-grained method to update the view.

To address the problems, we propose an effective view maintenance mechanism on distributed LSM-tree. Our contributions can be summarized as follows: (1) we examine the design space and conclude several design features for implementation of view on LSM-tree. (2) We decouple the procedure of view maintenance with transaction process, which reduces its impact on transaction processing. Meanwhile we utilize an asynchronous approach to update the view and propose two optimizing techniques that delay unnecessary updates. (3) We separate the storage and maintenance of view table from the server of memtable, which significantly reduces its overhead. (4) We also ensure query consistency under the asynchronous updates. (5) Experiments on popular benchmark show our method achieves a better performance than straightforward techniques.

The rest of paper is organized as follows. Section 2 gives the design consideration. Section 3 presents the overall structure. Section 4 describes the detailed query and update method. Section 5 shows the experiment results. Related works are discussed in Sect. 6 and the work is concluded in Sect. 7.

2   Background

2.1   LSM-Tree Model

In this paper, we study asynchronous view maintenance over a large-scale, distributed LSM-tree. A typical structure is illustrated in Fig. 1. Basically, it consists of an update server, several chunk-servers and multiple p-nodes.

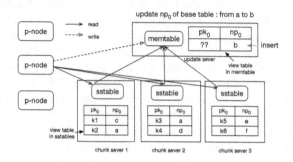

Fig. 1. LSM-tree model and UPSV

Memtable and Update Server. Memtable is the in-memory structure which keeps **incremental data**. In the LSM-tree, it resides in main memory to facilitate write performance. Data write (update, insert and delete) is only allowed to perform on the memtable. Data in memtableis managed through its primary key. It can be implemented as any index structure optimized for main memory access, such as bw-tree [14]. Memtable is placed on an **update server**, which usually has high-performance CPUs and large capacity of memory.

SSTables and Chunk Servers. SStable is the on-disk and immutable structure where **static data** is stored in lexicographic order based on its primary key. SStable is generated by freezing an active memtable. The frozen memtable is transferred into the distributed file system and becomes the sstable. In short, the log-structured storage firstly keeps written data in the memtable. It then freezes and merges the memtable into durable storage (sstables) when the memtable reaches a certain size. Thus, sstable is a **read-only** data structure. A memtable can correspond to several sstables, which are horizontally partitioned over the primary keys and each sstable corresponds to part of the memtable. Disk based indexes (such as block index), can be utilized in each sstable. SStables live in several **chunk-servers** according to a certain data distribution.

Data Write and Access. In LSM-tree, one primary key corresponds to one data record. We utilize **multi-version** for data write and access in our work. If a record is inserted, an entry (with primary key as its entry) is created and directly added into the memtable with an assigned version number; if a record is updated, its new value is appended into the memtable with a new assigned version number. For a read operation (a point read or a scan request), a process node (**p-node**) has to go through both the memtable and sstable to access records. From memtable, it obtains the latest state of record. Data (with same primary key) from memtable and sstable is fused on chunk-server, and then transferred to p-node for further processing.

Data Compaction. To reduce the memory space of update server, data compaction can be conducted in back-end to merge sstables and memtable together. It first freezes the old memtable and starts a new memtable to replace the old one for servicing further write requests. Then a write process would read entries from multiple sstables and merge the old memtable, i.e. if a record has a new state (via data writes) on the memtable, the new state will be merged into corresponding sstable based on its primary key. With the completion of the data compaction, the old memtable becomes expired.

2.2   View Table

The view table is created based on the base tables. In the distributed system, we divide the view table into different partitions (according to a range or a hash partitioning) and manage them separately on different chunk-servers. For example, recall the TPC-H benchmark [5]. It has total eight tables, where one of them named *lineitem* is used to store the purchasing records. We can create the view on *lineitem* according to the following statement:

CREAT VIEW *ShipDate*
SELECT   *L_shipdate, L_linenumber, L_orderkey,*   *L_discount,*   *L_quantity,*
L_extendedprice
FROM *lineitem*
WHERE *L_quantity* $>= 11$ AND *L_quantity* $<= 21$
PARTITION BY RANGE (*L_shipdate*);

A projection view table named *ShipDate* is distributed according to a composite primary key. It is constructed by combining the attribute of *L_shipdate* and the primary key of the base table (i.e. *L_orderkey* and *L_linenumber* from *lineitem*). Next, we formally define the base table and view table for further description. For simplicity, we only define the projection view, while other types of view tables are analogous. The schema of base table and view table are defined as:

Definition 1 (Base table $\mathcal{B}$). *A base table contains two type of columns (attributes): columns of primary key (PK) $PK_{\mathcal{B}} = \{pk_{b_1}, pk_{b_2}, \cdots, pk_{b_{|PK_{\mathcal{B}}|}}\}$; and columns of non-primary key (NP) $NP_{\mathcal{B}} = \{np_{b_1}, np_{b_2}, \cdots np_{b_{|NP_{\mathcal{B}}|}}\}$. pk_{b_i}/np_{b_i} is a PK/NP column respectively.*

For instance, *lineitem* is a base table, its has two PK columns $PK_{\mathcal{B}} = \{l\_orderykey, l\_partkey\}$ and it has total fourteen NP columns $NP_{\mathcal{B}} = \{l\_shipdate, l\_discount, \cdots\}$.

Definition 2 (View table $\mathcal{V}$). *A view table also contains two type of columns:* PK *columns* $PK_\mathcal{V} = \{pk_{v_1}, pk_{v_2}, \cdots, pk_{v_{|PK_\mathcal{V}|}}\}$ *(*$PK_\mathcal{B} \subset PK_\mathcal{V}$*) and* NP *columns* $NP_\mathcal{V} = \{np_{v_1}, np_{v_2}, \cdots np_{v_{|NP_\mathcal{V}|}}\}$ *(*$NP_\mathcal{V} \subset NP_\mathcal{B}$*), where* pk_{v_i}/np_{v_i} *is a* PK/NP *column of* $\mathcal{V}$ *respectively.*

Since $\mathcal{V}$ is constructed from $\mathcal{B}$, in projection view, we select several columns from $NP_\mathcal{B}$ and all columns from $PK_\mathcal{B}$ together as the PKs of $\mathcal{V}$, thus we have $PK_\mathcal{B} \subset PK_\mathcal{V} \subset \mathcal{B}$. Meanwhile, np_{v_i} is selected from NP columns of base table, which means $NP_\mathcal{V} \subset NP_\mathcal{B}$. For example, in the above view named *ShipDate*, its has three PK columns $PK_\mathcal{V} = \{l\_shipdate, l\_orderykey, l\_partkey\}$, where $l\_orderykey$, $l\_partkey$ are the PKs of base table *lineitem* and $l\_shipdate$ is a NP columns of base table. Three NP columns, $\{l\_discount, l\_quantity, l\_extendedprice\}$ from *lineitem* are selected for projection.

2.3 A Straightforward View Maintenance Design

We first introduce a straightforward design of view structure following the existing update and data access procedure on LSM-tree. We name it update server based view design (UPSV).

UPSV treats materialized views in the same way as the base tables, where records in the view table are ordered and managed through primary keys. View tables are also horizontally partitioned and the system can merge them from update server to chunk-servers through data compaction. For instance, Fig. 1 illustrates a view table with two columns (pk_0 and np_0), it is distributed on chunk-servers and partitioned by the PK column pk_0. If an update of the view happens, a new state will be added into the memtable of update server.

It relies on transaction processing to maintain the increment of view table. To ensure the consistency of the view table and the base table (see Sect. 2.4), view tables are updated synchronously in the transaction processing thread. To this end, an additional step of view maintenance will be added to the transaction if the base table updates are related to the view. Records of view tables associated with the base table updates will be updated/inserted/deleted in the memtable. Since the view table is same as the base table, updates of them only appear on the update server. A critical step is to identify those records in view tables to be updated. It is non-trivial on LSM-tree. For example, in Fig. 1, if a transaction updates a record of base table on attribute np_0 from a to b, then the view table should also be updated by inserting its new state into memtable. However, the update server do not know which record (since the base table and view table leverages different primary keys) within view table has np_0 with value a, so it has to visit all chunk-servers to find k_2 and k_3. Thus an **identifying step** must perform on the chunk-servers to find those updates and then give feedback to the update server for further movement.

UPSV is not efficient since it is cost-consuming in view maintenance, thus we need to reexamine its design space on LSM-tree and propose a new design.

2.4   Design Features

System for Hybrid Workload. We target for the mixed read-write workload scenario. The system can support one set of workloads focused on transaction processing (e.g. insert, update and delete on base tables) and another set configured to run analytical queries (e.g. SPJ query) on view tables without blocking the ongoing operational workloads. In term of materialized views, as known to all, it can be applied to query optimization to benefit query performance. For instance, a simple projection view can significantly facilitate data scan since it re-adjust the PK columns in the view table to make it better to use the index (e.g. block index, B+tree) for data scan. Nevertheless, to optimize query with materialized view, we must carefully design the strategy of view maintenance.

Decoupling of Transaction Processing and View Maintenance. First, view maintenance severely affects the performance of transaction processing. Take UPSV into consideration, which binds the view update to the transaction, it can directly prolong the locking time and increase the delay of a transaction. Worse still, the cost of view update is usually much higher than that of an update on the base table, so the design will significantly reduce the performance of transaction processing. Therefore, we need to decouple the procedure of view maintenance from transaction processing. In theory, we can put this step at any time point after the transaction and before the query, which is commonly known as lazy maintenance or asynchronous maintenance [6,20].

Reducing the Overhead of Update Server. It is not sufficient to simply split transactions processing and view updates. An important consideration is the overhead of update server. In a LSM-tree system, update server is the most resource consuming as its memory is limited, its CPU and networks are not only used to process transactions, but also to deal with data compaction. For UPSV, view updates are written in memtable. It greatly increases the use of CPUs and memory since the scale of view update is far beyond that of the base table, which also aggravates the burden of data compaction. To this end, we shift the storage of view table and update it on chunk-servers instead of update server.

View Consistency. Since view updates are no longer bound to transaction processing, the consistency that it provides for the query needs to be considered. We aim to provide a **consistent snapshot** [8] for data access on the view table. A query must read a unified version on all accessed data and this version can correspond to the latest updates on data. The problem is difficult since we update view tables asynchronously on remote chunk-servers.

Summary. We process the mixed workload and leverage materialized view to optimize query processing. To prevent a substantial decline in the performance of transaction processing, we decouple view maintenance from the transaction. In order not to bring additional overhead on update server, we choose to update views on chunk-servers instead of on update server. We also provide a way to query the view to ensure that we access a consistent snapshot. Next, we introduce a new design for view maintenance on LSM-tree.

3   System Architecture

In this section we describe a new design of view maintenance on LSM-tree. The core idea is to maintain view tables on chunk-servers. We name the method CSV.

3.1   View Storage

To reduce the overhead of update server, we do not keep any memtable of view table on update server, instead, we maintain them separately on different chunk-servers. There are four structures to store records of base tables and view tables:

- B-sstable stores the statistic data of base table. B-sstables are distributed on different chunk-servers according to the data partitioning algorithms.
- B-memtable is utilized to maintain all the incremental update of base table. B-memtable leverages a memory B+tree structure. Every leaf node stores a pair of $\langle PK_B, list \rangle$, where the entry is the PKs (PK_B) of a record. The list keeps all data writes on the record. An item in the list represents one update[1] with an assigned version number. Record locking [3] are employed to handle concurrent updates on the same list. Updates are only allowed in NP columns. As described in Sect. 2, B-memtable lives on update server. It becomes B-sstables through data compaction.
- V-sstable is the sstable which stores the static data of view table on disk. Similar to B-sstables, V-sstables are also distributed on chunk-servers.
- V-memtable is a memtable resides on a chunk-server. Each partition of view table contains a V-memtable and a V-sstable, V-memtable and its corresponding V-sstable live on a same chunk-server. V-memtable stores incremental data of view table. Similar to B-memtable, data writes are added into a B+tree, where the entry of list is the PK_V of the record in view table.

Table Partition. Note that both base table and view table are divided into several partitions on chunk-servers for parallel processing. Tables are partitioned over their PKs (PK_B or PK_V) by partitioning algorithms. Each partition has a specific **partition id** (denoted as *pid*), and for each record, we can compute its partition id based on its values of PK_B or PK_V.

3.2   Update on Base Table and View Table

Recall that UPSV updates the base table and the view table in the same transaction. It takes a cost-consuming identifying step to read remote data from chunk-server. In CSV, we decouple transaction processing (which updates the base table) and view maintenance. One of our contributions is that we propose a lightweight and asynchronous mechanism to avoid this identifying step in transaction processing.

---

[1] Delete as well as update, since the record with index will not be actually deleted but just marked for deletion.

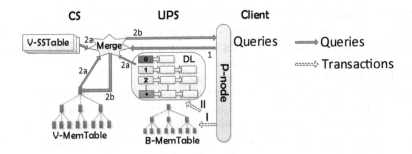

Fig. 2. Framework of CSV

When an update of record happens on a base table, we can obtain several lightweight *delta records* with only a few overhead. A delta record can be treated as a **corresponding modification of the view table** when an insert/update/delete happens on the base table. Formally it is defined as follows:

Definition 3 (Delta record δ_r). *Suppose an insert/update/delete happens on $\mathcal{B}$ and it generates a modification on column np_{v_i} of a record $r \in \mathcal{V}$, then it produces the corresponding δ_r including three parts: (i) its values on columns of $PK_{\mathcal{V}}$ and np_{v_i} (denoted as $\sqcap(PK_{\mathcal{V}} \cup np_{v_i})(r)$), which records the update on r; (ii) a partition id pid_r, which indicates the partition of r, (iii) an operation flag, which signs the operation type (insert, delete or update) of this record.*

It is worth noting that for the above first part, we may not be able to know the complete values on PK columns of r (i.e. $PK_{\mathcal{V}}$ of r). Since $PK_{\mathcal{B}} \subset PK_{\mathcal{V}}$, its values on columns $PK_{\mathcal{B}}$ are known by parsing the SQL which updates the base table. The rest columns in $PK_{\mathcal{V}}$-$PK_{\mathcal{B}}$ may not be known. To this end, we cannot compute the partition id of r without its exact values on $PK_{\mathcal{V}}$. In this case, we use a dummy mark $*$ to represent the partition id of a delta record.

We adopt an assisted structure called delta lists to store delta records.

Definition 4 (Delta lists $\mathcal{DL}$). *Suppose that the view table $\mathcal{V}$ is partitioned into p_n parts over the primary key, then $\mathcal{DL}$ contains $p_n + 1$ lists $\mathcal{DL} = \{\mathcal{DL}_1, \mathcal{DL}_2, \cdots, \mathcal{DL}_{p_n}, \mathcal{DL}_*\}$, where $\mathcal{DL}_i$ ($1 <= i <= pn$) stores delta records of whose partition id is i, and $\mathcal{DL}_*$ stores delta records that we can't get the partition id.*

Update Base Table. In transaction processing, we only append δ_r into $\mathcal{DL}$ for view maintenance. It takes the following two steps (as illustrated in Fig. 2):

(1) In p-node, we parse the SQL statement. There are three operations:

– Insert. There are all values of columns in a insert statement. P-node can calculate pid_r since all values of $PK_{\mathcal{V}}$ are known. Then it adds a δ_r of $(\sqcap(PK_{\mathcal{V}} \cup np_{v_i})(r), pid_r, insert)$ into $\mathcal{DL}_{pid_r}$.

- Delete. When deleting the record via $PK_\mathcal{B}$ of $\mathcal{B}$, since we do not access chunk-server, the value of corresponding $PK_\mathcal{V}$ is unknown, and we cannot compute the partition id. It adds a δ_r of $(\sqcap(PK_\mathcal{B})(r), *, delete)$ into $\mathcal{DL}_*$, where $\sqcap(PK_\mathcal{B})(r)$ is the values on columns of $PK_\mathcal{B}$.
- Update. There are further two situations:
  - (a) $PK_\mathcal{V} = PK_\mathcal{B} \cup np_{b_i}{}^2$, it means we need to update the PK of a record within the view table. It requires to delete the old record and insert a new one. For deletion, the value of $PK_\mathcal{V}$ is unknown without requesting chunk-server. p-node adds a δ_r of $(\sqcap(PK_\mathcal{B})(r), *, delete)$ into $\mathcal{DL}_*$ for deletion. Since the value of column after the update is known, P-node can calculate pid_r based on its values of $PK_\mathcal{V}$ in r. Thus, for insertion, it adds a δ_r of $(\sqcap(PK_\mathcal{V} \cup np_{v_i})(r), pid_r, insert)$ into $\mathcal{DL}_{pid_r}$.
  - (b) $PK_\mathcal{V} \neq PK_\mathcal{B} \cup np_{b_i}$, which means we would update a NP column of record of the view table. P-node constructs a δ_r of $(\sqcap(PK_\mathcal{B} \cup np_{v_i})(r), *, update)$, which will be inserted into $\mathcal{DL}_*$ due to lack of $PK_\mathcal{V}$.
- (2) The p-node requests update server with δ_r. Update server updates B-memtable as usual and appends δ_r to the corresponding $\mathcal{DL}$ according to the partition id of δ_r. Then the transaction processing is over, which does not involve any additional communication between update server and chunk-servers.

Update View Table. View maintenance becomes no-trivial since view table is no longer on the update server. Here we explain our main strategy first, and the detailed implements are introduced in Sect. 4. In essence, we utilize asynchronous update in CSV:

- Asynchronous update. Rather than update the view table in a transaction, CSV only appends the modification of view table (i.e. delta records) into $\mathcal{DL}$. It actually updates the view table by merging delta records into V-memtable **when a query requests relevant data**. This naturally reduces the latency of transaction processing.

Notice that there exists plenty of delta records generated by transactions to be updated into V-memtable. We only update those related to the query results instead of updating all of them directly. We propose two optimizing techniques to achieve this:

- Accurate Update. Only delta records within the query range will be updated while the rest will be left to an unprocessed list and wait for the next related queries. As it cause overhead to identify these related delta records, a segment tree is used to facilitate this progress.
- Merging Fresh. Note $\mathcal{DL}$ stores delta records over a period, and many of them are repeated modification on the same row. Since users only need to know the latest data at query time, CSV just update the most recent ("fresh") delta record into V-memtable, while the historical ones will be left to the unprocessed list.

---

[2] This is the most commonly used view schema for projection.

3.3 Query on View Tables

Next, we describe the procedure for the query on view table. A SQL query is first processed on the p-node, then rewritten as a query plan for leveraging the view through a SQL parser and a query optimizer. Ultimately, it generates several sub-plans to access the view table. As shown in Fig. 2, it takes three steps to run those sub-plans in CSV:

(1) P-node distributes all sub-plans to corresponding chunk-servers. The sub-plans are executed on different chunk-servers in parallel.
(2) For each chunk-server, (a) a merging thread pulls the corresponding delta records from update server, then it fuses the static data in V-sstable, the V-memtable and delta records together to obtain the merging results. (b) The results are returned to p-node. Meanwhile, the delta records are updated into V-memtable for view maintenance.
(3) Finally, p-node merges the results returned by all the chunk-servers and responds to the client.
    The second step is the most critical since it includes both data access and view maintenance at the same time. In Sect. 4, we introduce its details.

3.4 Extension to Join View

For the view of join ($\mathcal{R} \bowtie \mathcal{S}$), it is important to design its schema and the corresponding delta record. If we simply set the PKs of $\mathcal{V}$ as the PKs of $\mathcal{R}$ and $\mathcal{S}$, then any lack of values in $\mathcal{R}$ and $\mathcal{S}$ results in the incomplete $PK_\mathcal{V}$, causing all updates added into the $\mathcal{DL}_*$. To make the method effective, we use the fact that one of the two tables is usually more frequently updated than the other. Suppose $\mathcal{R}$ is updated more regularly, then $PK_\mathcal{V}$ is designed as $PK_\mathcal{R} \cup np_i$, where np_i is the join column. To this end, if the transaction updates np_i of $\mathcal{R}$, we can generate a delta record whose pid can be computed as the $PK_\mathcal{V}$ is known. On the other hand, the update on $\mathcal{S}$ generates a delata record into $\mathcal{DL}_*$. For the query on the view, the process is the same as mentioned in Sect. 3.3.

4 Incremental View Maintenance

In this section, we introduce our solution to ensure transaction-level read consistency under asynchronous updates. Section 4.1 describes how to support snapshot isolation for query on the view. Section 4.2 introduces the detailed process of updating V-memtable. Section 4.3 gives two optimizing techniques to facilitate this process.

4.1 Version Control

Asynchronous update causes the problem of consistency. We leverage the multi-version model to provide strong consistency by ensuring **snapshot isolation** [8]. For each query, it first assigns a snapshot point, i.e. determines a unified version for each accessed record, then it returns their (i.e. all accessed records) latest states smaller than this version.

In CSV, as the transaction node, update server assigns the unique global version for all transactions, so does a query request. To keep consistent, it should access the latest snapshot before the global version. There will be two cases.

Query on a Single Chunk Server. We first discuss the query which only accesses view table on one chunk-server. For each query, we should assign a snapshot point. In CSV, we specify the point at the first time when the query accesses update server (denoted as v_{ups}). Besides, we maintain a v_{cs} for each V-memtable, which keeps the latest updated version of the V-memtable. Every time the view maintenance is done by merging delta records and V-memtable, v_{cs} is updated to v_{ups}, i.e. the data before v_{cs} has been synchronized between this chunk-server and update server. For a query falls on a chunk-server, the detailed steps are as follows.

(1) When the query initiates, p-node accesses the specific chunk-server based on $PK_\mathcal{V}$. Then the chunk-server requests the latest delta records from update server with v_{cs}.

(2) Update server provides chunk-server with a global ordered version v_{ups}. Then the delta records δ_r between v_{cs} and v_{ups} in $\mathcal{DL}_{cs\_id}$ and $\mathcal{DL}_*$ (where cs_id indicates the id of the specific chunk-server) are identified and pulled from the update server to the chunk-server.

(3) A query access several data records in view table. It probes the V-sstable and V-memtable to retrieve the data. After fusing the data from $\mathcal{DL}$, V-memtable and V-sstable, it returns the results to p-node and writes them into V-memtable. Then it updates v_{cs} to v_{ups} for the next query requests.

Query Across Multiple Chunk Servers. Next we introduce our solution to guarantee reading the consistent data when the query access view tables across multiple chunk-servers. The problem is that all participated chunk-servers visit update server asynchronously, however, they should access $\mathcal{DL}$ with a unified version number. To solve the problem, we maintain a query ID for every sub-plans. When chunk-server pulls the incremental data, update server assigns the latest version to the sub-request that arrives first. Then update server stores this version number in a hashmap, and the following sub-requests get it from the hashmap according to the query ID before they access the delta lists.

4.2   Update of V-Memtable

Next we introduce how to merge delta records into V-memtable. Recall a delta record has three types of operation flags:

(1) Insert. As $(\sqcap(PK_\mathcal{V} \cup np_{v_i})(r), pid_r, insert) \in \mathcal{DL}_{pid_r}$. It represents a view table row which needs insertion. Since its values on $PK_\mathcal{V}$ are known, we directly add the new row into the corresponding list.

(2) Delete. As $(\sqcap(PK_\mathcal{B})(r), *, delete) \in \mathcal{DL}_*$. It means the record in view table which has $\sqcap(PK_\mathcal{B})(r) \in PK_\mathcal{V}$ needs to be deleted. We should mark these rows with delete flag in V-memtable. Since only $PK_\mathcal{B}$ is known, we need to check records in view table. If $PK_\mathcal{B}$ matches, we get its $PK_\mathcal{V}$ and apply the delete into the corresponding list of V-memtable.

(3) Update. As $(\sqcap(PK_\mathcal{B} \cup np_{v_i})(r), *, update) \in \mathcal{DL}_*$. For the record in the view table which has $\sqcap(PK_\mathcal{B})(r) \in PK_\mathcal{V}$, its value on column np_{v_i} should be updated. Similar to situation (2), since only $PK_\mathcal{B}$ is known, we need to check records in view table, if $PK_\mathcal{B}$ matches, we construct the new updated row and add it into V-memtable.

For all the data in $\mathcal{DL}_*$, it requires to check the view table for update, which is cost-consuming. We propose accurate update to optimize this problem.

Concurrent Updates. Several query requests will modify V-memtable at the same time in the condition of concurrency, i.e. several updates will be added to the list of a same entry simultaneously. We use an optimistic concurrency control method to handle conflict. Each thread makes its changes at local first. Before the change is mounted into the list, it inspects whether another modification has been made to the list with a self-validation which checks if the last item of the list has been changed. In case of conflicts, the update will be aborted.

4.3   Two Optimizations

Accurate Update. Since there are many delta records to be updated into V-memtable, it brings many overheads for a query if all of them are updated at once. To this end, we only update those related to the query results. For the rest delta records, we postpone updating and put them into an unprocessed list until the next queries require them. Figure 3 illustrates the procedure.

In particular, for $\mathcal{DL}_*$, it is the most critical performance bottleneck since all queries need to scan it and check if there are any related updates. To this end, we utilize a segment tree [4] to improve its efficiency. Figure 3 illustrates the above procedure. We name the segment tree update control tree. As shown in Fig. 3, we build the tree based on the partition range of view table. Each node of the tree stores a sub-range with a list containing temporary delta records. Given a query, its input is a single value of $PK_\mathcal{V}$ (a point query) or a range of $PK_\mathcal{V}$ (a scan request), it returns all the related delta records by recursively traversing the tree. When the query initiates, the root node will copy all δ_r pulled from $\mathcal{DL}_*$ into its list. Then we probe the tree from root to leaves. For the query with range $\langle l, r \rangle$, if the range of currently traversed node intersects with $\langle l, r \rangle$, we push down the delta records in the list to the lists of its children. The process is repeated until it reaches the leaf nodes. Then the delta records whose keys belong to $\langle l, r \rangle$ are merged into V-memtable. The remaining delta records in $\mathcal{DL}_*$ are left in tree and wait for the next queries.

In this way, we update the records of view table related to the query. The remaining delta records are left in the unprocessed list or the update control tree, and wait for the next queries.

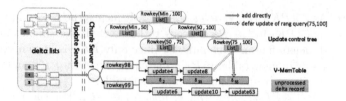

Fig. 3. Procedure of accurate update

Merging Fresh. Since users only care about the latest update at query time rather than the previous modifications, we also leverage this property to optimize. For different delta records in $\mathcal{DL}$, if they are modified on the same row, only the latest version of δ_r in $\mathcal{DL}$ will be calculated and added into the V-memtable, while other will be directly added into an unprocessed list.

Note that this strategy brings risks for concurrent queries which cross multiple chunk-servers. For instance, for a sub-plan $qsub_1$ of query q_1 on a V-memtable, if it obtains the $v_{ups} < v_{cs}$, it means the V-memtable has already been updated by other concurrent query q_2. Due to the fact that other sub-query of q_1 arrives at update server earlier than that of q_2, $qsub_1$ actually uses the smaller snapshot version than v_{cs}, which means the record in V-memtable is too "new" for $qsub_1$ that other sub-plans of q_1 are still using the old versions. Therefore, under workload of queries which cross multiple nodes, the updates in unprocessed list will still be computed to generate the query result.

Space Cost Analysis. The storage overheads of static data on disk for UPSV and CSV are identical, so we compare their overheads in memory. The total cost of V-memtable on all chunk-servers in CSV is similar to that of the memtable in UPSV. For CSV, there are additional delta records which occupy memory. A delta record only costs several bits as described. And the amount of delta records is proportional to the number of updates. When the delta record is merged into V-memtable, its memory is released in time. To this end, the memory overhead to store delta records is rather low. In addition, the space of the utilized segment tree is very small since it only stores a fixed number of nodes representing the sub-ranges of the partition. Overall, the total space cost of CSV is very close to UPSV. Besides, CSV occupies memory of chunk-servers while the memory cost of UPSV is mainly on update server.

5   Experiment

5.1   Experiment Setup

Setup. We conduct experiments to evaluate the effectiveness and cost of the CSV mechanism in an open source database Oceanbase [1]. The architecture of Oceanbase is a typical distributed LSM-tree structure consisting of update server, chunk-servers and p-nodes. We conduct our experiments on a database cluster of four Linux servers. Each server has two Intel Xeon E5-2620@2.00 GHZ

Table 1. Table schema

	View table	$PK_\mathcal{V}$	$NP_\mathcal{V}$
1UPSV/1CSV	V_1	L_shipdate,L_orderykey,L_partkey	L_quantity
3UPSV/3CSV	V_1	L_shipdate,L_orderykey,L_partkey	L_quantity
	V_2	L_quantity,L_orderykey,L_partkey	L_shipdate
	V_3	L_extendedprice,L_orderykey,L_partkey	L_quantity
5UPSV/5CSV	V_1	L_shipdate,L_orderykey,L_partkey	L_quantity
	V_2	L_quantity,L_orderykey,L_partkey	L_shipdate
	V_3	L_extendedprice,L_orderykey,L_partkey	L_quantity
	V_4	L_suppkey,L_orderykey,L_partkey	L_extendedprice
	V_5	L_linenumber,L_orderykey,L_partkey	L_tax

processors (each with 10 physical cores), 256 GB main memories and an Intel SATA SSD. In the experiment, an update server is deployed on one machine, and three chunk-servers and p-nodes are deployed on rest machines respectively. Servers in the cluster are connected via 1 Gb Ethernet.

Benchmarks. CSV aims to support analytical queries at the pressure of transaction processing. In our experiment, all queries are executed on a 10 GB version (SF = 10) of TPC-H [5]. We utilize *lineitem* as base table and generate five view tables with the similar schema defined in Sect. 2.2. The schemas of view tables are listed in Table 1. In the experiment, we may use different numbers of view tables. xUPSV/xCSV means we use x view tables for UPSV and CSV accordingly. To evaluate the interaction between view queries and transaction processing, we utilize two transactions. Both of them modify the *lineitem*, where one inserts a row and another updates a row.

5.2 Effect of View Query on Transaction Processing

We first measure the impact of view queries on transactions at different levels of data overlap for read and write requests. We record the peak throughput by increasing client threads from 1 to 300.

Queries and Transactions Fall on Different Data. Some applications tend to update the most recent data but query the historical data for analysis. In the experiment, the view table is range partitioned by l_shipdate in *lineitem*, where l_shipdate is an attribute recording time. First, we let the update occur only in the last partition which records the most recent data. While view queries are performed on all the other partitions, which only access historical data. In this case, in update server, UPSV needs to maintain the view tables as transactions process, while CSV would only append some lightweight updates to $\mathcal{DL}$.

Figure 4(a) shows the latency of transaction processing under different update throughput. Notice that CSV beats UPSV in every case because rather than update view table within the same transaction in UPSV, CSV just appends

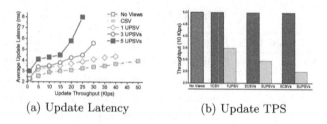

(a) Update Latency   (b) Update TPS

Fig. 4. Update performance

the modification of view table in $\mathcal{DL}$ for the asynchronous update, which has a lower impact on transaction processing. Also, with the increase of throughput, the average latency in UPSV increases significantly. Figure 4(b) plots the peak throughput by varying the amounts of view tables for UPSV and CSV. Our main observation is that the transaction process performs better in CSV than that of UPSV due to the decoupling of view maintenance and transaction processing in update server. Besides, as the number of UPSV increases, the peak throughput drops significantly, while the number of CSV hardly influences it, which is attributed to the fact that lightweight updates of view table have little influence on the performance of transaction.

Queries and Transactions Access the Same Data. Next we examine the effect of queries on transaction processing where the query and transaction access the same data, which means the updates of view tables need to be obtained.

Point Queries. First we set up a fixed point query with 1K throughput on view tables. Then we change the proportion of read and write requests in the transactions. Figure 5 shows the peak throughput in different ratios of reads and writes. We can observe that compared to UPSV, the peak throughput in CSV is higher and similar to that of no views, which means transactions are less affected by point queries on views. This is a result of the lower view maintenance cost of CSV through pulling δ_r in $\mathcal{DL}$ rather than requesting the whole memtable, which would not bring serious burden on update server. Note that with the increasing proportion of data writes, the overall throughput decreases. This is because in contrast to data reads, writes give more pressure on update server.

Range Queries. Then we turn to the effect of range queries for view tables on transaction processing with a mix of 50%/50% reads and writes. Every 30 s a range query on view tables will be initiated. We set the range to 100,000 and 10,000. As shown in Fig. 6(a) and (b), it shows that range queries have less effect on the peak throughput in CSV than UPSV. The reason is that CSV only accesses the incremental updates between v_{cs} and v_{ups} in $\mathcal{DL}$, while UPSV needs to request and merge the whole updates before v_{ups} in memtable, which makes a stronger impact on transaction processing in update server. Besides, the larger the query range is, the more updates UPSV needs to merge. That means more network cost and CPU resources in update server are occupied, which leads to the lower throughput and worse transaction performance.

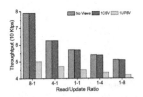

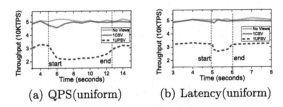

(a) QPS(uniform)   (b) Latency(uniform)

Fig. 5. Read/update ratio   **Fig. 6.** Scan performance

5.3   Effect of Transaction Processing on View Query

Our next experiment studies the impact of update transactions on view queries under different data distributions.

Point Queries. First we compare the peak throughput of point queries when the update throughput increases. Figure 7(a) shows the results. For UPSV, owing to the synchronous update of view table, view queries just simply require appropriate information in memtable, which provide better query throughput with a small amount of update loads. However, with the increase of update operations, CSV catches up with UPSV. Figure 7(b) plots the change of query latency under different update throughput. As the update load increases, the greater burden of transactions will be posed on update server, which causes lower query latency. In CSV, random point queries require a great deal of calculation for view maintenance, and this will bring about higher query latency.

Figure 7(c) and (d) show the case under zipfian data distribution. The performance of CSV is close to UPSV. Note that processing zipfian data leads to more lock contention and conflicts with more transactions aborted. In contrast to UPSV, CSV allows view updates to complete faster so locks are released sooner, which provides better query performance with high update loads. Thus with the rise of the throughput, CSV performs better than UPSV.

Range Queries. Then we measure the effect of update transactions on range queries via query latency. First, we set the queried range to 100,000 and vary the update throughput. As shown in Fig. 8(a), CSV performs better on range queries than UPSV. This is because in update server, CSV just pulls a little lightweight information from $\mathcal{DL}$ and leaves a great deal of computation and merging in chunk-servers, while UPSV consumes much more network and CPU resources of update server. More concrete analysis can be analyzed from Fig. 8(b), where we fix the update throughput to 30 kTPS and vary the query range. With the increase of query range, the advantages of CSV becomes more significant. Both small range and large range need to request the information of $\mathcal{DL}$ and process it. Processing small ranges of CSV is expensive compared to getting the updates using PKs in UPSV. While for large ranges, UPSV needs to request the full memtable for incremental data, which becomes cost-consuming.

From Figs. 7 and 8, we can also see that with the increase of update frequency, query throughput decreases and scan latency increases. The performance

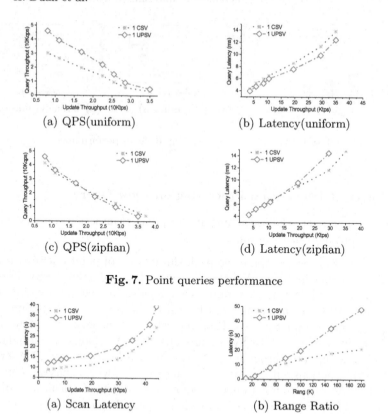

(a) QPS(uniform)

(b) Latency(uniform)

(c) QPS(zipfian)

(d) Latency(zipfian)

Fig. 7. Point queries performance

(a) Scan Latency

(b) Range Ratio

Fig. 8. Range query performance

decreases because $\mathcal{DL}$ stores more delta records as more updates happen. For the query, each chunk-server needs to pull the delta lists and update the related delta records, which results in the lower query throughput and higher scan latency. For UPSV, the performance reduction is more significant, this is because UPSV needs to maintain the view table for all the updates, which is more cost-consuming.

6   Related Work

Materialized View. Over the past decades, materialized views have been widely used as an effective optimization method in many commercial systems, such as Oracle [7], IBM DB2 [16], and Microsoft SQL Server [11]. With its improvement in query performance, materialized view also needs maintenance.

Asynchronous Maintenance. In database systems, while eager maintenance approaches usually update the view and base table synchronously within the same transaction, asynchronous view maintenance can reduce transaction latency and improve processing performance. The view is allowed to be inconsistent with the base table and brought up to date as necessary when queried [2,6,20]. [2,6] introduce the asynchronous maintenance mode in NoSQL databases without complete ACID characteristics. For the centralized database, [12,20] give us a lot of inspiration for lazy maintenance. In our system, distributed chunk-servers holds the view table partitions. They obtain the latest lightweight updates by requesting update server to maintain the view.

LSM-Tree. LSM [17] model was introduced in 1996 and received a reviving interest after Google released BigTable [9]. It prevails in workloads with a high rate of inserts and deletes. bLSM [18] improves LSM's read performance by using a geared scheduler and BloomFilters. [19] constructs the index on LSM at different consistency requirements.

7 Conclusion

In this paper, we propose an incremental strategy to maintain materialized views in LSM-tree. The view maintenance is decoupled with update transaction. Under the condition of ensuring data consistency, we use asynchronous method to identify the view update via lightweight information and propose an efficient algorithm for precise maintenance which postpones unnecessary updates. Experiment results show our method achieves good performance.

Acknowledgements. This work is supported by National Science Foundation of China under grant numbers 61702189, 61432006 and 61672232, and Youth Science and Technology - "Yang Fan" Program(17YF1427800). Huiqi Hu is the corresponding author.

References

1. Alibaba: OceanBase. https://github.com/alibaba/oceanbase/
2. Couchdb. https://couchdb.apache.org/
3. Record locking. https://en.wikipedia.org/wiki/Record_locking
4. Segment Tree. https://en.wikipedia.org/wiki/Segment_tree
5. TPC-H. http://www.tpc.org/tpch/
6. Agrawal, P., Silberstein, A., Cooper, B.F., et al. Asynchronous view maintenance for VLSD databases. In: SIGMOD, pp. 179–192. ACM (2009)
7. Bello, R.G., Dias, K., et al.: Materialized views in Oracle. VLDB, pp. 659–664. Morgan Kaufmann Publishers Inc., San Francisco (1998)
8. Berenson, H., Bernstein, P., Gray, J., Melton, J., O'Neil, E., O'Neil, P.: A critique of ANSI SQL isolation levels. In: Proceedings of the 1995 ACM SIGMOD International Conference on Management of Data, SIGMOD 1995, pp. 1–10. ACM, New York (1995)
9. Chang, F., Dean, J., Ghemawat, S., et al.: Bigtable: a distributed storage system for structured data. ACM Trans. Comput. Syst. **26**(2), 4 (2008)
10. Ghemawat, S., Gobioff, H., Leung, S.-T.: The Google file system. In: SOSP, vol. 37, pp. 29–43 (2003)

11. Goldstein, J., Larson, P.-Å.: Optimizing queries using materialized views: a practical, scalable solution. In: ACM SIGMOD Record, vol. 30, pp. 331–342. ACM (2001)
12. Katsis, Y., Ong, K.W., Papakonstantinou, Y., et al.: Utilizing IDs to accelerate incremental view maintenance. In: SIGMOD, pp. 1985–2000. ACM (2015)
13. Lakshman, A., Malik, P.: Cassandra: a decentralized structured storage system. Oper. Syst. Rev. **44**(2), 35–40 (2010)
14. Levandoski, J.J., Lomet, D.B., Sengupta, S.: The BW-Tree: a B-tree for new hardware platforms. In: ICDE, pp. 302–313 (2013)
15. LevelDB: LevelDB. http://leveldb.org/
16. Zaharioudakis, M., Cochrane, R., et al.: Answering complex SQL queries using automatic summary tables. ACM SIGMOD Rec. **29**, 105–116 (2000)
17. O'Neil, P., Cheng, E., Gawlick, D., O'Neil, E.: The log-structured merge-tree (LSM-tree). Acta Inf. **33**(4), 351–385 (1996)
18. Sears, R., Ramakrishnan, R.: BLSM: a general purpose log structured merge tree. In: SIGMOD, pp. 217–228. ACM (2012)
19. Tan, W., Tata, S., Tang, Y., Fong, L.L.: Diff-index: differentiated index in distributed log-structured data stores. In: EDBT, pp. 700–711 (2014)
20. Zhou, J., Larson, P.-A., Elmongui, H.G.: Lazy maintenance of materialized views. pp. 231–242. VLDB Endowment (2007)

CDSFM: A Circular Distributed SGLD-Based Factorization Machines

Kankan Zhao[1,2], Jing Zhang[1,2], Liangfu Zhang[1,2], Cuiping Li[1,2(✉)], and Hong Chen[1,2]

[1] School of Information, Renmin University of China, Beijing, China
{zhaokankan,zhang-jing,liangfu_zhang,licuiping,chong}@ruc.edu.cn
[2] Key Laboratory of Data Engineering and Knowledge Engineering, Beijing, China

Abstract. Factorization Machines (FMs) offers attractive performance by combining low-rank data vectors and heuristic features. However, it suffers from the growth of the dataset and the model complexity. Although much efforts have been made to distribute FMs over multiple machines, the computation efficiency is still limited by the foundational master-slave framework. In this paper, we propose CDSFM, which leverages Stochastic Gradient Langevin Dynamics (SGLD) to optimize FMs, and is distributed into a completely new circular framework. Experiments on two genres of datasets show that CDSFM can achieves a 2.3–4.7× speed-up over the comparison methods while obtains better performance.

1 Introduction

Factorization Machines (FMs) has been proposed by Stefen Rendle and successfully applied to recommendation and prediction tasks [10]. Despite much research on studying Factorization Machines [4,12,13], the problem remains largely unsolved in industry's real practice. The first challenge is when the well-adopted optimization algorithms such as Stochastic Gradient Descent (SGD) and Alternating Least square (ALS), are leveraged to solve FMs, it is difficult to escape a local optimum when it arrives. Secondly, due to the growth of the dataset and the model complexity, it is infeasible to solve problems using FMs in a single machine. Thus, distributed optimization is becoming a possible solution.

To solve the problem of local optimum, we propose to leverage Stochastic Gradient Langevin Dynamics (SGLD) to optimize FMs. To further adapt to large-scale dataset, we propose a fast distributed SGLD algorithm to solve FMs. We name the algorithm as Circular Distributed SGLD-based Factorization Machines (CDSFM).

Finally, we apply the proposed CDSFM algorithm into two genres of datasets. The results show that CDSFM clearly achieves better performance than FMs model that are inferred by SGD. To evaluate the efficiency of the proposed distributed framework, we compare CDSFM with the same algorithm implemented under Master-Slave (MS) framework. The results show that CDSFM achieves a 2.3–3.3× speed-up over the comparison methods, while obtains better accuracy performance.

J. Pei et al. (Eds.): DASFAA 2018, LNCS 10828, pp. 701–709, 2018.
https://doi.org/10.1007/978-3-319-91458-9_43

2   Circular Distributed SGLD-Based FM

Factorization Machines. Suppose the dataset of a prediction problem is formulated by a matrix $X \in \mathbb{R}^{n \times p}$, where the i-th row $\mathbf{x}_i \in \mathbb{R}^p$ of X represents a p-dimension data sample, and n is the data size. Based on the above defined notations, FMs model of order $d = 2$ can be defined as:

$$\widehat{y}(\mathbf{x}) = w_0 + \sum_{i=1}^{p} w_i x_i + \sum_{i=1}^{p} \sum_{j=i+1}^{p} \langle \mathbf{v}_i, \mathbf{v}_j \rangle x_i x_j \tag{1}$$

Stochastic Gradient Langevin Dynamics. Let θ denote a parameter vector. Suppose the dataset is denoted by $X = \{\mathbf{x}_i\}_{i=1}^{N}$, the posterior distribution is described as $p(\theta|X) \propto p(\theta) \prod_{i=1}^{N} p(\mathbf{x}_i|\theta)$. To learn the parameter vector θ, Stochastic Gradient Langevin Dynamics (SGLD) can be described as:

$$\theta_{t+1} \rightarrow \theta_t + \frac{\epsilon_t}{2} \left[\nabla \log p(\theta_t) + \frac{N}{n} \sum_{i=1}^{n} \nabla \log p(\mathbf{x}_{ti}|\theta) \right] + \nu_t \quad \nu_t \sim N(0, \epsilon_t I), \tag{2}$$

where ϵ_t is a sequence of step sizes, $\frac{N}{n} \sum_{i=1}^{n} \nabla \log p(\mathbf{x}_{ti}|\theta_t)$ is the mean score computed from multiple iterations of batches. SGLD [14] injects the gaussian noise into the parameter updates so that the variance of the data samples in a batch matches that of the posterior distribution.

Inferring FMs Using SGLD. Taking binary classification as an example, the parameter update rules of SGLD for Eq. (1) can be viewed as:

$$\theta^{t+1} \leftarrow \theta^t + \frac{\epsilon_t}{2} \{-\lambda^\theta \theta^t + \frac{N}{n} \sum_{i=1}^{N} \sigma(-\widehat{y}y)y \frac{\partial \widehat{y}}{\partial \theta}\} + \nu_t \tag{3}$$

where $\nu_t \sim N(0, \epsilon_t I)$, $\sigma(z) = \frac{1}{1+exp(-z)}$ and we abbreviate $\widehat{y}(X|\theta)$ as $\widehat{y}$.

Circular Distributed Framework. We propose to optimize FMs in a completely new distributed framework, named as Circular Distributed framework (CD). Instead of exchanging models between server and clients, CD discards the server and allows the communications between clients which release the bandwidth pressure of the server thoroughly.

In model training process, CD keeps one copy of the whole model in each worker, and allows each worker to learn the model parameters based on the partial dataset assigned to it. Thus, we will learn the same number of models as the number of workers in the framework. We also keep the scheduler node like the master-slave framework does to record the global variables such as the worker identifications. During the model prediction process, CD leverages all the models to predict the same test data independently, and then adopts the bagging method to merge the predictive results together as the final results.

Following the above idea, each model is learned by a worker independently. However, as each worker only captures partial sparse dataset, the model learned by each worker is not sufficient and cannot achieve good predictive performance. To obtain sufficient learning for each model, we design a *model scheduling* strategy. The basic idea is to allow each worker to be able to update not only the model being located at it, but also the models that are located at other workers.

Scheduling Strategy. We propose an efficient circular scheduling strategy to determine the next model to be updated by a worker. When a model is being updated by a worker, it can not be updated by any other workers. Thus, we keep a vacant model list in the scheduler node, to record all the models that are not being updated by any worker. In addition, we also keep an updated model list for each worker in the scheduler node, to allow each worker to only select the models that are not updated by it before. This principle makes a model to be learned in sufficient datasets as much as possible. The scheduler node should be accessed whenever a worker finishes current work and plans to select a next model to update. Our proposed circular scheduling strategy is to arrange all the workers in a ring and select the worker adjacent to the current worker in the ring, and then allow current worker to pull the model of the selected worker. When the selected model has been in the updated model list of the current worker, another model will be selected. The circular scheduling strategy can avoid the model conflict and thus reduce the checking time in updated model list and rescheduling time. Moreover, we select the model located at the same machine in priority. Specifically, we first select the workers located at the same machine as that of the current worker being located at (i.e., select the CPU cores with the same IP address), and then adopt the proposed strategy to select a worker within those workers (i.e., select one CPU core from the cores with the same IP address but different ports). Other workers in different machines will be tried when no available workers in the same machine can be selected. Then current worker pulls the model from the selected worker. The adopted strategy further avoids redundant communications between different machines. Figure 1 illustrates the general idea of model scheduling strategy in our method. From the figure, we can see that All the registered workers are arranged in a ring. When worker w_4 finishes updating the model m_3, it selects model m_4 on the ring to be next one. However, m_4 has already been updated by w_4, thus w_4 requests the next adjacent worker w_5 that is located at the adjacent IP address.

Other Update Strategies. First, we leverage a multiple-iteration update strategy to reduce the communication cost between the workers. In this way, the communication cost will be reduced to $\frac{1}{\tau}$ of that of one-time update. Moreover, we only need to transmit the necessary partial model parameters between workers, as the sparsity of high-dimensional dataset in each worker. Finally, to address the problem of delay caused by the imbalanced computational time, we randomly assign a key ranging from one to the number of the workers to each data sample, and aggregate the data samples with the same key into the

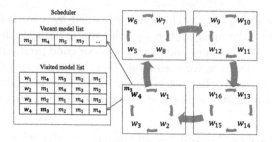

Fig. 1. Scheduling strategy of CDSFM.

same partition. By this method, the instances can be equally distributed to each worker which ensures the computational capabilities of the workers will be close to each other.

Theoretical Analysis of Model Prediction. As introduced before, we leverage all the learned models to predict the same test instance independently, and then adopt the bagging method [1] to merge all the results of different models. The bagging method actually follows the principle of majority voting. According to the theory of majority voting [6], we know that if multiple models are trained independently and all the models have the same individual accuracy, the result of majority voting is guaranteed to improve on the individual performance.

Suppose that we have learned S models independently and the models have the same individual accuracy p. A model with accuracy p indicates that the probability of a data sample being predicted correctly by the model is p. Thus, following the principle of majority voting, the probability of a data sample being predicted correctly by multiple models can be represented as:

$$p_c(S) = \sum_{i=k}^{S} \binom{S}{i} p^i (1-p)^{S-i},$$

(4)

where k is the smallest number of the models that composes the majority. We take k being odd as example to present the theoretical analysis as follows. The analysis is the same when k is even.

The recursive formula derived from Eq. (4) is given by [6] as Eq. (5). From Eq. (5), we can see that if each model's accuracy $p > 0.5$, $2p - 1 > 0$, then $p_c(2S + 1) > p_c(2S - 1)$. Thus we can get that $p_c(S)$ increases monotonically with the number of models, S. According to the above recursive formula, we can further derive the specific improvement of the predictive accuracy of multiple models on that of a single model as Eq. (6).

$$p_c(2S + 1) - p_c(2S - 1) = p^S (1-p)^S \binom{2S - 1}{S} (2p - 1).$$

(5)

$$p_c(S) - p_c(1) = (2p - 1) \sum_{i=1}^{\frac{S-1}{2}} p^i (1-p)^i \binom{2i - 1}{i},$$

(6)

3   Experiments Setting

Datasets. We perform experiments on two different genres of datasets for CTR prediction: Criteo and Avazu. The number of features in each data sample are 46,811 and 4,036, respectively. And the total number of data instances are 22,917,906 and 20,214,983, respectively.

Comparison Methods. We compare the following methods for classification:

FM: FMs optimized by SGLD and implemented in a single machine.
MSFM: FMs optimized by SGLD and implemented in MS framework.
CDFM-SGD: FMs optimized by SGD and implemented in CD framework.
CDSFM: FMs optimized by SGLD and implemented in CD framework.

Evaluation Measures and Platform. In accuracy performance, we use accuracy as the evaluation metric to compare the performance of different methods. In efficiency performance, we use the elapsed time of model learning to show the speedup and scalability of CDSFM comparing with other methods. All distributed methods are performed on a platform containing 15 machines, of which each machine contains 4 CPU cores (2.0 GHz) and 16 G memory. FM is performed in one machine of the platform.

Accuracy and Efficiency Performance. In Fig. 2, we compare the accuracy performance and efficiency performance of all the comparison methods. For each method, we conduct maximal 300 iterations, and record the accuracy and elapsed time when conducting 1/3, 1/2, 2/3 and the whole of the maximal iterations. From the results, we can see that first, the model performance can be improved by conducting more training iterations. Second, the proposed CDSFM can achieve better accuracy performance with less training time than the traditional MS framework. This may due to the reason that we reduce the communication cost, and keep the accuracy performance by the model bagging strategy. Finally, on all the datasets, CDFM-SGD performs a little worse than the proposed CDSFM, which indicates that SGLD can achieve better performance by injecting Gaussian noises. Note that on the dataset Criteo, the result of MSFM is not presented, as the method fails on the large dataset with large model size. The phenomenon is also shown in Fig. 4(a). We further vary the the number of maximal iterations, T, as 100, 200, and 300, and compare the final accuracy and elapsed time of different methods. The results in Fig. 3 show that CDSFM achieves a 3.3–4.7× speed up over FM, and a 2.3–3.3× speed up over MSFM, while obtains better accuracy than the two comparison methods.

Scalability. In Fig. 4, we increase the size of the dataset by replicating the original dataset as its 1.2, 1.5, 2.0, 2.4 and 3.0 times, and check the running time of different methods on each dataset. In these experiments, S is set as 29, T is set

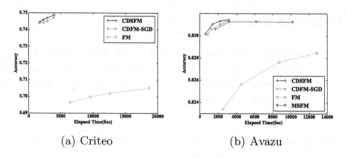

(a) Criteo   (b) Avazu

Fig. 2. Accuracy and efficiency performance of all the comparison methods.

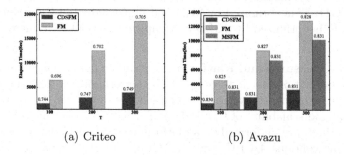

(a) Criteo   (b) Avazu

Fig. 3. Accuracy and efficiency performance under different number of iterations.

as 150, and τ is set as 100. We can see from the results that the increasing rate of the running time of CDSFM is much slower than that of FM, as the dataset is divided into multiple workers in CDSFM and the whole running time is saved. Meanwhile, on the largest dataset Criteo, after the dataset is larger than 2.4 times of the original dataset, FM fails to give the result, as the memory of a single machine is not large enough to hold the large dataset. In addition, MSFM fails even on the original Criteo dataset. It may be that with the increasing of the dataset, leading to the increment of the model size in each worker. The total communications may exceed the bandwidth limitation of the server.

In Fig. 5, we reduce the size of the model parameters by different proportions, and check the running time of different methods on each model size. We can see that the increasing rate of the running time of CDSFM is much slower than that of FM, as the model is divided into multiple workers in CDSFM and the whole running time is saved. The increasing rate of the running time of CDSFM is also slower than that of MSFM, as MSFM requires much more communication cost between the workers and the server. In addition, on the largest dataset Criteo, after the model size is larger than 83% of the original model size, MSFM fails to give the result, as the size of models exchanged between all the workers and the server exceeds the bandwidth limitation of the server.

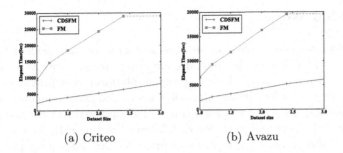

(a) Criteo   (b) Avazu

Fig. 4. Speedup when increasing the data size.

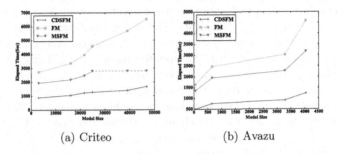

(a) Criteo   (b) Avazu

Fig. 5. Speedup when increasing the model size.

Parameter Analysis. We study how the number of worker number, S, affects the accuracy and efficiency performance of the proposed CDSFM. In these experiments, T is set as 150, and τ is set as 100. Figure 6 shows the accuracy and efficiency performance of CDSFM by varying the number of workers as 11, 15, 21, 25, and 29 on the datasets of Criteo. From the figure, we can see that first, the accuracy increases with the number of workers; second, the increment becomes slow and the accuracy becomes stable when there are more than 21 workers. The result is consistent with the theoretical analysis in Sect. 2.

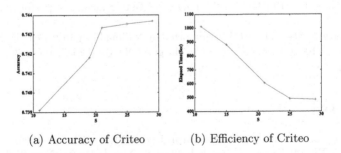

(a) Accuracy of Criteo   (b) Efficiency of Criteo

Fig. 6. Effect of worker number S on the performance of CDSFM.

4   Related Work

Distributed Architectures. In order to deal with the large-scale data mining problems, a great deal of distributed computing architectures were proposed. Hadoop [11] is a popular and easy to program platform, that provides easy programming interfaces for efficient data-flow operators, such as map, reduce and shuffle. The MapReduce implementation of Hadoop needs to write the program states into disk every iteration. Spark [15] improves upon Hadoop by keeping machine learning program states in memory, and thus gains large performance improvement. Graph-centric platforms such as GraphLab [8] and Pregel [9] efficiently partition graph-based models, but it is not clear whether the asynchronous graph-based models can produce correct machine learning programs, due to the lack of theoretical analysis. The above mentioned architectures follow the master-slave framework, where the clients always need to pull the latest models from the server and push the updated models back. Thus, the network bandwidth between the server and all the clients becomes a bottleneck. Recently, several systems were built upon the framework of Parameter Server [2,3,5,7]. Parameter Severs allow delayed asynchronous updates and also exchange partial model between the server and the clients, and thus can handle bigger models with billions of parameters. The conceptual architectural design of Parameter Server was introduced in [7] and the formalization was summarized in [16]. Based on the framework of Parameter Server, Dai et al. proposed Petuum [2], which allows programmable operation over global parameters.

5   Conclusions

In this paper, we describe CDSFM, a novel method to optimize and distribute Factorization Machines (FMs). Specifically, in CDSFM, we first optimize FMs using SGLD algorithm to alleviate the problem of local optimum. Then, we distribute SGLD-based FMs in a circular distributed framework. Furthermore, we adopt model bagging strategy when predicting to improve model performance. The experimental results show that CDSFM can achieves a 2.3–4.7× speed-up over the comparison methods while achieves better accuracy performance.

Acknowledgments. This work is supported by National Key Research&Develop Plan (No. 2016YFB 100702), and NSFC under the grant No. (61772537, 61772536, 61702522, 61532021).

References

1. Breiman, L.: Bagging predictors. Mach. Learn. **24**(2), 123–140 (1996)
2. Dai, W., Wei, J., Zheng, X., Jin, K.K., Lee, S., Yin, J., Ho, Q., Xing, E.P.: Petuum: a framework for iterative-convergent distributed ML. Eprint Arxiv (2013)
3. Dean, J., Corrado, G.S., Monga, R.: Large scale distributed deep networks. In: NIPS 2012, pp. 1223–1231 (2012)

4. Hirata, A., Komachi, M.: Sparse named entity classification using factorization machines. Eprint Arxiv (2017)
5. Jiang, J., Yu, L., Jiang, J., Liu, Y., Cui, B.: Angel: a new large-scale machine learning system. Natl. Sci. Rev. **5**, 216–236 (2017)
6. Lam, L., Suen, S.: Application of majority voting to pattern recognition: an analysis of its behavior and performance. IEEE Trans. Syst. Man Cybern. Part A Syst. Hum. **27**(5), 553–568 (1997)
7. Li, M., Andersen, D.G., Park, J.W.: Scaling distributed machine learning with the parameter server. In: OSDI 2014, vol. 14, pp. 583–598 (2014)
8. Low, Y., Gonzalez, J., Kyrola, A., Bickson, D., Guestrin, C.: GraphLab: a distributed framework for machine learning in the cloud. Eprint Arxiv (2011)
9. Malewicz, G., Austern, M.H., Bik, A.J., Dehnert, J.C., Horn, I., Leiser, N.: Pregel: a system for large-scale graph processing. In: SIGMOD 2010, pp. 135–146 (2010)
10. Rendle, S.: Factorization machines. In: ICDM 2010, pp. 995–1000 (2010)
11. Sun, H., Wang, W., Shi, Z.: Parallel factorization machine recommended algorithm based on MapReduce. In: SKG 2014, pp. 120–123 (2014)
12. Tsai, M.F., Wang, C.J., Lin, Z.L.: Social influencer analysis with factorization machines. In: WebSci 2015, p. 50. ACM (2015)
13. Wang, S., Du, C., Zhao, K., Li, C., Li, Y.: Random partition factorization machines for context-aware recommendations. In: WAIM 2016. pp. 219–230 (2016)
14. Welling, M., Teh, Y.W.: Bayesian learning via stochastic gradient Langevin dynamics. In: ICML 2011, pp. 681–688 (2011)
15. Zaharia, M., Chowdhury, M., Franklin, M.J., Shenker, S., Stoica, I.: Spark: cluster computing with working sets. In: HotCloud 2010, p. 10 (2010)
16. Zhong, E., Shi, Y., Liu, N., Rajan, S.: Scaling factorization machines with parameter server. In: CIKM 2016, pp. 1583–1592 (2016)

Industrial Track

An Industrial-Scale System for Heterogeneous Information Card Ranking in Alipay

Zhiqiang Zhang(✉), Chaochao Chen(✉), Jun Zhou(✉), and Xiaolong Li(✉)

Ant Financial Services Group, Hangzhou, China
{lingyao.zzq,chaochao.ccc,jun.zhoujun,xl.li}@antfin.com

Abstract. Alipay (https://global.alipay.com/), one of the world's largest mobile and online payment platforms, provides not only payment services but also business about many aspects of our daily lives (finance, insurance, credit, express, news, social contact, etc.). The homepage in Alipay app (https://render.alipay.com/p/s/download) integrates massive heterogeneous information cards, which need to be ranked in appropriate order for better user experience. This paper demonstrates an industrial-scale system for heterogeneous information card ranking. We implement an ensemble ranking model, blending online and chunked-based learning algorithms which are developed on parameter server mechanism and able to handle industrial-scale data. Moreover, we propose efficient and effective factor embedding methods, which aim to reduce high-dimensional heterogenous factor features to low-dimensional embedding vectors by subtly revealing feature interactions. Offline experimental as well as online A/B testing results illustrate the efficiency and effectiveness of our proposals.

Keywords: Ranking system · Industrial application · Embedding

1 Introduction

Alipay, which is known as one of the world's largest mobile and online payment platforms, also provides services on many aspects of our lives. For example, trading funds, purchasing insurances, and tracing express delivery can be done easily within the app. Users may also receive recommendations of news, movies, restaurants, promotions, or friends' updated posts based on their own interest. Moreover, one can even plant trees for public interest while using Alipay. Figure 1 demonstrates screenshots of the homepage in Alipay app. Massive heterogeneous information are integrated here, and each of which is demonstrated as 'card' (called **information card**, or **card** for short) with pinterest-style layout. Users scroll up and down to read them and click which they prefer to open a detail page for further operations. Massive cards can lead to information overload problem, hence personalized information card ranking becomes very important for better user experience.

© Springer International Publishing AG, part of Springer Nature 2018
J. Pei et al. (Eds.): DASFAA 2018, LNCS 10828, pp. 713–724, 2018.
https://doi.org/10.1007/978-3-319-91458-9_44

Fig. 1. Screenshots of the homepage in Alipay app. The left one is the first page, then the right one (i.e., the second page) comes after scrolling.

Information card ranking system for the homepage in Alipay app is challenging mainly from three aspects:

Heterogeneity. The ranking problem here is quite different from other industrial ranking or recommendation problems due to the *heterogeneity* of ranking targets (i.e., information cards). Most of industrial ranking/recommendation systems have a commonality that their target items are homogeneous (e.g., Movies in Netflix [4], user-generated videos in YouTube [3], and news feeds in Facebook [1]). However, in our scenario, ranking targets (i.e., information cards) can be quite different from each other. Heterogeneity of information cards means complicated feature engineering and very-large-scale feature space. Our system requires an efficient and effective method to alleviate this problem and reveal the subtle interaction between features.

Time-Awareness. Plenty of information cards are generated every second. The ranking system requires *time-awareness*, which means the ability to model newly generated information cards and the latest user operations. Our system should implement online feature extraction, real-time user feedbacks collection as well as online learning model to ensure time-awareness. Moreover, balancing new information cards with those have been well-established is also important for our system.

Scalability. Similar to other industrial scenarios, the homepage in Alipay app serves huge magnitude of users with massive information cards. Many state-of-the-art ranking or recommendation algorithms perform excellent on smaller datasets but fail to handle such a massive problem. Industrial-scale distributed learning system and algorithms become the foundation of our system. Moreover, efficient online services and real-time computation systems are also essential for scalability.

Matrix factorization [12] and its variants [7,9] play an importance role in recommendation, and achieve exciting performance in many real-world rating prediction problems. However, they fail to handle large-scale auxiliary features of heterogeneous business. Recently, deep learning draws more attention of the recommendation community [16] and has been applied successfully in real-world recommender system [3]. But it is difficult to be applied in online learning settings and thus fails to satisfy the time-awareness demand in our scenario. Recently, some proactive systems such as Google Now and Apple Siri also present information cards to their users. Some works propose zero-query ranking method [13] and proactive search/recommendation systems [14] for card ranking for these applications. But the variety of information cards in our problem is much greater, as well as the size of user feedbacks. That means methods worked well on those situations may fail in our scenario due to the heterogeneity and scalability challenges.

Considering the three challenges mentioned above, we formalize the information card ranking problem as **C**lick-**T**hrough **R**ate (CTR) prediction to predict whether users will click cards or not. When a user request occurs, its contained cards are ranked by the predicted CTR and sent back to the app. In this paper, we demonstrate an industrial-scale system with online/chunk-based learning ensemble model based on elaborate online/offline feature engineering. Online learning model (i.e., distributed bound delay FTRL-Proximal) captures latest user operations and newly generated information cards, while chunk-based learning model (i.e., a modified version of boosted trees plus logistic regression) is responsible for long-term user interest and well-established cards. The ensemble of them is able to balance the time-awareness demand between short-term variations and long-term characteristics. All of these algorithms are implemented on our industrial-scale parameter server based distributed learning system, and are able to handle billions of training samples and features. Moreover, inspired by word/document embedding methods, we propose efficient and effective factor embedding methods to reduce user's massive heterogenous features to a real-valued, low-dimensional vector representing his/her latent characteristics, which also reveals subtle interactions between features.

The remainder of this work is organized as follow: Sect. 2 presents a brief system overview. Sections 3 and 4 describe feature engineering and the ensemble model respectively. In Sect. 5, offline experiments and online A/B testings illustrate the performance of proposed factor embedding method as well as the online/chunk-based learning ensemble model. Finally, Sect. 6 presents the conclusion and future work.

2 System Architecture

Figure 2 demonstrates the architecture of our system. When users launch the Alipay app, our system will accomplish the online personalized card ranking procedure within several hundred milliseconds (red lines in Fig. 2). After that, users with different interests receive cards in different orders. In Fig. 1, cards about fund income and finance product promotion rank higher since the user heavily involved in finance business with Alipay. Users' feedbacks (e.g., click after

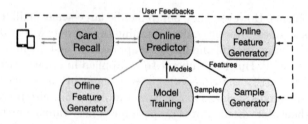

Fig. 2. Architecture of the heterogenous information card ranking system. Dashed lines indicate data flow of user feedbacks after receiving cards. (Color figure online)

exposure) will be recorded by clients or servers and sent back to our system for training sample construction and online feature generation (dash lines in Fig. 2).

Card Recall module receives user request from client, and then gathers all candidate cards from various business systems. Different from other industrial recommender systems [3], our system only focuses on ranking and leaves candidate generation to various business system. Specifically, cards come from different sources in different ways. For example, some are triggered by particular operations of user or state change of business (e.g., notification cards), while some are delivered in real time (e.g., news and friends' updated posts). Hence it's impractical to build an overall system containing candidate cards generation.

Feature Generator module performs feature engineering, which includes two sub-modules. *Online Feature Generator* is deployed on a distributed, high-efficiency real-time computation system. It collects the latest user feedbacks and provides up-to-the-minute features. *Offline Feature Generator*, based on the MaxCompute platform in Alibaba Cloud[1], gathers massive features of users and cards from various business systems and performs factor embedding technique (will introduce in Sect. 3.2). Especially, the proposed factor embedding procedure reduces user's high-dimensional heterogeneous business factor features into a low-dimensional embedding vector for model training.

Sample Generator is also deployed on the real-time computation system, similar like *Online Feature Generator*. It constructs real-time training samples, by joining features from online predictor with the corresponding user feedbacks (i.e., click as positive label while not click as negative). Samples are generated sequentially for online learning model, and also stored in data warehouse for chunk-based learning model.

Model Training module, implemented in parameter server mechanism on KunPeng platform [18], blends online learning model and chunk-based learning model. The online learning algorithm processes tens of thousand training sample and updates model in second, while chunk-based learning model is trained on billions of training samples and features within hours. Trained models are pushed periodically to *Online Predictor* for scoring.

---

[1] https://intl.aliyun.com/product/maxcompute.

Online Predictor is the central module of the whole system, which is responsible for two things: feature processing and scoring. After receiving candidate cards from *Card Recall* module, it queries both *Online Feature Generator* and *Offline Feature Generator* to gather features of the target user and his/her cards, and then perform the simple and effective feature interaction. Scoring is processed after feature processing, with the latest online/chunk-based learning ensemble model. Cards are ranked according to the personalized predicted CTR scores and returned to *Card Recall* module, and then back to the app. Another data flow of *Online Predictor* is to immediately deliver the whole feature vectors to *Sample Generator*, which is used to construct training samples for *Model Training* module.

Moreover, offline evaluation based on AUC metric helps to guide our system development, especially for training chunk-based model and factor embedding. Nevertheless, online A/B testing results hold the final determination of effectiveness of the system. This is because online A/B testing results may be inconsistent with offline evaluation.

3   Feature Generator

In this section, we will briefly describe the feature engineering framework and mainly focus on the proposed factor embedding methods.

3.1   Feature Generator Overview

Online Feature Extraction. Online or real-time feature extraction requires timeliness and is quite expensive. Hence we only extract simple but effective features which reflect temporary states of users and cards. For users, the latest actions greatly affect their following operations. For example, a user just reviewed funds information within the app is more likely to be attracted by a fund promotion card rather than other promotions or movie recommendation cards. To do this, we use the k latest interactions between users and our servers as online user features. On the other hand, the amount and ratio of click in recent minutes can describe current popularity of cards and are used as online card features.

Offline Feature Extraction. Our data scientists develop plenty of useful factors to comprehensively describe different perspectives of each user within his/her business (e.g., online shopping behaviors on Taobao/Tmall platform, offline payments with alipay, investments of funds on Ant Fortune platform, flight and hotel booking on fliggy.com, and so on). Thanks to the MaxCompute platform, aggregating users' heterogenous business factors can be done easily and efficiently. Then, the factor embedding module (will be presented in Sect. 3.2) reduces the original high-dimensional business factor space into low-dimensional embedding space, which represents long-term characteristics of users. On the other hand, when cards are published, their profiles (publisher, key words, relative business, etc.) will be tagged on them as card features.

Feature Interaction. The ranking problem in our scenario involves two roles: users and cards. To perform personalized ranking, user features must interact with card features. Considering that feature interaction must be done in *Online Predictor* which requires low latency, we construct interactive features by applying the efficient Cartesian product of user and card feature sets. Factor embedding procedure, which reduces the dimension of user's business factor space significantly, also prevents the curse of dimensionality in Cartesian product.

3.2 Factor Embedding

Users' business factors are collected from various heterogeneous business systems. Factors derived from one business rarely overlap with the others. Therefore, the dimension of business factor space is too high to perform real-time feature interaction. Moreover, communication traffic, storage, and the failure rate of feature queries are likely to increase as the feature size becomes bigger. Some unsupervised dimensionality reduction methods, such as PCA [15] and autoencoder [6], can help to transform high-dimensional vector into lower-dimensional latent vector. However, a small portion of factors update daily in our scenario, which require re-training PCA or autoencoder to update latent vectors.

In order to alleviate this problem, we propose to learn distributed representation of business factor and user in low-dimensional space, which is inspired by neural language models. Word2vec models [11] (e.g. continuous bag-of-words and skip-gram) learn low-dimensional distributed embedding of words by utilizing the dependence of target word and its local context. Inspired by word2vec, [8] involves vectors to represent paragraphs and learn word and paragraph vectors together. [2] introduces corruption to capture information of global context for better representation of paragraphs.

In our scenario, user's latent characteristics determine his/her behaviors in various business, which are characterized by massive business factors. The goal of our proposal is to learn low-dimensional vector to represent latent characteristics of business factors and users. In analogy with word or paragraph embedding methods, the proposal factor embedding methods regard a single business factor as a word, while the whole factor set of certain user as a paragraph. We first define some notations as follow:

- $\mathcal{U}$: training user set of size n, in which each user u contains a set of heterogenous business factors $\{f_u^1, \ldots, f_u^{t_u}\}$;
- $\mathcal{F}$: a set of m factors;
- $\mathbf{x} \in \mathbb{R}^m$: Bag of **W**ords (BoW) of a user, where $x_i = 1$ iff the i-th factor appears in the user's factor set;
- $\mathbf{c}^f \in \mathbb{R}^m$: BoW of the sampled factor subset w.r.t to factor f, $c_j^f = 1$ iff the j-th factor is sampled in current iteration;
- $\mathbf{U} \in \mathbb{R}^{d \times m}$: matrix of users' distributed representation, in which each column $\mathbf{U}_u$ denotes a d-dimensional embedding vector of user u;
- $\mathbf{W}, \mathbf{W}' \in \mathbb{R}^{d \times m}$: projection matrices from input space to hidden space and from hidden space to output space. We use $\mathbf{w}_f$ and $\mathbf{w}'_f$ to denote column in $\mathbf{W}$ and $\mathbf{W}'$ for factor f.

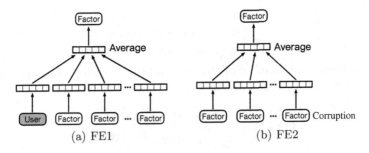

Fig. 3. Architectures of the factor embedding models.

Figure 3(a) shows the architecture of the first proposed model (called FE1). Similar to word/paragraph embedding, our proposal consists of an input layer, a hidden layer, and an output layer for target factor prediction. Target factor is predicted by an average of other t factors sampled randomly from the same user. Inspired by [8], user vector also contributes to target factor prediction. We define the probability of observing a target factor f of a user u given t sampled factors of the same user as:

$$P(f|\mathbf{c}^f, \mathbf{U}_u) = \frac{\exp(\mathbf{w}_f'^T(\frac{1}{1+t}(\mathbf{W}\mathbf{c}^f + \mathbf{U}_u)))}{\sum_{f' \in \mathcal{F}} \exp(\mathbf{w}_{f'}'^T(\frac{1}{1+t}(\mathbf{W}\mathbf{c}^f + \mathbf{U}_u)))}. \tag{1}$$

Matrices $\mathbf{U}$, $\mathbf{W}$, and $\mathbf{W}'$ are then learnt by minimizing the total negative log-likelihood on the whole training set as follow:

$$\mathbf{U}, \mathbf{W}, \mathbf{W}' = \underset{\mathbf{U}, \mathbf{W}, \mathbf{W}'}{\arg\min} - \sum_{u=1}^{n} \sum_{f \in \mathcal{F}_u} \log P(f|\mathbf{c}_u^f, \mathbf{U}_u), \tag{2}$$

where $\mathcal{F}_u$ represent the factor set of user u.

Each time, we choose a factor of certain user and randomly sample other t factors from his/her factor set to update the corresponding parameters. Exactly computing the probability in Eq. 1 is impractical, two approximation techniques (i.e., negative sampling and hierarchical softmax [11]) can help to approximate it efficiently. To generate embedding for a new user, matrices $\mathbf{W}$ and $\mathbf{W}'$ are fixed and only update the user vector until convergence.

To further improve efficiency and effectiveness, we propose another factor embedding method (called FE2, shown in Fig. 3(b)), inspired by [2], to represent user vector as the average of his/her factor vectors. [2] introduces *corruption* to capture information of global context for better representations of paragraphs. In our scenario, information of global context (i.e., user's latent characteristics) is essential for learning representations of factors and users, rather than local context. To perform unbiased corruption to user's factor set, we remove significant portion of factors with probability q, and represent the target factor using $\frac{1}{1-q}$ times of embedding vectors of the remained factors. We define the BoW of corrupt sampled factor subset $\tilde{\mathbf{c}}$ as follow:

$$\tilde{c}_i = \begin{cases} 0, & \text{with probability } q \\ \frac{c_i}{1-q}, & \text{otherwise} \end{cases}. \tag{3}$$

Then we define the probability of observing a target factor f of a user u given a corrupt factor subset of the same user as:

$$P(f|\tilde{\mathbf{c}}^f) = \frac{\exp(\mathbf{w}'_f{}^{\mathrm{T}}(\frac{1}{t_u}\mathbf{W}\tilde{\mathbf{c}}^f))}{\sum_{f' \in \mathcal{F}} \exp(\mathbf{w}'_{f'}{}^{\mathrm{T}}(\frac{1}{t_u}\mathbf{W}\tilde{\mathbf{c}}^f))}, \tag{4}$$

where t_u denotes the size of the corrupt factor set. Matices $\mathbf{W}$ and $\mathbf{W}'$ are learnt similarly with FE1.

[2] proves that corruption is equivalent to regularization on factors, which help to avoid overfitting. Factor that appears frequently will get larger penalty than a rare one. Moreover, regularization will also diminished as a factor is critical to a confident prediction. Compared with FE1, the inference of user vector is more efficient since it represents each user as an average of the embeddings of his/her factors: $\mathbf{U}_u = \frac{1}{t_u}\sum_{f \in \mathcal{F}_u}\mathbf{w}_f$.

4   Model Training

Based on feature engineering proposed in Sect. 3, our system ensembles online learning and chunk-based learning models. As mentioned above, we model the ranking problem as click-through rate prediction, which is a binary classification problem to distinguish whether a user will click a card or not. Figure 4 demonstrates the framework of the online/chunk-based learning ensemble model. All of the learning algorithms are implemented on KunPeng platform [18], an industrial parameter server based distributed learning system, and able to handle real-world datasets with billions of samples and features.

Online Learning Model. To perform online learning, we implement online FTRL-Proximal algorithm [10] (Online-FTRL for short) on Kunpeng. Online-FTRL can quickly adjust model according to user feedbacks (click information card or not), thus enable the model to reflect user's intentions and preferences in real time and further improve online prediction accuracy. Suppose $g_t \in \mathbb{R}^d$ is the d-dimensional gradient of Online-FTRL at time t with each element corresponding to the gradient of each model parameter, and $g_{t,i}$ is the i-th element in vector g_t. Similarly, $\eta_t \in \mathbb{R}^d$ is the d-dimensional learning rate vector of Online-FTRL at time t with each element corresponding to the learning rate of each model parameter, and $\eta_{t,i}$ is the i-th element in vector η_t. Then, the model w of Online-FTRL algorithm is iteratively updated by:

$$w_{t+1} = \arg\min_{w}(g_{1:t}w + \frac{1}{2}\sum_{s=1}^{t}\delta_s \parallel w - w_s \parallel_2^2 + \lambda_1 \parallel w \parallel_1 + \lambda_2 \parallel w - w_s \parallel_2^2), \tag{5}$$

where $g_{1:t} = \sum_{s=1}^{t} g_s$ and $\delta_t = \frac{1}{\eta_t} - \frac{1}{\eta_{t-1}}$. Typically, $\eta_{t,i}$ is designed as

$$\eta_{t,i} = \frac{\alpha}{\beta + \sqrt{\left(\sum_{s=1}^{t} g_{s,i}{}^2\right)}}, \tag{6}$$

where $\beta = 1$ is usually good enough and α depends on the features and datasets. For more details about FTRL-Proximal algorithm, please refer to [10].

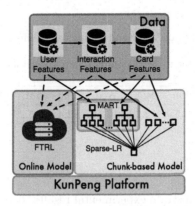

Fig. 4. Architecture of the online/chunk-based learning ensemble model.

Chunk-Based Learning Model. For better utilization of longer historical data to depict long-term characteristic of users/cards, chunk-based learning model needs to contribute for the final prediction. Furthermore, model learnt from massive historical data can alleviate instability of online learning model. As shown in the yellow box of Fig. 4, a modified version of the *boosted trees plus logistic regression* [5] is applied as chunk-based learning model. Distributed **M**ultiple **A**dditive **R**egression **T**rees (MART) [17] learns directly from user and card features (including offline user factor embeddings/card tag features and online user/card features). The outputs (i.e., one-hot encoding vector of leaf nodes) of each tree are concatenated and treated as inputs for a distributed **S**parse **L**ogistic **R**egression (Sparse-LR). Additionally, card features, as well as interactive features generated by Cartesian product of user factor embeddings and card features, are also used as input of Sparse-LR. The MART model learns more complex feature interactions inherently and achieves better generalization performance, while the Sparse-LR part, especially the additional wide component, achieves memorization and avoid over-generalization.

Since online prediction requires high efficiency, we make final prediction by simply applying weighted average on outputs of Online-FTRL model and chunk-based learning model. Weights of each model are adjusted according to online A/B testing results.

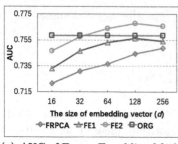 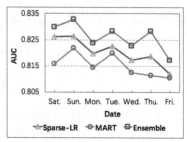

(a) AUC of Factor Emedding Methods

(b) AUC of Different Classifiers

Fig. 5. AUC performance of factor embeddings and different chunk-based model.

Table 1. CTR improvements of the online/chunk-based learning ensemble model.

Date	Mon.	Tue.	Wed.	Thu.	Fri.	Sat.	Sun.
Improvement	12.5%	12.7%	13.5%	12.4%	15.2%	15.1%	13.3%

5   Experiments

In this section, we demonstrate experimental results to show the effectiveness of proposed factor embedding methods, as well as the online/chunk-based learning ensemble model. Notice that in the following experiments, the hyper-parameters of all the models are set according to the best cross-validation results.

5.1   Effectiveness of Factor Embedding

Two datasets are used to show the effectiveness of factor embedding methods. Factor embeddings are learnt from the first dataset (called D1), while the effectiveness is evaluated on the second one (called D2).

We first present statistical information of D1 and describe how to generate user factor embedding on it. D1 contains tens of million of users with more than 10000 heterogenous business factors of them. Users in D1 are randomly chosen from all active users in the last week of March, 2017. A certain user's factor set consists of the server-side interfaces which he/she interacted during that time. It's a portion of the whole factor set in our scenario. It can approximately represents the user's recent behaviors in Alipay app. FRPCA, FE1 and FE2 are trained on D1. Specifically, FRPCA is performed on user-factor matrix, and each row of the transformed matrix represents user's feature for D2. For FE1 and FE2, we infer the user embeddings (as user feature for D2) based on factor embeddings learnt on D1.

We then present how to compare the performance of difference factor embedding methods on D2. D2 contains hundreds of millions of user feedbacks sampled from the first week of April, 2017 (next week from D1), and is divided into training set (the former 4 days) and validation set (the latter 3 days). We take user feature generated from D1 and card tag feature as the input of our model. Then, we train a MART on training set to predict which cards users will click on validation set. Finally, we calculate AUC (**A**rea **U**nder Receiver Operating Characteristic **C**urve) based on the predictions and the ground truth feedbacks.

According to Fig. 5(a), FE2 outperforms other methods with $d \geq 64$ (d is the size of embedding vector). FE1 with $d = 128$ is nearly the same as the original features, while FRPCA seems hard to achieve good performance in this setting. Our proposed factor embedding methods (especially FE2) can not only reduce high-dimensional factor space to low-dimensional space (more than 5000 original factors are reduced to embeddings of size 128 or less), but also reveal subtly interactions between factors for further improvement of the CTR prediction.

5.2   Effectiveness of Online/Chunk-Based Learning Ensemble Model

Figure 5(b) shows the AUC performance of chunk-based ensemble model, compared with single Sparse-LR and MART, during the first week of April, 2017. Card features and interactive features (generated by Cartesian product of user and card features) are applied to Sparse-LR, while user and card features are directly applied to MART since tree-based model can learn feature interactions inherently. All the models are trained on the entire user feedbacks during the last seven days. We report AUC score based on their predictions of user feedbacks in the next day. Obviously, the chunk-based ensemble model significantly outperforms the other two due to its ability to achieve both memorization and generalization. Remarkably, Sparse-LR outperforms MART in this scenario because of the effectiveness of feature interaction by Cartesian product, as well as its ability to handle larger-scale feature space. Notice that the AUC scores in Fig. 5(b) are greater than those in Fig. 5(a) since models in Fig. 5(b) are trained on the entire user feedbacks (i.e., billions of samples) rather than a sampled dataset.

Moreover, Table 1 shows the online A/B testing results of CTR improvement of the online/chunk-based learning ensemble model, compared with ranking by chunk-based model only, during the first week of April, 2017. Thanks to the ability to subtly capture user and card's short-term variations, more than 12% CTR improvement is achieved by the online/chunk-based learning ensemble model, which means tens of millions of clicks increase every day.

6   Conclusions

In this paper, we have demonstrated an industrial-scale system for heterogenous information card ranking in Alipay. Offline experiments illustrate the effectiveness of the proposed factor embedding methods and the online/chunk-based learning ensemble model, while online A/B testing result shows a significant CTR improvement achieved by the system. For further improvement of user experience, how to model the staying time of user to each card needs a deeper exploration.

References

1. Backstrom, L.: News Feed FYI: A Window Into News Feed. https://www.facebook.com/business/news/News-Feed-FYI-A-Window-Into-News-Feed (2013). Accessed 01 05 2017

2. Chen, M.: Efficient vector representation for documents through corruption. In: ICLR 2017 (2017)

3. Covington, P., Adams, J., Sargin, E.: Deep neural networks for Youtube recommendations. In: RecSys 2016, pp. 191–198 (2016)

4. Gomez-Uribe, C.A., Hunt, N.: The Netflix recommender system algorithms, business value, and innovation. ACM Trans. Manag. Inf. Syst. (TMIS) **6**, 13 (2016)

5. He, X., Pan, J., Jin, O., Xu, T., Liu, B., Xu, T., Shi, Y., Atallah, A., Herbrich, R., Bowers, S., Candela, J.Q.: Practical lessons from predicting clicks on ads at facebook. In: ADKDD 2014, pp. 1–9 (2014)

6. Hinton, G., Salakhutdinov, R.: Reducing the dimensionality of data with neural networks. Science **313**(5786), 504–507 (2006)

7. Koren, Y.: Factorization meets the neighborhood: a multifaceted collaborative filtering model. In: SIGKDD 2008, pp. 426–434 (2008)

8. Le, Q., Mikolov, T.: Distributed representations of sentences and documents. In: ICML 2014, pp. 1188–1196 (2014)

9. Lee, J., Kim, S., Lebanon, G., Singer, Y., Bengio, S.: LLORMA: local low-rank matrix approximation. J. Mach. Learn. Res. (JMLR) **17**(15), 1–24 (2016)

10. McMahan, H.B., Holt, G., Sculley, D., Young, M., Ebner, D., Grady, J., Nie, L., Phillips, T., Davydov, E., Golovin, D., et al.: Ad click prediction: a view from the trenches. In: SIGKDD 2013, pp. 1222–1230. ACM (2013)

11. Mikolov, T., Sutskever, I., Chen, K., Corrado, G., Dean, J.: Distributed representations of words and phrases and their compositionality. In: NIPS 2013, pp. 3111–3119 (2013)

12. Salakhutdinov, R., Mnih, A.: Probabilistic matrix factorization. In: Proceedings of the 20th International Conference on Neural Information Processing Systems (NIPS 2007), pp. 1257–1264 (2008)

13. Shokouhi, M., Guo, Q.: From queries to cards: re-ranking proactive card recommendations based on reactive search history. In: SIGIR 2015, pp. 695–704 (2015)

14. Song, Y., Guo, Q.: Query-less: predicting task repetition for nextgen proactive search and recommendation engines. In: WWW 2016, pp. 543–553 (2016)

15. Tulloch, A.: Fast Randomized SVD (2014). https://research.fb.com/fast-randomized-svd/. Accessed 01 05 2017

16. Wang, H., Wang, N., Yeung, D.Y.: Collaborative deep learning for recommender systems. In: SIGKDD 2015, pp. 1235–1244 (2015)

17. Zhou, J., Cui, Q., Li, X., Zhao, P., Qu, S., Huang, J.: PSMART: parameter server based multiple additive regression trees system. In: WWW 2017, pp. 879–880 (2017)

18. Zhou, J., Li, X., Zhao, P., Chen, C., Li, L., Yang, X., Cui, Q., Yu, J., Chen, X., Ding, Y., Qi, Y.A.: KunPeng: parameter server based distributed learning systems and its applications in Alibaba and ant financial. In: SIGKDD 2017 (2017)

A Twin-Buffer Scheme
for High-Throughput Logging

Qingzhong Meng[1](✉), Xuan Zhou[2], Shan Wang[1],
Haiyan Huang[3], and Xiaoli Liu[3]

[1] MOE Key Laboratory of DEKE, Renmin University of China, Beijing, China
mqz@ruc.edu.cn
[2] School of Data Science & Engineering, East China Normal University,
Shanghai, China
[3] Huawei Technologies Co., Ltd., Shenzhen, China

Abstract. For a transactional database system, the efficiency of logging is usually crucial to its performance. The emergence of new hardware, such as NVM and SSD, eliminated the traditional I/O bottleneck of logging and released the potential of multi-core CPUs. As a result, the parallelism of logging becomes important. We propose a parallel logging subsystem called TwinBuf and implemented it in PostgreSQL. This solution can make better use of multi-core CPUs, and is generally applicable to all kinds of storage devices, such as hard disk, SSD and NVM. TwinBuf adopts per-thread logging slots to parallelize logging, and a twin-log-buffer mechanism to make sure that logging can be performed in a non-stop manner. It performs group commit to minimize the persistence overheads. Experimental evaluation was conducted to demonstrate its advantages.

1 Introduction

Durability of transactions is the ability to protect data validity from software and hardware failure. Most transactional database systems utilize logging to ensure durability. When database systems shutdown abnormally due to loss of electrical power or errors, logs on the permanent storage guarantee that all modifications made by committed transactions are not lost. ARIES (Algorithms for Recovery and Isolation Exploiting Semantics) [10], a particular type of WAL (write-ahead logging) mechanism, is the most widely adopted approach, which dominates the majority of modern database systems. In ARIES, each change of a data object must be first written to a log and flushed to permanent storage before the change becomes persistent. That is, the log must persist before the modified data object. If this constraint is met, modified data objects can be recovered from the logs when the system shutdowns abnormally, by redoing committed transactions and undoing aborted transactions.

Most traditional database systems manage logs in a centralized way, in which all worker threads/processes insert logs into the same log buffer. Before a transaction commits, it must flush the log buffer, which contains its log records, to the

J. Pei et al. (Eds.): DASFAA 2018, LNCS 10828, pp. 725–737, 2018.
https://doi.org/10.1007/978-3-319-91458-9_45

hard disk. Upon the flush, the buffer may contain logs created by other transactions. This centralized mechanism is designed for hard disks. As disk I/Os are expensive, it is desirable to write as many logs as possible to the disks through a single I/O operation. Some systems even employ group commit [7,8,12,14], so the I/O bandwidth of hard disks can be fully exploited. However, as new storage devices emerge, such as NVM (Non-Volatile Memory) and SSD (Solid State Disk), the centralized design becomes the potential bottleneck of transaction processing. Especially when multicore CPUs are employed, locks on the single buffer can be congested, before the I/O bandwidth and the CPU cores are saturated. The bottleneck becomes increasingly serious when the number of CPU cores increases. Developers of PostgreSQL realized this problem and made several patches to improve the scalability of WAL insertions [1,2] . However, as discussed subsequently, these patches still suffer from lock contention. In this paper, we present a once-for-all solution for PostgreSQL.

With the emergence of SSD and NVM, researchers in both academia and industry starts to investigate how to deploy SSD and NVM in database systems. Both SSD and NVM offers much higher read/write throughput that disks. Intel released Optante^{TM} SSD DC P4800X Series whose read/write latency is less than $10\,\mu s$ and $4\,KB$ random read/write performance can achieve 500K IOPS. NVM's transfer rates is faster than hard disks by several orders of magnitude. If we use SSD or NVM as the persistent storage for logs, I/O may no longer be the performance bottleneck. Thus, the bottleneck of centralized logging will be more evident. In recent years, a lot of research work [4,8,9,14] have studied how to build specialized logging systems for SSD and NVM and SSD. However, they are not general solutions, as each of them only aims at one type of hardware. For a generic DBMS, such as PostgreSQL, a single logging scheme that applies to different types of storage devices, including hard disks, SSD and NVM, is desired.

In this paper, we designed and implemented a generic logging scheme for PostgreSQL, called TwinBuf. While TwinBuf was originally designed for NVMs and SSDs, it is also effective for disk equipped systems. In its essence, TwinBuf use two buffers, which alternatively receive logs from transactions and flush logs to persistent storage. This design allows logging to be performed in a non-stop manner. It is thus superior to the original single buffer approach of PostgreSQL, in which log insertion and log flush sometimes have to block each other. In each buffer, TwinBuf assigns each worker thread/process an exclusive log slot, so that they do not need to contend for buffer space. This allows us to maximally utilize the parallelism of multi-core CPUs. Moreover, TwinBuf applies group commit, to minimize the overheads of direct storage access. We conducted extensive experiments, whose results show that on all storage devices TwinBuf outperforms the centralized logging scheme in the original PostgreSQL.

The rest of the paper is organized as follows. Section 2 reviews the related work. Section 3 presents some backgrounds of PostgreSQL and NVM. Section 4 introduces the design of TwinBuf. Section 5 introduces the implementation of the system. Section 6 evaluates TwinBuf. Section 7 concludes the paper and discusses directions for future work.

2   Related Work

Several studies [4, 9] have investigated how to perform logging efficiently on SSD. Because the write latency of SSD is significantly longer than its read latency, the existing approaches all attempted to minimize the cost of write. For instances, the approach of [4] uses multiple SSDs; when a write request takes longer than expected, the system re-issues the request to other SSD devices. The approach of [15] tried to avoid doing redundant rewrite by using old version of the data on SSD. In [6], the authors try to integrate buffering and logging. In contrast to these approaches, TwinBuf attempts to be more general w.r.t. storage devices. It should not only work on SSD, but also work on newly emerging hardware such as NVM.

Using NVM as log storage is an obvious way to speedup an OLTP database. It thus has been studied extensively in recent years.

The authors of [8] developed a NV-Logging module in Shore-MT. NV-Logging adopts decentralized logging, which allocates each worker a private log space on NVM. Each worker writes logs to NVM directly. NV-logging uses the *cflush* instruction to flush specified cache lines to NVM to ensure the persistence of logs. As frequent invocation of *cflush* has negative impact on performance, group commit is used.

In [14], the authors proposed passive group commit. In this approach, every worker writes logs to NVM independently. To ensure correctness, each worker tracks the LSN (Log Sequence Number) of its last log record that is persistent on NVM (by issuing *mfence* instructions), and inserts this LSN into a queue. A daemon thread periodically examines the queue to decide which transactions can commit successfully. In this approach, whenever a transaction commits, a *mfence* will be issued. Frequent *mfence* may hurt performance.

There are many other work [5, 6, 11, 13, 16] attempting to perform logging on NVM. They all managed to parallelize the logging module, to enhance the throughput of a DBMS. However, as they all directly write logs to NVM, they cannot be used on slower storage devices such as SSD and HDD. As a general logging mechanism, TwinBuf utilizes DRAM buffers to amortize the latency of slow devices.

3   Background

PostgreSQL's logging subsystem is a well polished module, and has not changed significantly in recent years. In this section, we present how this logging subsystem works. We also give a brief introduction of NVM's features and performance characteristics.

3.1   Logging Subsystem of PostgreSQL

PostgreSQL uses a single WAL buffer. It allows multiple workers to insert WAL records to the same buffer concurrently. When inserting a log record, a worker takes the following steps:

1. Pack the update information into a log record;
2. Reserve a slot of the right size from the log buffer. The start address of of the next available slot is maintained by the variable *Insert->CurrBytePos*, which is protected by a single latch *Insert->insertpos_lck*.
3. Copy the log record to the reserved slot. A write latch is used to protect the buffer being written, to ensure that the buffer will not be flushed out before the write is completed.

The critical section in Step 2 is relatively short, while the one in Step 3 can be quite long. To improve the degree of concurrency, PostgreSQL 9.4 started to use 8 write latches in Step 3. Each worker only needs to get one of those latches before the write operation. While this method solves the concurrency issue temporarily, when the number of workers increases, the critical section can still be a bottleneck.

When a transaction is about to commit, it must flush all the log records it created to the disk. As the same log buffer is shared by all the workers, the log records created by different transactions interleave with each other. When a worker flushes the buffer, the logs of other transactions are flushed to the disk too. PostgreSQL intentionally utilizes this effect to improve I/O efficiency. When a transaction is about to commit, PostgreSQL will usually wait for a short period, so that other transactions can put more log records into the buffer. This allows more log records to be flushed to the disk through a single I/O operation.

As we can see, a centralized log buffer is able to combine several small I/Os to a bigger one, so as the improve the efficiency of disk I/O. However, such a centralized logging scheme can also be a potential bottleneck on platforms with high degree of parallelism, especially those equipped with multi-core CPUs and high-speed storage, such as SSD and NVM. To exploit the parallelism of multi-core CPUs and the features of SSD and NVM, we need to break the critical sections of the logging procedure and parallelize the logging scheme.

3.2   Non-Volatile Memories

NVM is a kind of byte-addressable storage, with very fast access speed. Thus, it can be used as RAM in computer systems. As NVMs have not entered mass production, we use NVDIMMs (Non-Volatile Dual In-line Memory Module) to emulate NVM in our platform. NVDIMM is actually an ordinary DRAM equipped with capacitors and SSDs. When encountering power outage, the capacitors provide sufficient electric power to flush the contents in the DRAM to the SSDs. Therefore, NVDIMM is non-volatile. The performance characteristics of NVDIMM is similar to that of DRAM. While NVM and NVDIMM may differ in many features, they are not relevant to our design of the logging scheme. Therefore, we believe that our logging scheme should be directly applicable to NVMs in the future.

Although NVM is non-volatile, data written to NVM will not be automatically persistent:

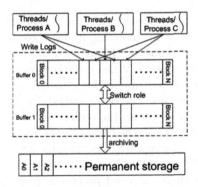

Fig. 1. Twin log buffer based log subsystem

1. Most modern CPUs have multiple levels of CPU caches and most systems adopt a write-back mode. In this mode, when a CPU writes to a NVM through the RAM interfaces, the data is placed in CPU caches first. The system needs to explicitly invoke the *cflush* or *mfence* instructions, to flush the data in CPU caches to NVMs. Otherwise, the data will be lost if the system crashes before the data is evoked from the cache. However, frequent invocation of *cflush* or *mfence* is harmful to the system's performance.
2. The latency of NVM's write operation is usually an order of magnitude longer than its read operation.

   When designing a logging scheme on NVM, we must take the above properties into consideration. To make the best of NVM, some prototypes [8] let workers directly write log records into NVM, without using a buffer. We believe that this design choice does not suit our case. First, frequent flushing CPU caches impacts the CPU performance seriously. Second, if we put data persistence operations in critical sections, it will impair the concurrency. Moreover, we expect that our scheme works for SSD and hard disk too. The performance of random access on SSDs and disks is however unacceptable for the logging subsystem.

4 Design of TwinBuf

4.1 Overview

TwinBuf has two log buffers, each of which is a segment of continuous space in DRAM. They take the following two roles, and switch their roles periodically.

1. Working buffer: when taking this role, the buffer is responsible for receiving log records generated by worker threads/processes.
2. Archiving buffer: when taking this role, the buffer flushes the log records it has received to the permanent storage.

A working buffer is partitioned into a large number of slots. Each worker is allocated an exclusive slot for inserting its log records. Thus, all workers can insert logs concurrently without interfering each other. This eliminates the critical section in the logging process. When a worker inserts its logs into a working buffer for the first time, it first applies for a slot from the buffer. Then all its log records will be inserted into the slot sequentially. When the slot is full, the worker applies for another slot from the buffer.

When the buffer's role is switched to an archiving buffer, it stops receiving logs and proceeds to flush the logs it has received to the permanent storage, which could be a hard disk, SSD or NVM. The complete set of logs in an archiving buffer is called a *log batch*. A single backend process is responsible for the archiving job. When certain conditions are met, TwinBuf triggers a role switch between the two log buffers: the archiving buffer that have finished archiving will be turned into an empty working buffer; the original working buffer will be turned into an archiving buffer and passed to the backend process. Figure 1 shows the architecture of TwinBuf.

As the backend process archives the buffers one by one, the log batches are written to the permanent storage in a serial order. However, the log records in each batch are ordered by slots, instead of following the temporal orders of the transactions. During the recovery process, the system must replay log records in the temporal order. In the original logging scheme of PostgreSQL, the temporal order of log records is exactly the order of insertions. Therefore, the offset of a log record in log file is exactly the order of the redo operation. As mentioned above, when using the parallel logging scheme of this paper, the offset of a log record is no longer its temporal order. Therefore, we assign each log record a global timestamp to represent the time it is inserted.

4.2   Group Commit

When applying a WAL method, transactions can not commit before its log records are permanent. If we flush log records each time a transaction commits, it will impair the performance of the system. The latency of disk and SSD is far lengthier than that of DRAM. Even if we use NVM as the permanent storage, we still need to flush all CPU caches to materialize the write to NVM, which is also harmful to the performance. Group commit is able to smooth out the overheads of log persistence, as it transforms the overheads of multiple commits into one.

The design of TwinBuf allows us to perform group commit naturally. In Twin-Buf, when a worker is ready to commit a transaction, it is suspended and waits for the working buffer containing the commit record to turn into an archiving buffer and have the logs flushed out. When the backend process finishes flushing the logs, it wakes up the suspended workers, which in turn will wrap up the transactions and inform the clients. When the system is crowded, multiple workers may wait for the same archiving step. In effect, their transactions will commit in a single group.

As we can see, the frequency of group commit is not determined by individual transactions, but by how frequent the two buffer switch their roles.

4.3   Switch of Log Buffers' Roles

In TwinBuf, only the backend process is responsible for switching the log buffers' roles. The switch occurs at a single point of time. After the point, the roles are exchanged, transactions start inserting logs to the new working buffer. The previous working buffer stops receiving logs and becomes the archiving buffer. The backend process starts flushing the archiving buffer to the permanent storage. In principle, the switch process does not involve any locking or blocking. Therefore, it can be very efficient and smooth.

The frequency of role switch is important to the system's performance. If it is too frequent, the effects of group commit cannot be achieved. If the frequency is too low, the latency of transaction will be unacceptable. Considering all the aspects, we trigger the role switch using the following three conditions:

1. Amount of logs. If the threshold of amount of logs in the working log buffer reaches a limit, we switch the roles.
2. Time interval. If the roles have not been switched for a certain period of time, we switch the roles.
3. Number of suspended transaction processes. If too many transactions are waiting for committing, we switch the roles.

4.4   Checkpointing

When performing a checkpoint, all the dirty pages in the buffer pool are written to the permanent storage and a checkpoint record is written to the log pointing to the last log record inserted before the checkpoint. This last log record is known as the checkpoint position. When performing recovery, redo only needs to start from the checkpoint position. After the checkpoint, the logs before the checkpoint position can be safely discarded and their space can be recycled. In TwinBuf, when a checkpoint is initiated, a role switch is immediately triggered. The checkpoint position will then be the end of the log batch flushed to the storage immediately after the role switch.

Most databases systems perform checkpoints automatically. The interval between checkpoints is usually tunable. If the interval is too short, too many checkpoints may hurt the performance. If the interval is too big, we need much more space to store logs and the recovery will take a longer time. In Twinbuf, we set a upper bound to the checkpoint interval, to make sure there is enough space on permanent storage for logs. If it is NVM, logs can be flushed faster. Then the upper bound will be smaller. If there are more CPU cores or more powerful CPUs, the upper bound will also be smaller.

4.5 Recovery

During recovery, the log space is backwardly traversed. Then, the last checkpoint record is located and checkpoint position is retrieved. TwinBuf copies the log records from the last checkpoint position forward into the memory, resorts them and perform the undo and redo procedures. As mentioned above, log records of TwinBuf do not follow their temporal order in the permanent storage. Therefore, they have to be sorted based on their timestamps before being replayed. Fortunately, log batches strictly follow the temporal order, so that resorting only needs to be conducted within each batch. Therefore, each time TwinBuf only needs to retrieve a batch of logs into the memory, resort and replay them. This extra resorting step will not incur significant overheads to the recovery process.

5   Detailed Implementation

In this section, we show how to implement TwinBuf in PostgreSQL so it can work efficiently. We focus on two processes – the process workers execute to insert logs into appropriate buffer slots, and the one the backend process executes to perform archiving.

We added several shared variables to the shared memory of PostgreSQL to coordinate the logging. They include *logswitchcount* and *xlogcount*[]. *logswitchcount* records the number of buffer role switches that have been conducted by TwinBuf. Its parity indicates which buffer is the working buffer and which one is the archiving buffer. To perform role switch, TwinBuf just needs to increment *logswitchcount* by 1. *xlogcount*[0] and *xlogcount*[1] records the number of log records in the two log buffers respectively. They also plays as locks – when the value is −1, it means that the corresponding log buffer is locked by the backend process.

Log insertion is performed in the following three steps:

1. Call Algorithm 1 to get the identifier of the current working buffer and increment the corresponding log counter.
2. If the worker is using the buffer for the first time or its current buffer slot is not enough, apply for a new buffer slot.
3. Insert the log record into the buffer slot.

Algorithm 1 is used for obtaining the current working buffer. It uses a CAS(compare-and-swap)[1] operation instead of an exclusive latch to protect *xlogcount*. This latch-free implementation proves to be much more efficient than an exclusive latch.

The backend process iteratively performs the following procedure to switch buffer roles and flush the buffers:

1. If the log buffer switch condition is not met, sleep for a while and return; otherwise, perform the following steps.

---

[1] https://en.wikipedia.org/wiki/Compare-and-swap.

Data:
logswitchcount: shared variable, used for getting the identifier of the working buffer;
xlogcount[2]: count of logs in log buffers, -1 indicates that the log buffer is locked;

Result: get the current working buffer identifier *workinglognumber* and increase the corresponding counter;

```
while true do
    while true do
        workinglognumber ← logswitchcount%2;
        t ← xlogcount[workinglognumber];
        if t == −1 then
            usleep(100L);
        end
        else
            break;
        end
    end
    /* CAS is the compare and swap operation */
    if CAS(&xlogcount[workinglognumber],t,t+1) then
        break;
    end
end
```

Algorithm 1. Increase log counter of the working log buffer

Data:
logswitchcount: shared variable, used for getting the identifier of the working buffer;
xlogcount[2]: count of logs in log buffers, -1 indicates that the log buffer is locked;
Result: lock the current working buffer;

```
while true do
    workinglognumber ← logswitchcount%2;
    t ← xlogcount[workinglognumber];
    /* CAS is the compare and swap operation */
    if CAS(&xlogcount[workinglognumber],t,-1) then
        break;
    end
end
```

Algorithm 2. The backend process locks the archiving buffer

2. Switch the roles of the two buffers by incrementing *logswitchcount* by 1.
3. Get the identifier of the current archiving buffer, and lock it by invoking Algorithm 2, which sets *xlogcount*[*buffno*] to −1.
4. Wait for all transactions that are currently inserting logs to the archiving buffer to finish. (The backend process counts the number of log records in the buffer, to know if the transactions have finish inserting.)
5. Write logs in the archiving log buffer to the permanent storage. If NVM is used, CPU cache is flushed to make sure that all logs are persistent.
6. Unlock this log buffer by resetting the corresponding counter(*xlogcount* [*buffno*]) to 0.
7. Notify the suspended worker waiting for group commit.

Algorithm 2 is invoked for locking the archiving buffer. Again, a latch-free implementation is adopted for improved efficiency.

6   Evaluation

We implemented TwinBuf in PostgreSQL 9.4 and conducted experiments to evaluate its performance. We evaluated TwinBuf's adaptivity to NVM, SSD and hard disks. We used three different benchmarks in our tests, which include Smallbank [3], TPC-C and our custom benchmark. We compared TwinBuf against the original PostgreSQL 9.4 to demonstrate its advantage. (The logging module in PostgreSQL 9.4 is the same as that of the newest version – PostgreSQL 10.1.)

The experiments were conducted on a HP Z820 workstation, equipped with two 2.60 GHz Intel Xeon E5-2670 processors and 256 GB DDR3 RAM. Its primary storage is a HDD of 10 TB. Its secondary storage is a PCIe SSD of 1 TB. We used 64 GB of DRAM to simulate NVM. The operating system installed on the workstation was CentOS 7.1. This size of each log buffer in TwinBuff is 512 MB.

Scalability on Multi-Core CPUs. Our first set of experiments aimed to evaluate the scalability of the logging modules on multi-core CPUs. We varied the number of physical cores used by the system, and observed how the throughput increases with the cores. In all the experiments, the number of workers was set to two times as many as the number of cores. Figures 2 and 3 show the results on Smallbank and TPC-C respectively.

As we can see, the logging module is indeed a bottleneck of PostgreSQL on multi-core platforms. TwinBuf's scalability is significantly better than that of the original PostgreSQL 9.4. This advantage is less visible on disks, as disk I/O seem to be a more significant bottleneck than centralized logging. Nevertheless, TwinBuf's advantage on NVM and SSD is quite obvious. When the number of cores is below 4, TwinBuf performs a bit worse than the original PostgreSQL 9.4. This is because of the application of group commit – when the degree of concurrency is low, the benefit of group commit may not compensate its overheads. As we did not consider think time in the experiments, this phenomenon was actually magnified. In real-world cases, it will be much rarer.

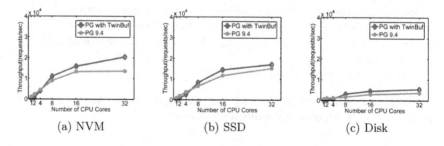

(a) NVM   (b) SSD   (c) Disk

Fig. 2. Scalability on smallbank. (Scale factor = 10)

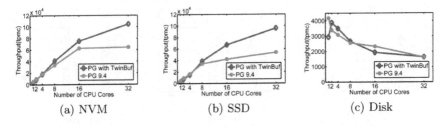

Fig. 3. Scalability on TPC-C. (Number of warehouses = 100)

The Case of Intensive Logging. To confirm that the performance difference in the standard benchmarks is caused by logging, we did an additional set of experiments on write-intensive transactions. In the experiments, each transaction is comprised of three random updates on a large table. This setting was intended to make sure that the other scalability bottleneck of PostgreSQL, such as that of the snapshot isolation mechanism, is not encountered during the experiments. The results are shown in Fig. 4. As we can see, in this write-intensive scenario TwinBuf can outperform the original PostgreSQL 9.4 by 60%. It can be predicted, if the number of cores increase further, the advantage of TwinBuf could be even more significant.

Recovery. In the final set of experiments, we tested the recovery time of PostgreSQL when using TwinBuf. In all the tests, the checkpoint interval was fixed to 60 s, during which the system could generate 1–2 GB of logs. We killed the running PostgreSQL using the command *killall -s 9 postgres*, and measured the average amount of time required to restart PostgreSQL. The results are shown in Fig. 5. As expected, the recovery on NVM and SSD is much faster than that on disks. As TwinBuf needs to perform sorting before redoing the transactions, it requires more time in recovery. Nevertheless, as sorting is performed within each log batch, this extra overhead is quite marginal most of the time.

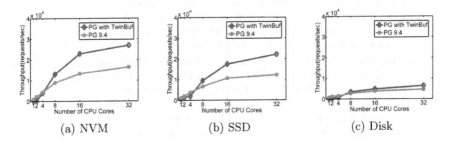

Fig. 4. Scalability on write-intensive transactions

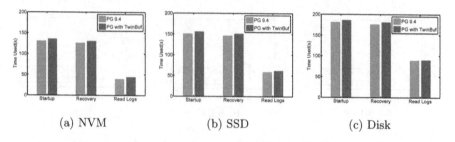

Fig. 5. Recovery Time (*Startup* – the entire amount of time required before the system is available for accepting request. *Recovery* – the amount of time for doing recovery. *ReadLog* – the amount of time required for retrieving log from the storage, including time on sorting.)

7   Conclusion

In this paper, we presented TwinBuf, a redesign of the logging module in PostgreSQL. We showed that centralized logging is a potential bottleneck of PostgreSQL on multi-core platforms. To parallelize the logging scheme, TwinBuf allocates each worker with an exclusive buffer slot. Two buffers are utilized to ensure that log insertion and log flushing do not block each other. Experimental evaluation showed that TwinBuf can speedup PostgreSQL significantly in intensive transaction processing.

References

1. Improve scalability of wal insertions. https://github.com/postgres/postgres/commit/9a20a9b21baa819df1760b36f3c36f25d11fc27b
2. Replace the xloginsert slots with regular lwlocks. https://github.com/postgres/postgres/commit/68a2e52bbaf98f136a96b3a0d734ca52ca440a95
3. Alomari, M., Cahill, M., Fekete, A., Rohm, U.: The cost of serializability on platforms that use snapshot isolation. In: IEEE 24th International Conference on Data Engineering, 2008. ICDE 2008, pp. 576–585. IEEE (2008)
4. Chen, S.: Flashlogging: exploiting flash devices for synchronous logging performance. In: Proceedings of the 2009 ACM SIGMOD International Conference on Management of Data, pp. 73–86. ACM (2009)
5. Fang, R., Hsiao, H.-I., He, B., Mohan, C., Wang, Y.: High performance database logging using storage class memory. In: 2011 IEEE 27th International Conference on Data Engineering (ICDE), pp. 1221–1231. IEEE (2011)
6. Gao, S., Xu, J., He, B., Choi, B., Hu, H.: PCMLogging: reducing transaction logging overhead with PCM. In: Proceedings of the 20th ACM International Conference on Information and Knowledge Management, pp. 2401–2404. ACM (2011)
7. Helland, P., Sammer, H., Lyon, J., Carr, R., Garrett, P., Reuter, A.: Group commit timers and high volume transaction systems. In: Gawlick, D., Haynie, M., Reuter, A. (eds.) HPTS 1987. LNCS, vol. 359, pp. 301–329. Springer, Heidelberg (1989). https://doi.org/10.1007/3-540-51085-0_52

8. Huang, J., Schwan, K., Qureshi, M.K.: NVRAM-aware logging in transaction systems. Proc. VLDB Endow. **8**(4), 389–400 (2014)
9. Lee, S.-W., Moon, B., Park, C., Kim, J.-M., Kim, S.-W.: A case for flash memory SSD in enterprise database applications. In: Proceedings of the 2008 ACM SIGMOD International Conference on Management of Data, pp. 1075–1086. ACM (2008)
10. Mohan, C., Haderle, D., Lindsay, B., Pirahesh, H., Schwarz, P.: Aries: a transaction recovery method supporting fine-granularity locking and partial rollbacks using write-ahead logging. ACM Trans. Database Syst. (TODS) **17**(1), 94–162 (1992)
11. Pelley, S., Wenisch, T.F., Gold, B.T., Bridge, B.: Storage management in the NVRAM era. Proc. VLDB Endow. **7**(2), 121–132 (2013)
12. Rafii, A., DuBois, D.: Performance tradeoffs of group commit logging. In: International CMG Conference, pp. 164–176 (1989)
13. Son, Y., Kang, H., Yeom, H.Y., Han, H.: A log-structured buffer for database systems using non-volatile memory. In: Proceedings of the Symposium on Applied Computing, pp. 880–886. ACM (2017)
14. Wang, T., Johnson, R.: Scalable logging through emerging non-volatile memory. Proc. VLDB Endow. **7**(10), 865–876 (2014)
15. Xiao-Feng, L.Z.-P.M., Da, Z.: HV-recovery: a high efficient recovery technique for flash-based database. Chin. J. Comput. **12**, 007 (2010)
16. Xu, J., Swanson, S.: NOVA: a log-structured file system for hybrid volatile/nonvolatile main memories. In: FAST, pp. 323–338 (2016)

Qualitative Instead of Quantitative: Towards Practical Data Analysis Under Differential Privacy

Xuanyu Bai[1(✉)], Jianguo Yao[1], Mingyuan Yuan[2], Jia Zeng[2],
and Haibing Guan[1]

[1] Shanghai Key Laboratory of Scalable Computing and Systems,
Shanghai Jiao Tong University, Shanghai, China
{xybjtu,jianguo.yao,hbguan}@sjtu.edu.cn
[2] Huawei Noah's Ark Lab, Hong Kong, China
{Yuan.Mingxuan,zeng.jia}@huawei.com

Abstract. Differential privacy (DP) has become the de facto standard in the academic and industrial communities. Although DP can provide strong privacy guarantee, it also brings a major of performance loss for data mining systems. Recently there has been a flood of research into the quantitative mining of DP based algorithms, which are designed to improve the performance of data mining systems. However, industrial applications demand accurate quantitative mining results. Results containing noise are actually difficult to use. This paper rethinks to apply DP in industrial big data from another perspective: qualitative analysis, which aims to dig the data about rank, pattern, important set, etc. It does not require accurate results and naturally has a greater ability to accommodate noise. We design a framework about DP data publication based attribute importance rank to support the qualitative analysis of DP, which assists data buyers to perform qualitative analysis tasks and to know the credibility of their results. We show the realization of this framework using two typical qualitative tasks. Experimental results on public data and industrial data show that making use of this framework, qualitative analysis tasks can be completed with a high confidence support even when privacy budget ϵ is very small (e.g., 0.05). Our observations suggest that qualitative analysis of DP has the potential ability to realize applying DP in industrial applications.

Keywords: Differential privacy · Qualitative analysis
Quantitative mining

1 Introduction

Differential privacy (DP) has become the state-of-the-art technology that protects privacy since it was proposed by Dwork [8]. In recent years, it develops very fast and a lot of research are proposed [12,13,18,19]. However, numerous

J. Pei et al. (Eds.): DASFAA 2018, LNCS 10828, pp. 738–751, 2018.
https://doi.org/10.1007/978-3-319-91458-9_46

studies focus on quantitative mining of DP based algorithms, which improve the performance of algorithms under DP. The quantitative mining tasks usually need accurate numerical results, so it is hard to use results containing noise. On the contrary, qualitative analysis focuses on digging data about rank, pattern, important set and so on, which does not need specific numerical values and has a greater ability to accommodate noise. Therefore we believe that qualitative analysis about strong privacy protection has a broader industrial application prospect. In this paper, we propose a framework to apply DP from a new prospective: qualitative analysis. The framework relies on one type of qualitative analysis approaches: attribute importance rank, i.e., attribute relationship analysis. For a dataset, depend on some measurement methods, e.g., mutual information or correlation coefficient, we are able to figure out how great an attribute is influential to the target value (class label). Therefore an attribute importance rank can be obtained, which represents the attribute relationship.

About DP, one general usage is data publication, which transforms raw data into synthetic data with similar characteristics. The synthetic data can be published to the public, which protects the privacy of raw data. In recent years, DP based data publication method has achieved great progress. In this paper, we design a data publication based qualitative analysis framework, which can use any data publication algorithms. We make use of the latest two DP data publication methods, i.e., PrivBayes [20] and DPTable [5] as examples to show how this framework works. Details about the methods are shown in Sect. 3.1.

In our work, we study the influence of data transformation from raw data to synthetic data on attribute relationship of raw data and explore the changes in two general aspects: coincidence rate of attribute rank (CRAR) and comparison of mining accuracy (CMA). Then on the other side, in industrial systems, data provider usually does not release raw data or relevant important information to the public directly, especially for sensitive data or information, so synthetic data will be put on sell. Therefore data buyers get nothing essential except the synthetic data that contains a lot of noise. They have no idea how great they can trust the model performance obtained from the synthetic data. If attribute rank of the raw data is accessible to data buyers, it must be beneficial for them to make better use of the data bought. However, the problem is that attribute importance rank belongs to sensitive information and will not be published directly from data provider. To solve this problem, we design a framework to assist data buyers to perform qualitative analysis tasks and to learn more about attribute relationship of raw data without leaking privacy. The tasks are represented by two stable classifiers, i.e., Belong to Top-K (BTK) and Be Larger (BL) classifiers, which answer two different questions. Stability means the classifier is in a convergent state and we conduct experiments on public and industrial data.

To summarize, we make the following contributions:

- Compared with quantitative mining, qualitative analysis has a greater ability to accommodate noise. We take the first attempt to apply DP in qualitative analysis and find a way to realize industrial application of DP.

- We propose a DP based qualitative analysis framework, which assists data buyers to perform qualitative analysis tasks and to know the credibility of their results. We make use of two typical qualitative tasks represented by two classifiers, i.e., BTK and BL as examples to show the application of the framework.
- We conduct experiments on public data and industrial data respectively. Experimental results show that making use of this framework, qualitative analysis tasks can be completed with a high confidence support even when ϵ is very small (e.g., 0.05), which has the potential ability to realize industrial application of DP.

2   Background and Motivation

2.1   Differential Privacy

Differential privacy technology takes a strong quantified control of privacy exposure. Insertion or deletion of any individual record has no influence on the output of queries or calculations on datasets.

Definition 1. A random function F provides ϵ–differential privacy if for any neighboring databases DB_1 and DB_2 ($DB_1 \triangle DB_2 = 1$), for any output $O \in Range(F)$, $Pr[F(DB_1) \in O] \leq e^\epsilon \times Pr[F(DB_2) \in O]$.

Neighboring databases DB_1 and DB_2 are two databases that contain only one individual record difference. ϵ is a parameter treated as privacy budget, which controls the grade of privacy protection. When ϵ decreases, more noise is added and stronger privacy protection will be provided.

2.2   Qualitative or Quantitative Mining Under DP

Previous work about DP usually focus on quantitative mining, which aims to get satisfying numerical results. In data publication algorithms, researchers hope to obtain better synthetic data with more similar characteristics of the original data. In other cases, researchers aim to construct better models that learn the properties of raw data. In general, the purpose of these improvements is to get better DP based model performance.

However, all these algorithms usually bring a major of performance loss for data mining systems, which is unacceptable for industrial applications. That is why DP still cannot be applied in real-life systems. It is difficult for business people to trust the results containing noise. In this paper, we study DP from another perspective: qualitative analysis, which focuses on digging data about rank, pattern, important set, etc. Qualitative analysis naturally has a greater ability to accommodate noise, so maybe it is a good new way to realize application of DP in industrial big data.

3   Qualitative Analysis of Attribute Relationship

Before proposing our own approaches about qualitative analysis, we discuss the influence of data transformation on attribute relationship through data publication algorithms. That is to say, how attribute relationship changes when a raw dataset is transformed into a synthetic one.

3.1   Data Publication with DP

For DP based data publication algorithm [17], it usually uses some methods to simulate the data distribution of the raw dataset and then under DP protection, generates a new different synthetic dataset, which contains similar characteristics. The synthetic dataset is privacy-insensitive, so any data mining applications can be applied on it directly without caring about privacy. However, the shortcoming is quite clear: since data mining algorithms run on the synthetic data instead of the original data, their performance is restricted by the method of how to generate the synthetic dataset seriously. In particular, the performance loss may not be acceptable for industrial big data applications.

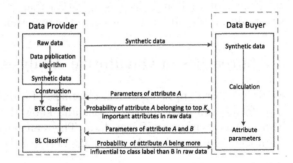

Fig. 1. Skeleton of how BTK and BL classifiers work.

In this paper, we make use of two latest data publication algorithms, i.e., PrivBayes [20] and DPTable [5]. PrivBayes is a differentially private method for releasing high-dimensional data using Bayesian network. DPTable develops a robust sampling-based framework to preserve the joint distribution of high-dimensional datasets. Both algorithms do not change the number or meaning of the original attributes in raw data. Due to space limit, we will not talk about the details here. Interested readers can learn more about them in [20] and [5]. Next we discuss the influence of data transformation on attribute relationship in two aspects. Both them show the effect of data transformation in some ways.

3.2   Metrics of Attribute Relationship

Coincidence Rate of Attribute Rank (CRAR). Through data publication algorithm, a raw dataset can be transformed into a synthetic one. Based on mutual information, we can get the attribute importance rank. Now assume attribute rank $Rank_r$ for raw data and $Rank_s$ for synthetic data have been obtained, we compare top K attributes of $Rank_r$ with top K of $Rank_s$. We figure out the percentage of the same attributes, i.e., coincidence rate of attribute rank (CRAR). Sometimes the value of CRAR is small for complex datasets, which does not mean top K of $Rank_r$ definitely has a loose connection with top K of $Rank_s$. With small CRAR, top K of $Rank_r$ may still have similar influence on the class label with top K of $Rank_s$. Therefore we propose another metric, i.e., comparison of mining accuracy (CMA).

Comparison of Mining Accuracy (CMA). To get results of CMA, obtaining $Rank_r$ and $Rank_s$, based on instances in raw data, we respectively use top K of $Rank_r$ and $Rank_s$ to train classifiers, e.g., random forests [3] or xgboost [6] and then compare their performance. It should be noted that all training instances in these models come from raw data instead of synthetic data. This is because our goal is to study the change of attribute relationship rather than how suitable the synthetic data is to build models. In Sect. 5, we give the experimental results.

4   Proposed Approaches of Qualitative Analysis

In this section, we propose two approaches (i.e., classifiers) for data provider to help data buyers perform qualitative analysis tasks and know the credibility of their results, without leaking privacy of raw data. This can help data buyers understand the data they buy more deeply and make better use of it.

4.1   Skeleton Design

The two approaches proposed are named as Belong to Top-K (BTK) and Be Larger (BL) classifiers respectively. Figure 1 shows the overall skeleton of how these two classifiers work. Being afraid of revealing privacy, data provider only can put synthetic data on sell, which is purchased by data buyers. However, since synthetic data usually contains a lot of noise, data buyers have no idea how great they can trust the model performance running on it. If data buyers have access to attribute importance rank about raw data, which is sensitive information and not released by data provider, they can make better use of the data they buy.

Since attribute rank is not accessible, we construct BTK and BL classifiers for data provider and provide their interfaces to data buyers. Owning synthetic data, data buyers are able to calculate some parameters about each attribute, e.g., mutual information, information gain, etc. For BTK, if data buyers calculate parameters of attribute A as input and invoke the interface of BTK. The classifier will return the probability of A belonging to top K attributes of raw

---

Algorithm 1. Belong to Top-K Classifier (BTK)

Input: $Data_r$ - raw dataset, K - number of attributes participating in the comparison;
Output: $model$ - BTK classifier, T - number of synthetic datasets used when convergent;
1: obtain attribute rank $Rank_r$ of $Data_r$;
2: $newInstances = \emptyset$, $saveScore = [-1, -1]$, $T = 0$, $index = 0$;
3: **while** $index \geq 0$ **do**
4:   generate synthetic dataset $Data_s$ from $Data_r$ using data publication algorithm;
5:   **for** $attr$ in attributes of $Data_s$ **do**
6:     calculate CS, CC, En, ED, IG, MI, VC of $attr$;
7:     **if** $attr \in$ top K of $Rank_r$ **then**
8:       $label = 1$;
9:     **else**
10:      $label = 0$;
11:    **end if**
12:    $newInstance = [attr_{CS}, attr_{CC}, attr_{En}, attr_{ED}, attr_{IG}, attr_{MI}, attr_{VC}, label]$;
13:    $newInstances = newInstances \cup newInstance$;
14:   **end for**
15:   $index++$;
16:   $isConvergent, T = ctc(index, newInstances, saveScore, T)$;
17:   **if** $isConvergent$ is true **then**
18:     break;
19:   **end if**
20: **end while**
21: $model \leftarrow$ random forests or xgboost built on $newInstances$;
22: return $model$ and T;

---

data. It needs to be emphasized that a probability will be returned not a definite prediction result, which protects the privacy. For instance, a relatively high probability (>0.5) means for A, the probability of belonging to top K is larger than the probability of not belonging to top K, rather than A definitely belongs to top K. High probability means high credibility of attribute relationship. Similarly, for BL, data buyers want to know whether the influence of attribute A on class label is larger than the influence of B. They can put parameters of A and B into the interface and get the corresponding probability. Both BTK and BL classifiers built are in a convergent state, which protects privacy.

In this framework, data buyers have access to two knowledge: synthetic data and the probability from the two classifiers. First, synthetic dataset is generated using DP based data publication algorithms, which satisfies DP. Second, during construction of classifiers, more and more synthetic datasets are used to train the classifier until it is converged. Each synthetic dataset is a sample, which is under the DP noise distribution. Therefore, when it is converged, the classifier provides no more information than the noise distribution of DP. In general, the framework can still guarantee the DP protection. Next we explain the details of how to construct BTK and BL classifiers.

4.2 Belong to Top-K Classifier (BTK)

To build BTK, many (T) pieces of different synthetic datasets will be generated. We transform each attribute in each synthetic dataset into a newly generated instance. For each attribute, based on class label, seven parameters are calculated, i.e., Chi Square (CS), Correlation Coefficient (CC), Entropy (En),

---

Algorithm 2. Belong Larger Classifier (BL)

Input: $Data_r$ - raw dataset;
Output: $model$ - BL classifier, T - number of synthetic datasets used when convergent;
1: obtain attribute rank $Rank_r$ of $Data_r$;
2: $newInstances = \emptyset$, $saveScore = [-1, -1]$, $T = 0$, $index = 0$;
3: **while** $index \geq 0$ **do**
4:     generate synthetic dataset $Data_s$ from $Data_r$ using data publication algorithm;
5:     $attrs \leftarrow$ array of attributes of $Data_s$;
6:     **for** A in $attrs$ **do**
7:         **for** B in $attrs$ **do**
8:             **if** $A == B$ **then**
9:                 continue;
10:            **end if**
11:            **if** $A > B$ according to $Rank_r$ **then**
12:                $label = 1$;
13:            **else**
14:                $label = 0$;
15:            **end if**
16:            $newInstance = [A_{CS}, A_{CC}, A_{En}, A_{ED}, A_{IG}, A_{MI}, A_{VC}, B_{CS}, B_{CC}, B_{En}, B_{ED},$ $B_{IG}, B_{MI}, B_{VC}, A_{CS} - B_{CS}, A_{CC} - B_{CC}, A_{En} - B_{En}, A_{ED} - B_{ED}, A_{IG} - B_{IG},$ $A_{MI} - B_{MI}, A_{VC} - B_{VC}, A_{CS}/B_{CS}, A_{CC}/B_{CC}, A_{En}/B_{En}, A_{ED}/B_{ED}, A_{IG}/B_{IG},$ $A_{MI}/B_{MI}, A_{VC}/B_{VC}, label]$;
17:            $newInstances = newInstances \cup newInstance$;
18:        **end for**
19:    **end for**
20:    $index$++;
21:    $isConvergent, T = ctc(index, newInstances, saveScore, T)$;
22:    **if** $isConvergent$ is true **then**
23:        break;
24:    **end if**
25: **end while**
26: $model \leftarrow$ random forests or xgboost built on $newInstances$;
27: return $model$ and T;

Euclidean Distance (ED), Information Gain (IG), Mutual Information (MI) and Vector Cosine (VC). Chi Square of attribute A is represented as A_{CS}. These seven parameters are deemed as attributes of the new instance, which describe the characteristics of A in different aspects. Due to space limit, we will not talk about the details of these parameters here. They are all general methods, so interested readers can learn about them by many ways. On the other hand, each attribute in synthetic data must belong to top K important attributes in raw data or not, which generating the class label of this new instance. An example is given here. Assume that a raw dataset with 14 attributes and 1 class label is used. K equals 4 and T equals 50. Eventually, 700 $(14 * 50)$ new instances are generated, including 200 $(4 * 50)$ positive (belong to top 4 influential attributes in raw dataset) and 500 $(10 * 50)$ negative instances (not belong to top 4 in raw dataset). Each new instance consists of 7 attributes and 1 class label. Finally these newly generated instances are used to train a model, i.e., random forests or xgboost, which returns data buyers probabilities (BTK classifier).

On the other side, to protect privacy, we want the classifier to be in a convergent state. This means how we set the value of T. Since $T * sizeOf(attributes)$ newly generated instances are used, convergence means that the values (V_{T-1} and V_T) of loss function of classifiers built on $(T - 1) * sizeOf(attributes)$ and

---

Algorithm 3. Check termination condition - ctc(*index*, *newInstances*, *saveScore*, *T*))

Input: *index* - number of synthetic datasets used, *newInstances* - set of newly generated instances, *saveScore* - array saving the latest two values, *T* - number of synthetic datasets used when convergent;

Output: *isConvergent* - true or false, which represents whether the classifier is convergent, *T* - number of synthetic datasets used when convergent;

1: *score* ← objective function value of random forests or xgboost built on *newInstances*;
2: **if** *saveScore*[0] == -1 **then**
3:     *saveScore*[0] = *score*;
4:     *saveScore*[1] = *score*;
5: **else**
6:     *saveScore*[0] = *saveScore*[1];
7:     *saveScore*[1] = *score*;
8: **end if**
9: *isConvergent* = false, T = -1;
10: **if** —*saveScore*[1] - *saveScore*[0]— ≤ *saveScore*[0] * 0.01% **then**
11:     *isConvergent* = true;
12:     T = *index*;
13: **end if**
14: **return** *isConvergent* and T;

$T * sizeOf(attributes)$ instances are very similar. In our work, if $|V_T - V_{T-1}| \leq V_{T-1} * 0.01\%$, we say the classifier built on $T * sizeOf(attributes)$ instances has been convergent. Using more and more newly generated instances, we can find the moment of convergence (the value of T). Finally we use $T * sizeOf(attributes)$ instances to train the classifier and provide its interface to the public. Algorithm 1 shows the whole process of how to build a BTK classifier.

4.3   Be Larger Classifier (BL)

Next we discuss BL classifier. Many (T) pieces of synthetic datasets will be generated. Different from BTK, for BL, every combination of two different attributes in synthetic dataset is used to generate two new instances, which consist of 28 attributes and 1 class label. The class label implies whether attribute A has greater influence on class label than attribute B ($B \mathrel{!=} A$) in raw dataset, which is decided by attribute rank (based on mutual information) of raw data. Besides, 28 attributes are respectively CS, CC, En, ED, IG, MI, VC of A and B, and their subtraction and division. It needs to be noted that two different instances are generated due to combination of A and B, i.e., A has greater influence than B (positive instance) and B has less influence than A (negative instance).

For instance, assume that a raw dataset with 14 attributes and 1 class label is used. T equals 50. Finally 9100 (14 * 13 * 50) new instances will be generated, including 4550 positive and 4550 negative instances. Based on them, classification models are built to return data buyers probabilities (BL classifier). Similar to BTK, we also need to use the same method to make the classifier built to be convergent. Algorithm 2 shows the whole process of how to build a BL classifier.

5   Experiments

5.1   Experimental Settings

Datasets. We make use of two datasets in our experiments: (i) *Adult* [1] (public), which is extracted from the 1994 census bureau database. It includes information of 45222 individuals and the prediction task is to determine whether a person makes over 50K a year. It contains 14 attributes and 1 class label. (ii) *Industry* (private), which is industrial data extracted from one of the biggest telecommunication operators in China. It includes information of more than one million users and the prediction task is to predict whether a customer will subscribe the service next month. It contains 53 attributes and 1 class label.

Parameters. To transform raw datasets, we use two data publication algorithms, i.e., PrivBayes [20] and DPTable [5] and build two classification models, i.e., random forests and xgboost. Due to space limit, we mainly demonstrate the results of PrivBayes and xgboost. Similar results can be observed by DPTable and random forests and we will only show two example results of them in Figs. 4(b) and 6. On the other side, we use the default settings in PrivBayes and DPTable. For PrivBayes, usefulness of each noisy marginal distribution in second phase θ equals 4 and the degree of Bayesian network k equals 3. For DPTable, marginals generated are all 2-way. Besides, depend on settings in PrivBayes and DPTable, we set privacy budget ϵ as 0.05, 0.1, 0.2, 0.4, 0.8 or 1.6 respectively. For *Adult*, K is set as 3, 6 or 9. For *Industry* that contains much more instances and attributes, K is set as 8, 16, or 24. For getting results of CRAR and CMA, we conduct experiments 50 times and get average results eventually. For constructing BTK and BL, the value of T is determined by the convergent state of classifiers. Every time we build a model, we split the dataset into two parts: 4/5 of instances as training data and 1/5 as test data.

5.2   Metrics of Attribute Relationship

Coincidence Rate of Attribute Rank (CRAR). About CRAR, Fig. 2(a) shows the experimental results for *Adult*. We can see that as budget ϵ gets larger, CRAR also becomes larger. This is because when ϵ increases, size of noise added into raw data is smaller. The distribution of synthetic data is more similar to that in raw data. Besides, it is easy to understand larger CRAR with larger K, which means that more attributes participate in the comparison.

Figure 2(b) shows the results on *Industry*. Since *Industry* is much more complex, the influence of ϵ on CRAR is smaller and CRAR increases slightly with larger ϵ. And CRAR is obviously lower than results for *Adult*. However, it does not mean top K of synthetic data has a loose connection with top K of raw data. We give the results about CMA.

Comparison of Mining Accuracy (CMA). Fig. 3 shows the results about CMA. For *Adult*, in Fig. 3(a), for the line *Top* 6, we use instances in raw dataset

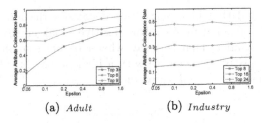

(a) *Adult*   (b) *Industry*

Fig. 2. Results of CRAR.

$Data_r$ to build models on the basis of top 6 influential attributes of synthetic dataset $Data_s$. For the line *Baseline Top* 6, the difference is that we use top 6 attributes of $Data_r$ instead of $Data_s$. We observe that model performance of *Top* 6 is worse than *Baseline Top* 6, which represents that top 6 attributes of $Data_s$ are not as good as top 6 of $Data_r$. However, as ϵ becomes larger, less noise is added and top 6 attributes of $Data_s$ are closer to top 6 of $Data_r$. Therefore, performance of models built becomes better.

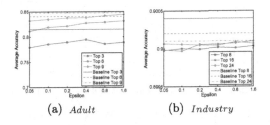

(a) *Adult*   (b) *Industry*

Fig. 3. Results of CMA.

For *Industry*, we observe the same phenomenon in Fig. 3(b). The two lines with the same color become closer as ϵ increases, which means top K of $Data_s$ has a strong connection with top K of $Data_r$. Besides, using more attributes to build models does not always lead to better model performance. For example, the line *Baseline Top* 24 is lower than the line *Baseline Top* 8.

5.3   Belong to Top K Classifier (BTK)

For *Adult*, experimental results about BTK are shown in Fig. 4(a)(b). As *Adult* is a public dataset, which meets relatively more uniform distribution, mostly accuracy of models is high (larger than 0.8) even when the privacy budget is very small ($\epsilon = 0.05$). When K equals 6 (not very small or large), models usually have the best performance. During the prediction, for each test instance, the classifier returns the probability of this instance being positive (attribute A belongs to top K attributes of raw data). Using these probabilities, we get cumulative distribution function (CDF) curves, which are shown in Fig. 5(a)(b)(c). From these curves, we find that in general, as ϵ decreases, the number of instances whose probability of being predicted as positive is larger than 0.8 or smaller than 0.2 becomes smaller. That is to say, when ϵ becomes smaller, less instances are predicted as positive or negative with a high probability. It becomes more difficult for the classifier to make a prediction.

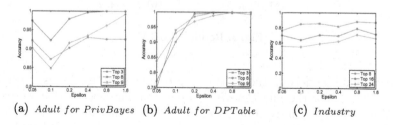

(a) *Adult for PrivBayes*   (b) *Adult for DPTable*   (c) *Industry*

Fig. 4. Accuracy of BTK classifier.

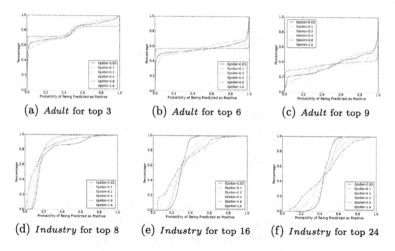

(a) *Adult* for top 3   (b) *Adult* for top 6   (c) *Adult* for top 9

(d) *Industry* for top 8   (e) *Industry* for top 16   (f) *Industry* for top 24

Fig. 5. CDF curve of BTK classifier.

For *Industry*, experimental results are shown in Fig. 4(c). ϵ has smaller influence on accuracy of classifier. But we can still find that as ϵ increases, classifier performance becomes better. When K equals 8, models have the best performance, i.e., the accuracy is larger than 0.8. Figure 5(d)(e)(f) show the corresponding CDF curves. We can still observe that as ϵ decreases, it becomes more difficult to predict an instance with a high probability (larger than 0.8 or smaller than 0.2). On the other side, compared with results on *Adult*, less instances can be predicted as positive or negative with a high probability for *Industry*. This is because *Industry* is a much more complex industrial dataset, its distribution is not uniform as *Adult*. Test instances seem harder to be predicted.

5.4   Be Larger Classifier (BL)

For *Adult*, about BL, Figs. 6(a) and 7(a) show the similar phenomena. Accuracy of BL classifier is larger than 0.8 even when ϵ is very small (e.g., 0.05) and with smaller ϵ, instances become more difficult to predict.

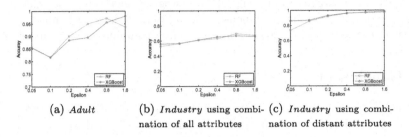

(a) *Adult*   (b) *Industry* using combi- (c) *Industry* using combi-
                    nation of all attributes   nation of distant attributes

Fig. 6. Accuracy of BL classifier.

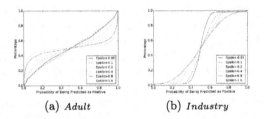

(a) *Adult*   (b) *Industry*

Fig. 7. CDF curve of BL classifier.

For *Industry*, Figs. 6(b) and 7(b) show the experimental results. The accuracy is not very high (between 0.5 and 0.75). We believe that this is because with 53 attributes, the number of combination of random two attributes is very large. Classifier becomes so complex that it is hard to make a very accurate prediction and we can still observe that less instances are predicted with a high probability for smaller ϵ. In order to get better model performance, we make an improvement on model construction: only combination of two distant attributes can be used to generate new instances. Specifically, since our standard measurement method is mutual information, if attributes A and B meets $|A_{MI} - B_{MI}|$ / $max\{A_{MI}, B_{MI}\} > 10\%$, they are deemed as distant attributes. This improves model performance very well in Fig. 6(c) and the accuracy is very high even when ϵ is very small (e.g., 0.05).

6   Related Work

Researchers prefer DP due to its strong control of privacy leakage. Unlike previous privacy protection techniques [7,14], DP [8,15,16] does not worry about how much attackers learn about the background knowledge (*quasi-identifiers*).

A lot of research work about different fields have been conducted on DP. Chaudhuri et al. [4] applied DP in regular logistic regression. Friedman et al. [11] studied how to apply DP into the decision tree model. Blum et al. [2] realized applying DP in K-means algorithm and [9] completely proposed the generalization about DP based histogram method. Dwork et al. [10] pointed out that the biggest advantage of applying DP in histogram method is that calculation of sensitivity is irrelevant to dimension of dataset. Besides, DP can be applied in other scenarios, such as graph dataset, item set mining, crowdsourcing, etc.

7   Conclusion

In this paper, we propose a DP based qualitative analysis framework, which assists data buyers to perform qualitative analysis tasks and to know the credibility of their results without leaking privacy. From a new perspective, i.e., qualitative analysis, we want to find a way to apply DP in industrial systems. We make use of two typical qualitative tasks to show how this framework works. Experimental results on public and industrial data show that even when ϵ is very small, qualitative analysis tasks can be completed with a high confidence support. Our observations suggest that qualitative analysis of DP is more applicable in large-scale industrial systems than quantitative mining. In future work, we would like to continue to study qualitative analysis under DP in some other different ways and design better strategies to apply DP in real-life applications.

Acknowledgment. This work was supported in part by the Program for NSFC (No. 61525204, 61772339), and the STCSM project (No. 16QA1402200). The corresponding author is Prof. Yao, Jianguo.

References

1. http://archive.ics.uci.edu/ml/datasets/Adult
2. Blum, A., et al.: Practical privacy: the SuLQ framework. In: PODS (2005)
3. Breiman, L.: Random forests. Mach. Learn. **45**, 5–32 (2001)
4. Chaudhuri, K., et al.: Privacy-preserving logistic regression. In: NIPS (2008)
5. Chen, R., et al.: Differentially private high-dimensional data publication via sampling-based inference. In: SIGKDD (2015)
6. Chen, T., et al.: XGBoost: a scalable tree boosting system. In: SIGKDD (2016)
7. Domingoferrer, J., et al.: A critique of k-anonymity and some of its enhancements. In: ARES (2008)
8. Dwork, C.: Differential privacy. In: ICALP (2006)
9. Dwork, C.: A firm foundation for private data analysis. In: CACM (2011)
10. Dwork, C., McSherry, F., Nissim, K., Smith, A.: Calibrating noise to sensitivity in private data analysis. In: Halevi, S., Rabin, T. (eds.) TCC 2006. LNCS, vol. 3876, pp. 265–284. Springer, Heidelberg (2006). https://doi.org/10.1007/11681878_14
11. Friedman, A., et al.: Data mining with differential privacy. In: SIGKDD (2010)
12. To, H., et al.: A framework for protecting worker location privacy in spatial crowdsourcing. Proc. VLDB Endow. **7**, 919–930 (2014)

13. Hu, X., et al.: Differential privacy in telco big data platform. In: VLDB (2015)
14. Machanavajjhala, A., et al.: L-diversity: privacy beyond k-anonymity. In: TKDD (2007)
15. McSherry, F.: Privacy integrated queries: an extensible platform for privacy-preserving data analysis. In: CACM (2010)
16. Bai, X., et al.: Embedding differential privacy in decision tree algorithm with different depths. SCIS **60**, 082104 (2017)
17. Mohammed, N., et al.: Differentially private data release for data mining. In: SIGKDD (2011)
18. Qardaji, W., et al.: Differentially private grids for geospatial data. In: ICDE (2013)
19. Xiao, Q., et al.: Differentially private network data release via structural inference. In: SIGKDD (2014)
20. Zhang, J., et al.: PrivBayes: private data release via Bayesian networks. In: SIGMOD (2014)

Client Churn Prediction with Call Log Analysis

Nhi N. Y. Vo[1], Shaowu Liu[1], James Brownlow[2], Charles Chu[2], Ben Culbert[2], and Guandong Xu[1(✉)]

[1] Advanced Analytics Institute, University of Technology Sydney, Ultimo, Australia
Nhi.Vo@student.uts.edu.au, {Shaowu.Liu,Guandong.Xu}@uts.edu.au
[2] Colonial First State, Sydney, Australia
{James.Brownlow,Charles.Chu,Ben.Culbert}@cba.com.au

Abstract. Client churn prediction is a classic business problem of retaining customers. Recently, machine learning algorithms have been applied to predict client churn and have shown promising performance comparing to traditional methods. Despite of its success, existing machine learning approach mainly focus on structured data such as demographic and transactional data, while unstructured data, such as emails and phone calls, have been largely overlooked. In this work, we propose to improve existing churn prediction models by analysing customer characteristics and behaviours from unstructured data, particularly, audio calls. To be specific, we developed a text mining model combined with gradient boosting tree to predict client churn. We collected and conducted extensive experiments on 900 thousand audio calls from 200 thousand customers, and experimental results show that our approach can significantly improve the previous model by exploiting the additional unstructured data.

Keywords: Text mining · Churn prediction · Call log analysis
Customer data analytics · Machine learning

1 Introduction

Customer data analytics has always been the core business function of any firm. Without a good insights and customer relationship management (CRM) strategies, there would be no success for any business. In the past decades, companies have been intensively analyze their customer data for different segmentations, churn prediction and marketing planning. Many of those business intelligence applications have been proven to be effective in customer attraction and retention. As the cost of gaining new customers is approximately three times higher than that of retaining the existing ones, churn prediction and customer retention is the most challenging task for any business. Every percentage increase in accurate churn prediction could result in million dollars saved in cost and revenue. Within financial services field where customer engagement and loyalty is low in general, this analytics problem has become vital to both long-term and short-term business plan, especially for Superannuation funds.

Most of current churn prediction models are using only structured data from customer, e.g. demographic (age, gender, educational background, etc.), transactional data (account balance, balance change, change ratios, etc.), network (advisor, connection with other accounts, etc.) and financial decisions (investment, churn, buying insurance, etc.). These structured data are easy to be collected and studied, which make it an efficient approach within customer analytics field. On the other, there are also some effort to incorporate communications data from call centre into customer analytics field. However, these are still structured features such as call frequency, intervals between calls, call lengths, which still slightly improve the prediction accuracy, but not significantly meaningful for the financial gains of the firms.

From methodology perspective, since researchers have exhausted all the feature engineering and stacking techniques, the performance lift of any predictive model using these features would be very minimal. Those table-formatting features from structured data are normally accounted for only 20% of business insights. The remaining 80% lies within unstructured data from daily customer interactions, e.g. emails, calls, chats, social media interactions, face-to-face meetings (Fig. 1). These data are often undocumented, unrecorded and unorganized, which make it harder to perform any analysis. Researches have performed opinion mining and sentimental analysis [14]. However, we believe there is more information in other types of text features rather than just sentiment scores. The application of text mining to extract further term-based features from these communicative data would further increase the churn prediction and provide deeper customer insights than basic account factors.

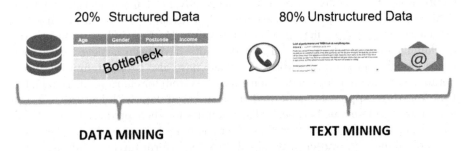

Fig. 1. Structured and unstructured data

Considering the recent advancement in text mining and machine learning, we now have more means to extract meaningful business insights from these unstructured data. Within the scope of this paper, we will utilize the textual information from transcriptions of customer call logs to predict whether they will churn or not. The forecast churn result will directly complement the prediction model using structured data, increasing the prediction accuracy by at least 4%. It will help the company in planning the suitable marketing and customer services strategies targeting at customers with high churn probability. By improving customer retention rate, the firm will save 4.5 millions Australian dollars in business annual revenue.

There would be three main contributions of our paper:

- To the best of our knowledge, this is the first attempt to incorporate text mining techniques for churn prediction model within Superannuation and Pension fund industry in Australia.
- Our prediction model was built using big unstructured data in combination with structured data. This is the initial foundation for the incorporation of more unstructured data into daily customer analytics in financial services industry.
- The final contribution is the improvement of churn prediction accuracy using customer call logs, which is vital for the financial success with on-time customer retention strategies for any type of business.

The rest of this paper is organized as follows. Section 1 introduces the current background of churn prediction application with limitations and our proposed text mining solution. In Sect. 2, we review the literature on customer churn prediction with focus on recent methodologies as the motivation for our research work. Section 3 is devoted to describe the technical details of our methodologies. In Sect. 4, the proposed text mining approaches are applied to private business datasets to perform prediction on customer call logs. Finally, conclusions are drawn in Sect. 5.

2   Preliminary

Churn prediction models have utilize many advanced machine learning algorithms to improve the prediction accuracy [2]. Current literature has been focusing more on churn prediction within telecommunication industry due to the availability of large database [11].Good results have been achieved with tree-based models and neural networks [7]. There have also been attempts to incorporated customer call logs in decision support system for churn prediction [5,18]. However, these approaches still used the structured data with features, e.g. call length as time intervals between calls, as inputs for their prediction models [8]. This is due to the high cost as well as other customer privacy and ethical concern when obtaining the transcription of call logs to perform further analysis.

Within financial services industry, there are already some churn prediction models using different approaches, especially more research effort has been done in private banking [1]. The most common methodology for these models is support vector machines, which achieves good results in general and particularly with unbalance dataset of credit card customers [6]. Other researchers try to improve further by combining the machine learning algorithm with fuzzy methodologies [9]. There also has been some effort to use more advance tree-based algorithms for electronic banking churn prediction [10]. A hybrid methodology combining k Reverse Nearest Neighborhood and One Class support vector machine (OCSVM) has been proposed to predict credit card churn rate [17]. As the common practice, all these models are using only demographic and financial transaction of customers to predict churn.

Little churn prediction research have been conducted on Superannuation and Pension funds, and those are mainly using qualitative methodologies rather than machine learning quantitative approaches. Some researchers have conduct a comprehensive experiments to build churn prediction models for Superannuation industry using multiple machine learning algorithms [4]. According to their results, random forest has the best average prediction accuracy across different datasets. Overall, the performances of these models are notable customer data analytics works. However, the prediction accuracy of these models cannot be significantly improved just by using a different machine learning algorithms on the same demographic features from structured dataset. We believe with the complement of unstructured data from customer communication would provide deeper insights into the financial behaviours, i.e. churn decision. The current literature on this particular research topic definitely lacks of text mining approaches and the usage of unstructured data.

On the other hand, researchers have also applied text mining methodologies for sentiment analysis based on customer feedback and social media posts [19]. There also has been a proposed hybrid model that comprises fuzzy formal concept analysis and concept-level sentiment analysis (FFCA + SA) for opinion mining on complaints from financial services customers [14]. However, the opinion mining task using written text did not go further to predict churn or other financial and business outcomes. We believe there are more important information lies in the words and sentences used by customers rather than those raw number or simple positive and negative feelings. Realizing this research gap, we will apply text mining techniques to analyse customer call logs and build a churn prediction model based on these extracted features in combination with structural data.

3   Methodology

Within the scope of this research project, we applied three text mining methodologies on the private customer call logs datasets to obtain three different sets of text features: Semantic Information, Word Importance and Word Embedding. After that, we will combine each and all three approaches with the current customer structured database to build the final churn prediction model with higher accuracy.

3.1   Semantic Information

For semantic information, the most common text mining approach is sentiment analysis with just positive or negative dimension. In our dataset, we want to have a more comprehensive semantic features. Therefore, we use Linguistic Inquiry and Word Count 2015 (LIWC) to extract semantic text features. The LIWC

2015 master dictionary is composed of almost 6,400 words, word stems, and selected emoticons. For each dictionary word, there is a corresponding dictionary entry that defines one or more word categories. For example, the word cried is part of five word categories: Sadness, Negative Emotion, Overall Affect, Verb, and Past Focus. Hence, if the word cried was found in the target text, each of these five subdictionary scale scores would be incremented. As in this example, many of the LIWC2015 categories are arranged hierarchically. All sadness words, by definition, will be categorized as negative emotion and overall affect words. The LIWC 2015 [13] dictionary covers many topic-related features (e.g. work, family, friend, money), sentiment (e.g. possemo, negemo) and even speech related features (non-fluent). A total 93 text features were extracted.

3.2   Word Importance

Despite the popularity of unigram or multi-gram Bag of Words models, these methodologies will produce a large number of features considering the amount of call logs in our dataset. We believe the most suitable "word importance" text mining technique in our context is the "term frequency inverse document frequency" (TF-IDF) [12,16]. This methodology reduces the dimension of features significantly by removing many common and unimportant terms. In total, we extracted around 10,000 TF-IDF text features (Fig. 2).

– Term Frequency (TF): measures of how often the term appears within the call
– Inverse Document Frequency (IDF): measures of how much information the word provides, i.e., whether the term is common or rare across all call logs

$$idf(terms, calls) = log \frac{number\ of\ calls}{numbers\ of\ calls\ contain\ the\ term}$$

– Term Frequency - Inverse Document Frequency (TF-IDF): TF-IDF = TF x IDF.

3.3   Word Embedding

In most cases, different combination of similar terms in English could result in a totally different meaning. It is not sufficient to build a text mining model using words as features alone. Therefore, we also look at the place of these terms in a sentence, their positions and their connections. Word embedding methodology would help us evaluate the relationships and interactions between these terms better. In this paper, we will use the Word2Vec model from python package Gensim [15] to extract total 50 word embedding features (Fig. 3).

10k terms

Call \ Term	close	transfer	hello	the	yes	no	Churn
Call 1	1.63	0.24	0.07	0	0.20	0.73	1
Call 2	0	2.19	0.07	0	0.27	0.15	1
Call 3	0.54	0.97	0.07	0	0.40	0	0
Call 4	0	0	0.07	0	0.27	0	0
Call 5	0	0.49	0.07	0	0.20	0.15	0
Call 6	0	0.24	0.07	0	0	0.15	0
Call 7	0	0	0	0	0	0.15	0

Fig. 2. Sample TFIDF features set

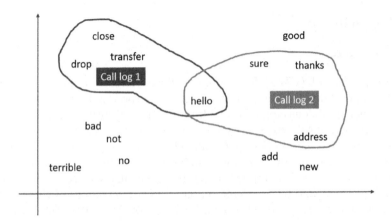

Fig. 3. Word embedding model captures relationships between terms

3.4 Combined Prediction Model

After extracting all three different set of text features from the customer call logs, we will then merge them with other demographic and account performance features from customer database to build a churn prediction model. We will use the "Extreme Gradient Boosting" (XGBoost) Classifier from Python package XGBoost [3] as our machine learning algorithm without any specific parameter tuning. From the predicted churn, the company can propose suitable customer retention strategy targeting at the right customer at the right time (Fig. 4).

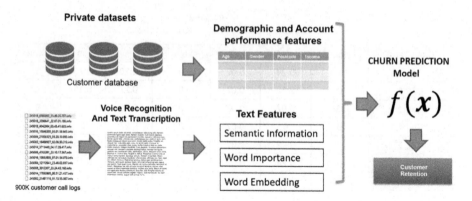

Fig. 4. Text mining for churn prediction methodology

4 Experiment

4.1 Datasets

Call Logs dataset (CL). The customer call logs dataset are privately provided by a Superannuation company for research purpose only. In total, there are more than 3 millions calls recorded from April 2011 until April 2017. The text transcription of calls were provided by a third-party service, where the words spoken by customer service agents and clients were separated with some confidence levels. Using the customer number recorded by the call center system, we were able to match approximately 900, 000 calls with the right customer from the client database. Therefore, only the transcriptions of these call logs will be used for the research within the scope of this paper. Furthermore, to avoid using call logs when customers specifically contact the agent regarding closing their accounts, we will exclude any calls happened within 14 days before the churn day. The final cleaned dataset contains 173, 000 calls in total.

Demographic and Account Performance dataset (DAP). We have four different customer databases: employer superannuation accounts with approximately 268, 000 clients, non-employer superannuation accounts with approximately 302, 000 clients, investment accounts with approximately 170, 000 clients and pension accounts with approximately 131, 000 clients. These databases contain key demographic features (e.g. age, sex, location) and account performance features (e.g. account balance, balance change ratio, balance change). Not all customer has made calls to the company, so will only use the sub-datasets with customers who have called the company since the beginning of 2014. The statistics on the final cleaned datasets are described in Table 1. All these customers have been labeled with binary coding with 1 for churn and 0 for not churn. These labels will be the ground truth labels for our churn prediction model. We achieve similar results for all four datasets, which prove our methodology is suitable in general.

Table 1. Statistics of DAP datasets

Dataset	Full datasets			Called customer datasets			Basic
	Churn	No churn	Total	Churn	No churn	Total	Features
Non-employer super	35, 239	266, 244	301, 483	2, 067	47, 929	49, 996	106
Employer super	18, 796	249, 203	276, 999	1, 959	34, 649	36, 608	133
Pension	2, 907	127, 772	130, 679	582	27, 464	28, 046	106
Investment	7, 596	161, 987	169, 583	2, 098	21, 324	23, 422	106

4.2 Evaluation

For churn prediction accuracy, even though the train and test label are binary, our predicted churn scores are ranging from 0 to 1 as the probability for churn. We will compare these values with the ground truth labels after running 10-fold cross validation experiment and evaluate our model performance using Area Under the Curve (AUC) scores and plot the results using the Receiver Operating Characteristic (ROC) curve.

4.3 Results

Firstly, we will build separate churn predictions using different set of features: (1) basic features from DAP dataset, (2) Semantic Information features from LIWC 2015, (3) Word Importance features from TF-IDF and (4) Word Embedding features from Word2Vec. The predicted churn values of each model will be evaluated against the ground truth labels provided by the company. Lastly, we combined all three text features sets and basic features set to build a churn prediction model. The AUC scores results are as in Table 2 and ROC curves are as in Fig. 5 for non-employer superannuation accounts, Fig. 7 for pension customers, Fig. 6 for employer superannuation accounts and Fig. 8 for investment customers (Table 2).

Table 2. AUC results on model prediction accuracy

Dataset	AUC scores			
	Basic features	Text features	Combined	Improvement
Non-employer super	0.7261	0.7306	0.7812	7.6%
Employer super	0.7980	0.7599	0.8346	6.4%
Pension	0.7843	0.7311	0.8169	4.2%
Investment	0.6847	0.7798	0.8108	18.4%

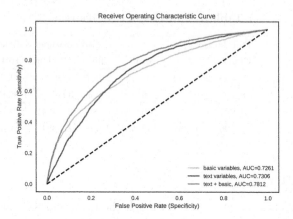

Fig. 5. Churn prediction for non-employer superannuation customers

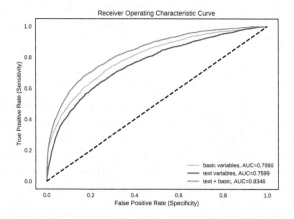

Fig. 6. Churn prediction for employer superannuation customers

The ROC plots show that the combination of both structured and unstructured data can increase the churn prediction accuracy by at least 4% comparing to the basic features only model. Especially in the Investment dataset, our prediction model achieve a significant increase of 18.4%. It confirms that our combined features approach are effective for churn prediction model. This result also proves our hypothesis that there is meaningful information in the words and sentences of customer, and these unstructured data should be incorporated in customer data analytics. The empirical result further implies the potential improvement of the model using other unstructured data from other customer interaction channels like social media.

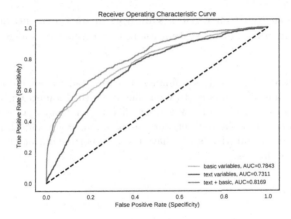

Fig. 7. Churn prediction for pension customers

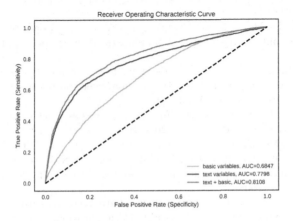

Fig. 8. Churn prediction for investment customers

5   Conclusions

The big data analytics field advances providing businesses more means to understand their customer in a broader quantitative basis using different type of data
rather than traditional ones. Within the scope of this paper, we have taken a new
approach when using unstructured data from customer call logs to build a churn
prediction model. This is the first research within financial services industry to
incorporate both structured and text features for such customer data analytics
problem. The results show that unstructured data contains vital information
which improves the accuracy of churn prediction by at least 5% on different customer datasets. It not only help the company on planning for customer retention
strategy and save million dollars in revenue but also lay an initial foundation
for the incorporation of more and more unstructured data in daily business
intelligence system. For future research, we would look into using these unstruc-

tured data for other customer insights analysis such as segmentation or building personalized recommendation system for financial services and Superannuation products.

Acknowledgment. This work was financially supported by our industry partner at UTS and was technically supported by our faculty with the infrastructures and computing power for empirical works. We would like to send our appreciation to fellow researchers who provided the datasets for this research. We would also like to thanks our colleagues, family and friends who have been supportive during the experimenting and paper writing process.

References

1. Ali, O.G., Arıtürk, U.: Dynamic churn prediction framework with more effective use of rare event data: the case of private banking. Expert Syst. Appl. **41**(17), 7889–7903 (2014)
2. Almana, A.M., Aksoy, M.S., Alzahrani, R.: A survey on data mining techniques in customer churn analysis for telecom industry. J. Eng. Res. Appl. **4**(5), 165–171 (2014)
3. Chen, T., Guestrin, C.: XGBoost: a scalable tree boosting system. In: Proceedings of the 22nd ACM SIGKDD International Conference on Knowledge Discovery and Data Mining, pp. 785–794. ACM (2016)
4. Chu, C., Xu, G., Brownlow, J., Fu, B.: Deployment of churn prediction model in financial services industry. In: 2016 International Conference on Behavioral, Economic and Socio-cultural Computing (BESC), pp. 1–2. IEEE (2016)
5. Coussement, K., Van den Poel, D.: Integrating the voice of customers through call center emails into a decision support system for churn prediction. Inf. Manag. **45**(3), 164–174 (2008)
6. Farquad, M.A.H., Ravi, V., Raju, S.B.: Churn prediction using comprehensible support vector machine: an analytical CRM application. Appl. Soft Comput. **19**, 31–40 (2014)
7. Hassouna, M., Tarhini, A., Elyas, T., AbouTrab, M.S.: Customer churn in mobile markets a comparison of techniques. arXiv preprint arXiv:1607.07792 (2016)
8. Huang, Y., Zhu, F., Yuan, M., Deng, K., Li, Y., Ni, B., Dai, W., Yang, Q., Zeng, J.: Telco churn prediction with big data. In: Proceedings of the 2015 ACM SIGMOD International Conference on Management of Data, pp. 607–618. ACM (2015)
9. Karahoca, A., Bilgen, O., Karahoca, D.: Churn management of e-banking customers by fuzzy AHP. In: Handbook of Research on Financial and Banking Crisis Prediction Through Early Warning Systems, pp. 155–172. IGI Global (2016)
10. Keramati, A., Ghaneei, H., Mirmohammadi, S.M.: Developing a prediction model for customer churn from electronic banking services using data mining. Financ. Innov. **2**(1), 10 (2016)
11. Linoff, G.S., Berry, M.J.A.: Data Mining Techniques: For Marketing, Sales, and Customer Relationship Management. Wiley, Hoboken (2011)
12. Luhn, H.P.: A statistical approach to mechanized encoding and searching of literary information. IBM J. Res. Dev. **1**(4), 309–317 (1957)
13. Pennebaker, J.W., Francis, M.E., Booth, R.J.: Linguistic Inquiry and Word Count: LIWC 2001, p. 71. Lawrence Erlbaum Associates, Mahway (2001)

14. Ravi, K., Ravi, V., Prasad, P.S.R.K.: Fuzzy formal concept analysis based opinion mining for CRM in financial services. Appl. Soft Comput. **60**, 786–807 (2017)
15. Řehůřek, R., Sojka, P.: Software framework for topic modelling with large corpora. In: Proceedings of the LREC 2010 Workshop on New Challenges for NLP Frameworks, Valletta, Malta, pp. 45–50. ELRA, May 2010. http://is.muni.cz/publication/884893/en
16. Sparck Jones, K.: A statistical interpretation of term specificity and its application in retrieval. J. Doc. **28**(1), 11–21 (1972)
17. Sundarkumar, G.G., Ravi, V.: A novel hybrid undersampling method for mining unbalanced datasets in banking and insurance. Eng. Appl. Artif. Intell. **37**, 368–377 (2015)
18. Wei, C.-P., Chiu, I.-T.: Turning telecommunications call details to churn prediction: a data mining approach. Expert Syst. Appl. **23**(2), 103–112 (2002)
19. Yee Liau, B., Pei Tan, P.: Gaining customer knowledge in low cost airlines through text mining. Ind. Manag. Data Syst. **114**(9), 1344–1359 (2014)

Unpack Local Model Interpretation for GBDT

Wenjing Fang[1(✉)], Jun Zhou[1], Xiaolong Li[1], and Kenny Q. Zhu[2]

[1] Ant Financial Services Group, Hangzhou, China
{bean.fwj,jun.zhoujun,xl.li}@antfin.com
[2] Shanghai Jiao Tong University, Shanghai, China
kzhu@cs.sjtu.edu.cn

Abstract. A gradient boosting decision tree (GBDT), which aggregates a collection of single weak learners (i.e. decision trees), is widely used for data mining tasks. Because GBDT inherits the good performance from its ensemble essence, much attention has been drawn to the optimization of this model. With its popularization, an increasing need for model interpretation arises. Besides the commonly used feature importance as a global interpretation, feature contribution is a local measure that reveals the relationship between a specific instance and the related output. This work focuses on the local interpretation and proposes an unified computation mechanism to get the instance-level feature contributions for GBDT in any version. Practicality of this mechanism is validated by the listed experiments as well as applications in real industry scenarios.

1 Introduction

Machine learning has great success in modeling data and making predictions automatically. In many real-world applications, we need an explanation rather than a black-box model. For example, when customers apply for a loan on credit, the loan officers will compute their credit scores based on their historical behaviors. In this case, it's far from enough to only show the customers the final scores, and the loan officers would better give some detailed reasons. While most efforts in data mining have been made on improving the accuracy and efficiency, which results in better models, little attention is paid to model interpretation for these models. Several common measures for the variable significance have been proposed. Gini importance is one of the commonly used importance measure for Random Forest, which is derived from the Gini index [2]. Gini is used to measure impurity between the parent node and two descendent nodes of samples after splitting. The final importance is accumulated from the Gini changes for each feature over all the trees in forest. This general feature importance(FI), also known as *global interpretation* , shows the important factors of the target, which unpacks the general information in the trained models. However, it doesn't take any feature values of an instance into consideration, which is insufficient sometimes. *Local interpretation,* on the other hand, places particular emphasis on a

ⓒ Springer International Publishing AG, part of Springer Nature 2018
J. Pei et al. (Eds.): DASFAA 2018, LNCS 10828, pp. 764–775, 2018.
https://doi.org/10.1007/978-3-319-91458-9_48

specific case and reveals the main causes of each record. This type of interpretation makes up for the shortages of the global one. One approach proposed to define the feature contributions(FC) [12] , which is accumulated from label distribution changes, as a measure of the feature impact on the output. The value of feature contribution reveals how much a feature contributes and the sign represents whether it's a positive impact or not.

GBDT [6] is an ensemble model built on top of a bunch of regression decision trees. It has some appealing characteristics. For example, GBDT can naturally handle nonlinearity and tolerate missing values. As a winning model in many data mining challenges [1,3,7], GBDT is a good option for regression, classification and ranking problems with well-known ability to generalize. Besides its wide range of applications, GBDT is also flexible in allowing users to define their own suitable loss functions. Furthermore, there are many implementations [4,8] and much work has been done to speed up the training process.

In most cases, GBDT outperforms linear models and random forest. Given the popularity and high quality of GBDT, it's important to uncover internals of the model. For GBDT, global feature importances calculation is widely used to do the feature selection. For example, Breiman proposed a method to estimate feature importance [6]. However, existing work has largely ignored the exploration of local interpretations, which will be the focus of this paper. Specifically, we will study feature contributions for GBDT. We starts from previous approaches of model interpretation for random forest [12] and update the definition of the feature contribution. The proposed mechanism is flexible enough to interpret all versions of GBDT. The original definition based on label distribution change is proved to be a special case of ours under a particular loss function.

The rest of the paper is organized as follows. Section 2 provides a brief review of related work on local interpretations. Section 3 gives out the formal definition of feature contribution as preliminary and presents the approach for calculating feature contributions for random forests. In Sect. 4, we describe the rationale behind as well as main actions in interpreting GBDT. Section 5 contains experiment settings and the process to examine the proposed methodology. At the end, Sect. 6 concludes our work.

2   Related Work

Local model interpretation provides convincing reasons to the model outputs. One type of interpretations prefer both the good performance of complex models and interpretability of simple models. The pipeline of this type will first make use of advanced models as a black-box and then extract useful information out of it with the help of a more interpretable model. For example, a novel approach in [5] formally treats the interpretation of additive tree models as extracting the optimal actionable plan. It models the optimization problem as an integer linear programming and utilizes existing toolkit as the solver. The constraints are based on both the output score and the objective function. Notice that, this kind of approaches need extra training process especially for the interpretation and bring new models or tasks to solve.

Some other researchers come up with model-independent local interpretations. They mainly make changes to feature value and test the chain effect to performance loss of predictions. The loss is then taken as the measure of local importance of feature [11]. This method only relies on the output evaluation and provides an unified way to check feature contribution for black-box models. By replacing the actual feature values with missing, zero or average values, the impact of a feature in predicting is then removed. The instance-level contributions of all the features can be calculated separately and compared with each other. Moreover, this method is also work for global feature importance.

As a derivative of decision tree, the random forest goes further on model interpretation than GBDT. The method in [10,12] computes the feature contributions so as to show informative results about the structure of model and provide valuable information for designing new compounds. This method makes full use of the information, not only the training data but also the model structure. It is natural to design the interpretations with the model structures to get a more reasonable result.

This work proposes an easy way to get the feature contributions on the instance-level. Generally, it can be applied to all versions of GBDT implementations with little preprocessing and modification to the prediction process.

3   Preliminary

Additive tree models are a powerful branch of machine learning but are often used as black boxes. Though they enjoy high accuracies, it's hard to explain their predictions from a feature based point of view. Different ensemble strategies bring out different models while sharing the tree structure as a basis. So the model interpretations for different addictive tree models share some key spirits and can spread out from one to another with appropriate adaptation. In this section, we first review a practical interpretation method for random forest (for the binary classification) and introduce the general definition of feature contribution to better illustrate the proposed model interpretation for GBDT.

3.1   Interpretation for Random Forest

Random forest is one of the most popular machine learning models due to its exordinary accuracy utilizing categorical or numerical features on regression and classification problems. A random forest is a bunch of decision trees that are generated respectively and vote together to get a final prediction. Every tree is trained on randomly sampled data and subsampling feature columns to introduce the diversity for better generalization, which is the key weakness of single decision tree models. Random forest is known as a typical bagging model and the bagging strategy works out by averaging the noises to get a lower variance model.

An instance starts a path from the root node all the way down to a leaf node according to its real feature value. All the instances in the training data will fall into several nodes and different nodes have quite different label distributions of the instances in them. Every step after passing a node, the probability of being the positive class changes with the label distributions. All the features along the path contribute to the final prediction of a single tree.

A practical way to evaluate feature contributions is explored [12]. The key idea is taking the distribution change values for the positive class as the feature contribution. Concretely, it takes four procedures to work:

1. Computing the percentage of positive class of every node in a tree;
2. Recording the percentage difference between every parent node and its children;
3. Accumulating the contributions for every feature on each tree;
4. Averaging the feature contribution among all the trees in the forest;

The method consists of an offline preparation embedded in training (steps 1–2) and an online computing with the prediction process (step 3–4). It is easy to record the local contribution (or local increment) and related split feature to every edge on a tree.

3.2   Gradient Boosting Decision Tree

GBDT is another type of ensemble model that consists of a collection of regression decision trees. However, the ensemble is based on gradient boosting which promotes the prediction gradually by reducing the residual. For every iteration, a new model is built up to fit the negative gradient of the loss function until it converges under an acceptable threshold. The final prediction is the summation of all stagewise model predictions. Gradient boosting is a general framework and different models are available to be embedded. GBDT introduces decision tree as the basic weak learner. When square error is chosen as the loss function, the residual between current prediction and target label is the negative gradient which is computational friendly.

From the above definition, we can see the differences between random forest and GBDT, some of which are the main obstacles that prevent us from adapting the model interpretation for random forest to GBDT:

1. Random forest aggregates trees by voting, while GBDT sums up the scores from all the trees. This means that the trees in GBDT are not equal and the trees have to be trained in sequential order. The interpretation should make proper adaptations to deal with this problem.
2. Decision tree in GBDT outputs a score instead of a majority class type for classification problems. Though we can get the label distribution changes as random forest interpretation, the output scores in GBDT should be wisely taken into consideration.

3.3 Problem Statement

Given a training dataset $D = \{x^{(i)}, y^{(i)}\}_{i=1}^{N}$, where N is the total number of training samples, $x = (x_1, x_2, ..., x_S)$ implies a S dimensional feature vector, $x^{(i)}$ is the feature vector for the i-th sample and $y^{(i)}$ is the related label. We can illustrate training process of GBDT as in Algorithm 1. r_{mi} is the residual for sample i in the m-th iteration.

Algorithm 1. Gradient Boosting Decision Tree

1: **function** TRAIN(D,M)
2:     Init $f_0(x) = 0$
3:     **for** $m = 1, 2, ..., M$ **do**
4:         Compute residual:
5:         $r_{mi} = y_i - f_{m-1}(x_i)$, $i = 1, 2, ..., N$
6:         Train a regression decision tree from residual:
7:         $T_m =$ BUILDTREE(D)
8:         Cumulated prediction sum:
9:         $f_m(x) = f_{m-1}(x) + T_m$
10:     **end for**
11:     Get finally boosting function:
12:     $f_M = \sum\limits_{m=1}^{M} T_m$
13:     **return** f_M
14: **end function**
15: **function** PREDICTINSTANCE(X_i, f_M)
16:     score $= \sum\limits_{m=1}^{M}$ TREEPREDICT(X_i, T_m)
17:     **return** score
18: **end function**

Besides the basics of model, the feature contribution(FC) , as the key concept for local interpretation, is clarified below. We introduce the notation of FC by denoting the model interpretation for random forest in Sect. 3.1 :

$$LI_f^c = \begin{cases} Y_{mean}^c - Y_{mean}^p & \text{if the split in the parent is performed over the feature } f; \\ 0, & \text{otherwise} \end{cases} \quad (1)$$

LI_f^n in Eq. 1 is the Local Increment(LI) of feature f for node n defined before. For binary classification, Y_{mean}^n represents the percentage of the instances belonging to the positive class in node n.

$$FC_{i,m}^f = \sum_{c \in path(i)} LI_f^c \quad (2)$$

$$FC_i^f = \frac{1}{M} \sum_{m=1}^{M} FC_{i,m}^f \quad (3)$$

On a single tree m, $FC_{i,m}^f$ in Eq. 2 cumulates the feature contribution of feature f for a specific instance i. Equation 3 later average all the feature contribution for feature f among all the trees.

4   Mechanism

Looking back at model interpretation for random forest, its central spirit is to establish the idea of feature contribution. By computing label distribution, a measure of the change is then obtained and associated with the split feature. In the case of GBDT, we can expand this computation with a slight modification. Because the targets of the latter trees are the residual, it should replace the instance label while computing label distribution. Nevertheless, the problem of this version is that the average of labels on a leaf node is not always equal to the score on it. So the valuable model information in these scores are not utilized and the method is not appropriate for different GBDT versions [4,6].

In fact, the loss function determines the optimal coefficient and Table 1 shows some common examples. LS and LAD stand for Least Square and Least Absolute Deviation respectively. $\tilde{y}_i$ is the residual updated after each iteration. $F_{m-1}(x_i)$ is the approximation on iteration $(m-1)$. g_i and h_i are the first and second order gradient statistics on the loss. Different from the numerical optimization essence to compute negative gradient (for LS and LAD), XGB [4] first approximates the loss function with its second order Taylor expansion and an analytic solution is then got. So it contains no negative gradient computation and the evaluation of leaf weights is far from the label average. Particularly, only if the LS loss function and traditional GBDT training process is used, the label averages meet the scores.

Table 1. Loss functions of GBDT

Settings	Loss function	Negative gradient	Leaf weight
LS	$\frac{1}{2}[y_i - f(x_i)]^2$	$y_i - f(x_i)$	$ave_{x_i \in R_{jm}} \tilde{y}_i$
LAD	$\mid y_i - f(x_i) \mid$	$sign[y_i - f(x_i)]$	$median_{x_i \in R_{jm}} \{y_i - F_{m-1}(x_i)\}$
XGB	$\sum_{i=1}^{n} [l((y_i, \hat{y}^{(t-1)})) + g_i f_t(x_i)$ $+ \frac{1}{2} h_i f_t^2(x_i))] + \Omega(f_t)$	/	$-\frac{\sum_{i \in I_j} g_i}{\sum_{i \in I_j} h_i + \lambda}$

Without loss of generality, the interpretation for GBDT needs to work on the leaf scores. Since the scores are only assigned to leaf nodes, we have to find a way to propagate them back all the way to the root. The left tree of Fig. 1 shows an example tree in a GBDT model, with split feature and split value marked on arcs. Observing the three nodes in the rounded rectangle, the instances in node 6 will get a score difference as: $S_{n11} - S_{n12} = 0.085 - 0.069 = 0.016$, where S_{nk} is the score on node k. Moreover, this difference is caused by splitting feature $feat5$ branching by a threshold of 1.5. We can allocate this difference to the two branches by assigning the average score of child nodes to their parent node. For instance, $S_{n6} = \frac{1}{2}(S_{n11} + S_{n12}) = \frac{1}{2} \times (0.085 + 0.069) = 0.0771$. Then, the local increment metrics could be calculated using the scores, $LI^{n11}_{feat5} = S_{n11} - S_{n6} = 0.085 - 0.0771 = 0.0079$. Similarly, the leaf scores as well as the local increment could be spread to the whole tree.

The interpretation process during predicting is the same as that of the random forest. On the right hand side of Fig. 1, all the node average scores and feature contributions on the tree are marked. Supposing an instance gets a final prediction on leaf node 14 of tree t, a cumulation through the path: $n0 \rightarrow n2 \rightarrow n5 \rightarrow n9 \rightarrow n14$ will be executed: $FC^t_{feat5} = LI^{n2}_{feat5} = -0.0201$, $FC^t_{feat2} = LI^{n5}_{feat2} = -0.0073$, $FC^t_{feat4} = LI^{n9}_{feat4} + LI^{n14}_{feat4} = -0.0015 + 0.0010 = 0.0025$.

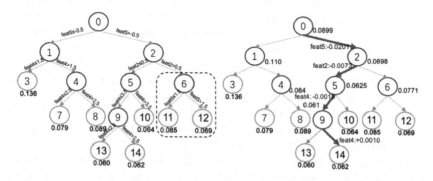

Fig. 1. Feature contribution example for GBDT

By the propagation strategy, the average score is assigned to the node 6 which assumes an instance falls into the left branch or the right with equal probability. So the expectation of intermediate nodes could be revised as in Eq. 4:

$$S_p = \frac{1}{2}(S_{c1} + S_{c2}) \rightarrow \frac{N_{c1} \times S_{c1} + N_{c2} \times S_{c2}}{N_{c1} + N_{c2}}, \qquad (4)$$

where the N_{c1} and N_{c2} is the number of the instances fall into child nodes node c1 and c2. These statistics need extra information from training process.

By viewing the computation in this brand new way, we get a flexible interpretation mechanism by only using the leaf node scores and instance distributions, regardless of the implement settings of GBDT. Under the setting of the LS loss function, we can see that not only the label distribution meets the prediction score on leaf node but also the label distribution of the intermediate node meets our back propagated score. That is to say, the label distribution method is a special case of our mechanism with this particular setting. Furthermore, this method also supports the multiple classification problems.

5   Experiment

In this section, we demonstrate the experiments on the proposed interpretation. In the first place, we show the mechanism is reliable and generally agrees with global feature importance. Then we compare our interpretations to those of random forest and find it accord with the global feature importance better. Finally, we study the interpretations of real cases in our scenario and get a satisfied analysis for them.

5.1 Experiment Setup

The GBDT version in our experiment is the Scalable Multiple Additive Regression Tree(SMART) [13], which is a distributed algorithm under the parameter server. Hundreds of billions of samples with thousands of features could be trained by the algorithm. Not only the storage usage but also the running time cost is optimized without the loss of the accuracy.

The training data is drawn from transactions under the scene of Fast Pay(FP) in Alipay[1]. A transaction is marked as a positive if it is reported as a fraud by the customer. To keep a balanced ratio between positive and negative cases, only 1% of normal transactions are retained by random sampling.

```
<Node id="0">
  <True/>
  <Node id="1">
    <SimplePredicate field="feat5" operator="lessOrEqual" value="-.5"/>
    <Node id="3" score="0.1366064239336732">
      <SimplePredicate field="feat4" operator="lessOrEqual" value="1.5"/>
    </Node>
    <Node id="4">
      <SimplePredicate field="feat4" operator="greaterThan" value="1.5"/>
      <Node id="7" score="0.07905535474853541">
        <SimplePredicate field="feat4" operator="lessOrEqual" value="2.5"/>
      </Node>
      <Node id="8" score="0.08930692524151827">
        <SimplePredicate field="feat4" operator="greaterThan" value="2.5"/>
      </Node>
    </Node>
  </Node>
  <Node id="2">
    <SimplePredicate field="feat5" operator="greaterThan" value="-.5"/>
    <Node id="5">
      <SimplePredicate field="feat2" operator="lessOrEqual" value=".5"/>
      <Node id="9">
        <SimplePredicate field="feat4" operator="lessOrEqual" value="3.5"/>
        <Node id="13" score="0.06043183878498103">
          <SimplePredicate field="feat4" operator="lessOrEqual" value="2.5"/>
        </Node>
        <Node id="14" score="0.06211634427742983">
          <SimplePredicate field="feat4" operator="greaterThan" value="2.5"/>
        </Node>
      </Node>
      <Node id="10" score="0.06457350478010203">
        <SimplePredicate field="feat4" operator="greaterThan" value="3.5"/>
      </Node>
    </Node>
    <Node id="6">
      <SimplePredicate field="feat2" operator="greaterThan" value=".5"/>
      <Node id="11" score="0.08556975499229456">
        <SimplePredicate field="feat5" operator="lessOrEqual" value="1.5"/>
      </Node>
      <Node id="12" score="0.0691611364527228">
        <SimplePredicate field="feat5" operator="greaterThan" value="1.5"/>
      </Node>
    </Node>
  </Node>
</Node>
```

Fig. 2. GBDT model in PMML format

---

[1] https://global.alipay.com/.

Figure 2 is a fraction of GBDT model in Predictive Model Markup Language (PMML) format[2] and the tree embedded in it can be translate as shown in Fig. 1. The element *Node* is an encapsulation for a tree node, which contains a predicative rule to choose itself or its siblings. The attribute *id* assigns a unique number to each node in a tree. The value of *score* in a *Node* is the predicted value for an instance falling into it. *SimplePredicate* is a simple Boolean expression indicating the split information. Our pre-trained model is stored as a PMML file. JPMML[3] is employed as the evaluator and we implement the proposed interpretation based on it.

5.2 Consistency Check

We implement the feature contribution as the previous description in [6]. In order to make the interpretation be independent of the training process of GBDT, the training algorithm is not changed in our experiment. In order to get the distribution of instances in Eq. 4, we use JPMML to predict the training instances and record instance distributions on every node. According to the tree structure in model and instance distributions, the pre-process is done by back propagating the local increments as shown in Sect. 4. With the local increments, the feature contributions of the new instances could be computed. After interpreting lots of instances, we can get a distribution of feature contributions among the instances. The median is a robust estimator for the expectation of the general feature contribution and should somehow keep accordance with the global feature importances metrics [12].

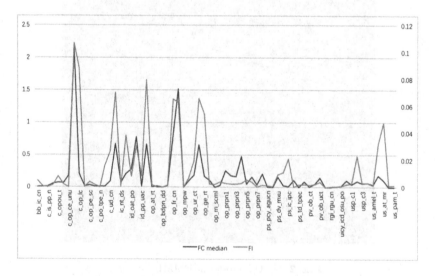

Fig. 3. Feature importance and feature contribution medians

---

[2] http://dmg.org/pmml/v4-3/GeneralStructure.html.
[3] https://github.com/jpmml/jpmml-evaluator.

Figure 3 plots the global Feature Importance(FI) for GBDT and Feature Contribution(FC) medians for every feature. As we can see, this two statistics have similar distributions and are in good agreement. It proves that the interpretation for GBDT is practical and reasonable.

5.3   Comparison to Random Forest

Following the experiment of last section, we get a ranking of the feature contribution median. This ranking is a measure of feature importance and reflects the quality of local interpretation. We implement the work for random forest in [12] and compare it with our ranking. For justice, we replace the GBDT Feature Importance with Information Value(IV) as the importance metric. IV is a concept from information theory and shows the predictive strength for the features [9].

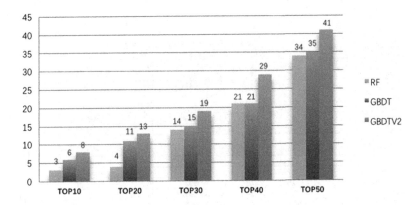

Fig. 4. Interpretation: GBDT v.s RF

In Fig. 4, we compute the intersection size on different variable coverage (i.e. Top 10–50 features of IV). *RF* implies the method explained in Sect. 3.1. *GBDT* is the simple average strategy with only the information in PMML file. *GBDTV2* is the revised version in Eq. 4. From the result, our interpretations capture the importance better and the revised version works best.

5.4   Case Study

Besides the general evaluation, we analysis the 300 specific instances in the test data. Figure 5 shows a case, we only list some representative fields and divide them into 4 parts. The variables are ranked by IV (general feature importance). Domain experts check the feature risk manually and draw the following conclusions:

- Part I: Variables in this section are with high IV, our interpretation is able to capture the features that are judged to be high risk(marked as blue fields). The feature with high IV but low risk (judging from the feature value) is assigned a lower score, so the interpretation is good for instance-level contributions.
- Part II: There are 2 variables(colored pink) with high IV and marked high risk is missed by the interpretation, which mainly due to its low occurrence in split features. The global importance of these two variables is also low and model interpretations are limit by the model quality.
- Part III: Variables with median or low IVs are not caught by mistake and is assigned a low feature contribution for that case.
- Part IV: Several variables are considered to be high risk for the particular instance, even the general IVs of them are low. Our interpretation finds them out, which shows the superiority of the local feature contribution over the global feature importance.

Variables	IV	IVrank	Feature Importance	Feature Value	Risk	
c_op_ud_lk	2.203	1	0.106791601	2	High	I
opcd_pn_cy	1.762	2	0.065620865	0	High	
id_pp_cy	1.525	3	0.030499262	0	High	
op_st_at_d	1.445	4	0.079317491	1	High	
op_fd_pob	1.315	5	0.063102101	-1	Low	
ic_dcn_dr	1.126	6	0.07008353	0.333333333	High	
op_fr_cn	1.065	7	0.065047636	1	High	
c_op_lc	0.92	8	0.088589539	0	High	
op_ur_ct	0.874	9	0.019405404	2	High	II
id_oat_po	0.842	10	0.008055243	1	High	
c_op_cr_y	0.811	11	0.002003042	2	High	
uicy_icdcy_ud	0.778	12	0.00064423	0.078567	Low	
c_opcd_cn	0.693	13	0.00193592	0	Median	III
uicy_icd_osu_po	0.68	14	0.000859095	0.117214	Low	
ps_pn_ct	0.651	15	0.00016962	15	Low	
op_prpn1	0.624	16	0.002749283	0	Low	
ic_ay_dcn	0.618	17	0.038145332	-1	Low	
us_bste_t	0.608	18	0.03190312	0	High	
usp_c1	0.343	29	0.002563547	4	High	
op_ge_rt	0.203	52	0.054343817	0	High	IV
us_at_mr	0.196	53	0.047708771	56.50257164	High	

Fig. 5. Case study for interpretation

Further more, if we conduct interpretations on a batch of fraud cases which are missed by the model, the local feature contributions will help analysts improve the model.

6   Conclusion

Employing models as a black-box is not enough. A measure for the impact of a feature on the prediction convinces analysts in an intuitive way. The local interpretation provides an explanation when necessary and contributes to the promotion of the models. We describe a method to unpack the interpretation for the advanced model GBDT. To the delight of analysts, the whole process is independent from the training details and technical optimizations. Only the tree structure and instance distribution are needed, which can be easily extracted by a post-processing after training. The label distribution based method of random forest is proved to be a special case of our method. We explore the distribution of local feature contributions and prove it to be in agreement with global feature importance. The method is applied to real case studies in different scenarios and serves as a good translator of our models.

References

1. Bennett, J., Lanning, S., et al.: The netflix prize. In: Proceedings of KDD Cup and Workshop, New York, NY, USA, vol. 2007, p. 35 (2007)
2. Breiman, L.: Random forests. Mach. Learn. **45**(1), 5–32 (2001)
3. Chapelle, O., Chang, Y.: Yahoo! learning to rank challenge overview. In: Proceedings of the Learning to Rank Challenge, pp. 1–24 (2011)
4. Chen, T., Guestrin, C.: XGBoost: a scalable tree boosting system. In: Proceedings of the 22nd ACM SIGKDD International Conference on Knowledge Discovery and Data Mining, pp. 785–794. ACM (2016)
5. Cui, Z., Chen, W., He, Y., Chen, Y.: Optimal action extraction for random forests and boosted trees. In: Proceedings of the 21th ACM SIGKDD International Conference on Knowledge Discovery and Data Mining, pp. 179–188. ACM (2015)
6. Friedman, J.H.: Greedy function approximation: a gradient boosting machine. Ann. Stat. **29**, 1189–1232 (2001)
7. He, X., Pan, J., Jin, O., Xu, T., Liu, B., Xu, T., Shi, Y., Atallah, A., Herbrich, R., Bowers, S., et al.: Practical lessons from predicting clicks on ads at Facebook. In: Proceedings of the Eighth International Workshop on Data Mining for Online Advertising, pp. 1–9. ACM (2014)
8. Ke, G., Meng, Q., Finley, T., Wang, T., Chen, W., Ma, W., Ye, Q., Liu, T.Y.: LightGBM: a highly efficient gradient boosting decision tree. In: Advances in Neural Information Processing Systems, pp. 3149–3157 (2017)
9. Kullback, S.: Information Theory and Statistics. Courier Corporation, Chelmsford (1997)
10. Kuz'min, V.E., Polishchuk, P.G., Artemenko, A.G., Andronati, S.A.: Interpretation of QSAR models based on random forest methods. Mol. Inform. **30**(6–7), 593–603 (2011)
11. Lei, J., G'Sell, M., Rinaldo, A., Tibshirani, R.J., Wasserman, L.: Distribution-free predictive inference for regression. J. Am. Stat. Assoc. (2017, accepted)
12. Palczewska, A., Palczewski, J., Robinson, R.M., Neagu, D.: Interpreting random forest models using a feature contribution method. In: 2013 IEEE 14th International Conference on Information Reuse and Integration (IRI), pp. 112–119. IEEE (2013)
13. Zhou, J., Cui, Q., Li, X., Zhao, P., Qu, S., Huang, J.: PSMART: parameter server based multiple additive regression trees system. In: Proceedings of the 26th International Conference on World Wide Web Companion, pp. 879–880. International World Wide Web Conferences Steering Committee (2017)

Cost-Sensitive Churn Prediction in Fund Management Services

James Brownlow[1,2], Charles Chu[1,2], Bin Fu[1], Guandong Xu[2(✉)],
Ben Culbert[1,2], and Qinxue Meng[1]

[1] Colonial First State, Sydney 2000, Australia
{James.Brownlow,Charles.Chu,Bin.Fu,Ben.Culbert,Qinxue.Meng}@cba.com.au
[2] Advanced Analytics Institute, UTS, Sydney 2007, Australia
Guandong.Xu@uts.edu.au

Abstract. Churn prediction is vital to companies as to identify potential churners and prevent losses in advance. Although it has been addressed as a classification task and a variety of models have been employed in practice, fund management services have presented several special challenges. One is that financial data is extremely imbalanced since only a tiny proportion of customers leave every year. Another is a unique cost-sensitive learning problem, i.e., costs of wrong predictions for churners should be related to their account balances, while costs of wrong predictions for non-churners should be the same. To address these issues, this paper proposes a new churn prediction model based on ensemble learning. In our model, multiple classifiers are built using sampled datasets to tackle the imbalanced data issue while exploiting data fully. Moreover, a novel sampling strategy is proposed to deal with the unique cost-sensitive issue. This model has been deployed in one of the leading fund management institutions in Australia, and its effectiveness has been fully validated in real applications.

Keywords: Customer retention · Churn prediction
Cost-sensitive classification · Imbalanced data

1 Introduction

Fund management services refer to the institutions that help customers achieve their wealth goals by providing them with a range of investment options, i.e., funds. Since a customer could have an investment of thousands or even millions of dollars, it is vital for them to retain their valuable customers. To this end, a practical approach is to predict which customers would quit, i.e., churners as soon as possible, then a retention campaign which targets these potential churners could be launched. Churn prediction can be viewed as a binary classification task, which is one of the fundamental concepts in data mining. Basically, a set of customers classified as *churner* or *non-churner* aka a training set is used to learn a predictive model, which is used to predict churn probabilities of customers

© Springer International Publishing AG, part of Springer Nature 2018
J. Pei et al. (Eds.): DASFAA 2018, LNCS 10828, pp. 776–788, 2018.
https://doi.org/10.1007/978-3-319-91458-9_49

whose classes are unknown. Nowadays, churn prediction is receiving increasing attention from both academia and industry. A multitude of methods such as boosting [1], random forest [2], and neural network [3] etc. have already been investigated and employed for churn prediction in various applications, including telecommunication [1,4], online community [5], and social game [6], and so forth.

Despite these achievements in other industries, the fund management industry has its own particular challenges, meaning that existing methods cannot be employed directly. One is that financial data is even more imbalanced compared with other industries. Sampling techniques like undersampling are commonly used to cope with this issue [7]. However, how to sample a set of informative and diverse subsets still needs further investigation. Another major challenge is that a unique cost-sensitive problem is presented. Costs in existing applications are either class-dependent or instance-dependent [8]. However, the cost of churn prediction in financial industry belongs to neither of them. On one hand, costs of wrong predictions for churners should be proportional to their account balance, so these costs are instance-dependent. One the other hand, wrong predictions for non-churners could be the same loss which should be less than the loss associated with any churner, so these costs are class-dependent. Thus the cost here actually is a hybrid of class-dependent cost and instance-dependent cost. To our knowledge, there are few approaches for dealing with this special type of cost at the moment.

To tackle these challenges, we propose a novel approach based on ensemble learning for churn prediction in this paper. Specifically, multiple balanced subsets are sampled from the original dataset, multiple classifiers are then learnt and combined using these subsets. Although similar paradigms have been used in [2,9], we introduce a new sampling strategy that consists of two separate sampling steps with different weighting mechanisms for two classes respectively. The advantages of our approach include: (1) this novel sampling strategy uses different weighting mechanisms for different classes, thus the special cost-sensitive issue can be handled properly; (2) sizes of the subsets are determined randomly, so they are varied instead of being fixed as in [9], this additional randomness could increase the diversity of classifiers and achieve better performance accordingly. Gradient boosting machine [10] is used in our approach to learn the classifiers to improve the performance further. To summarize, this paper makes the following three main contributions. (1) A new weighting mechanism and sampling strategy is proposed to deal with the imbalanced data and special cost-sensitive problem. (2) The concrete process of how this approach has been deployed in real production is introduced. (3) Extensive experiments with real-world data have been conducted to validate the effectiveness of our approach.

The rest of this paper is organised as follows. Section 2 previews related work. Section 3 gives the notations used throughout this paper as well as a formal formulation of the learning task. Section 4 introduces the specific implementation of our proposed model in the real scenario. Section 5 presents the experimental results, followed by conclusions in Sect. 6.

2   Related Work

2.1   Churn Prediction

Over the last decade, churn prediction has been applied in various fields, e.g., telecommunication, social networks, and mobile application [4,6,11]. In most cases, it is solved as a classification problem through learning a predictive model using a set of customers whose classes are known. A customer is usually represented as a vector of features, and the relationship between a customer's features and class could be captured by the model. Generally, there are two keys to learn a good model, one is how to define a set of discriminative features that could cover underlying factors, and the other is how to determine the form of model that is suitable for current data. Every particular application has its distinctive data from which features can be derived. For example, business data and operation data are exploited in the telecom industry [4], question and comment data are analysed in online question answering services [12], etc. Although each application has its unique features, existing classification methods can be used in these applications commonly. Popular methods such as boosting, random forest and logistic regression have already been employed [1,4,12], and a comprehensive review of methods used in the telecom industry is also given in [13].

2.2   Imbalanced Data and Cost-Sensitive Learning

Imbalanced data must be carefully handled otherwise the learning process would be skewed towards the majority class while the minority class is ignored. Two primary strategies can be employed to cope with imbalanced data, i.e., method transformation and data transformation [14]. The former adapts learning methods to enable them to handle imbalanced data directly. For instance, a skew-insensitive splitting criteria is adopted in decision tree [15]. By contrast, the latter aims to obtain balanced datasets, so existing methods can be used without adaptation. For example, oversampling and undersampling techniques obtain balanced data via varying the size of data of one particular class [7]. Since useful information could be missed in undersampling, ensemble learning based methods have become popular recently. These methods follow the same paradigm in which multiple subsets are sampled, but differ from each other in terms of weighting mechanisms in the sampling process [2,9,16]. Cost-sensitive learning is closely related to imbalanced data and has been used as a weighting mechanism to make data balanced [17]. As stated above, there could be a class-dependent cost or an instance-dependent cost. A classical strategy of dealing with class-dependent cost is to define a cost matrix and determine predictions using Bayes optimal rule [18]. In addition, weighting instances according to their relevant costs is another typical strategy of encoding costs into the learning process [19].

From aforementioned work, it can be observed that assigning appropriate weights to instances is a critical way of dealing with imbalanced data as well as the cost-sensitive learning issue. Inspired by the EasyEnsemble method [9], our approach also adopts the ensemble learning paradigm to obtain balanced subsets as well as take full advantage of available data. The key difference is that a novel weighting mechanism based on customers' balance is designed in our approach to handle the special cost-sensitive issue.

3   Problem Formulation

Let $P = \{(x_i, 1)\}_{1 \leq i \leq |P|}$ be a dataset of minority class in which x_i denotes the ith customer whose class is 1, i.e., churner. Similarly, Let $N = \{(x_i, 0)\}_{1 \leq i \leq |N|}$ be a dataset of majority class in which every customer x_i' class is 0, i.e., non-churner. The size of N should be considerably larger than the size of P, i.e., $|P| \ll |N|$. A customer x is represented as a feature vector $x = \langle x_1, x_2, \ldots, x_n \rangle$, and these features could be demographic information and behavioural patterns extracted from historical transactions.

The task of classification is to learn a predictive model f based on a training set $D = P \bigcup N$. Essentially, a model f is a function that establishes a mapping from instance space to class space, i.e.,

$$f(x) \rightarrow c, c \in \{0, 1\} \tag{1}$$

Given an instance x, c is the class predicted for it by f. The output f could also be a real value $y(0 \leq y \leq 1)$ which indicates the probability of $c = 1$.

In order to learn a good model, aforementioned imbalanced data and the cost-sensitive learning issue must be handled properly. An effective strategy to handle imbalanced data is undersampling. Specifically, a subset N' is sampled from N, and a model is then learnt based on training set $D' = N' \bigcup P$. Usually we choose $|N'| = |P|$, so D' is balanced. One issue of undersampling is that the majority of N is excluded, resulting in much useful information being unexploited. Hence, recent methods often follow the paradigm of integrating ensemble learning with sampling as shown in Algorithm 1.

Algorithm 1. Ensemble of multiple samplings

Data: Training set N and P, iteration number t
Result: Multiple classifiers $f = (f_1, f_2, \ldots f_t)$

1 **for** $k \leftarrow 1$ **to** t **do**
2   sample P_i from P according to weights of instances in P;
3   sample $N_i(|N_i| = |P_i|)$ from N according to weights of instances in N;
4   learn a model f_i using $D_i = N_i \bigcup P_i$
5 return $f = (f_1, f_2, \ldots f_t)$;

As shown in Algorithm 1, every classifier f_i is learnt using a balanced dataset D_i, and N is fully exploited through multiple samplings. Methods that follow this paradigm differ mainly on: (1) how to set the weights of instances in N and P, (2) the size of N_i and P_i, and (3) the method used to learn classifiers. For example, in the EasyEnsemble method, N_i is sampled evenly from N with every instance having the same weight, P_i is simply set as P so that $|N_i| = |P_i| = |P|$, and AdaBoost [9] is used to learn classifiers. Our approach also adopts this paradigm, and the remaining problem is how to design weighting mechanisms, determine sizes of subsets, and combine multiple classifiers to deal with the special cost-sensitive issue. The solution is introduced in following section.

4   Model Design and Implementation

In this section, our proposed approach is introduced. Particularly, the framework and steps of its implementation in practice are also presented.

4.1   Our Learning Approach

Two types of wrong predictions could possibly happen, i.e., predicting a churner as a non-churner and predicting a non-churner as a churner. The former is costly because failing to identify a churner could lead to loss of all his or her money. The more money he or she has, the greater the cost will be. Consequently, the cost of a wrong prediction for churners should be proportional to their account balance. However, the latter would not incur much loss and has nothing to do with customers' account balance. Hence it is reasonable to set the cost of wrong predictions for non-churners as a fixed value. With this assumption, the weight w_i assigned to every instance x_i in dataset N and P is set according to Eq. (2) in our approach.

$$w_i = \begin{cases} \frac{1}{|N|} & \text{if } x_i \in N \\ \frac{b_i}{\sum_{x \in P} b_x} & \text{if } x_i \in P \end{cases} \tag{2}$$

Here b_i is x_i's account balance. It can be seen from Eq. (2) that weights assigned to churners are proportional to their individual account balance, while weights assigned to non-churners are the same which is a class level value.

Next, instances should be sampled from N and P according to their weights to take costs into consideration when learning models. Instances with greater weights would appear more times in the new training set, thus the likelihood of making wrong predictions for them is reduced. Here a key point is how to determine the sizes of sampled subsets. Instead of setting $|N_i|$ and $|P_i|$ always as $|P|$, we use a straightforward method to introduce randomness in the sizes of subsets. Specifically, when sampling a subset P_i from P, a subset P' of size $|P|$ is sampled according to Eq. (2) firstly, then these instances which exist in P but not in P' will also be added into P' to form P_i. In this way, the size of P_i is a random value which ranges from $|P|$ to $2|P| - 1$. A subset N_i of size of $|P_i|$ is then sampled from N, so that $(P_i \bigcup N_i)$ is a balanced dataset. We can see that now the imbalanced data and the cost-sensitive issue are well addressed in this way.

Algorithm 2. Our proposed approach

Data: Training set N and P, iteration number t
Result: Multiple classifiers $f = (f_1, f_2, \ldots f_t)$

1 for $i \leftarrow 1$ **to** t **do**
2   sample $P'(|P'| = |P|)$ from P using weights according to Equation (2);
3 $P_i = P' \bigcup (P \setminus P')$;
4   sample $N_i(|N_i| = |P_i|)$ from N using weights according to Equation (2);
5 $f_i \leftarrow Xgboost(N_i \bigcup P_i)$
6 return $f = (f_1, f_2, \ldots f_t)$;

Furthermore, Xgboost [10], which is a popular implementation of the gradient boosting machine model, is employed in our approach to learn models. It is an additive model which consists of multiple submodels, and every submodel is obtained through minimizing the residuals produced by previous models. Now, all the key issues are solved, and the details of our approach are specified in Algorithm 2.

The output of Xgboost for binary classification is a real value in $[0, 1]$ which denotes the probability of being a churner. After obtaining multiple models, we simply use the average of their outputs as the final prediction for x as shown in Eq. (3).

$$f(x) = \frac{1}{t} \sum_{i=1}^{t} f_i(x) \tag{3}$$

Here f_i is the ith model learnt in the ith iteration.

It can be observed that our approach has several advantages: (1) weights based on account balance are introduced, so it is less likely to make wrong predictions for high value churners; (2) line 3 of Algorithm 2 indicates the size of every subset is randomly determined, so models learnt using these subsets would be more diverse and the performance could be improved via reducing variance accordingly; (3) the size of subset N_i is larger than $|P|$, so more information about the majority class could be exploited when learning models compared with other methods like EasyEnsemble.

4.2 Model Implementation

Our approach has been applied in a fund service company in Australia. In this section, how to prepare data and define features in practice is introduced.

Data Sources. Multiple sources of data regarding various entities exist in reality, and data from heterogeneous sources should be integrated to get a comprehensive understanding of customers. In our implementation, the primary types of data that have been exploited are: (1) customer demographic information, (2) customer behaviour, such as call log and online system login, (3) account status, (4) transaction records, (5) fund performance such as daily records of fund price, (6) insurance records, and (7) interaction with advisers such as records of adviser fees etc.

Feature Engineering. Six types of features as below are extracted.

(1) **Customer demographic features.** These features provide information regarding customers' profiles, such as *gender, age,* and *occupation* etc.
(2) **Customer behavioural features.** Customers' past behaviours or interactions with a company contain some useful clues for their future behaviours. Typical features of this type includes *call frequency, survey rating,* and so on.
(3) **Account level features.** Two types of account level features are extracted. The first one relates to an account's current status, such as *tenure* and *balance.* the other describes an account's changing trend in the past, i.e., *balance change,* and *option change.*
(4) **Fund performance.** Customers are usually sensitive to their investment returns. Therefore, we also extract features like *fund performance* to measure the growth rate of a customer's investment in the past year.
(5) **Adviser and dealer features.** Although we do not have much data about advisers and dealers, we can infer their features through customers associated with them. Features such as *number of customers,* and *number of churn customers* are constructed under the assumption that if many customers who belong to an adviser have left, other customers belonging to the same adviser are also likely to leave in the near future.
(6) **Employer features.** We also extract a set of features regarding employers. Features like *number of employees, number of churn employees* are extracted to measure the impact of an employer on its employees.

Around 120 features are defined totally. For every customer, his or her final feature vector is the combination of features of all above 6 types. The overall framework of model implementation is outlined in Fig. 1.

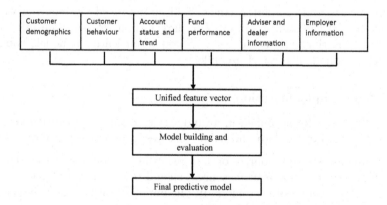

Fig. 1. Framework of model implementation

As shown in Fig. 1, for every customer, multiple sets of features are extracted from different perspectives. These features are then combined into a single feature vector. Therefore, a unified view which covers the influential factors as much as possible is obtained, increasing the probability of building a reliable model.

5   Experiments

In this section, extensive experiments using data from real applications are conducted to validate our approach's effectiveness.

5.1   Datasets

Datasets of four different funds are used in experiments. They are *retail super-annuation, corporate superannuation, pension,* and *investment.* To generate these datasets, data between Jan 2016 and Dec 2016 (observation window) are extracted to generate features that are introduced in previous section, and data between Jan 2017 and Jun 2017 (label window) are extracted to determine classes. A customer is classified as a churner if his or her account is closed in the specified label window, otherwise is classified as a non-churner. The purpose here is to use a customer's information in the past one year to predict his decision in the next six months.

After excluding outliers and customers whose accounts are opened within the observation window because they do not have sufficient historical data, Table 1 gives the summary of the four datasets in detail.

Table 1. Description of datasets

Dataset	$|D|$	$|N|$	$|P|$	*churn ratio*
Retail super	220000	210320	9680	4.4%
Corporate super	260000	243300	16640	6.4%
Pension	135000	128925	6075	4.5%
Investment	160000	152480	7502	4.7%

In Table 1, $|D|$, $|N|$, and $|P|$ is the size of the whole population, non-churners, and churners respectively, and *churn ratio* is the ratio of churners in the population, i.e., $|P|/|D|$. It can be observed that all of these datasets are extremely imbalanced.

5.2   Evaluation Metrics

In practice, churn prediction models are used to predict the churn probabilities or attrition scores of existing customers. These scores are then ordered descendingly, so a retention campaign could focus on the most likely churners, i.e., the top

K customers. In this case, a model can be evaluated in two manners. One is the number of true churners in top K customers, and the other is the sum of true churners' account balance in top K customers. Accordingly, two evaluation criteria are used in our experiments.

The first one is *recall*, and its definition is given in Eq. (4)

$$R@k = \frac{\sum_{x \in Top(k)} c_x}{|P|} \tag{4}$$

Here $Top(k)$ denotes the top K customers. c_x is customer x's label, and it could be 1 or 0, 1 indicates x is a churner and 0 indicates the opposite.

The second one is *balance recall*, which is defined in Eq. (5)

$$BR@k = \frac{\sum_{x \in Top(k)} c_x * b_x}{\sum_{x \in P} b_x} \tag{5}$$

Here $Top(k)$ and c_x have the same meaning as above, and b_x is customer x's account balance. For both of these two criteria, a greater value means a better model performance.

5.3   Baselines and Settings

We compare our proposed method with three classical methods of coping with imbalanced or cost-sensitive data. These methods are:

- Balanced random forest [2]. In its ith iteration of learning a decision tree, a subset P_i and $N_i(|P_i| = |N_i| = |P|)$ is evenly sampled from P and N.
- WeightGBM. It is Xgboost with class-dependent weights [10]. Weights of instances in P are set as $|N|/|P|$ in this method.
- EasyEnsemble [9]. In its ith iteration, only a subset N_i is sampled from N, and $P_i = P$.
- CostGBM, our proposed approach.

The purpose of comparing our approach with these baselines is to validate the effectiveness of the weighting mechanism designed in this paper, especially in terms of the criterion *balance recall*. To make the comparison fair and convincing, Xgboost is also used in EasyEnsemble instead of Adaboost. The size of Balance random forest, i.e., number of trees is set as 200. In all other 3 methods, the number of iterations is 10 and a Xgboost model with 200 trees is learnt in every iteration. When learning Xgboost model, 'binary:logistic' is chosen as the objective function, and the optimal learning rate is chosen from 0.05–0.3 through multiple trials. All these methods are implemented in R environment, and the R package *Xgboost* is used to learn Xgboost models.

5.4   Results and Analysis

Datasets are split into training set (80%) and test set (20%). Models are then built using the training sets and evaluated using the test sets. All these methods

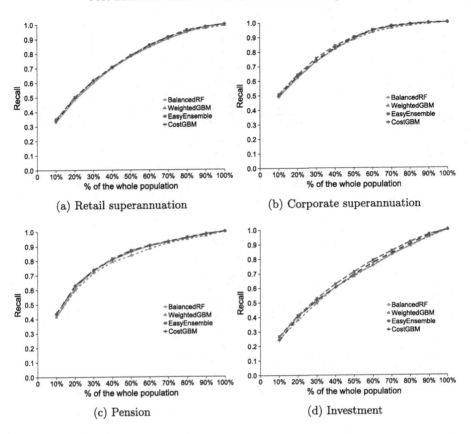

Fig. 2. Model performance in terms of recall

generate numeric predictions as churn probabilities, and the population as in test sets are ranked in terms of their predictions in a descending order.

To begin with, these methods are evaluated and compared in terms of recall, and the results on the four datasets are depicted in Fig. 2. For any point in Fig. 2, its x value is the top percentage of the whole population, and its y value is the recall. We can see that while EasyEnsemble performs slightly better on these datasets, our proposed method also shows competitive performance in terms of recall, even it gives more focus on high value customers.

When it comes to balance recall, our proposed method outperforms the other three methods significantly on all datasets as shown in Fig. 3. Take the results on corporate superannuation as example, when we look at the top 10% percent of the population, the balance recall of Balanced random forest, WeightedGBM, EasyEnsemble, and CostGBM is around 0.1, 0.1, 0.15, and 0.35 respectively. We can see that the total balance of true churners identified by our method is around 2 times greater than those identified by other methods. Given the volume of corporate superannuation, it means the improvement gained by our method could be millions of dollars.

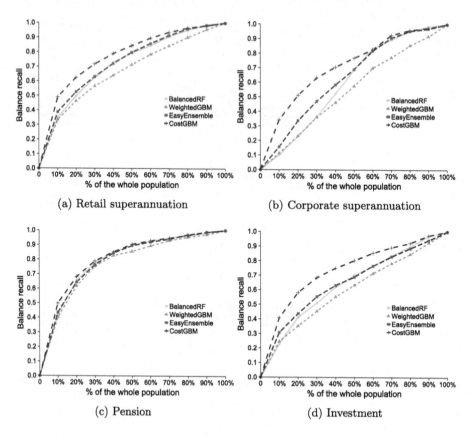

Fig. 3. Model performance in terms of balance recall

6   Conclusions

This paper introduces a novel method for churn prediction in fund management services and its implementation in a fund management company in Australia. A sampling framework based on ensemble learning and a new weighting mechanism based on account balance are proposed to deal with imbalanced and cost-sensitive issues with financial data. The practical steps of model implementation are also introduced, especially how various data from heterogeneous sources are exploited and integrated to gain a unified view of customers. Evaluation using real word data validates our model's superiority in capturing high value churners compared with traditional methods. Moreover, our method has been applied in real applications and assists the marketing team to narrow down their campaign target. In future work, strategies of incorporating account balance based cost into other advanced models will be investigated, and more features will also be extracted to enhance learning performance.

References

1. Lu, N., Lin, H., Lu, J., Zhang, G.: A customer churn prediction model in telecom industry using boosting. IEEE Trans. Ind. Inform. **10**(2), 1659–1665 (2014)
2. Chen, C., Liaw, A., Breiman, L.: Using random forest to learn imbalanced data. Technical report, University of California, Berkeley (2004)
3. Ismail, M.R., Awang, M.K., Rahman, M.N.A., Makhtar, M.: A multi-layer perceptron approach for customer churn prediction. Int. J. Multimed. Ubiquitous Eng. **10**(7), 213–222 (2015)
4. Huang, Y., Zhu, F., Yuan, M., Deng, K., Li, Y., Ni, B., Dai, W., Yang, Q., Zeng, J.: Telco churn prediction with big data. In: Proceedings of the 2015 ACM International Conference on Management of Data, pp. 607–618 (2015)
5. Rowe, M.: Mining user lifecycles from online community platforms and their application to churn prediction. In: Proceedings of the 13th IEEE International Conference on Data Mining, pp. 637–646 (2013)
6. Runge, J., Gao, P., Garcin, F., Faltings, B.: Churn prediction for high-value players in casual social games. In: Proceedings of the 2014 IEEE Conference on Computational Intelligence and Games, pp. 1–8 (2014)
7. He, H., Garcia, E.A.: Learning from imbalanced data. IEEE Trans. Knowl. Data Eng. **21**(9), 1263–1284 (2009)
8. Zhang, Y., Zhou, Z.H.: Cost-sensitive face recognition. IEEE Trans. Pattern Anal. Mach. Intell. **32**(10), 1758–1769 (2010)
9. Liu, X.Y., Wu, J., Zhou, Z.H.: Exploratory undersampling for class-imbalance learning. IEEE Trans. Syst. Man. Cybern. Part B (Cybern.) **39**(2), 539–550 (2009)
10. Chen, T., Guestrin, C.: Xgboost: a scalable tree boosting system. In: Proceedings of the 22nd ACM SIGKDD International Conference on Knowledge Discovery and Data Mining, pp. 785–794. ACM (2016)
11. Rothenbuehler, P., Runge, J., Garcin, F., Faltings, B.: Hidden Markov models for churn prediction. In: Proceedings of the SAI Intelligent Systems Conference, pp. 723–730 (2015)
12. Dror, G., Pelleg, D., Rokhlenko, O., Szpektor, I.: Churn prediction in new users of Yahoo! answers. In: Proceedings of the 21st International Conference Companion on World Wide Web, pp. 829–834 (2012)
13. Mahajan, V., Misra, R., Mahajan, R.: Review of data mining techniques for churn prediction in telecom. J. Inf. Organ. Sci. **39**(2), 183–197 (2015)
14. Galar, M., Fernandez, A., Barrenechea, E., Bustince, H., Herrera, F.: A review on ensembles for the class imbalance problem: bagging-, boosting-, and hybrid-based approaches. IEEE Trans. Syst. Man. Cybern. Part C **42**(4), 463–484 (2012)
15. Cieslak, D.A., Chawla, N.V.: Learning decision trees for unbalanced data. In: Daelemans, W., Goethals, B., Morik, K. (eds.) ECML PKDD 2008. LNCS (LNAI), vol. 5211, pp. 241–256. Springer, Heidelberg (2008). https://doi.org/10.1007/978-3-540-87479-9_34
16. Galar, M., Fernández, A., Barrenechea, E., Herrera, F.: Eusboost: enhancing ensembles for highly imbalanced data-sets by evolutionary undersampling. Pattern Recognit. **46**(12), 3460–3471 (2013)
17. Sun, Y., Kamel, M.S., Wong, A.K., Wang, Y.: Cost-sensitive boosting for classification of imbalanced data. Pattern Recognit. **40**(12), 3358–3378 (2007)

18. Domingos, P.: Metacost: a general method for making classifiers cost-sensitive. In: Proceedings of the Fifth ACM SIGKDD International Conference on Knowledge Discovery and Data Mining, pp. 155–164. ACM (1999)
19. Zadrozny, B., Langford, J., Abe, N.: Cost-sensitive learning by cost-proportionate example weighting. In: Proceedings of the Third IEEE International Conference on Data Mining, pp. 435–442. IEEE (2003)

Demonstration Track

A Movie Search System with Natural Language Queries

Xin Wang[1,2(✉)], Huayi Zhan[2], Lan Yang[2], Zonghai Li[2], Jiying Zhong[2],
Liang Zhao[2], Rui Sun[2], and Bin Tan[2]

[1] Southwest Jiaotong University, Chengdu, China
`xinwang@swjtu.cn`
[2] ChangHong Inc., Mianyang, China
{`huayi.zhan,lan.yang,zonghai.li,jiying.zhong,liang6.zhao,`
`rui1.sun,bin.tan`}`@changhong.com`

Abstract. In this demo, we present MSeeker, a user-friendly movie search system with following characteristics: it (1) transforms natural language queries (nlq) into graph pattern queries Q with a special node u_o as "query focus"; (2) identifies diversified top-k matches of u_o by early termination algorithm; and (3) provides graphical interface to help users interact with the system.

1 Introduction

Recently, knowledge graphs have attracted a lot of attentions in academia and industry, since they organize rich information with structured data, and hence are able to efficiently provide answers to users' queries. Figure 1(a) depicts a sample knowledge graph G. Each node in G either denotes a person, labeled by *name*; or a movie (m), with attributes *title*, *genres*, *rating* and *year*. Each directed edge labeled by "P", "D" or "P&D" indicates the person *played in*, *directed* or *played in and directed* the movie. On knowledge graphs, queries are typically evaluated with graph pattern matching, *i.e.*, given a pattern query Q and knowledge graph G, it is to find all the matches of Q in G.

Key issues for querying knowledge graphs are query understanding and evaluation. (1) Users' queries are often expressed with natural languages, which can not be evaluated directly on knowledge graphs, and need to be properly transformed into pattern queries Q. (2) Knowledge graphs are often very big, and query semantic is typically defined in terms of subgraph isomorphism, which is an NP-complete problem [2], these together bring following challenges: (a) query evaluation is cost prohibitive, (b) it is a daughting task to understand query results, as there may exist excessive matches of Q in G, and (c) users are often interested in top-k matches of the "query focus" u_o of Q, that are not only relevant to u_o, but are also as diverse as possible, simultaneously.

To tackle the issues, we demonstrate MSeeker, a prototype system for movie search on knowledge graphs with Chinese. MSeeker has the following two main features.

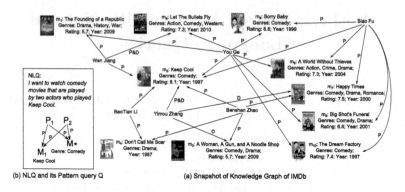

(b) NLQ and its Pattern query Q   (a) Snapshot of Knowledge Graph of IMDb

Fig. 1. Knowledge graph and queries

Querying understanding. MSeeker takes natural language queries (Chinese) nlq as input, and transforms nlq into pattern queries with output node u_o as "query focus".

Diversified top-k matching. MSeeker proposes metrics to measure relevance and distance of matches, identifies diversified top-k matches with *early termination* algorithm.

The prototype of MSeeker was deployed and tested by one of our industrial collaborators, and shows its performance in result diversification and high efficiency.

Demo Overview. We demonstrate the functionality of MSeeker in two parts. (1) We introduce how natural language queries are transformed into pattern queries with *query focus*. (2) We illustrate how diversified top-k matches are efficiently identified.

Below, we first present the foundation (Sect. 2) and the functional components (Sect. 3) of MSeeker. We then propose a detailed demonstration plan (Sect. 4).

2   Preliminary

We start with a review of natural language query understanding, and diversified top-k graph pattern matching [3], which are the foundations of MSeeker.

Query Understanding. To understand a natural language query nlq, it is necessary to (1) identify named entities and their relationship from nlq, (2) recognize the *query focus* of nlq, and (3) generate a pattern query with entities as nodes, their relationship as edges, and a designated node, referred to as the "output node", as *query focus*.

Named entity recognition (NER). NER is a typical application of sequence label-ing, where the sequence is a sentence and the labels are the classes that a word can take on. Sequence labeling problem is usually solved by Conditional Ran-dom Fields (CRF) [4]. Thus, MSeeker applies CRF++ [1] to identify entities in a query sentence.

Entity relationship identification. After named entities are recognized, each pair of entities needs to be examined to decide whether they have task-specific rela-tions using classification model. To this end, MSeeker applies a strategy for relationship identification. It first extracted several typical rules by mining query logs. For example, *"played in"*, *"directed"* are typical relationships between entity pairs in query sentences. It then adopts these rules to determine the relationship of each entity pair.

Pattern query construction. Given a natural language query nlq, MSeeker con-structs a pattern query Q based on entities and their relationships in nlq. MSeeker also extends Q by specifying a node u_o, referred to as "output node", to indicate *query focus*.

Example 1. A natural language query and its corresponding pattern query Q is shown in Fig. 1(b). Observe that each node in Q represents an entity, and each edge is marked with "P" indicating the *played in* relationship between two entities. In particular, the *query focus* M is marked with "*" as "output node". □

Diversified top-k Matching. Considering that (1) it is expensive to conduct graph pattern matching with subgraph isomorphism on large graphs; (2) there may exist excessive matches of Q in a large graph G, which makes understanding very difficult; (3) users are often only interested in top-k matches of the "query focus" u_o; and (4) result diversification has been proven effective in improving users' satisfaction, MSeeker identifies diversified top-k matches of u_o with *early termination* algorithm.

Matching semantic. Given a pattern query Q with output node u_o and a knowl-edge graph G, the matches of u_o in G is defined to be $M_u(Q, G, u_o) = \{v|h(u_o) = v, v \in V_s, G_s = (V_s, E_s) \in M(Q, G)\}$, *i.e.,* all the matches of the output node u_o, where $h(\cdot)$ is the bijective function that maps each node of Q to each node of the subgraph G_s in G, and $M(Q, G)$ is the match set of Q in G.

Result diversification. To measure the diversification of a match set $\mathcal{S} = \{v_1, v_2, \cdots, v_k\}$, a function $F(\cdot)$ is defined as $F(\mathcal{S}) = (1 - \lambda) \sum_{v_i \in \mathcal{S}} w(v_i) + \frac{2 \cdot \lambda}{k-1} \sum_{v_i \in \mathcal{S}, v_j \in \mathcal{S}, i<j} d(v_i, v_j)$, where $w(\cdot)$ and $d(\cdot)$ are the relevance and distance functions, respectively, and $\lambda \in [0, 1]$ is a parameter set by users. The diversity metric is scaled down with $\frac{2 \cdot \lambda}{k-1}$, since there are $\frac{k(k-1)}{2}$ numbers for the difference sum, while only k numbers for the relevance sum.

One may define $w(\cdot)$ and $d(\cdot)$ by using a variety of functions. While in movie search application, given a match v_i of u_o, its relevance $w(v_i)$ can be simply defined as its *rating*; and for a pair of matches (v_i, v_j), their difference $d(v_i, v_j)$

can be defined to be the total distance on attributes that are used to measure their difference.

Example 2. Consider knowledge graph G and pattern query Q in Fig. 1(a) and (b). With subgraph isomorphism, the match set S is $\{m_5, m_6, m_7, m_8, m_{10}\}$. If only attribute *year* is used to measure the distance, then $d(m_5, m_6) = \frac{2}{13}$, since the distance 2 is normalized by the largest distance 13 among the match set. One may further verify that (a) when $\lambda < \frac{13}{153}$, a top-2 set is $\{m_7, m_{10}\}$; (b) when $\frac{13}{153} < \lambda < \frac{13}{73}$, a top-2 set is $\{m_7, m_8\}$; and (c) when $\lambda > \frac{13}{73}$, $\{m_8, m_{10}\}$ makes the best result.

3   The System Overview

The architecture of the system, shown in Fig. 2, consists of the following components.

(1) A *Query Interpreter* (QI) for understanding natural language queries nlq, and transforming them into pattern queries. (2) A *Query Engine* (QE) that evaluates pattern queries Q and identifies diversified top-k matches of the "output node" u_o.

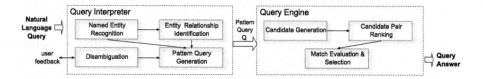

Fig. 2. Architecture of MSeeker

Query Interpreter. QI takes natural language queries nlq as input, recognizes entities, their relationships, and the query focus from nlq, and constructs pattern queries Q. As entities and their relationship may not be recognized correctly, which may lead to incorrect pattern queries, QI hence allows users to adjust the pattern queries after generation.

Query Engine. QE performs query evaluation with *early termination* algorithm.

Query evaluation. Upon receiving k and Q, QE (1) identifies a set of candidate matches v_o of u_o; (2) ranks diversification value $F(\cdot)$ for each pair of candidates; and (3) iteratively verifies whether a pair of candidate are valid matches starting from candidate pairs with highest $F(\cdot)$. Specifically, given a candidate match v_o, QE starts depth first search from u_o and v_o, simultaneously, following the topological structure of Q. After verification, if the pair of candidate matches are both true matches, they are included in the match set $\mathcal{S}$. When k is odd, after $\lfloor k/2 \rfloor$ rounds, $|\mathcal{S}| = k - 1$, QE only picks a "true" match that is not in $\mathcal{S}$ and can maximize $F(\cdot)$. Once the set $\mathcal{S}$ of k matches are identified, QE terminates search immediately, and returns $\mathcal{S}$ as final result.

Following the strategy, the diversification value $F(\cdot)$ of top-k matches identified is no less than $\frac{1}{2} F(\mathcal{S}_{OPT})$, where $\mathcal{S}_{OPT}$ is the optimal solution of the given instance.

4 Demonstration Overview

The demonstration shows: (1) how QI understands a natural language query nlq, transforms nlq into a pattern query Q with output node u_o; and (2) how QE identifies diversified top-k matches of u_o with *early termination* algorithm. The back-end of the system is implemented in Java and deployed on a machine with 2.9 GHz CPU, 8 GB Memory.

Performance of QI. We aim to show (a) how a pattern query is transformed from a natural language query, and how it is modified to eliminate ambiguation. **Performance of QE.** As shown in Fig. 3, we will show effectiveness of result diversification. We will also show the efficiency of our *early termination* algorithm.

搜索结果

Fig. 3. Query results of "I want to watch action movies that are played by Jackie Chan"

5 Summary

This demonstration aims to show the key idea and performance of the system MSeeker. From our industry collaborator's feedback, MSeeker can understand users' query intention, and efficiently find top-k diversified movies from large knowledge graphs, hence we contend that MSeeker can serve as a promising searching tool on knowledge graphs.

Acknowledgement. Xin Wang is supported in part by the NSFC 61402383 and 71490722, Sichuan Provincial Science and Technology Project 2014JY0207, and Fundamental Research Funds for the Central Universities, China.

References

1. Crf++. https://taku910.github.io/crfpp/
2. Cordella, L.P., Foggia, P., Sansone, C., Vento, M.: A (sub) graph isomorphism algorithm for matching large graphs. TPAMI **26**(10), 1367–1372 (2004)
3. Fan, W., Wang, X., Wu, Y.: Diversified top-k graph pattern matching. PVLDB **6**(13), 1510–1521 (2013)
4. Lafferty, J.D., McCallum, A., Pereira, F.C.N.: Conditional random fields: probabilistic models for segmenting and labeling sequence data. In: ICML, USA, pp. 282–289 (2001)

EventSys: Tracking Event Evolution on Microblogging Platforms

Lin Mu[1], Peiquan Jin[1(✉)], Lizhou Zheng[1], and En-Hong Chen[1,2]

[1] University of Science and Technology of China, Hefei, China
jpq@ustc.edu.cn
[2] Anhui Province Key Laboratory of Big Data Analysis and Application,
Hefei, China

Abstract. In this paper, we demonstrate a prototype system named EventSys, which provides efficient monitoring services for detecting and tracking event evolution on microblogging platforms. The major features of EventSys are: (1) It describes the lifecycle of an event by a staged model, and provides effective algorithms for detecting the stages of an event. (2) It offers emotional analysis over the stages of an event, through which people are able to know the public emotional tendency over a specific event at different time. (3) It provides a novel event-type-driven method to extract event tuples, which forms the foundation for event evolution analysis. After a brief introduction to the architecture and key technologies of EventSys, we present a case study to demonstrate the working process of EventSys.

Keywords: Event evolution · Emotional evolution · Tracking
Microblog · Detection

1 Introduction

Microblog platforms have been one of the major sources for new events detection and spreading. Motivated by the massive fresh information generated by microblog users, many works on event detection and analysis on microblogs have been conducted in recent years [1–4]. However, previous studies mainly focused on extracting structural tuples of events, e.g., extracting the 5W1H (who, where, when, what, whom, how) information [2]. In addition to event tuple extraction, some studies paid attention to the evolution analysis of events [5], but they cannot grasp the development process of events. On the other hand, an event usually has a developing process in the real world, i.e., from birth to death, which is similar to the lifecycle of people. The lifecycle information of an event is very useful in information mining and decision making. For example, company managers can make specific decisions according to the developing stage of the events related to products.

In this paper, we propose to extract the lifecycle of events from microblogs. Basically, the lifecycle of an event can be defined as a five-stage process including a budding stage, a developing stage, a peak stage, a recession stage, and a pacification stage. Although there are some previous studies focusing on event evolution [6], to the best of our knowledge, they are not able to extract the lifecycle of events.

© Springer International Publishing AG, part of Springer Nature 2018
J. Pei et al. (Eds.): DASFAA 2018, LNCS 10828, pp. 797–801, 2018.
https://doi.org/10.1007/978-3-319-91458-9_51

Particularly, we propose a prototype system called EventSys for detecting and analyzing the lifecycle of events from microblogs. The major features of EventSys are as follows:

(1) *Microblog Event Tuple Extraction and Semantic Element Extraction.* Given an event keyword, one problem is how to effectively extract the keyword-related events from the microblog set. We propose to incorporate event type into the event tuple extraction. Inspired by the studies in the news-report area that describe an event based on the news features [1], i.e., when, where, who, whom, what, and how, we consider to detect the news features of events from microblogs.

(2) *Microblog Event Evolution Stage Detection.* In order to grasp the lifecycle of events, we describe the lifecycle of events based on a five-stage model that consists of five stages: budding, development, peek, recession, and pacification.

(3) *Emotional Evolution Analysis.* The public emotional tendency to an event varies with time. Based on the extracted stages of an event, we develop a visual interface to monitor the public emotional evolution for each stage of specific events.

2 Architecture and Key Technologies of EventSys

Figure 1 shows the architecture of EventSys. The modules of EventSys include event tuple extraction, event tuple linking, event lifecycle detection, and emotional evolution analysis.

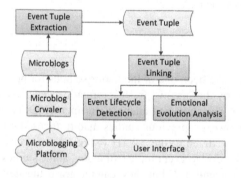

Fig. 1. Architecture of EventSys

Event Tuple Extraction. The extraction of event tuples is based on event type, which is defined as follows.

Definition 1. *Event Type. Given a collection of microblog post T which is obtained by one event query word, the event type is defined as a* quadruple $< p_l, p_n, p_o, p_t >$, *in which p_l, p_n, p_o, p_t represent the importance of location, person name, organization and time entity in the collection respectively, and $p_l + p_n + p_o + p_t = 1$.*

Given a microblog post collection, we represent it as a feature vector x and then employ the Multinomial Logistic Regression method to train the model [1]. The result $p_i = p(y = i|x^{(i)}, w)$ where $i = l, p, o$ and t for different named entity categories is used as the probabilistic distribution. Then, we use the quadruple $< p_l, p_n, p_o, p_t>$ to represent the event type, based on which we perform event-type-based clustering for microblogs by calculating the similarity among microblogs. We use the named entity probability distribution to adjust the similarity of the named entity of the microblog text to enhance the extraction effect, and finally get several events microblogging clusters. Each cluster of microblogging describes the same event. Next, for each cluster, we extract the 5W1H information [2], and finally get event tuples.

Event Tuple Linking. After extracting event tuples, we need to link the event tuples that describes the same event. This is mainly because an event will evolve with time. Given the microblogging data set at time t_i, we first get the set of event tuples, represented by eventTupleSet$_i$. For each event tuple in eventTupleSet$_i$, we calculate the similarity of the event tuple to previous events, find the most similar event, and link the event tuple to that event. If there is no similar event, we create a new event and add attach the event tuple to the newly created event.

Event Lifecycle Detection. Figure 2 shows the representation framework of an event. Each event has a unique ID and a set of event attributes. It also has a unique lifecycle that is five-bit structure indicating the current evolution process of the event. An event has a list of event tuples that are linked by the event tuple linking algorithm. All event tuples are arranged along the timeline and each tuple has an indicator describing what stage it belongs to.

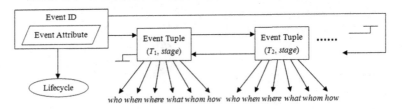

Fig. 2. Event representation

The key issue of event evolution analysis is to determine the right stage of an event. Sometimes we need to predict the stage of an event in future. In our system, we use the popularity of events to detect the event lifecycle. We define the popularity of an event in terms of the following features: (1) The forwarding number and commenting number as well as the total number of related microblog posts are used to measure the popularity of an event. (2) If users' emotional tendency towards an event changes dramatically, it implies that the evolutional stage of the event may change. (3) People cannot always focus on one specific event. When a new and interesting event happens, it will attract user attention and change the evolutional stage of current events.

(4) When the locations that are embedded in the event tuples change, it usually implies the change of the evolutional stage of the event.

2.1 Emotional Evolution Analysis

The public emotional tendency over a certain event will change with time. Thus, it is helpful to track the emotional evolution of events in decision making. In EventSys, we take three steps to extract the emotional evolution information of an event: (1) First, We extract the microblogging event tuples from each microblogging slices and extract the emotional tendencies of the event tuple based on a given sentiment dictionary. (2) Second, we map the event tuple to a developing stage of the event, as shown in the event representation framework in Fig. 2. (3) Finally, we compute the overall emotional polarity for each stage of an event based on the emotional tendencies of the event tuples linked within the stage. We use a weighted sum to aggregate the emotional tendencies of the event tuples in a stage, in which we put high weights for recent event tuples.

3 Demonstration

Figure 3 shows a screenshot of *EventSys*. Users are first required to select the time interval and event keywords as well as other parameters. The system will extract all events related to the selected event keywords. For example, Fig. 3 shows the output

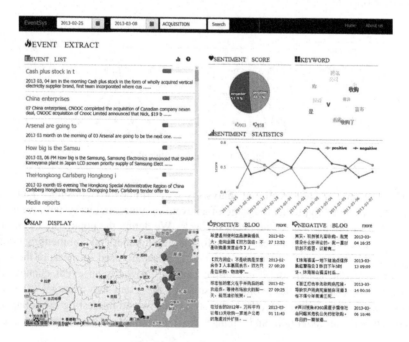

Fig. 3. Screenshot of EventSys

after inputting the event keyword *"acquisition"*, Zone A shows a list of all the events associated with the keyword. The event is sorted by the number of related microblogs. In zone B, we can see that the acquisition events occurred mostly in the eastern coastal areas of China. The right part of Fig. 3 shows the emotional evolution of the event, where different kinds of emotional information are presented, including the static statistical emotion, the dynamic emotional tendency, and the supported microblog posts.

Acknowledgements. This work is supported by the National Science Foundation of China (61672479 and 71273010).

References

1. Jin, P., Mu, L., et al.: News feature extraction for events on social network platforms. In: WWW, pp. 69–78 (2017)
2. Zheng, L., Jin, P., Zhao, J., Yue, L.: A fine-grained approach for extracting events on microblogs. In: Decker, H., Lhotská, L., Link, S., Spies, M., Wagner, Roland R. (eds.) DEXA 2014. LNCS, vol. 8644, pp. 275–283. Springer, Cham (2014). https://doi.org/10.1007/978-3-319-10073-9_22
3. Sakaki, T., Okazaki, M., et al.: Earthquake shakes twitter users: real-time event detection by social sensors. In: WWW, pp. 851–860 (2010)
4. Zhao, J., Wang, X., et al.: Feature selection for event discovery in social media: a comparative study. Comput. Hum. Behav. **51**, 903–909 (2015)
5. Cai, H., Huang, Z., et al.: Indexing evolving events from tweet streams. In: ICDE, pp. 1538–1539 (2016)
6. Huang, J., Peng, M., et al.: A probabilistic method for emerging topic tracking in microblog stream. World Wide Web **20**(2), 325–350 (2017)

AdaptMX: Flexible Join-Matrix Streaming System for Distributed Theta-Joins

Xiaotong Wang[1], Cheng Jiang[1], Junhua Fang[3],
Xiangfeng Wang[1], and Rong Zhang[1,2(✉)]

[1] School of Computer Science and Software Engineering,
East China Normal University, Shanghai, China
{xtwang,cj}@stu.ecnu.edu.cn, {xfwang,rzhang}@sei.ecnu.edu.cn
[2] School of Data Science and Engineering, East China Normal University,
Shanghai, China
[3] Soochow University, Suzhou, China
jhfang@suda.edu.cn

Abstract. Stream join is a fundamental and important processing in many real-world applications. Due to the complexity of join operation and the inherent characteristic of streaming data (e.g., skewed distribution and dynamics), though massive research has been conducted, adaptivity and load-balancing are still urgent problems. In this paper, an enhanced adaptive join-matrix system AdaptMX for stream theta-join is presented, which combines the key-based and tuple-based join approaches well: (i) at outer level, it modifies the well-known join-matrix model to allocate resource on demand, improving the adaptivity of tuple-based parititoning scheme; (ii) at inner level, it adopts a key-based routing policy among grouped processing tasks to maintain the join semantics and cost-effective load balancing strategies to remove the stragglers. For demonstration, we present a transparent processing of distributed stream theta-join and compare the performance of our AdaptMX system with other baselines, with $3\times$ higher throughput.

1 Introduction

Stream computation is another important form of big data processing apart from batch computation, which is tailored for continuously processing with requirements of low latency and high throughput. Among various streaming computations, stream join processing is a fundamental online real-time operation on data from different streams, and has been the focus of much research in many applications, such as stock trading, mobile and network information management systems.

Although a considerable quantity of research has been conducted on stream join processing, there are still urgent problems to tackle with such as adaptivity and load balancing: (i) previous work on join-matrix, a high-performance tuple-based stream join model [3], is subject to the inflexible "The power of two" data

J. Pei et al. (Eds.): DASFAA 2018, LNCS 10828, pp. 802–806, 2018.
https://doi.org/10.1007/978-3-319-91458-9_52

partitioning scheme and scales out/down in a costly way [1]; (ii) the naive key-based *KeyGrouping* approach is easy to implement by sending the tuples with same key to the same tasks, but leads to severe load imbalance due to data skew and slows down the overall performance.

In this paper, we propose AdaptMX, a novel stream join processing system which mixes together well the key-based and tuple-based join approaches. At outer level, we modify the join-matrix model by designing a more flexible tuple-based partitioning scheme which adaptively scales out/down and allocates resource on demand. At inner level, after partitioning the join-matrix into processing units, we group several joiner tasks within each unit and adopt the key-based join approach: (i) we maintain a routing table (RT) and a basic hash function (HF) for each processing unit to route tuples; (ii) we propose cost-effective load balancing strategies to remove the straggler tasks within a processing unit based on the idea of "Split keys on demand and merge keys as far as possible".

2   System Overview and Key Techniques

As Fig. 1(a) shows, AdaptMX consists of two parts. The **data source** contains applications which continuously generates data, and an input adapter. We adopt Kafka as input adapter. The **join processing** is built on the top of Storm, a widespread distributed real-time computation engine, and contains three components, which coordinate to work as a topology: (i) The **spout** subscribes to Kafka and routes original tuples to the downstream **joiner** tasks (instances of Storm component). Under the current $\alpha \times \beta$-partition scheme, each stream corresponds to one side of the matrix. If stream R is for row side, it first randomly selects a row and then sends the tuples to all units along the chosen row. It is the same for stream S corresponding to column side. Then inside a processing unit, it routes a tuple to its destination task by a RT or HF. (ii) The **joiner** does the actual join computation. Several joiner tasks are grouped into processing units which are organized as a join-matrix and each task periodically reports its load statistics to the **controller**. (iii) The **controller** is responsible for constructing the partition scheme of join-matrix model, generating unit mapping, building migration plan to schedule the data migration and conducting load balancing adjustment among **joiner** tasks. Four key techniques of AdaptMX's implementation including partition scheme, unit mapping, data migration and load balance are discussed as followings.

Constructing Partition Scheme. As Fig. 1(b) shows, a join operation between two data streams R and S can be modeled as a join-matrix model, the calculation area of which equals a rectangular of $|R| \cdot |S|$. A partition scheme on the matrix model splits the stream join $R \bowtie S$ into $\alpha \times \beta$ smaller join processing units. Each unit holds partial subset data of two streams and does the local join $R_i \bowtie S_j (0 \leq i < \alpha, 0 \leq j < \beta)$. Figure 1(c) gives an example of a 2×3 - partition scheme on Fig. 1(b)'s join-matrix. Since those subsets are replicated along rows or columns, α and β decide the memory and cpu consumption, which are proportional to the semi-perimeter $(|R_i| + |S_j|)$ and area $(|R_i| \cdot |S_j|)$ of each

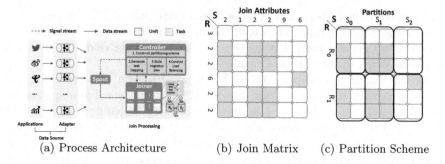

(a) Process Architecture   (b) Join Matrix   (c) Partition Scheme

Fig. 1. Processing Example: (a) Architecture of AdaptMX; (b) Join matrix for the predicate =; (c) A 2×3 - partition scheme.

units respectively. Given the predefined memory resource, we aim to minimize the number of used units $\alpha \cdot \beta$ by adopting two well known theories [2]: (i) given the area, the square has the smallest perimeter among all the rectangles; (ii) given the perimeter, the square has the biggest area among all the rectangles. Current partition scheme are proved to have superfluous units along the last row or last column [2]. We aim to remove the redundant units for the sake of resource utilization. And then we may generate irregular partition scheme compared to previous one along the last row and column, but still promise load balancing.

Generating Unit Mapping Pairs. Before building migration plan, we need to determine the unit mapping pairs from the old (regular) scheme to the new (irregular) scheme, with the purpose to minimize the data migration (including tuples and corresponding states) during matrix transformation. We define a correlation coefficient λ to measure the data overlap between two units, which are from old and new scheme respectively. The naive mapping pairs generation can be divided into two steps: (i) we enumerate all possible mapping pairs; (ii) we select the ones with the biggest λ. Likewise, we can conduct a start-point-alignment method to further optimize the eventual mapping pairs. We align the range start point of data for unit p in the new scheme with unit q in the old scheme. It is proved in [2] that the total migration volume is smaller than that in the previous naive method.

Building Migration Plan. To make it comprehensible, we first introduce how to build migration plan for a regular scheme, and discuss the details for an irregular one later. We define two actions for data migration: duplicating and moving. Apart from duplicating actions which occur to units along the same row or column, the remaining cases are moving actions. Building migration plan is divided into three steps: (i) we first get the whole dataset of stream R or S by combining the data from the first row or column of the old scheme; (ii) then we fill each unit in the new scheme by duplicating or moving actions; (iii) we finally delete data which has been migrated from the unit of the old scheme by moving action. Since the new scheme may be irregular, we need to randomly reassign the data of the deleted units to the remaining ones in the last row or column, and set the deleted units inactive.

Conducting Load Balancing within Each Processing Unit. We adopt the mix routing policy to do join processing: (i) As each processing unit maintains a RT, when a tuple is incoming, it first checks whether the key of the tuple exists in RT. If it exists, the tuple is sent to some joiner task corresponding to that routing entry. Otherwise, the tuple is routed by a HF. (ii) Based on the real-time load statistics, we apply the cost-effective load balancing strategies among joiner tasks within a processing unit independently using the idea of "Split keys on demand and merge keys as far as possible" [4] with the purpose of minimizing RT size but guaranteeing load balancing.

3 Demonstration

Our system runs on top of Storm[1], and the front-end system is deployed on github[2]. The control signal data of AdaptMX is managed by Redis. Screenshots of front-end system are shown in Fig. 2: (i) Fig. 2(a) displays the skew distribution and the fluctuation of stream; (ii) Fig. 2(b) presents the load statistics of each joiner task in AdaptMX; (iii) Fig. 2(d) demonstrates the data migration among joiner tasks during matrix transformation; (iv) To demonstrate the combination of key-based and tuple-based techniques, Fig. 2(e) shows the tuple routing both among and inside units; (v) Fig. 2(c) compares the performance among our AdaptMX, DYNAMIC and Bi [3] where AdaptMX has 3× higher throughput.

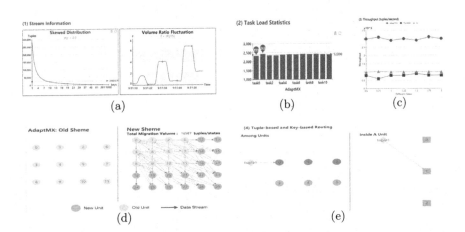

Fig. 2. Demonstration: (a) Stream information; (b) Task load statistics; (c) Throughput of AdaptMX, DYNAMIC and Bi under different; (d) Matrix transformation of AdaptMX compared with DYNAMIC; (e) Data routing.

---

[1] http://storm.apache.org/.
[2] Web link of the demonstration: https://github.com/CJECNU/AdaptMX.

Acknowledgements. The work is partially supported by the Key Program of National Natural Science Foundation of China (Grant No. 61672233, No. 61572194 and No. 61702113).

References

1. Elseidy, M., Elguindy, A., Vitorovic, A., Koch, C.: Scalable and adaptive online joins. PVLDB **7**(6), 441–452 (2014)
2. Fang, J., Zhang, R., Wang, X., Fu, T.Z.J., Zhang, Z., Zhou, A.: Cost-effective stream join algorithm on cloud system. In: CIKM, pp. 1773–1782 (2016)
3. Okcan A., Riedewald, M.: Processing theta-joins using MapReduce. In: SIGMOD, pp. 949–960 (2011)
4. Wang, X., Fang, J., Li, Y., Zhang, R., Zhou, A.: Cost-effective data partition for distributed stream processing system. In: Candan, S., Chen, L., Pedersen, T.B., Chang, L., Hua, W. (eds.) DASFAA 2017. LNCS, vol. 10178, pp. 623–635. Springer, Cham (2017). https://doi.org/10.1007/978-3-319-55699-4_39

A System for Spatial-Temporal Trajectory Data Integration and Representation

Douglas Alves Peixoto[1(✉)], Xiaofang Zhou[1], Nguyen Quoc Viet Hung[2], Dan He[1], and Bela Stantic[2]

[1] The University of Queensland, Brisbane, Australia
{d.alvespeixoto,zxf,d.he}@uq.edu.au
[2] Griffith University, Gold Coast, Australia
{henry.nguyen,b.stantic}@griffith.edu.au

Abstract. Different GPS devices and transportation companies record and store their data using various formats. Even though GPS data often contains the same spatial-temporal and semantic attributes, describing the moving object's trajectory, the integration of these datasets into a single format and storage platform is yet an issue. Therefore, we deliver a data integration system for simplified loading and preprocessing of trajectory data into a standard text platform; this facilitates data access and processing by any trajectory application using multiple and heterogeneous datasets.

1 Introduction

With the increasing of GPS trajectory data volume and sources, large amount of spatial-temporal trajectory data formats have emerged. Therefore, spatial-temporal trajectory data integration is significant to combine data from different sources into a unified format and platform for trajectory data-based applications [3,4]. We introduce a novel system to represent and integrate spatial-temporal trajectory data from different sources and formats. This system targets researchers and professionals working on trajectory data-driven systems and applications, which often demands the collection of data from several sources in order to perform experiments and trajectory-based analytics. The application parses the input data to a predefined output and compressed CSV format, and stores the formatted data into any of the provided primary storage platforms, i.e., MongoDB[1], HBase[2], VoltDB[3], or Local directory. This allows any trajectory-based system to process data from multiple heterogeneous datasets in a user-provided storage platform, without the need of re-implementation.

Current spatial-temporal trajectory data sources generate and store data in a semi-structured textual format, containing the latitude, longitude, and timestamp of the trajectory coordinates points, along with additional semantic infor-

---

[1] MongoDB. https://www.mongodb.com/.

[2] HBase. https://hbase.apache.org/.

[3] VoltDB. https://www.voltdb.com/.

© Springer International Publishing AG, part of Springer Nature 2018
J. Pei et al. (Eds.): DASFAA 2018, LNCS 10828, pp. 807–812, 2018.
https://doi.org/10.1007/978-3-319-91458-9_53

mation, which varies from one dataset to another. Furthermore, several independent sensors may be used in different circumstances to collect data [1]. However, it is challenging to interpret and integrate trajectory data from the multitude of textual formats and sensors available, and it is still an issue [2]. Therefore, in order to represent and integrate data from different formats, we firstly introduce the *Trajectory Data Description Format (TDDF)*, a data description format for spatial-temporal trajectory data. The TDDF was designed based on a survey on several real GPS trajectory datasets, both public and private, accessible by our research groups. Then, based on the user-provided TDDF, our application loads and parses the input data into the selected output data format using lossless Delta compression, in order to reduce the size of the stored data. Our system also generates statistical information (Metadata) about the input datasets. A data parser was built to convert each data record from the input datasets to the output format provided.

2 System Design

Figure 1(a) introduces the system workflow. Briefly, raw trajectory data is read and parsed based on a user-provided input data format (Input TDDF). The parser identifies trajectory records and attributes from the raw data, and parse the raw data to any of the provided output data formats, along with the metadata and the description of the output data format (Output TDDF). The parsed data can be stored into any of the primary storage platforms provided.

Spatial-temporal trajectory datasets available are organized in basically three manners, (1) each document in the dataset contains one trajectory record, (2) each document contains several records, one per line, (3) each document contains several records in multiple lines separated by a delimiter. Attribute values in a record are separated by a delimiter (such as a comma or semicolon). Attributes are either atomic or multi-valued (i.e., list). We overcome the problem of reading different formats by telling the parser how the records are organized in the dataset, that is, the format, type, and order of each attribute in the trajectory records.

2.1 Trajectory Data Description Format: TDDF

We introduce a set of data description keywords to describe the input data format. The format (fields/attributes) of the input data must be provided as they appear in the source files. The *TDDF* is a user-specified script containing the descriptions of the input data files, similar to a *Data Description Language (DDL)*, assisting the parser to identify trajectory records and attributes. The TDDF scope contains both *attribute* declarations, and *commands* to be executed while parsing the data. We introduce a set of declarative keywords to the TDDF, for both attributes' (*Data Definition Keywords*) and command's (*Data Control Keywords*) declarations. Identifiers and spatial-temporal attributes have a special tag since they represent the core of trajectory data. The scope of the

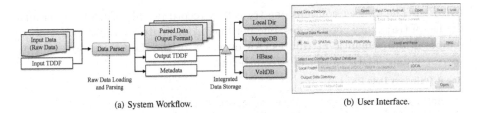

(a) System Workflow.   (b) User Interface.

Fig. 1. Trajectory data loader workflow and GUI.

TDDF was designed based on a survey of existing spatial-temporal trajectory formats, in order to cover a wide range of trajectory datasets. Due to space limitations, however, we omit the TDDF grammar. The complete list of TDDF keywords and their meanings, with further usage examples, can be found at the system's repository [4].

For each *attribute* of the data record, one must provide the attributes' NAME, TYPE and DELIMITER, separated by space or tab. When providing the TDDF script, the user must declare one attribute per line in the exact order they appear in the input file. Commands, on the other hand, are declared in the form NAME, and VALUE. Three different output formats are provided, namely SPATIAL, SPATIAL-TEMPORAL, and ALL. In SPATIAL format the output records contain the trajectory ID and the list of spatial attributes of the coordinates only; the SPATIAL-TEMPORAL format adds the temporal information of every coordinate; the ALL format contains the complete set of attributes declared in the TDDF. The output formats follow a CSV (comma separated values) style. Attribute values are separated by semicolon, and array items are separated by comma. The output documents contain one trajectory record per line. Documents can also be output as BSON [5] documents in MongoDB. Furthermore, to reduce storage consumption, the spatial-temporal attributes in the list of coordinates are delta-compressed. The records attributes are always in the order: ID;LIST_OF_COORDINATES;SEMANTIC_ATTRIBUTES.

3   Case Study and Demonstration Outline

Figure 1(b) shows the main application GUI. We present a set of case studies using real spatial-temporal trajectory datasets. We demonstrate how our application can be used to integrate data from different sources and formats into a unique format. For each case, we provide an overview of the input raw data, as well as the Input and Output TDDF scripts, and the parsed data. The sources of the data files, as well as some attribute values, will be omitted for privacy reasons. For the sake of simplicity, and to demonstrate how our system can be used to integrate datasets into a standard format, we output all datasets using the SPATIAL_TEMPORAL output format.

---

[4] https://github.com/douglasapeixoto/trajectory-data-loader.
[5] https://www.mongodb.com/json-and-bson.

CASE 1: The dataset contains one trajectory record per file. Each line contains the list of trajectory coordinates, one per line. Coordinates contain both spatial-temporal and semantic attributes. An overview of the dataset records and its corresponding TDDF script are given below.

Input Trajectory Data 1:

```
40.008304,116.319876,0,492,39745.0902662037,2008-10-24,02:09:59
40.008413,116.319962,0,491,39745.0903240741,2008-10-24,02:10:04
. . .
```

Input TDDF Script 1:

```
_RECORDS_DELIM    EOF
_COORD_SYSTEM     GEOGRAPHIC
_IGNORE_LINES     [1-6]
_AUTO_ID          db1_t
_COORDINATES      ARRAY(_LAT        DECIMAL      ,
                        _LON        DECIMAL      ,
                        zeroVal     INTEGER      ,
                        alt         INTEGER      ,
                        timeFrac    DECIMAL      ,
                        _TIME       DATETIME["yyyy-MM-dd,HH:mm:ss"] LN) EOF
```

CASE 2: The dataset contains several trajectory records per file, delimited by the character #. The first line of each record contains a set of semantic attributes of the trajectory, followed by the list of coordinates, one per line. Coordinates contain both spatial-temporal and semantic attributes.

Input Trajectory Data 2:

```
#,1,3/2/2009 9:23:12 AM,3/2/2009 10:02:17 AM,10.4217737338017 km
3/2/2009 9:23:12 AM,39.929961,116.355872,23570
3/2/2009 9:23:42 AM,39.926785,116.356007,23526
. . .
#,2,3/2/2009 10:04:14 AM,3/2/2009 10:56:23 AM,13.1721183785493 km
3/2/2009 10:04:14 AM,39.969738,116.288209,32482
3/2/2009 10:04:44 AM,39.973138,116.288661,13208
. . .
```

Input TDDF Script 2:

```
_RECORDS_DELIM    #
_COORD_SYSTEM     GEOGRAPHIC
_AUTO_ID          db2_t
_IGNORE_ATTR      ,
_IGNORE_ATTR      ,
timeIni           STRING   ,
timeEnd           STRING   ,
length            STRING   LN
_COORDINATES      ARRAY(_TIME    DATETIME["M/d/yyyy H:mm:ss a"]  ,
                        _LAT     DECIMAL                          ,
                        _LON     DECIMAL                          ,
                        alt      INTEGER                         LN) #
```

CASE 3: The dataset contains several records per file, one per line. The dataset contains a list of coordinates, and a set of semantic attributes. This dataset had been used for map-matching, hence the coordinate points also contain semantic attributes regarding map-matching. A record in this dataset, corresponding to a single line in the input file, and its corresponding TDDF are given below.

Input Trajectory Data 3:

```
1018_1450,1018,1,27|27|27|19,3639865:0:57:114.33708:30.50130:1427933750|
3639862:6:59:114.33715:30.50128:1427933759
```

Input TDDF Script 3:

```
_RECORDS_DELIM    LN
_COORD_SYSTEM     GEOGRAPHIC
_ID               STRING    ,
sourceId          INTEGER   ,
carType           INTEGER   ,
citySequence      ARRAY(cityId      INTEGER   |)    ,
_COORDINATES      ARRAY(linkID      INTEGER   :
                        oDistance   INTEGER   :
                        mDistance   INTEGER   :
                        _LON        DECIMAL   :
                        _LAT        DECIMAL   :
                        _TIME       INTEGER   |)   LN
```

For all three cases, the output TDDF is the following, since in all cases the input datasets have been parsed to the same output format.

Output TDDF Script:

```
_OUTPUT_FORMAT    SPATIAL_TEMPORAL
_COORD_SYSTEM     GEOGRAPHIC
_DECIMAL_PREC     5
_ID               STRING
_COORDINATES      ARRAY(_LON DECIMAL _LAT DECIMAL _TIME INTEGER)
```

The output formated data, in CSV and BSON documents, for the three datasets is the following. Notice that now all datasets are in the same format SPATIAL_TEMPORAL.

Output Trajectory Data (.csv):

```
db1_t_1;11631987,4000830,1224814199000,8,10,5000
db2_t_1;11635587,3992996,1235985792000,13,-318,30000
db2_t_2;11628820,3996973,1235988254000,46,340,30000
1018_1450,11433708,3050130,1427933750,7,-2,9
```

Output Trajectory Data (.bson):

```
{_id : "db1_t_1", _coordinates : [11631987,4000830,1224814199000,8,10,5000]}
{_id : "db2_t_1", _coordinates : [11635587,3992996,1235985792000,13,-317,30000]}
{_id : "db2_t_2", _coordinates : [11628820,3996973,1235988254000,46,340,30000]}
{_id : "1018_1450", _coordinates : [11433708,3050130,1427933750,7,-2,9]}
```

CASE 4 (Unsupported formats): Despite our efforts to provide a universal parser, some data formats may still not fit perfectly in our parser. However, some pain-less preprocessing can be done in the raw data in order to fit it in our model. For instance, we have access to a dataset collected by a private bus company, they collect the GPS locations of all their buses after certain time interval, and store all GPS coordinates collected at the same time together in a text file. Consequently, the GPS coordinates for a given bus trip were spread across multiple files. Since the GPS records also contained the buses IDs and trip IDs, we simply had to perform a quick sort-and-aggregate algorithm to group coordinates of a same bus and trip into the same file sorted by time-stamp. After that, the trajectory records could be easily parsed by our application.

4   Conclusions

In this demonstration we introduced a novel system for spatial-temporal trajectory data integration and representation. Our application interprets and integrates trajectory data from several textual formats into a standard format, using a novel Trajectory Data Description Format (TDDF), and outputs the integrated data into a user-specified storage platform, in order to assist researchers and developers working on trajectory data-driven applications.

Acknowledgments. This research is partially supported by the Brazilian National Council for Scientific and Technological Development (CNPq).

References

1. Jo, J., Tsunoda, Y., Stantic, B., Liew, A.W.-C.: A likelihood-based data fusion model for the integration of multiple sensor data: a case study with vision and lidar sensors. In: Kim, J.-H., Karray, F., Jo, J., Sincak, P., Myung, H. (eds.) Robot Intelligence Technology and Applications 4. AISC, vol. 447, pp. 489–500. Springer, Cham (2017). https://doi.org/10.1007/978-3-319-31293-4_39
2. Spaccapietra, S., Parent, C., Damiani, M.L., de Macedo, J.A., Porto, F., Vangenot, C.: A conceptual view on trajectories. Data Knowl. Eng. **65**(1), 126–146 (2008)
3. Zheng, Y.: Trajectory data mining: an overview. ACM Trans. Intell. Syst. Tech. (TIST) **6**, 29 (2015)
4. Zheng, Y., Zhou, X.: Computing with Spatial Trajectories. Springer Science and Business Media, Heidelberg (2011). https://doi.org/10.1007/978-1-4614-1629-6

SLIND: Identifying Stable Links in Online Social Networks

Ji Zhang[1(✉)], Leonard Tan[1], Xiaohui Tao[1], Xiaoyao Zheng[2], Yonglong Luo[2(✉)], and Jerry Chun-Wei Lin[3]

[1] The University of Southern Queensland, Toowoomba, Australia
{Ji.Zhang,Leonard.Tan,Xiaohui.Tao}@usq.edu.au
[2] Anhui Normal University, Wuhu, China
zxiaoyao_2000@163.com, ylluo@ustc.edu.cn
[3] Harbin Institute of Technology Shenzhen Graduate School, Shenzhen, China
jerrylin@ieee.org

Abstract. Link stability detection has been an important and long-standing problem in the link prediction domain. However, it is often easily overlooked as being trivial and has not been adequately dealt with in link prediction [1]. In this demo, we introduce an innovative link stability detection system, called SLIND (Stable LINk Detection), that adopts a Multi-Variate Vector Autoregression analysis (MVVA) approach using link dynamics to establish stability confidence scores of links within a clique of nodes in online social networks (OSN) to improve detection accuracy and the representation of stable links. SLIND is also able to determine stable links through the use of partial feature information and potentially scales well to much larger datasets with very little accuracy to performance trade-offs using random walk Monte-Carlo estimates.

Keywords: Link stability · Graph theory · Online social networks
Hamiltonian Monte Carlo (HMC)

1 Introduction

Links in Online Social Network models represent complex relationships between individuals in real life. Link prediction is the likelihood estimation that unobserved relationships exist in a future time space. Several methods of predicting potential relational ties center around either node, topology or social theory based techniques. Link stability detection shares the same stochastic based approaches as link predictive methods. However, instead of predicting unobserved links, it detects and ranks social relations that score a high likelihood of stable future occurences from past observations of information transaction activity.

Link stability detection plays an important role to identify key structural framework for many social and industry applications such as transport, communication networks, engineering, science, business, governments, etc. [2]. For

© Springer International Publishing AG, part of Springer Nature 2018
J. Pei et al. (Eds.): DASFAA 2018, LNCS 10828, pp. 813–816, 2018.
https://doi.org/10.1007/978-3-319-91458-9_54

example, transient protein interactions in a peptide network mirror important properties of cellular function. Recommendation systems and influence mechanisms require stable properties in a social structure to function effectively.

Existing methods of link stability detection include feature similarity based techniques (e.g. CN, JC, Katz, AA, etc.) [1], clustering ensemble methods [6], signed relations [5] and graph-embedding approaches [4]. However, the main limitations of these methods are a bias approach towards a single feature (often similarity based) in the decision process [1]. This flaw often leads to detection inaccuracies and data misrepresentations [2].

In order to mitigate the limitation of the existing methods for stable link detection in OSNs, an innovative link stability detection system, called SLIND (Stable LINk Detection), is introduced in this demo paper. SLIND serves a good system platform to label stable links for OSNs. The novel scientific contribution of our work involves bridging the gap between temporality and stability of links in any online social network by using dynamic link features instead of conventional static features. The innovative feature of SLIND are summarized as follows:

1. SLIND implements a novel idea of running a Multi-Variate Autoregressive model based on dynamic correlated link-based features. This yields highly accurate results when detecting and representing link stability in OSNs;
2. SLIND features a user-immersive interface that allows for manipulation of various link-based feature inputs into the analysis of both univariate and multivariate regression models. It also provides a rich set of visualization modes to display the final results at either a fixed or continuous time duration through different 3D presentations;
3. SLIND accepts large-scale, high dimensional datasets such as the crawled information from real-life OSNs (e.g., Facebook) and can also perform stability analysis on small tightly knitted cliques as well. The system model is computationally efficient and versatile;
4. In comparison to traditional structural and attribute based univariate methods, SLIND is a major advancement for detecting stable links within OSNs featuring much better computational performance, data representation and intelligent predictive accuracy.

2   Method and Architecture of SLIND

In SLIND, the MVVA (Multi-Variate Vector Auto-regression Analysis) method provides the core capability to scale towards the problem complexity. SLIND runs efficiently on small scale networks. However, MVVA itself is insufficient to tackle the problem of large-scaled partially observable networks. SLIND assimilates the Hamiltonian Monte Carlo to build a probabilistic chain of states through time that converges to the actual Stability Index (SI) distribution [3].

The overview of the system architecture of SLIND is presented in Fig. 1. In SLIND, crawled data is first decoded through a de-anonymization module to produce the features of interest. An optional 2-D topology plot of the dataset can be subsequently produced. The feature selection module, next handles the de-anonymized data (e.g. sentiment, trust, frequency, etc.) and a univariate / multivariate stability index (SI) score is calculated from the UV/MV modules. A 3D node coordinate set is generated, and the links are subsequently tagged with SI scores. A 3D universe module establishes and transforms the labeled social structure onto the euclidean space. Finally, a 3D canvas module is called to instantiate and draw the corresponding OSN architecture for visual representation to users.

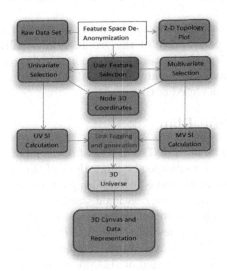

Fig. 1. System architecture of SLIND

The dataset chosen for this study, as well as for the demo, was crawled from Facebook and obtained from the repositories of the Common Crawl (August 2016)[1]. It is de-anonymized to reveal the following relational features in the wall posts: the Cumulative Frequency, Sentiment, Similarity, Trust and Number of Posts at corresponding Unix time samples.

The links are tagged based on their SI scores with higher scores denoting more stable links.

3   Human Computer Interaction

SLIND allows for user interactions when selecting features for regression analysis. Additionally, SLIND also allows users to specify tuning parameters, L (Number of steps) and ϵ (Stepsize) of our developed HMC model. In addition, users can also specify the length of the HMC sampling chain. Furthermore, SLIND provides the option for users to specify initial state values (if known), of the feature

---

[1] http://commoncrawl.org/2016/09/august-2016-crawl-archive-now-available/.

array as parameterization constraints. Alternatively, users can choose to generate initial values for their model randomly. Finally, users can choose the sample sizes, number of parallel repetitions and Markov Chain lengths for the burn-in phase of the HMC.

There are two presentation modes supported by SLIND to display the link stability analysis results, namely the snapshot mode and the timeline mode. In the snapshot mode, the link stability analysis results are presented at a fixed time frame, while the timeline mode supports the continuous real-time presentation of all the results.

The HMC stochastic distribution model can also be toggled to display the evolution of the linked stability of the graph. Additionally, SLIND is able to transform between views for better visualization. Furthermore, users are able to generate plots, run real-time analysis simulations, define labels for link SI and fill in the markers of detected outliers.

4   Demonstration Plan

The demonstration plan for SLIND will consist of the following four major parts. First, we describe to the audience the limitations of the existing link predictive methods and the main motivations behind using a time-series Mulivariate model supported by HMC for predicting relational stability in socio-network links. Second, we will showcase the system architecture of SLIND. Third, a demonstration on the interactive interfaces developed on SLIND will be given. Fourth, an on-site demo of SLIND will be played to the audience. Audience interaction with the software platform is encouraged at this stage. On-site assistance will also be available upon request.

Acknowledgements. This research is partially supported by National Science Foundation of China (No. 61672039, No. 61772034, No. 61503092) and Guangxi Key Laboratory of Trusted Software (No. kx201615).

References

1. Ozcan, A., Oguducu, S.G.: Multivariate temporal link prediction in evolving social networks. In: International Conference on Information Systems (ICIS-2015), pp. 113–118 (2015)
2. Mengshoel, O.J., Desai, R., Chen, A., Tran, B.: Will we connect again? Machine learning for link prediction in mobile social networks. In: Eleventh Workshop on Mining and Learning with Graphs, Chicago, Illinois, pp. 1–6 (2013)
3. Sol-Dickstein, J., Mudigonda, M., DeWeese, M.R.: Hamiltonian Monte Carlo without detailed balance. In: Proceedings of the 31st International Conference on Machine Learning (JMLR), vol. 32 (2014)
4. Ryohei, R.: Semi-supervised graph embedding approach to dynamic link prediction. arXiv preprint arXiv:1610.04351 (2016)
5. Song, D., Meyer, D.A.: Link sign prediction and ranking in signed directed social networks. Soc. Netw. Anal. Min. **5**(1), 52 (2015)
6. Feng, X., Zhao, J.C., Xu, K.: Link prediction in complex networks: a clustering perspective. Eur. Phys. J. B **85**(1), 3 (2012)

MusicRoBot: Towards Conversational Context-Aware Music Recommender System

Chunyi Zhou[1], Yuanyuan Jin[1], Kai Zhang[2], Jiahao Yuan[1], Shengyuan Li[1],
and Xiaoling Wang[1(✉)]

[1] Shanghai Key Laboratory of Trustworthy Computing,
MOE International Joint Lab of Trustworthy Software,
East China Normal University, Shanghai, China
cyzhou@stu.ecnu.edu.cn, xlwang@sei.ecnu.edu.cn
[2] Shenzhen Gowild Robotics Co. Ltd., Shenzhen, China
zhangkai_ai@gowild.cn

Abstract. Traditional recommendation approaches work well on depicting users' long-term music preference. However, in the conversational applications, it is unable to capture users' real time music taste, which are dynamic and depend on user context including users' emotion, current activities or sites. To meet users' real time music preferences, we have developed a conversational music recommender system based on music knowledge graph, *MusicRoBot* (Music RecOmmendation Bot). We embed the music recommendation into a chatbot, integrating both the advantages of dialogue system and recommender system. In our system, conversational interaction helps capture more real-time and richer requirements. Users can receive real time recommendation and give feedbacks by conversation. Besides, *MusicRoBot* also provides the music Q&A function to answer several types of musical question by the music knowledge graph. A WeChat based service has been deployed piloted for volunteers already.

Keywords: Music recommendation · Online recommendation
Dialogue system · Recommender system

1 Introduction

Listening to music has been common during many people's leisure time. Generally, the recommended content includes hit songs, daily playlist and music radio, but these kind of interactive ways limit the expression of requirements. In this work, we have implemented a conversational music recommender system, *MusicRoBot* (Music RecOmmendation Bot), which embeds music recommendation

X. Wang—This work was supported by NSFC grants (No. 61472141), Shanghai Knowledge Service Platform Project (No. ZF1213) SHEITC and Shanghai Agriculture Applied Technology Development Program (No. G20160201).

into a ordinary chatbot innovatively. Comparing with traditional recommendation scenarios, there're many differences in the conversational scenario: (1) it emphasizes more on online interactions; (2) this scenario is more context-sensitive; (3) dialogues carry richer but more complex information. Obviously, conversational recommendation is significant but challenging. In fact, conversational recommendation has been already studied. Christakopoulou et al. [1] proposes a conversational recommender system for restaurant recommendation by asking user absolute or relative questions. Sun et al. [2] demonstrates a conversational products recommendation agent based on deep learning technologies, but this demo seems like a virtual sales agent using a task-oriented dialog system. Besides, different from most existing chatbot, we focus on music-domain rather than open-domain, and we also construct *Music Knowledge Graph* (MKG) in support of musical entity recognition and recommendation.

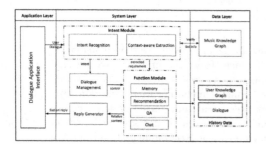

Fig. 1. Architecture of MusicRoBot

Fig. 2. Entities and relationships in *MKG*

2   System Design

Our system can be is divided into three layers: data layer, system layer and application layer. Figure 1 shows the architecture of *MusicRoBot*.

2.1   Music Knowledge Graph (MKG)

In support to better recognition and recommendation, we construct *Music Knowledge Graph*, raw data of which comes from *Xiami*[1]. We organize all musical entities as a graph in consideration of advantages in inference and analysis. Figure 2 shows entities and their relationships, the entity is represented as node and the relationship as edge. We define four types of entities: song, album, artist and genre, and our genres include both professional genres and common tags. *MKG* is stored in neo4j[2], and there're currently over 6 million songs, 600 thousand albums, 130 thousand artists, nearly 500 genres and still increasing.

---

[1] xiami's homepage: http://www.xiami.com/.
[2] neo4j's homepage: https://neo4j.com/.

2.2   Scenario Design

In our demo, we have designed four scenarios as follow:

Memory-based User Portrait Construction. In Fig. 1, it's short for *Memory*. System can capture users' basic properties and preferences on music during dialogues and store in *User Knowledge Graph (UKG)*. These properties can be used as explanation for recommendations and relieve the cold-start problem. In this implementation, basic property contains age, gender and current emotion. Preference contains all kinds of music entities in *MKG*. In addition, we conduct collision detection on user's basic properties.

Q&A. Q&A consists of user properties' Q&A and music knowledge's Q&A. This function is mainly designed for enhancing the interactivity between user and system. It may help discover useful user preferences for our future research.

Recommendation. It's the core module in this system. There're three kinds of recommendation scenarios: specific-query based recommendation, free recommendation and emotion-based recommendation. The recommended items include song, album and artist. Besides, in coping with online and interactive recommendation, we adopt a bandit-based recommendation algorithm, C^2UCB [3]. We compare C^2UCB with most popular strategy using *Xiami* user's listened song list and show the average reward (AR) for each user in Table 1. The result shows that user prefer less popular songs and prove the advantage of C^2UCB.

Chat. This function is in charge of the other scenarios which don't match the above situations. It is essential but not our focus, we employ the existed implementation by *emotibot*[3].

2.3   Intent Recognition and Dialogue Management

In *Intent Module*, system recognizes user intent and extracts useful constraints from input. We summarize this task into *Intent Recognition* and *Realtime Requirements Extraction*. This module is implemented by *Gowild*[4], applying both template matching and classifier. *Dialogue Management* is in charge of making next system action, which plays a role as a central controller. Both users' current intents and the context of previous dialog are considered during decision making.

3   Demonstration

Our demo is published as a *WeChat Service* and it currently supports only Chinese text input. Figure 3 shows the representative example motion-based recommendation scenario: a new user comes in and expresses his negative emotion, in this case, system inquires user preferences under this emotion, then recommends song as normal. Currently, system provides multi-turn recommendations at most three times, when user doesn't accept recommendations.

---

[3] *emotibot*'s homepage: http://www.emotibot.com.
[4] *Gowild*'s homepage: http://www.gowild.cn.

Table 1. Comparison with Most-popular recommendation

#round	#user	AR for MP	AR for C^2UCB based	Promotion (%)
1	645	0.40	0.41	2.5
10	645	4.09	4.5	10.02
20	645	8.51	9.24	8.58
50	645	20.09	23.87	18.82
100	645	39.84	48.27	21.16
200	504	80.21	98.87	23.26
500	277	200.46	261.77	30.58

Fig. 3. WeChat Service Demonstration for the recommendation scenario

References

1. Christakopoulou, K., Radlinski, F., Hofmann, K.: Towards conversational recommender systems. In: Proceedings of the 22nd ACM SIGKDD International Conference on Knowledge Discovery and Data Mining, pp. 815–824. ACM (2016)
2. Sun, Y., Zhang, Y., Chen, Y., et al.: Conversational recommendation system with unsupervised learning. In: Proceedings of the 10th ACM Conference on Recommender Systems, pp. 397–398. ACM (2016)
3. Qin, L., Chen, S., Zhu, X.: Contextual combinatorial bandit and its application on diversified online recommendation. In: Proceedings of the 2014 SIAM International Conference on Data Mining, pp. 461–469. Society for Industrial and Applied Mathematics (2014)

HDUMP: A Data Recovery Tool for Hadoop

Zhongsheng Li[1], Qiuhong Li[2(✉)], Wei Wang[2], Qitong Wang[2], Fengbin Qi[1], Yimin Liu[3], and Peng Wang[2]

[1] JiangNan Institute of Computing Technology, Wuxi, China
lizhsh@yean.net, qifb118@sina.com
[2] School of Computer Science, Fudan University, Shanghai, China
{qhli09,weiwang1,qtwang16,pengwang5}@fudan.edu.cn
[3] Third Affiliated Hospital of Second Military Medical University, Chongqing, China
liuyiminzsh@aliyun.com

Abstract. Hadoop is a popular distributed framework for massive data processing. HDFS is the underlying file system of Hadoop. More and more companies use Hadoop as data processing platform. Once Hadoop crashes, the data stored in HDFS can not be accessed directly. We present HDUMP, a light-weight bypassing file system, which aims to recover the data stored in HDFS when Hadoop crashes.

1 Introduction

MapReduce [3] is a popular parallel programming model and Hadoop [1] is its open-source implementation. Many internet companies are dependent on Hadoop for their massive datasets. HDFS stores meta data on a master node, called Name Node. Application data are stored on other servers called Data Nodes. For Hadoop 1.x, there is only one Name node. In [4], the authors proposed the second Name Node for Hadoop. However, even with two Name Nodes, it is possible that the two Name Nodes crash at the same time. The recovery of the whole Hadoop clusters needs professional engineers. Sometimes, users can not wait a long time to get the urgent data from HDFS. We present HDUMP, an off-line data recovery tool, which can bypass Hadoop and HDFS to fetch the data directly. We know that HDFS is made up of the local directories of the Data nodes. The data is stored in the data blocks, which are distributed in the local directories of the data nodes. Thus, by acquiring the mapping relationships between the files and the data blocks, we can fetch the files from the local directories directly.

In this demo, we introduce HDUMP, a light-weight data recovery tool for HDFS. The total installation package is only about 25 MB. The main idea of

The work is supported by the Ministry of Science and Technology of China, National Key Research and Development Program (No. 2016YFE0100300, No. 2016YFB1000700), National Key Basic Research Program of China (No. 2015CB358800), NSFC (61672163, U1509213), Shanghai Innovation Action Project (No.16DZ1100200).

© Springer International Publishing AG, part of Springer Nature 2018
J. Pei et al. (Eds.): DASFAA 2018, LNCS 10828, pp. 821–824, 2018.
https://doi.org/10.1007/978-3-319-91458-9_56

HDUMP is to restore the information from the meta data of HDFS. The meta file *fsimage* includes the file and directory properties and the mapping relationships between the file and the data blocks. However, a data node includes many data blocks, to speed up the searching, we build a BTree [2] index for the block ids and the node ids. HDUMP implements the following functions without the support of Hadoop:

1. Show the information of HDFS including the users, files and the mapping relationships between HDFS files and HDFS data blocks.
2. A similar file system as HDFS including common commands, such as *ls,cd,pwd,list,cp,cat*, and so on.
3. Scan and download files from HDFS in an off-line way.

The rest of paper is as organized as follows. Section 2 introduces HDUMP. Section 3 presents the demonstration and evaluation of HDUMP. Section 4 gives the conclusion.

2   HDUMP

2.1   HDUMP Overview

The main idea of HDUMP is to restore the directory hierarchy, the HDFS file properties and the mapping relationship between the HDFS files and the data blocks according to the meta data, named as *fsimage*. We implement a build-in file system in HDUMP, which is a similar file system as HDFS.

2.2   HDUMP Architecture

HDUMP is comprised of two main components. One is *fsimage* analyzer and the other is the HDUMP file system. *Fsimage* analyzer transforms the meta data of HDFS to an xml format. HDUMP file system uses the xml to restore the HDFS file system. The architecture of HDUMP is illustrated in Fig. 1. HDUMP is implemented by a client/server architecture. The server includes four components, which are fsimage analyzer, HDUMP file system, file fetcher and BTree index respectively. The client is a window which supports *SSH* protocol (Table 1).

2.3   Fsimage Analyzer

We analyze the meta file *fsimage* to get the data hierarchy of HDFS. HDUMP supports Hadoop 1.x and Hadoop 2.x. The *fsimage* of Hadoop 1.x is in the format of xml. We interpret the xml format to get the requisite information to restore HDUMP file system. The *fsimage* of Hadoop 2.x is in a binary format. We transform it to an xml format.

Table 1. HDUMP commands

Name	Description
load	Load the *fileimage* of HDFS
head	Show the info of *fsimage*
list	List the directories and files included in the *fsimage*
print	Print all information in the *fsimage*
cat	Show the contents of the file in HDFS
cp	Copy the file in HDFS to the local machine
find	Find file in HDFS
ls	Show the files and the directories in the current HDFS directory
shell	Execute shell command

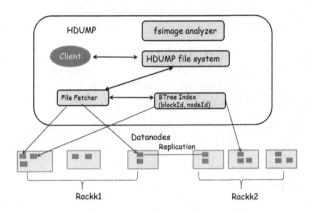

Fig. 1. HDUMP

2.4 Data Recovery Utilizing HDUMP

HDUMP provides a off-line way to fetch the files stored in HDFS. When Hadoop crashes, we can use HDUMP to restore the files. For HDFS, the data blocks are stored in the data directory specified in the hdfs-site.xml in data nodes. There are a large amount of blocks in a data node. To speed up the time efficiency, we use a BTree [2] index to implement a quick searching for the data blocks.

When HDUMP starts, we use *load* command to load the *fsimage*. The *fsimage* analyzer of HDUMP restore the file and directory properties from *fsimage* and HDUMP provides a simple file system for users to interact.

3 Demonstration and Evaluation

As illustrated in Fig. 2, HDUMP serves as a file system similar with HDFS. When Hadoop crashes, we use HDUMP to scan and fetch the data stored in HDFS. The steps fetching the files on HDFS are listed below:

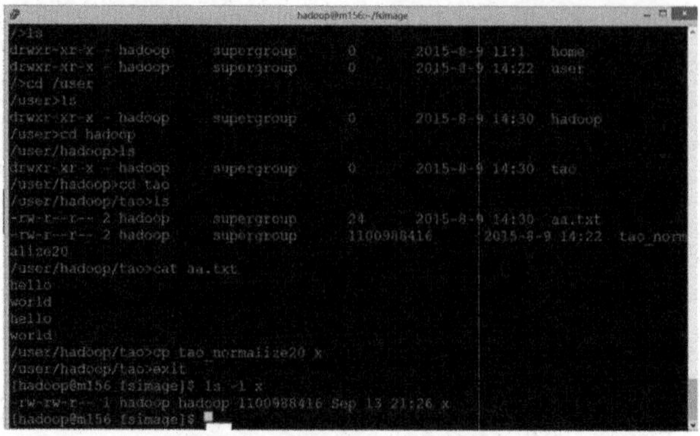

Fig. 2. HDUMP interactive interface

1. Step1: Load $fsimage$ into HDUMP
2. Step2: Find the files to be fetched using $find$ command
3. Step3: Enter the specified directory using cd command
4. Step4: Download the files on HDFS using cp command.

The file transferring speed utilizing HDUMP is dependent on the network speed and the disk speed. We copy a file with size of 500 MB and take about 5 s.

4   Conclusions

In the demo, We present HDUMP, a data recovery tool for Hadoop. When Hadoop crashes, using HDUMP, we can browse and fetch files from HDFS directly without the recovery of Hadoop.

References

1. http://hadoop.apache.org/
2. Comer, D.: The ubiquitous B-tree. ACM Comput. Surv. **11**, 121–137 (1979)
3. Dean, J., Ghemawat, S.: MapReduce: simplified data processing on large clusters. Commun. ACM **51**, 107–113 (2008)
4. Deshpande, P., Bora, D.: The recovery system for Hadoop cluster. In: The 20th International Conference on Distributed Multimedia Systems: Research Papers on Distributed Multimedia Systems, Distance Education Technologies and Visual Languages and Computing, Pittsburgh, PA, USA, 27–29 August 2014, pp. 416–420 (2014)

Modeling and Evaluating MID1 ICAL Pipeline on Spark

Zhongsheng Li[1], Qiuhong Li[2(✉)], Yimin Liu[3], Wei Wang[2], Fengbin Qi[1], Mingmin Chi[2], and Yitong Wang[2]

[1] JiangNan Institute of Computing Technology, Wuxi, China
lizhsh@yean.net, qifb118@sina.com
[2] School of Computer Science, Fudan University, Shanghai, China
{qhli09,weiwang1,mmchi,ytwang}@fudan.edu.cn
[3] Third Affiliated Hospital of Second Military Medical University, Chongqing, China
liuyiminzsh@aliyun.com

Abstract. Squire Kilometre Array (SKA) project generates almost the hugest data volume in the world. SKA data flow pipelines need almost real-time processing ability, which brings huge challenges to the execution frameworks (EF for short). We propose a cost model for a typical SKA data flow pipeline named as MID1 ICAL pipeline on Spark. By simulating the I/O of MID1 ICAL pipeline with a reduced SKA data, we evaluate several different implementations of MID1 ICAL pipeline and conclude the optimized method for this pipeline on Spark.

1 Introduction

Squire Kilometre Array [2] is the next generation telescope producing almost the hugest data volume in the world. The SKA science data processing (SDP for short) pipelines in SKA include imaging pipelines and calibration pipelines mainly. Both of these two kind of pipelines are iterative, data-intensive and computing-intensive. Spark [4] is reported as a popular distributed in-memory computing framework for large scale data analytics, especially suitable for iterative data processing.

MID1 ICAL pipeline is an astronomical calibration pipeline for SKA. In [1], the authors propose an implementation of the IO for MID1 ICAL on Spark, which is treated as the baseline. We do not need the astronomical background by only considering the nodes as a collection of data and the edges as the data dependency between nodes. As illustrated in Figs. 1 and 2, the edges between nodes represent the data dependency. The main bottleneck of the baseline is the shuffle caused by the "cogroup" and "flatMap". We propose two methods to improve it. First, we use Spark partitioning to replace "cogroup" and "flatMap".

The work is supported by the Ministry of Science and Technology of China, National Key Research and Development Program (No. 2016YFE0100300), National Key Basic Research Program of China (No. 2015CB358800), NSFC (61672163, U1509213), Shanghai Innovation Action Project (No. 16DZ1100200).

© Springer International Publishing AG, part of Springer Nature 2018
J. Pei et al. (Eds.): DASFAA 2018, LNCS 10828, pp. 825–828, 2018.
https://doi.org/10.1007/978-3-319-91458-9_57

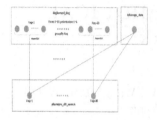

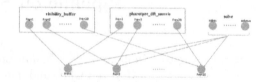

Fig. 1. Pharotpre_dft_sumvis Stage   **Fig. 2.** Solve Stage

Secondly, we use Alluxio [3] to replace these two operations. The key contributions of this demo are as follows.

1. We propose a cost model for MID1 ICAL pipeline on Spark. We model the execution of MID1 ICAL concerning of partition granularity, shuffle amount and task overheads.
2. We use Alluxio to solve the performance bottleneck caused by Spark shuffles.

2   Modeling MID1 ICAL Pipeline on Spark

2.1   MID1 ICAL Pipeline Overview

The smallest independent processing unit of MID1 ICAL pipeline generates about 62.3 TB data with one loop, one island worth of frequencies (20 frequencies) and one snapshot. Figures 1 and 2 illustrate the data dependencies for Stage Pharotpre_dft_sumvis and Stage solve. Each stage includes a collection of <key, value> pairs. The key represents the identifier and the value represents the data. The key of Stage degkerupd_deg is marked as (beam, major_loop, frequency, time, facet, polarisation). Here beam, time and major_loop are set to 1. The frequency is from 1 to 20 and facet is from 1 to 81. The polarisation is from 1 to 4. Thus there are 6480 <key, value> pairs for Stage degkerupd_deg.

2.2   Cost Model

We propose a cost model for MID1 ICAL pipeline as the following:

$$COST_{I/O} = MEM_{cost} + Shuffle_{cost} + NET_{cost} + Task_{overheads} \qquad (1)$$

$$MEM_{cost} = RDD_{memory} + Broadcast_{memory} + GC_{cost} \qquad (2)$$

$$Shuffle_{cost} = Sort_{cost} + DISK_{cost} + SER_{cost} + DESER_{cost} \qquad (3)$$

$$NET_{cost} = TRANS_{RDD} + TRANS_{Broadcast} + TRANS_{shuffle} \qquad (4)$$

Spark task overheads per task is less than 3 ms. We reduce the data scale to a very small value and treat the execution time as the task overheads. We use

3 ms as the average Spark overheads for a task (not include the data processing time). For the auto-generated version, 12964 tasks are launched. The time for extra task overheads is about $12964 * 3\,ms = 38.9\,s$. The number of Spark tasks is related with the number of the partitions of RDDs. Thus, coarse granularity of RDD partitions can decrease the extra Spark task overheads.

The number of Spark nodes affects the cost mainly by the amount of network IOs caused by both the RDD processing and Spark Broadcast. Both intra-RDD and inter-RDD operations need to consider the network cost. The processing speed for shuffle is quite slow, less than 50 MB/s for our cluster because of sort operations and disk IOs are concerned. The Spark partitioning does not generate shuffle.

3 Comparisons of Different MID1 ICAL Pipeline on Spark

We compare two proposed implementations with the baseline method.

The Auto-partitioning is the Spark implementation of the MID1 ICAL pipeline mentioned in [1]. The main idea is to utilize "flatMap" operation to copy the data and use "cogroup" to generate an RDD from two or three RDDs. To match the keys, many "flatMap" operations are used to copy the RDD items to generate new items with different keys.

We solve the cogroup problem by utilizing Spark partitioning and Spark broadcast. Take pharotpre_dft_sumvis stage as an example. We use mapPartitions function to partition the RDD degrid by frequency. This operation does not cause shuffle operations. By the combination of broadcasting small datasets and partitioning big datasets, we can avoid most of the previous cogroup operations.

We use Alluxio, an in-memory distributed file system to replace the "cogroup" operations and the effect is significant. Fortunately, Alluxio is also a product from AMPLab, the same as Spark. Alluxio can provide data sharing across different jobs and different systems with in-memory speed. By the same way, the "cogroup" operations can be replaced by the combination of Spark and Alluxio.

4 Evaluation

4.1 Evaluation Environments and Baselines

We conduct the experiments to compare different implementations of the reduced program for MID1 ICAL pipeline on Spark. We use two experiment environments. The first includes one machine, which has 1.5 TB memory, 80 CPU cores of 2.2 GHZ. The second includes five machines, each with 64 GB memory, 8 CPU cores, each with 1.8 GHZ.

4.2   Comparisons of Different Spark Implementations of MID1 ICAL Pipeline

We use data scale $= 1/8$ to evaluate three implementations of auto-partitioning, partitioning and Alluxio + Spark. The results are illustrated in Table. 1. We verify the data by checking both the number and the size for the data in Alluxio. The system resources are illustrated in Figs. 3 and 4.

Table 1. Comparisons of Different Spark Implementations (data scale $= 1/8$)

Version name	Stages number	Time (min)	Shuffle (Read)	Shuffle (Write)
Auto-partitioning	12964	40	18.0 GB	15.8 GB
Partitioning	19	4.4	84.7 MB	107.9 MB
Spark + Alluxio	22	2.6	75.7 MB	108.1 MB

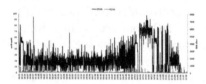

Fig. 3. System summary (Auto-partitioning)

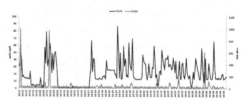

Fig. 4. System summary (Partitioning)

5   Conclusions

Traditional astronomical applications use MPI to implement parallelism. In this demo, we present two efficient implementations for MID1 ICAL pipeline on Spark.

References

1. https://confluence.ska-sdp.org/display/arch/evaluation+using+generated+pipeline
2. https://skatelescope.org/
3. Li, H., Ghodsi, A., Zaharia, M., Shenker, S., Stoica, I.: Tachyon: reliable, memory speed storage for cluster computing frameworks. In: Proceedings of the ACM Symposium on Cloud Computing, Seattle, WA, USA, 03–05 November 2014, pp. 6:1–6:15 (2014)
4. Zaharia, M., Chowdhury, M., Das, T., Dave, A., Ma, J., McCauly, M., Franklin, M.J., Shenker, S., Stoica, I.: Resilient distributed datasets: a fault-tolerant abstraction for in-memory cluster computing. In: Proceedings of the 9th USENIX Symposium on Networked Systems Design and Implementation, NSDI 2012, San Jose, CA, USA, 25–27 April 2012, pp. 15–28 (2012)

Correction to: Coverage-Oriented Diversification of Keyword Search Results on Graphs

Ming Zhong, Ying Wang, and Yuanyuan Zhu

Correction to:
Chapter "Coverage-Oriented Diversification of Keyword Search Results on Graphs" in: J. Pei et al. (Eds.):
Database Systems for Advanced Applications,
LNCS 10828, https://doi.org/10.1007/978-3-319-91458-9_10

In the originally published version of chapter 10 the funding information in the acknowledgement section was incomplete. This has now been corrected.

The updated version of this chapter can be found at
https://doi.org/10.1007/978-3-319-91458-9_10

Author Index

Printed in the United States
By Bookmasters